HANDBOOK OF RENEWABLE ENERGY TECHNOLOGY & SYSTEMS

Highly Recommended Titles

Renewable Energy Finance: Funding the Future of Energy
Second Edition
edited by Charles W. Donovan
ISBN: 978-1-78634-859-3
ISBN: 978-1-78634-862-3 (pbk)

Solar Power Finance Without the Jargon
by Jenny Chase
ISBN: 978-1-78634-739-8
ISBN: 978-1-78634-745-9 (pbk)

Handbook of Energy Finance: Theories, Practices and Simulations
edited by Stéphane Goutte and Duc Khuong Nguyen
ISBN: 978-981-3278-37-0

Chemistry and Technology of Alternate Fuels
by James G. Speight
ISBN: 978-981-120-364-0

Energy Efficiency
Innovations: Driving Prosperity, Slashing Emissions
edited by Henry Kelly
ISBN: 978-981-121-785-2

Wind, Water and Fire: The Other Renewable Energy Resources
edited by Gerard M. Crawley
ISBN: 978-981-122-591-8

Electricity: Humanity's Low-carbon Future
Safeguarding Our Ecological Niche
by Hans B. (Teddy) Püttgen and Yves Bamberger
ISBN: 978-981-122-435-5
ISBN: 978-981-122-930-5 (pbk)

HANDBOOK OF RENEWABLE ENERGY TECHNOLOGY & SYSTEMS

Editors

Ramesh C Bansal

University of Sharjah, United Arab Emirates

University of Pretoria, South Africa

Ahmed F Zobaa

Brunel University London, UK

World Scientific

NEW JERSEY • LONDON • SINGAPORE • BEIJING • SHANGHAI • HONG KONG • TAIPEI • CHENNAI • TOKYO

Published by

World Scientific Publishing Europe Ltd.

57 Shelton Street, Covent Garden, London WC2H 9HE

Head office: 5 Toh Tuck Link, Singapore 596224

USA office: 27 Warren Street, Suite 401-402, Hackensack, NJ 07601

Library of Congress Cataloging-in-Publication Data

Names: Bansal, Ramesh C., editor. | Zobaa, Ahmed F., editor.

Title: Handbook of renewable energy technology and systems / Ramesh C, Bansal, Department of Electrical Engineering, University of Sharjah, Sharjah, United Arab Emirates, Ahmed F Zobaa, College of Engineering, Design and Physical Sciences, Brunel University London, Uxbridge, United Kingdom.

Description: First edition. | London ; Singapore ; hackensack, NJ : World Scientific, 2022. | Includes bibliographical references and index.

Identifiers: LCCN 2020051963 | ISBN 9781786349026 (hardcover) | ISBN 9781786349033 (ebook for institutions) | ISBN 9781786349040 (ebook for individuals)

Subjects: LCSH: Renewable energy sources.

Classification: LCC TJ808 .H358 2022 | DDC 621.042--dc23

LC record available at https://lccn.loc.gov/2020051963

British Library Cataloguing-in-Publication Data

A catalogue record for this book is available from the British Library.

For any available supplementary material, please visit
https://www.worldscientific.com/worldscibooks/10.1142/Q0264#t=suppl

Desk Editors: Ramya Vengaiyan/Britta Ramaraj/Amanda Yun

Typeset by Diacritech Technologies Pvt. Ltd.
Chennai - 600106, India

Contents

Preface

Worldwide, the effects of Global warming, pollution due to power generation from fossil fuels, and its depletion have led to the rapid deployment of renewable energy-based power generation. The leading renewable technologies are wind and photovoltaic (PV) systems. The incorporation of these generation technologies has led to the development of a broad array of new methods and tools to integrate renewable generation into the power system networks. This book composes of 22 chapters contributed by the authors from different countries. The book is arranged into four sections, which present a comprehensive analysis of various renewable energy-based distributed generation (DG) technologies (wind, PV, microgrid/hybrid) briefly described as follows:

Section 1 composes of five chapters on wind power systems. Chapter 1 discusses the steady-state modelling of DFIG-based wind energy system for unbalanced operation. Chapter 2 presents the mathematical modelling and small-signal stability analysis of a DFIG-based wind system using various control techniques. Chapter 3 presents small-disturbance, short-, and long-term voltage stability analysis of a DFIG-based wind system using Nordic test system.

A comparative analysis of 10 commercially available wind turbines integration in radial distribution systems for various seasons of the year based on practical data of wind farm located in Southern India is presented in Chapter 4. Chapter 5 analyzes the services for current and future power systems with respect to active power, reactive power, frequency, and voltage services. Different grid code requirements of several countries are discussed to understand the requirements of the wind power plants.

Section 2 contains the next eight chapters on solar PV systems. Concentrating solar power (CSP) technologies and power tower configuration can provide the choice for large-scale electricity generation in the near future. Both technologies can have energy storage and auxiliary heat production during sunlight unavailability. Other CSP technologies examined in Chapter 6 include dish–Stirling engines, solar chimneys, and solar ponds.

Chapter 7 presents the calculations of the position of the sun as a function of location, date, and time. Direct, diffuse, and reflected components of solar insolation on a tilted surface are determined. These calculations can be used to determine the angles of single and two-axis tracking systems to maximize the utilization of solar energy. Chapter 8 shows the importance of parameter estimation of PV cells and their impact on the performance characteristics of the PV array. The performance of PV module under different environmental conditions at two different locations is critically evaluated.

Chapter 9 focuses on the application of the different types of DC-to-DC converters and inverters employed for DC–AC power conversion. In this conversion, single-stage and two-stage converters, that is, DC–DC converter for DC load system and DC–AC converter for AC load system is presented to make the efficient output of PV as per the requirement of grid/load. In Chapter 10, various design aspects of PV system based on consumer requirements and load patterns are presented. Sizing of PV panels for applications with and without batteries, selection of batteries, sizing of cables, and the design of protection system against overcurrent, overvoltage, and lightning surges are presented.

Chapter 11 analyzes the performance of a PV system connected to a single-phase grid supply and nonlinear behavior of the system operating under varying irradiance levels. The response of the maximum power point tracking (MPPT) controller is analyzed. Chapter 12 discusses three main inverter configurations namely central inverter, string inverter, and microinverter. The effects of shading on the performance of central inverter and microinverter connected to low power solar-based domestic or rural applications are presented. A case study has been included where in the comparison of microinverter and the central inverter has been made on the basis of experimental results. Techno-economical analysis of microinverter is presented to show the energy-saving with the use of microinverter technology.

Chapter 13 focuses on the protection of feeders supplying residential customers with rooftop PV systems. The chapter illustrates that time and sequence of the PVs' isolation following the fault will vary depending on the impedance and location of fault, PV generation–demand ratio, and network's earthing type.

Section 3 contains six chapters on hybrid powers/microgrid/DG. Chapter 14 iterates a generalized comprehensive trilevel framework combining the synergies of system modelling, multi-objective optimization, and multi-criteria decision analysis models to design microgrids based not only on technical and economic factors, but also on social, environmental, and institutional/political traits as well.

Chapter 15 presents an overview of various optimization techniques and its applications for optimal sizing of hybrid renewable energy systems consisting of multiple energy generation and energy storage sources. The purpose of Chapter 16 is to optimally size an off-grid hybrid renewable energy system for a university campus in Nigeria and conducts a techno-economic analysis. In Chapter 17, distributed resources (DRs) allocation problem in radial distribution systems is modelled by considering more realistic load profile of the system and the consequences of ignoring load diversity among distribution buses on the performance of the distribution system are investigated.

In Chapter 18, optimum sizing of DGs at optimum power factor and shunt capacitor sizing at optimum locations are obtained using genetic algorithm considering the objectives of minimizing the burden on the substation, reducing real and reactive power losses, and improving the voltage profile while maintaining branch current constraints.

Chapter 19 discusses various control system techniques for the primary and secondary control of microgrids. Active and reactive power-sharing among the distributed energy resources (DERs) are illustrated under the primary control along with voltage and frequency control, which are discussed under the secondary control. The small-signal stability of microgrids is also detailed focusing on the states of the system, which have a dominating factor in making the system unstable.

Section 4 comprises three chapters on power quality, the economic incentive for renewable energy expansion planning, and the smart grid. Chapter 20 presents a review of power quality issues of grid-connected renewable energy systems and mitigation techniques using active power filters (APFs) and flexible AC transmission systems (FACTS). In Chapter 21, economic incentive schemes for improving the effective utilization of renewable energy sources in composite expansion planning are evaluated. The optimal expansion planning strategy is achieved by minimizing the investment cost for the renewable energy sources, minimizing the operational cost and emissions of the fossil fuel generators, as well as maximizing the incentive schemes while satisfying constraints. This problem is solved using advanced interactive multidimensional modelling system (AIMMS) and it is tested on IEEE 24-bus and real-world data of the Nigerian power system.

A coordinated dispatch of both supply and demand resources is presented in Chapter 22. The mathematical model of the operational dynamics of a prosumer smart grid is developed and the problem of power balance as a constrained optimization problem is formulated using Ontario's grid network data. The problem is solved using AIMMS that minimizes the operating cost of the grid and demand dispatch curtails the adverse effects of the randomness of renewable energy generation.

We are very grateful to a number of individuals who have made contributions to this book. In particular, we would like to appreciate all the authors for their contributions and reviewers for reviewing the book chapters and thus improving the quality of this handbook. Editors would also like to thank their university authorities and staff members for maintaining a cordial atmosphere and providing the facilities for completion of the book. We would also like to thank the World Scientific Publishing Europe specially to mention Amanda Yun, Britta Ramaraj, and Ramya Vengaiyan for being very supportive and timely bringing out this handbook. Finally, we would like to express our gratitude and sincere regard to our family members who have provided us great support during the preparation of this handbook.

Ramesh C. Bansal
Ahmed F. Zobaa

About the Editors

Ramesh Bansal has more than 25 years of diversified experience in research, scholarship of teaching and learning, accreditation, industrial, and academic leadership in several countries. Currently, he is a professor in the Department of Electrical Engineering at University of Sharjah, UAE, and an Extraordinary professor in the Department of Electrical, Electronic and Computer Engineering at University of Pretoria (UP), South Africa. Previously, he was professor and group head (Power) in the Department of Electrical and Computer Engineering at University of Pretoria (UP), South Africa. Prior to his appointment at UP, he was employed by the University of Queensland, Australia; University of the South Pacific, Fiji; BITS Pilani, India; and Civil Construction Wing, All India Radio.

Prof. Bansal has significant experience in collaborating with industry and government organizations. These utilities include NTPC (a 60 GW Indian power generation company), Powerlink, and ESKOM. During sabbaticals, he has worked with Powerlink, an Australian government-owned corporation responsible for Queensland's high-voltage electricity transmission network. He has made significant contributions to the development and delivery of BS and ME programs for utilities. He has extensive experience in the design and delivery of CPD programs for professional engineers. He has carried out research and consultancy and attracted significant funding from industry and government organizations.

Professor Bansal has published over 325 journal articles, presented papers at conferences, and wrote books and chapters in books. He has Google citations of over 12000 and an h-index of 50. He has supervised 25 PhDs, 4 postdocs, and is currently supervising several PhD students. His diversified research interests are in the areas of renewable energy (wind, PV, DG, microgrid) and smart grids. Professor Bansal is an editor of several highly regarded journals: *IET-RPG, IEEE Systems Journal, Electric Power Components and Systems,* and *Technology and Economics of Smart Grids and Sustainable Energy*. He is a fellow and chartered engineer at IET-UK; a fellow of Institution of Engineers (India); and senior member of IEEE-USA.

Ahmed Faheem Zobaa received his BSc (Hons), MSc, and PhD degrees in Electrical Power & Machines from Cairo University, Egypt, in 1992, 1997, and 2002, respectively. He received his Doctoral of Science from Brunel University, UK in 2017. He is currently a Reader in electrical and power engineering, an MSc Course Director and a Full Member of the Institute of Energy Futures at Brunel University, UK. His main areas of expertise include power quality (marine) renewable energy, smart grids, energy efficiency, and lighting applications.

Dr Zobaa is an Executive Editor for *International Journal of Renewable Energy Technology*, an Executive Editor-in-Chief for *Technology and Economics of Smart Grids and Sustainable Energy*, and an Editor-in-Chief for *International Journal of Electrical Engineering Education*. He is a registered Chartered Engineer, Chartered Energy Engineer, European Engineer, and International Professional Engineer. He is also a registered member of the Engineering Council, UK; the Egypt Syndicate of Engineers; and the Egyptian Society of Engineers. He is a Senior Fellow of Higher Education Academy, UK; Fellow of the Institution of Engineering and Technology, Energy Institute, UK, Chartered Institution of Building Services Engineers, UK, Institution of Mechanical Engineers, UK, The Royal Society of Arts, UK, The African Academy of Sciences, and Chartered Institute of Educational Assessors, UK. Finally, he is a senior member of IEEE.

Section I: Wind Power Systems

CHAPTER 1

Steady-State Modeling of DFIG-Based Wind Energy System for Unbalanced Operation

T. Datta[1] and P. Bajpai[2]

[1]Department of Electrical Engineering, Techno India University, West Bengal, India. ayanatdatta@gmail.com
[2]Department of Electrical Engineering, Indian Institute of Technology Kharagpur, Kharagpur, India. pbajpai@ee.iitkgp.ernet.in

Abstract Steady-state performance analysis of DFIG-based wind energy system requires accurate mathematical representation of the induction generator considering its operational limits and network unbalance. As distribution systems, generally, have on-load tap changers for voltage regulation, the choice of power-flow algorithm is governed by its performance in presence of these regulators. This chapter compares the efficiency of different load-flow algorithms available in literature in terms of its computational requirement and convergence characteristics. Also the current state of research carried out to establish a steady-state model for DFIG is discussed here. Although some algorithms can serve to initialize the dynamic equations, steady-state representation requires a proper strategy for application of operating limits in the mathematical model with representation of change in operating point. Further simultaneous solution of steady-state operating point for DFIG with network power flow is necessary in order to eliminate the necessity of an estimate. This chapter compares two steady-state mathematical models in terms of their performance for different operating conditions of the DFIG. The performance of these algorithms in normal operating conditions as well as in cases of violation of limit is evaluated here.

Keywords: DFIG, Operating limits, Power flow analysis, Voltage regulator, Wind generation, Unbalanced distribution system.

1.1 Introduction

Depletion of conventional nonrenewable energy sources and detrimental effects of greenhouse gases produced during generation of electricity from such sources has shifted the focus toward increased utilization of renewable sources. Wind is one such renewable energy source that is being efficiently tapped. The global total installed capacity for wind-power generation has reached 487 GW in the year 2016 [1]. Innovations in power electronics technology have ensured maximum power extraction by wind turbines using conversion technologies such as the Doubly Fed Induction Generator (DFIG) [2]. The DFIG systems employ back-to-back power converters of only one-third of the generator rating, that is, the Rotor Side Converter (RSC) and the Grid Side Converter (GSC), which connect the rotor to the grid. Thus, the investment in power electronics, the loss incurred through switching of converters as well as the harmonic distortion introduced due to switching, is much less in case of DFIG-based wind-conversion system [3]. The back-to-back converters are capable of controlling the reactive power injection through stator and GSC, respectively [4]. DFIG-based wind-power units have a share of over 50% of the total generation capacity all over the world [5].

Use of DFIGs for stand-alone systems as well as small units serving as distributed resource for the modern distribution system raises concern over second harmonic oscillations in electromagnetic torque caused by the unbalanced distribution network [6]. Increasing installations of DFIG-based wind generators in the existing distribution system has led to rigorous research in the area of control of DFIG to mitigate the detrimental effects of an unbalanced system voltage [7, 8]. These oscillations not only affect the DFIG operation by overheating the machine but also reduce the life of moving parts, drawing increased cost on the maintenance of the project [9].

Power-flow analysis serves as a useful tool in establishing the steady-state response of the DFIG-based wind-generation unit. Moreover, it provides an initial state for studying the dynamic response of the system. Unbalance in the distribution system is high due to the presence of un-transposed lines, unequally distributed loads, and small single-phase distributed resources. Thus, in order to study the effect of this unbalance factor on the operation of the DFIG, a three-phase analysis becomes necessary. The DFIG may operate under restricted conditions due to overloading or large unbalance in the distribution system. Because the DFIG can reach its operating limits, it becomes necessary to monitor the internal parameters of the DFIG.

Power-flow studies with DG units are generally conducted considering them as constant active and reactive power sources (PQ mode) or constant active power and terminal voltage sources (PV mode) [10]. The steady-state representation of DFIG in power flow is also conducted in PQ or PV modes [11, 12]. As also for islanded micro-grids, simple representation in PQ or PV mode with active power updated according to droop is practiced [13]. Steady-state calculations of DFIG internal states as in [14] for reduced model and [15] for full-order model, which use direct methods for solution of a set of linear equations and one quadratic equation, have the prerequisite of a converged power flow and operation at a fixed slip. This is only possible if:

- Power-flow analysis of the rest of the system is carried out previously with an estimate of the active power output.
- Fixed slip operation is considered and reaching of limit by positive-sequence component of any state is not considered.

However, the maximum amount of power extracted by the DFIG for a particular value of wind speed (it can be calculated from the turbine-operating characteristic [16]) is equal to the active power available at the DFIG terminal and internal losses (copper losses and switching plus conduction losses in the converters and the grid-side transformer). Internal losses are not known prior to power-flow analysis. Thus, representation as in [11] and [12] also needs approximate estimates of active power output. The three-phase terminal voltage and current injections at the terminal of DFIG become unknown quantities if the power flow is to be solved simultaneously with the steady-state calculations of DFIG and to eliminate the necessity of an erroneous estimate. Thus, the previously considered set of linear equations and one quadratic equation becomes a set of mostly quadratic equations. An added disadvantage of nonlinearity is attributed to steady-state calculations of internal parameters if slip is taken to be unknown on consideration of nonoptimal cases of operation.

In [17] and [18], the authors have proposed a mathematical model for representing the steady-state operation of DFIG, considering power extracted from wind turbine, and in [17] have also discussed cases where the limit on the stator current is violated. But operation in unbalanced condition has not been considered. Also, the effect of violation of limits, for positive-sequence components, on DFIG rotor speed is not represented. In [11, 12, 19–21], unbalanced operation with negative-sequence controller action is modeled. Although [20, 21] do not consider limits on DFIG states, both [12] and [19] have modeled the effect of these limits. However, variation of slip on violation of limit is not considered in any of the literature mentioned above. Although this may serve to initialize dynamic simulation, it does not provide the steady-state system states for the given conditions.

Moreover, the strategy used to represent violation of limit while solving these algorithms affect their convergence characteristics as well as accuracy, for example, in [12], the control objectives of positive- and negative-sequence controllers of RSC and GSC are readjusted on violation of a limit and fixed at those values until the next case of violation occurs. This strategy of treating the cases of violation of limits may result in pessimistic values, as the parameters may deviate from the limit in the following iterations. However, enforcement of any limit after convergence will increase the number of iterations required for convergence.

The steady-state model of DFIG can be used for initializing dynamic simulations, as mathematical model in planning algorithms such as DFIG sizing, placement and network reconfiguration, or steady-state performance analysis such as voltage-stability, eigenvalue analysis.

Steady-state analysis of DFIG-based wind generator requires accurate representation in load-flow studies. A comparative study of the various methods, used for representation of unbalanced operation of DFIG-based wind generator, has been presented in this chapter. Section 1.2 highlights the different load-flow algorithms used for unbalanced distribution systems. This section also compares the efficiency of the load-flow algorithms in terms of convergence and computational burden. Section 1.3 presents the mathematical models for representing DFIG in the load-flow algorithms discussed in Section 1.2. Moreover, the strategy used for representing the violation of limit while solving these algorithms is also discussed as it affects their convergence characteristics as well as accuracy. Section 1.4 compares the models presented in Section 1.3 using different case studies conducted on IEEE 123-bus distribution system. Finally, Section 1.5 summarizes the content presented in this chapter.

1.2 Load-Flow Algorithms for Unbalanced Distribution Systems with Voltage Regulators

The voltage profile in a distribution network is generally regulated using On-Load Tap Changers (OLTCs) and capacitor banks. OLTC-based voltage regulators are generally designed to either maintain the voltage at a particular node or provide line load compensation. Load-flow algorithms for unbalanced distribution system should have the ability to mathematically represent the effect of these voltage regulators. In this section we will briefly discuss the various algorithms used for unbalanced load-flow analysis, their strategy for representing voltage regulators, and its effect on the convergence of these algorithms. The distribution systems are generally radial in nature. Linear power-flow algorithms provide additional advantage of less computational time in comparison to nonlinear methods. Thus, computational time for [22] is much more than that for semi-decoupled algorithm of [23], and linear algorithms of [24–26]. However, distribution systems also means the presence of unbalanced loads and feeders and OLTCs for load compensation or voltage regulation. All linear and nonlinear methods are not capable of handling multiple three-phase OLTCs without affecting the convergence characteristics of the algorithm.

1.2.1 Newton–Raphson

Newton–Raphson (NR)-based three-phase unbalanced load flow [22] is used to solve the equations for active (P_{im}) and reactive power injection (Q_{im}) of all the phases "m" of each bus "i" in terms of the state variables, that is, the real ($\Re$) and imaginary ($\Im$) components of voltage (V_{im}) at these phases of each bus. Thus, the matrix equation for solution of NR algorithm is given by (1.1) for all buses "i" and phases "m".

$$\begin{bmatrix} \Delta P_{im} \\ \Delta Q_{im} \end{bmatrix} = [J]^{-1} \begin{bmatrix} \Delta \Re(V_{im}) \\ \Delta \Im(V_{im}) \end{bmatrix} \tag{1.1}$$

The Jacobian matrix, J, comprises the partial derivatives of active and reactive power injection of all phases of each bus in terms of the state variables. The algorithm is iterative in nature due to the nonlinearity of the power-flow equations. Convergence is achieved when the precision of the correction vector goes below a target value. This algorithm is mainly used for meshed transmission systems. The equivalent circuit representation of three-phase transformers for regulators as given in [27] is utilized to obtain the equivalent admittance matrix as in (1.2), where V_p, V_q, I_p, I_q are the three-phase voltage and current vectors at the two terminals "p" and "q" of the voltage regulator and all admittance matrices, that is, Y_{pp}, Y_{pq}, Y_{qp}, Y_{qq}, are functions of off-nominal tap-ratio a_n (n = a, b, c, ab, bc, ca as per the voltage regulator objective and design).

$$\begin{bmatrix} I_p \\ I_q \end{bmatrix} = \begin{bmatrix} Y_{pp} & Y_{pq} \\ Y_{qp} & Y_{qq} \end{bmatrix} \begin{bmatrix} V_p \\ V_q \end{bmatrix} \tag{1.2}$$

The off-nominal tap-ratio, a_n, can be written in terms of integer tap number Tap_n as $1/(1 + Tap_n V_{step})$, where V_{step} is the change in voltage per regulator tap (5/8% of set voltage) [28]. The objective of the line-compensator regulators or the voltage setting vector V_{set} can be represented as in (1.3), where Z_{set} is a matrix created using the R and X settings of the OLTC [29].

$$V_{set} = V_q + Z_{set} I_q \tag{1.3}$$

The error vector incorporated in NR load flow is given as $\Delta|V_{setn}|^2$ (n = a, b, c, ab, bc, ca as per the voltage regulator objective and design). This error vector is incorporated with the main error vector for active and reactive power injection at each phase of each node, if at a particular iteration, the voltage regulator setting goes beyond the allowable bandwidth. V_{setn} is expressed in terms of Tap_n, which is the state variable to be calculated, using (1.3). Thus, the Jacobian matrix for load-flow calculation incorporates the partial derivatives corresponding to the voltage regulator. As a result, the system state variables, that is, magnitude of voltage and phase angle, are calculated simultaneously along with the state variables for the voltage regulator, that is, the integral tap positions. The system admittance matrix is updated in each iteration as per the tap position number.

1.2.2 Backward and Forward Sweep

The Backward and Forward Sweep (BF) Algorithm [24] was developed to utilize the radial nature of distribution systems. Thus this algorithm reduces the computational requirement for load-flow calculations for the systems of radial or weakly meshed nature. The backward sweep of the algorithm is an application of Kirchhoff's Current Law (KCL) at each bus starting from the terminal nodes toward the substation node of the distribution feeder. This sweep requires the calculation of current injected by the loads using the initialized voltage profile or the voltages obtained from forward sweep of previous iteration. The backward sweep results in obtaining the branch current of all branches of the network.

Forward sweep is then carried out by applying Kirchhoff's Voltage Law (KVL) for all branches starting from the substation node where magnitude and angle of voltage are known. This forward sweep uses the branch voltage drops calculated using the branch currents obtained from backward sweep to obtain the magnitude and phase angle of voltages at each phase of each node, that is, the state variables. BF algorithm is iterative in nature for systems with constant power loads. The iteration continues till the difference between state variables of successive iterations go below a target precision. Weakly meshed systems are treated in BF algorithm as pseudo radial systems with connecting links cut open to achieve the radial nature.

Voltage regulator taps are calculated from the R and X setting for line compensation-based OLTCs during the forward sweep. Thus, OLTCs located at upstream nodes get updated earlier and influence the calculation of taps of OLTCs downstream. Thus, effectively simultaneous solutions for regulator tap positions are obtained along with solution for system state variables.

1.2.3 Sequential Load Flow

Sequential load flow (SQ) [23] attempts to decrease the computational requirement of NR three-phase load flow. The voltages of positive-sequence network is obtained as in NR algorithm using (1.1), utilizing a Jacobian matrix 1/9 times the size of the one used in NR algorithm. The negative-sequence and zero-sequence current injections ($I_{0/2}$) are calculated from the unbalanced load and initial values or values obtained from the previous iteration of negative-sequence and zero-sequence voltages. Then magnitude and phase angles of zero-sequence and negative-sequence voltages ($V_{0/2}$) for the current iteration is obtained by solving $V_{0/2} = Y_{0/2}^{-1} I_{0/2}$, where the zero-sequence and negative-sequence admittance matrices ($Y_{0/2}$) have a size of almost 1/9 times the size of Jacobian matrix for NR algorithm. Also, in the absence of voltage regulators, $Y_{0/2}$ is constant throughout the solution procedure and hence inversion operation is necessary only once. State variables for single-phase and two-phase feeders need to be solved for using BF algorithm.

Representation of unbalanced voltage regulator is very tricky in this type of load-flow algorithms where each sequence network is treated separately. Hence, simultaneous solution is not possible. The tap positions are updated separately for each OLTC based on the compensation or control objective after state variables are obtained for each iteration.

1.2.4 Case Study: Comparison of Computational Requirement and Convergence

The standard IEEE test distribution systems available in [30] have OLTC or voltage regulators of various configurations and controller objectives. The effect of different representations of voltage regulator in different algorithms in the presence of single and multiple voltage regulators are studied in this section.

1.2.4.1 Single Voltage Regulator

The state variables of the standard IEEE 13-bus system are calculated using NR, BF, and SQ-based power-flow algorithms. All the algorithms converge to a solution for a target precision of 10^{-4}. Fig. 1.1(a) depicts execution time required for calculating the state variables for each iteration, whereas Fig. 1.1(b) represents the convergence of the algorithms in terms of the precision achieved following each iteration. It can be clearly observed from Fig. 1.1(a) that NR-based three-phase load-flow algorithm requires more time mainly due to inversion of Jacobian matrix, which is nine times the size of the three matrices inverted in SQ-based power flow. BF-based methods have no requirement for matrix inversion and solution is achieved by using simple KCL and KVL. Hence it takes the minimum time for calculation per iteration.

Faster convergence of NR-based algorithm is observed (Fig. 1.1(b)) as the solutions converge to target precision within three iterations. BF load-flow algorithm has a low convergence rate near target precision taking seven iterations to converge as can be observed from the magnified inset. Thus, the number of iterations required for convergence can be greatly reduced by degrading the target precision. SQ load-flow employs NR-based inversion of Jacobian matrix for positive-sequence circuit and BF-based calculation for single-phase and two-phase feeders. Thus, the convergence rate of SQ-based algorithm falls in between NR and BF algorithms (six iterations for convergence).

The tap positions calculated by NR, BF, and SQ algorithms for the three phases of the voltage regulator (RG) located at bus 650 of IEEE 13-bus system is displayed in Fig. 1.2. All three algorithms converge to the same tap position. The sudden changes in precision (Fig. 1.1(b)) are coincident with the changes in tap position (Fig. 1.2).

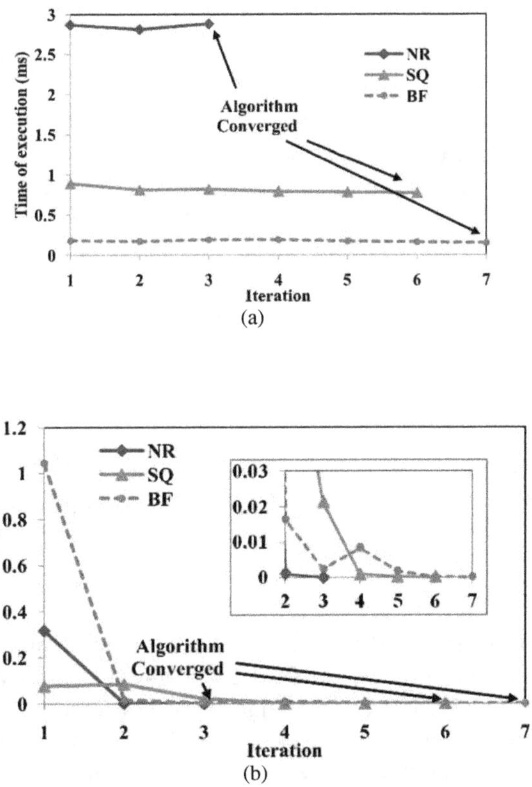

Figure 1.1. IEEE 13-bus system with OLTC (RG) at bus 650: (a) Time for execution per iteration in ms and (b) precision per iteration in pu.

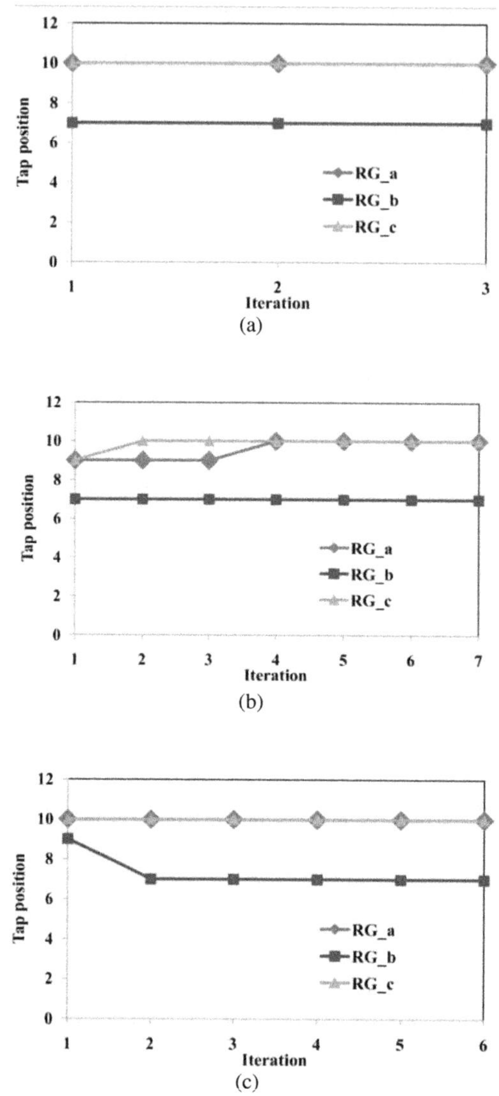

Figure 1.2. Calculated tap positions of each phase of OLTC (RG) at bus 650 of IEEE 13-bus system: (a) NR Algorithm; (b) BF Algorithm; and (c) SQ Algorithm.

1.2.4.2 Multiple Voltage Regulator

The state variables of the standard IEEE 34-bus system are calculated using NR, BF, and SQ-based power-flow algorithms. The IEEE 34-bus system has two OLTCs (RG1 and RG2) at buses 814 and 852. NR and BF algorithms converge to a solution for a target precision of 10^{-4}, while SQ algorithm diverges. Fig. 1.3(a) depicts execution time required for calculating the state variables for each Iteration, while Fig. 1.3(b) represents the convergence of the algorithms in terms of precision achieved following each iteration. As stated earlier, Fig. 1.3(a) highlights the fact that NR-based three-phase load-flow algorithm

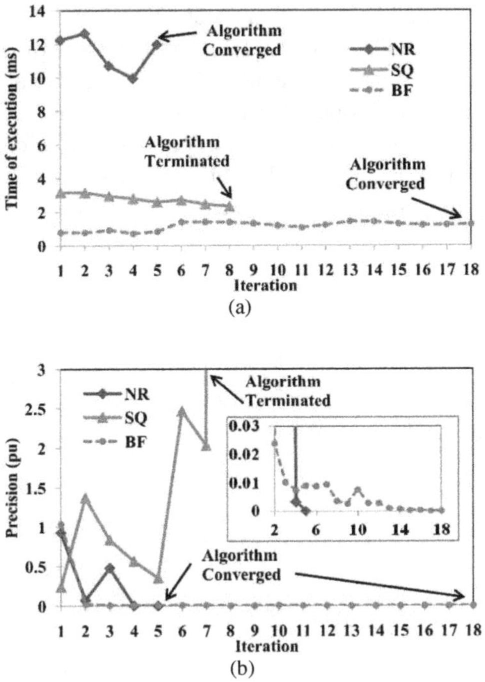

Figure 1.3. IEEE 34-bus system with two OLTCs (RG1 and RG2) at buses 814 and 852: (a) Time for execution per iteration in ms and (b) precision per iteration in pu.

requires more time per iteration as compared to SQ algorithm due to the inversion of Jacobian matrix whose size increases with the system size. Thus, time required for NR algorithm is more for 34-bus system as compared to 13-bus system (Fig. 1.1(a)). BF-based method takes the minimum time for calculation per iteration.

Faster convergence of NR-based algorithm is depicted in Fig. 1.3(b) as the solutions converge to target precision within five iterations. The low convergence rate of BF algorithm near target precision is depicted by the small changes in the precision in subsequent iterations taking 18 iterations to converge. This can be observed from the magnified inset of Fig. 1.3(b). The number of iterations required for convergence can be reduced by seven iterations if the target precision is degraded. SQ load flow solves for OLTCs after a solution for voltage magnitude and phase angle is obtained in each iteration. Adjustment of tap position in OLTCs one after the other leads to oscillation of tap changes between the two OLTCs. This leads to increasing precision and the algorithm is terminated following the blowing up of state variables as the precision becomes higher than a threshold. Solution of single OLTC for some iterations succeeded by solution of the other regulator yields similar case of divergence. This observance asserts the fact that the simultaneous solution for voltage regulator tap position is necessary as in NR and BF algorithms. Thus, SQ-based algorithm diverges and is terminated after eight iterations. If the algorithm is not terminated due to high value of adjustment to be made (precision), the solution is seen to oscillate till the maximum number of iterations.

The tap positions calculated by NR, BF, and SQ algorithms for the three phases of the voltage regulators (RG1 and RG2) located at buses 814 and 852 of IEEE 34-bus system are displayed in Fig. 1.4. The tap positions calculated by NR and BF algorithms converge to a similar solution. The difference of one integral tap position for RG2_a can be ignored due to the bandwidth allowed for voltage regulation as per data given for the IEEE test systems. The tap positions calculated by the SQ algorithm are seen to oscillate as stated earlier. The sudden changes in precision (Fig. 1.3(b)) are coincident with changes in tap position (Fig. 1.4).

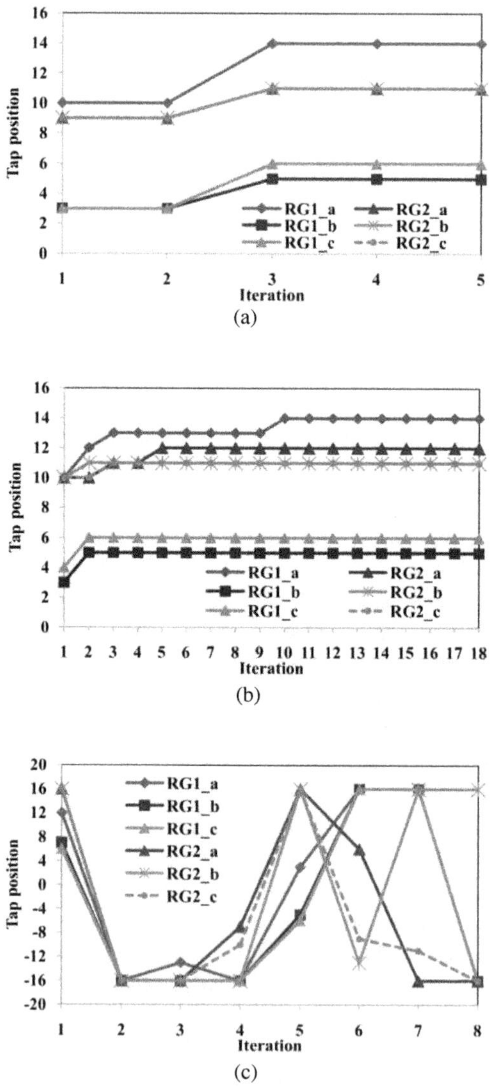

Figure 1.4. Calculated tap positions of each phase of the two OLTCs (RG1 and RG2) at buses 814 and 852 of IEEE 34-bus system: (a) NR Algorithm; (b) BF Algorithm; and (c) SQ Algorithm.

1.3 Steady-State Representation of Type-III Wind Energy System

1.3.1 Representation of Wind Turbine

The power extracted by the wind turbine is given by [31]

$$P_t = 0.5\rho\pi C_p\left(\lambda, \beta_w\right) R^2 V_w^3 \tag{1.4}$$

where ρ is the density of air, R is the radius of the turbine, and V_w is the wind speed. The power coefficient C_p is a function of tip-speed ratio λ and pitch angle β_w, where λ is given by (1.5).

$$\lambda = \frac{\omega_r R}{V_w Gear\, ratio} \tag{1.5}$$

where ω_r is the angular speed of the rotor. Manufacturers of wind turbine generally provide an empirical formula for the power coefficient of the turbine as functions of pitch and tip-speed ratio. One such sample of empirical formula used to express the performance coefficient *Cp* for the wind turbine in the studies carried out in this chapter is given by (1.6) as in [32].

$$\begin{aligned} C_p\left(\lambda, \beta_w\right) = c_1 & \left(\frac{c_2}{\lambda + c_8\beta_w} - \frac{c_2 c_9}{\beta_w^3 + 1} - c_3\beta_w - c_4\beta_w^{c_5} - c_6\right) \\ & * exp\left(-\frac{c_7}{\lambda + c_8\beta_w} + \frac{c_7 c_9}{\beta_w^3 + 1}\right) + c_{10}\lambda \end{aligned} \tag{1.6}$$

where the values of constants c_1–c_{10} are summarized in Table 1.1.

1.3.2 DFIG Sequence Model

A type-III wind generator (DFIG) is capable of operating as an active/reactive power source (PQ mode: generally when connected to grid) and also as a voltage-controlled power source (PV mode: when islanded with a small system) [33]. The DFIG is modeled in sequence components for calculating the internal parameters as unbalanced control of DFIG is done in sequence d-q reference frame so that a decoupled control scheme can be implemented [34]. The DFIG windings are generally connected in delta or ungrounded star, preventing the flow of zero-sequence currents [35]. Hence, zero-sequence components of DFIG parameters are not considered. The converter conduction and switching losses are accounted for by using a resistance in series with the back-to-back converters as in [36]. Mathematical equations are expressed using the equivalent sequence component circuit of DFIG (Fig. 1.5) as given in [37] based on the different control objectives of RSC and GSC. Here R_p and X_p are the resistance and reactance of winding respectively where subscript p = r (rotor) or p = s (stator) or p = g (transformer interfacing Grid Side Converter i.e. GSC).

Table 1.1. Constants for calculation of performance coefficient of wind turbine.

c_1	c_2	c_3	c_4	c_5	c_6	c_7	c_8	c_9	c_{10}
0.22	116	0.4	0	0	5	12.5	0.08	0.035	0

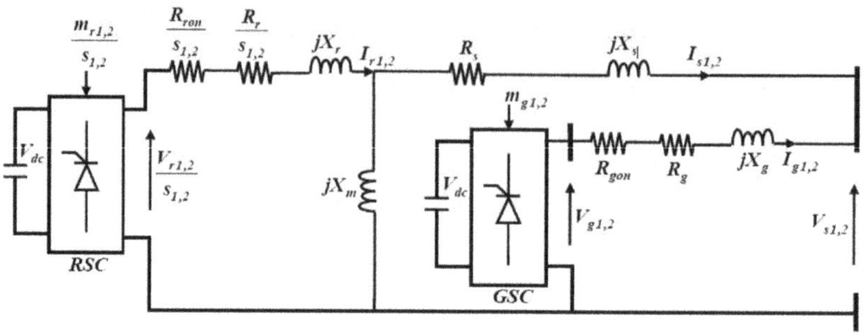

Figure 1.5. DFIG equivalent circuit representation in sequence components.

R_{pon} is the equivalent resistance to represent switching and conduction losses in converters p = r (RSC) or p = g (GSC). I_{pn}, V_{pn} represents positive/negative (subscript n = 1/2) sequence components of current, voltage respectively for subcript p = r (rotor) or p = s (stator) or p = g (Grid Side Converter i.e. GSC). m_{pn} is the modulation index of converter terminal-voltage where p = r (RSC) or p = g (GSC). The positive/negative-sequence slips are represented by $s_n$where n = 1/2 respectively.

1.3.3 Operating Limits

The rotor of the DFIG is connected via RSC and GSC to the grid. Thus, apart from the rating of the induction generator, the rating of these converters will also affect the overall operating limit of the DFIG. The maximum value of phase modulation index is 1 for Sinusoidal Pulse Width Modulation (SPWM) and 1.1547 for balanced Space Vector Pulse Width Modulation (SVPWM). Voltage unbalance factor (VUF) is defined as the ratio of the negative-sequence voltage component to the positive-sequence voltage component [38]. Thus, the limits on positive- and negative-sequence voltages produced at the converter terminals is calculated based on the choice of maximum allowable unbalance factor at the rated condition. In per-unit system, the maximum value of phase modulation index for RSC is set at 0.5 for SPWM and 0.5773 for SVPWM. A VUF of 6.4% and 25% is chosen for the GSC and RSC, respectively, for the study carried out in this chapter.

The maximum magnitude of current in each phase is limited by the converter rating for the grid side as well as the current rating of the machine itself. Generally, the GSC and RSC rating is about 30%–40% of the DFIG rating. Nowadays, DFIG-based distributed generators are designed to operate in fault-ride-through mode. Hence, the equipments are designed to withstand this condition. The limiting value of maximum current supplied by the converters play a major role in deciding the contributed fault current from the machine. The phase current magnitude limit is fixed at about 20% overload on rated value for the study carried out in this chapter. Because in many cases the negative-sequence controller of the GSC is kept inactive, large load unbalances in the system result in high negative-sequence current.

The stator current rating depends on the designed rated speed of the DFIG rotor. The stator as a result carries about 70%–80% of the total active power output of the DFIG. The phase current magnitude limit is fixed at about 20% overload on rated value. The unbalance factor is limited to 10%. All the limits imposed are actually on phase currents. Thus, the boundary values on sequence components are calculated based on the worst-case assumption of collinear sequence components [39].

1.3.4 Representation with NR Load-Flow Algorithm

The rest of the three-phase distribution system is expressed using phasor components, as NR in phasor components has better convergence. Also, in presence of OLTCs (On Load Tap Changer) in the distribution network, NR in phasor components is superior to other algorithms. This is because equations representing network power flow and objectives of control for OLTCs are simultaneously solved using the Jacobian matrix.

1.3.4.1 Representation of RSC Control Strategy

The RSC is generally controlled with the ultimate objective to extract maximum power available from the wind turbine, which is dependent on the turbine efficiency, while operating within limits [32]. The total active power output (positive and negative sequence of both rotor and stator) of the wind generator in addition to its copper losses and the converter switching plus conduction losses is equal to the maximum power available from wind turbine (P_{t_max}) as given in (1.7).

$$\begin{aligned}&\Re\left[V_{s1}\left(I_{s1}+I_{g1}\right)^{*}+V_{s2}\left(I_{s2}+I_{g2}\right)^{*}\right]+\left(|I_{r1}|^{2}+|I_{r2}|^{2}\right)\left(R_{r}+R_{ron}\right)\\&+\left(|I_{s1}|^{2}+|I_{s2}|^{2}\right)R_{s}+\left(|I_{g1}|^{2}+|I_{g2}|^{2}\right)\left(R_{g}+R_{gon}\right)=P_{t\_max}\end{aligned} \tag{1.7}$$

In earlier published works [11, 12], the copper losses and the negative-sequence active power injections are either neglected or an approximate estimate is required to specify the net active power output. However, like in [17], (1.7) incorporates losses without the necessity of any estimate. The maximum output extracted from the wind turbine can be accurately calculated using the turbine characteristic and current speed of wind. This fulfills the first objective of this work and hence improves the accuracy of results. The positive-sequence stator reactive power injection is controlled for the PQ (constant power) mode of operation, whereas the positive-sequence stator voltage is controlled for PV (constant voltage) mode of operation. PQ and PV modes of operation are represented by (1.8) and (1.9).

$$\Im\left[V_{s1}I_{s1}^{*}\right]=Q_{s\_sp_{1}} \tag{1.8}$$

$$|V_{s1}|=V_{s\_sp_{1}} \tag{1.9}$$

where Q_{s_sp1} and V_{s_sp1} are the specified values of positive-sequence stator reactive power injection and positive-sequence stator voltage respectively.

1.3.4.2 Representation of GSC Control Strategy

The GSC controller, usually, maintains the voltage of the DC-link between the back-to-back converters [32], thus balancing the average active power exchanged across the DC-link. The average active power flow through the DC-link has components due to both positive- and negative-sequence networks. The contribution of negative-sequence voltage and current in power balance is neglected in [12]. The loss in the DC-link is due to capacitor leakage conductance, which is very low during steady-state operation [32] and hence is neglected here.

In this model, expression of the active power balance across the back-to-back converters through the DC-link is given by (1.10).

$$\Re\left[V_{g1}I_{g1}^{*} + V_{g2}I_{g2}^{*} + V_{r1}I_{r1}^{*} + V_{r2}I_{r2}^{*}\right] = 0 \tag{1.10}$$

The positive-sequence component of reactive power injection is also controlled at its specified value (Q_{g_sp1}) through the GSC as presented in (1.11).

$$\Im\left[V_{s1}I_{g1}^{*}\right] = Q_{g\_sp_1} \tag{1.11}$$

The GSC is generally operated at unity power factor to reduce its rating and hence the cost. The positive- and negative-sequence slips ($s_n$where n = 1/2) are given as in (1.12) and (1.13), respectively.

$$s_1 = \frac{\omega_s - \omega_r}{\omega_s} \tag{1.12}$$

$$s_2 = 2 - s_1 \tag{1.13}$$

where ω_s and ω_r are the synchronous angular speed of the stator field and the angular speed of the rotor respectively.

1.3.4.3 Power Balance at the DFIG Terminal

As stated earlier, the rest of the distribution system, including the coupling transformer, is represented in phasor components. Hence, (1.14) and (1.15) representing net injection of active (P_{DF_injm}) and reactive (Q_{DF_injm}) power at the DFIG terminal (ideally zero in the absence of any local load) are also included for all three phases.

$$P_{DF\_injm} = P_{PCCm} + \Re\left[V_{sm}\left(I_{sm} + I_{gm}\right)^{*}\right] \tag{1.14}$$

$$Q_{DF\_injm} = Q_{PCCm} + \Im\left[V_{sm}\left(I_{sm} + I_{gm}\right)^{*}\right] \tag{1.15}$$

where P_{PCCm} and Q_{PCCm} are net active and reactive power injected by the distribution network into *PCC* bus, *m* corresponds to the phases *a, b*, and *c* and *s* represents stator.

1.3.4.4 Representation of Negative Sequence Control Modes of RSC and GSC

The GSC and RSC, generally, inject negative-sequence current to achieve various control modes [34, 40]. They are represented in NR algorithm as:

- **Balanced rotor or stator currents via RSC**: The RSC is controlled to prevent the flow of negative-sequence currents in the stator or rotor [40], that is, unbalanced voltage is generated at the RSC terminal. The mathematical equation used to represent this mode of operation is

$$\Re\left[I_{r2,s2}\right] = 0 \quad \text{and} \quad \Im\left[I_{r2,s2}\right] = 0 \tag{1.16}$$

The GSC operates to allow or prevent negative-sequence current exchange with the grid. If the negative-sequence current exchange is prevented via the GSC, then

$$\Re\left[I_{r2,s2}\right] = 0 \quad \text{and} \quad \Im\left[I_{r2,s2}\right] = 0 \tag{1.17}$$

- **Balanced stator currents via GSC**: The GSC is controlled to prevent the flow of negative-sequence currents in the stator by adding a negative-sequence voltage component at its output [12]. The negative-sequence controller of the RSC is disabled, hence, rotor voltage remains balanced. For steady-state calculations this control mode can be represented using (1.16).
- **Reduction of double-frequency component of stator active power**: This component of stator power output can be removed by the control action of RSC [40] and is represented by (1.18) and (1.19). The GSC operation is similar to the control modes where rotor or stator currents were balanced via RSC, that is, it allows or prevents the flow of negative-sequence currents. Prevention of exchange of negative-sequence currents through GSC is represented by (1.17).

$$\Re\left[V_{s1}I_{s2}^{*} + V_{s2}I_{s1}^{*}\right] = 0 \tag{1.18}$$

$$\Im\left[V_{s1}I_{s2}^{*} - V_{s2}I_{s1}^{*}\right] = 0 \tag{1.19}$$

- **Reduction of double-frequency component of electromagnetic torque**: The RSC is controlled to remove stator reactive power double-frequency component while GSC control is similar to the previous mode. This negative-sequence control strategy can be expressed by

$$\Re\left[V_{s1}I_{s2}^{*} - V_{s2}I_{s1}^{*}\right] = 0 \tag{1.20}$$

$$\Im\left[V_{s1}I_{s2}^{*} + V_{s2}I_{s1}^{*}\right] = 0 \tag{1.21}$$

- **Coordinated control**: GSC and RSC are controlled to remove the double-frequency component of electromagnetic torque, as discussed earlier, as well as to remove the double-frequency components of the active power output of the DFIG through stator and GSC [37]. This mode can be modeled using (1.22) and (1.23) in addition to the ones used for representing control mode for mitigating double-frequency electromagnetic torque component, that is, (1.20) and (1.21).

$$\Re\left[V_{s1}I_{s2}^{*} + V_{s2}I_{s1}^{*} + V_{s1}I_{g2}^{*} + V_{s2}I_{g1}^{*}\right] = 0 \tag{1.22}$$

$$\Im\left[V_{s1}I_{s2}^{*} - V_{s2}I_{s1}^{*} + V_{s1}I_{g2}^{*} - V_{s2}I_{g1}^{*}\right] = 0 \tag{1.23}$$

Equations (1.7)–(1.23), corresponding to a particular operating mode, are included in NR algorithm. The error matrix Δf_{DFIG}, incorporated in the original NR algorithm, for PQ mode of operation of DFIG is given as

$$\left[\Delta P_{DF\_injm}\ \Delta Q_{DF\_injm}\ \Delta P_{t\_\max}\ \Delta Q_{s\_sp_1}\ \Delta P_{dc}\ \Delta Q_{g\_sp_1}\ \Delta Eq_1\ \Delta Eq_2\ \Delta Eq_3\ \Delta Eq_4\right]$$

where m = a, b, c. The error ΔQ_{s_sp1} is replaced by ΔV_{s_sp1} for PV mode of operation of the DFIG. Eq_1, Eq_2, Eq_3,and Eq_4 are optional and are meant to represent the different negative-sequence control modes. The NR load-flow is solved to calculate the voltage magnitude and phase angles for all phases of each bus of the distribution system, the voltage regulator tap-settings and the positive- and negative-sequence voltage magnitude $|V_{pn}|$ and angle θ_{pn} at the terminals of RSC ($p = r$) and GSC ($p = g$), that is, the state variables of the DFIG. Total fourteen state variables of DFIG, where m = a, b, c, as obtained from the NR algorithm-based mathematical model, are given as

$$\left[\Delta\theta_{sm}\ \Delta|V_{sm}|\ \Delta\theta_{r1}\ \Delta|V_{r1}|\ \Delta\theta_{g1}\ \Delta|V_{g1}|\ \Delta\theta_{r2}\ \Delta|V_{r2}|\ \Delta\theta_{g2}\ \Delta|V_{g2}|\right]$$

where $\Delta\theta_{r2}$, $\Delta|V_{r2}|$, $\Delta\theta_{g2}$, $\Delta|V_{g2}|$ are required, wherever necessary, for representing various negative-sequence control modes.

1.3.4.5 Limit Handling Strategy

This section highlights the second objective of this work. The flowchart (Fig. 1.6) depicts the steps for solving the NR algorithm in order to calculate the state variables of the DFIG. As shown in Fig. 1.6, parameters of DFIG are checked for cases of violation of limits. This

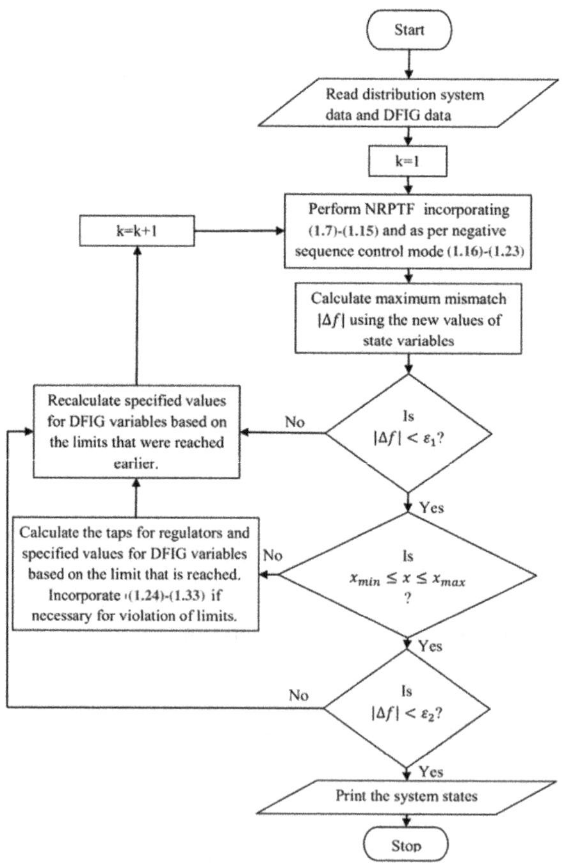

Figure 1.6. Flowchart of power-flow analysis.

condition checking is carried out only if the maximum value of mismatch $|\Delta f| < \varepsilon_1$. ε_1 is chosen such that the solution is near convergence. This strategy is adopted to prevent any unnecessary enforcement of limits. This approach also reduces the time for computation and the number of iterations that is required for convergence. The algorithm converges to a solution when $|\Delta f| < \varepsilon_2$. Here $\varepsilon_1 = 0.01$ and $\varepsilon_2 = 0.0001$. These values are adopted considering quadratic nature of the convergence characteristics of NR method. For all iterations, k, after a state reaches its limit, the specified values of the control objectives are recalculated. This procedure is adopted to avoid erroneous fixed values based on non-converged machine parameters of an intermediate iteration. Only one quantity, which satisfies the limits on all others, is used for objective readjustment, if several quantities reach their respective limits simultaneously.

1.3.4.6 Negative-sequence Limits

Violation of limits on negative-sequence state variables is taken care of in the following way:

- **Negative-sequence GSC modulation index limit m_{g2_max}**: If in the k-th iteration, this limit is surpassed, then it is fixed at its maximum value. The control of GSC is achieved by maintaining the $\Re/\Im$ components of I_{g2} at the values calculated from (1.24).

$$I_{g2}^{k+1} = \left[\frac{m_{g2\_max} K_{inv} V_{dc} \mathrm{V}_{g2}^{k}}{|\mathrm{V}_{g2}^{k}|} - \mathrm{V}_{s2}^{k}\right] / \left(R_g + R_{gon} + jX_g\right) \tag{1.24}$$

- **Negative-sequence GSC current magnitude limit $|I_{g2}|_{max}$**: If a case of violation of this limit is observed, then operation of the GSC controller is represented by maintaining the $\Re/\Im$ components of I_{g2} at the values calculated from (1.25).

$$I_{g2}^{k+1} = \frac{I_{g2}^{k}}{|I_{g2}^{k}|}|I_{g2}|_{max} \tag{1.25}$$

- **Negative-sequence RSC modulation index limit m_{r2_max}**: If in the k-th iteration, this limit is exceeded, then it is limited at its maximum value. The RSC control is expressed by maintaining the $\Re/\Im$ components of I_{s2} at the values calculated from (1.26).

$$I_{s2}^{k+1} = \frac{\dfrac{m_{r2\_max} K_{inv} V_{dc} V_{r2}^{k}}{s_2 |\mathrm{V}_{r2}^{k}|} - \mathrm{V}_{s2}^{k}\left(1 + \dfrac{\dfrac{R_r + R_{ron}}{s_2} + jX_r}{jX_m}\right)}{R_s + jX_s + \left(\dfrac{R_s + j(X_m + X_r)}{jX_m}\right)\left(\dfrac{R_r + R_{ron}}{s_2} + jX_r\right)} \tag{1.26}$$

- **Negative-sequence stator current magnitude limit $|I_{s2}|_{max}$**: If a case of violation of this limit is observed, then the operation of the RSC controller is represented by maintaining the $\Re/\Im$ components of I_{s2} at the values obtained from (1.27).

$$I_{s2}^{k+1} = \frac{I_{s2}^{k}}{|I_{s2}^{k}|}|I_{s2}|_{max} \tag{1.27}$$

1.3.4.7 Positive-sequence Limits

The parameters specified in (1.8) and (1.11) are replaced by readjusted values calculated from (1.30)–(1.33) and the state variables of current iteration are recalculated after the corresponding violation of limit of a positive-sequence component of system state variable. Equation (1.10) representing the balance of active power between the back-to-back converters, that is, across the DC-link, remains unchanged. The balance between the turbine output and the reduced active power output of the DFIG is achieved by changing the speed (slip) of the DFIG. This change is incorporated in the mathematical expressions by replacing P_{t_max} in (1.7) with P_t [32] (1.4). The general expression for $C_p(\lambda, \beta_w)$ is given in (1.6). Also, (1.28) and (1.29) specify the restricted value of positive-sequence active powers for stator (P_{s1}) and GSC (P_{g1}), respectively. Either (1.28) or (1.29) is needed to be solved as per the corresponding limit, which is violated, that is,

$$\Re\left[V_{s1}I_{s1}^*\right] = P_{s1} \tag{1.28}$$

$$\text{or} \quad \Re\left[V_{s1}I_{g1}^*\right] = P_{g1} \tag{1.29}$$

Slip becomes a calculated system state following this type of violation of limit. The additional equation (1.28) or (1.29) is necessary for obtaining the new state variable, that is, slip. During these adjustments, if the DFIG rotor speed ω_r reaches its maximum limit, that is, 30% above synchronous speed, then slip is fixed at the limit. Thus, λ is fixed at the value calculated for maximum value of ω_r, as λ is a function of ω_r (1.5). The output power at restricted condition with a constant λ is given by P_t from (1.4). P_{t_max} is replaced by P_t in (1.7) and is solved for the state variable β_w, that is, pitch angle of the wind turbine. This change, which is incorporated as power coefficient C_p, is a function of β_w in (1.4).

- **Positive-sequence GSC modulation index limit m_{g1_max}**: If in the k-th iteration, this type of violation occurs, then the value of the modulation index is fixed at the maximum value, and the restricted value of positive-sequence GSC current is found using (1.30). The limited values of active and reactive power injected by the GSC into the grid, that is, P_{g1}, Q_{g1}, are calculated using this readjusted value of current in (1.11) and (1.29).

$$I_{g1}^{k+1} = \left[\frac{m_{g1\_max}K_{inv}V_{dc}V_{g1}^k}{|V_{g1}^k|} - V_{s1}^k\right] / \left(R_g + R_{gon} + jX_g\right) \tag{1.30}$$

- **Positive-sequence GSC current magnitude limit $|I_{g1}|_{max}$**: The positive-sequence component of GSC current is set at its maximum value obtained from (1.31). P_{g1}, Q_{g1} are calculated using this readjusted value of current.

$$I_{g1}^{k+1} = \frac{I_{g1}^k}{|I_{g1}^k|}|I_{g1}|_{max}. \tag{1.31}$$

- **Positive-sequence RSC modulation index limit m_{r1_max}**: If this limit is exceeded, then the modulation index is restricted at the maximum value. Equation (1.32) is used to obtain the readjusted value of I_{s1} for PQ mode. This value is used to calculate reduced values of P_{s1}, Q_{s1} in (1.8) and (1.28).

$$I_{s1}^{k+1} = \frac{\dfrac{m_{r1\_max}K_{inv}V_{dc}V_{r1}^{k}}{s_1|V_{r1}^{k}|} - V_{s1}^{k}\left(1 + \dfrac{\dfrac{R_r + R_{ron}}{s_1} + jX_r}{jX_m}\right)}{R_s + jX_s + \left(\dfrac{R_s + j\left(X_m + X_r\right)}{jX_m}\right)\left(\dfrac{R_r + R_{ron}}{s_1} + jX_r\right)} \tag{1.32}$$

For PV mode of operation, the limiting value of voltage magnitude at stator terminal is calculated by rearranging (1.32) for implementing (1.9).

- **Positive-sequence stator current magnitude limit $|I_{s1}|_{max}$**: If this limit is violated, then the stator current is set at the maximum value obtained from (1.33). For PQ or PV modes of operation, the restricted values of P_{s1}, Q_{s1}, or $|V_{s1}|$ are recalculated similarly as in case of violation of limit on positive-sequence component of modulation index of RSC (given in this section).

$$I_{s1}^{k+1} = \frac{I_{s1}^{k}}{|I_{s1}^{k}|}|I_{s1}|_{max} \tag{1.33}$$

1.3.5 Representation with SQ Load-Flow Algorithm

The representation of DFIG-based wind energy system in SQ-based algorithm is carried out in [12]. The sequence model given in Figure 1.5 is utilized to represent the DFIG in three separate sequences.

1.3.5.1 Positive-sequence Network

As discussed earlier, the positive-sequence network is solved using Newton–Raphson-based Jacobian inversion method. [12] proposes to specify the DFIG positive-sequence active and reactive power ($Q_{s\_sp1} + Q_{g\_sp1}$) output at the DFIG terminal for PQ mode, or specify the positive-sequence terminal voltage magnitude (V_{s_sp1}) along with the active power output. This model is dependent on the availability of initial estimates for slip and active power output at stator terminal, P_{s_sp1}. Thus, the positive-sequence active power output at DFIG terminal is specified as $(1 - s)P_{s\_sp1}$. This representation ignores internal losses of DFIG and the contribution of negative-sequence voltage and current components to active power balance across DC-link because rotor power is approximated as $(-s)P_{s\_sp1}$.

1.3.5.2 Negative-sequence Network

The negative-sequence network model of DFIG (Fig. 1.5) along with the control strategies of the DFIG, which injects negative-sequence voltage at the terminals of RSC and GSC, are represented in terms of Norton's equivalent circuit, that is, a negative-sequence current source with a negative-sequence impedance in parallel to the source (Fig. 1.7) in [12]. As stated earlier, in SQ load-flow algorithm, negative-sequence network is solved as $V_2 = Y_2^{-1}I_2$. The Norton's equivalent current source adds to the vector I_2, whereas the equivalent impedance in parallel modifies Y_2 in every successive iteration.

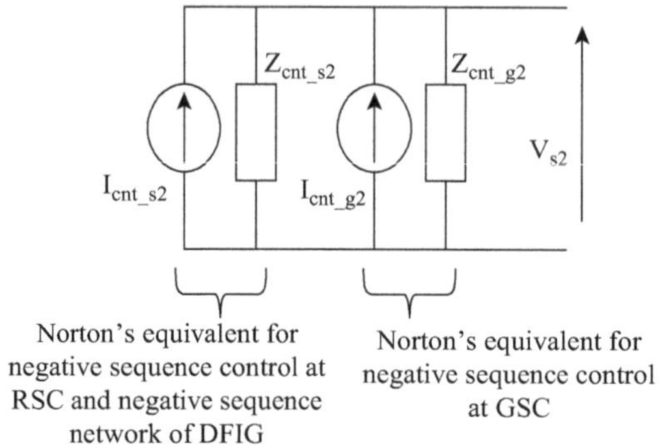

Figure 1.7. Norton's equivalent representation of negative-sequence circuit and control strategy of DFIG.

1.3.5.3 Negative-sequence Control Modes of RSC and GSC

The negative-sequence network and control modes are represented in SQ algorithm as discussed below:

- **Balanced rotor or stator currents via RSC**: The parameters of the approximated Norton's equivalent circuit used to represent this mode is given by (1.34) for balanced rotor current and (1.35) for balanced stator current.

$$Z_{cnt\_s2} = R_s + j\left(X_s + X_m\right); \quad I_{cnt\_s2} = 0 \tag{1.34}$$

$$Z_{cnt\_s2} = R_s + j(X_s + X_m); \quad I_{cnt\_s2} = V_{s2}/\left(R_s + j(X_s + X_m)\right) \tag{1.35}$$

If the GSC prevents negative-sequence current exchange with the grid then

$$I_{cnt\_g2} = 0 \quad \text{and} \quad Z_{cnt\_g2} = \infty, \quad \text{else} \quad Z_{cnt\_g2} = R_g + R_{gon} + jX_g \tag{1.36}$$

- **Balanced stator currents via GSC**: The parameters used to represent this mode is given by $Z_{cnt\_s2} = \infty$; $I_{cnt\_s2} = 0$; $Z_{cnt\_g2} = \infty$; I_{cnt_g2} = negative-sequence current injected into load and distribution network at DFIG terminal.
- **Reduction of double-frequency component of stator active power**: The parameters of Norton's equivalent circuit used to represent this mode are given by (1.37). The GSC operation is denoted by (1.36).

$$Z_{cnt\_s2} = R_s + j(X_s + X_m); \quad I_{cnt\_s2} = \frac{V_{s2}}{R_s + j(X_s + X_m)} - \frac{V_{s2}I_{s1}^*}{V_{s1}^*} \tag{1.37}$$

- **Reduction of double-frequency component of electromagnetic torque**: The parameters of Norton's equivalent circuit used to represent this mode are given by (1.38). The GSC operation is denoted by (1.36).

$$Z_{cnt\_s2} = R_s + j(X_s + X_m); \quad I_{cnt\_s2} = \frac{V_{s2}}{R_s + j(X_s + X_m)} + \frac{V_{s2}I_{s1}^*}{V_{s1}^*} \tag{1.38}$$

- **Coordinated control**: The parameters of Norton's equivalent circuit used to represent this mode are given by (1.38). The GSC operation is denoted by

$$Z_{cnt\_s2} = \infty; \quad I_{cnt\_s2} = -\frac{V_{s2}\left(I^*_{s1} + I^*_{g1}\right)}{V^*_{s1}} \tag{1.39}$$

1.3.5.4 Limit-handling Strategy: Positive Sequence

The algorithm checks for violation of any of the operating limits discussed earlier in each iteration. Violation of limit on positive-sequence modulation index of GSC or magnitude of positive-sequence current through GSC is treated by updating the values of $(-s)P_{s\_sp1}$ and Q_{g_sp1}, based on the restricted parameter, while keeping slip s constant in that particular iteration where violation is observed. Violation of limit on positive-sequence modulation index of RSC or magnitude of positive-sequence current through stator is treated by updating the values of P_{s_sp1} and Q_{s_sp1}, for PQ mode of operation and P_{s_sp1} and V_{s_sp1}, for PV mode of operation based on the restricted parameter, while keeping slip, s, constant in the iteration where violation is observed. In case of violation of multiple limits, only the minimum value that satisfies all operating limits is updated. Slip at which the machine operates is not changed on violation of any limit. The updated values of specified power serve as the target set-point in all successive iteration till another case of violation of limit is observed. Thus, the DFIG does not operate any more in MPPT mode.

1.3.5.5 Limit-handling Strategy: Negative Sequence

Violation of limit on negative-sequence modulation index of GSC or magnitude of negative-sequence current through GSC is treated by updating the value of I_{cnt_g2}, while violation of limit on negative-sequence modulation index of RSC or magnitude of negative-sequence current through stator is treated by updating the value of I_{cnt_s2} based on the restricted parameter in the iteration where violation is observed. In cases of violation of multiple limits only the minimum value that satisfies all operating limits is updated. As stated earlier, these updated values serve as target set-points in all succeeding iterations and the control objective are not achieved.

1.4 Comparison between Models Represented with NR and SQ Algorithms

The model represented in NR and SQ algorithms is tested on IEEE standard 13-, 34-, and 123-bus systems. Data of these unbalanced distribution systems are available in [30]. The data used here for the wind turbine and a 50-Hz, 4-pole DFIG set are summarized in Tables 1.2 and 1.3, respectively. The converter loss resistances for RSC and GSC are chosen as 0.0061 pu and 0.0087 pu. These values are obtained by average fitting of data available in [41]. The limit on the magnitude of phase currents is set at 20% overload on rated value. Voltage at the DC-link is taken as 2.20 times of peak phase voltage of GSC. The type of modulation adopted for the converters is SPWM for a DFIG unit of rating 1 MW, 0.69 kV.

The bi-factorization method [42], used for inverting a sparse matrix, is modified for block sub-matrices to take advantage of the block-sparsity structure of the Jacobian matrix of this three-phase power-flow technique. A precision of 10–4 is chosen for convergence.

Table 1.2. Parameters of wind turbine.

Parameter	Value	Parameter	Value
Rated wind speed	12.5 m/s	Cut-in wind speed	3 m/s
Rated tip speed	79.0625 m/s	R	24.64 m
ρ	1.225 kg/m3	Gear ratio	64
Maximum Cp	0.48	Optimal λ	8.1

Table 1.3. Parameters of DFIG system.

Parameter	Value	Parameter	Value
Rated Voltage	690 V	DFIG rating	1 MW
R_s	0.005 pu	R_r	0.0055 pu
L_m	4 pu	L_s	0.04 pu
L_r	0.0602 pu	Rated ω_r	1.3073 pu
L_g	0.032 pu	R_g	0.00189 pu

The model represented with NR algorithm in [43] is compared with the model using SQ algorithm presented in [12], as [12] encompasses the topic of discussion presented in this chapter. The positive-sequence active power output at the point of common coupling (PCC) for [12] is fixed at an approximate estimate considering an internal loss of 1.15%. This is necessary as the accurate output is unavailable or unknown without a prior knowledge of the total internal losses. These losses, if ignored, will introduce a difference of 1%–2% in active power output between the two models.

Table 1.2 summarizes the values of different parameters of the 1-MW wind turbine chosen for the study conducted in this chapter. Table 1.3 summarizes the values of different parameters of a 1-MW, 4-pole, 50-Hz DFIG-based wind generator.

The converter loss resistances for RSC (R_{ron}) and GSC (R_{gon}) are taken as 0.0061 pu and 0.0087 pu. These values are obtained by average fitting of the data available in [41].

1.4.1 Results of Load-Flow Algorithms for Comparing Various Operating Modes

Results for calculated state variables are obtained for three DFIG units connected to buses 7, 18, and 47 of the IEEE 123-bus distribution system (without voltage regulators) via 0.69 kV/4.16 kV transformers. The system states are calculated for the NR-based model (Model II) and compared with the SQ-based model (Model I). Table 1.4 summarizes the values of Bus Voltage Magnitudes (BVM), Bus Voltage Phase Angles (BVPA) of point of coupling, internal currents, modulation indices, slip, active and reactive power output of DFIG connected at buses 7, 18, and 47 for both models. The steady-state operating point of the system without connecting any DFIG for the two algorithms under consideration have been found to be identical and the positive- and negative-sequence voltages in pu at buses 7, 18, and 47 are 0.9860, 0.0045, 0.9698, 0.0075, 0.9633, and 0.0082, respectively, for this condition.

The DFIG connected to buses 7 and 18 operate in voltage-controlled mode while extracting maximum power at a rated wind speed of 12.5 m/s (i.e., super-synchronous rotor speed). Operation at rated wind speed is indicated by rated slip in Table 1.4. The RSC is controlled to maintain positive-sequence voltage at the DFIG terminal at 0.99 pu. The GSC of the DFIG is operated at unity power factor. Negative-sequence control is not activated for DFIG connected to bus 7 (m_{g2} and m_{r2} in Table 1.4 are zero as negative-sequence voltage is not injected by RSC and GSC), while DFIG at 18 is operated in coordinated control mode. The DFIG parameters are observed to differ slightly mainly in negative sequence. As DFIG at bus 7 does not have any control in negative sequence, the difference of 0.28% and 0.09% in magnitude, observed for currents I_{g2} and $I_{s2,}$ respectively, is due to approximations in output power and inability to achieve power-balance across the DC-link in model I as discussed earlier. The positive- and negative-sequence voltages in pu at buses 7 and 18 are 0.9911, 0.0024 and 0.9823, 0.0046 for model I, and 0.9911, 0.0025 and 0.9823, 0.0050 for model II. The maximum difference in values of BVM and BVPA for DFIGs at buses 7 and 18 is 0.04% (BVM of phase a of bus 18 in Table 1.4). Both models show a reduction in negative-sequence voltage at PCC on connecting a DFIG as it acts as a sink.

The DFIG connected to bus 47 is used to verify the sub-synchronous operation of DFIG in PQ control mode for a wind speed of 8 m/s (i.e., sub-synchronous rotor speed indicated by positive slip in Table 1.4). This DFIG injects 90 kVAr as positive-sequence stator reactive power at the terminal. The negative-sequence controller of RSC is used to balance stator current as shown by results from both models ($|I_{s2}|$ is zero in Table 1.4) and negative-sequence current exchange through GSC is allowed (m_{g2}, i.e., negative-sequence GSC modulation index is zero in Table 1.4). The positive- and negative-sequence voltages in pu at bus 47 in this case is 0.9778 and 0.0042 for model I and 0.9778 and 0.0047 for model II. The maximum difference in values of BVM and BVPA for DFIG at bus 47 is 0.06% (BVM of Phase a of bus 47 in Table 1.4). Because DFIG at bus 47 does not have any control in negative sequence for GSC, the difference of 0.59% in magnitude observed for current I_{g2} in Table 1.4 is due to the inability of model I to achieve power balance at the DC-link.

In these results, limit violation does not occur and, hence, negligible difference (maximum 0.59%) is observed between the two models.

1.4.2 Limit Violation

A single type-III wind generator rated 1 MW, 0.69 kV, is interfaced with bus 18 of 123-bus system, which is considered for case studies on violation of limit. The limiting magnitudes of I_{s1} and I_{g1} are set at 0.1636 pu and 0.0738 pu, respectively, according to boundary values of state variables as discussed earlier in Section 1.3.3. Four different case studies are presented where violation of limit on a DFIG parameter is observed. All cases considered in this section are briefly outlined in Table 1.5.

1.4.2.1 Case Study I

The steady-state wind speed is 12.4 m/s. The DFIG connected to bus 18 of IEEE 123-bus system (with an OLTC at bus 149) operates in PQ mode. Q_{s_sp1}, specified at the stator terminal, is increased in steps. Operating conditions are generated where violation of the prescribed limit of one or more state variables occur due to the unbalance in the system.

Table 1.4. Comparison of Models I and II for BVM in pu, BVPA in degrees of point of coupling, various internal parameters of DFIG connected at buses 7, 18, and 47.

	Phase a				Phase b				Phase c			
Model	I	II	I	II	I	II	I	II	I	II	I	II
Bus	BVM		BVPA		BVM		BVPA		BVM		BVPA	
7	0.9853	0.9851	−0.49	−0.49	0.9989	0.9991	−120.05	−120.04	0.9892	0.9893	120.06	120.05
18	0.9701	0.9697	−1.07	−1.07	0.9973	0.9975	−120.15	−120.13	0.9796	0.9798	120.18	120.17
47	0.9654	0.9648	−1.16	−1.16	0.9921	0.9924	−120.34	−120.31	0.9761	0.9764	120.13	120.10
	DFIG Parameters											
Model	I	II	I	II	I	II	I	II	I	II	I	II
Bus	$\lvert I_{g1} \rvert$ (pu)		$\lvert I_{s1} \rvert$ (pu)		m_{g1}		m_{r1}		$\lvert I_{g2} \rvert$ (pu)		$\lvert I_{s2} \rvert$ (pu)	
7	0.0469	0.0460	0.1596	0.1607	0.9032	0.9031	0.2683	0.2683	0.0047	0.0075	0.0016	0.0025
18	0.0469	0.0456	0.1546	0.1557	0.9032	0.9031	0.2785	0.2785	0.0016	0.0018	0.0007	0.0008
47	0.0103	0.0109	0.0657	0.0663	0.8926	0.8926	0.1530	0.1531	0.0099	0.0158	0	0
Bus	m_{g2}		m_{r2}		P_{total} (kW)		Q_{s1} (kVAr)		Q_{g1} (kVAr)		slip	
7	0	0	0	0	988.35	989.04	−229.22	−229.58	0	0	−0.3073	−0.3073
18	0.0041	0.0045	0.0097	0.0105	988.34	987.69	118.88	118.60	0	0	−0.3073	−0.3073
47	0	0	0.0028	0.0045	259.06	258.93	90.000	90.000	0	0	0.1633	0.1633

Table 1.5. Case studies on violation of limit with DFIG connected at bus 18.

Case Study	Operating Mode of DFIG	Voltage Regulator	Wind Speed (m/s)	Specification
I	PQ	Bus 149	12.4	Q_{s_spl} is varied in steps of 2.5 kVAr from 352.5 kVAr to 390 kVAr
II	PQ	-	12.4	Q_{s_spl} is varied in steps of 2.5 kVAr from 352.5 kVAr to 390 kVAr
III	PV	-	Varied in steps of 0.1 from 12 to 12.4	$V_{s\_spl} = 0.995$ pu
IV	PV	Bus 149	Varied in steps of 0.1 from 12 to 12.4	$V_{s\_spl} = 1.02$ pu

The limit on $|I_{s1}|$ is violated in model of [12] for all operating points. However, in NR-based model, violation is observed for Q_{s_sp1}>367 kVAr. The NR-based model is able to track this limit (indicated by constant magnitude of I_{s1} at 0.1636 pu in Fig. 1.8(a)). The model given in [12] is unable to do so and operates at a much lower active and reactive power output than is actually necessary as indicated in Fig. 1.8(b) and (c). Moreover, model of [12] operates in restricted condition due to the limit violated by $|I_{s1}|$ for Q_{s_sp1} <360 kVAr, whereas if limits were not assessed, then the steady-state value of current obtained as final solution would have remained within the designed limits. This behavior is observed due to the fact that variables reaching the limit might revert back in succeeding iterations. Thus, the limit-handling strategy of the NR-based model avoids both unnecessary enforcement of limit and reduction of output power to erroneous fixed values.

The active and reactive power outputs of the DFIG for model of [12] are set at lower values as displayed in Fig. 1.8(b) and (c), but the DFIG is considered to operate with same slip (Fig. 1.8(d)). The reduction in power output at the DFIG terminal, due to the violation of limit observed in this case, has to be met by a similar reduction in the power extracted from the wind turbine. This reduction in power extracted from the turbine can only be obtained by varying either the slip or the pitch angle of the turbine blades or both. One of these variables can remain constant during a case of reduced output if the other is varied. However, the pitch controller comes into operation, that is, β_w is varied only if extracted real power or rotor speed surpasses its rated value, which is not observed for operating points Q_{s_sp1} <380 kVAr of this case study. Thus, the operating state calculated by [12] is not a steady equilibrium point; instead, it initializes the dynamic equations representing motion of the DFIG rotor. These dynamic equations when solved would yield the steady equilibrium point. The enforcement of limit on $|I_{s1}|$ for the NR-based model causes a reduction of P_{s1} and is followed by a change in ω_r as indicated in Fig. 1.8(d) by the value of slip. The operating speed increases on restricting the stator active power output as $P_{g1}/P_{s1} \approx -s_1$. Thus, the NR-based method predicts the steady equilibrium point. At operating points Q_{s_sp1} >380 kVAr, the rotor speed exceeds the rated value in the NR-based model due to change in rotor slip. The machine is restricted to operate at rated speed (Fig. 1.8(d)) by the pitch controller, and hence, the active power output of the NR-based model is further reduced for these operating points (Fig. 1.8(b)).

1.4.2.2 Case Study II

The positive-sequence reactive power output Q_{s_sp1}, specified at the stator terminal, is increased in steps for the DFIG connected to bus 18 of IEEE 123-bus system. Voltage regulators or OLTCs are not considered for this case. The models of [43] and of [12]

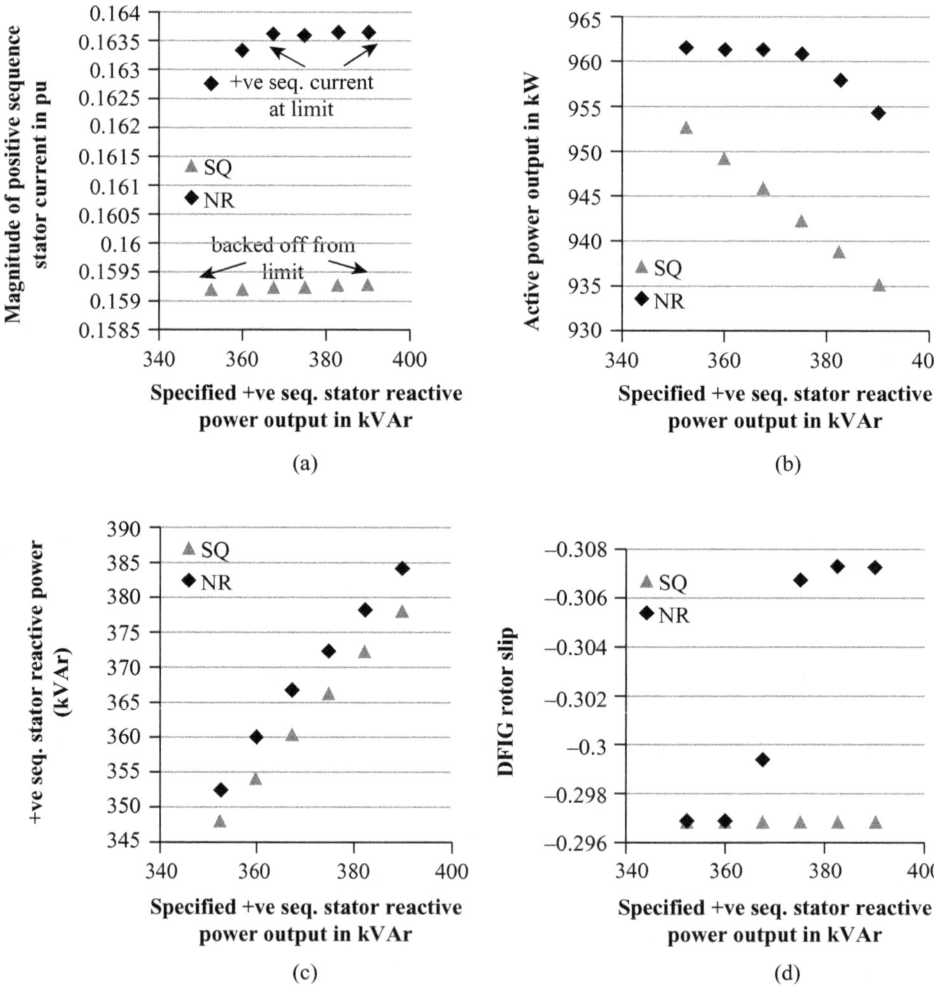

Figure 1.8. System states of IEEE 123-bus system (Case study I): (a) Magnitude of positive-sequence stator current $|I_{s1}|$; (b) Total active power output of DFIG; (c) Positive-sequence stator reactive power output; and (d) Rotor slip of DFIG.

showcase the fact that $|I_{s1}|$ has reached its limit (Fig. 1.9(a)) for all operating conditions. The DFIG active and reactive power output (Fig. 1.9(b) and (c)) show similar values for both models. These results are obtained for the system without any OLTC. Comparison of the results, obtained in Case studies I and II, indicate that model of [12] gives better results in the absence of OLTC. This inference is drawn from the fact that the stator current does not revert back from its limit in succeeding iterations (Fig. 1.9(a)) for Case study II. Similar to Case study I, Fig. 1.9(b) and (c) highlights that the active and reactive power outputs of the DFIG for model of [12] are set at lower values. However, the DFIG operates with the slip required to extract maximum power from the turbine (Fig. 1.9(d)), which does not present an equilibrium condition. As the DFIG is restricted to operate at rated speed by the pitch controller following variation of slip, the total active power output of model in [43] is slightly less as compared to that of the model in [12].

1.4.2.3 Case Study III

Case study III is conducted to understand the effect of violation of limit on $|I_{s1}|$ while DFIG is operating in PV mode. Voltage regulators or OLTCs are not considered in the distribution network. Fig. 1.10 shows the calculated states of the 123-bus system (without regulator) for different values of wind speed. Operation in PV mode is obtained by maintaining the voltage at the terminal of DFIG unit at 0.9950 pu. Both models show that the calculated states lie within their operating limits for wind speeds of 12 m/s and 12.1 m/s. A difference of 0.0009 pu in magnitude of the positive-sequence stator current is observed (Fig. 1.10(a)) between the two models. This causes a corresponding difference of 3.9 kW in the total active power output (Fig. 1.10(b)) for these two wind speeds. This difference is observed because of average estimation of loss in model of [12]. Violation of limit on $|I_{s1}|$ is observed in both SQ- and NR-based models for operating wind speeds of 12.2 m/s to 12.4 m/s. As $|I_{s1}|$ reaches its limit, the DFIG cannot maintain the terminal voltage at the previously specified value of 0.995 pu. The reactive power output (Fig. 1.10(c)) is fixed at a value obtained due to the corresponding enforcement of limit on $|I_{s1}|$. Thus, the DFIG operates with a terminal voltage different from the specified value. The rotor slip (Fig. 1.10(d)) changes due to this violation of limit for NR model at these values of wind speeds. As discussed earlier, the model of [12] indicates the change in output power (Fig. 1.10(b)), but the value of slip does not change. This is not a steady equilibrium point. Further, the NR-based model also shows the limit on rotor speed (Fig. 1.10(d)) being handled by adjusting the pitch angle at wind speeds 12.3 m/s and 12.4 m/s.

1.4.2.4 Case Study IV

Case study IV is conducted to observe the efficiency of the algorithms in handling PV mode of operation in presence of voltage regulator/OLTC. Fig. 1.11 highlights the calculated states of the system (with regulator at bus 149) for varying values of wind speed. The voltage at the stator terminal is maintained at 1.02 pu. This case illustrates conditions where violation of limit for PV mode occurs in the presence of a voltage regulator in the system. Increasing active power output causes the positive-sequence stator current (Fig. 1.11(a)) to subsequently reach its limit for NR-based model. The voltage at the stator terminal gets readjusted to meet this violation of limit. Also the total active power output (Fig. 1.11(b)), positive-sequence reactive power output (Fig. 1.11(c)), and slip (Fig. 1.11(d)) get readjusted

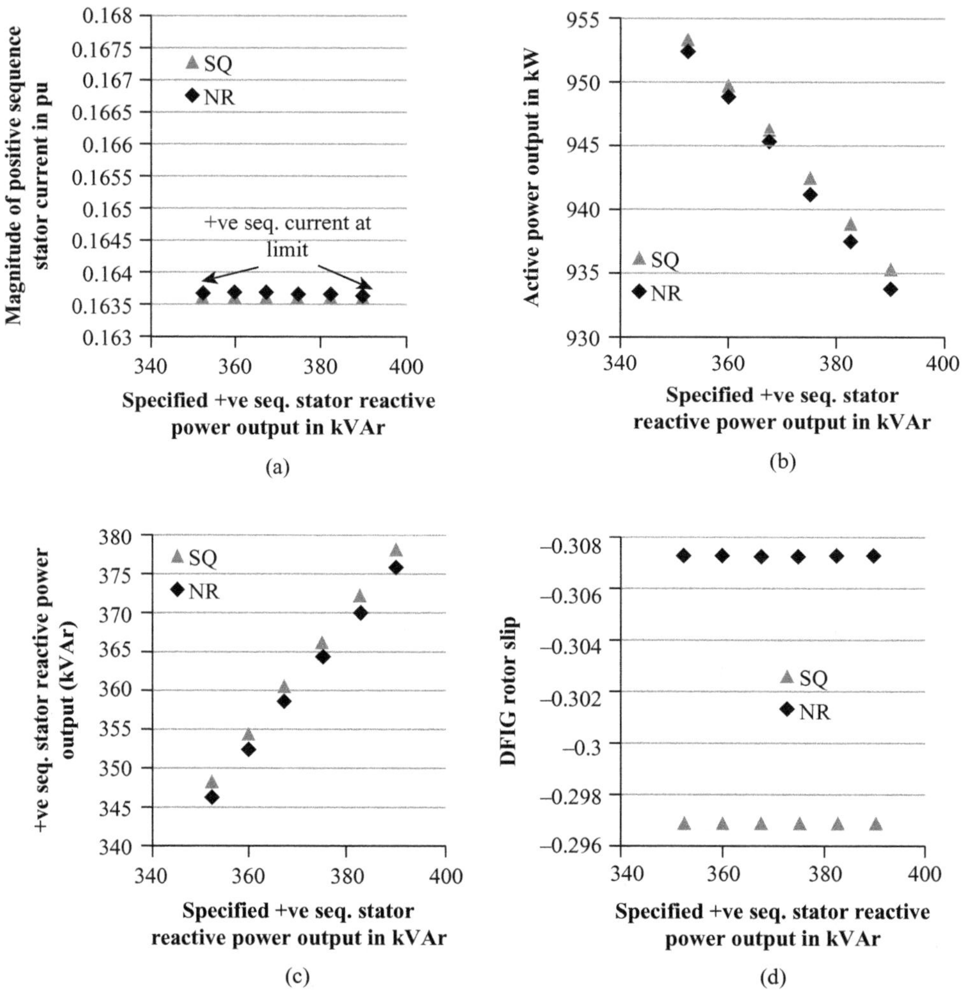

Figure 1.9. System states of IEEE 123-bus system (Case study II): (a) Magnitude of positive-sequence stator current $|I_{s1}|$; (b) Total active power output of DFIG; (c) Positive-sequence stator reactive power output; and (d) Rotor slip of DFIG.

to satisfy the limit on $|I_{s1}|$. States for model presented in [12] are not displayed in Fig. 1.11. This is due to the fact that the algorithm of [12] does not converge for these operating conditions. The main reason behind this inability is the nonsimultaneous calculation of voltage regulator tap-settings and the readjustment of voltage at the stator terminal caused by the violation of the limit on $|I_{s1}|$.

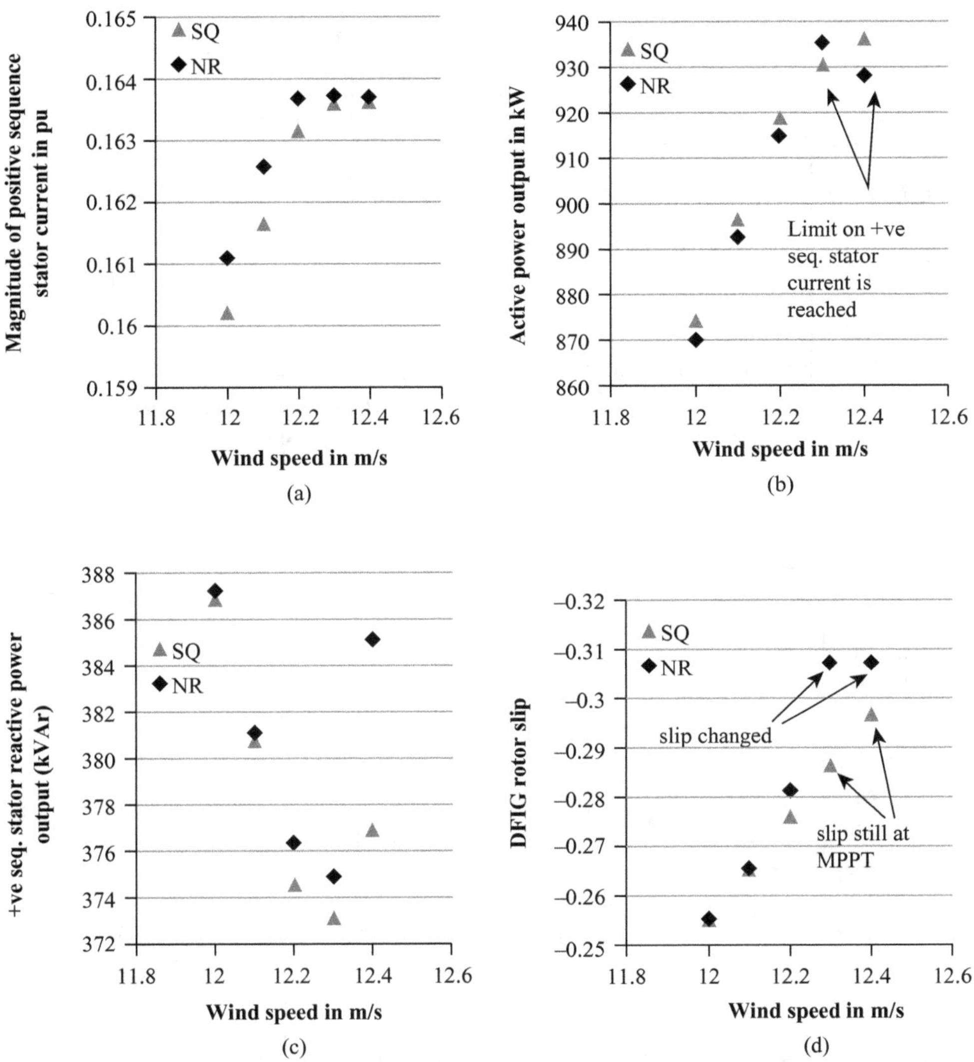

Figure 1.10. System states of IEEE 123-bus system (Case study III): (a) Magnitude of positive-sequence stator current $|I_{s1}|$; (b) Total active power output of DFIG; (c) Positive-sequence stator reactive power output; and (d) Rotor slip of DFIG.

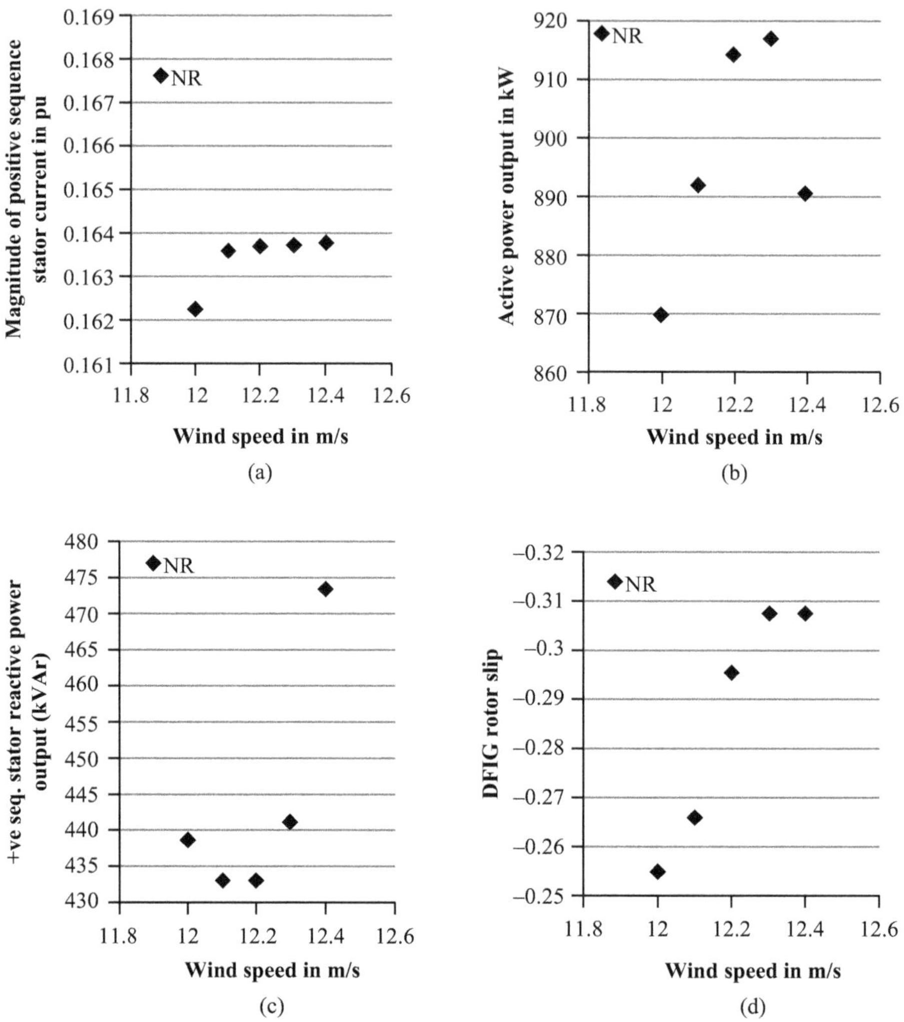

Figure 1.11. System states of IEEE 123-bus system (Case study IV): (a) Magnitude of positive-sequence stator current $|I_{s1}|$; (b) Total active power output of DFIG; (c) Positive-sequence stator reactive power output; and (d) Rotor slip of DFIG.

1.5 Summary

Steady-state performance analysis of DFIG-based wind generator requires accurate mathematical representation of the induction generator considering its operational limits and network unbalance. Different power-flow algorithms are compared in terms of their convergence characteristics and computational requirement using IEEE standard 13- and 34-bus distribution systems. While NR algorithm depicts rapid convergence, BF algorithm takes minimum time per iteration. The time required per iteration as well as the convergence rate of SQ-based algorithm falls in between NR and BF algorithms. The effect of multiple voltage regulators on convergence of different load-flow algorithms are investigated. Adjustment of tap position in OLTCs one after the other leads to oscillation of tap changes between multiple OLTCs. SQ-based algorithm is prone to divergence due to these oscillations. Simultaneous solution for voltage regulator tap position is necessary as in NR and BF algorithms.

A model for three-phase steady-state analysis of DFIG using NR algorithm is compared with a model using SQ algorithm. Results for various operating modes without violation of limits are compared in Section 1.4.1. A negligible difference (maximum 0.59%) is observed between NR and SQ-based models. SQ-based model achieves power-balance at the DC-link in terms of positive-sequence parameters only. Thus, maximum mismatches are observed in magnitudes of negative-sequence currents.

Results for IEEE 123-bus distribution system are effectively presented in Section 1.4.2 showing the advantage of the NR-based model for enforcing limits on DFIG internal parameters as it prevents unnecessary enforcement. SQ-based model gives better results in the absence of regulators as shown in Case study II as the stator current does not revert back from the limit in successive iterations. Also, DFIG operating state at restricted condition, with variation in rotor speed, is represented for NR-based model, which does not reflect in the SQ-based model. This is highlighted in all four case studies where after violation of limit the rotor speed changes to extract a reduced amount of active power from the wind. Again from Case study IV, it is established that the SQ-based model tends to diverge from solution in presence of OLTC and voltage-regulated buses. However, the computational requirement of the NR-based model makes it more useful for planning or off-line analytical studies while SQ-based model can be used for real-time analysis.

1.6 References

1. REN21, "Advancing the Global Renewable Energy Transition." 2017. Available online: http://www.ren21.net/wpcontent/uploads/2017/06/170607_GSR_2017_Highlights.pdf, (Renewable Energy Policy Network for the 21st Century).
2. A. F. Zobaa, and R. C. Bansal. (ed.), *Handbook of Renewable Energy Technology*, World Scientific Publishers Singapore, 2011.
3. M. Tazil, V. Kumar, R. C. Bansal, S. Kong, Z. Y. Dong, W. Freitas, and H. D. Mathur, "Three-phase doubly-fed induction generators: An overview," *IET-Electric Power Applications*, vol. 4, no. 2, pp. 75–89, 2010.
4. M. Rahimi, "Coordinated control of rotor and grid sides converters in DFIG based wind turbines for providing optimal reactive power support and voltage regulation," *Sustainable Energy Technologies and Assessments,* vol. 20, pp. 47–57, 2017.
5. V. Yaramasu, B. Wu, P. C. Sen, S. Kouro, and M. Narimani, "High-power wind energy conversion systems: State-of-the-art and emerging technologies," *Proceedings of the IEEE*, vol. 103, no. 5, pp. 740–788, 2015. doi:10.1109/JPROC.2014.2378692
6. L. Fan, and Z. Miao, "Chapter 4—Analysis of DFIG with unbalanced stator voltage" in *Modeling and Analysis of Doubly Fed Induction Generator Wind Energy Systems,* pp. 55–73, Academic Press, 2015. ISBN 978-0-12-802969-5.
7. M. J. Zandzadeh, and A. Vahedi, "Modeling and Improvement of Direct Power Control of DFIG Under Unbalanced Grid Voltage Condition", *International Journal of Electrical Power & Energy Systems*, vol. 59, pp. 58–65, 2014.
8. M. Farshadnia, and S. A. Taher, "Current-based direct power control of a DFIG under unbalanced grid voltage," *International Journal of Electrical Power and Energy Systems,* vol. 62, pp. 571–582, 2014.
9. R. Pena, R. Cardenas, E. Escobar, J. Clare, and P. Wheeler, "Control strategy for a doubly-fed induction generator feeding an unbalanced grid or stand-alone load," *Electric Power Systems Research,* vol. 79, no. 2, pp. 355–364, 2009.

10. H. E. Farag, E. F. El-Saadany, R. E. Shatshat, and A. Zidan, "A generalized power flow analysis for distribution systems with high penetration of distributed generation," *Electric Power Systems Research,* vol. 81, no. 7, pp. 1499–1506, 2011.
11. Y. Ju, W. Wu, B. Zhang, and H. Sun, "Three-phase DFIG steady model and fast three-phase load flow algorithm for distribution power systems" in *IEEE International Conference on Power System Technology (POWERCON)*, pp. 1–6, 2010. doi:10.1109/POWERCON.2010.5666718
12. M. Z. Kamh, and R. Iravani, "Three-phase steady-state model of type-3 wind generation unit part I: Mathematical models," *IEEE Transactions on Sustainable Energy,* vol. 2, no. 4, pp. 477–486, 2011. doi:10.1109/TSTE.2011.2156821.
13. M. H. Moradi, V. B. Foroutan, and M. Abedini, "Power flow analysis in islanded micro-grids via modeling different operational modes of DGs: A review and a new approach," *Renewable and Sustainable Energy Reviews,* vol. 69, pp. 248–262, 2017.
14. L. Holdsworth, X. G. Wu, J. B. Ekanayake, and N. Jenkins, "Direct solution method for initialising doubly-fed induction wind turbines in power system dynamic models," *IEE Proceedings—Generation, Transmission and Distribution,* vol. 150, no. 3, pp. 334–342, 2003. doi:10.1049/ip-gtd: 20030245.
15. M. Wu, and L. Xie, "Calculating steady-state operating conditions for DFIG-based wind turbines," *IEEE Transactions on Sustainable Energy,* vol. 9, no. 1, pp. 293–301, 2018. doi:10.1109/TSTE.2017.2731661.
16. J. G. Slootweg, H. Polinder, and W. L. Kling, "Dynamic modelling of a wind turbine with doubly fed induction generator" in *IEEE Power Engineering Society Summer Meeting*, vol. 1, pp. 644–649, 2001. doi:10.1109/PESS.2001.970114.
17. K. C. Divya, and P. S. N. Rao, "Models for wind turbine generating systems and their application in load flow studies," *Electric Power Systems Research,* vol. 76, no. 910, pp. 844–856, 2006. doi:10.1016/j.epsr.2005.10.012.
18. J. P. A. Vieira, M. V. A. Nunes, and U. H. Bezerra, "Analysis of steady-state operation of DFIG-based wind turbines in power systems" in *IET Conference on Renewable Power Generation (RPG)*, pp. 1–5, 2011. doi:10.1049/cp.2011.0237.
19. Y. He, and J. Hu, "DFIG wind power generation systems during unbalanced network fault: Analysis and operation" in *IEEE International Conference on Electrical Machines and Systems,* pp. 2331–2336, 2008.
20. Y. Ju, F. Ge, W. Wu, Y. Lin, and J. Wang, "Three-phase steady-state model of doubly fed induction generator considering various rotor speeds," *IEEE Access,* vol. 4, pp. 9479–9488, 2016. doi:10.1109/ACCESS.2016.2646683.
21. A. Dadhania, B. Venkatesh, A. B. Nassif, V. K. and Sood, "Modeling of doubly fed induction generators for distribution system power flow analysis," *International Journal of Electrical Power and Energy Systems,* vol. 53, pp. 576–583, 2013.
22. R. G. Wasley, and M. A. Shlash, "Newton–Raphson algorithm for 3-phase load flow," *Proceedings of the Institution of Electrical Engineers,* vol. 121, no. 7, pp. 630–638, 1974. doi:10.1049/piee.1974.0145.
23. M. A. Akher, K. M. Nor, and A. H. A. Rashid, "Improved three-phase power-flow methods using sequence components," *IEEE Transactions on Power Systems,* vol. 20, no. 3, pp. 1389–1397, 2005. doi:10.1109/TPWRS.2005.851933.
24. C. S. Cheng, and D. Shirmohammadi, "A three-phase power flow method for real-time distribution system analysis," *IEEE Transactions on Power Systems,* vol. 10, no. 2, pp. 671–679, 1995. doi:10.1109/59.387902.
25. J.-H. Teng, "A direct approach for distribution system load flow solutions," *IEEE Transactions on Power Delivery,* vol. 18, no. 3, pp. 882–887, 2003.
26. J.-H. Teng, "Modelling distributed generations in three-phase distribution load flow," *IET Generation, Transmission and Distribution,* vol. 2, no. 3, pp. 330–340, 2008.

27. J. Arrillaga, and N. R. Watson, *Computer Modelling of Electrical Power Systems*, 2nd edn. John Wiley and Sons, 2001. ISBN 0 471 81249 0.
28. Siemens, "Distribution Voltage Regulators." 2017. Available online: https://www.energy.siemens.com/us/pool/us/power-distribution/voltage-regulators/voltage-regulator-brochure.pdf.
29. P. Ortmann, "Understanding Power Quality, University of Idaho, College of Engineering." 2010. Available online: http://www.ece.uidaho.edu/ee/power/ECE528/.
30. PES, "Resources: 1992 Test Feeder Cases, IEEE Power and Energy Society." 1992. Available online: http://sites.ieee.org/pes-testfeeders/resources/.
31. G. M. Masters, *Renewable and Efficient Electric Power Systems,* John Wiley & Sons, Inc., 2004.
32. F. Mei, "Small signal modelling and analysis of doubly fed induction generators in wind power applications," Ph.D. thesis, Department of Electrical and Electronic Engineering, Imperial College London, University of London, 2007.
33. F. A. Bhuiyan, and A. Yazdani, "Multimode control of a DFIG-based wind-power unit for remote applications," *IEEE Transactions on Power Delivery,* vol. 24, no. 4, pp. 2079–2089, 2009. doi:10.1109/TPWRD.2009.2021030.
34. Y. Zhou, P. Bauer, J. A. Ferreira, and J. Pierik, "Operation of grid-connected DFIG under unbalanced grid voltage condition," *IEEE Transactions on Energy Conversion,* vol. 24, no. 1, pp. 240–246, 2009. doi:10.1109/TEC.2008.2011833.
35. W. Kersting, *Distribution System Modeling and Analysis*, 2nd edn. Taylor & Francis, New York, 2007.
36. A. Yazdani, and R. Iravani, *Voltage-Sourced Converters in Power Systems: Modeling, Control and Applications*, John Wiley & Sons, Inc., Hoboken, New Jersey, 2010.
37. L. Xu, "Coordinated control of DFIG's rotor and grid side converters during network unbalance," *IEEE Transactions on Power Electronics,* vol. 23, no. 3, pp. 1041–1049, 2008. doi:10. 1109/TPEL.2008.921157.
38. P. Pillay, and M. Manyage, "Definitions of voltage unbalance," *IEEE Power Engineering Review,* vol. 21, no. 5, pp. 49–51, 2001. doi:10.1109/MPER.2001.4311362.
39. M. I. Martinez, G. Tapia, A. Susperregui, and H. Camblong, "DFIG power generation capability and feasibility regions under unbalanced grid voltage conditions," *IEEE Transactions on Energy Conversion,* vol. 26, no. 4, pp. 1051–1062, 2011.
40. J. Hu, and Y. He, "Modeling and enhanced control of DFIG under unbalanced grid voltage conditions," *Electric Power Systems Research,* vol. 79, no. 2, pp. 273–281, 2009. doi:10.1016/j.epsr.2008.06.017.
41. S. Meier, "Losses in Power Electronic Converters." 2005. Available online: www.elkraft.ntnu.no/smola2005/Topics/3.pdf, pp. 1–7
42. K. Zollenkopf, "Bifactorization-basic computational algorithm and programming techniques" in *Conference on Large Sparse Sets of Linear Equations*, pp. 75–95, Oxford, 1971.
43. T. Datta, A. K. Sinha, and P. Bajpai, "Three-phase steady state model for unbalanced operation of grid-connected wind generation unit" in *IEEE Annual India Conference (INDICON),* pp. 1–6, 2016. doi:10.1109/INDICON.2016.7839157.

CHAPTER 2

Study of Mathematical Modeling and Small-Signal Stability of a Wind-Driven DFIG-Based System Using Different Types of Control Approach

Ravi Bhushan[1] and Kalyan Chatterjee[2]

[1]*Electrical Engineering Department, NIT Srinagar, J&K, India 190006.*
ravibhushaniitbhu@gmail.com
[2]*Electrical Engineering Department, IIT (ISM) Dhanbad, India 826004.*
kalyanchatterjee@iitism.ac.in

Abstract This chapter discusses the different types of the modern controller such as linear quadratic regulator (LQR) and observer designs, which are based on the state-space model for a wind-driven doubly fed induction generator (DFIG)-based system. The control approaches for these controllers are based on the states/outputs feedback system in which the controllers guarantee the closed-loop stability of the system. In this concern, first a comprehensive nonlinear, as well as a linearized mathematical model for each component of the wind-driven DFIG system, is formulated in the *d–q* axes synchronous frame of reference. Further, the concept of small-signal analysis has been discussed for the linearized model by using eigenvalues method. The simulation studies of the proposed controller performances have been done on the MATLAB/SIMULINK software.

Keywords: Mathematical model, Wind, Doubly fed induction generator (DFIG), Linear quadratic regulator (LQR), Observer, Small-signal stability.

List of Acronyms and Symbols

DFIG	:	Doubly Fed Induction Generator
DPC	:	Direct Power Control
DTC	:	Direct Torque Control
FSWT	:	Fixed Speed Wind Turbine
GSC	:	Grid-Side Converter
LQR	:	Linear Quadratic Regulator
LVRT	:	Low Voltage Ride-Through
MIMO	:	Multi-Input-Multi-Output
MPPT	:	Maximum Power Point Tracking
PCC	:	Point of Common Coupling
PI	:	Proportional-Integral
PID	:	Proportional-Integral-Derivative
PSO	:	Particle Swarm Optimization
PV	:	Peak value
RSC	:	Rotor-Side Converter
SISO	:	Single-Input-Single-Output
TSR	:	Tip Speed Ratio
VC	:	Vector Control
VSWT	:	Variable Speed Wind Turbine
WECS	:	Wind Energy Conversion System
α	:	Tip speed ratio
α_{opt}	:	Tip speed ratio corresponding to the C_{p_max}
Δ	:	Small deviation

θ_{tw}	:	Shaft twist angle
ρ	:	Air density
ψ_{qr}, ψ_{dr}	:	Rotor flux linkages of the *d–q* axes
ψ_{qs}, ψ_{ds}	:	Stator flux linkages of the *d–q* axes
ω_B	:	Base speed
ω_r	:	Generator-angular speed
ω_s	:	Synchronous angular speed
ω_t	:	Turbine-angular speed
C_{dc}	:	Capacitance of the DC-link
C_p	:	Power coefficient
C_{p_max}	:	Maximum power coefficient
H	:	Total equivalent inertia constant (Wind turbine + generator)
H_t, H_g	:	Wind turbine and generator inertia constant
i_{dc}	:	Current flowing through capacitor of the DC-link
i_{qg}, i_{dg}	:	GSC currents of the *d–q* axes
i_{qr}, i_{dr}	:	Rotor currents of the *d–q* es
i_{qs}, i_{ds}	:	Stator currents of the *d–q* axes
K_{sh}, D_{sh}	:	Shaft stiffness and damping coefficients
L_m	:	Mutual inductance
L_{rr}	:	Rotor self-inductance
L_{ss}	:	Stator self-inductance
N_{s2r}	:	Stator-to-rotor turns ratio
P_{dc}	:	DC-link power
P_g, Q_g	:	Active and reactive power at the AC terminal of the GSC
P_r, Q_r	:	Active and reactive power at the AC terminal of the RSC
P_s, Q_s	:	Stator active and reactive power
P_{total}, Q_{total}	:	Total active and reactive power at the PCC
R_b	:	Rotor radius of the wind turbine
R_g, L_g	:	GSC filter resistance and inductance
R_r, L_r	:	Rotor resistance and leakage inductance
R_s, L_s	:	Stator resistance and leakage inductance
s	:	Slip speed
T'_0	:	Rotor circuit time constant

T_e	:	Electromagnetic torque
T_m	:	Turbine input mechanical torque
T_{sh}	:	Shaft torque
V_b	:	Slack bus voltage
V_r	:	Rotor voltage
V_s	:	Stator voltage
V_w	:	Wind speed
v_{dc}	:	Voltage across the capacitor of the DC-link
v'_q, v'_d	:	Voltage behind the transient reactance of *d–q* axes
v_{qg}, v_{dg}	:	*d–q* axes voltages at the GSC terminal
v_{qr}, v_{dr}	:	Rotor voltages of the *d–q* axes
v_{qs}, v_{ds}	:	Stator terminal voltages of the *d–q* axes
X_s	:	Stator reactance
X'_s	:	Stator transient reactance

2.1 Introduction

The repeated reduction of fossil fuels, worrying uncertainties in climate, and growing pollution levels have enforced many countries to explore new, nonpolluting sources of energy. In the last few decades, lack of power generation has been a major problem in achieving the growing energy requirement, and therefore, awareness in many researchers, concerning the search for an alternative way, has gained momentum. Distributed power generation is increasingly being utilized to the fulfillment of energy requirements. In recent years, wind power has stood out among the normally utilized renewable energy sources as it is clean, yields no greenhouse gasses, abundantly available in nature (both on land (onshore) and at sea (offshore)), reliable, fastest-growing, and less expenditure with respect to the other renewable energy sources (e.g., compared to solar photovoltaic) [1, 2]. Simultaneously, inclusive studies have been carried out for the advancement of different kinds of turbines, which are an integral part of the wind energy production. The prime objective of the wind turbine electrical generator is transforming the mechanical torque/power from the wind rotor turbine into electric power to the grid/load. The variable speed wind turbine (VSWT) is recommended over the fixed speed wind turbine (FSWT) owing to its larger operating range. Currently, the idea of DFIG, which is driven by a VSWT, has received additional importance because of its notable advantages over other wind turbine generating systems [3–10].

Mehta et al. [3] present the state–space modeling of DFIG for small-signal stability assessment. Wu et al. [4] developed a complete DFIG-based wind turbine model including the grid-side converter (GSC) current dynamics in the modeling process. Further, they analyzed the small-signal stability of the reduced system model with the help of eigenvalues and participation factor methods. Pal and Mei [5] operated the DFIG system at different rotor speeds (i.e., subsynchronous, synchronous, and super-synchronous) and analyzed by obtaining the eigenvalues and participation factors of the system state matrix. Mei and Pal

[7] have formulated different models mathematically, such as one- or two-mass drive train as well as with or without stator transients of the DFIG-based wind energy conversion system (WECS) and established the nonlinear differential equations for each of them.

Vicatos and Teqopoulos [11] observed the global performance ability of the DFIG system in the steady-state condition under synchronous operation. They have demonstrated mathematically the mechanical power, electromagnetic torque, currents, and powers flowing through the DFIG in terms of rotor excitation voltage, angle, slip, and DFIG parameters. While Slootweg et al. [12] proposed a VSWT model driving DFIG and also established a set of mathematical equations that provided the behavior of the wind turbine.

Increasing penetration of wind energy into the power grids create many technical issues, such as grid integration, voltage instability, small-signal instability, and transient instability, which are major concerns of power system operators [13–16]. The nonlinearity and the stochastic behavior of the wind also provide several complications in the design and control of the wind power system.

Yang et al. [17] developed an accurate mathematical model of the wind-driven DFIG system and a Jacobian matrix, which help in the investigation of the small-signal stability. Due to the improper tuning of the controller, the system loses its stability via a Hopf bifurcation. Further, they performed eigenvalue sensitivity analysis toward the DFIG and control parameters and identified those critical parameters that affected the system stability. Moreover, to ensure the stable operation, they used this information as a tuning guide of the DFIG controller.

It is found from the above literature that the structures and operation of the DFIG is more complex than the typical induction machine. Consequently, the control of DFIG is also more complicated in comparison to another induction machine. To overcome these technical challenges, many of the researchers worldwide have presented different types of control techniques. Tremblay et al. [18] used three types of control techniques namely direct power control (DPC), direct torque control (DTC), and vector control (VC) for the DFIG-based wind turbine control. Yamamoto and Motoyoshi [19] developed a control system for the DFIG active and reactive power, which is controlled independently through a stator-flux oriented method that relies on the *d–q* axes frame of reference. Pena et al. [20] discussed a vector control scheme that has been applied to both rotor- and grid-side converters to control the active and reactive power of the grid-connected DFIG system independently. Mishra et al. [10, 21, 22] presented a modified PI controller scheme using bacterial foraging techniques to enhance the damping of the overall system. Mehta et al. [3] designed a PI controller with the use of particle swarm optimization (PSO) technique for rotor-side converter (RSC) of the DFIG system, which is based on torque and voltage control scheme. Wu et al. [4] derived a small-signal stability model of the wind-driven DFIG system and used a novel tuning method to optimize the PI controller parameters, which are based on PSO. Yang et al. [23] analyzed the small-signal stability analysis of an islanding system with DFIG wind farm in which authors designed a self-tuning frequency control by using PSO techniques. Zhou et al. [24] proposed a robust dynamic equivalent model of the DFIG-based wind farm for control and stability analysis. The authors used standard IEEE-9 bus test system to demonstrate the effectiveness and robustness of the proposed system.

A modern control system can be categorized on the basis of the physical nature of the rules followed by the system and the mathematical nature of the governing differential equations [25]. The fundamental concept in modern control is to define the system's state.

The robust controller may play an important role to overcome the above mentioned technical issues [26–36]. Ali et al. [29] proposed an LQR controller for the wind-driven DFIG system to reduce the impact of line resistance initiated as a noise constraint in the stability of the distribution system. The weighting matrices of the LQR were chosen by the authors in diagonal form. Barros et al. [30] presented an optimal feedback control strategy for the stator terminal voltage along with the active power supplied to the grid control. The rotor current controls have been included in this new control strategy. Hossain et al. [31] presented a new robust control topology based on output-feedback to boost the low-voltage ride-through (LVRT) ability of the wind-driven DFIG system. They applied this controller on the GSC to moderate the deviations in DC-link voltage. In addition, they simultaneously applied the strategy on the RSC to moderate the effects of the voltage dip on the system transients. Qin and Sun [32] proposed an RSC control of DFIG system, which was based on state-dependent Riccati equation to enhance the LVRT capability. Li et al. [33] proposed the feedback linearization and sliding mode control to mitigate the sub-synchronous problems, which appear in the DFIG system because of the series compensated system. Further, the small-signal stability analysis is used to know the performance and effectiveness of the designed controller. Li et al. [34] estimated the wind speed, which is very difficult to measure accurately in short term, by using the physically measurable signal such as pitch angle, rotor speed, and electrical power. In this concern, they adopted a genetic algorithm-based support vector machine model. In addition, to accommodate the external disturbances and system uncertainty, they used the neural network concept with an adaptive algorithm. Ghafouri et al. [35] presented an LQR for damping of sub-synchronous problems in DFIG-based wind farms. Further, LQR-based observer controller is designed to estimate the DFIG state variables. To fulfill the new grid code for fault ride through, they include the nonlinearities and the essential transient functions in the DFIG modelling. Sahoo et al. [36] designed a new control named dual mode fuzzy, which is based on switching logic, and also have the capability to improve the frequency profile of DFIG-based wind-integrated system.

The outline of the chapter is as follows: After the Introduction, the mathematical model for each component of the system is discussed in Section 2.2. In Section 2.3, small-signal analysis concept based on the state-space approach is used for designing control systems. Section 2.4 introduces the control strategies for DFIG-based WECS, which leads to meet all the desired outputs more efficiently. The results and discussions have been presented in Section 2.5. Finally, the chapter concluded with a summary in Section 2.6 followed by references in Section 2.7.

2.2 Modelling of DFIG-based Wind Systems

This part considers an inclusive mathematical modelling of each component of the wind-driven DFIG system. For the analysis of small-signal stability along with the accurate and precise design of a controller in subsequent sections of this chapter, it is necessary to know the exact mathematical modelling and dynamics of the system [28, 37, 38].

In doing so, the wind turbine model is described in which the concept of WECS, different types of drive-train model and the concept of maximum power curve have been discussed. The turbine generator model is simplified by using equivalent circuits and d–q axes frame reference in order to bring the system into a state from which it is easy to obtain its dynamic model.

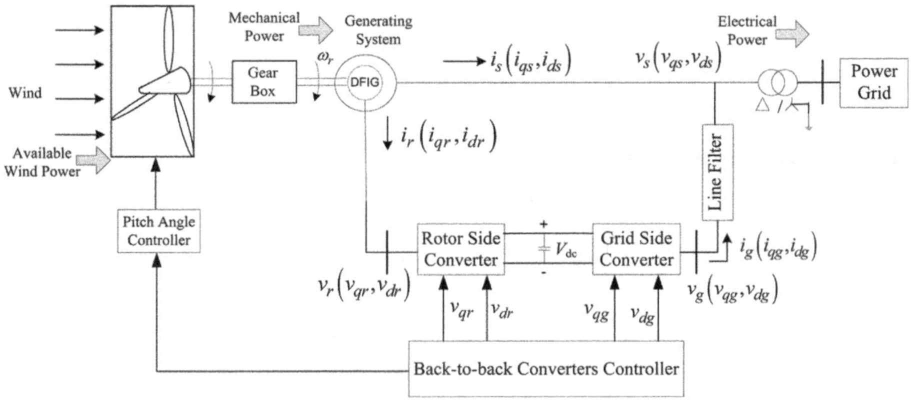

Figure 2.1. Block diagram of WECS [7].

2.2.1 Wind Turbine Model

2.2.1.1 Concept of WECS

Figure 2.1 signifies the single line diagram of the wind-driven DFIG system [7]. The wind-turbine model transforms the kinetic energy of the wind into the useful mechanical energy, which can be used as a prime mover of the DFIG and the DFIG finally converts to electrical energy [6]. The mechanical and the electrical part are the two main components of the WECS. The mechanical part consists with the wind turbine and associated control, including the gearbox, whereas the electrical part consists with the DFIG comprising back-to-back converters and related control, the grid-side filter model, and the infinite bus (grid). Here, the DFIG is coupled to the wind turbine through a gearbox.

The total power available in the wind (P_w) for extraction by the turbine blade can be defined as follows [6]:

$$P_w = \frac{1}{2}\rho a V_w^3 = \frac{1}{2}\rho\pi R_b^2 V_w^3 \tag{2.1}$$

where ρ is the air density; a is the turbine blade swept area through which power is extracted; R_b is the rotor radius of the turbine blade; and V_w is the wind speed.

From Equation (2.1), it can be seen that for a constant air density and wind speed, power availability for the extraction is greater when the turbine blade is larger. However, the wind turbine cannot convert all the available wind power to mechanical power (P_m). In general, it can extract only a fraction of that power (P_w), which is defined as follows [6]:

$$P_m = C_p P_w = \frac{1}{2}\rho\pi R_b^2 C_p V_w^3 \tag{2.2}$$

where C_p is the power coefficient or the coefficient of performance, a dimensionless variable quantity that signifies the effectiveness of the wind turbine in terms of the transformation of wind power into mechanical power. The value of C_p basically depends upon the wind speed, turbine angular speed (ω_t), and the blade pitch angle (β). Generally, C_p is a function

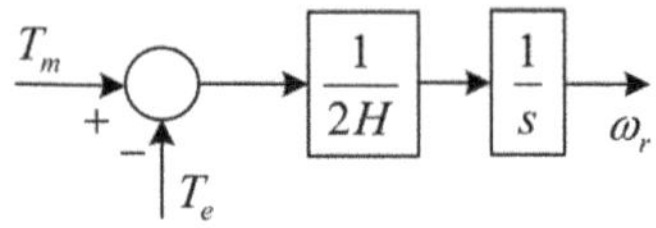

Figure 2.2. Single-mass drive-train model.

of tip speed ratio (TSR) (α) and the blade pitch angle as follows [6]:

$$C_p(\alpha, \beta) = 0.5176\left(\frac{116}{\alpha_i} - 0.4\beta - 5\right)e^{\frac{-21}{\alpha_i}} + 0.0068\alpha \tag{2.3}$$

where α is defined as the ratio of turbine blade speed to the wind speed as follows [39]:

$$\alpha = \frac{R\omega_t}{V_w} \tag{2.4}$$

and α_i is also the function of α and β, which is given by [39]

$$\frac{1}{\alpha_i} = \frac{1}{\alpha + 0.08\beta} - \frac{0.035}{\beta^3 + 1} \tag{2.5}$$

2.2.2 Drive-Train Model

2.2.2.1 Single-Mass Drive Train Model

In this case, the generator-angular speed must be identical to turbine speed as the shaft and gearbox effects have not been considered into account. This is the simplified form of a VSWT model as a single-mass model in which the system states are available for the feedback to control design purpose. The dynamic behavior is expressed by the differential equation (or swing equation) as follows [4, 7, 28]:

$$p(\omega_r) = \frac{1}{2H}T_m - \frac{1}{2H}T_e \tag{2.6}$$

Equation (2.6) is used to compute the generator-angular speed by using mechanical and electromagnetic torque. The structure of the single-mass drive-train model is shown in Figure 2.2.

2.2.2.2 Two-Mass Drive Train Model

This is a more generalized form and gives a more accurate response during the large perturbations, such as the grid faults and the conditions of heavy wind fluctuations.

In this case, both the turbine and generator angular speeds are different, which depends on the amount of shaft twist angle and shaft torque as follows [7, 21, 38, 40]:

$$2H_t p(\omega_t) = T_m - T_{sh} \tag{2.7}$$

$$2H_g p(\omega_r) = T_{sh} - T_e \tag{2.8}$$

$$p(\theta_{tw}) = \omega_B(\omega_t - \omega_r) \tag{2.9}$$

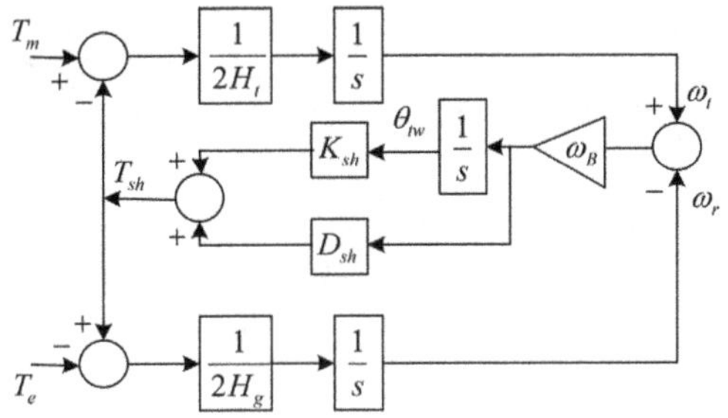

Figure 2.3. Two-mass drive-train model.

$$T_{sh} = K_{sh}\theta_{tw} + D_{sh}p\left(\theta_{tw}\right) \tag{2.10}$$

The effect of the two-mass model on the turbine shaft has been minimized by using the controllers and therefore use of such a model is not necessary for the control design process in the closed-loop system.

2.2.3 Concept of Maximum Power Curve

The top priority of the DFIG system is extracting the maximum available energy from the wind. This is accomplished when the power coefficient is maximized. The value of the power coefficient C_p increases when the pitch angle β decreases and achieves its extreme value when the pitch angle is fixed at zero.

In this work, the dynamics of the pitch angle controller has not been taken into consideration because of its sluggish response with respect to the electrical control action. The instability happens before dynamics of the pitch angle control becomes significant enough and the turbine power extracted, and hence the mechanical torque developed can be defined as follows [6, 28, 41]:

$$P_m = k_{opt}\omega_r^3 \tag{2.11}$$

$$T_m = k_{opt}\omega_r^2 \tag{2.12}$$

The constant $k_{opt} = \frac{1}{2}\rho\pi\frac{R^5}{\alpha_{opt}^3}C_{p\_max}$ rely on the turbine properties and surrounding circumstances, where $\alpha_{opt} = \frac{R\omega_r}{V_w}$ is the TSR with respect to the maximum power coefficient C_{p_max}.

This concept is an example of indirect speed control and the wind turbine mechanical output power follows the maximum power curve by adjusting the DFIG output powers, that is, disturbances in rotor speed will affect the wind turbines mechanical output and adjusts its speed again around steady-state value by adjusting the electromagnetic torque (T_e) of the DFIG.

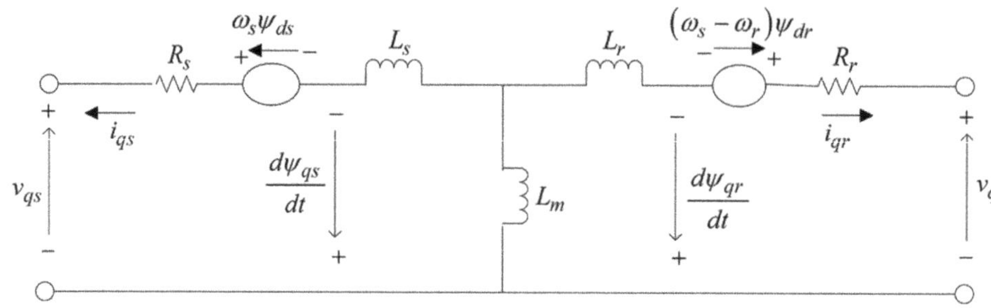

Figure 2.4. Inner circuit configuration of the DFIG for *q*-axis in the synchronously rotating frame of reference.

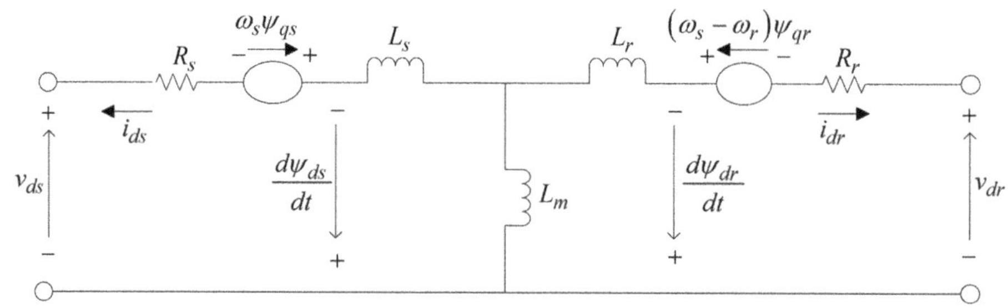

Figure 2.5. Inner circuit configuration of the DFIG for *d*-axis in the synchronously rotating frame of reference.

2.2.4 Wind Turbine Generator Model

The prime objective of the generator is to convert the mechanical torque or power from the wind rotor turbine into electric power to the grid or load. Nowadays, DFIG is one of the most broadly usable devices for wind energy production. Since in the conventional structure of the DFIG, the stator coil winding is directly coupled with the grid. Therefore, it has the ability to keep the terminal voltage essentially constant at grid values, despite the speed of the rotor. Whereas the rotor winding is coupled via the AC–DC–AC converters. The capacity of the converter required is only 20%–30% of the generator rating. This makes the wind turbine cheaper and lighter and also it has the ability to control the generator power factor and to inject/absorb the reactive power to/from the grid [29, 37]. The advantage of this configuration is to widen the generator speed range up to ±30% and therefore it can work in sub-synchronous, synchronous, and super-synchronous mode. The general assumptions have been taken throughout the thesis as follows:

a) The *d*-axis is supposed to be 90° forward of the *q*-axis in the direction of rotation.
b) The sign convention for the current and the self-flux linkage produced by a current in the generating mode are assumed as positive for outward flow.
c) Neglecting magnetic saturation, hysteresis, and eddy currents, and assuming the purely sinusoidal distribution of flux waves.

Figures 2.4 and 2.5 represent the inner circuit configuration of the DFIG system for *d*–*q* axes in the synchronously rotating frame of reference [6, 38]. The elementary equations of the proposed system, which are derived from its equivalent circuit diagrams of Figures 2.4 and 2.5, are as follows [4–6, 8, 28, 37, 38]:

$$\psi_{qs} = L_{ss}i_{qs} + L_m i_{qr} \tag{2.13}$$

$$\psi_{ds} = L_{ss}i_{ds} + L_m i_{dr} \tag{2.14}$$

$$\psi_{qr} = L_m i_{qs} + L_{rr}i_{qr} \tag{2.15}$$

$$\psi_{dr} = L_m i_{ds} + L_{rr}i_{dr} \tag{2.16}$$

$$L_{ss} = L_s + L_m \tag{2.17}$$

$$L_{rr} = L_r + L_m \tag{2.18}$$

$$v_{qs} = -R_s i_{qs} + \omega_s \psi_{ds} - \frac{1}{\omega_B} p\left(\psi_{qs}\right) \tag{2.19}$$

$$v_{ds} = -R_s i_{ds} - \omega_s \psi_{qs} - \frac{1}{\omega_B} p\left(\psi_{ds}\right) \tag{2.20}$$

$$v_{qr} = -R_r i_{qr} + \left(\omega_s - \omega_r\right) \psi_{dr} - \frac{1}{\omega_B} p\left(\psi_{qr}\right) \tag{2.21}$$

$$v_{dr} = -R_r i_{dr} - \left(\omega_s - \omega_r\right) \psi_{qr} - \frac{1}{\omega_B} p\left(\psi_{dr}\right) \tag{2.22}$$

where $p = d/dt$ represents the derivative operator.

The DFIG model can be represented in two importantly different forms on the basis of the rearrangements of the dynamic equations (2.13)–(2.22), which are commonly applied in the literature for the stability study of the overall system.

2.2.4.1 Model-I

The DFIG, in this model, relies on the elementary flux linkage and is represented in terms of *d–q* axes components of the voltage behind transient reactance $\left(v'_q, v'_d\right)$ by ignoring the stator transients and shorting the rotor windings [37, 38]. Therefore, equations (2.19)–(2.22) become:

$$v_{qs} = -R_s i_{qs} + \omega_s \psi_{ds} \tag{2.23}$$

$$v_{ds} = -R_s i_{ds} - \omega_s \psi_{qs} \tag{2.24}$$

$$0 = -R_r i_{qr} + \left(\omega_s - \omega_r\right) \psi_{dr} - \frac{1}{\omega_B} p\left(\psi_{qr}\right) \tag{2.25}$$

$$0 = -R_r i_{dr} - \left(\omega_s - \omega_r\right) \psi_{qr} - \frac{1}{\omega_B} p\left(\psi_{dr}\right) \tag{2.26}$$

This model is derived by eliminating the rotor currents of equations (2.15) and (2.16) in terms of the rotor flux linkage as follows:

$$i_{qr} = \frac{\psi_{qr} - L_m i_{qs}}{L_{rr}} \tag{2.27}$$

Therefore, from equations (2.13) and (2.27), it can be written as follows:

$$\psi_{qs} = L_{ss} i_{qs} + \frac{L_m\left(\psi_{qr} - L_m i_{qs}\right)}{L_{rr}} = \frac{L_m}{L_{rr}}\psi_{qr} + \left(L_{ss} - \frac{L_m^2}{L_{rr}}\right) i_{qs} \tag{2.28}$$

Similarly,

$$\psi_{ds} = \frac{L_m}{L_{rr}}\psi_{dr} + \left(L_{ss} - \frac{L_m^2}{L_{rr}}\right) i_{ds} \tag{2.29}$$

Substituting the value of ψ_{ds} from equation (2.29) in equation (2.23) and v_{qs} can be written as follows:

$$v_{qs} = -R_s i_{qs} + \omega_s\left(\frac{L_m}{L_{rr}}\psi_{dr} + \left(L_{ss} - \frac{L_m^2}{L_{rr}}\right) i_{ds}\right) = -R_s i_{qs} + \frac{\omega_s L_m}{L_{rr}}\psi_{dr} + \omega_s\left(L_{ss} - \frac{L_m^2}{L_{rr}}\right) i_{ds} \tag{2.30}$$

And therefore,

$$v_{qs} = -R_s i_{qs} + X_s' i_{ds} + v_q' \tag{2.31}$$

Similarly,

$$v_{ds} = -R_s i_{ds} - X_s' i_{qs} + v_d' \tag{2.32}$$

where

$$v_q' = \left(\frac{\omega_s L_m}{L_{rr}}\right)\psi_{dr} \tag{2.33}$$

$$v_d' = -\left(\frac{\omega_s L_m}{L_{rr}}\right)\psi_{qr} \tag{2.34}$$

And the stator transient reactance,

$$X_s' = \omega_s\left(L_{ss} - \frac{L_m^2}{L_{rr}}\right) \tag{2.35}$$

The rotor currents can be written as follows:

$$i_{qr} = -\frac{L_m}{L_{rr}} i_{qs} + \frac{1}{\omega_s L_m} v_d' \tag{2.36}$$

$$i_{dr} = -\frac{L_m}{L_{rr}} i_{ds} - \frac{1}{\omega_s L_m} v_q' \tag{2.37}$$

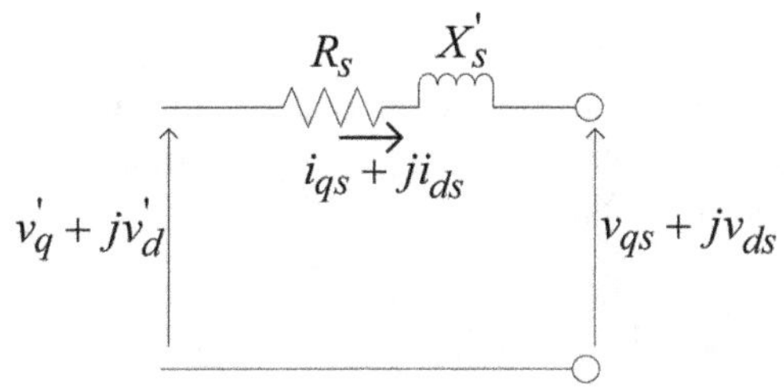

Figure 2.6. Transient-equivalent circuit of the DFIG [37].

The combination of stator voltage equations (2.31) and (2.32) make it possible to represent the DFIG as a transient-equivalent circuit of Figure 2.6 [37].

Finally, with the help of above-derived equations, the differential equations can be rearranged as follows [21, 38]:

$$p\left(i_{qs}\right) = -\frac{\omega_s\omega_B}{X'_s}\left\{R_s + \frac{1}{\omega_s T'_0}\left(X_s - X'_s\right)\right\} i_{qs} + \omega_s\omega_B i_{ds} + \frac{\omega_r\omega_B}{X'_s}v'_q - \frac{\omega_B}{X'_s T'_0}v'_d$$
$$-\frac{\omega_s\omega_B}{X'_s}v_{qs} + \frac{\omega_s\omega_B L_m}{X'_s L_{rr}}v_{qr} \tag{2.38}$$

$$p\left(i_{ds}\right) = -\omega_s\omega_B i_{qs} - \frac{\omega_s\omega_B}{X'_s}\left\{R_s + \frac{1}{\omega_s T'_0}\left(X_s - X'_s\left(\right.\right.\right\} i_{ds} + \frac{\omega_B}{X'_s T'_0}v'_q + \frac{\omega_r\omega_B}{X'_s}v'_d$$
$$-\frac{\omega_s\omega_B}{X'_s}v_{ds} + \frac{\omega_s\omega_B L_m}{X'_s L_{rr}}v_{dr} \tag{2.39}$$

$$p\left(v'_q\right) = \frac{\omega_B}{T'_0}\left(X_s - X'\right)_s i_{ds} - \frac{\omega_B}{T'_0}v'_q + \omega_B\left(\omega_s - \omega_r\right)v'_d - \frac{\omega_s\omega_B L_m}{L_{rr}}v_{dr} \tag{2.40}$$

$$p\left(v'_d\right) = -\frac{\omega_B}{T'_0}\left(X_s - X'_s\right)i_{qs} - \omega_B\left(\omega_s - \omega_r\right)v'_q - \frac{\omega_B}{T'_0}v'_d + \frac{\omega_s\omega_B L_m}{L_{rr}}v_{qr} \tag{2.41}$$

where $X_s = \omega_s L_{ss}$ is the stator reactance; $X_s - X'_s = \dfrac{\omega_s L_m^2}{L_{rr}}$, and the constant $T'_0 = \dfrac{L_{rr}}{R_r}$ is the rotor circuit time constant.

The electromagnetic torque T_e can be defined mathematically as follows [38, 42].

$$T_e = \left(\frac{v'_q}{\omega_s}\right)i_{qs} + \left(\frac{v'_d}{\omega_s}\right)i_{ds} \tag{2.42}$$

2.2.4.2 Model-II

In this case, the DFIG model relies on the generalized flux linkage of equations (2.13)–(2.22) and then readjusted in terms of the differential equations of d–q axes stator and rotor currents, which are as follows [28]:

$$p\left(i_{qs}\right) = \left(\frac{\omega_B R_s L_{rr}}{L_m^2 - L_{ss}L_{rr}}\right) i_{qs} + \left(\frac{\omega_B}{L_m^2 - L_{ss}L_{rr}}\right)\left\{\left(\omega_s - \omega_r\right)L_m^2 - \omega_s L_{ss}L_{rr}\right\} i_{ds} - \left(\frac{\omega_B R_r L_m}{L_m^2 - L_{ss}L_{rr}}\right) i_{qr} + \left(\frac{\omega_B}{L_m^2 - L_{ss}L_{rr}}\right)\left\{\left(\omega_s - \omega_r\right)L_m L_{rr} - \omega_s L_m L_{rr}\right\} i_{dr} - \left(\frac{\omega_B L_m}{L_m^2 - L_{ss}L_{rr}}\right) v_{qr} + \left(\frac{\omega_B L_{rr}}{L_m^2 - L_{ss}L_{rr}}\right) v_{qs} \tag{2.43}$$

$$p\left(i_{ds}\right) = \left(\frac{\omega_B}{L_m^2 - L_{ss}L_{rr}}\right)\left\{\omega_s L_{ss}L_{rr} - \left(\omega_s - \omega_r\right)L_m^2\right\} i_{qs} + \left(\frac{\omega_B R_s L_{rr}}{L_m^2 - L_{ss}L_{rr}}\right) i_{ds} + \left(\frac{\omega_B}{L_m^2 - L_{ss}L_{rr}}\right)\left\{\omega_s L_m L_{rr} - \left(\omega_s - \omega_r\right)L_m L_{rr}\right\} i_{qr} - \left(\frac{\omega_B R_r L_m}{L_m^2 - L_{ss}L_{rr}}\right) i_{dr} - \left(\frac{\omega_B L_m}{L_m^2 - L_{ss}L_{rr}}\right) v_{dr} + \left(\frac{\omega_B L_{rr}}{L_m^2 - L_{ss}L_{rr}}\right) v_{ds} \tag{2.44}$$

$$p\left(i_{qr}\right) = -\left(\frac{\omega_B R_s L_m}{L_m^2 - L_{ss}L_{rr}}\right) i_{qs} + \left(\frac{\omega_B}{L_m^2 - L_{ss}L_{rr}}\right)\left\{\omega_s L_m L_{ss} - \left(\omega_s - \omega_r\right)L_m L_{ss}\right\} i_{ds} + \left(\frac{\omega_B R_r L_{ss}}{L_m^2 - L_{ss}L_{rr}}\right) i_{qr} + \left(\frac{\omega_B}{L_m^2 - L_{ss}L_{rr}}\right)\left\{\omega_s L_m^2 - \left(\omega_s - \omega_r\right)L_{ss}L_{rr}\right\} i_{dr} + \left(\frac{\omega_B L_{ss}}{L_m^2 - L_{ss}L_{rr}}\right) v_{qr} - \left(\frac{\omega_B L_m}{L_m^2 - L_{ss}L_{rr}}\right) v_{qs} \tag{2.45}$$

$$p\left(i_{dr}\right) = \left(\frac{\omega_B}{L_m^2 - L_{ss}L_{rr}}\right)\left\{\left(\omega_s - \omega_r\right)L_m L_{ss} - \omega_s L_m L_{ss}\right\} i_{qs} - \left(\frac{\omega_B R_s L_m}{L_m^2 - L_{ss}L_{rr}}\right) i_{ds} + \left(\frac{\omega_B}{L_m^2 - L_{ss}L_{rr}}\right)\left\{\left(\omega_s - \omega_r\right)L_{ss}L_{rr} - \omega_s L_m^2\right\} i_{qr} + \left(\frac{\omega_B R_r L_{ss}}{L_m^2 - L_{ss}L_{rr}}\right) i_{dr} + \left(\frac{\omega_B L_{ss}}{L_m^2 - L_{ss}L_{rr}}\right) v_{dr} - \left(\frac{\omega_B L_m}{L_m^2 - L_{ss}L_{rr}}\right) v_{ds} \tag{2.46}$$

Equation (2.42) can also be written in terms of the d–q axes stator and rotor currents as follows [28, 43].

$$T_e = L_m i_{qs} i_{dr} - L_m i_{ds} i_{qr} \tag{2.47}$$

The structure of the electromagnetic torque model is depicted in Figure 2.7.

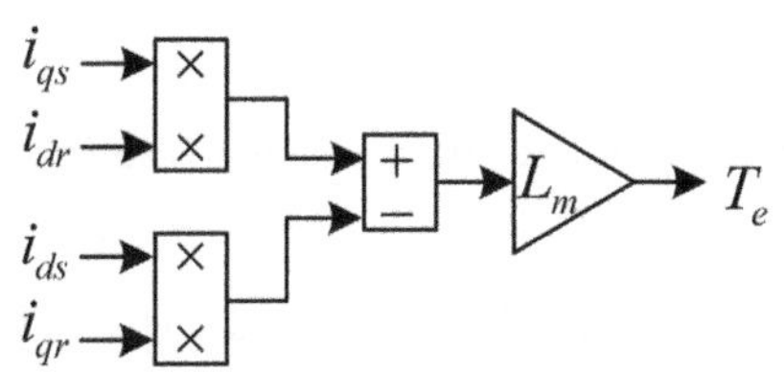

Figure 2.7. Electromagnetic or generator torque model.

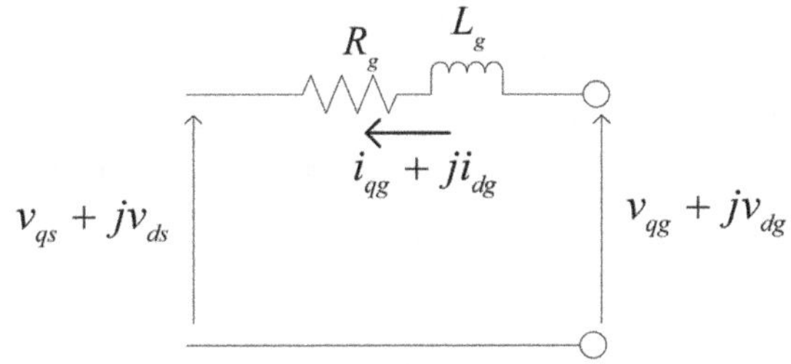

Figure 2.8. Inner circuitry of the GSC filter [28].

2.2.5 Grid-Side Converter Filter Model

The adequate grid-side filter is necessary between the GSC and the point of common coupling (PCC). It is installed at the output terminal of the GSC, which amends the amplitude and phase of the voltage introduced by the GSC in addition to attenuating the harmonics content. The inner circuitry of the GSC filter is illustrated in Figure 2.8 [28, 38].

The dynamics of the GSC filter (here, *d–q* axes inductor current) have also been taken into consideration in the modeling process to secure more accurate results. From Figure 2.8, the GSC voltage in complex form can be written as follows:

$$v_{qg} + jv_{dg} = R_g\left(i_{qg} + ji_{dg}\right) + j\omega_s L_g\left(i_{qg} + ji_{dg}\right) + \frac{1}{\omega_B}p\left(i_{qg} + ji_{dg}\right) + \left(v_{qs} + jv_{ds}\right) \quad (2.48)$$

By equating real and imaginary parts of equation (2.48) from both the sides, the GSC voltages (*d–q* axes components) can be written as follows:

$$v_{qg} = R_g i_{qg} - \omega_s L_g i_{dg} + \frac{1}{\omega_B}p\left(i_{qg}\right) + v_{qs} \quad (2.49)$$

$$v_{dg} = R_g i_{dg} + \omega_s L_g i_{qg} + \frac{1}{\omega_B}p\left(i_{dg}\right) + v_{ds} \quad (2.50)$$

By rearranging the equations (2.49) and (2.50), the GSC current (*d–q* axes components) in terms of a differential equation can be written as follows:

$$p\left(i_{qg}\right) = -\frac{\omega_B R_g}{L_g}i_{qg} + \omega_s\omega_B i_{dg} + \frac{\omega_B}{L_g}v_{qg} - \frac{\omega_B}{L_g}v_{qs} \quad (2.51)$$

$$p\left(i_{dg}\right) = -\omega_s\omega_B i_{qg} - \frac{\omega_B R_g}{L_g}i_{dg} + \frac{\omega_B}{L_g}v_{dg} - \frac{\omega_B}{L_g}v_{ds} \quad (2.52)$$

2.2.6 DFIG Powers Model

Let P_s and Q_s, respectively, symbolize the stator active and reactive powers, which are defined as follows [28]:

$$P_s + jQ_s = v_{qds} i_{qds}^* = (v_{qs} + jv_{ds})(i_{qs} + ji_{ds})^* \tag{2.53}$$

$$P_s = v_{qs} i_{qs} + v_{ds} i_{ds} \tag{2.54}$$

$$Q_s = v_{ds} i_{qs} - v_{qs} i_{ds} \tag{2.55}$$

The structure of the stator power model is illustrated in Figure 2.9.

Similarly, the rotor active and reactive powers can be determined as follows:

$$P_r = v_{qr} i_{qr} + v_{dr} i_{dr} \tag{2.56}$$

$$Q_r = v_{dr} i_{qr} - v_{qr} i_{dr} \tag{2.57}$$

The structure of the rotor power model is illustrated in Figure 2.10.

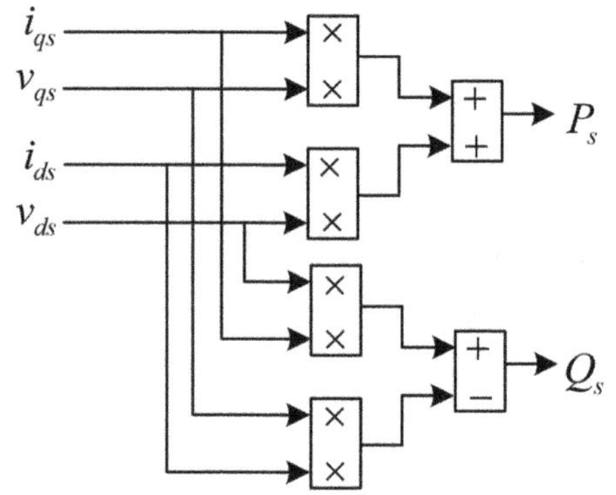

Figure 2.9. Stator powers model.

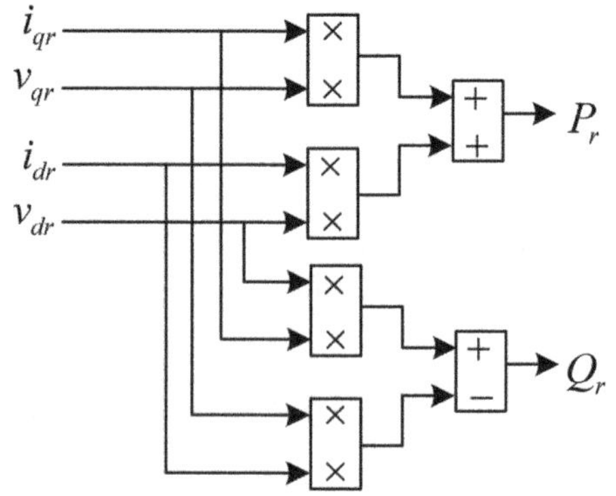

Figure 2.10. Rotor powers model.

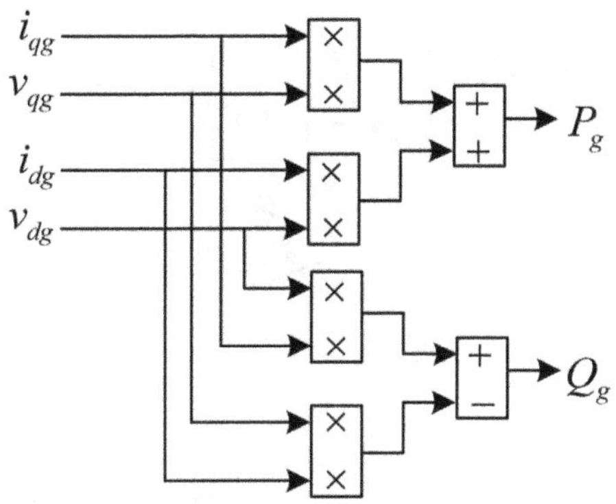

Figure 2.11. Powers model at the GSC.

The active and reactive powers at the GSC can be written in a similar manner of the stator and rotor powers as follows:

$$P_g = v_{qg}i_{qg} + v_{dg}i_{dg} \tag{2.58}$$

$$Q_g = v_{dg}i_{qg} - v_{qg}i_{dg} \tag{2.59}$$

The structure of the GSC power model is represented in Figure 2.11. The total active and reactive powers at the PCC can be defined as follows:

$$P_{total} = P_s + P_g \tag{2.60}$$

$$Q_{total} = Q_s + Q_g \tag{2.61}$$

2.2.7 DC-Link Model

The DFIG contains a rectifier-inverter pair in which power can flow in both the directions. A DC-link capacitor C_{dc} is installed between the rectifier-inverter, which serves as a reactive power source that can control the DFIG stator-side power factor. In this work, the losses in the converters and the switching dynamics are assumed to be neglected. In addition, the transfer functions of the converters are considered to be unity because of its high switching frequency [5, 6]. The DC-link power (P_{dc}) can be formulated as follows [4, 7, 28]:

$$P_{dc} = P_r - P_g \tag{2.62}$$

The expressions for P_{dc}, P_r and P_g are, respectively, specified in equations (2.63), (2.56) and (2.58). Then, equation (2.62) can be written as follows:

$$P_{dc} = v_{dc}i_{dc} = C_{dc}v_{dc}\frac{dv_{dc}}{dt} = P_r - P_g \tag{2.63}$$

The DC-link voltage in terms of differential equation relies on P_r and P_g as follows:

$$p\left(v_{dc}\right) = \frac{1}{C_{dc}v_{dc}}\left(P_r - P_g\right) \tag{2.64}$$

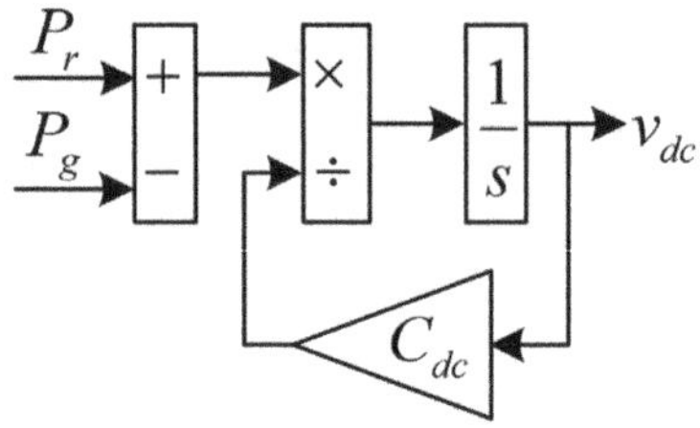

Figure 2.12. DC-link voltage model.

By using equations (2.56) and (2.58), equation (2.64) can be written as follows:

$$p\left(v_{dc}\right)=\frac{1}{C_{dc}v_{dc}}\left(v_{qr}i_{qr}+v_{dr}i_{dr}-v_{qg}i_{qg}-v_{dg}i_{dg}\right) \quad (2.65)$$

The structure of the DC-link voltage model is illustrated in Figure 2.12.

The choice of DC-link voltage is influenced by the rotor and the PCC voltage. However, generally the DFIG works on the slip of range ±0.3 and hence the rotor voltage is very small in comparison to the PCC voltage. Therefore, the choice of DC-link voltage has been estimated by considering only PCC voltage with the help of the following equation [44]:

$$v_{dc0}\geq\frac{2\sqrt{2}}{\sqrt{3}\times(\text{Modulation index})}\times(\text{Line voltage at the PCC}) \quad (2.66)$$

For the linear range, the maximum modulation is assumed as unity. From the various literatures [45–47], it can be observed that the permissible limit of the DC-link voltage during disturbances is lying between 1.2 and 1.35 pu.

The capacitor of the DC-link is estimated as follows [48]:

$$C_{dc}\geq\frac{0.9\times(\text{Permissible peak value of ac line current})}{4\sqrt{2}\times\pi\times(\text{Grid frequency})\times(\text{Permissible ripple DC}-\text{link voltage})} \quad (2.67)$$

2.3 Small-Signal Analysis Concept of the DFIG-based Wind Systems

The small-signal stability is the ability of power system to regain its original steady point after being subjected to a small disturbance, which leads to a corresponding incremental change in one or more of the system state variables [31]. The presence of nonlinearity in the study system, which was discussed in Section 2.2, gives worse dynamic performances. Hence, linearization of the nonlinear model around a nominal operating point is needed.

In order to linearize, the state-space approach is used, which facilitates a compact mathematical methodology to examine the dynamic behavior of the system [3, 29, 49]. The accurate initialization of such a model is essential to reach quickly the steady-state condition if any disturbance has occurred in the system. The nominal operating point is obtained by load flow analysis.

2.3.1 Concept of Model Linearization

The discussed mathematical model in Section 2.2 can be characterized as a set of generalized first-order nonlinear ordinary differential equations as follows [14, 28, 37, 38, 50]:

$$\dot{x} = f(x, u) \tag{2.68}$$

$$y = g(x, u) \tag{2.69}$$

where $f = [f_1; f_2; \cdots; f_n]$ and $g = [g_1; g_2; \cdots; g_m]$ represent the nonlinear functions;

$x = [x_1; x_2; \cdots; x_n]$ is the "n" number of state variables;

$u = [u_1; u_2; \cdots; u_r]$ is the "r" number of input variables, and

$y = [y_1; y_2; \cdots; y_m]$ is the "m" number of input variables, and

In order to develop the small-signal model, a small perturbation is considered in the input variables (i.e., $u = u_0 + \Delta u$), consequently the respective variation is noted in the state variables (i.e., $x = x_0 + \Delta x$) and thus in the output variables (i.e., $y = y_0 + \Delta y$). The nonlinear equations, with respect to small deviation around the equilibrium point, have been linearized and that can be arranged in the state-space form as given below [14, 28, 37, 50]:

$$\dot{\Delta x} = A\Delta x + B\Delta u \tag{2.70}$$

$$\Delta y = C\Delta x + D\Delta u \tag{2.71}$$

where Δ denotes the small deviation and the coefficient A, B, C, and D represent the system or state matrix of dimension $n \times n$, the control or input matrix of dimension $n \times r$, the output matrix of dimension $m \times n$, and the feed-forward matrix of dimension $m \times r$, respectively, whose analytical values are calculated as follows [14, 37, 50]:

$$A = \begin{bmatrix} \frac{\partial f_1}{\partial x_1} & \cdots & \frac{\partial f_1}{\partial x_n} \\ \vdots & \cdots & \vdots \\ \frac{\partial f_n}{\partial x_1} & \cdots & \frac{\partial f_n}{\partial x_n} \end{bmatrix} \tag{2.72}$$

$$B = \begin{bmatrix} \frac{\partial f_1}{\partial u_1} & \cdots & \frac{\partial f_1}{\partial u_r} \\ \vdots & \cdots & \vdots \\ \frac{\partial f_n}{\partial u_1} & \cdots & \frac{\partial f_n}{\partial u_r} \end{bmatrix} \tag{2.73}$$

$$C = \begin{bmatrix} \frac{\partial g_1}{\partial x_1} & \cdots & \frac{\partial g_1}{\partial x_n} \\ \vdots & \cdots & \vdots \\ \frac{\partial g_m}{\partial x_1} & \cdots & \frac{\partial g_m}{\partial x_n} \end{bmatrix} \tag{2.74}$$

$$D = \begin{bmatrix} \frac{\partial g_1}{\partial u_1} & \cdots & \frac{\partial g_1}{\partial u_r} \\ \vdots & \cdots & \vdots \\ \frac{\partial g_m}{\partial u_1} & \cdots & \frac{\partial g_m}{\partial u_r} \end{bmatrix} \tag{2.75}$$

The above partial derivatives are calculated around their operating point about which the small-signal stability is to be analyzed.

2.3.2 Eigenvalues Analysis

System eigenvalues analysis plays a prominent role in deciding the small-signal stability and in knowing the system dynamic behavior. The eigenvalues of the system matrix exhibit necessary information concerning the system response corresponding to small perturbations and hence characterize the system stability. The state matrix A provides the system eigenvalues by using the characteristic equations as follows [37, 50, 51]:

$$|A - \lambda I| = 0 \tag{2.76}$$

where λ represents the eigenvalues of the system and the numbers of eigenvalues are the same as the numbers of first-order differential equations considered for the linearization, whereas I represents the identity matrix of size A.

The values of λ may be real or complex. A positive real eigenvalue represents an unstable system of exponentially growing behavior, whereas a negative real eigenvalue indicates a stable system with non-oscillatory mode. However, the complex eigenvalues always occur in conjugates pairs, which are oscillatory in nature and can be written as follows:

$$\lambda = \sigma \pm j\omega \tag{2.77}$$

where σ is the real part of the eigenvalue, which must be negative for a stable system. In case of a positive value of σ, a small amount of disturbance would lead to an increase in the system oscillations exponentially and the system becomes unstable. The system damping is also dependent on the values of σ. The ω represents the imaginary part of the eigenvalues, which gives the frequency of oscillation and is written as follows [37, 50, 51]:

$$\text{Damped Frequency or Frequency of Oscillation} = \frac{\omega}{2\pi}\text{Hz} \tag{2.78}$$

The rate of decay of system oscillation magnitude is obtained by the damping ratio and expressed as follows [37, 50, 51]:

$$\text{Damping ratio} = \frac{-\sigma}{\sqrt{\sigma^2 + \omega^2}} \tag{2.79}$$

2.4 Control Strategies for DFIG-Based WECS

Recently, the prompt growth of wind power penetration significantly affects the static and dynamic behavior of the power system. The variations in load are leading to the voltage instabilities at stator terminal of the DFIG-based WECS. Moreover, the stochastic behavior

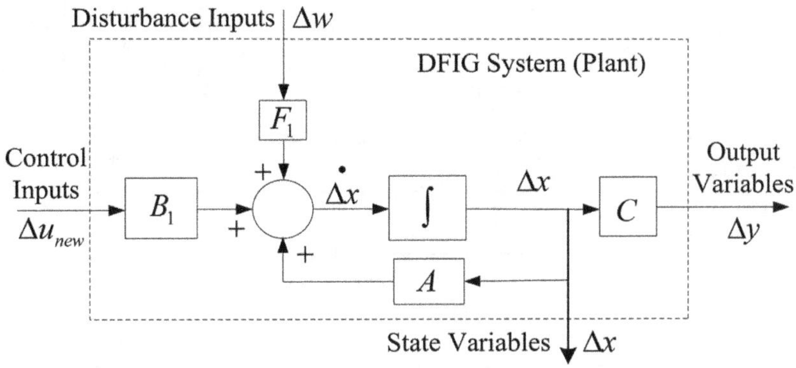

Figure 2.13. State-space based block diagram approach of the DFIG system [28].

of the wind speed creates many issues in the modelling and control of the DFIG-based WECS. In order to overcome these technical issues, an appropriate controller is needed, which improves the small as well as the large signal stability of the studied system and hence makes the system more stable and robust.

In Figure 2.13, the DFIG system inputs are bifurcated into two parts such as control inputs (Δu_{new}) and disturbance inputs (Δw) [28]. Therefore, the corresponding change in matrices B and D are as $\left(B = \left[B_1 \ \vdots \ F_1\right]\right)$ and $\left(D = \left[D_1 \ \vdots \ F_2\right]\right)$, respectively. Therefore, equation (2.70) becomes:

$$\Delta\dot{x} = A\Delta x + B_1\Delta u_{new} + F_1\Delta w \tag{2.80}$$

In this work, all the state variables are assumed to be the output variables of the system. Thus, equation (2.71) can be written as follows:

$$\Delta y = C\Delta x \tag{2.81}$$

The deviations in the states, control inputs, disturbance inputs, and the outputs variables of the DFIG system are considered as follows:

$$\Delta x = \left[\Delta i_{qs}\ \Delta i_{ds}\ \Delta i_{qr}\ \Delta i_{dr}\ \Delta i_{qg}\ \Delta i_{dg}\ \Delta\omega_r\ v_{dc}\right] \tag{2.82}$$

$$\Delta u_{new} = \left[\Delta v_{qr}\ \Delta v_{dr}\ \Delta v_{qg}\ \Delta v_{dg}\right] \tag{2.83}$$

$$\Delta w = \left[\Delta v_{qs}\ \Delta v_{ds}\right] \tag{2.84}$$

$$\Delta y = \left[\Delta i_{qs}\ \Delta i_{ds}\ \Delta i_{qr}\ \Delta i_{dr}\ \Delta i_{qg}\ \Delta i_{dg}\ \Delta\omega_r\ v_{dc}\right] \tag{2.85}$$

2.4.1 PI Controller

Generally, PI controllers provide adequate outcomes in the DFIG single-input-single-output (SISO) system but face troubles in the investigation of multi-input-multi-output (MIMO)

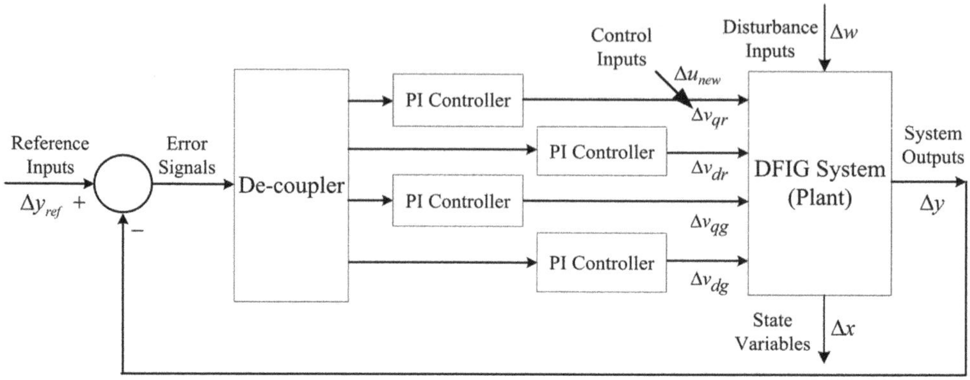

Figure 2.14. Structure of the PI control approach of the DFIG system.

systems. Furthermore, PI controllers are very successful in the decoupled control of DFIG rotor and grid side converter, but face difficulties in controlling the rotor and grid side converter as a whole [52]. If all the output is accessible for feedback, then, the PI control mechanism can be employed to diminish the output deviations. The tunable control structure of the PI controller is illustrated in Figure 2.14. The installed de-coupler in Figure 2.14 provides the signal according to the number of control variables essential for controlling the DFIG system. In this work, the MATLAB programming (using loop-tune method) is utilized to tune the controller gains for the MIMO feedback system. The controller gains are adjusted automatically and generate the control signals according to the errors between the reference signals and the measured values of output signals.

The major steps that are followed in tuning the PI controllers are given as follows:

Step 1: To tune the control system, first of all, create a state-space model of the plant.
Step 2: Then create tunable models of the controller elements and interconnect them to build a controller model.
Step 3: Then use a ***"looptune"*** MATLAB command to tune the free parameters of the controller model.
Step 4: Finally, examine the performance of the tuned system with the help of success indicator to confirm that the tuned controller yields desirable performance.

The PI controller is not enough for getting the desired results because of its limited operation.

Recently, the DFIG system utilized the concept of the LQR [29, 53, 54], observer-based [53, 54] control action to overcome the complications faced by the PI/PID controller. The LQR design approach guarantees the optimality conditions such as minimize the state deviations as well as the control input energy. To design this controller, the state-space of the DFIG system must be defined first. The state-space concept of the DFIG has already been discussed in the previous section 2.3, can handle these concerns in a better way because it facilitates a compact mathematical methodology to study the dynamic behavior of the system [3, 38, 49]. This chapter focuses on the control designs of the wind-driven DFIG system, which rely on the state-space methodology.

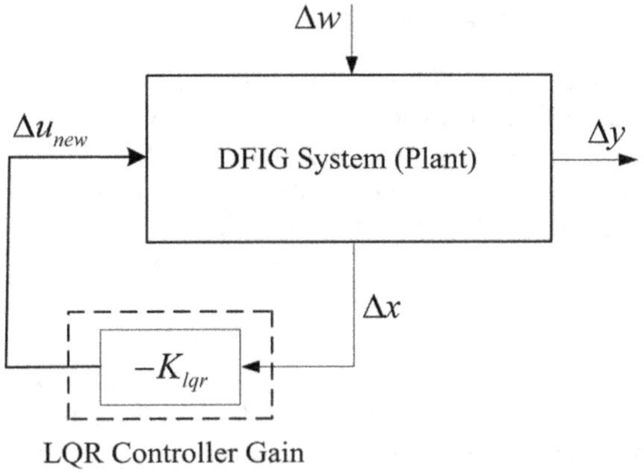

Figure 2.15. Structure of the LQR control approach of the DFIG system [27].

2.4.2 LQR Controller

It offers an alternate control design approach for the DFIG-based WECS. In this case, it is assumed that all the states and the outputs of the linearized system are measurable and accessible for the feedback. Thus, the full state variable feedback control through LQR method can be implemented. Figure 2.15 demonstrates the structure of the conventional LQR controller. LQR control offers a proficient way of calculating unconstrained optimization problems with linear dynamics and quadratic cost functions. The LQR offers an efficient approach of calculating the state feedback gain matrix and sets all the states to zero [3]. The performance index used in the design of the LQR controller is assumed as follows [53, 54]:

$$J_{lqr} = \int_0^{\infty} [\Delta x]^T Q [\Delta x] + \left[\Delta u_{new}\right]^T R \left[\Delta u_{new}\right] dt \tag{2.86}$$

where Q is a real positive semi-definite and R is a positive-definite symmetric matrix. To obtain the optimal feedback gain matrix ($K_{lqr} = R^{-1}B_1^T P_{dlqr}$), first, choose the proper weight matrices (Q and R) and then obtained the positive-definite matrix (P_{dlqr}) by using the following Riccati equation:

$$A^T P_{dlqr} + P_{dlqr} A - P_{dlqr} B_1 R^{-1} B_1^T P_{dlqr} + Q = 0 \tag{2.87}$$

The value of K_{lqr} is acceptable only when the system remains stable, that is, all the eigenvalues of the matrix is $A_{lqr} = A - B_1 K_{lqr}$ must have negative real parts. The closed-loop system is asymptotically stable if $\left(A, B_1\right)$ and (Q, A) are stabilizable and observable, respectively.

However, the assumption that all the state and the output signals are measurable and available for the feedback is often unrealistic because of physical or economic constraints. This type of issue is typically handled by estimating the states using an observer, which is discussed in Section 2.3.

2.4.3 Observer-based Controller

This section presents the concept of observer design for the DFIG system. When state variables are not available for feedback because of nonavailability of sensors, expensive sensors, and unacceptable sensors (because of high noise and high power consumption), then observer design is needed to estimate unavailable state variables. The state-space representation is utilized as a basic model for the observer design and the full/reduced order model can be designed on the basis of the desired observer model order.

The plant is defined by the previously discussed state-space equations (2.70) and (2.71). Consider $\widehat{\Delta x}$ is the observed (or estimated) state, where symbol "∧" denotes estimation. The observer mathematical model is the same as that of the considered plant, except the additional estimation error (or observer error) term. This error is actually the difference between the measured and observed output. Therefore, the observer mathematical model can be defined as follows [35, 53, 54]:

$$\dot{\widehat{\Delta x}} = A\widehat{\Delta x} + B_1 \Delta u_{new} + F_1 \Delta w + K_{ob}\left(\Delta y - \widehat{\Delta y}\right) \tag{2.88}$$

$$\widehat{\Delta y} = C\widehat{\Delta x} \tag{2.89}$$

where K_{ob} is known as observer gain.

From equations (2.88) and (2.89), it can be written as follows:

$$\dot{\widehat{\Delta x}} = \left(A - K_{ob}C\right)\widehat{\Delta x} + B_1 \Delta u_{new} + F_1 \Delta w + K_{ob}\Delta y \tag{2.90}$$

The observer error dynamics can be obtained by subtracting equation (2.90) from equation (2.80) as follows [35, 53, 54]:

$$\Delta\dot{x} - \dot{\widehat{\Delta x}} = \left(A - K_{ob}C\right)\left(\Delta x - \widehat{\Delta x}\right) \tag{2.91}$$

If the error vector is assumed as $e = x - \widehat{x}$, therefore

$$\Delta\dot{e} = \Delta\dot{x} - \dot{\widehat{\Delta x}} = \left(A - K_{ob}C\right)e \tag{2.92}$$

The state-feedback control based on the observed state $\widehat{x}$ can be written as follows:

$$\Delta u_{new} = -K\widehat{\Delta x} \tag{2.93}$$

where K is known as the state-feedback control gain.

Therefore, the state equation becomes:

$$\Delta\dot{x} = \left(A - B_1 K\right)\Delta x + B_1 K \Delta e + F_1 \Delta w \tag{2.94}$$

The augmented state equation, which describes the dynamics of the observed state-feedback control system, can be obtained by combining equations (2.92) and (2.94) as follows [53, 54]:

$$\begin{bmatrix} \Delta\dot{x} \\ \Delta\dot{e} \end{bmatrix} = \begin{bmatrix} A - B_1 K & B_1 K \\ 0 & A - CK_{ob} \end{bmatrix} \begin{bmatrix} \Delta x \\ \Delta e \end{bmatrix} + \begin{bmatrix} F_1 \\ 0 \end{bmatrix} [\Delta w] \tag{2.95}$$

Here, from equation (2.95), the augmented system matrix and then the characteristic equation for the same system can be obtained as follows:

$$A_{aug} = \begin{bmatrix} A - B_1 K & B_1 K \\ 0 & A - CK_{ob} \end{bmatrix} \tag{2.96}$$

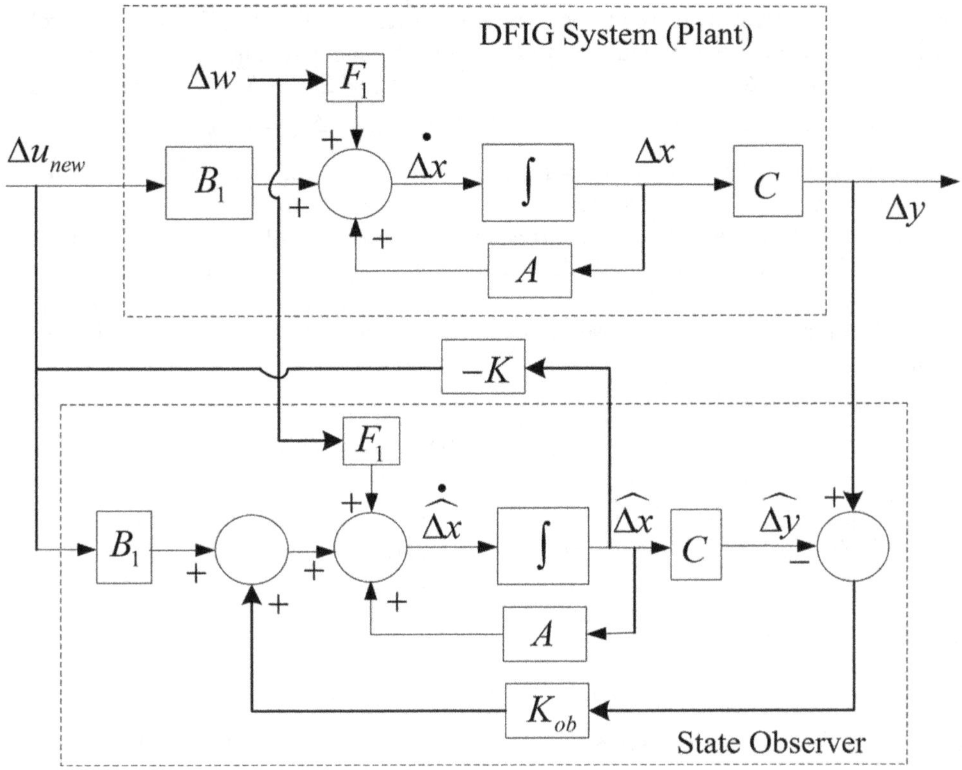

Figure 2.16. Block diagram of the observed state-feedback control system [53].

$$|sI - A_{aug}| = 0 \tag{2.97}$$

$$|sI - A + B_1 K||sI - A + CK_{ob}| = 0 \tag{2.98}$$

From equation (2.98), it can be observed that both the feedback design and the observer design are independent of each other because the poles of the augmented system (i.e., observed state-feedback system) comprises the poles because of the feedback design $\left[sI - \left(A - B_1 K\right)\right]$ and the poles because of the observer design $\left[sI - \left(A - CK_{ob}\right)\right]$. Thus, it can be designed separately and joined to form the observed state-feedback control system. The block diagram for this system is shown in Figure 2.16 [53].

2.5 Results and Discussions

The small-signal stability analysis was done with the help of eigenvalues of the system (*for model-II*). The system matrix was evaluated in the case of different controllers, which are used in determining the eigenvalues of the system. System stability has been decided on the basis of eigenvalues (i.e., poles location).

The state-feedback gain matrix (K) of the observer model is designed using the LQR control method (i.e., assume $K = K_{lqr}$) and the observer gain matrix (K_{ob}) is designed by choosing the observer poles (L) to be two times faster than the state-feedback controller poles, which ensures the estimation error (Δe) converges to zero quickly.

Suppose $\Delta e_1 = \Delta i_{qs} - \widehat{\Delta i_{qs}}$; $\Delta e_2 = \Delta i_{ds} - \widehat{\Delta i_{ds}}$; $\Delta e_3 = \Delta i_{qr} - \widehat{\Delta i_{qr}}$; $\Delta e_4 = \Delta i_{dr} - \widehat{\Delta i_{dr}}$; $\Delta e_5 = \Delta i_{qg} - \widehat{\Delta i_{qg}}$; $\Delta e_6 = \Delta i_{dg} - \widehat{\Delta i_{dg}}$; $\Delta e_7 = \Delta \omega_r - \widehat{\Delta \omega_r}$; $\Delta e_8 = \Delta v_{dc} - \widehat{\Delta v_{dc}}$

MATLAB/SIMULINK software was used for the simulation of the system considered in this chapter. The system parameters are given in the Appendix-A.

The obtained PI controller gains corresponding to the control signal as in Figure 2.14 is shown in Table 2.1. To find out the LQR controller gain, the weighting matrices Q and R are assumed as diag([1 0.0001 100 0.0001 1 1 1 1]) and diag([0.1 0.1 0.001 0.01]), respectively. Therefore, LQR and the observer-based controller gains are obtained, which are as follows:

$$K_{lqr} = \begin{bmatrix} 0.29 & 5.19 & -31.48 & 5.17 & 3.9\times 10^{-06} & -8.3\times 10^{-06} & 0.05 & -0.21 \\ 1.56 & -4.17 & 1.57 & -4.19 & -8.3\times 10^{-06} & 1.8\times 10^{-05} & 1.14 & 0.44 \\ 0.15 & -0.05 & 0.15 & -0.05 & 31.61 & -0.04 & -1.29 & -31.24 \\ 0.0002 & -0.00005 & 0.0002 & -0.0006 & -0.004 & 9.88 & -0.001 & -0.03 \end{bmatrix}$$

$$K_{ob} = \begin{bmatrix} 13.15 & 1.6\times 10^{04} & 17.14 & 1.7\times 10^{04} & 0 & 0 & 152.29 & 0 \\ -1.6\times 10^{04} & 13.15 & -1.7\times 10^{04} & 17.14 & 0 & 0 & -1.3\times 10^{04} & 0 \\ 15.58 & -1.6\times 10^{04} & 2.0\times 10^{05} & -1.6\times 10^{04} & 0 & 0 & -153.81 & 0 \\ 1.6\times 10^{04} & 15.58 & 1.6\times 10^{04} & 8.8\times 10^{05} & 0 & 0 & 1.3\times 10^{04} & 0 \\ 0 & 0 & 0 & 0 & 2.8\times 10^{05} & -314.16 & 0 & 0 \\ 0 & 0 & 0 & 0 & -314.16 & 63.91 & 0 & 0 \\ -0.031 & 0.00007 & -0.03 & -0.5 & 0 & 0 & 126.13 & 0 \\ 0 & 0 & 5.56 & 0 & -8.33 & -0.007 & 0 & 16.97 \end{bmatrix}$$

The above controller parameters are used to find out the controlled DFIG system matrix, which helps in evaluating the corresponding eigenvalues and tabulated in Table 2.2. From Table 2.2, it is observed that the eigenvalues in case of the robust controller are significantly shifted toward the left-hand side of the imaginary axis and also reduces the number of oscillatory modes. Both reasons lead to enhance system stability.

Table 2.1. PI controller gains.

Signal	K_P	K_I
Corresponding to Δv_{qr}	-1.23×10^{-05}	0.013
Corresponding to Δv_{dr}	9.59×10^{-05}	−0.116
Corresponding to Δv_{qg}	1.56×10^{-04}	0.585
Corresponding to Δv_{dg}	-2.97×10^{-05}	−0.0201

Table 2.2. Eigenvalues of the DFIG system with different controllers.

PI Controller	LQR Controller	Observer-based Controller
$-16.18 \pm j313.51$	−100,027.60	−441,536.90
$-167.13 \pm j314.07$	−441,536.90	−100,027.60
−4.62	−139,626.95	−139,626.95
$-14.52 \pm j120.08$	$-14.48 \pm j313.87$	$-14.48 \pm j313.87$
−0.76	−115.73	−115.73
−0.12	−62.98	−62.98
−0.03	−8.49	−8.49
$-0.01 \pm j0.00032$		$-28.97 \pm j627.74$
		−200,055.21
		−883,073.81
		−279,253.90
		−231.46
		−125.96
		−16.97

The response curve of the errors between actual states and estimated states to the initial condition can be determined and shown in Figures 2.17–2.24. Figures 2.17–2.24 represent the natural response (*not forced response, where input must be applied*) of the errors between actual and observed states with respect to the initial condition of every system state, which itself acts as a disturbance by setting the input vectors to zero [53]. In order to design the observer, the matrices A, B_1, F_1, and C in the observer are considered to be exactly the same as of the original physical plant.

From Figures 2.17–2.24, it is seen that the magnitude of errors in all the cases are approached zero, which indicate that the observer is stable and well designed. In other words, it can also conclude that there are no discrepancies in the observer matrices A, B_1, F_1, and C and in the physical plant.

2.6 Summary

In this chapter, different mathematical models for generator (in model-I, stator transient neglected whereas, in model-II, stator transient is considered) and for drive-train (single-mass and two-mass) of DFIG-based system are addressed in detail. The model is established for small-signal stability analysis. In addition, the modern control approaches of LQR and observer design for the wind-driven system have also been discussed. This study can offer a good understanding of developing the mathematical model for SISO and MIMO systems. Further, the response to the initial condition of the error between the actual state and observed state have been presented. From the results, it can be concluded that system damping is improved in case of modern control approaches because of the significant poles shifting toward the left-hand side in s-plane. Results also indicated that the estimation errors decay very fast and converge to zero very quickly (within 0.3 s), and hence, it can be said that the overall performance of the wind-driven DFIG system is enhanced.

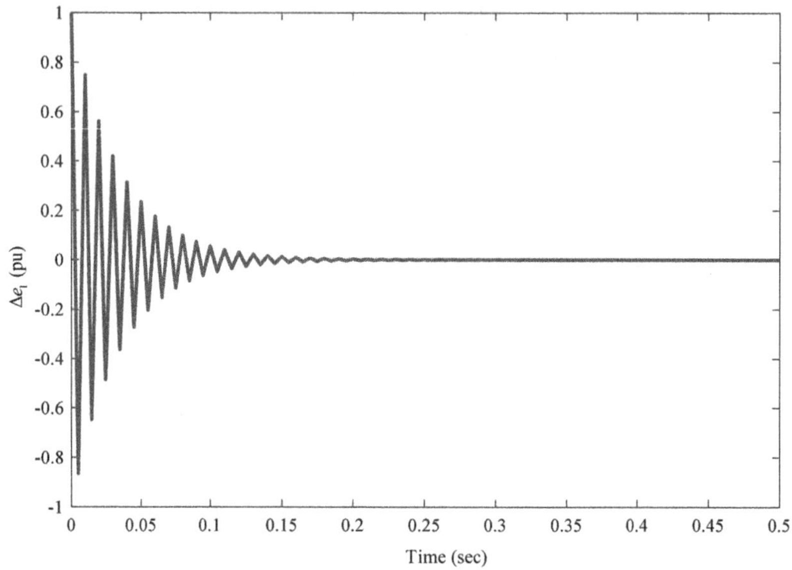

Figure 2.17. Response curve of the errors between actual q-axis stator current and observed q-axis stator current to the initial condition.

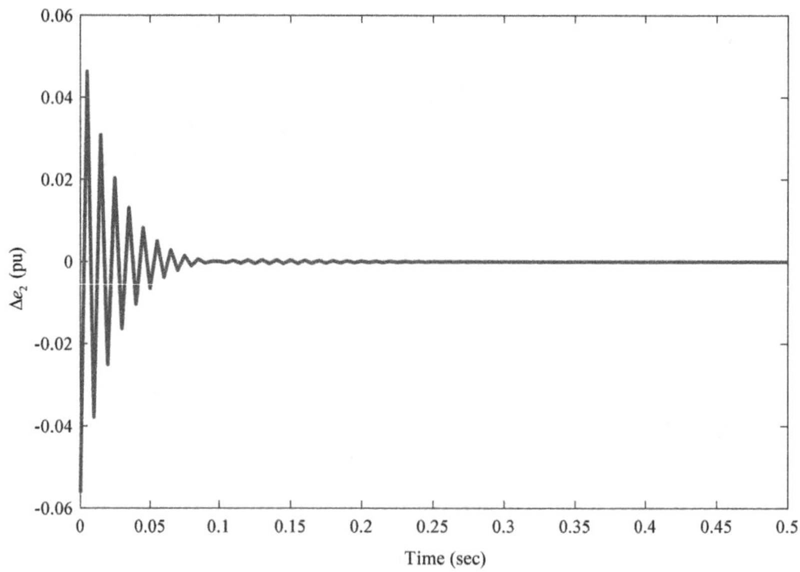

Figure 2.18. Response curve of the errors between actual d-axis stator current and observed d-axis stator current to the initial condition.

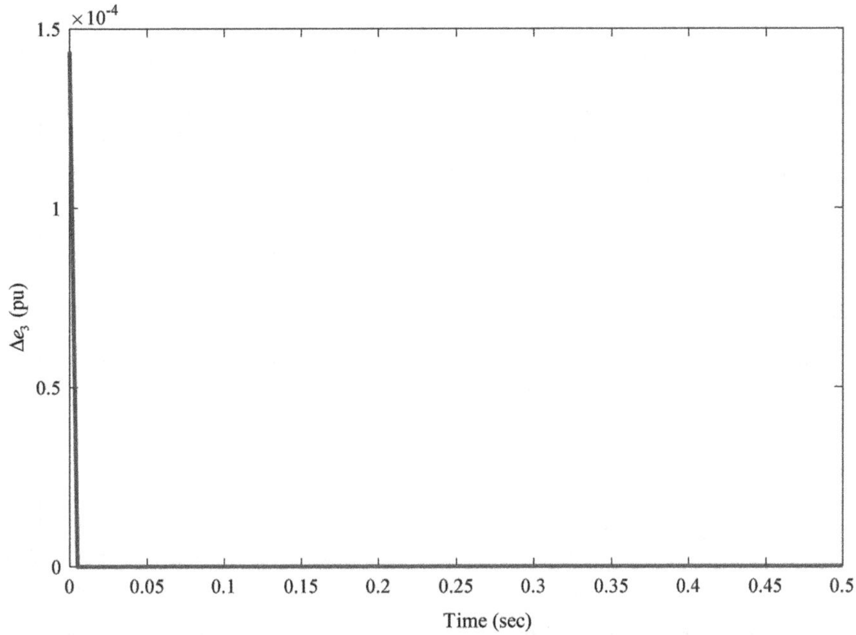

Figure 2.19. Response curve of the errors between actual *q*-axis rotor current and observed *q*-axis rotor current to the initial condition.

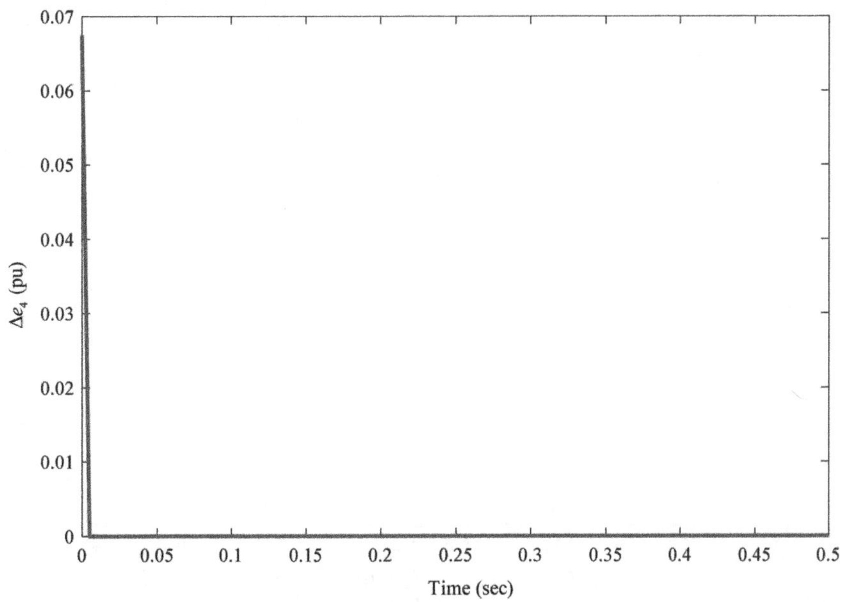

Figure 2.20. Response curve of the errors between actual *d*-axis rotor current and observed *d*-axis rotor current to the initial condition.

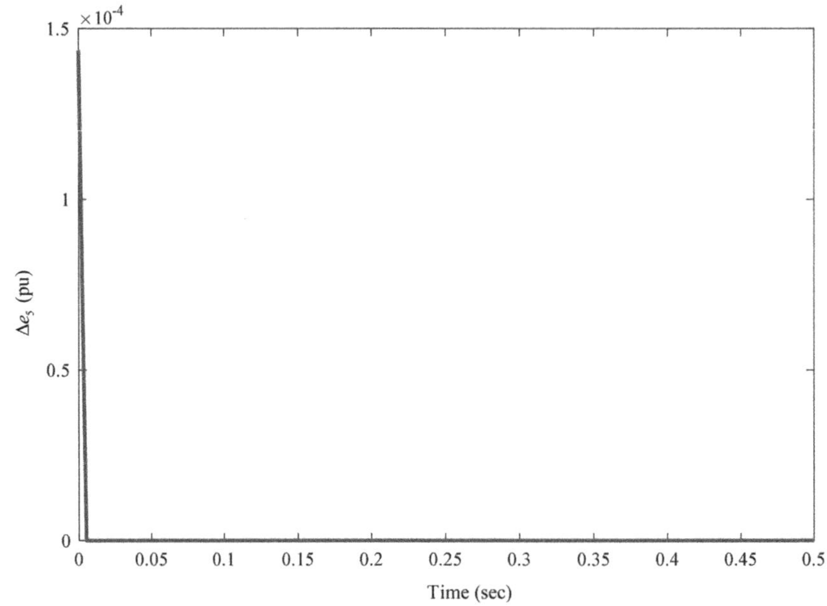

Figure 2.21. Response curve of the errors between actual *q*-axis GSC current and observed *q*-axis GSC current to the initial condition.

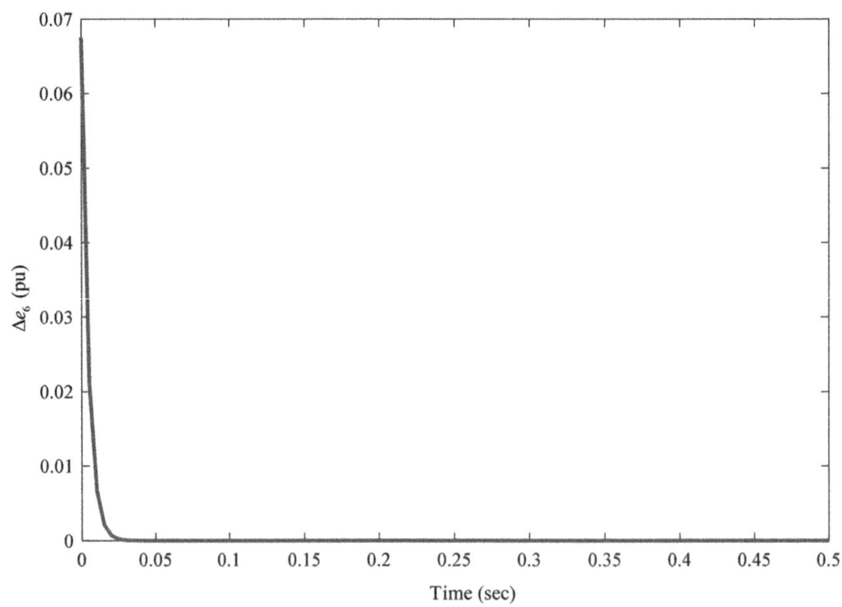

Figure 2.22. Response curve of the errors between actual *d*-axis GSC current and observed *d*-axis GSC current to the initial condition.

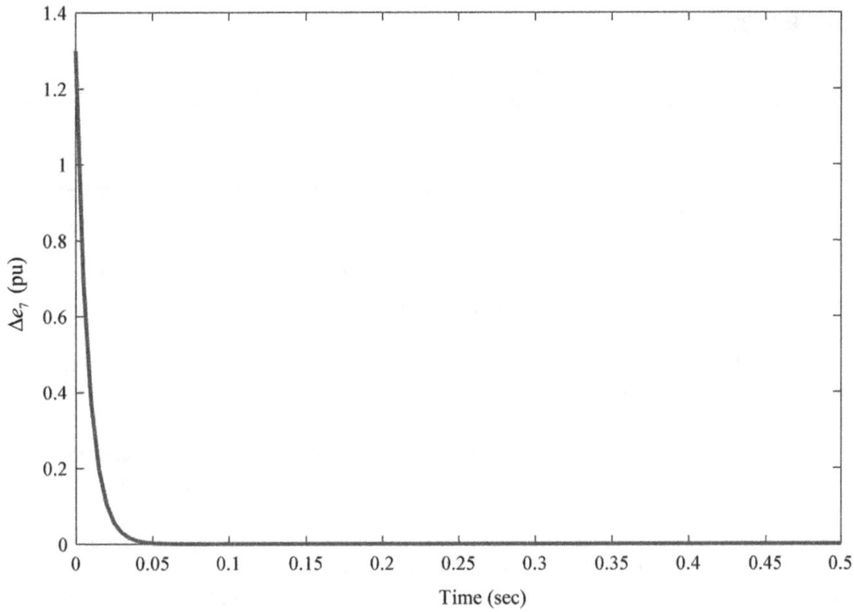

Figure 2.23. Response curve of the errors between actual generator-angular speed and observed generator-angular speed to the initial condition.

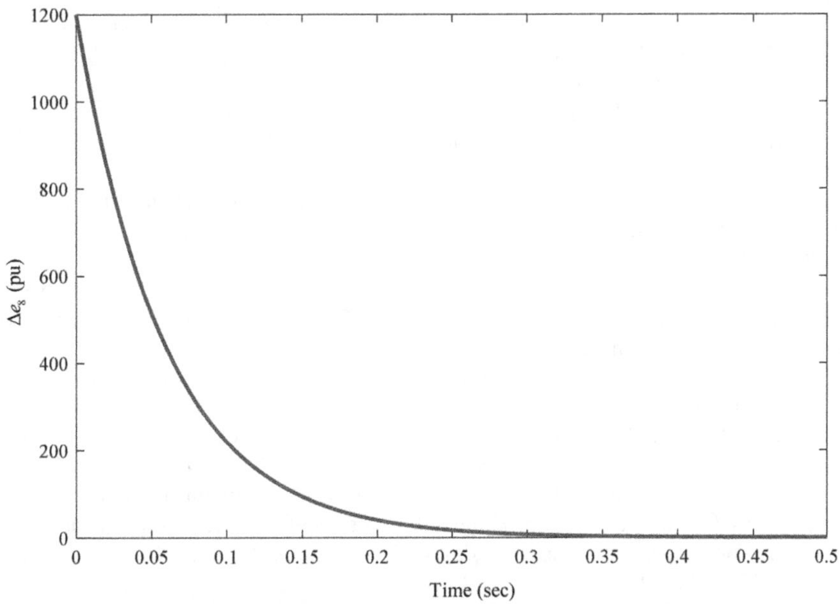

Figure 2.24. Response curve of the errors between actual DC-link voltage and observed DC-link voltage to the initial condition.

2.7 References

1. M. Ahmed, M. EL-Shimy, and M. A. Badr, "Advanced modeling and analysis of the loading capability limits of doubly-fed induction generators," *Sustainable Energy Technologies Assessments*, vol. 7, pp. 79–90, 2014.
2. S. Ravichandran, R. P. Kumudinidevi, S. G. Bharathidasan, and V. Evangelin Jeba, "Coordinated controller design of grid connected DFIG based wind turbine using response surface methodology and NSGA II," *Sustainable Energy Technologies Assessments*, vol. 8, pp. 120–130, 2014.
3. B. Mehta, P. Bhatt, and V. Pandya, "Small signal stability enhancement of DFIG based wind power system using optimized controllers parameters," *International Journal Electrical Power Energy Systems*, vol. 70, pp. 70–82, 2015.
4. F. Wu, X.-P. Zhang, K. Godfrey, and P. Ju, "Small signal stability analysis and optimal control of a wind turbine with doubly fed induction generator," *IET Generation Transmission Distribution*, vol. 1, no. 5, pp. 751–760, 2007.
5. B. C. Pal, and F. Mei, "Modelling adequacy of the doubly fed induction generator for small-signal stability studies in power systems," *IET Renewable Power Generation*, vol. 2, no. 3, pp. 181–190, 2008.
6. G. Abad, J. Lopez, M. Rodríguez, L. Marroyo, and G. Iwanski, *Doubly Fed Induction Machine: Modeling and Control for Wind Energy Generation*, John Wiley & Sons, 2011.
7. F. Mei, and B. Pal, "Modelling and small-signal analysis of a grid connected doubly-fed induction generator," *IEEE Power Engineering Society General Meeting*, pp. 2101–2108, 2005.
8. A. Oudah, I. I. Mohd, and A. Hameed, "Wind turbines control: Features and trends," *Modern Applied Science*, vol. 8, no. 6, pp. 272–295, 2014.
9. M. Tazil, V. Kumar, R. C. Bansal, S. Kong, Z. Y. Dong, and W. Freitas, "Three-phase doubly fed induction generators: an overview," *IET Electric Power Applications*, vol. 4, no. 2, pp. 75–89, 2010.
10. Y. Mishra, S. Mishra, L. Fangxing, Z. Y. Dong, and R. C. Bansal, "Small-signal stability analysis of a DFIG-based wind power system under different modes of operation," *IEEE Transactions Energy Conversion*, vol. 24, no. 4, pp. 972–982, 2009.
11. M. S. Vicatos, and J. A. Teqopoulos, "Steady state analysis of a doubly-fed induction generator under synchronous operation," *IEEE Transactions Energy Conversion*, vol. 4, no. 3, pp. 495–501, 1989.
12. J. G. Slootweg, H. Polinder, and W. L. Kling, "Dynamic modelling of a wind turbine with doubly fed induction generator," *Power Engineering Society Summer Meeting, Conference Proceedings Cat No01CH37262*, vol. 1, pp. 644–649, 2001.
13. D. Gautam, V. Vittal, and T. Harbour, "Impact of increased penetration of DFIG-based wind turbine generators on transient and small signal stability of power systems," *IEEE Transactions Power Systems*, vol. 24, no. 3, pp. 1426–1434, 2009.
14. J. G. Slootweg, and W. L. Kling, "The impact of large scale wind power generation on power system oscillations," *Electric Power Systems Research*, vol. 67, no. 1, pp. 9–20, 2003.
15. R. D. Fernández, R. J. Mantz, and P. E. Battaiotto, "Impact of wind farms on a power system. An eigenvalue analysis approach," *Renewable Energy*, vol. 32, no. 10, pp. 1676–1688, 2007.
16. A. Kunwar, and R. C. Bansal, "Effect of high-voltage direct current link on small signal stability of a power system with different penetration level of doubly fed induction generator and direct drive synchronous generator based wind farms," *International Journal Green Energy*, vol. 13, no. 4, pp. 335–343, 2016.
17. L. Yang, Z. Xu, J. Østergaard, Z. Y. Dong, K. P. Wong, and X. Ma, "Oscillatory stability and eigenvalue sensitivity analysis of a DFIG wind turbine system," *IEEE Transactions Energy Conversion*, vol. 26, no. 1, pp. 328–339, 2011.

18. E. Tremblay, S. Atayde, and A. Chandra, "Comparative study of control strategies for the doubly fed induction generator in wind energy conversion systems: A DSP-based implementation approach," *IEEE Transactions Sustainable Energy*, vol. 2, no. 3, pp. 288–299, 2011.
19. M. Yamamoto, and O. Motoyoshi, "Active and reactive power control for doubly-fed wound rotor induction generator," *IEEE Transactions Power Electronics*, vol. 6, no. 4, pp. 624–629, 1991.
20. R. Pena, J. C. Clare, and G. M. Asher, "Doubly fed induction generator using back-to-back PWM converters and its application to variable-speed wind-energy generation," *IEE Proceedings Electric Power Applications*, vol. 143, no. 3, pp. 231–241, 1996.
21. Y. Mishra, S. Mishra, M. Tripathy, N. Senroy, and Z. Y. Dong, "Improving stability of a DFIG-based wind power system with tuned damping controller," *IEEE Transactions Energy Conversion*, vol. 24, no. 3, pp. 650–660, 2009.
22. Y. Mishra, S. Mishra, and F. Li, "Coordinated tuning of DFIG-based wind turbines and batteries using bacteria foraging technique for maintaining constant grid power output," *IEEE Systems Journal*, vol. 6, no. 1, pp. 16–26, 2012.
23. J. Yang, Y. Chen, and Y. Hsu, "Small-signal stability analysis and particle swarm optimisation self-tuning frequency control for an islanding system with DFIG wind farm," *IET Generation Transmission Distribution*, vol. 13, no. 4, pp. 563–574, 2019.
24. Y. Zhou, L. Zhao, and W. Lee, "Robustness analysis of dynamic equivalent model of DFIG wind farm for stability study," *IEEE Transactions Industry Applications*, vol. 54, no. 6, pp. 5682–5690, 2018.
25. A. Tewari, *Modern Control Design With MATLAB and SIMULINK*, John Wiley & Sons, Ltd., 2002.
26. N. K. Roy, H. R. Pota, M. A. Mahmud, and M. J. Hossain, "Voltage control of emerging distribution systems with induction motor loads using robust LQG approach," *International Transactions Electrical Energy Systems*, vol. 24, no. 7, pp. 927–943, 2014.
27. G. Sharma, Ibraheem, and R. C. Bansal, "DFIG Based AGC of power system using robust methodology," *Energy Procedia*, vol. 105, pp. 590–595, 2017.
28. R. Bhushan, and K. Chatterjee, "Mathematical modeling and control of DFIG-based wind energy system by using optimized linear quadratic regulator weight matrices," *International Transactions Electrical Energy Systems*, vol. 27, no. 11, p. e2416, 2017.
29. M. S. Ali, H. R. Pota, M. A. Mahmud, and M. J. Hossain, "Critical parameter analysis and LQR control for the distribution system with DFIG," *Power and Energy Society General Meeting*, no. 2012, pp. 1–6, 2012.
30. L. Barros Sales, W. Santos Mota, J. J. Da Silva, and C. M. Vital Barros, "An optimal control strategy for DFIG," *IEEE International Conference on Industrial Technology (ICIT)*, pp. 1727–1732, 2010.
31. M. J. Hossain, T. K. Saha, N. Mithulananthan, and H. R. Pota, "Control strategies for augmenting LVRT capability of DFIGs in interconnected power systems," *IEEE Transactions Industrial Electronics*, vol. 60, no. 6, pp. 2510–2522, 2013.
32. B. Qin, and H. Sun, "State dependent riccati equation based rotor-side converter control for doubly fed wind generator," *IEEE Access*, vol. 6, pp. 27853–27863, 2018.
33. P. Li, J. I. E. Wang, L. Xiong, and M. Ma, "Robust nonlinear controller design for damping of sub-synchronous control interaction in DFIG-based wind farms," *IEEE Access*, vol. 7, pp. 16626–16637, 2019.
34. D. Li, W. Cai, P. Li, Z. Jia, and H. Chen, "Neuroadaptive variable speed control of wind turbine with wind speed estimation," *IEEE Transactions Industrial Electronics*, vol. 63, no. 12, pp. 7754–7764, 2016.
35. M. Ghafouri, U. Karaagac, H. Karimi, S. Jensen, J. Mahseredjian, and S. O. Faried, "An LQR controller for damping of subsynchronous interaction in DFIG-based wind farms," *IEEE Transactions Power Systems*, vol. 32, no. 6, pp. 4934–4942, 2017.

36. S. K. Sahoo, G. Sharma, A. Panwar, and R. C. Bansal, "Frequency regulation of wind integrated power system using dual mode the fuzzy," *Energy Procedia*, vol. 158, pp. 6321–6327, 2019.
37. P. Kundur, *Power System Control and Stability*, Mcgraw-Hill, Inc., 1994.
38. R. Bhushan, and K. Chatterjee, "Effects of parameter variation in DFIG-based grid connected system with a FACTS device for small-signal stability analysis," IET Generation Transmission Distribution, vol. 11, no. 11, pp. 2762–2777, 2017.
39. E. B. Muhando, T. Senjyu, A. Uehara, T. Funabashi, and C. H. Kim, "LQG design for megawatt-class WECS with DFIG based on functional models' fidelity prerequisites," *IEEE Transactions Energy Conversion*, vol. 24, no. 4, pp. 893–904, 2009.
40. A. Moharana, and R. K. Varma, "Subsynchronous resonance in single-cage self-excited-induction-generator-based wind farm connected to series-compensated lines," *IET Generation Transmission Distribution*, vol. 5, no. 12, pp. 1221–1232, 2011.
41. S. G. Varzaneh, H. Rastegar, and G. B. Gharehpetian, "A new three-mode maximum power point tracking algorithm for doubly fed induction generator based wind energy conversion system," Electric Power Components Systems, vol. 42, no. 1, pp. 45–59, 2014.
42. F. Mei, and B. Pal, "Modal analysis of grid-connected doubly fed induction generators," *IEEE Transactions Energy Conversion*, vol. 22, no. 3, pp. 728–736, 2007.
43. H. A. Mohammadpour, E. Santi, and A. Ghaderi, "Analysis of sub-synchronous resonance in doubly-fed induction generator-based wind farms interfaced with gate – controlled series capacitor," *IET Generation Transmission Distribution*, vol. 8, no. 12, pp. 1998–2011, 2014.
44. K. Swami Naidu, and B. Singh, "Doubly fed induction generator for wind energy conversion systems with integrated active filter capabilities," *IEEE Transactions Industrial Informatics*, vol. 11, no. 4, pp. 923–933, 2015.
45. J.J. Justo, F. Mwasilu, and J. W. Jung, "Doubly fed induction generator wind turbines: A novel integrated protection circuit for low-voltage ride-through strategy," *Journal Renewable Sustainable Energy*, vol. 6, no. 5, pp. 3129–3144, 2014.
46. L. Huchel, M. S. El Moursi, and H. H. Zeineldin, "A parallel capacitor control strategy for enhanced FRT capability of DFIG," *IEEE Transactions Sustainable Energy*, vol. 6, no. 2, pp. 303–312, 2015.
47. S. Swain, and P. K. Ray, "Short circuit fault analysis in a grid connected DFIG based wind energy system with active crowbar protection circuit for ridethrough capability and power quality improvement," *International Journal Electrical Power Energy Systems*, vol. 84, pp. 64–75, 2017.
48. S. M. Tripathi, A. N. Tiwari, and D. Singh, "Optimum design of proportional-integral controllers in grid-integrated PMSG-based wind energy conversion system," *International Transactions Electrical Energy Systems*, vol. 26, pp. 1006–1031, 2016.
49. G. Díaz, C. González-Morán, and C. Viescas, "State-space representation of DFIG-based wind power plants," *IET Renewable Power Generation*, vol. 7, no. 3, pp. 254–264, 2013.
50. S. Y. Kong, R. C. Bansal, and Z. Y. Dong, "Comparative small-signal stability analyses of PMSG-, DFIG- and SCIG-based wind farms," *International Journal Ambient Energy*, vol. 33, no. 2, pp. 87–97, 2012.
51. B. Mehta, P. Bhatt, and V. Pandya, "Small signal stability analysis of power systems with DFIG based wind power penetration," *International Journal Electrical Power Energy Systems*, vol. 58, pp. 64–74, 2014.
52. B. Kedjar, and K. Al-Haddad, "DSP-based implementation of an LQR with integral action for a three-phase three-wire shunt active power filter," *IEEE Transactions Industrial Electronics*, vol. 56, no. 8, pp. 2821–2828, 2009.
53. O. Katsuhiko, *Modern Control Engineering*, 4th ed. Prentice Hall, Aeeizh, 1972.
54. R. C. Dorf, R. H. Bishop, and P. Hall, *Modern Control Systems*, Pearson Education, Inc., 2011.

2.8 Appendix-A: System Parameters

$P_{total} = 1$ pu

$N_{s2r} = 1.5$

$V_b = 1$ pu

$\omega_{r0} = 1.03$ pu

$R_r = 0.0055$ pu

$L_{rr} = 2.2839$ pu

$X_g = 0.0225$ pu

$i_{qr0} = 1.44 \times 10^{-4}$ pu

$v_{dc0} = 1,200$ V

$V_s = 1$ pu

$V_r = \dfrac{V_s}{N_{s2r}}$ pu

$\omega_s = 1$ pu

$k_{opt} = 0.5787$

$R_g = 0.012$ pu

$L_m = 2.25$ pu

$C_{dc} = 0.0001$ pu

$i_{dr0} = 0.0675$ pu

$v_{qg0} = 1.0000017244$ pu

$v_{qs0} = 1$ pu

$v_{qr0} = 0.6667$ pu

$\omega_B = 314$ rad/ sec

$H = 3.5$ sec

$L_{ss} = 2.2725$ pu

$X_m = 2.25$ pu

$i_{qs0} = 0.9999$ pu

$i_{qg0} = 1.44 \times 10^{-4}$ pu

$v_{ds0} = 0$

$v_{dr0} = 0$

$s = -0.03$

$R_s = 0.005$ pu

$X_s = 2.2725$ pu

$L_g = 0.0225$ pu

$i_{ds0} = -0.0563$ pu

$i_{dg0} = 0.0675$ pu

$v_{dg0} = 0.0008$ pu

CHAPTER 3

Voltage Stability with Wind Power Generation

Lasantha Meegahapola[1], Enkhtsetseg Munkhchuluun[1], and Kanchana Amarasekara[2]

[1]*Electrical and Biomedical Engineering, School of Engineering, RMIT University, Melbourne, Australia. lasantha.meegahapola@rmit.edu.au and s3554528@student.rmit.edu.au*

[2]*School of Electrical, Computer and Telecommunications Engineering, University of Wollongong, Wollongong, Australia. hwkma481@uowmail.edu.au*

Abstract Voltage stability is an imperative issue for power networks with significant wind penetration as synchronous machines that provide voltage support are being replaced by wind generation. This chapter comprehensively examines different types of voltage stability issues with wind power generation with special emphasis on reactive power capability of different wind generator types. More specifically, small-disturbance voltage stability and short- and long-term voltage stability are investigated with doubly fed induction generator (DFIG)-based wind generation using different test systems. Small-disturbance voltage stability is investigated by considering different wind penetration levels and control strategies for wind generators by using the IEEE-14 bus test system. The short-term voltage stability is investigated by using a test network with induction machine load, and the Nordic test system was used to assess the long-term voltage stability. According to the investigation, each voltage stability type is affected by DFIG wind generation and reactive power capability, and the dynamic response of DFIG wind farms have a direct influence on voltage stability.

Keywords: Doubly fed induction generator, Full-converter wind generator, Voltage stability, QV-curves, Wind power generation, Reactive power control.

3.1 Introduction

Wind power penetration levels are increasing rapidly, inflicting various challenges for power grid operators. Power system stability is one of the major concerns as the poorly managed high wind power penetration levels could potentially lead to catastrophic stability issues in power networks resulting in power blackouts [1,2]. Such blackout incidents have been reported in the past decade, and one of the first notable catastrophic incidents occurred in Europe in 2006 [1]. The poor fault ride-through (FRT) performance of wind farms was one of the critical factors for that major blackout [1]. Most recently, South Australian (SA) power network has also experienced a similar blackout incident. That incident has also occurred due to inadequate low-voltage ride-through (LVRT) settings at wind farms and voltage instability within the wind-rich region of the network following multiple contingency events in the SA network [2]. These are good examples of poor planning and operation of the power grid with high penetration of wind generation. Therefore, it is imperative to manage high wind power penetration levels in power networks effectively; particularly it is essential to analyze and manage potential instability issues in the power network.

This chapter primarily focuses on the voltage stability issues that could arise due to high wind power penetration levels in power networks. The wind power industry utilizes different technologies for wind power generation, hence each wind generation technology has different effects on voltage stability [3]. Also, voltage stability has different classifications based on the type of the disturbance and time frame [4], hence each type of voltage stability should be analyzed separately.

This chapter is organized as follows: voltage stability classification is presented in Section 3.2. Wind power generation technologies and their reactive power characteristics are described in Section 3.3. Voltage stability studies with wind power generation, reported in the literature, are succinctly discussed in Section 3.4. Small-disturbance stability (or static voltage stability) with wind power generation is investigated in Section 3.5. Short- and long-term voltage stability issues are investigated in Sections 3.6 and 3.7, respectively. Chapter conclusions are presented in Section 3.8.

3.2 Classification of Voltage Stability

Voltage stability issues can be categorized into different types: either based on the disturbance type (i.e., small-disturbance voltage stability and large-disturbance voltage stability) or based on the timescale (i.e., short- or long-term voltage stability) [4]. Figure 3.1 illustrates the classification for voltage stability.

The small-disturbance voltage stability refers to the ability of the power grid to maintain steady-state voltage levels after being subjected to small disturbances, such as small load variations. In some literature, it is also called static voltage stability. The voltage could be influenced by either active or reactive power in the network. For example, if the power system is heavily loaded (i.e., high active power demand), then system voltages may reach the steady-state stability limit. The voltage stability with respect to active power demand is typically assessed through PV curves. Similar to active power, reactive power also strongly influences the steady-state voltage. In particular, the transmission networks have a high reactance to resistance ratio (i.e., X/R ratio), hence reactive power has a strong coupling with the transmission system voltage. Typically, the reactive power sensitivity of voltage is assessed through QV curves. Therefore, power grid operators should maintain a sufficient reactive power reserve to maintain network voltage stability. The conventional synchronous generators fulfill a large portion of the reactive power requirement of the power network.

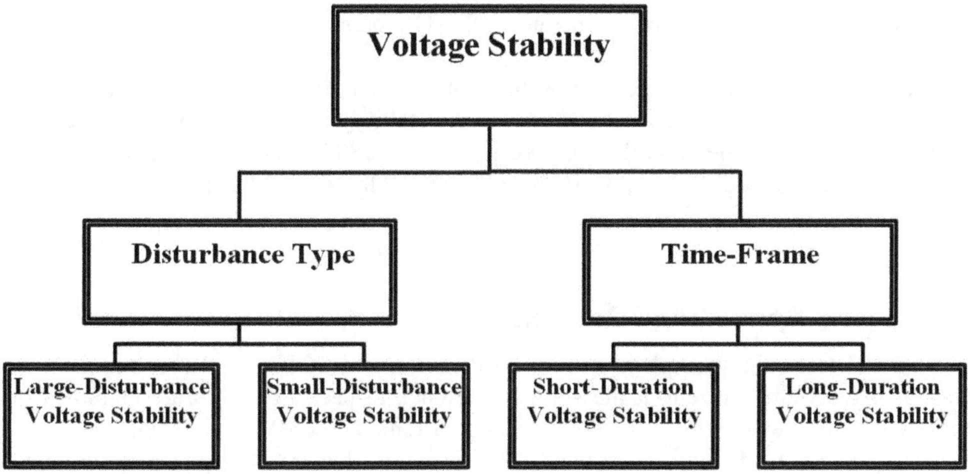

Figure 3.1. Classification of voltage stability.

Also, capacitor banks and static VAR compensators (SVCs) also contribute to the reactive power requirement of the power grid. Other than the PV and QV curve-based methods, modal analysis is also used to assess the small-disturbance voltage stability of the power network [4]. In addition, various indices have also been proposed for steady-state voltage stability assessment [5,6]. Although PV curves, QV curves, and modal analysis could be used for steady-state or small-disturbance stability assessment, large-disturbance voltage stability assessment cannot be performed using these tools, because during large disturbances control actions of various power system devices (e.g., synchronous generators, tap-changing transformers) influence voltage stability. Hence, dynamic simulations should be performed to assess the large-disturbance voltage stability of the power network.

The fast-acting devices, such as power electronic converters (PECs), and various types of loads, such as induction motors (IMs), could trigger voltage stability issues within short-term periods following disturbances in the network. Therefore, the voltage collapse/instability is not apparent at the instance of the fault, and voltages will progressively decrease to low values after a short-time period of the fault occurring. Therefore, these voltage stability issues are categorized as short-term voltage stability, and the time period could vary from 2 to 30 s. In contrast, the slow acting devices, such as tap-changing transformers, over-excitation limiters (OELs) in synchronous generators could also trigger voltage stability issues after a relatively long time period of the fault in the network. Therefore the slow acting devices driven voltage stability issues are categorized as long-term voltage, stability and period could vary from 1 min to several minutes. With the increased penetration of wind power penetration in the power system, the synchronous generators are phasing out from generation portfolio, hence wind generation would have a distinct impact on each of the voltage stability issue mentioned above.

3.3 Wind Power Generation Technologies

The wind generator technologies can be broadly categorized into four types: (1) *Type-1*: Squirrel-cage induction generator (SCIG)–based fixed-speed wind generator (FSWG) technology, (2) *Type-2*: Wound rotor induction machine (WRIM)–based limited variable-speed wind generation technology, (3) *Type-3*: WRIM-based doubly fed induction generator technology (DFIG), and (4) *Type-4*: Full-converter wind generator (FCWG) (based on WRIG, permanent-magnet synchronous generator [PMSG], or separately excited synchronous generator).

Each of the above wind generation technologies has distinct characteristics, hence have a different impact on power system voltage stability. The important parameter that influences voltage stability is the reactive power, hence basic architecture and reactive power characteristics of various wind generator technologies are discussed in subsequent subsections.

3.3.1 Fixed-Speed Wind Generator

The FSWG technology employs directly grid- connected SCIGs, and they require substantial magnetizing current at the start, hence back-to-back thyristors are employed in these wind turbines to reduce the high starting current. The FSWGs also require reactive power during its steady-state operation and also have a no-load reactive power requirement. Therefore, capacitor banks are employed in this type of wind generators, and typically the capacitor banks are rated at its no-load reactive power requirement [7]. A schematic and the reactive power characteristic of the FSWG are shown in Figure 3.2(a) and (b), respectively.

As shown in Figure 3.2(b), the reactive power requirement is increasing with the active power generation of the FSWG, hence adequate reactive power reserve must be available in power networks to achieve high fixed-speed wind generation-based wind power penetration in power networks.

3.3.2 Limited Variable-Speed Wind Generator

The limited variable-speed wind generator technology is similar to the FSWG technology, as the induction generator stator is directly connected to the grid. However, in limited variable-speed wind generator, it utilizes a WRIG, where rotor is connected to an external resistance, hence rotor resistance can be varied enabling limited variable-speed operation. As the variable rotor resistors offer limited variability in operating speed, it offers improved wind power extraction capability. A schematic of the limited variable-speed wind generator is shown in Figure 3.3.

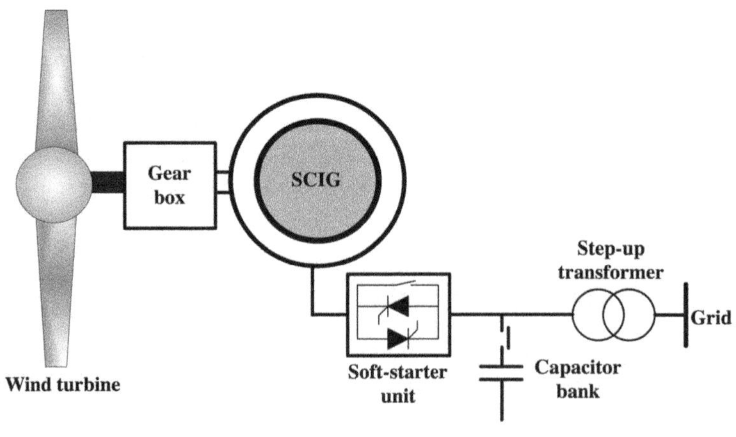

(a)

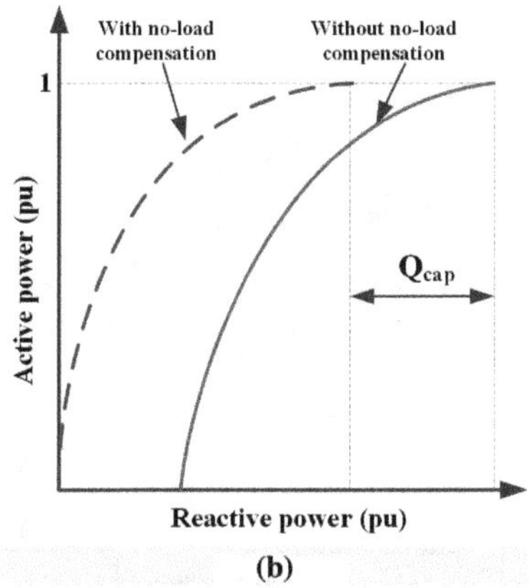

Figure 3.2. (a) Fixed-speed wind generator architecture and (b) fixed-speed wind generator reactive power characteristics.

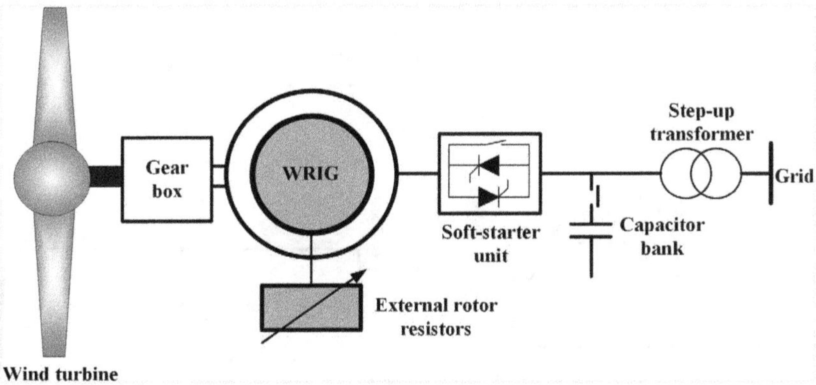

Figure 3.3. Limited variable-speed wind generator architecture.

3.3.3 Doubly Fed Induction generator

The DFIG utilizes the WRIG, and the stator terminals are directly connected to the grid, whereas the rotor terminals are connected to the grid via back-to-back converter system. The rotor-connected converter (commonly known as the rotor-side converter [RSC]) enables variable-speed operation (typically between 0.7 and 1.2 pu), whereas the grid-side converter (GSC) maintains the DC-link voltage constant. The RSC is also capable of independently controlling the active and reactive of the stator terminals, hence active and reactive power characteristics are governed by the RSC. In addition, GSC can also independently act as a reactive power source based on the available capacity after transferring active power requirements of the RSC. A schematic of the DFIG and its reactive power characteristics is shown in Figure 3.4(a)–(c).

As shown in Figure 3.4(b), the DFIG reactive power capability varies with the active power generation and operating slip of the machine. Aforementioned, the GSC also has substantial reactive power capability and this capability could be used for stability improvement.

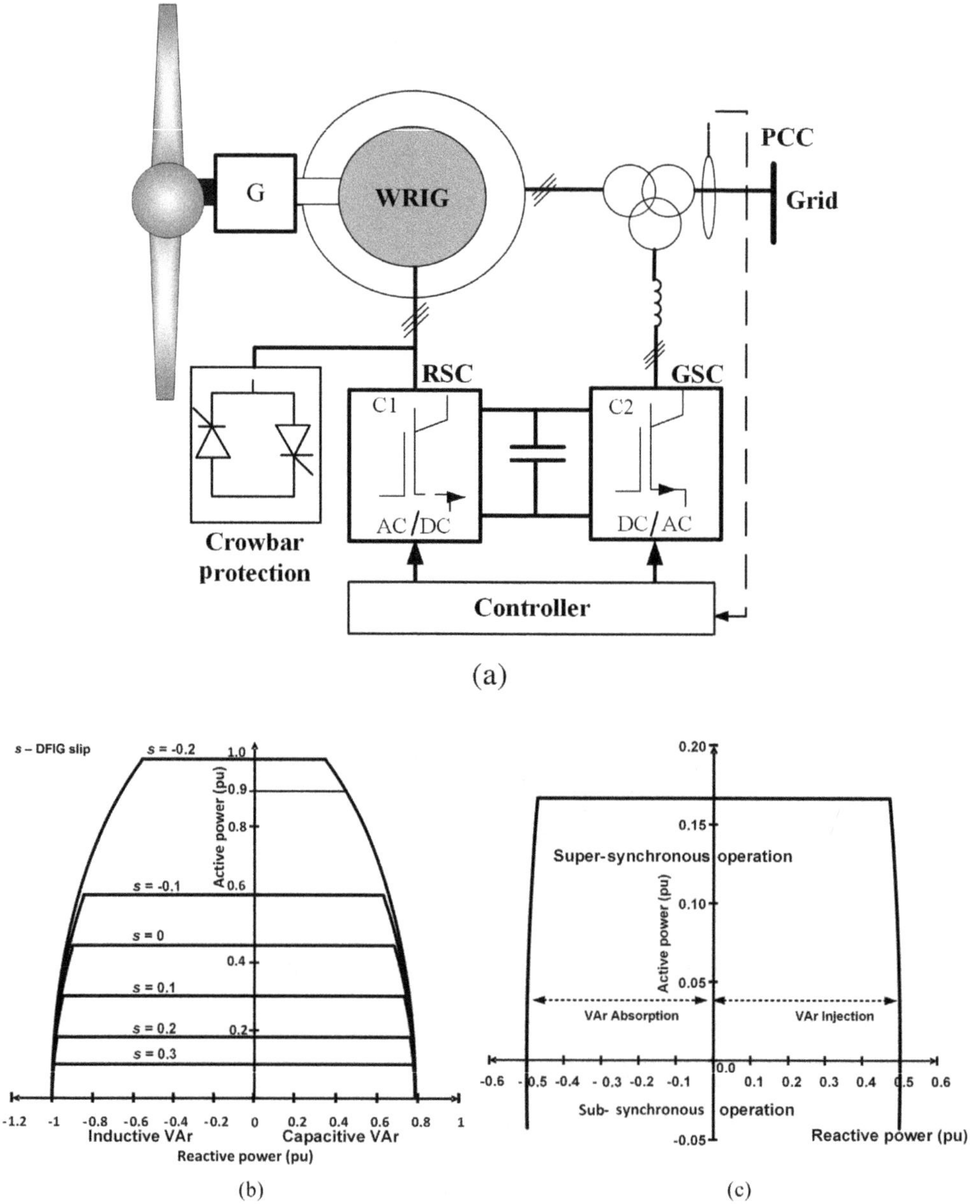

Figure 3.4. (a) DFIG architecture; (b) DFIG-RSC capability curve; and (c) GSC capability curve [8].

3.3.4 Full-Converter Wind Generator

The FCWG utilizes a fully rated PEC system. However, several variants in FCWGs are based on the generator type. In some FCWGs, asynchronous generators (i.e., induction generators) are used, whereas in many FCWGs either permanent magnet synchronous generators (PMSGs) or electrically excited synchronous generators are used. A schematic of the FCWG and its reactive power characteristics are shown in Figure 3.5(a) and (b), respectively.

As shown in Figure 3.5, reactive power characteristics of the FCWG depend on the GSC (i.e., inverter characteristics), and the GSC is typically rated at 1.1 pu of the active power rating, hence at the rated active power, still inverter has some reactive power capability.

3.4 Voltage Stability Studies with Wind Generation

Many research studies have been conducted during the last 15 years to characterize the impact of wind power generation on voltage stability. The early studies are mainly focused on large-disturbance voltage stability with the FSWGs and DFIGs [9–19]. These studies have used test systems and actual power networks for voltage stability analysis. The studies conducted by Akhmatov et al. [9,10] analyzed large-disturbance instability in the Danish power grid with FSWG- and DFIG-based wind farms. These studies illustrated that risk of voltage collapse in the system could be reduced by disconnecting the FSWG-based wind generators, as the disconnecting of FSWGs could reduce the dynamic reactive power

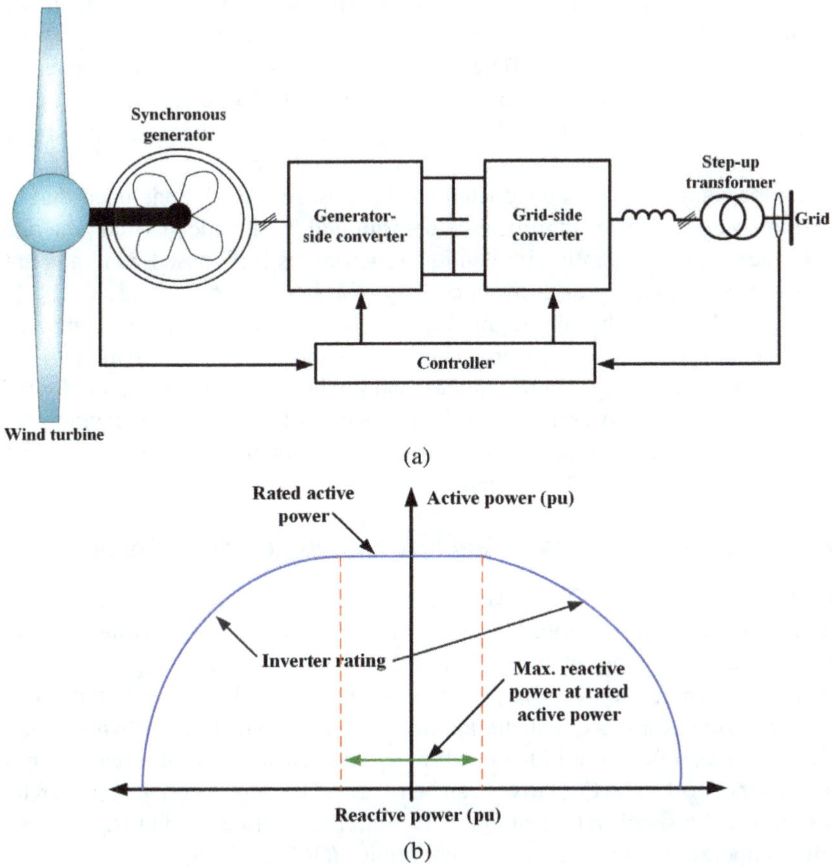

Figure 3.5. (a) Full-converter wind generator architecture and (b) full-converter wind generator capability curve.

demand during network faults (because the FSWG absorbs inductive reactive power during short-circuit faults). Authors have further emphasized the fact that improved FRT capabilities of wind generators could reduce the risk of voltage collapse during a network fault.

Furthermore, some research studies have shown that dynamic reactive power compensation devices installed at specific nodes in the network could improve the voltage stability in the network. The research study conducted by Rodriguez et al. [11] and Usala et al. [12] have examined voltage stability issues in the Spanish power system with FSWGs and DFIGs in the network. A comparison between both wind generation technologies showed that DFIGs have assisted in improving the voltage stability performance in comparison to FSWGs. The Nordic system study conducted with the FSWG-based wind generation has also analyzed the voltage stability during grid faults [13] and shown that the impact of FSWGs is local, but may result in sustained power oscillations in the power network.

Stability studies presented in [14] illustrated that transient and voltage stability improved by high-voltage network connection of wind farms, and a study on wind technology mixes have shown that optimal wind technology mix [15] can improve overall system stability by using complementary characteristics of different wind generators. However, some stability studies conducted on variable-speed wind generation technology (i.e., FCWGs and DFIGs) are limited to simplified network structures and few fault scenarios [16–17]. In [18], impact of wind integration approaches to stability has been presented and shown that stability outcomes are different based on the way in which wind power generation is integrated into the power network. The impact of DFIG operating mode (i.e., voltage control and power factor control modes) on voltage stability [19] has been investigated and concluded that voltage control mode is beneficial for stability.

The voltage stability studies conducted by Amarasekara et al. concluded that both the short- and long-term voltage stabilities are deteriorated by the added wind generation (by replacing synchronous generation by wind generation), as PEC-based wind generators do not have overload capability as conventional synchronous generators [3,20,21]. Furthermore, voltage stability studies highlighted that wind farm location (i.e., local or remote), wind farm reactive power capability characteristics, wind farm operating mode (i.e., power factor control mode or voltage control mode), and loading level also significantly influence the voltage stability of the system [3]. As the majority of the newly added wind generators are power electronic-based variable-speed wind generators, the voltage stability investigations are conducted using the DFIG wind generator.

3.5 Small-Disturbance Voltage Stability with Wind Power Generation

Aforementioned, the PV and QV curves can be used to assess the small-disturbance voltage stability of the power system. However, the PV curve approach is more suitable for analyzing load-driven voltage stability phenomena. As the wind power generation replaces the conventional generation, reactive power is more affected than the active power, hence, QV curve approach is a more realistic approach to assess small-disturbance voltage stability assessment as QV curve illustrates the voltage sensitivity to the reactive power on location in the network. The QV curves can be obtained through a series of power-flow calculations. Figure 3.6 shows a typical QV curve and the minimum point (L_2) at the bottom is the critical operating point, because at that point $(dQ/dV) = 0$ [5].

The points that are located at the right side of the point L2 are stable, whereas left-side operating points are unstable. Therefore, the relative movement of the critical point could be used as an indicator for small-disturbance voltage instability of the power network. The small-disturbance voltage stability with DFIG-based wind generation is evaluated using QV curves using the IEEE-14 bus network in Figure 3.7.

When the DFIG-based wind generation is added to the power network, synchronous generators are progressively withdrawn from the network. The DFIG-based wind farms were added to busbars 4, 5, 13, and 14, and wind generation at each busbar was increased from 5 to 35 MW with 10-MW increments; subsequently, a QV curve was generated for each busbar. All wind farms were operated at unity power factor mode, and synchronous machines at busbars 3, 6, and 8 were operated in synchronous condenser mode. The generation portfolio for each wind scenario is summarized in Table 3.1.

Figure 3.8 (a) and (b) illustrate the QV curves generated for busbars 3 and 12, respectively.

As illustrated in Figure 3.8, integration of wind generation has impacted the power network stability in two ways. Regarding the high-voltage network, bus voltage stability has decreased with the increased wind generation in the network. For example, the voltage instability occurs at −232 MVAr consumption for bus 3 with no wind generation, whereas with a 140-MW wind generation (i.e., scenario-5) the instability point is observed at −183.33 MVAr. Therefore, at the HV network, busbar QV instability margin has reached with low

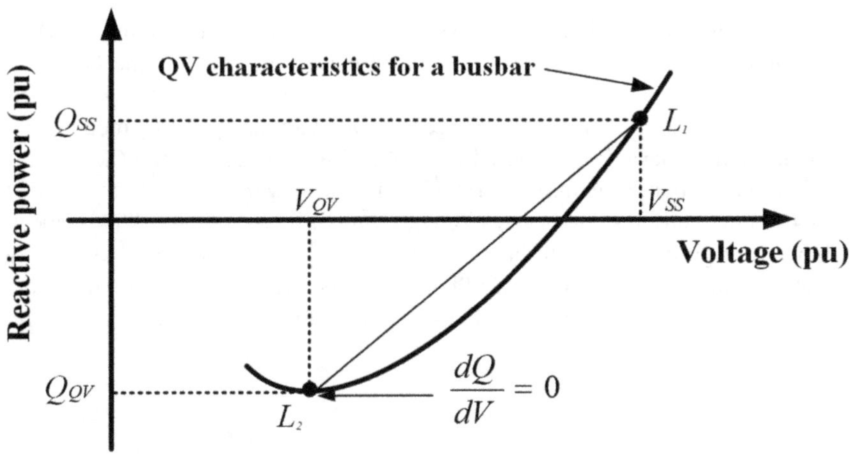

Figure 3.6. QV curve-based voltage stability index calculation.

Table 3.1. Generation portfolio for DFIG wind generation scenarios.

Scenario	Active Power Dispatch (MW)					
	Bus 1 (SG)	Bus 2 (SG)	Bus 4 (DFIG)	Bus 5 (DFIG)	Bus 13 (DFIG)	Bus 14 (DFIG)
1	232.9	40	0	0	0	0
2	210.3	40	5	5	5	5
3	187.8	20	15	15	15	15
4	168.0	0	25	25	25	25
5	128.1	0	35	35	35	35

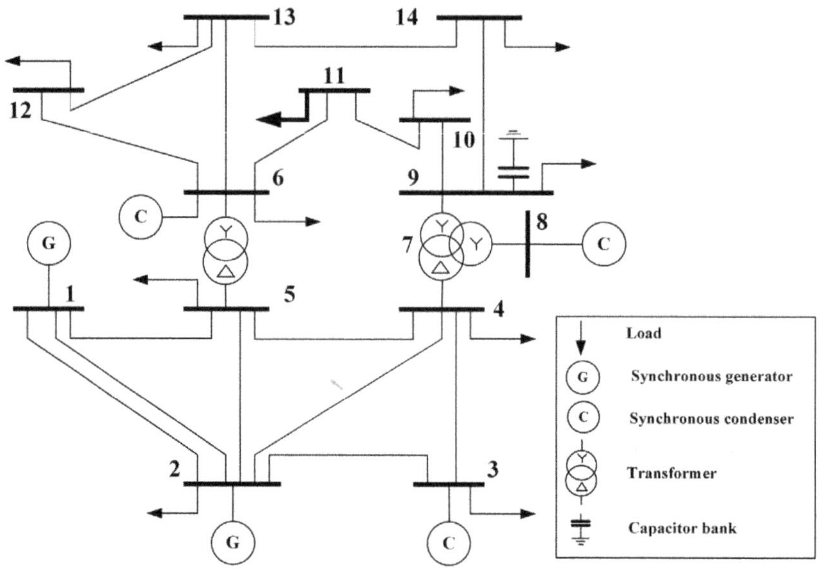

Figure 3.7. The IEEE-14 bus model.

reactive power consumption, hence static voltage stability has deteriorated in HV busbar with wind integration. Conversely, for the LV network (i.e., bus 12), the QV instability point has slightly improved with the integration of wind generation in some scenarios.

To further investigate the impact on voltage stability with the operating mode of the wind farm, an additional scenario is considered with 140-MW wind generation (i.e., Scenario-5) in the IEEE-14 bus network. In this scenario, all wind farms were operated at voltage control mode. The QV curves for busbars 3 and 12 are shown in Figure 3.9(a) and (b), respectively.

As shown in Figure 3.9, at both high- and low-voltage busbars, static-voltage stability has substantially improved when the wind farms are operated at voltage control mode. For example, for the low-voltage busbar (i.e., bus 12), the voltage instability point has reached

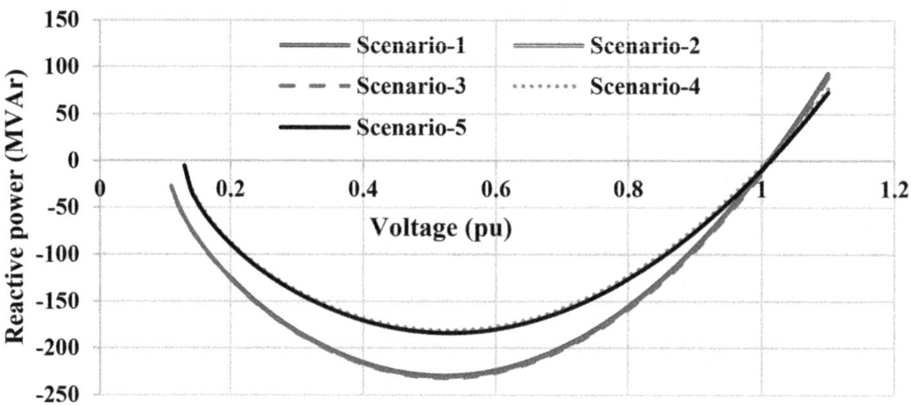

(a)

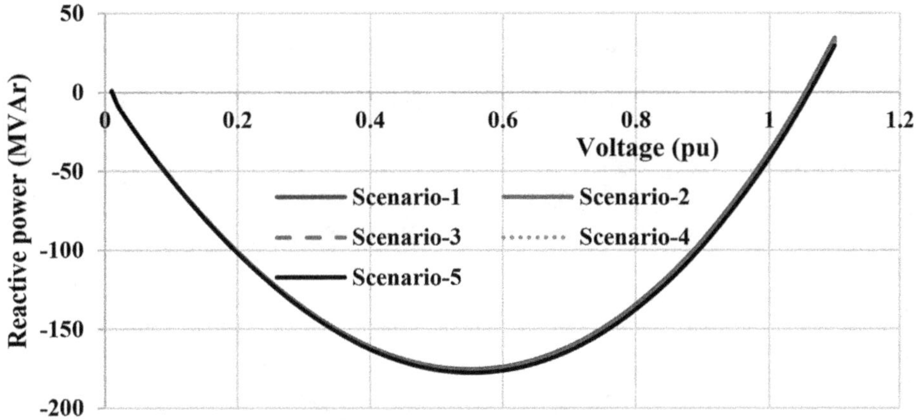

(b)

Figure 3.8. (a) QV curves for bus 3 of the IEEE-14 bus system under wind scenarios and (b) QV curves for bus 12 of the IEEE-14 bus system under wind scenarios.

the operating point where the reactive power consumption becomes −185 MVAr at voltage control mode, however in power factor control mode, voltage instability occurs when the reactive power consumption becomes −178 MVAr. This clearly indicates that voltage control mode is more beneficial for voltage stability of the network.

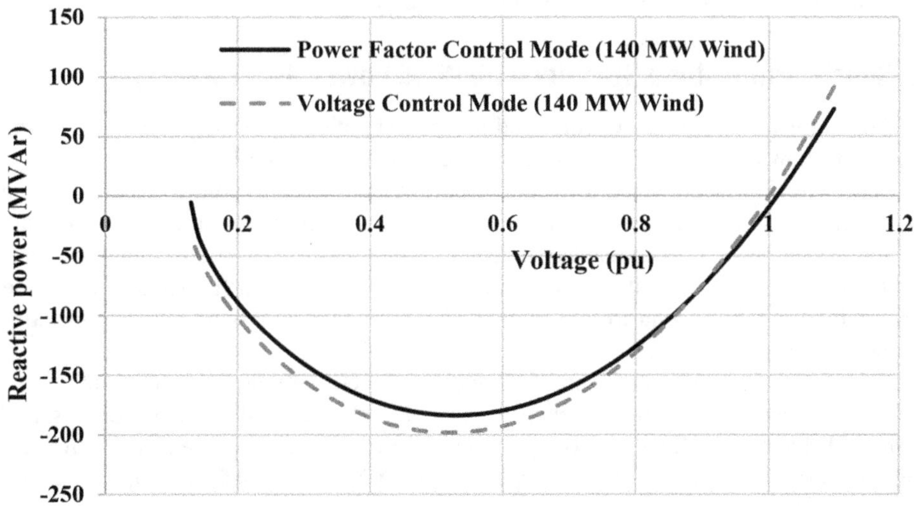

(a)

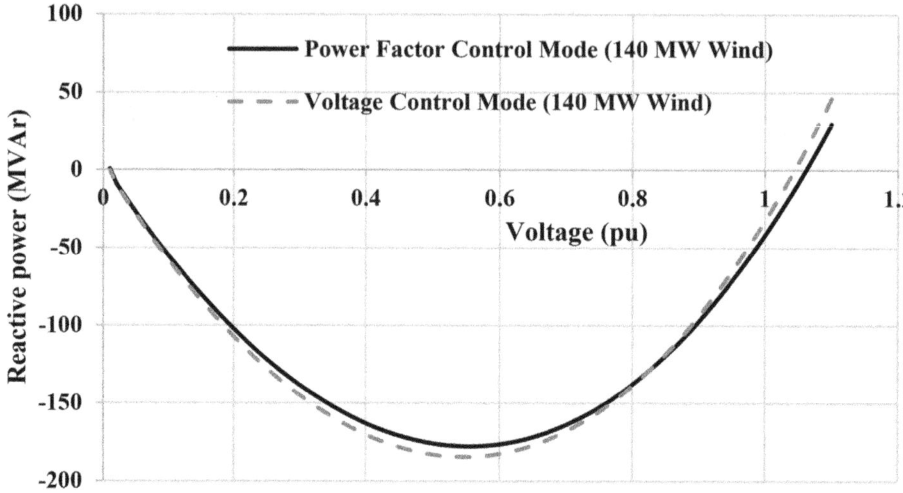

(b)

Figure 3.9. Impact of wind farm operating mode on voltage stability: (a) bus 3 and (b) bus 12.

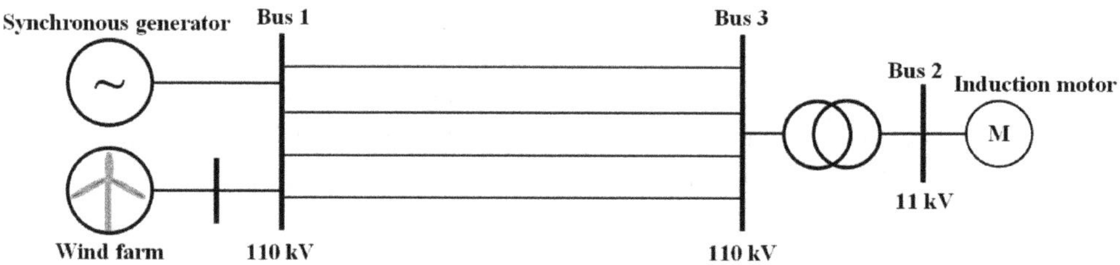

Figure 3.10. Test system—short-term voltage stability analysis.

3.6 Short-Term Voltage Stability with Wind Power Generation

The short-term voltage stability with wind power generation is investigated using a simple test system shown in Figure 3.10, modeled in DIgSILENT Power Factory.

Test system consists of a single synchronous generator and IM connected by four transmission lines. As the main driving factor for short-term voltage instability is the dynamic load, the load at bus 2 was modeled as an IM. To investigate the short-term voltage instability, a three-phase short-circuit fault is simulated in one of the four transmission lines near busbar 3 at $t = 1$ s. Two scenarios are studied: Cases A and B with two fault clearance times 110 and 120 ms, respectively.

As shown in the results in Figure 3.11, with extended fault clearance time, the voltage does not recover back to pre-disturbance value. For this case, the reason for this is the stalling of IM load, which then absorbs a large quantity of reactive power. The DFIG-based wind generator with a capacity of 39 MW and 97.5 MW (Cases C and D, respectively) is added to busbar 1 of the test system to study the impact of wind generation on voltage stability. The total generation capacity at busbar 1 is maintained the same by reducing the synchronous generation capacity at bus 1 by an equivalent amount. The same fault scenario was simulated with a fault clearance time of 110 ms.

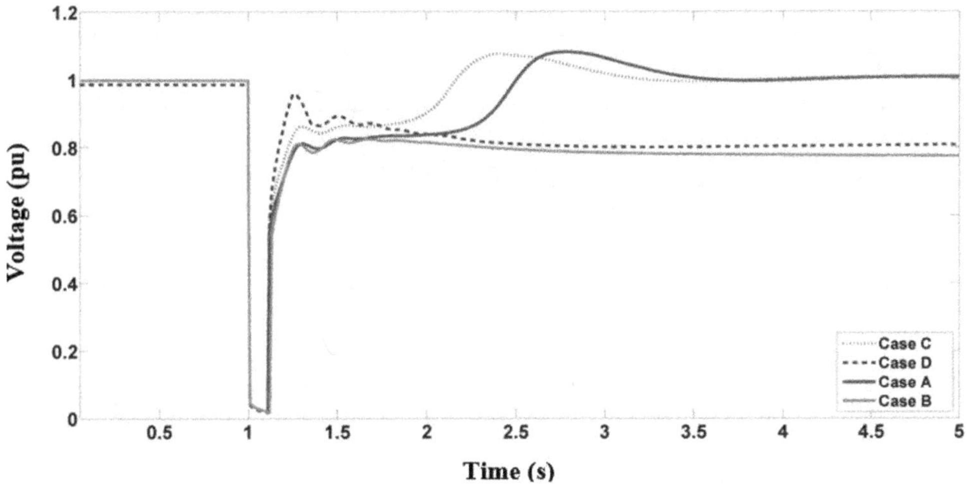

Figure 3.11. Voltage profile at busbar 2.

As shown in the results in Figure 3.11, if the DFIG wind capacity is only about 39 MW, then the short-term voltage stability of the network can be maintained. However, when the DFIG generation capacity at bus 1 increases, it shows a detrimental impact on short-term voltage stability.

3.7 Long-Term Voltage Stability with Wind Power Generation

The Nordic test network [22] was used for assessing the long-term voltage stability with wind generation. The Nordic test system is consisted of 74 buses, 20 generators, 102 transmission lines, 22 loads, 9 shunt capacitors, 5 series compensators, and 2 shunt reactors. Nordic test system consists of four areas: North, Central, Equiv, and South. All synchronous generators are equipped with a governor, an automatic voltage regulator (AVR), power system stabilizer (PSS), and OEL, whereas the transformers are equipped with on-load tap changers (OLTCs). The exponential load model was used to represent the loads in the Nordic test network. Figure 3.12 illustrates the Nordic test system architecture.

According to Table 3.2, the system is heavily loaded due to large power transfers from North to Central region. For example, the Central region has an active power load of 6190 MW, and has only 2850 MW active power generation, hence the active power is imported to the Central region.

Previous studies conducted on the Nordic test system concluded that the test system has five heavily stressed synchronous generators (e.g., g6, g7, g14, g15, and g16) [23]. Therefore, to investigate the impact on long-term voltage stability with wind generation, these five heavily stressed synchronous generators in the Nordic test system are methodically replaced by the DFIG-based wind farms. The following conditions were maintained when replacing the synchronous generators with wind generators, to ensure an accurate judgment on long-term voltage stability impact from wind generation.

- Replaced the synchronous generators with DFIG-based wind farms with same mega-volt ampere (MVA) capacity.

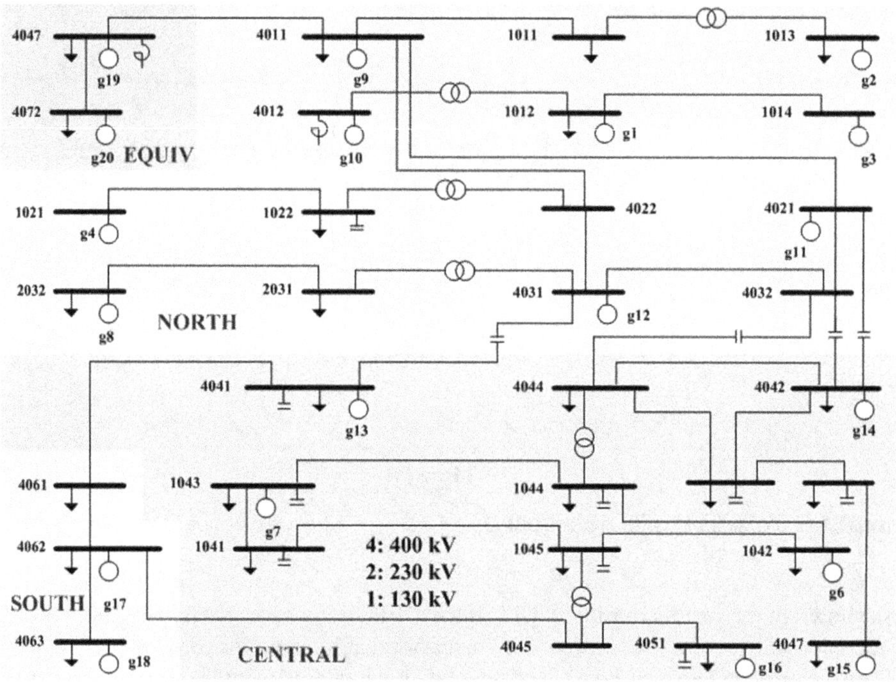

Figure 3.12. Nordic test system architecture.

- Steady-state active and reactive power levels of DFIG wind farms were maintained same as replaced synchronous generators.

Also, a range of loading levels (i.e., 70%, 50%, and 25%) was considered for each wind generation scenario in this study. Bus 1041 has been identified as the most critical busbar of the Nordic test system as its voltage collapses first during a contingency in the network. Table 3.3 illustrates the voltage of bus 1041 under various wind generation scenarios, and their loading conditions when the same disturbance (i.e., three-phase short circuit at line 4032-4044) is applied in the Nordic test system.

Table 3.2. The generation and active power load of each region in the Nordic test system.

Regions	Active Power Generation (MW)	Active Power Load (MW)
North	4628.5	1180
Central	2850.0	6190
South	1590.0	1390
Equiv	2437.4	2300
Total	11,505.9	11,060

Table 3.3. Critical bus (bus 1041) voltage for various wind generation scenarios.

No	Scenarios	Penetration Level (%)	No Wind	DFIG Loading Level 70%	50%	25%
1	g 7	2	0.732	0.736	0.738	0.741
2	g 6	3	0.732	0.730	0.731	0.733
3	g 14	5	0.732	0.728	0.730	0.733
4	g 16	5	0.732	0.727	0.730	0.735
5	g 15	9	0.732	0.727	0.730	0.735
6	g 15, g 16	14	0.732	0.724	0.729	0.737
7	g 14, g 15, g 16	20	0.732	0.722	0.728	0.738
8	g 6, g 14, g 15, g 16	23	0.732	0.720	0.727	0.740
9	g 6, g 7, g 14, g 15, g 16	25	0.732	0.727	0.736	0.749

Table 3.4. The improvement of reactive power capability.

No	Generator-Replaced Scenarios	Penetration Level (%)	No Wind	DFIG Loading Level 70%	50%	25%
1	g 6	3	175	174	174	174
2	g 7	2	96	97	102	116
3	g 14	5	288	233	250	284
4	g 16	6	345	304	303	306
5	g 15	9	525	556	558	560

According to Table 3.3, the critical busbar (bus 1041) voltage has improved with the increased wind power penetration in the system under low loading scenarios. For example, at 25% wind penetration (i.e., scenario 9), when the DFIG wind generators are 25% loaded, the critical busbar voltage improves to 0.749 pu from the base case, which had a critical voltage level of 0.732 pu. This improvement has resulted due to the improved reactive power reserve in the network during the partially loaded wind generation scenarios. The improvement in the reactive power capability of the system when wind generators replace the synchronous generators is summarized in Table 3.4.

According to Table 3.4, when the wind generator replaces the synchronous generator g7, reactive power reserve has improved by 97 MVAr, 102 MVAr, and 116 MVAr for 70%, 50%, and 25% wind generator loading levels, respectively. Subsequently, when the synchronous generator g7 is replaced by wind generation and loaded by 70%, the critical bus voltage has improved to 0.736 pu, whereas at 50% loading level it has improved to 0.738 pu.

The g15 wind farm has 560 MVAr reactive power capability at 25% loading level, whereas synchronous generator's reactive power output was 525 MVAr. Thus, the partially loaded wind generation (25% loading level) provides an additional 35 MVAr to the system.

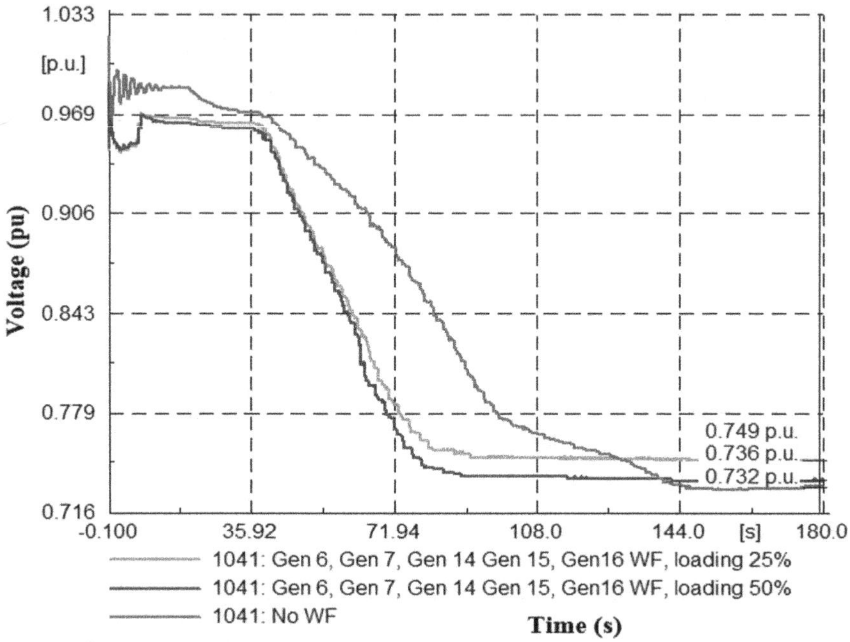

Figure 3.13. The voltage profile at the critical busbar (bus 1041).

As a result, the critical busbar voltage has increased from 0.732 to 0.735 pu, even though it is not sufficient to arrest the voltage collapse in the network.

Furthermore, after the initial disturbance, the reactive power output of wind farms is comparatively low compared with the synchronous generators. Once a significant voltage drop occurs (<0.90 pu) in the network, wind farms increase their reactive power output depending on their loading level. Although the synchronous generator also improves its reactive power output, the operation of the OEL substantially decreases the reactive power output of the synchronous generator and ultimately results in a voltage collapse. However, as illustrated in Figure 3.13, improved reactive power capability under low loading levels improves the final voltage level. For example, with no wind farms, the voltage at t = 144 s is 0.732 pu, and that has been improved to 0.749 pu with a 25% loading level.

3.8 Conclusions

This chapter investigated the impact of wind power generation on power system voltage stability. The reactive power capability of wind generators is one of the critical factors influencing voltage stability of the power system, hence the reactive power capability of different wind generator types is thoroughly examined in the chapter. Different voltage stability aspects are investigated with DFIG-based wind generation, such as small-disturbance stability, short-term voltage stability, and long-term voltage stability. The small-disturbance voltage stability analysis revealed that voltage stability would deteriorate at high wind power penetration levels. However, by operating DFIG wind farms at voltage control mode, voltage stability could be substantially improved in comparison to the power factor control mode.

The short-term voltage stability results demonstrate that at high wind power penetration levels, short-term voltage stability would deteriorate due to less reactive power provided by the DFIG wind generators, although low wind power penetration has less impact on short-term voltage stability. With regard to the long-term voltage stability, it has indicated that DFIG wind generation has adversely affected the voltage stability at their high loading levels. However, DFIG wind farms could improve long-term voltage stability as they could improve the reactive power reserve during low loading conditions. This outcome highlights an essential fact that reactive power capability of wind generators plays an imperative role in every aspect of voltage stability. Therefore, if FCWGs are employed, then the voltage stability could be improved in comparison to the DFIG-based wind generation, as the FCWG has much better reactive power capability than a typical DFigure However, if DFIG GSC capability is taken into consideration, then the voltage stability could be improved, and that improvement would be much better than the improvement expected from the FCWG. Therefore, power grid operators should encourage wind farm operators to utilize their maximum reactive power capability to improve the voltage stability of the network.

3.9 References

1. Union for the Co-ordination of Transmission of Electricity. "Final Report on the Disturbances on 4 November 2006 [Internet]. UCTE: Brussels – Belgium." Nov. 2006. Available online: http ://entsoe.eu/.
2. Australian Energy Market Operator (AEMO). "Black System South Australia 28 September 2016 [Internet]. AEMO: Melbourne, Australia." Mar. 2017. Available Online: https://www.ae mo.com.au.
3. K. Amarasekara, L. G. Meegahapola, A. P. Agalgaonkar, and S. Perera, "Characterisation of long-term voltage stability with variable-speed wind power generation," *IET Generation, Transmission & Distribution*, vol. 11, no. 7, pp. 1848–1855, 2017.
4. P. Kundur et al, "Definition and classification of power system stability IEEE/CIGRE joint task force on stability terms and definitions," *IEEE Transactions on Power Systems*, vol. 19, no. 3, pp. 1387–1401, 2014.
5. L. Meegahapola, and T. Littler, "Characterisation of large disturbance rotor angle and voltage stability in interconnected power networks with distributed wind generation," *IET Renewable Power Generation*, vol. 9, no. 3, pp. 272–283, 2015.
6. F. Gubina, and B. Strmcnik, "Voltage collapse proximity index determination using voltage phasors approach," *IEEE Transactions on Power Systems*, vol. 10, no. 2, pp. 788–794, 1995.
7. L. Meegahapola, S. Durairaj, D. Flynn, and B. Fox, "Coordinated utilisation of wind farm reactive power capability for system loss optimisation," *European Transactions on Electrical Power*, vol. 21, no. 1, pp. 40–51, 2011.
8. L. Meegahapola, T. Littler, and S. Perera, "Capability curve based enhanced reactive power control strategy for stability enhancement and network voltage management," *International Journal of Electrical Power and Energy Systems*, vol. 52, pp. 96–106, 2013.
9. V. Akhmatov, and P. B. Erilkson, "A large power system in almost island operation-a Danish case study," *IEEE Transactions Power Systems*, vol. 22, no. 3, pp. 937–947, 2007.
10. V. Akhmatov, and H. Kundsen, "Large penetration of wind and dispersed generation into Danish power grid," *Electric Power Systems Research*, vol. 77, no. 9, pp. 1228–1238, 2007.
11. J. M. Rodriguez, J. L. Fernandez, D. Beato, R. Iurbr, J. Usala, P. Ledesma, and J. R. Wilhelmi, "Incidence of power system dynamics of high penetration of fixed speed and doubly-fed wind energy systems: Study of Spanish case," *IEEE Transactions on Power Systems*, vol. 17, no. 4, pp. 1089–1095, 2002.

12. J. Usala, P. 1Ledesma, J. M. Rodriguez, J. L. Fernandez, D. Beato, R. Iturbe, and J. R. Wilhelmi, "Transient stability studies in grids with great wind power penetration-modeling issues and operational requirements," IEEE PES GM, Toronto, Canada, 2003.
13. C. Jauch, P. Sorensen, I. Norheim, and C. Rasmussen, "Simulation of the impact of wind power on the transient fault behaviour of the Nordic power system," *Electric Power System Research*, vol. 77, pp. 143–136, 2006.
14. L. Meegahapola, D. Flynn, and T. Littler, "Transient stability analysis of a power system with high wind penetration," UPEC, Padova, Italy, 2008.
15. L. Meegahapola, D. Flynn, J. Kennedy, and T. Littler, "Impact of wind generation mix on transient stability for an interconnected power system," EWEC, Marseille, France, 2009.
16. E. Muljadi, C. P. Butterfield, B. Parsons, and A. Ellis, "Effect of variable speed wind turbine generator on stability of a weak grid," *IEEE Transactions Energy Conversion*, vol. 22, no. 1, pp. 29–36, 2007.
17. M. V. A. Nunes, J. A. P. Lopes, H. H. Zurn, U. H. Bezerra, and R. G. Almeida, "Influence of the variable speed wind generators in transient stability margin of the conventional generators integrated in electrical grids," *IEEE Transactions Energy Conversion*, vol. 19, no. 4, pp. 692–701, 2004.
18. L. Meegahapola, and D. Flynn, "Impact on transient and frequency stability for a power system at very high wind penetration," IEEE PES GM, Minneapolis, USA, 2010.
19. E. Vittal, A. Keane, and M. O'Mally, "A steady-state voltage stability analysis of power systems with high penetrations of wind," *IEEE Transactions Power Systems*, vol. 25, no. 1, pp. 433–442, 2010.
20. K. Amarasekara, L. Meegahapola, A. P. Agalgaonkar, and S. Perera, "Impact of variable speed wind power generators on short-term and long- term voltage stability," Wind Integration Workshop, Brussels, Belgium, 2015.
21. K. Amarasekara, L. Meegahapola, A. P. Agalgaonkar, S. and Perera, "Investigation of load driven long-term voltage instability phenomenon with wind generation," Australasian Power Engineering Conference (AUPEC), Wollongong, Australia, 2015.
22. Reliability Test System Task Force of the Voltage stability and security Assessment, "IEEE Reliability Test System," IEEE Trans. on PES-TR19, 2015, New York. USA.
23. L. Meegahapola, and E. Munkhchuluun, "Impact of the solar photovoltaic (PV) generation on long-term voltage stability of a power network," IEEE PES Innovative Smart Grid Technologies Asia (ISGT-ASIA), Auckland, New Zealand, 2017.

CHAPTER 4

Seasonal Impact of Wind Turbine Integration on the Performance of Radial Distribution Networks: A Comparative Case Study with Practical Wind Data of Southern India

Ashwani Kumar and Sooraj Narayan

Department of Electrical Engineering, National Institute of Technology, Kurukshetra, Haryana, India. ashwani.k.sharma@nitkkr.ac.in; soorajn14@gmail.com

Abstract This chapter presents a comparative case study to investigate the impacts of wind turbine integration in radial distribution systems for various seasons of the year. A total of 10 commercially available wind turbines are selected and integrated to the test system. Hourly wind speed data for an entire year is obtained for a wind turbine site located in Southern India. Seasonal impacts of the wind turbine placement are studied on the basis of various performance indices of the distribution system. The main contribution of the chapter is (1) wind speed data for a practical wind farm located in Southern India for a year and determination of its Weibull distribution parameter for four seasons, (2) capacity factor determination for different wind turbines based on Weibull parameters, (3) integration of different wind turbines in radial distribution system considering the probabilistic load model, and (4) comparative study of different wind turbines to observe the impact on performance of distribution system. In this study, the substation voltage has also been considered a random variable. Monte Carlo Simulation (MCS) is used to perform the probabilistic load flow (PLF) of the distribution system. The comparative analysis has been carried out for the distribution system based on the determination of performance indices such as Voltage Deviation Index (VDI), Active Power Loss Index (APLI), Reactive Power Loss Index (RPLI), and Line Loading Index (LLI) for various seasons of the year. Cape Comorin in Southern India is selected as the site to take the wind speed data for the study. The results are obtained for IEEE 33 bus radial distribution system with wind power integration. This study can help extensively in planning the integration of wind turbines in distribution systems.

Keywords: Active and reactive power loss; Radial distribution system; Line loading; Monte Carlo Simulation; Wind power integration.

4.1 Introduction

Decentralization and deregulation of the electric power industry along with the ever-changing market scenario has led to a new era of flexibility and dynamism in the power sector [1]. Distributed generation (DG) has acquired an increased interest in the past decade or so because of the techno-socio-economical benefits provided by it. The strategies and techniques adopted for the planning and operation of the traditional unidirectional power systems have to be modified and reconstructed with the integration of DGs into the distribution systems. This is mainly because of the fact that DGs convert the formerly passive distribution system into an active one. The various advantages of integrating DG into the power network have been well studied and researched in the past [2–4]. Among the possible realms of the DG sector, the renewable energy sources have the inherent advantage of being environment friendly and inexhaustible. Wind power is widely considered as one of the fastest growing power sources of the current era [5]. According to the global wind energy statistics report published by the Global Wind Energy Council (GWEC) in 2021, the global installed wind power capacity was 586 GW by the end of 2019 [6]. The enviable exponential growth of the wind power sector has been ably aided by the development in the scientific know-how and technology boom that has helped strategize new energy efficient methods of trapping the air mass for the production of electricity.

India currently ranks 4th in the world in terms of wind power generation, behind China, United States, Germany. The installed capacity in India was 39.1 GW by the end of March 2020 [6]. About two-thirds of India's renewable energy capacity comprises wind power. As a major player in the world energy market scenario, wind energy planning and policies play a very major role in India's power sector. A known fact is that as the wind speed increases, more energy can be trapped in the turbine blades and hence the efficiency of the turbine increases, and the costs involved decreases. With a coastline of more than 7000 km, India has a huge potential of wind energy. The average wind speed for a major part of India's landmass is less than 5.6 m/s, which is comparatively low. This is evident from the wind power density map of India shown in Fig. 4.1 [7]. The Ministry of New and Renewable Energy (MNRE) of India, which is the primal body of the Government of India for all topics related to renewable energy, has announced the setting up of a National Wind Energy Mission (NWEM) [8]. From 2014 onward, the country is planning to add an amount of 3000 MW of wind capacity every year [9]. Considering all this, it is necessary to have a study on the possible impacts of wind turbine integration, under various seasons of the year, on the distribution system based on local wind speed data.

Many studies have been conducted in the past with regard to the various benefits of wind turbine integration in distribution networks. In [10], the techno-economic benefits of a small wind electric generator (SWEG) for providing deregulated power supply in rural India are explained. Environmental issues with regard to the SWEGs are also discussed in the referenced paper. The effect of wind turbine integration on the voltage profile of distribution networks is discussed in [11]. Certain concepts for the enhanced integration of wind turbines into the distribution system in Germany are discussed in [12]. A new method for voltage fluctuation analysis on the integration of wind turbines into distribution systems is provided in [13]. In this referenced paper, the probabilistic load flow (PLF) method is used to incorporate the variability of the considered parameters. In [13], a comparison is done between two methods, namely deterministic and statistical, in order to observe the effect of wind turbines on the voltage profile of distribution networks. In [14], a benefit analysis of the distribution system on wind turbine integration is done. In the referenced paper, it is shown that wind turbine integration will lead to reduced system losses and improved voltage profile in distribution networks. In [15], the impact of wind turbine performance on electrical grid connection for a wind farm in Jordan is studied. In [16] integrating high levels of wind power systems in Maui island in Hawaii is discussed. The literature review for this work has revealed that even though a lot of research has been done on determining the benefits of wind turbine placement on distribution networks, a seasonal case study investigating the impacts on radial distribution networks for commercially available turbines based on hourly wind data for a site in India has not been conducted.

Owing to the intermittent nature of the wind power sources, integrating these dispersed generating units brings forward some technical difficulties. Because the wind speed is a random variable, the output of the wind turbine, which traps the kinetic energy in the air, is also a variable. Analytical methods have been used to model these renewable resources as nondispatchable sources of power [17]. A probabilistic approach for modelling the wind speed may incorporate various probability functions such as Weibull, Rayleigh, or Lognormal distribution functions. With regard to the radial test distribution system, a probabilistic

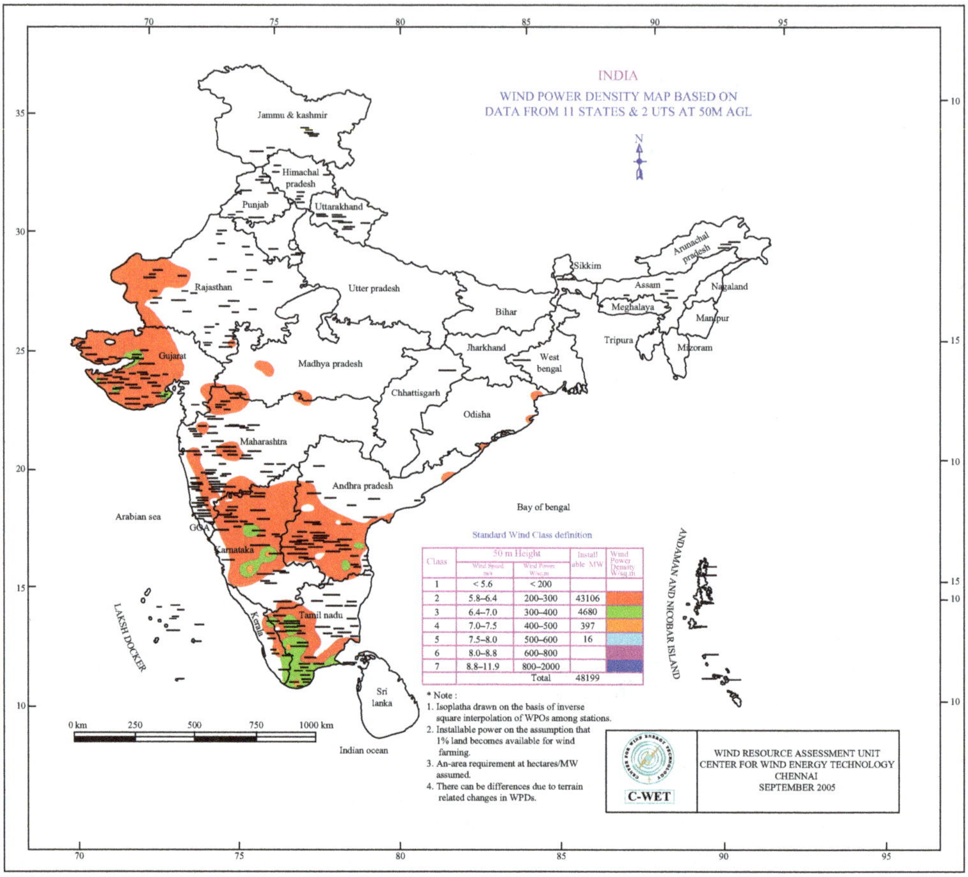

Figure 4.1. Wind power density map of India.

approach of performing the load flow is utilized in this chapter. Even though deterministic methods for distribution load flow produce precise results, they assume that the distribution system is in steady state. Methods that take into consideration the various uncertainties in the system are bound to be more realistic. In PLF calculations, only the load is normally taken as a random variable. Both the load and the substation voltage are assumed to be random variables in [18]. MCS is used to run the PLF calculations. However, the impact of seasonal probabilistic load variation along with the seasonal wind power generation needs to be addressed for assessment of performance of the distribution system.

In this chapter, 1-h wind speed data for a whole year is obtained for a particular site selected for the study. The data obtained are then grouped into 24-h data for 365 days of the year. Then, the wind speed is divided into four seasons of the year, namely winter, spring, summer, and fall. In each season, the mean and standard deviation of the wind speed data have been obtained for each hour of the day. Weibull distribution function is used to model the wind speed in this work. For each hour of each season, the Weibull parameters are determined. The parameters are different for each turbine because of the difference in hub heights. Then, for this particular site, the capacity factor (C_f) of 10 commercially available

Table 4.1. Details of the site under study.

Site	Location	Period of data
Cape Comorin	$8°4'33''N\ 77°32'53''E$	January 1, 2011–December 31, 2011

wind turbines are determined for each hour of each season. The C_f values are determined according to the quadratic model, which is discussed in the further sections of this chapter. The C_f values obtained are utilized to determine the average power of each wind turbine, which is to be applied to the test distribution system. The major contribution of the chapter is (1) wind speed data for a practical wind farm located in Southern India for a year and determination of its Weibull distribution parameter for four seasons, (2) capacity factor determination for different wind turbines based on Weibull parameters, (3) integration of different wind turbines in radial distribution system considering the probabilistic load model, and (4) comparative study of different wind turbines to observe the impact on performance of distribution system. Monte Carlo Simulation (MCS) is used to perform the PLF of the distribution system. The comparative analysis has been carried out for the distribution system based on determination of performance indices such a Voltage Deviation Index (VDI), Active Power Loss Index (APLI), Reactive Power Loss Index (RPLI), and Line Loading Index (LLI) for various seasons of the year. Cape Comorin in Southern India is selected as the site to take the wind speed data for the study. The results are obtained for IEEE 33 bus radial distribution system with wind power integration. This study can help extensively in planning the integration of wind turbines in distribution systems. Four impact indices are selected for showing the impacts of the wind turbines on the test system. The system load is considered a probabilistic load that follows probability distribution function. The variation in the load pattern is assumed to follow the pattern as for the IEEE-RTS system [25]. The average power values obtained from the capacity factor calculations of each turbine are integrated to the test system at the optimal bus to carry out the load flow studies using MCS.

The rest of the chapter is as follows: Section 2 details the site and wind turbines selected for the study. Section 3 discusses the modelling of load power demands, substation voltages, and the wind energy source along with the capacity factor model. Section 4 discusses the stability index-based method used for finding the optimum wind turbine location. Section 5 defines the various impact indices used for comparison study. Section 6 details the computational procedure. The various case studies and simulation results are discussed in Section 7.

4.2 Site Description and Wind Turbine Specifications

Hourly wind data for the year of 2011 at Cape Comorin in Southern India is collected from [19] and is used for obtaining the performance of different manufacturers wind turbine integration into the radial distribution system. The details about the location of the site are provided in Table 4.1 with longitude and latitude of the site and the site is also shown in Fig. 4.1 with higher wind speed marked in green color. The selected site has comparatively higher average wind speeds throughout the year. Based on the wind speed data obtained

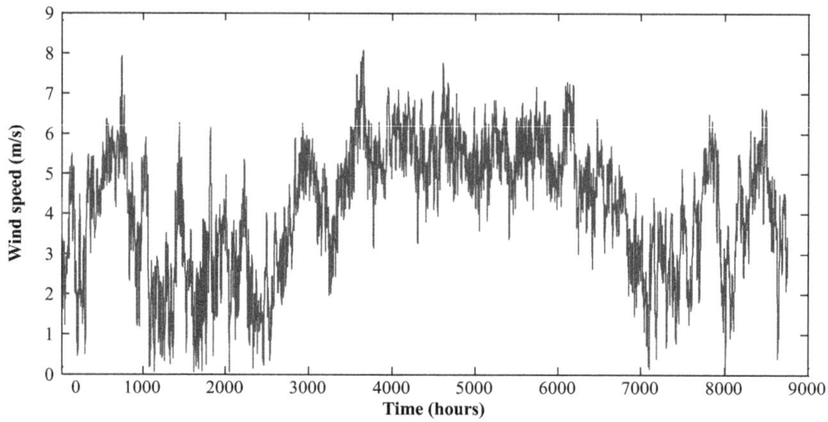

Figure 4.2. Hourly wind speed variation for Cape Comorin for an entire year at 10-m height.

from [19], the hourly wind speed variation is obtained for the entire period and is shown in Fig. 4.2. Figure 4.2 shows the hourly wind speed in m/s for the wind speed data for a year. As observed in Fig. 4.2, there are a total of 8760 data points. The wind speed data measured is assumed to be at 10.0 m height. The mean wind speed measured for the site is 4.19 m/s for the period of study. The wind turbines selected for this study and their specifications are provided in Table 4.2 [20–23]. The details provided include the rated power in kW, hub height in m, rotor diameter in m, and wind speed specifications. The range of rated power of the wind turbines selected is 50 kW–3000 kW. In the further sections of this chapter, the turbines will be referred according to their respective numbers as summarized in Table 4.2. The manufacturers of the wind turbines mentioned in the Table 4.2 are Wind Energy Solutions (WES), Enercon, FUHRLAENDER, GMBH, VESTAS V, General Electric (GE), and Siemens Wind Turbines (SWT). The wind turbines are chosen based on their cut-in speed as available in India at the site. In the table, the data are provided for rated power in kW, rotor diameter, turbines operating speed range in m/s, and the hub height in m. The practical wind speed data are utilized for the different wind turbines to obtain their power output over the entire period.

4.3 Mathematical Model of Load, Substation Voltage, Wind Speed, and Wind Turbine

In this study, we have considered probabilistic load model as this load model can provide better planning of the distribution system connected with the wind sources. The substation voltage does not remain constant during the entire day as it is dependent on the load pattern, and therefore, in this study we have considered the substation voltage as a random variable, which follows normal distribution [18]. Because the wind speed is stochastic in nature, it follows the Weibull distribution function and can be utilized to obtain the k and c parameter known as shape and scale parameters for wind, which characterizes its shape and wind speed.

Table 4.2. Wind turbine specifications from various manufacturers.

Turbine Number	Turbine Manufacturer	Rated Power (kW)	Rotor Diameter (m)	Turbine Operating Speed Range (m/s)			Hub Height (m)
				v_{ci}	v_r	v_{co}	
1	WES-50	50	20.3	2.9	10	25	30
2	WES-250	250	30	2.5	13	25	36
3	ENERCON E-40	600	44	2.5	13	25	46
4	FUHRLAENDER, GMBH	1300	62	2.22	15	26.9	50
5	VESTAS V-82	1650	82	3	13	20	70
6	VESTAS V-100	2000	100	3	12	22	95
7	VESTAS V-110	2000	110	3	11.5	20	120
8	GE 2.3	2300	94	3	14	25	100
9	SWT-2.3-108	2300	108	3	11	25	80
10	SWT-3-101	3000	101	3	12	25	99.5

4.3.1 Load Modelling

The load model may be time independent or time dependent. In this study, we have considered both time-independent and time-dependent load models. The time independency of the load may be because of unscheduled outages, errors in measurement, or unknown load power values. The time dependency of the load model may be because of the location and different seasons of the year.

4.3.2 Time-Independent Load Modelling Using Probability Distribution Function

The electric power load may vary because of unscheduled outages, errors in measurement, or unknown load power values. This type of variation is instantaneous and hence time independent. In this case, the load can be modelled by using probability distribution functions. The probabilistic nature of load at each bus in a distribution system can be incorporated into load flow studies by visualizing the loads as random variables distributed with a variance about a mean value. In this chapter, the load demands at each bus are assumed to be random variables with Gaussian or normal distribution [24].

$$f\left(P_{L,i}\right) = \left(\frac{1}{\sigma\sqrt{2\pi}}\right)\exp-\frac{\left(P_{L,i} - \overline{P}_{L,i}\right)^2}{2\sigma^2} \tag{4.1}$$

where $P_{L,i}$ is the active load demand at bus number i and $\overline{P}_{L,i}$, σ are the mean and standard deviation values of each load power, respectively.

4.3.3 Time-Dependent Load Modelling Using Statistical Data

Using statistical data available for a location, the load variation for various seasons of the year are calculated. The percentage of the load with respect to the peak value is calculated for seasons, months, weeks, and hours of the year. For example, Fig. 4.3 shows the hourly peak load in percentage of daily peak for the four seasons of the year for the IEEE-RTS [25]. In this chapter, both the modelling techniques are merged together to model the load. This type of load modelling is more realistic in nature as it incorporates both the instantaneous dynamism and the seasonal variability of the load in conjunction. In the new model, (4.1) becomes:

$$f\left(P_{L,i}(t)\right) = \left(\frac{1}{\sigma\sqrt{2\pi}}\right)\exp-\frac{\left(P_{L,i}S(t) - \overline{P_{L,i}S(t)}\right)^2}{2\sigma^2} \tag{4.2}$$

where $S(t)$ is the percentage of hourly peak load in terms of daily peak at hour t of the day.

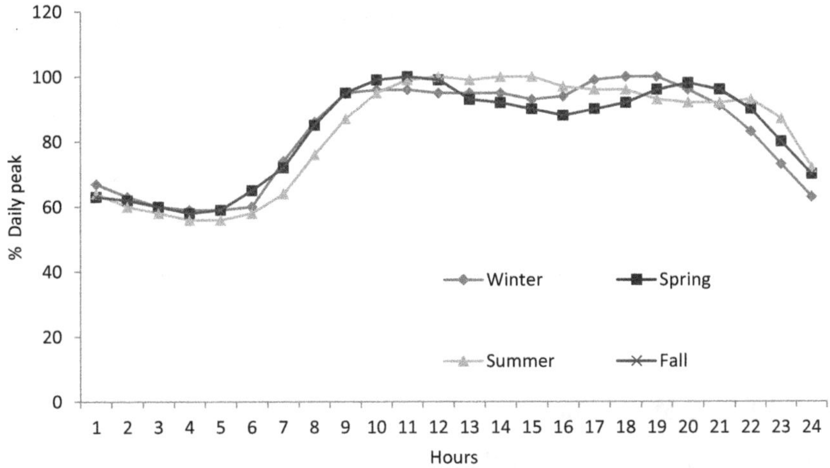

Figure 4.3. Hourly peak load in percentage of daily peak for IEEE RTS.

4.3.4 Substation Voltage Modelling

In this chapter, the substation voltage is also considered to be a random variable following normal distribution [18].

$$f(V_s) = \left(\frac{1}{\sigma\sqrt{2\pi}}\right)\exp - \frac{\left(V_s - \overline{V}_s\right)^2}{2\sigma^2} \tag{4.3}$$

where V_s is the substation bus voltage and $\overline{V}_s$, σ are the mean and standard deviation values of substation voltage, respectively.

4.3.5 Wind Speed Model

The intermittency of wind speed and air density at any given location results in the wind turbine power output also being variable. Because of the frequent variation of wind speed, it is assumed to be a random variable. Many probability distribution functions can be used to model the wind speed stochastically. Weibull, Rayleigh, and Lognormal distributions are usually utilized to simulate the wind speed characteristics [20]. This chapter uses the Weibull distribution function to sample the wind speed. The Weibull distribution function is a two-parameter function, which is used to describe wind speed mathematically as:

$$f(v) = \frac{k}{c}\left(\frac{v}{c}\right)^{k-1} \exp\left(\frac{-v}{c}\right)^k, 0 \le v \le \infty \tag{4.4}$$

where v is the wind speed, k is the shape parameter, and c is the scale parameter [26]. k is a dimensionless parameter, whereas c has the units of m/s. The Rayleigh distribution is a special case of Weibull distribution, where $k = 2$.

Several numerical methods are available in literature to determine the Weibull parameters k and c from statistical wind speed data. In this chapter, the empirical method [27] is used to determine the values of k and c. The following equations describe the method:

$$k = \left(\frac{\sigma}{\bar{v}}\right)^{-1.086} \tag{4.5}$$

$$c = \frac{\bar{v}}{\Gamma\left(1 + \frac{1}{k}\right)} \tag{4.6}$$

where $\bar{v}$ is the mean wind speed in m/s and σ is the standard deviation of the wind speed in m/s.

4.4 Wind Turbine Modelling

4.4.1 Wind Turbine Power Output

The wind turbine produces electrical power by trapping the kinetic energy present in the moving mass of air. The power output from a wind turbine can be written mathematically as:

$$P(v) = \frac{1}{2} C_p \rho A v^3 \tag{4.7}$$

where C_p is the power coefficient of the turbine, A is the area swept by the rotor blades of the turbine, and ρ is the air density. The value of C_p is limited to 0.59 known as Betz limit. This means that a turbine can never extract more than 59% of the power present in the air mass. For a pitch-controlled turbine, the turbine power output is reduced to zero at very high wind speeds. This is done in order to reduce excess stress and damage to the rotor blades at high wind speeds. The power output for a pitch-controlled wind turbine can be expressed using the linear model, quadratic model, or the cubic model [28]. In this chapter, the quadratic model is used to express the power output from the turbine.

$$P_w = \begin{cases} 0, v \leq v_{ci} \\ P_r \dfrac{v^2 - {v_{ci}}^2}{{v_r}^2 - {v_{ci}}^2}, v_{ci} < v < v_r \\ P_r, v_r < v < v_{co} \\ 0, v > v_{co} \end{cases} \tag{4.8}$$

where P_w is the power output of wind turbine, v is the wind velocity in m/s, v_{ci} is the cut-in speed of the wind turbine in m/s, v_{co} is the cut-out speed of the wind turbine in m/s, v_r is the rated speed of the wind turbine in m/s, and P_r is the rated power output of the wind turbine.

4.4.2 Wind Turbine Capacity Factor

The capacity factor of a wind turbine is defined as the ratio of average power of the turbine over a period of time and the rated power of the turbine.

$$C_f = \frac{P_{avg}}{P_r} \tag{4.9}$$

where P_{avg} is the average power of the wind turbine. The capacity factor of the wind turbine is a very important parameter as it indicates how much of the rated capacity of the turbine is actually utilized. The power output of a particular wind turbine depends on two factors: the design characteristics of the turbine and the wind speed parameters of the selected site. Hence, the capacity factor of a wind turbine can be approximated from the wind resource data available. The C_f value of a turbine varies from site to site. Mathematically, C_f can be expressed in accordance with the Weibull distribution function and the quadratic power output model as [28]:

$$C_f = \left(\frac{2c^2}{{v_r}^2 - {v_{ci}}^2}\right)\frac{\Gamma\left(\frac{2}{k}\right)}{k} * \left(\gamma\left(\left(\frac{v_r}{c}\right)^k, \frac{2}{k}\right) - \left(\left(\frac{v_{ci}}{c}\right)^k, \frac{2}{k}\right)\right) - e^{-\left(\frac{v_{co}}{c}\right)^k} \tag{4.10}$$

where Γ (.) is the gamma function and γ (.) is the lower incomplete gamma function.

4.4.3 Wind Speed Variation with Hub Height

Wind speed varies significantly with height. Most of the commercially available wind turbines have different hub heights than the height at which wind speed is measured at a site. Hence, the measured wind speed has to be converted into a new value in accordance with the hub height of the turbine. The power law [29] is used to express the variation of wind speed with hub height.

$$\frac{v_2}{v_1} = \left(\frac{h_2}{h_1}\right)^m \tag{4.11}$$

where v_2 and v_1 are wind speeds at heights h_2 and h_1, respectively. m is an exponential factor that depends upon atmospheric stability and surface roughness. The value of m lies in the range 0.05–0.5. Usually the value is approximated to be 0.14 [28].

4.5 Optimum Location for Wind Turbine Placement

Stability-based indices have been used in the recent past to obtain the optimum location for placing the DGs in the distribution system [30]. In this chapter, the voltage stability index (VSI) has been obtained for the test system and the bus, where the VSI is minimum is the weakest bus from voltage stability point of view. The bus with lowest VSI has been selected as the candidate bus for the wind generator placement. This index identifies the most voltage sensitive bus in the system.

The bus with the least value of VSI is the most sensitive bus and the DG is to be placed on that bus for voltage profile improvement. This index is obtained from the two-node distribution system shown in Fig. 4.4.

$$SI(r) = 2V_s^2V_r^2 - V_r^4 - 2V_r^2(PR + QX) - |Z|^2\left(P^2 - Q^2\right) \tag{4.12}$$

where s and r are the sending and receiving end nodes, respectively. V_s and V_r are the sending and receiving end voltages in per unit, respectively. P and Q are the active and reactive power load at node r. R, X, and Z are the resistance, reactance, and impedance of the line connecting nodes s and r, respectively.

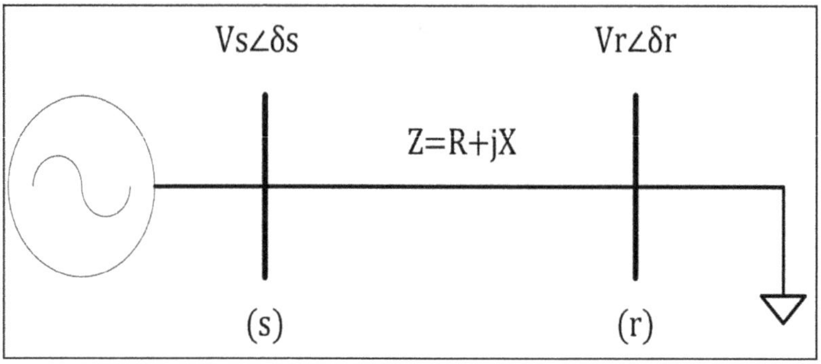

Figure 4.4. One-line diagram of a two bus distribution system.

4.6 Performance Indices

In this study, performance indices have been evaluated to analyze the impact of different wind turbine integration into the distribution system taking into consideration seasonal and probabilistic load patterns. These are summarized below.

4.6.1 Voltage Deviation Index

The *VDI* is a measure of how much the voltage value of the system deviates from the nominal substation voltage. This index provides important information about the voltage stability of the system. For an ideal DG interconnection, the *VDI* value should decrease with DG placement.

$$VDI = \sum_{i=1}^{N} \left(\frac{V_{nom} - V_i}{V_{nom}} \right)^2 \tag{4.13}$$

where V_{nom} is the nominal substation voltage (1.00 pu), V_i is the voltage in pu at i^{th} bus of the system, and N is the total number of nodes in the system.

4.6.2 Active Power Loss Index

The *APLI* is a measure of how much the total active power loss has changed with the interconnection of wind turbine-based DG.

$$APLI = \frac{TPL_{WT}}{TPL_{woWT}} \tag{4.14}$$

where TPL_{WT} and TPL_{woWT} are the total active power loss with and without wind turbine placement, respectively.

4.6.3 Reactive Power Loss Index

Similar to the *APLI*, the *RPLI* is a measure of how much the total reactive power loss has changed with the placement of the wind turbine in the system. The lower the reactive power loss in the system, the better will be the voltage profile and active power losses will also be lower.

$$RPLI = \frac{TQL_{WT}}{TQL_{woWT}} \tag{4.15}$$

where TQL_{WT} and TQL_{woWT} are the total reactive power loss with and without wind turbine placement, respectively.

4.6.4 Line Loading Index

The *LLI* is the ratio of the maximum line loading in the distribution system without and with the wind turbine.

$$LLI = \max\left(\frac{S_j}{S_{j-woWT}}\right) \forall j, j \in (1, N_b) \tag{4.16}$$

where S_j and S_{j-woWT} are the complex power flows in line j with and without wind turbine, respectively. N_b denotes the total number of branches/lines in the radial distribution system.

4.7 Computational Procedure

A computational procedure is developed for analyzing the seasonal impacts of wind turbine placement on the distribution system. Firstly, the data of 1-h wind speeds for an entire year are divided into 24-h data for each day of the month and for all the months of the year. Then, each month is represented by one set of 24-h wind speed data. The wind speed of each of these hours is the mean wind speed of that hour for that entire month. The standard deviations of the wind speeds are also determined for each of these hours. Then, the months are grouped into four seasons: winter, summer, spring, and fall. This leads to each season being represented by one set of mean and standard deviation of 24-h wind speed data. Table 4.2 summarizes the technical specifications of the commercially available wind turbines selected. The following are the analytical steps followed:

Step 1: Select the season of the year: winter, spring, summer, or fall.
Step 2: Select the wind turbine, denoted by the corresponding no. T, from Table 4.1
Step 3: Start time counter $t = 0$. While ($t \leq 24$), perform the following steps:

a) Input wind speed data set, network line data, and load data.
b) Find mean and variance of wind speed for hour t.
c) Find the Weibull parameters k and c according to (4.5) and (4.6), respectively.
d) Find C_f of the turbine T for hour t using (4.10).
e) From the C_f value and the rated power of the turbine from Table 4.1, find the P_{avg} of turbine T using (4.9).
f) Perform PLF using MCS and obtain the indices viz. *VDI*, *APLI*, *RPLI*, and *LLI* using (4.13), (4.14), (4.15), and (4.16), respectively.
g) Increase the time counter, $t = t + 1$

Step 4: Repeat steps (2) and (3) until all wind turbines have been selected individually.
Step 5: Repeat steps (2) to (4) until all seasons have been selected.
Step 6: Plot all indices for all wind turbines for four seasons and all days

Each season of the year is represented by a day. Even though the study is conducted for 24 h for the 4 days representing their respective seasons, because of the complexity of the results, the observations and results of only 2 h are analyzed in this chapter. The wind turbines are assumed to be operating at unity power factor. Only one wind turbine is placed at a time.

4.8 Case Study

The study of wind integration into the distribution system has been conducted on an IEEE 33 bus radial distribution test system shown in Fig. 4.5 and the data of the system are summarized in [31]. The base power of the system is 100 MVA and the base voltage is 12.66 kV. The total connected active power load is 3.72 MW and reactive power load is 2.30 MVAR. The analysis has been carried in MATLAB 7.6. The results are obtained for all seasons in the year and 24 h a day. However, the results have been presented for four seasons and 7–8 a.m. in the day and 7–8 p.m. in the night.

For the analysis, the four seasons in a year have been considered. The results have been obtained for four seasons of the year and 24 h in a day, however, the results are presented for 2 h in a day and nighttime. Within each season of study, 1 h is selected from daytime and 1 h is selected from nighttime. The selected hours are 8th hour (7–8 a.m.) and 20th hour (7–8 p.m.). In summary, the following scenarios are considered:

(i) Winter: Daytime (7–8 a.m.) and Nighttime (7–8 p.m.)
(ii) Spring: Daytime (7–8 a.m.) and Nighttime (7–8 p.m.)
(iii) Summer: Daytime (7–8 a.m.) and Nighttime (7–8 p.m.)
(iv) Fall: Daytime (7–8 a.m.) and Nighttime (7–8 p.m.)

Figure 4.6 shows the percentage of hourly load in terms of the daily peak load for the abovementioned scenarios. From the figure, it is evident that summer daytime has the minimum load demand, whereas spring nighttime and fall nighttime have the highest demands. The load pattern has been considered to follow the load change as for IEEE-RTS system [25]. Figure 4.7 shows the seasonal load variation considered for the study of the IEEE 33

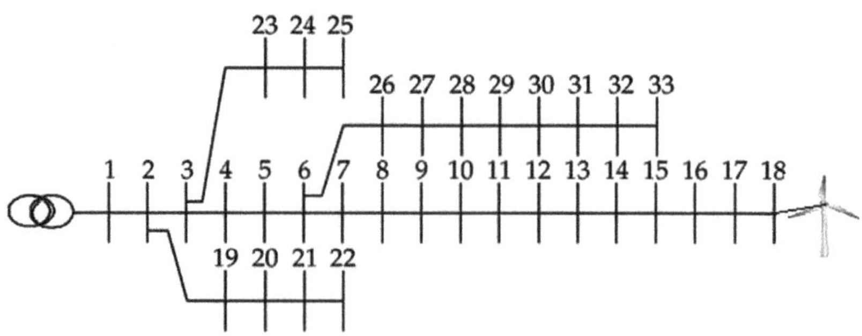

Figure 4.5. IEEE 33 bus radial distribution system.

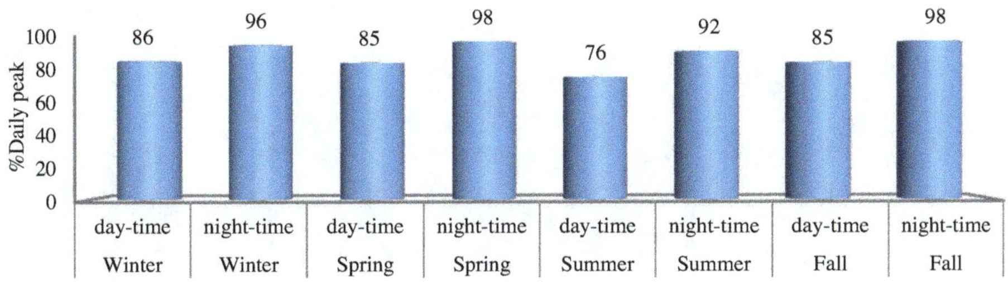

Figure 4.6. Seasonal load variation of the test system.

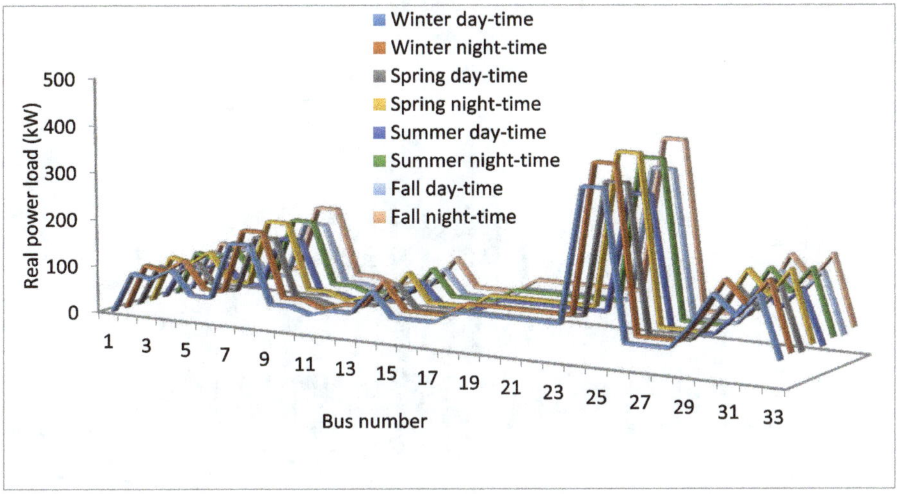

Figure 4.7. Seasonal load variation of the test system following IEEE-RTS.

bus test system. Then, using (2), the load is modelled as probabilistic load using normal distribution function. The number of samples for MCS-based PLF is set to 15,000. The data values of load at each load bus are assumed to be their respective mean values. A standard deviation of 10% is set for each load bus. The substation voltage mean and standard deviation values are assumed to be 1.0 pu and 1.5%, respectively.

For the test system in consideration, VSI values are calculated for the test system without any wind integration. Figure 4.8 shows the variation of VSI values with bus number. The bus with the least value of VSI is selected to be the location for the wind turbine placement. The lowest VSI value of 0.66804233 is obtained for bus 18. This bus is the most voltage-sensitive bus in the system for wind generation system. The different manufacturers wind turbines are placed one at a time at bus 18 and are shown in Fig. 4.5.

4.9 Numerical Results

Because the wind turbines have different hub heights, the Weibull parameters for each hour in each season will be different for them. The wind speed data from Fig. 4.2 are adjusted to match the respective wind turbine hub heights using the power law mentioned in (11).

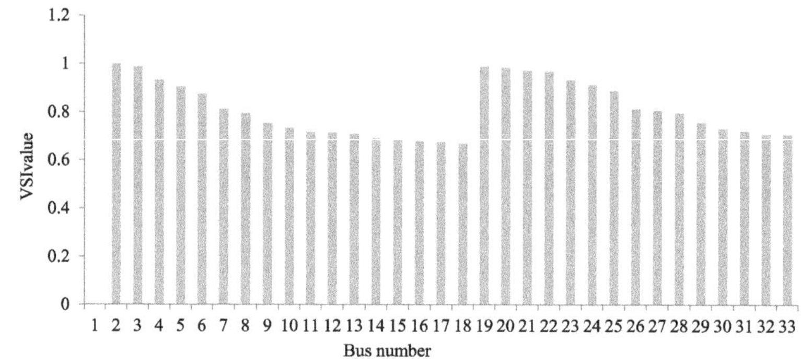

Figure 4.8. VSI value for each bus of the test system.

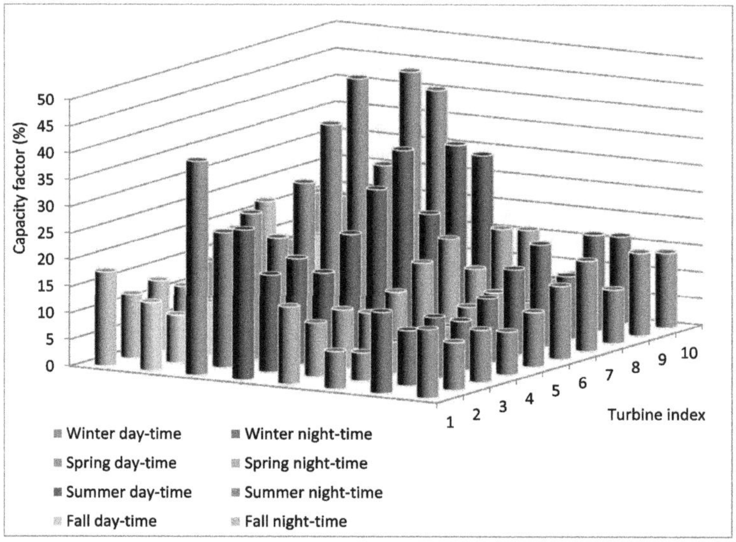

Figure 4.9. Variation of C_f values for various turbines.

After that, using (5) and (6), the values of Weibull parameters are obtained as summarized in Table 4.3. Using the Weibull parameter values obtained for each hour of each season, (10) is used to determine the C_f values of different turbines. After determining the C_f values, the P_{avg} values are also determined using (9). The C_f and P_{avg} values thus obtained are summarized in Table 4.3. The variations of the C_f and P_{avg} values for each turbine are shown in Figs. 4.9 and 4.10, respectively. It is observed from Fig. 4.8 that the capacity factor C_f is higher for summer nighttime due to the higher wind speed at this time. The capacity factor for other hours are observed lower compared to summer nighttime. In Fig. 4.10, it can be observed that maximum average power is produced by the turbines during summer nighttime. This is because during this period the wind speed is maximum during the year. Similarly, spring daytime is observed to be the time when minimum average power is produced by the turbines due to the lower wind speed during this period.

Table 4.3. Weibull shape parameter (***k***), Weibull scale parameter (***c*** in m/s), capacity factor (***C_f*** in %), and average power output (***P_avg*** in kW).

Season	Parameter	Various Wind Turbines with Different Hub Heights									
		1	2	3	4	5	6	7	8	9	10
Winter (Day-time)	k	3.24	3.24	3.24	3.24	3.24	3.24	3.24	3.24	3.24	3.24
	c	4.69	4.81	4.97	5.03	5.28	5.51	5.75	5.55	5.38	5.54
	C_f	13.02	9.07	9.96	8.12	10.31	13.82	17.02	10.19	15.56	14.08
	P_{avg}	6.51	22.67	59.73	105.92	170.13	276.48	10340.33	10234.26	10357.77	10422.41
Winter (Night-time)	k	2.89	2.89	2.89	2.89	2.89	2.89	2.89	2.89	2.89	2.89
	c	4.90	5.03	5.21	5.27	5.52	5.76	5.96	5.81	5.63	5.80
	C_f	15.41	10.47	11.45	9.27	12.04	16.05	19.18	11.81	18.10	16.33
	P_{avg}	7.70	26.18	68.73	120.58	198.65	320.95	10383.66	10271.72	10416.39	10489.94
Spring (Day-time)	k	3.16	3.16	3.16	3.16	3.16	3.16	3.16	3.16	3.16	3.16
	c	3.92	4.03	4.16	4.21	4.42	4.61	4.68	4.64	4.50	4.64
	C_f	7.13	5.40	6.01	5.12	5.91	8.08	9.35	5.97	8.99	8.25
	P_{avg}	3.56	13.51	36.05	66.49	97.53	161.66	10186.96	10137.39	10206.91	10242.67
Spring (Night-time)	k	4.06	4.06	4.06	4.06	4.06	4.06	4.06	4.06	4.06	4.06
	c	4.29	5.05	5.23	5.29	5.55	5.79	5.98	5.84	5.65	5.83
	C_f	14.67	10.15	11.13	9.07	11.58	15.51	18.59	11.42	17.46	15.79
	P_{avg}	7.33	1025.38	66.80	10117.92	10191.15	10310.16	10371.94	10262.72	10401.66	10473.74

(Continued)

Table 4.3. Weibull shape parameter (***k***), Weibull scale parameter (***c*** in m/s), capacity factor (C_f in %), and average power output (P_{avg} in kW). *(Continued)*

Season	Parameter	Various Wind Turbines with Different Hub Heights									
		1	2	3	4	5	6	7	8	9	10
Summer (Day-time)	k	8.75	8.75	8.75	8.75	8.75	8.75	8.75	8.75	8.75	8.75
	c	6.14	6.29	6.52	6.59	6.90	7.21	7.49	7.26	7.04	7.25
	C_f	28.28	18.34	19.92	15.75	21.56	28.43	34.26	20.89	32.28	28.89
	P_{avg}	14.14	25.86	119.52	204.71	355.73	568.55	685.27	480.44	742.39	866.55
Summer (Night-time)	k	11.68	11.68	11.68	11.68	11.68	11.68	11.68	11.68	11.68	11.68
	c	6.98	7.17	7.42	7.51	7.87	8.21	8.48	8.27	8.02	8.26
	C_f	40.23	25.43	27.50	21.48	30.23	39.63	46.83	29.09	45.14	40.23
	P_{avg}	20.11	63.55	165.01	279.33	498.85	792.52	936.63	669.08	1038.27	1206.9
Fall (Day-time)	k	3.76	3.76	3.76	3.76	3.76	3.76	3.76	3.76	3.76	3.76
	c	4.71	4.84	5.01	5.06	5.31	5.54	5.73	5.58	5.41	5.58
	C_f	12.84	9.03	9.92	8.14	10.23	13.74	16.55	10.13	15.44	13.99
	P_{avg}	6.42	22.56	59.52	105.80	186.78	274.78	331.04	232.88	335.18	419.92
Fall (Night-time)	k	4.85	4.85	4.85	4.85	4.85	4.85	4.85	4.85	4.85	4.85
	c	5.26	5.39	5.58	5.65	5.92	6.18	6.28	6.23	6.03	6.22
	C_f	17.75	12.05	13.17	10.63	13.84	18.48	22.09	13.61	20.87	18.81
	P_{avg}	8.87	30.13	79.04	138.18	228.32	369.72	441.95	312.94	480.01	564.35

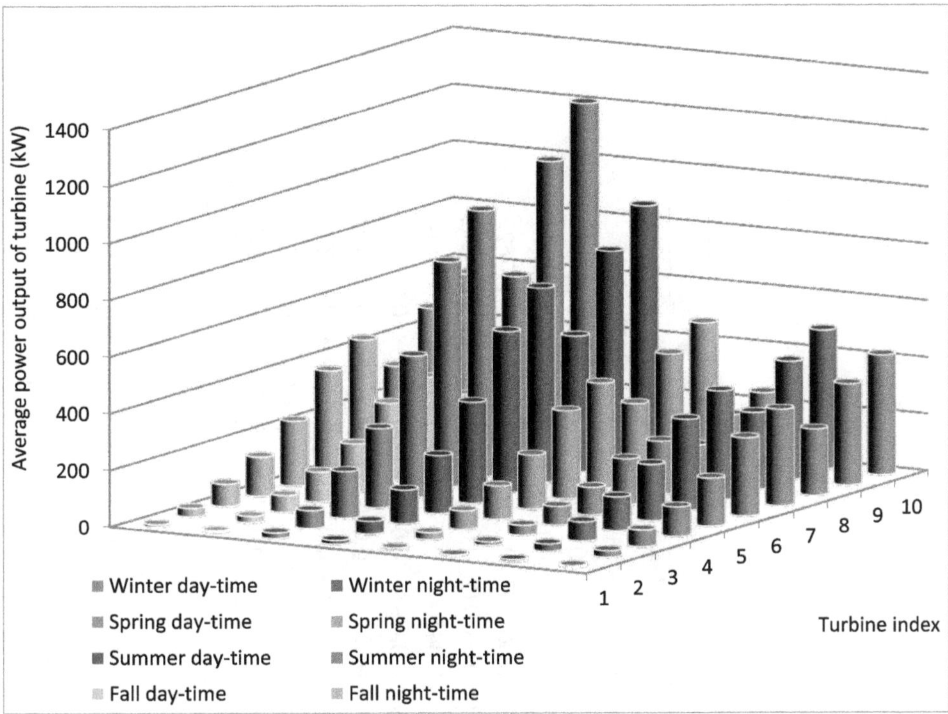

Figure 4.10. Variation of P_{avg} values for various turbines.

4.9.1 Impact of Wind Turbine Placement on *VDI*

The *VDI* values were obtained for the test system with integration of all types of wind turbines. Table 4.4 summarizes the *VDI* values obtained on placing the 10 wind turbines for various scenarios. The *VDI* values are shown graphically in Fig. 4.11. From Table 4.4 and Fig. 4.11, it can be observed that for all the 10 turbines, minimum *VDI* is obtained for summer daytime. Because *VDI* is an indicator of how much the voltage of the nodes deviated from the nominal value, lesser *VDI* is always better for the system. As the *VDI* values are reduced, the system voltage becomes closer to unity, which is ideal for voltage stability. The *VDI* values are also calculated for the test system before placing the wind turbines for comparison and are shown in Fig. 4.12. Figure 4.12 shows the *VDI* values of the system for different seasons. It is evident from Table 4.4 that the *VDI* values for day and night hours in all the seasons have reduced after placing wind turbines, which shows the impact of wind integration on the improvement of voltage stability. The voltage has considerable improved after the wind generation integration into the distribution network.

From Fig. 4.12, it can be concluded that the minimum *VDI* among all the cases is obtained when placing turbine 10 during summer daytime. The *VDI* value obtained here is 0.021665. Before placing the wind turbine, the *VDI* value was 0.0741 for summer daytime. This means that on placing turbine 10 during summer daytime, there is a reduction of 70.7% in *VDI*. After summer daytime, summer nighttime is generally observed to have reduced *VDI* values. The maximum values of *VDI* are obtained during spring nighttime. On placing turbine 1 during spring nighttime, the *VDI* value is 0.127133. Before placing

Table 4.4. *VDI* values for different turbines.

Turbin Index	Winter		Spring		Summer		Fall	
	Day-time	Night-time	Day-time	Night-time	Day-time	Night-time	Day-time	Night-time
1	0.09578	0.121389	0.093733	0.127133	0.072717	0.109112	0.093406	0.12681
2	0.093964	0.119001	0.092637	0.124624	0.069683	0.103871	0.091619	0.123989
3	0.089846	0.113575	0.090132	0.119189	0.062992	0.092405	0.087573	0.11762
4	0.084903	0.107225	0.086813	0.112737	0.055912	0.080802	0.082694	0.110272
5	0.078384	0.098213	0.083498	0.103988	0.045007	0.062114	0.076396	0.099814
6	0.06862	0.085547	0.077097	0.091169	0.032957	0.04365	0.066808	0.085299
7	0.063322	0.07969	0.07473	0.085144	0.027827	0.036965	0.062197	0.078706
8	0.072364	0.09046	0.079472	0.096101	0.03751	0.050582	0.070468	0.090882
9	0.061895	0.076712	0.072835	0.082308	0.025634	0.033034	0.060257	0.075357
10	0.057008	0.070529	0.069203	0.075938	0.021665	0.027976	0.055438	0.068547

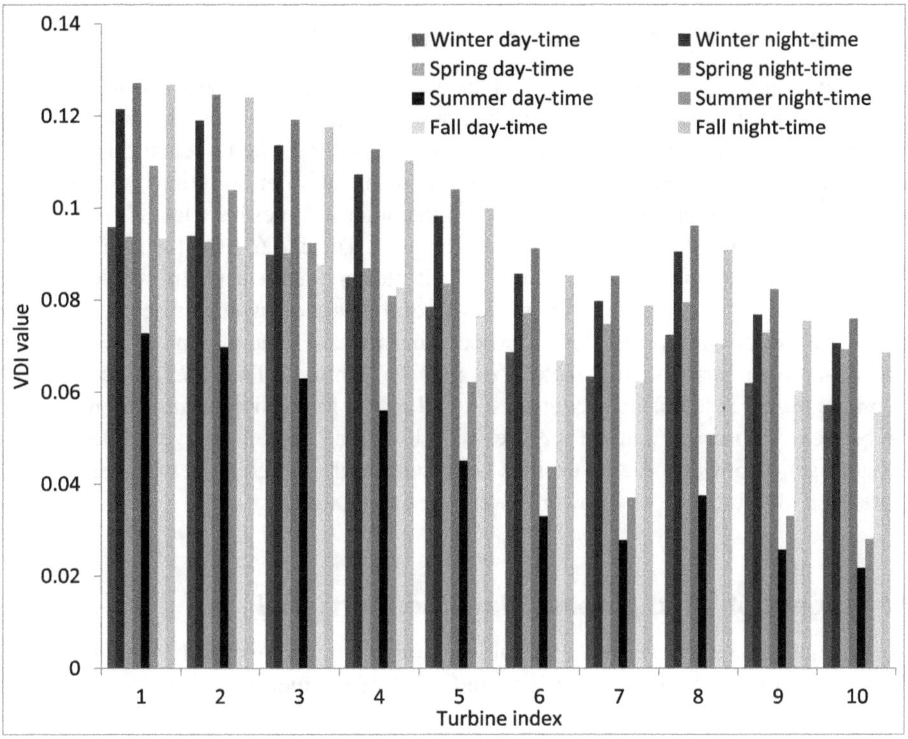

Figure 4.11. Variation of *VDI* values for different turbines.

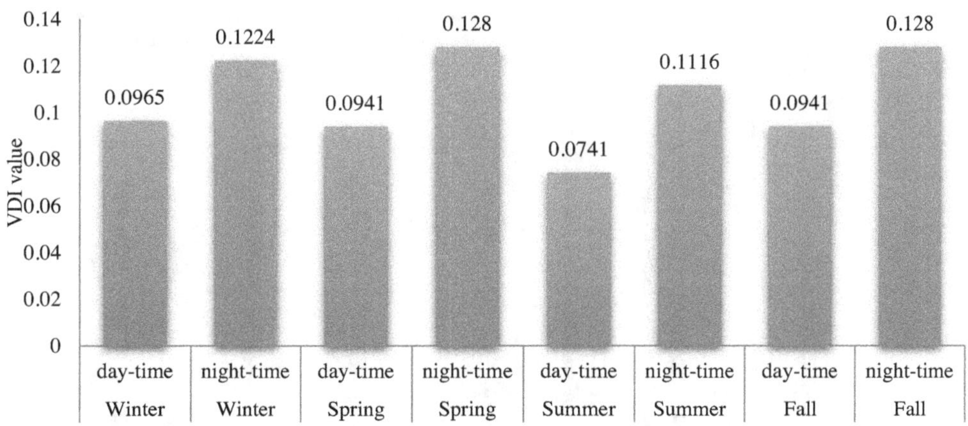

Figure 4.12. Variation of *VDI* values before wind turbine placement

the wind turbine, the *VDI* value was 0.128. This means that there is a reduction of 0.67% in *VDI* for this case, which is the minimum reduction in *VDI* for all the cases considered. In general, the *VDI* values are observed to be higher during spring nighttime and fall nighttime. *VDI* values are at their minimum during the summer season. Also, turbine 10, which has the highest capacity factor value among all the other turbines, is seen to provide better *VDI* values.

4.9.2 Impact of Wind Turbine Placement on *APLI*

Table 4.5 and Fig. 4.13 show the *APLI* values obtained on placing the 10 wind turbines for various scenarios. Because *APLI* is an indicator of the measure of active power loss reduction on wind turbine placement, the lesser the value of *APLI* the better it is for the system. In general, *APLI* value is observed to be minimum during summer nighttime compared to all other time periods. Except turbines 7 and 9, all the other turbines have less *APLI* value during summer nighttime compared to summer daytime. Turbine 6 gives the minimum value of *APLI* of all cases, that value being 0.6934. This means that the active power loss on placing turbine 6 during summer daytime is 69.34% of the active power loss before placing the turbine. The reduction in power loss is 30.66% in this case. The maximum *APLI* value of 0.9968 is obtained during spring daytime on placing turbine 1. The *APLI* value is observed to be on the higher side during spring season and on the lower side during summer season. It can be clearly observed that the turbines with low rated power does very little to reduce the system active power losses. Turbines 6–10 with higher rated powers, more than 2000 kW, are seen to reduce the real power losses to a great extent.

4.9.3 Impact of Wind Turbine Placement on *RPLI*

Table 4.6 and Fig. 4.14 show the *RPLI* values obtained on placing the 10 wind turbines for various scenarios. Like *APLI*, *RPLI* is an indicator of the measure of reactive power loss

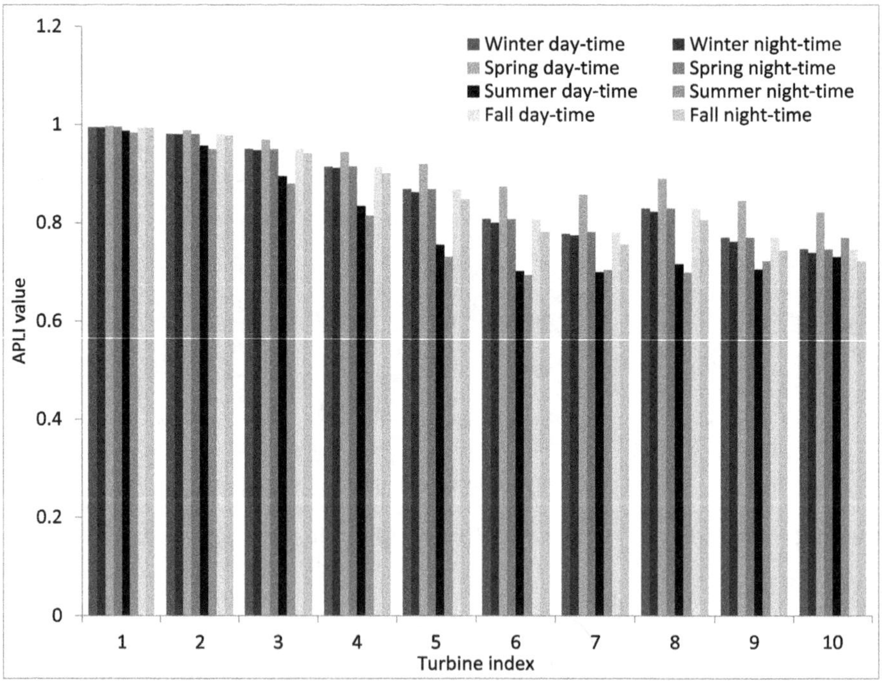

Figure 4.13. Variation of *APLI* values for different turbines

Table 4.5. *APLI* values for different turbines.

Turbin Index	Winter		Spring		Summer		Fall	
	Day-time	Night-time	Day-time	Night-time	Day-time	Night-time	Day-time	Night-time
1	0.9942	0.9938	0.9968	0.9948	0.9861	0.9834	0.9943	0.993
2	0.9807	0.9797	0.9885	0.9807	0.9568	0.9497	0.9806	0.977
3	0.9501	0.9479	0.9692	0.9501	0.8947	0.8794	0.9497	0.9414
4	0.9143	0.9116	0.9439	0.9147	0.8339	0.8146	0.9136	0.9016
5	0.8691	0.8624	0.9192	0.8688	0.7547	0.7304	0.8689	0.8479
6	0.8073	0.7996	0.8737	0.8071	0.7014	0.6934	0.8068	0.7815
7	0.7776	0.7738	0.8577	0.781	0.6991	0.7039	0.7801	0.7555
8	0.83	0.8228	0.8901	0.8298	0.7154	0.6989	0.8294	0.8056
9	0.7699	0.7614	0.845	0.7693	0.704	0.7212	0.7694	0.7432
10	0.7461	0.7383	0.8218	0.7455	0.7298	0.7687	0.7455	0.7217

Table 4.6. *RPLI* values for different turbines.

Turbin Index	Winter		Spring		Summer		Fall	
	Day-time	Night-time	Day-time	Night-time	Day-time	Night-time	Day-time	Night-time
1	0.9937	0.9932	0.9965	0.9944	0.9849	0.9819	0.9937	0.9924
2	0.9789	0.9779	0.9874	0.9789	0.9531	0.9455	0.9788	0.975
3	0.9458	0.9435	0.9665	0.9459	0.8866	0.8706	0.9455	0.9365
4	0.9075	0.9046	0.9392	0.908	0.8233	0.8036	0.9067	0.894
5	0.8598	0.8528	0.9128	0.8594	0.7457	0.7239	0.8595	0.8377
6	0.7964	0.7886	0.8646	0.796	0.7071	0.7084	0.7958	0.7707
7	0.7671	0.7634	0.8479	0.7703	0.7187	0.7355	0.7696	0.746
8	0.8193	0.812	0.8819	0.8191	0.7133	0.7028	0.8187	0.7946
9	0.7597	0.7515	0.8347	0.7589	0.7317	0.7667	0.7593	0.7347
10	0.7377	0.7305	0.811	0.7368	0.7783	0.841	0.7372	0.7166

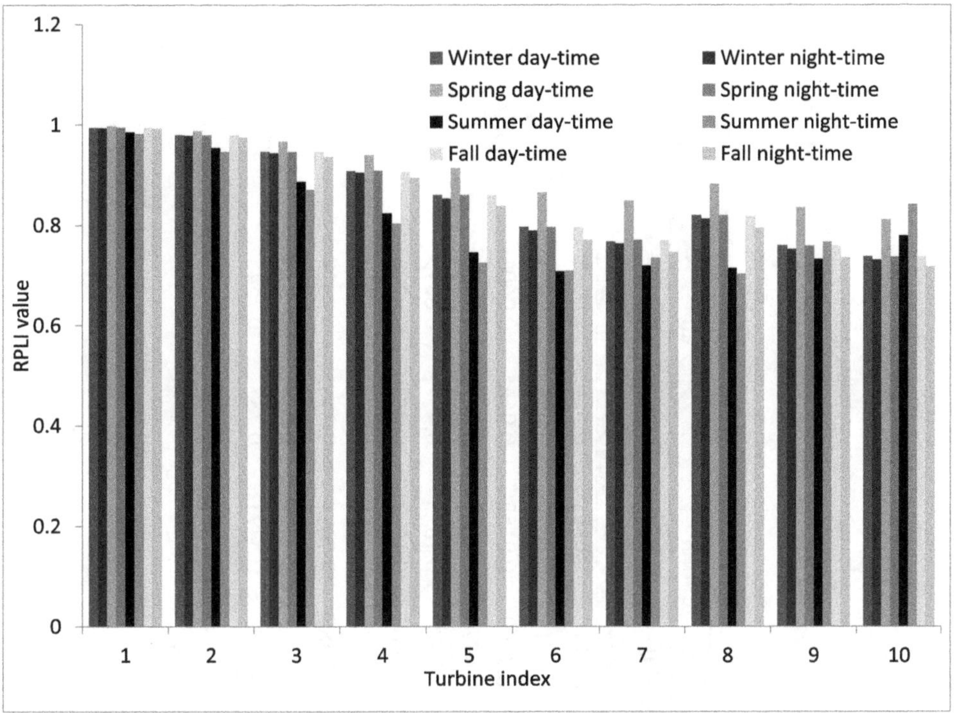

Figure 4.14. Voltage PDF plot at bus 18 during winter day-time.

reduction on wind turbine placement. Among all the cases, turbine 8 produces the minimum *RPLI* value of 0.7028 during summer nighttime. Also, turbine 8 produces an *RPLI* value of 0.7071 during summer daytime. A maximum *RPLI* value of 0.9965 is produced by turbine 3 during spring daytime. Similarly, turbine 1 produces an *RPLI* value of 0.9944 during spring nighttime. It can be observed that summer season provides comparatively lower values of *RPLI*, whereas spring season provides higher values of *RPLI*.

4.9.4 Impact of Wind Turbine Placement on *LLI*

Table 4.7 and Fig. 4.15 show the various *LLI* values obtained after placing the different types of the wind turbines considering various scenarios. It can be observed that summer nighttime produces the minimum values of *LLI*, whereas spring day and nighttime produce very high *LLI* values. The minimum value of *LLI* is observed to be 0.5471 for summer nighttime on placing turbine 10. Turbine 9 also produces a low value, 0.5897, of *LLI* during summer nighttime. Like the variation of *APLI*, it can be observed that turbines 6–10 produce very low *LLI* as compared to turbines 1–5. Turbine 1 produces the highest *LLI* value of 0.9979 during spring daytime. This implies that there is only a 0.21% decrease in line loading when turbine 1 is placed in spring daytime. The same turbine 1 when placed during summer nighttime produces an *LLI* value of 0.9891, indicating a decrease in line loading of 1.09%. Similarly, the *LLI* value of turbine 10 increases from 0.588 in summer daytime to 0.8673 in spring daytime.

Table 4.7. *LLI* values for different turbines.

Turbin Index	Winter		Spring		Summer		Fall	
	Day-time	Night-time	Day-time	Night-time	Day-time	Night-time	Day-time	Night-time
1	0.9963	0.996	0.9979	0.9966	0.991	0.9891	0.9963	0.9954
2	0.9874	0.9867	0.9925	0.9874	0.9714	0.9665	0.9873	0.985
3	0.9667	0.9651	0.9797	0.9667	0.9269	0.9152	0.9665	0.9606
4	0.9412	0.9391	0.9623	0.9414	0.8779	0.8609	0.9407	0.9317
5	0.9072	0.9018	0.9449	0.9068	0.7982	0.7675	0.907	0.8899
6	0.8544	0.8473	0.9109	0.8543	0.7003	0.6624	0.8539	0.8294
7	0.8248	0.8213	0.8981	0.8289	0.6533	0.6185	0.8274	0.8008
8	0.8747	0.8685	0.9234	0.8746	0.7387	0.7038	0.8742	0.8529
9	0.8166	0.8077	0.8877	0.8166	0.6314	0.5897	0.8161	0.7858
10	0.7879	0.7787	0.8673	0.7883	0.588	0.5471	0.787	0.7545

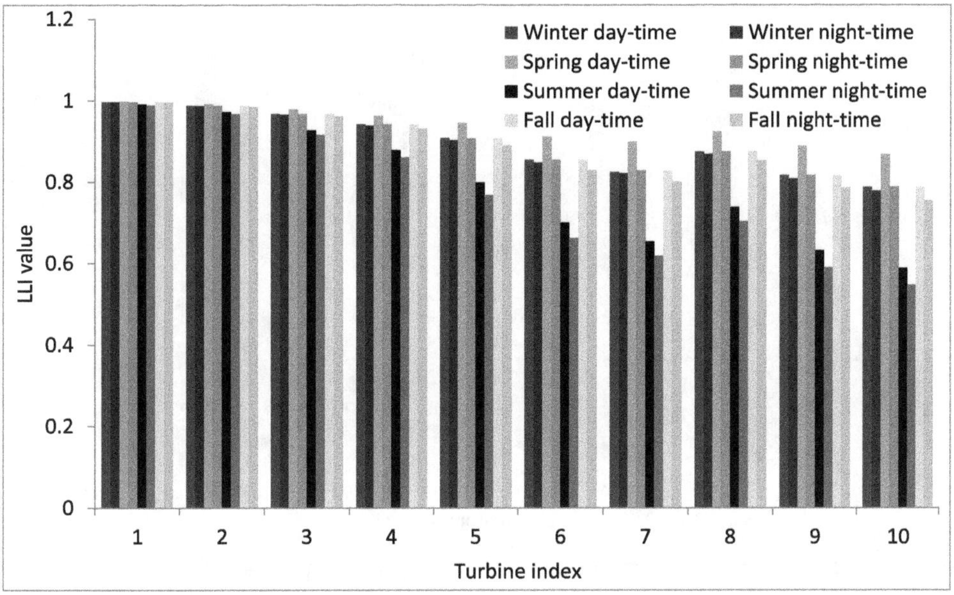

Figure 4.15. Voltage PDF plot at bus 18 during winter night-time.

4.9.5 Voltage Probability Density Function Plots

The probability density function (PDF) plots are a good indicator of the voltage profile improvement of the system with wind turbine integration. Figures 4.16–4.23 show the PDF plots at bus 18 with various types of wind turbines for the different scenarios under study. The more nearer the PDF plot is to unity, the better is the voltage stability of the system. From these figures, it is evident that the turbines with higher rated powers leads to more enhanced voltage profile improvement of the system. Also, it can be observed that the PDF plots are nearer to unity during summer day and nighttimes. The maximum voltage profile improvement for spring daytime leads to the PDF plot being grouped around 0.94, as evident from Fig. 4.18. Similarly, the maximum voltage profile improvement for fall daytime leads to the PDF plot being grouped around 0.95, as evident from Fig. 4.22. Winter and fall are observed to produce better results compared to the spring. The voltage profile of the system is observed to be significantly enhanced by the addition of the wind turbines. A huge variation in the magnitude of voltage profile improvement is caused by placing the turbines in daytime and nighttime in different seasons of the year. The rating and capacity factor of the turbine is also observed to vary the voltage profile of the system.

4.10 Summary

In this study, an overall analysis of impacts of placing wind turbines with different specifications on the radial distribution network is studied. Hourly wind speed data for a whole year is obtained for Cape Comorin, located in Southern India. Ten wind turbines of different manufacturers are included in the study. The impacts on the test system are observed

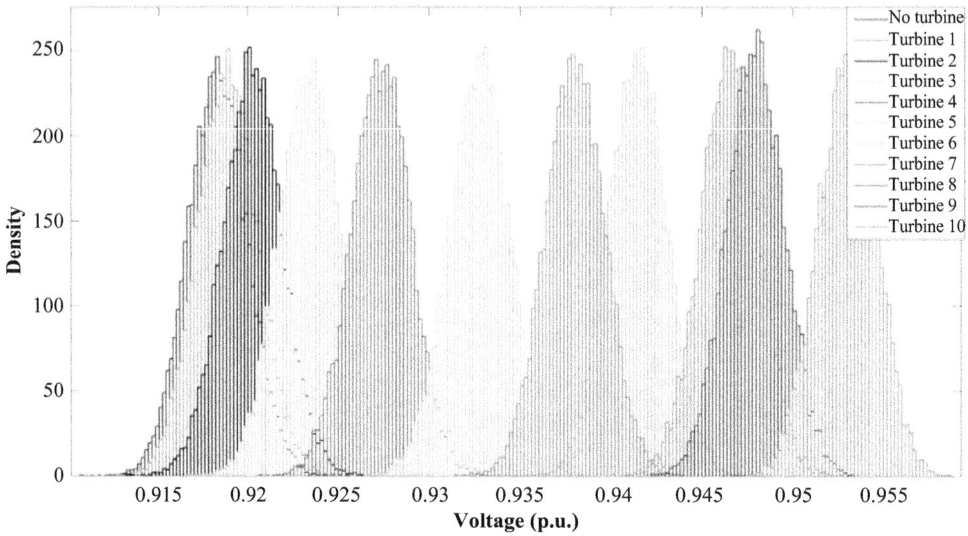

Figure 4.16. Voltage PDF plot at bus 18 during spring day-time.

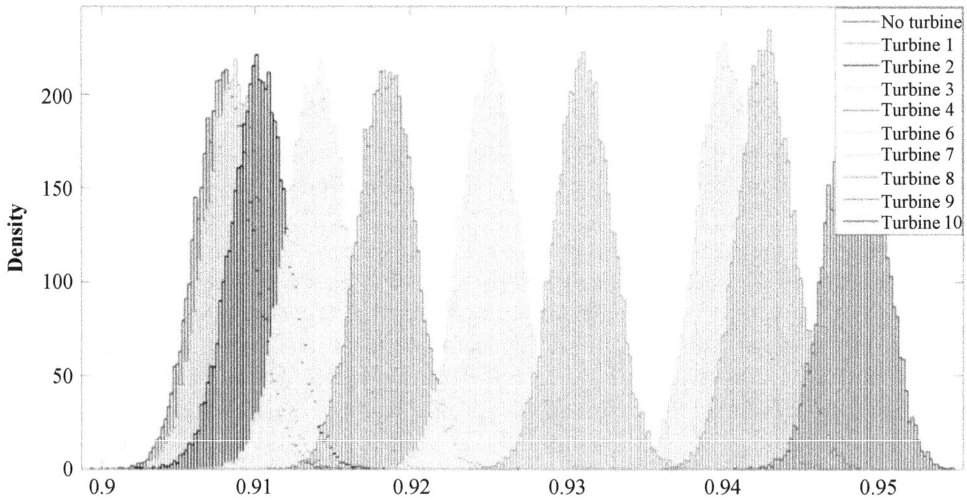

Figure 4.17. Voltage PDF plot at bus 18 during spring night-time.

for each season of the year. The results are obtained for all seasons and for each hour, however, the results are presented for each season and 2 h: a daytime hour and a nighttime hour. The capacity factor and average power output of each of the turbines are determined for each hour of each season. To analyze the effects of the wind turbines on the test system, a set of four performance indices are obtained. The results reveal that there is a significant difference in the performance of the distribution system from hour to hour, season to season, and turbine to turbine. In general, it is observed that turbines with higher rated power

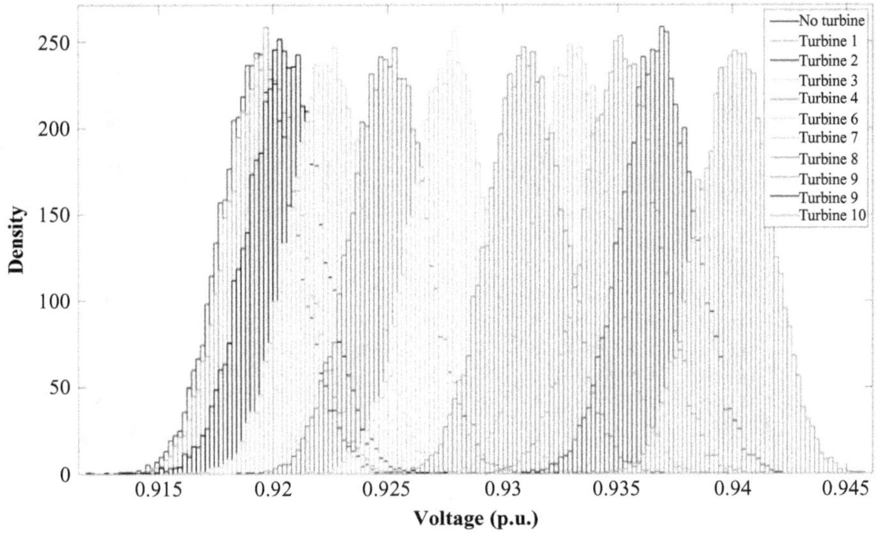

Figure 4.18. Voltage PDF plot at bus 18 during summer day-time.

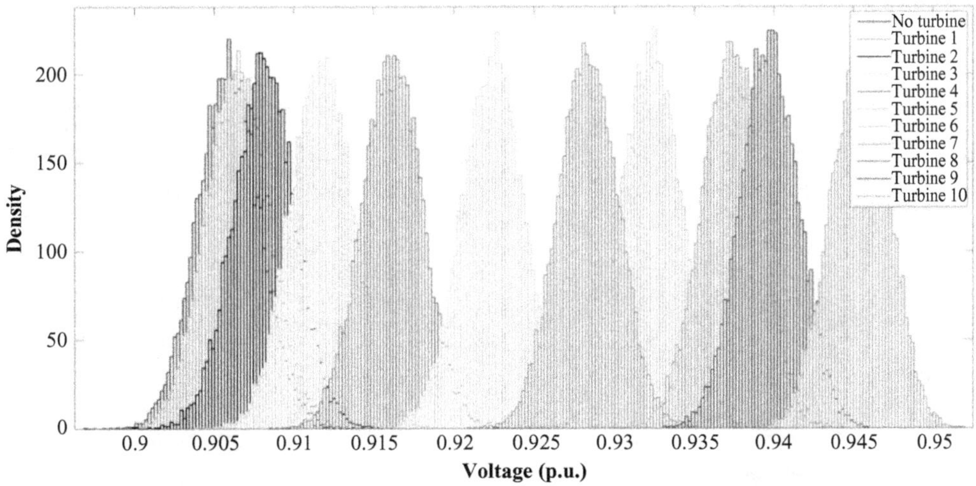

Figure 4.19. Voltage PDF plot at bus 18 during summer night-time.

and higher capacity factor provide better results. The wind turbines when placed during the summer season of the year produce much better voltage profile, reduced real and reactive power losses and line loading when compared to other seasons of the year. Similarly, the spring season is observed to produce the least amount of enhancement in the system performance with wind turbine placement. Of all the turbines considered for the study, turbines with rated power of more than 2000 kW, hub heights more than 70 m, and rotor diameter more than 65–70 m are observed to be better suitable for the site under study.

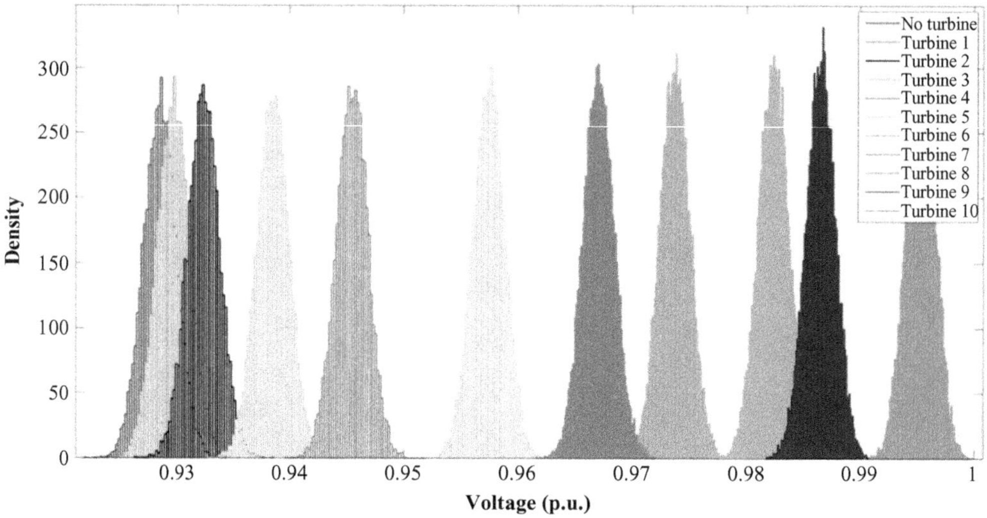

Figure 4.20. Voltage PDF plot at bus 18 during fall day-time.

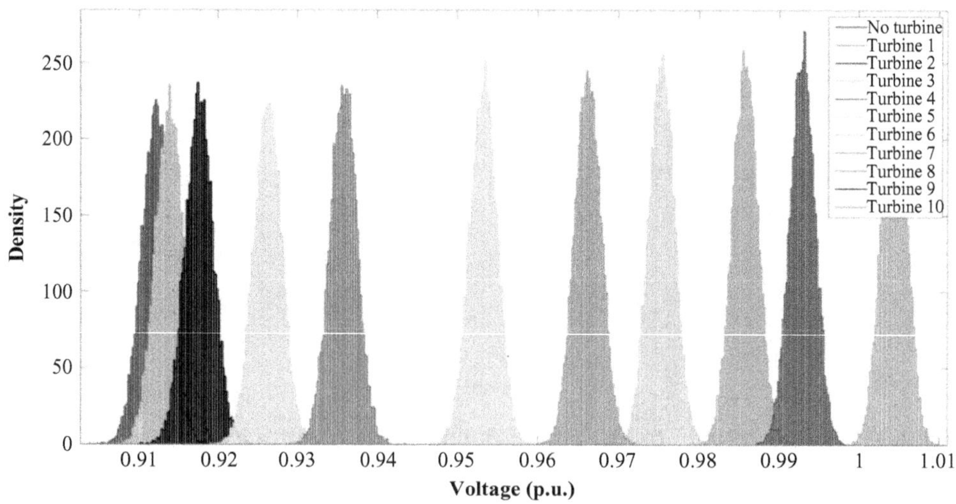

Figure 4.21. Voltage PDF plot at bus 18 during fall night-time.

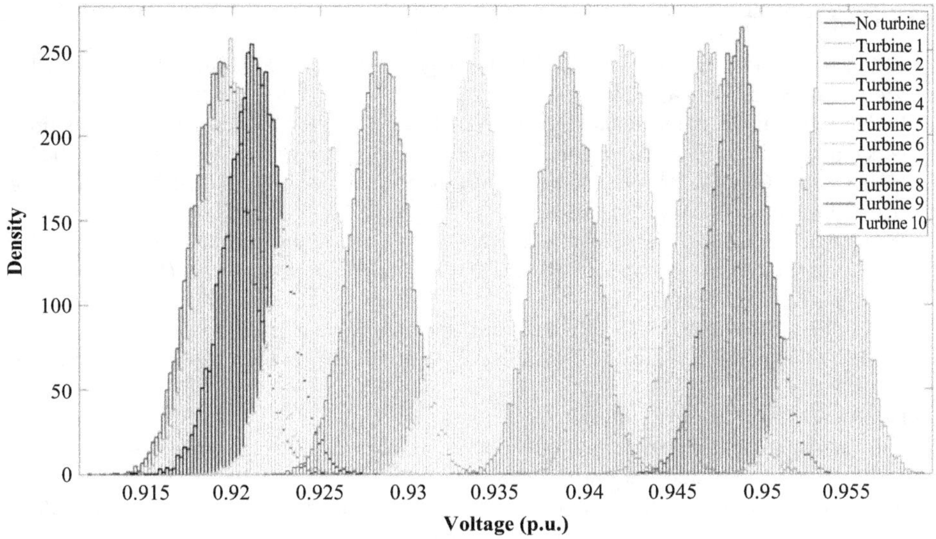

Figure 4.22. Voltage PDF plot at bus 18 during fall day-time.

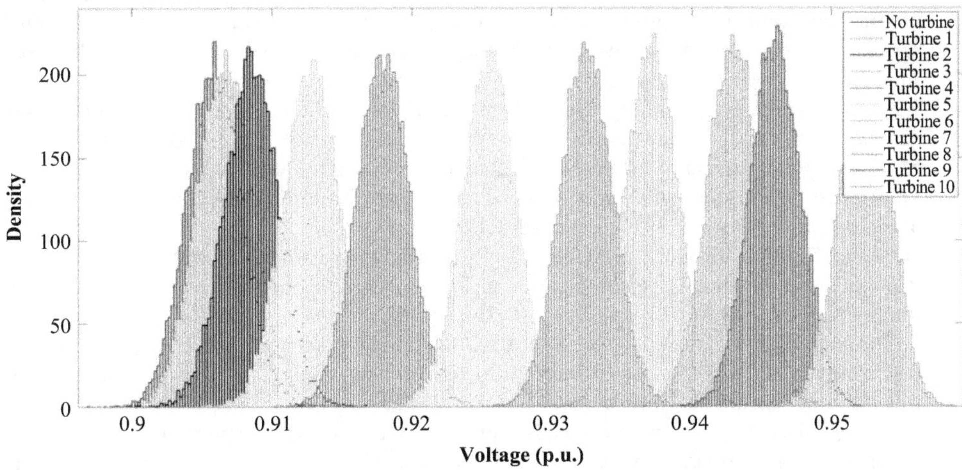

Figure 4.23. Voltage PDF plot at bus 18 during fall night-time.

4.11 References

1. R. C. Dugan, and T. E. Mcdermont, "Distributed generation," *IEEE Industry Applications Magazine*, pp. 19–25, Mar./Apr. 2002.
2. P. Chiradeja, and R. Ramakumar, "An approach to quantify the technical benefits of distributed generation," *IEEE Transactions Energy Conversion,* vol. 19, no.4, pp. 764–773, Dec. 2004
3. H. A. Gil, and G. Joos, "Models for quantifying the economic benefits of distributed generation," *IEEE Transactions Power Systems,* vol. 23, no. 2, pp. 327–335, May. 2008.
4. P. Siano, L. F. Ochoa, G. P. Harrison, and A. Piccolo, "Assessing the strategic benefits of distributed generation ownership for DNOs," *IET-Generation, Transmission & Distribution,* vol. 3, no. 3, pp. 225–236, Mar. 2009.
5. Renewables 2014 Global Status Report. Available online: www.ren21.net
6. GWEC, Global Wind Energy Statistics. 2021. Available online: https://gwec.net/global-wind-report-2021/
7. http://www.inwea.org/maps.htm
8. http://www.nrdc.org/international/india/files/renewable-energy-wind-financing-IP.pdf
9. http://www.ibef.org/industry/power-sector-india.aspx
10. M. R. Nouni, S. C. Mullick, and T. C. Kandpal, "Techno-economics of small wind electric generator projects for decentralized power supply in India," *International Journal of Energy Policy System,* vol. 35, pp. 2491–2506, April. 2007.
11. N. G. Boulaxis, S. A. Papathanassiou, and M. P. Papadopoulos, "Wind turbine effect on the voltage profile of distribution networks," *International Journal Renewable Energy*, vol. 25, pp. 401–415, March. 2002.
12. P. Siemes, H., Haubrich, H. Vennegeerts, and S. Ohrem, "Concepts for the improved integration of wind power into the German interconnected system," *IET - Renewable Power Generation,* vol. 2, no. 1, pp. 26–33, March. 2008.
13. Z. Bie, G. Li, H. Liu, X. Wang, and X. Wang, "Studies on voltage fluctuation in the integration of wind power plants using probabilistic load flow," *IEEE Power and Energy Society General Meeting - Conversion and Delivery of Electrical Energy in the 21st Century,* pp. 1–7, 20–24 July. 2008.
14. J. N. Sheen, M. T. Tsai, and S. W. Wu, "A benefit analysis for wind turbine allocation in a power distribution system," *Energy Conversion and Management*, vol. 68, pp. 305–312, April. 2013.
15. M. H. Abderrazzaq, and O. Aloquilli, "Evaluating the impact of electrical grid connection on the wind turbine performance for Hofa wind farm scheme in Jordan," *Energy Conversion and Management*, vol. 49, pp. 3376–3380, Nov. 2008.
16. N. Miller, D. Manz, H. Johal, S. Achilles, L. Roose, and J. P. Griffin, "Integrating high levels of wind in Island systems: Lessons from Hawaii," *IEEE International Conference on Sustainable Energy Technologies (ICSET),* pp.1–8, 6–9, Dec. 2010.
17. D. Q. Hung, N. Mithulananthan, and K. Y. Lee, "Optimal placement of dispatchable and nondispatchable renewable DG units in distribution networks for minimizing energy loss," *International Journal Electrical Power Energy Systems,* vol. 55, pp. 179–186, Feb. 2014.
18. V. Hengsritawat, T. Tayjasanant, and N. Nimpitiwan, "Optimal sizing of photovoltaic distributed generators in a distribution system with consideration of solar radiation and harmonic distortion," *International Journal Electrical Power Energy Systems,* vol. 39, pp. 36–47, July. 2012.
19. http://rredc.nrel.gov/solar/new_data/India/about.html
20. T. R. Ayodele, A. A. Jimoh, J. L Munda, and J. T. Agee, "Wind distribution and capacity factor estimation for wind turbines in the coastal region of South Africa," *Energy Conversion and Management*, vol. 64, pp. 614–625, Dec. 2012
21. http://www.windenergysolutions.nl/
22. http://www.vestas.com/products_and_services/turbines
23. http://www.energy.siemens.com/hq/en/renewable-energy/wind-power/

24. M. Aien, R. Ramezani, and S. M. Ghavami, "Probabilistic load flow considering wind generation uncertainty," *Engineering, Technology & Applied Science Research*, vol. 1, pp. 126–132, 2011.
25. P. Wong, P. Albrecht, R. Allan, R. Billinton, Q. Chen, C. Fong, S. Haddad, W. Li, R. Mukerji, D. Patton, A. Schneider, M. Shahidehpour, and C. Singh, "The IEEE reliability test system-1996. a report prepared by the reliability test system task force of the application of probability methods subcommittee," *IEEE Transactions Power Systems*, vol.14, no. 3, pp. 1010–1020, Aug. 1999.
26. C. Chen, B. Zhang, and W. Wu, "A fast probabilistic voltage assessment method for distribution system integrated with wind power generation," *IEEE Region 10 Conference, TENCON*, pp.1–6, 19–22, Nov. 2012.
27. P. A. C. Rocha, R. C. de Sousa, C. F. de Andrade, and M. E. V. da Silva, "Comparison of seven numerical methods for determining Weibull parameters for energy generation in northeast region of Brazil," *Applied Energy*, vol. 89, pp. 395–400, Jan. 2012.
28. T. P. Chang, F. J. Liu, H. H. Ko, S. P. Cheng, L. C. Sun, and S. C. Kuo, "Comparative analysis on power curve models of wind turbine generator in estimating capacity factor," *Energy*, vol. 73, pp. 88–95, Aug. 2014.
29. E. K. Akpinar, S. Akpinar, "An assessment on seasonal analysis of wind energy characteristics and wind turbine characteristics," *Energy Conversion and Management*, vol. 46, pp. 1848–1867, July. 2005
30. M. M. Aman, G. B. Jasmon, H. Mokhlis, and A. H. A. Bakar, "Optimal placement and sizing of a DG based on a new power stability index and line losses," *International Journal Electrical Power Energy Systems*, vol. 43, pp. 1296–1304, Dec. 2012.
31. R. Ranjan & Das, "Simple and efficient computer algorithm to solve radial distribution networks," *Electric Power Components and Systems,* vol. 31, no.1, pp. 95–107, June. 2010

CHAPTER 5

Wind Power Plant System Services

Kaushik Das[1], Müfit Altin[2], and Anca Daniela Hansen[1]

[1]Department of Wind Energy, Technical University of Denmark (DTU), Risø, Denmark. kdas@dtu.dk and anca@dtu.dk
[2]Energy Systems Engineering Department, Izmir Institute of Technology, Izmir, Turkey. mufitaltin@iyte.edu.tr

Abstract This chapter focuses on system service capability of wind power plants. Needs for system services for current and future power system with respect to active power, reactive power, frequency, and voltage services are analyzed. Different grid code requirements from countries all over the world are discussed to understand the requirements from the wind power plants. Wind power plant capabilities and models are explained for different operating conditions of the power system, namely – normal condition, alert operation, and emergency operation.

Keywords: Wind power plant, Wind turbine, Ancillary services, System services, Frequency control, Voltage control, Active power control, Reactive power control.

5.1 Introduction

Growing concerns for environment, energy security, sustainable development goals, and technological development have led the way for increasing share of renewable generations, mainly wind and solar power in power systems all over the world. Increased installations of wind turbines (WTs) and continuous improvement of technology have made wind power one of the cheapest sources of electricity generation as of in 2018 [1]. Major milestones were reached for offshore wind power in 2017 when new offshore wind power plants (WPPs) were planned to be built without government subsidies [2]. These improvements indicate optimistic future for wind power with even higher share of wind power in power systems all over the world. According to Global Wind Energy Council (GWEC) report,

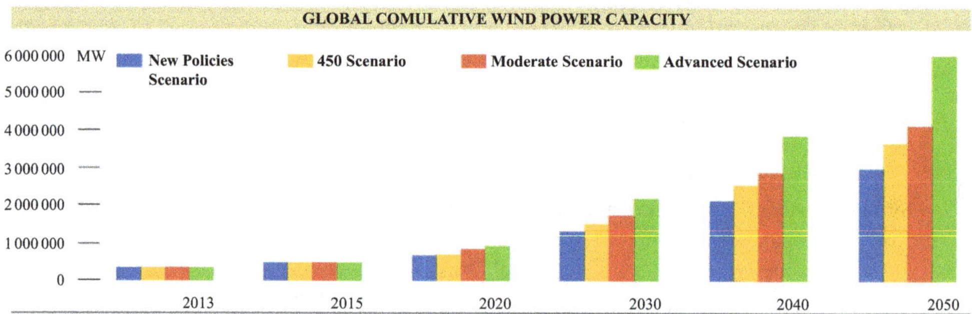

	2013	2015	2020	2030	2040	2050
New Policies Scenario						
MW	318 354	432 656	639 478	1 259 974	2 052 583	2 869 611
TWh/a	714	868	1 569	3 311	5 394	7 541
450 Scenario						
MW	318 354	432 656	658 009	1 454 395	2 458 757	3 545 595
TWh/a	714	868	1 614	3 822	6 462	9 318
Moderate Scenario						
MW	318 354	432 656	797 028	1 675 624	2 767 351	3 983 995
TWh/a	714	868	1 955	4 404	7 273	10 470
Advanced Scenario						
MW	318 354	432 656	879 446	2 110 161	3 720 919	5 805 882
TWh/a	714	868	2 157	5 546	9 779	15 258

Figure 5.1. GWEC's scenario forecasts of global wind installations [3].

total installed capacity of wind power all over world has reached around 540 GW by the end of 2017. Based on different forecasted scenarios by GWEC as shown in Figure 5.1, installed capacity of wind power may reach as high as 5.8 TW by 2050 [3].

5.2 Wind Power Plant Characteristics

The modern WTs are connected to power systems through power electronic devices. These power electronic devices provide fast and improved controllability. However, they also separate the WTs from the power systems; consequently, WTs cannot provide inherent support toward system inertia and short-circuit power. Additional controls are needed to provide the system support to power system from the WPPs. Traditionally, system support has been provided by conventional generations. Frequency support has been provided by the governor or load frequency controller (LFC), whereas voltage support is provided by exciter-based automatic voltage regulator (AVR) among others. However, with more and more traditional generations being replaced by power electronics converter–connected renewable generations, these modern generations should also step up in providing system supports instead of conventional generations [4–7].

WTs have come a long way in terms of capabilities to provide system supports. Modern WPPs have capabilities to control their power production during normal operation. According to several national grid codes, WPPs must have the capability to regulate their power production to an imposed reference value, set remotely or locally thereby controlling active and reactive power independently. Grid codes also require support from WPPs in case of grid faults like low-voltage ride-through (LVRT) capability. Modern WTs have abilities to withstand low voltages during short-circuit faults, stay connected and provide reactive power support during the fault, and resume their pre-fault normal operation shortly following the fault clearance.

Details of capabilities of WTs to provide support during normal operations, fault conditions, and emergencies are discussed in detail in later sections.

5.3 Need for System Services

To operate the power system in secured and reliable manner, certain system services are required. Such services are additional to the transmission and distribution of electricity and determine the quality and reliability of the electricity supply. These system services also called as ancillary services mainly include the following [8, 9].

5.3.1 Frequency Reserves

It is the responsibility of each transmission system operator (TSO) to balance the active power in their own control area while honoring the inter-area power transfer commitment. Any imbalance in generation and load in the individual control area is handled using operating reserves. Operating reserves can be classified as automatic or manual reserves based on the method of activation. Based on the cause of imbalance, reserves can be classified as frequency-controlled normal operation reserve and frequency-controlled disturbance reserve. These reserves are generally termed as primary, secondary, and tertiary reserves. Primary reserves (also called frequency containment reserves in European power

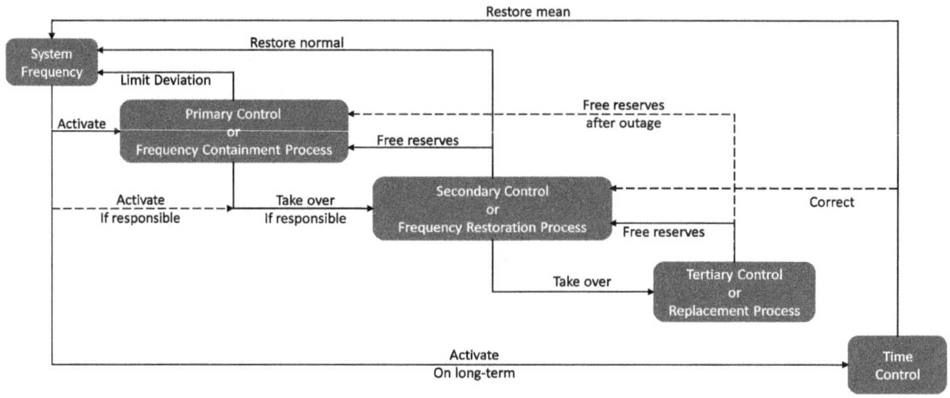

Figure 5.2. Different types of reserve activation by different controls [11].

systems) are automatic reserves required to be activated within seconds following the frequency change is detected. Secondary reserves (also called frequency restoration reserves in Europe) are activated through secondary control (or frequency restoration process) in minutes range. The purpose of secondary control is to restore the frequency back within the normal range as well as relinquishing the already activated primary reserve. Tertiary reserves (or replacement reserve) are activated in time frame from minutes to hour through tertiary control to relinquish activated secondary reserves. Primary reserve requirement for the power system is generally quantified based on security criteria of the power system. Typically, the security criterion is maximum generation/load that may get disconnected when any one of the power system equipment goes down (sometimes, referred as dimensioning fault). Secondary and tertiary reserves as discussed before (can be seen from Figure 5.2) are used to free primary and secondary reserves, respectively, to sustain further imbalances and faults [10]. Figure 5.2 [11] explains the activation of different reserves as discussed earlier.

Notice that traditionally all the reserves are deployed from conventional generators. However, in future power systems with large share of renewables, reserves can also be deployed from WPPs. Although WPPs may need to be operated in downregulated mode to provide this service, which incurs revenue loss for power plant owners (since maximum available power is not produced in downregulated mode) and therefore must be financially compensated.

5.3.2 Voltage Support

The voltage at any busbar in transmission network can be affected by the exchange of reactive power. It is not advisable and feasible to transmit reactive power over large distances due to increased losses. Reactive power should either be generated or consumed locally at the vicinity of the node to either increase or reduce the voltage, respectively. Generally, it is required that each power plants provide reactive power volume to support the voltage at the point of connection. There can be other kinds of services such as power oscillation damping, synchronizing power, LVRT, emergency defense plans, black-start support, and so on.

5.4 Overview of Grid Code Requirements

In this section, a brief overview of grid codes is given for WTs and WPPs. Integration studies have been conducted by power system operators, and thus the grid code requirements have been published when there is a need due to the result of these studies to maintain the reliability and the stability of power systems. Since it is very difficult and not feasible to track the last status of each grid code in various countries, general considerations of the grid code requirements are presented in this section. The focus of this overview is on the grid codes published by the TSOs, not the distribution system operators (DSOs).

Grid codes are viable in power systems' planning and operation stages with regard to the following statements:

- Stability and reliability of power system must always be maintained regardless of the energy mix and the generation technologies.
- Technical communication between power system operators and power plant operators must be clear, transparent, and reduced as much as possible.
- Power plant developers must design equipment and controllers considering the grid codes to be able to connect to power system, and accordingly they should not make changes without the TSO's permission to maintain power system stability and security.

When the grid codes first appeared, they were focused on the conventional power plants that consisted of synchronous generators. Generally, conventional power plants provide support to the transmission network in terms of inertial response, synchronizing power, power oscillation damping, voltage support, short-circuit capacity, and black-start capability. Some of the features are inherent (i.e., characteristic property) and allow the conventional power plants to support the grid, thus today TSOs have stable and reliable power system operation worldwide [12].

To sustain the integration of WPPs and the stable operation of power systems, TSOs published grid code requirements specific to WPPs and updated them regularly depending on integration studies and needs. These specific requirements are introduced due to WT generators' technical characteristics, which are fixed speed WT (i.e., Types 1 and 2) as older technology and variable-speed turbines with partial- or full-scale power electronic interfaces (Types 3 and 4) as modern technology. The idea behind these requirements is that WPPs should behave in similar manner to the conventional power plants for better integrity and easy transition. Simultaneously, the WT manufacturers have been challenged by these new requirements of grid codes as they must adapt their technology to meet these requirements. After upgrading the capability in the WT technology, power system operators and WPP developers revise the grid codes together to attain stable WPP operation without destabilizing the transmission system [13]. This is an iterative process between power system operators, WPP developers, and manufacturers. The recent grid codes of some countries that have large-scale wind power are listed in Table 5.1.

Table 5.1. Grid codes in countries with large share of wind power

Country	Power System Operator	Release Date	Reference
Denmark	Energinet	July 2016	[14]
United Kingdom	National Grid Electricity Transmission (NGET)	March 2017	[15]
Netherland, Germany	TenneT	November 2015	[16]
Ireland	EIRGRID	July 2015	[17]
United States	Federal Energy Regulatory Commission (FERC), Western Electricity Coordinating Council (WECC)	June 2016 May 2013	[18] [19]
Australia	Australian Energy Market Operator (AEMO)	July 2016	[20]
Europe	European Commission	April 2016	[21]

According to the grid codes given in Table 5.1, the technical requirements are defined for the grid connection and operation of WPPs in power systems for various countries. These requirements are mainly categorized into two groups with regard to the normal operation and the operation under grid disturbances/faults as follows:

- Normal operation
  - Maintaining frequency and voltage ranges
  - Active power control
  - Reactive power control
  - Voltage control
  - Frequency regulation
- Operation under grid disturbances/faults/extreme conditions
  - Frequency control
  - Fault ride-through capability
  - Reactive current injection
  - Post-fault active power recovery
  - Synthetic inertia
  - Power oscillation damping
  - Active power control during storm conditions

5.4.1 Maintaining Frequency and Voltage Ranges for Normal Operation

The voltage–frequency limitations of WPPs in grid codes are graphically given in Figure 5.3 as a generic representation. This requirement can be described typically as frequency/voltage operation zones as continuous operation in a limited range, time-limited operation with possible reduced output, and immediate disconnection.

5.4.2 Active Power Control

Active power control capability of WPPs is to provide the active power set point within a specified period and a tolerance defined by TSOs. This control capability is defined by the minimum and maximum ramp rates for both upregulation and downregulation [21]. This requirement is illustrated in Figure 5.4. Since the active power set point can be scheduled with this capability by the TSO externally, this requirement aims at ensuring that WPP can regulate their active power output to respond to changes in scheduled output and wind speed changes. Maximum and minimum ramp rates are specified that may depend on WT characteristics and limitations. These maximum and minimum values are given as follows for different countries:

- Germany, with an upper ramp rate limit of 10% of grid connection capacity per minute [16].
- Ireland, with a ramp rate of 1–30 MW/min [17].
- Nordic countries, with an upper ramp rate limit of 600 MW/h [22].
- For Denmark, the maximum standard value for the ramp rate constraint is 100 kW/s [14].

There are additional requirements, which are absolute power and delta power constraints [16], with regard to how to define the active power set point. The absolute power constraint is used to limit the active power output at a maximum power limit. Additionally, the delta

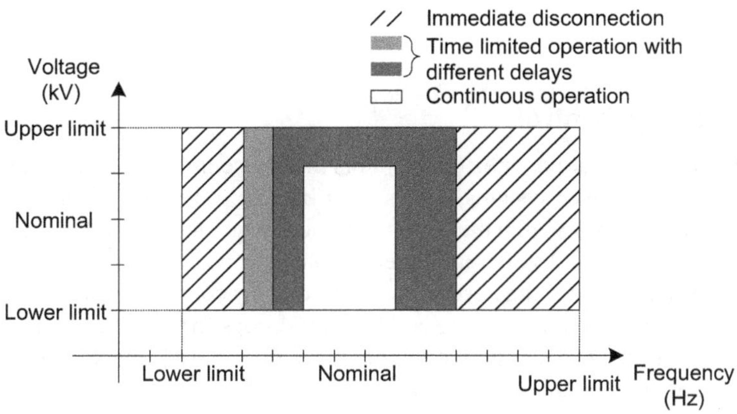

Figure 5.3. Generic representation of voltage–frequency limitations for WPPs.

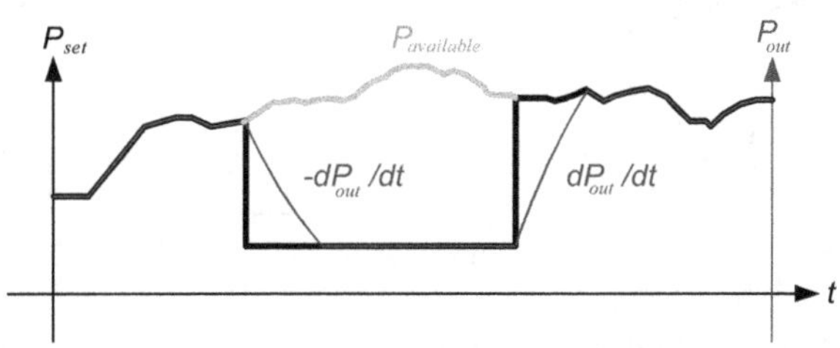

Figure 5.4. Illustration of the rate of change of active power output requirement.

power constraint is used to constrain the active power output in proportion to the available active power. For these two constraints, the control must be commenced within 2 s and completed no later than 10 s after receipt of an order to change the parameter [14].

5.4.3 Reactive Power/Voltage Control

Similar to the active power control, reactive power/voltage control requirements are specified in grid codes as a function of voltage at nominal active power and a function of active power for the cases, where WPPs operate at various active power levels for different ranges of voltages. Generic profiles for this requirement are shown in Figure 5.5. For the performance of the reactive power set point change, in [21] the starting time required is 2 s and the ending time is no later than 30 s after receipt with an accuracy of 1 kVAr.

5.4.4 Frequency Regulation

This requirement is defined as secondary frequency control, frequency restoration reserve, or tertiary reserve in the literature. For instance, frequency restoration reserve must be activated within 15 min [21] and a delay of at most 30 s [14].

5.4.5 Frequency Control

In case of frequency deviations (i.e., due to faults that cause imbalance between generation and consumption) in power systems, WPPs must be able to provide frequency control to stabilize the grid frequency within the aforementioned frequency limits. The frequency control must start no later than 2 s after a frequency change is detected and must be completed within 15 s [14]. The droop characteristic for the frequency control of WPPs is illustrated in Figure 5.6. It can be observed that there are different droops for different frequency ranges, illustrating different control setting requirements for different frequency regimes, namely normal, alert, and emergency operation.

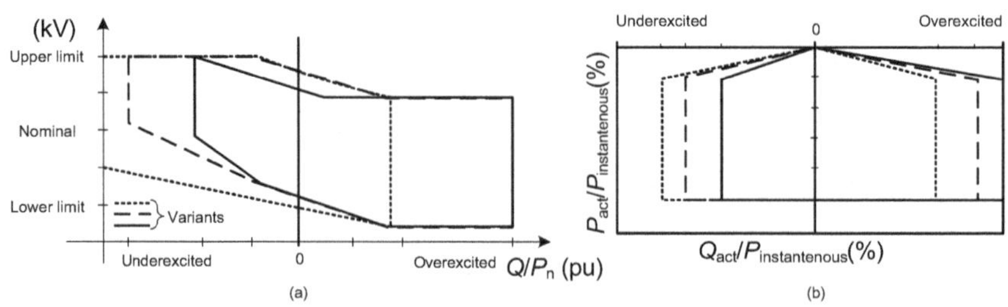

Figure 5.5. Variants of Q-V (a) and P-Q dependencies (b) for reactive power/voltage control.

5.4.6 Fault Ride-Through Capability

The objective of fault ride-through capability is to keep WTs and WPPs stay connected to the power system during short-circuit faults, so substantial amount of power is not disconnected in a securely cleared fault case. It should be noted that the requirement allows the WPPs to disconnect unless the grid fault is cleared (grid voltage does not come back to normal) after a certain fault time duration. To represent this explanation and behavior, the fault ride-through curves (Figure 5.7) are implemented in grid codes. The modern variable-speed WTs (Types 3 and 4) can successfully survive through the short-circuit faults, owing to their power electronics converters. In Figure 5.7, within the dashed lines WPPs must stay connected even if the voltage of point of common coupling (PCC) point is very low (i.e., 0). The operating time of relays in general is assumed as 150 ms.

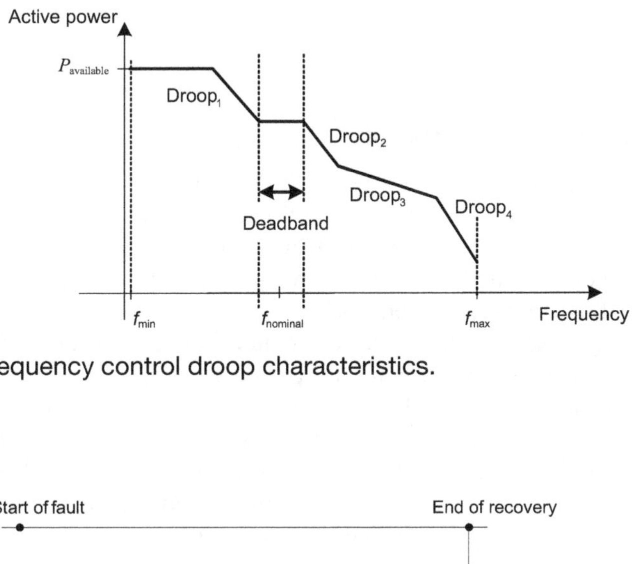

Figure 5.6. Frequency control droop characteristics.

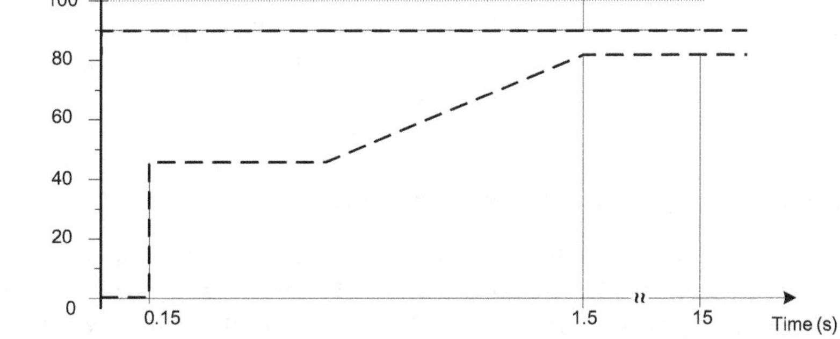

Figure 5.7. Fault ride-through capability of WPPs.

5.4.7 Reactive Current Injection

As described in the fault ride-through capability, WPPs must be connected during the faults, and with the reactive current injection requirement, they should provide the reactive current proportional to the voltage deviation. This requirement aims to support the short-term voltage stability. The reactive current injection should be responded within maximum 20 ms. The reactive current should be controlled within −10% to +20% of the rated current [16].

5.4.8 Post-Fault Active Power Recovery

After the fault ride-through, the active power must be recovered to pre-fault level within a certain amount of time post-fault. The objective of this requirement is to limit the short-term loss of active power infeed and to supply the load and balance the system after the fault has been cleared to prevent frequency collapse. During and after the recovery, active power oscillations should be adequately damped. Additionally, the total active energy delivered during the period of the oscillations is at least that which would have been delivered if active power was constant [15].

5.4.9 Synthetic Inertia Capability

Synthetic inertia is required in power systems with low inertia to contain the rate of change of frequency (RoCoF) less than certain value. If RoCoF is high, it can trigger many RoCoF-based relays, resulting in possible instability. Conventional controller settings of frequency control might also not be sufficient if RoCoF is very high [23]. An example of the synthetic inertia capability of the requirement follows from the Hydro-Quebec grid code [24]; WPPs with a rated output greater than 10 MW have to emulate inertial response of a conventional synchronous generator whose inertia constant (H) equals to 3.5 s. The active power of WPP varies dynamically and rapidly by at least 5% of active power for about 10 s, when a large, short-duration frequency deviation occurs on the power system. Although this requirement is also mentioned in [21], there is not a standard way to implement and tune its parameters. The studies are still going on in academia, industry, and operators' side. Similar to the frequency control, the WT set point can be modified using a frequency measurement with proper parameters in order to provide the synthetic inertia response. The overloading capability of the WT is employed. In this overloading case, the WT provides the active power set point above the available power by using the kinetic energy of the rotor following a recovery period, resulting in deceleration of the WT rotor. Another approach to provide synthetic inertia can be allocation of a reserve power in the WT as in the case of the frequency control.

5.4.10 Power Oscillations Damping Control

Different grid codes require different minimum and maximum damping levels. For example, in the National grid requirements [15], the tuning of the power system stabilizer is adequate if the corresponding active power response shows improved damping with the power system stabilizer functionality together with the automatic voltage regular (AVR) compared with the AVR alone over the frequency range 0.3–2 Hz.

5.4.11 Active Power Control during Storm Conditions

Storm conditions are normally not considered as grid disturbances but have high impacts on power system operations. WT's control system is usually designed to stop the turbine through the mechanical brake inside the nacelle when stormy wind speed occurs, which is often referred as high wind speed shutdown (HWSD) control. However, advanced storm control mechanism called high wind ride through (HWRT) can be explained by the power curve of a single turbine, as shown in Figure 5.8. This situation can be implemented as a test case, where the input wind speed can be increased until 39 m/s and later reduced below 20 m/s.

5.5 WPP Services

5.5.1 Hierarchical Control Architecture

A WPP typically consists of several WTs connected through a substation transformer to the WPP's PCC.

Figure 5.9 illustrates a typical configuration and control architecture for a WPP. It consists of a WPP, composed of several WTs connected to the main grid at the PCC through a main transformer, a WPP controller, communication system and measurement devices for voltage, current, grid frequency, and power at the PCC. The WPP controller gets external power set points and commands via SCADA system. WPP controller (WPPC) also uses a wind power forecast algorithm to estimate the future available wind power of the WTs in the WPP.

A WPP controller controls the power production by sending power set point signals to each individual WT, whereas a local WT control level ensures that the set points sent by the WPP controller are reached [25]. The WPP controller communicates back and forth with every individual WT controller through a set of signals to control the power production from the whole WPP and to provide if required grid support at PCC. The WPP controllers interact with the WT controllers to provide system services in such a way that the individual WTs maintain the ability to maximize power production, limit mechanical loads and minimize life time reduction.

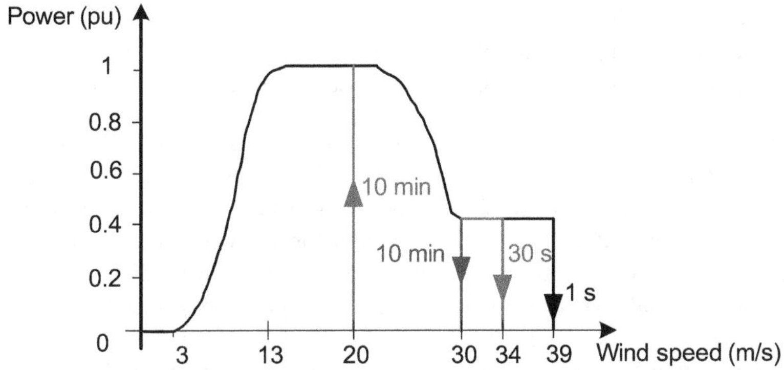

Figure 5.8. Siemens High Wind Ride–Through™ control algorithm [18].

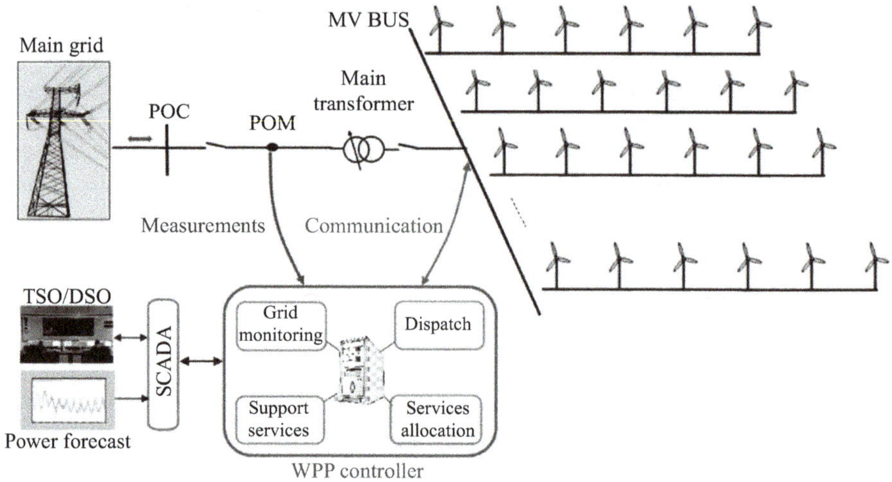

Figure 5.9. WPP controlling in the point of connection.

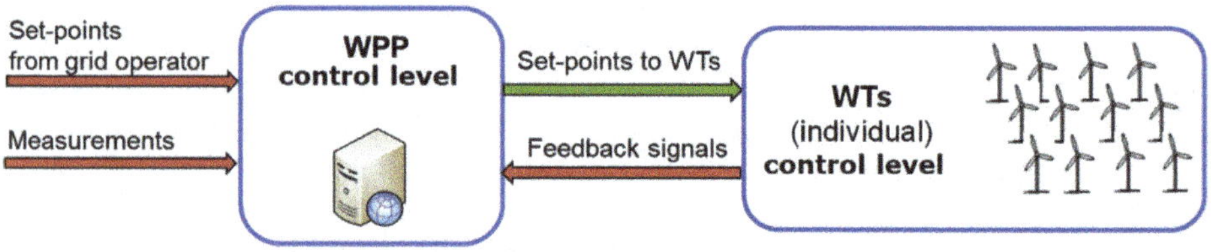

Figure 5.10. WPP hierarchical control architecture.

Inside the WPP controller there is -

- **Grid monitoring:** delivers info about the grid condition, that is, the grid frequency and frequency gradient estimation, active and reactive power calculation based on voltage and current measurements.
- **Dispatch block**: distributes the individual power set points to each WT set points, which are generated based on the measurements, set points from TSO, and WTG's conditions.
- **Support services:** contains different technical control functionalities such as balance, delta, frequency, or voltage control.
- **Services allocation:** makes possible to select/activate and even combine sequential or simultaneously multiple control functionalities with a specific weighting.

As depicted in Figure 5.10, WPP control architecture consists of a WPP control level and a WT control level. WPP level interacts with the WT control level through a set of signals and status feedback signals to provide the relevant system service. The WPP operators control the production of the WPP all the time and for any wind conditions namely either as maximum production or as controlled production, according to different control tasks, similar to a conventional power plant. Depending on the grid conditions, the system operator can issue specific demands to the WPP control level.

WPP control level behaves as a single central unit. It prepares and sends further reference signals to each individual WT control, based on several measurements in the PCC and available power from each individual WT. The individual WT controller then ensures that the power references sent from the WPP control level to the WT connection point are reached.

5.5.1.1 Wind Power Plant Control Level

The WPP control level is responsible for housing all control functionalities (i.e., active power control, frequency control, reactive power control, and voltage control) and selecting which controllers to enable, as well as measuring relevant quantities at PCC, and performing closed-loop control on active/reactive power. Notice that the different control functions for active and reactive power are simultaneously active. For all the ancillary service functionalities, active power (and reactive power) set points should be within the plant's capabilities. The dispatching block distributes the individual power references to the WTs based on a predefined criterion. The services' allocation block makes possible to select/activate and even combine sequential or simultaneously multiple control functionalities with a specific weighting.

As can be seen from Figure 5.11, there are two concomitant control loops in the WPP control level: active power control and reactive power control. Both the control loops consist of an anti-wind-up proportional integral (PI) controller. This controller computes a power error and sends the power reference set point for the whole WPP. Notice that the WPP power set points obtained from TSO can be, if necessary, adjusted to support frequency and voltage in the power system, respectively.

- The active power set point is defined based on the needed control mode by TSO, which can be either maximum production or limited production (i.e., balance control, delta control, and ramp rate control). In case there is any deviation in the frequency, the **subordinated frequency control loop** can be activated. It can be implemented as frequency droop control or inertia control.
- The reactive power set point can be defined by the system operator requiring a certain amount of reactive power from WPP. In case of voltage deviation in the power system, the subordinated voltage control loop will be activated.

Notice that in the case when control of WPP's active and reactive power production is in focus, then the frequency and voltage control loops can be neglected.

5.5.1.2 Wind Turbine Control Level

The WT control level addresses the local control system of every single WT and ensures that the power set points sent from the WPP control level, depicted in Figure 5.10, are reached. WT control level adjusts the WT production to the power set point obtained from WPP control system, according to the set points imposed by TSO. In case of normal operation, WTs aim to produce maximum power, whereas in power limitation operation mode, the WTs need to limit their production owing to generator limits to the power set point received from the WPP control center. The feedback signals sent back to the WPP control level contain typically information about WTs' power capability.

As described in [25] in detail, each local WT control system is also built up with a hierarchical control architecture, containing a slow dynamic control level (control of speed and power) and a fast dynamic control level (electrical control of the generator currents) as depicted in Figure 5.12.

Slow dynamic control has the responsibility to provide current references required for faster electrical converter dynamic control [26] as well as pitch angle reference to slower pitch control. These slow dynamic controls are referred to as power controller and speed controller, respectively. The power controller computes the current signal making sure the power reference obtained from WPP controller is maintained. The aim of the speed controller is to control the generator speed by controlling pitch angle (θ).

The reference for the speed controller can be locally computed inside the WT control level using "optimum speed" look-up table. This look-up table, assumed known beforehand, is function of the wind speed and described in detail in [25].

5.6 System Services during Different States of Power Systems

5.6.1 Normal Operation

The objective of the WT control traditionally has been to maximize energy production while minimizing cost. This is generally achieved by aiming to generate maximum power from available wind energy while minimizing the structural loads on the mechanical components. As share of wind power has increased in different power systems drastically over the years, the power system has become more dependent on weather-based wind power production and this is a serious concern for the system operators. This has led to the situation that today WPPs need to be controlled actively to support the power system in the same manner

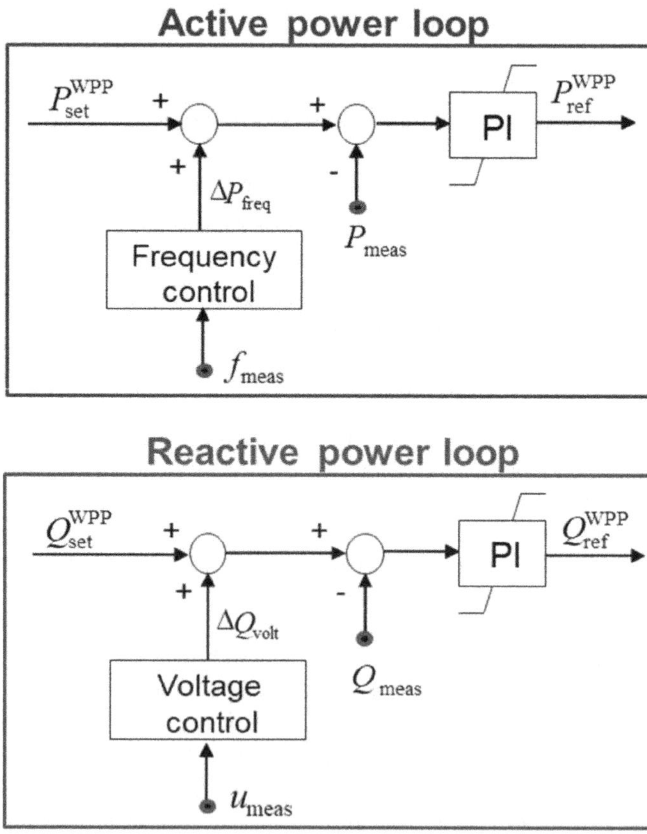

Figure 5.11. WPP control level.

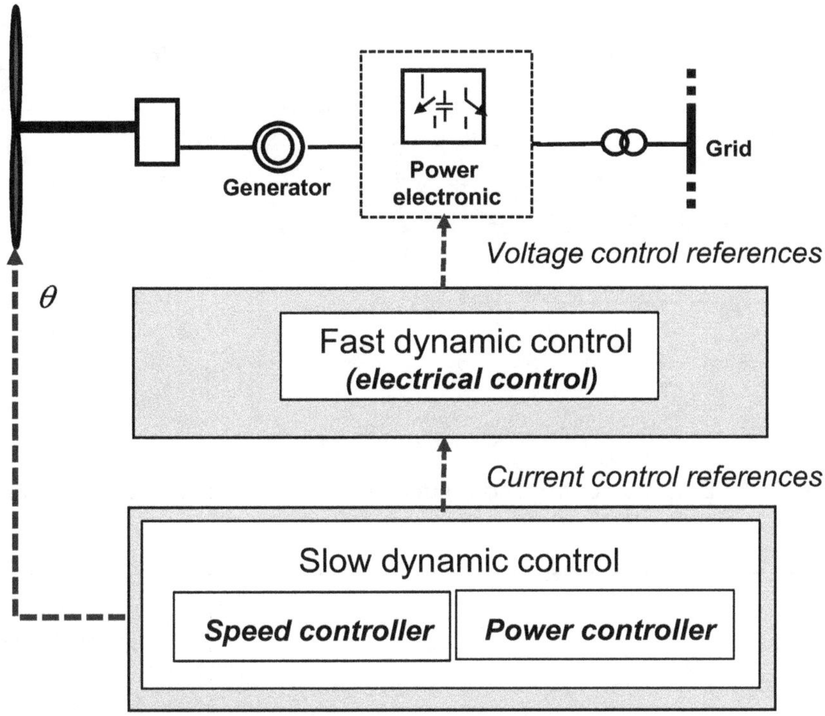

Figure 5.12 WT control level.

as conventional power plants do, namely controlling active power and reactive power, and providing frequency support and voltage control in the grid.

5.6.1.1 Active Power Control and Frequency Regulation

Active power control capabilities of modern WPPs have been many times addressed in the literature [25, 27–31].

A set of simulations showing the active power control capability of modern WPPs is illustrated in Figure 5.13(a) and (b) respectively, namely

- **balance control:** whereby the WPP is ordered to adjust downward or upward its active power production. These active power set points are sent as constant step inputs.
- **delta control:** WPP is sent set points to operate in downregulated mode with certain constant reserve with respect to possible power production capacity. This reserve power capacity can be used for frequency control.
- **power gradient limiter:** limits the ramp rate of WPP production and can be adjusted to either ramp up or ramp down.

For each control action, both the available and the actual active power measured in the PCC of the WPP are illustrated. In both the cases, the WPP produces maximum power in the first 100 s. Actual power can be seen to follow the available power since the ramp rate is less than the ramp limiter. In the time period between 100 and 300 s, the WPP operates under balance control as illustrated in Figure 5.13(a) and under delta control as illustrated in Figure 5.13(b). As shown in the figure, in balance control the power production of WPP is ordered down to 60 MW, whereas in delta control, WPP is ordered to operate with

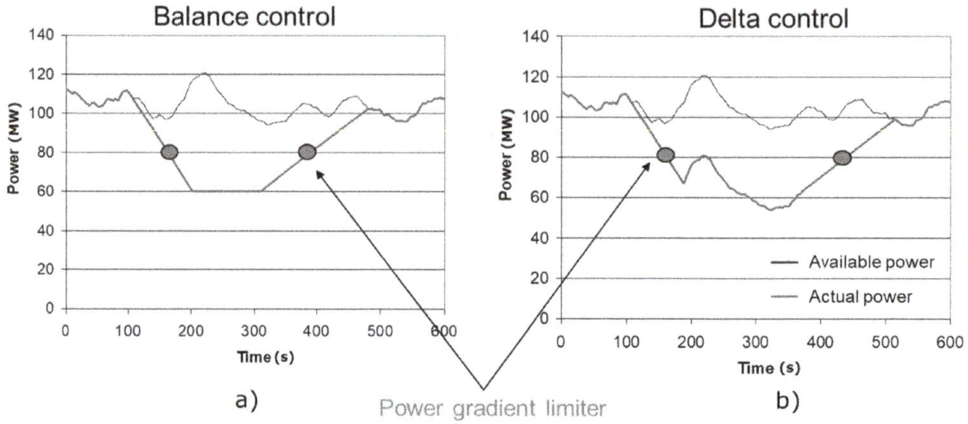

Figure 5.13. (a) Balance control and (b) delta control combined with power gradient limiter.

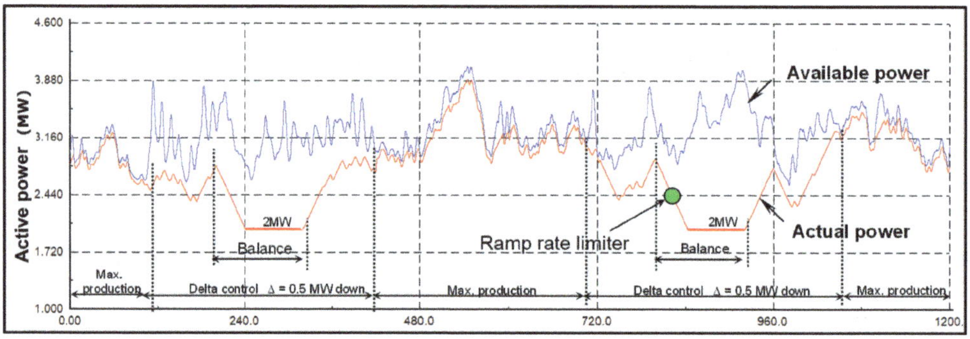

Figure 5.14. Simulation of activation of different control functions.

40 MW reserve power. Notice that the WPP power production adjustment upward and downward is performed with a power gradient limiter, which prevents the power from changing immediately.

Different control functions can be activated either consecutively or simultaneously. As an example, Figure 5.14 shows that both available and actual power production in the PCC for a 9-m/s wind speed and 10% turbulence intensity when different control actions such as balance control, delta control, and power gradient limiter are activated according to the following sequence:

- WPP is ordered to produce maximum power in the first 100 s. Actual power can be seen to follow the available power because the ramp rate is less than the ramp limiter.
- WPP is ordered to operate with a delta control (i.e., 0.5 MW power reserve) in the time period between 100 and 420 s.
- WPP is ordered to operate with balance control (i.e., downregulation to 2 MW) in the time period between 200 and 320 s. Both delta and balance control are activated simultaneously during this period. The upregulation and downregulation of the WPP power is performed with a power gradient limit about ±1.2 MW/min.

As presented by Hansen et al. [25] and Hansen and Michalke [32], modern WPP technologies, consisting of doubly-fed induction generator (DFIG)-based WTs and permanent magnet synchronous generators (PMSGs) with full-scale converter-based WTs, can independently control their active and reactive power production. This means that WPPs can respond to both active power demands (like balance control, delta control) and reactive power demands, which may be imposed by power system operators.

5.6.1.2 Reactive Power Control and Voltage Regulation

The aim of voltage regulation is to maintain the voltage profile within certain predefined tolerance band in time frame of hours. Analogous to traditional conventional generator [33], this control is achieved by injecting or absorbing reactive power by WPP. However, if the reactive power support is still not sufficient, additional auxiliary reactive power equipment, such as shunt capacitors and flexible alternating current transmission systems (FACTs) devices and synchronous condensers, can be used.

Reactive power is generally required to be controllable to specific power factor (typically unity power factor) in the grid codes [14]. However, modern WPPs using power converters can provide advanced system services. Depending on the technology and the electrical design, WPPs normally have additional reactive power capacity depending on the active power production. This additional reactive power capacity can be used for power factor control/reactive power control/automatic voltage control (i.e., combined droop and dead-band control), as illustrated in Figure 5.15.

It is important to underline that the automatic voltage control is performed on the WPP control level to avoid voltage instability. If WTs individually aim to control the voltage in the WT connection points, there is a risk for current flow between the WTs causing voltage instability.

Figure 5.16 illustrates the WPP automatic voltage control capability to a set of emulated steps in the voltage in the PCC. The figure illustrates the voltage in the PCC and both the actual reactive power and the reactive power reference of WPP. Notice that if the voltage

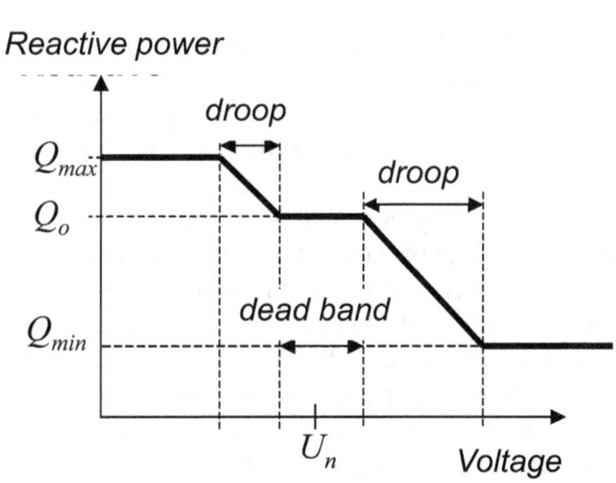

Figure 5.15. Automatic voltage control requirements with adjustable droop and dead band.

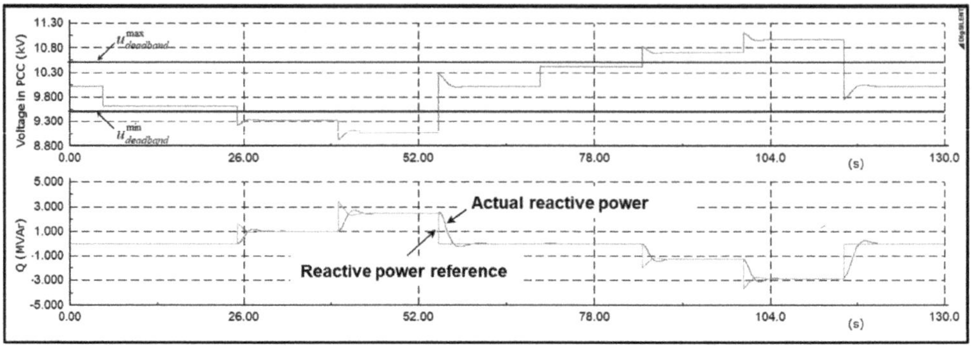

Figure 5.16. Emulated steps in voltage showing the automatic voltage control capability of WPP.

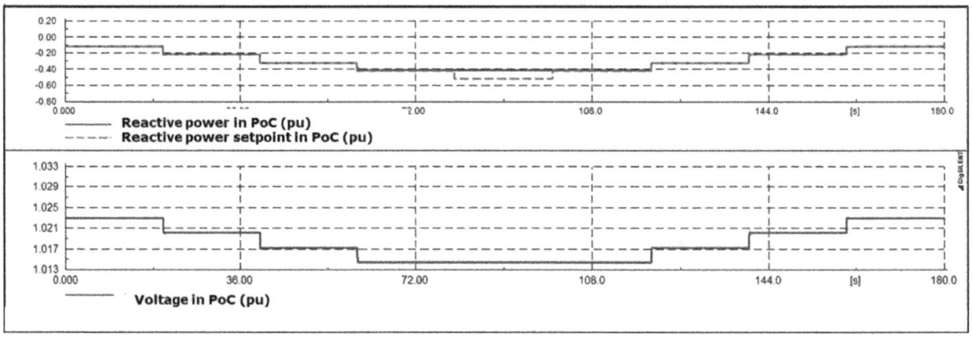

Figure 5.17. WPP reactive power control: Impact of WPP reactive power control on the local voltage.

is within the dead-band, the WPP is in normal operation with a certain reactive power production. When the voltage decreases lower or increases higher than the minimum or maximum dead-band voltage value, the WPP starts increasing or decreasing its reactive power production, respectively. Notice that as the specified reactive power reference is achieved properly, the control system performs well.

Reactive power control capability of a WPP is a control functionality, whereby the WPP is required to produce or absorb a constant specific amount of reactive power. To illustrate the impact of WPP reactive power control capability on the power system voltage, a set of reactive power references (downward and upward) for the WPP are emulated, as depicted in Figure 5.17. The first graph illustrates both the reactive power reference and the reactive power generated by the WPP. Notice that the reactive power of WPP follows very well its reference if the controller does not reach its maximum or minimum limit.

As illustrated in the second graph of Figure 5.17, the steps in the WPP reactive power production lead to changes in the voltage in PCC of the WPP. Notice that when WPP reactive power level is reduced or increased, the voltage levels are reduced or increased as well, respectively.

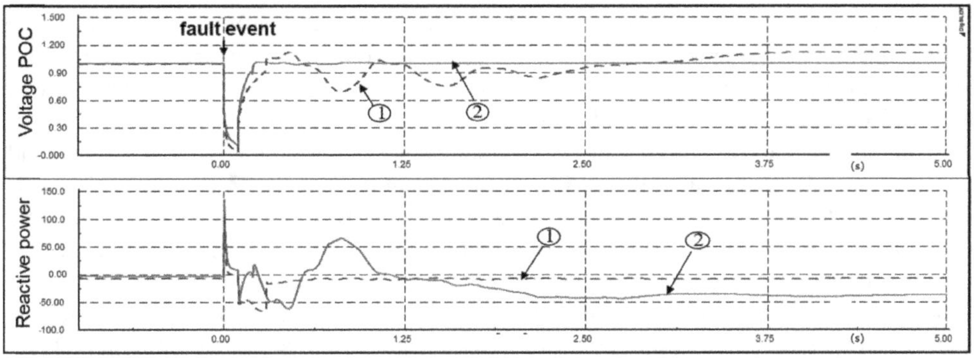

1) Without WPP voltage control 2) With WPP voltage control

Figure 5.18. WPP with and without voltage control: Impact on the power system voltage stability.

An example for the impact of the voltage control of modern WPP on the power system voltage stability during a severe three-phase short-circuit grid fault, described in detail by Hansen et al. [34], is shown in Figure 5.18. It should be noted that both modern WPP equipped with advanced control equipment and old-fashion WTs absorbing reactive power are connected to the considered power system. During the grid fault, the old-fashion WTs are disconnected from the grid, due to their protection and grid connection requirements. The figure illustrates the voltage in the point of connection (POC) of WPP and its reactive power for two situations:

- WPP without voltage control—WPP maintains a power factor of 1.
- WPP with voltage control capability.

Notice that the impact of voltage control can be observed both during the fault and post fault condition. Without voltage control, the grid voltage has higher oscillation and stabilizes to a higher voltage level postfault condition. The reason for such phenomenon can be explained by the existence of sudden reactive power surplus in the system as a result of the old-fashion WTs' disconnection during the grid fault. This reactive power surplus can be seen to be absorbed by the WPP when voltage control is enabled. Notice that the WPP with voltage control reestablishes the grid voltage to 1 pu quickly without any oscillation.

5.6.2 Alert Operation

5.6.2.1 Low-Voltage Ride-Through

WTs connected to the grid are subjected to grid faults, which can reduce the grid voltage to very low values. The faults are usually characterized by voltage drop for certain amount of time.

Up to two decades ago, WTs were only required to disconnect from the grid when a grid fault was detected, to avoid large inrush currents when the voltage recovered. However, with the increased capacity of wind power in the power system over the years, disconnection of large number of WTs could generate frequency and voltage instability in the system.

Besides power control capabilities, grid code requirements for WPPs also address the LVRT capability requirements today. The first grid codes only required that WTs were capable of "riding through" faults, that is, to stay connected to the network when voltage dip was detected. However, today most grid codes include specific requirements to supply a specified amount of reactive current during and immediately after voltage dips to control and stabilize the voltage during faults. Grid support capability requires WTs to provide ancillary services to the power system such as voltage control to ensure a secured and reliable network. WTs with power electronics connected to generators can support with these grid code requirements. These requirements normally ensure enough injection of reactive power to support the voltage during fault/voltage dip and for few seconds immediately after it is cleared. However, certain grid codes prioritize injection of active current rather than reactive current. This is particularly important to ensure that the frequency instability (which is controlled by active current) does not occur due to the fault that caused the voltage drop. If any WPP does not have enough reactive power capacity to support the voltage, auxiliary equipment is used. Reactive power capacity of shunt capacitors and SVCs is proportional to the square of the voltage. Therefore, these are not useful when voltage drop occurs. Static synchronous compensators (STATCOMs) are more suitable component to be connected at the PCC of WPPs.

Generally, LVRT requirements define WT dynamic support in three continuous periods when a grid fault occurs: fault transient, fault continuous, and fault recovery [35]. Dynamic voltage requirements are usually defined with typical response times in the range of 30–50 ms at the WT/WPP level [21].

The LVRT requirement, imposed to avoid significant loss of wind generation during grid fault events, addresses challenges primarily with regard to design of the WT controller rather than the WPP controller with the aim that the WT can remain grid-connected during grid faults and to support the grid to recover the voltage during and after a grid fault. The LVRT capability depends on the size of power electronic converters, which vary for different WT technology. The early generation of WTs would disconnect from the grid even during quite small disturbances as they did not have any LVRT capability owing to unavailability of power electronics in the WTs. Modern variable-speed WTs, for example, those equipped with power electronics (DFIG with partial-scale converter and PMSGs with full-scale converter), have capability not only to remain connected during and after the fault, but also provide support in terms of active and reactive current, if they are equipped with a frequency converter protection system.

The LVRT capability of these types of variable-speed WTs and their interaction with the power system especially during grid faults have been investigated extensively over the years in the literature [31,34,36]. Furthermore, they are able to provide system services to the power system and even improve the behavior of old WTs with conventional technology, if appropriate control is implemented for grid support. The modeling and control principles of these modern WT concepts are similar but differ inside due to the different system configuration and their advanced control strategies, which enhance their grid support capability.

In the work by Hansen and Michalke [31], it is described and analyzed in detail the dynamic interaction between variable-speed DFIG WTs with power transmission system during and immediately after grid faults. A test transmission network model [37] has been implemented for this purpose as depicted in Figure 5.19. A severe three-phase short-circuit grid fault of 100 ms is simulated at the end of the line connecting the WPP.

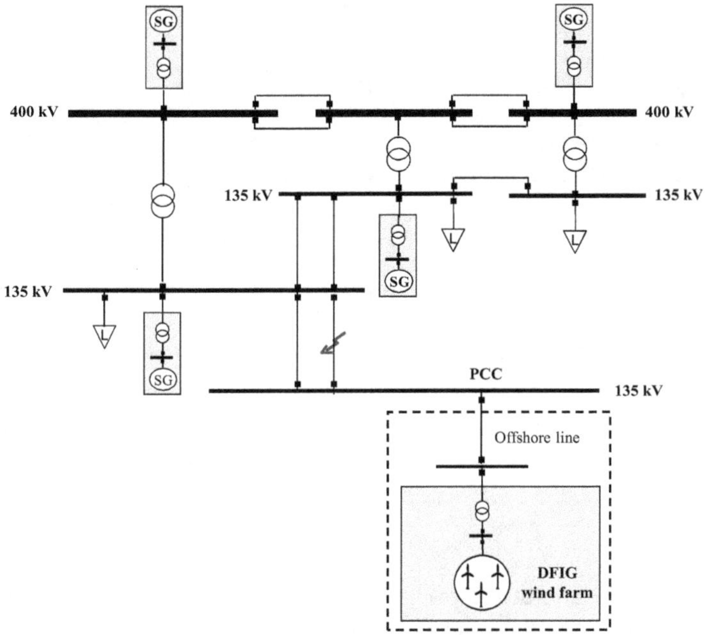

Figure 5.19. Simplified power system layout.

Dynamic interaction of DFIG WTs with power system during and shortly after the short-circuit fault is explained in detail by Hansen and Michalke [31], analyzing the impact of the grid fault on both the mechanical (i.e., drive train) and electrical (i.e., power converter) components of the WT. The converter control and protection are particularly essential for fault ride-through capability. Although providing LVRT support, torsional oscillations can be observed in the drive train, which might lead to self-excitation and high mechanical stress of the drive train. Controller should be properly designed to actively damp the torsional oscillations.

As shown in Figure 5.20(a), in the fault instant, the voltage at the DFIG generator terminal drops and it leads to a corresponding decrease of the stator and rotor flux in the generator. This results in a reduction in the active power and demagnetization of the generator over the stator, depicted by the fact that the reactive power peaks in the moment of the fault. In the fault moment, as the stator voltage decreases significantly, high-current transients appear in the stator and rotor windings, as shown in Figure 5.20(b). Note that the rotor current resembles the stator current.

The rotor voltage reference is increased in the rotor side converter to compensate for the increasing rotor current. However, the grid-side converter is not able to transfer the whole power from the rotor through the converter further to the grid, because the grid voltage has dropped during fault. The grid-side converter's control of the DC voltage reaches its limitation. Consequently, the energy that cannot be transferred to the grid charges the DC-link capacitor, resulting in an increase of DC voltage, as shown in Figure 5.20(b). Stator voltage gets restored when the fault is cleared allowing the electromagnetic torque and active power to increase. As the grid voltage and the flux increases, the demagnetized stator

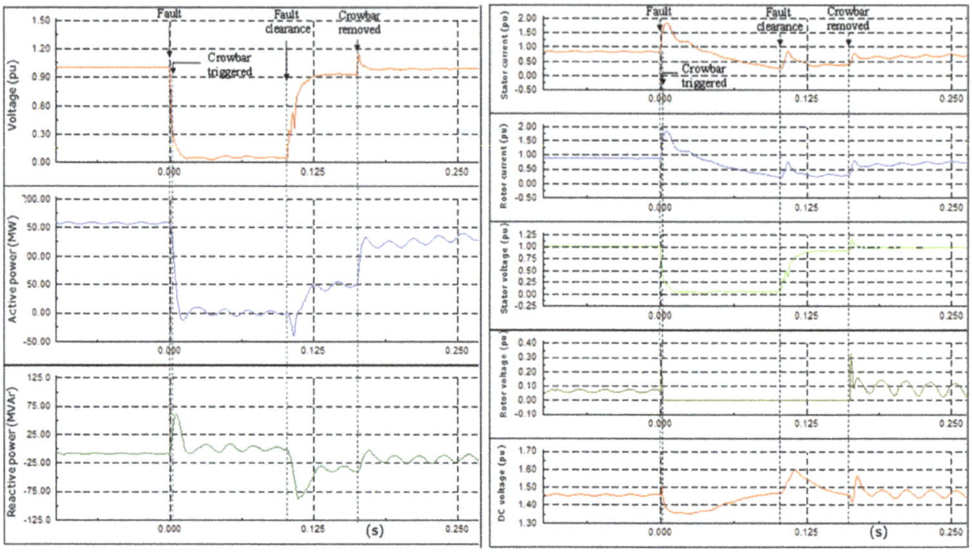

a) b)

Figure 5.20. (a) DFIG generator terminal voltage, active power, reactive power, (b) rotor and stator currents, and voltages.

and rotor oppose this change in flux resulting in increase in rotor and stator currents, as depicted in Figure 5.20(b).

The bigger fault ride-through and grid support capability of fully rated converter-based WTs compared with DFIG WTs have been discussed and presented in detail in the literature over the years [37–39]. Nevertheless, it has been shown by Hansen and Michalke [36] and Michalke et al. [32] that both novel control strategies and the use of a chopper in the DC-link can reduce the WT mechanical stress during fault ride-through (i.e., smaller oscillations in the shaft torque) and enhance even further the fully rated converter-based WT's fault ride-through capability. This is illustrated in Figure 5.21, where a short-circuit of 400 ms happens at 0 s. The PMSG WT is connected to a grid modeled as a Thevenin equivalent and the wind speed is kept constant during the simulation. Before the fault incident, the turbine operates at rated power.

Owing to voltage dip during fault, the grid-side converter can only transfer limited amount of active power to the grid. If there is no chopper, the generator-side converter must reduce active power to keep the DC voltage constant, leading to power imbalance between the reduced generator power and unchanged input aerodynamic wind power. This makes the generator to accelerate and oscillate because the drive train is stiff, as depicted in Figure 5.21. When chopper is used, the surplus power in the DC-link is burned in the chopper; therefore, reduction of active power is not required. The chopper thus effectively reduces the impact of fault on the generator and turbine.

As described in detail by Michalke et al. [40], PMSG WTs with full-scale frequency converter have a better grid support capability than DFIG WTs with partial-scale frequency converter. Due to the full converter system, PMSG WTs can provide a higher amount of reactive power to the grid. This yields that the voltage level is supported to a higher level

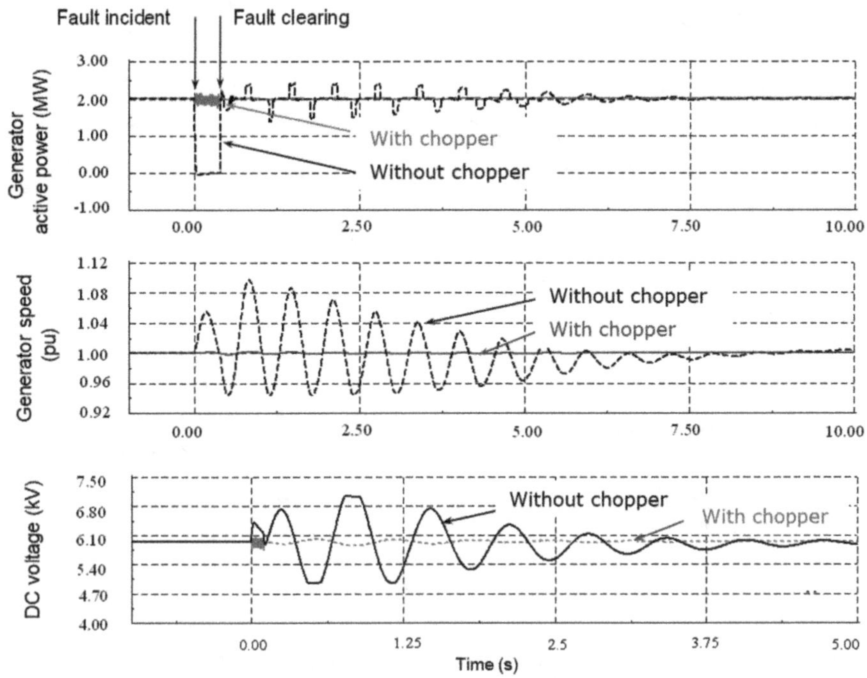

Figure 5.21. PMSG WT's active power, speed, and DC voltage.

and recovers faster. The PMSG WTs are furthermore fully decoupled from the grid by the converter and are thus less subjected to the impact of a grid fault.

5.6.2.2 Frequency Control

According to the Danish grid code [14], the frequency control must include droop and dead band as shown in Figure 5.22. This shows how a WPP is capable of increasing or reducing active power production to compensate for frequency fluctuation measured at the PCC.

In normal operation (without activated balance or delta control), the power set point P_o will be equal to the maximum available power P_{max}, and the power–frequency characteristic in Figure 5.22 will only have reserve for overfrequencies, because there is no reserve power available for underfrequencies.

To illustrate how the WPP reacts to changes in frequency, namely the automatic frequency control capability of a modern WPP, a set of emulated steps in frequency (downward and upward) are simulated and illustrated in Figure 5.23, for a WPP operating with a delta control, that is, it has reserve power available for underfrequencies. The figure illustrates the frequency and both the available and the actual active power of WPP. Notice that if the frequency is within the dead-band, the WPP is in normal operation with a power reserve. When the frequency decreases lower or increases higher than the minimum or maximum dead-band frequency value, the WPP starts increasing or decreasing its active power

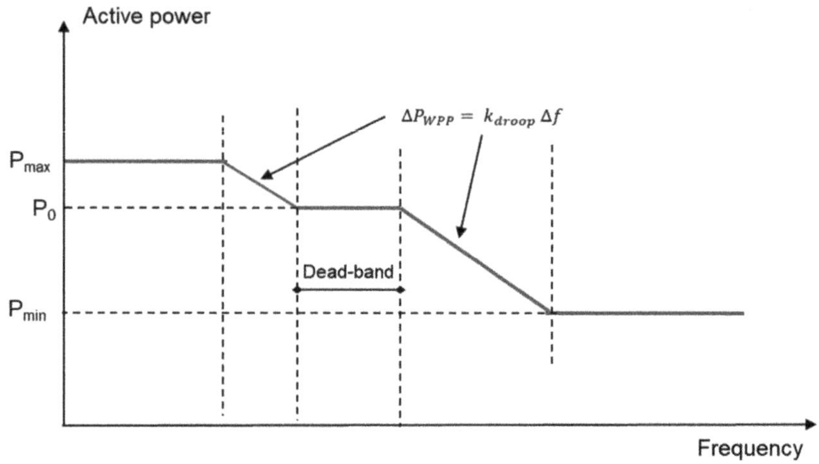

Figure 5.22. Requirement for automatic frequency control with adjustable droop and dead band.

production, respectively. The simulation results show a good performance of the control system. The specified reference for the active is achieved properly.

Figure 5.24 illustrates both the frequency excursion and the WPP frequency control contribution during the loss of a large generation unit in the power system at the time = 0 s for different values of the droop gains for the WPP. The loss of the generator unit yields to an unbalance between power generation and power consumption and therefore to a frequency drop. Notice that WPP operates with reserve power of 0.4 MW and therefore WPP is expected to be able to increase its production following a frequency event. For a zero-droop gain, WPP does not provide any additional power into the power system and this yields to a large frequency drop. Figure 5.24 shows that the higher the droop gain k_{droop}, higher is the WPP production with reduction in frequency, and this yields to smaller frequency drops in the power system following the loss of a generation unit.

5.6.3 Emergency Operation

Although the power system is designed to operate in stable conditions, however, it might be inevitable to prevent unforeseen events leading the power system to an emergency state. Defense plans are designed as ultimate preventive steps to prevent blackout in an emergency state [41,42]. Overfrequency emergency is one such situation occurring due to excess generation than load. During overfrequency situations, conventional generations decrease their power outputs based on their governor. However, in high wind power generation scenarios, conventional generators may be operated closer to their minimum power set points and therefore have limited ramping down capabilities. Furthermore, inertia of the power system also can be low, especially because the power system is generally islanded during frequency emergencies. Faster ramping capabilities of power electronic converter-based WTs can have major support in such scenarios.

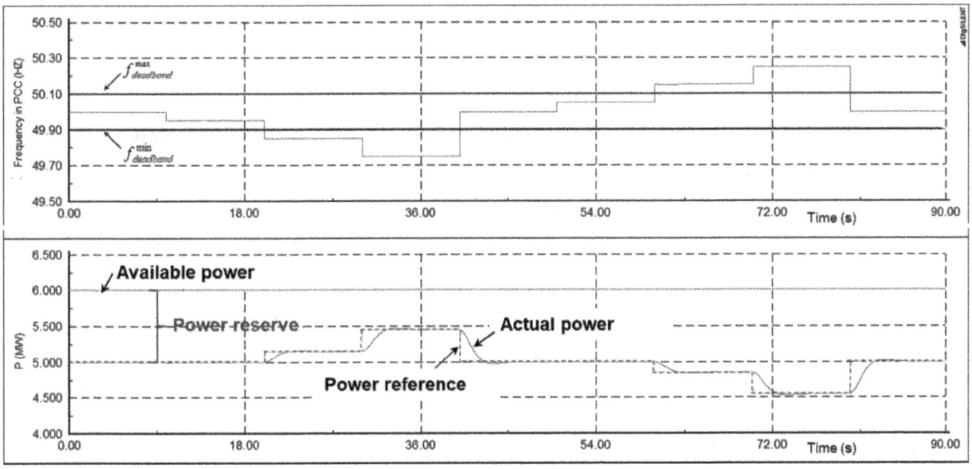

Figure 5.23. Emulated steps in frequency showing the automatic frequency control capability of WPP.

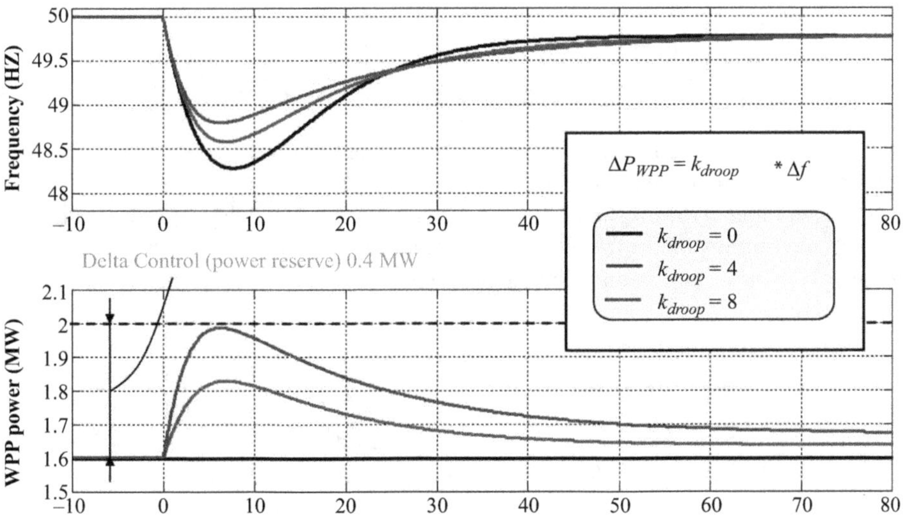

Figure 5.24. WPP automatic frequency control capability for different droop gains.

The disturbance on November 4, 2006 [43], at the "Union for the Co-ordination of Transmission of Electricity" (UCTE) network, is an event where frequency emergencies were observed. Cascaded disconnections of lines led to the separation of the entire UCTE network into three islands as shown in Figure 5.25. Schematic map of UCTE area is split into three areas [43].

The frequency excursions in these three islands after islanding are shown in Figure 5.26. The countries in the northeastern part of the UCTE network encountered a surplus of generation leading to frequency rise over 50.5 Hz in most of the cases while peaking at 51.4 Hz. This area had high proportion of wind power generation, which was being disconnected and reconnected arbitrarily, thereby, demonstrating the need for improved control and protection settings of wind power generations.

5.6.3.1 Frequency Control Capabilities of WT during Overfrequency Emergency

The model for frequency control is shown in Figure 5.27. Overfrequency responses from converter-connected WT generators mainly depend on ramping rate, droop settings, and delay.

- $\Delta\omega$: change in frequency in pu with respect to reference frequency (50 Hz)
- $T_{delayMeas}$: frequency measurement delay
- T_{delayP}: power activation delay
- $droop_{WT}$: droop settings in percentage
- $RampRate$: ramp rate in pu/s
- $Pmax_{WT}$: maximum power output from WT in pu
- $Pmin_{WT}$: minimum power output from WT in pu

Total delay consists of two parts:

1. measurement delay $T_{delayMeas}$ and
2. power activation delay T_{delayP},

Studies are performed where $T_{delayMeas}$ is assumed as 100 ms, whereas T_{delayP} is varied from 100 ms to 1 s. Droop settings $droop_{WT}$ for WT is varied from 2% to 10%. *RampRate* is varied between 0.1 and 0.5 pu/s. $Pmax_{WT}$ is assumed as 1 pu. $Pmin_{WT}$ is chosen as 0.1 pu based on the recommended value of Energinet's Technical Regulation 3.2.5 [44]. Frequency thresholds are chosen as 49.8 and 50.2 Hz.

The recommended $droop_{WT}$ settings for different activation delay and ramp rate limitation to get maximum support from WT are summarized in Table 5.2. Table 5.2 depicts that ramp rate plays more important role than activation delay in deciding the droop settings.

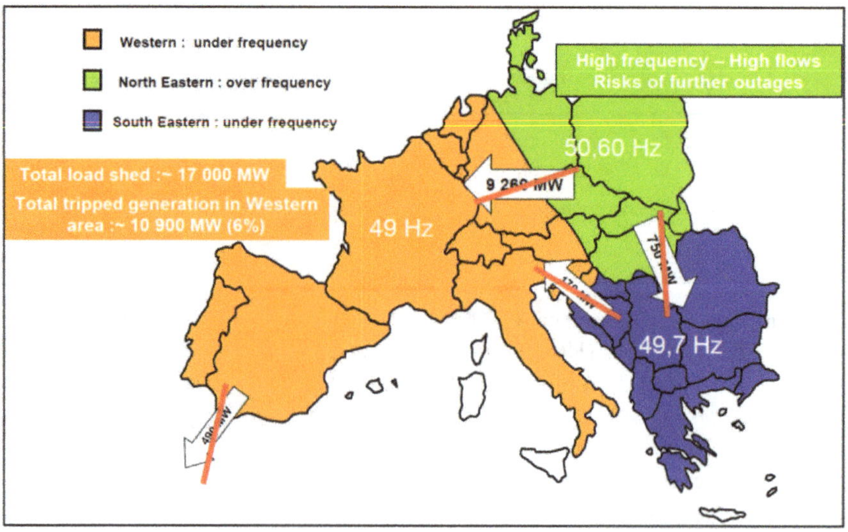

Figure 5.25. Schematic map of UCTE area split into three areas [43].

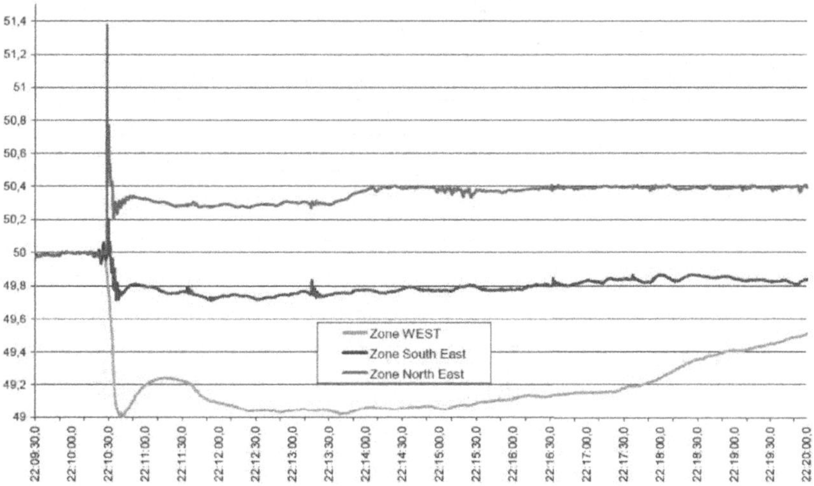

Figure 5.26. Frequency excursions after islanding [43].

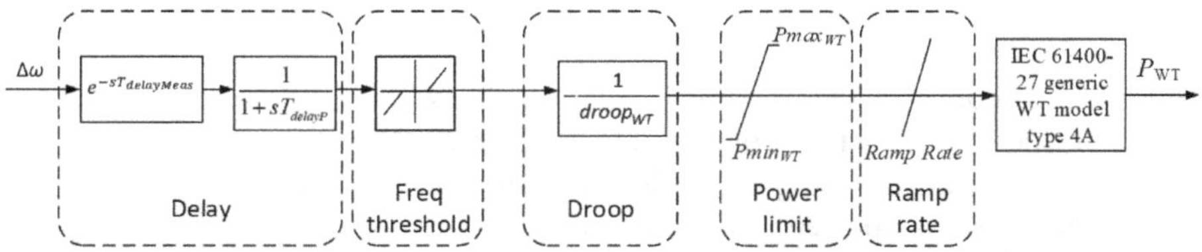

Figure 5.27. Frequency control model for overfrequency emergency studies.

Table 5.2. Recommended droop settings for different activation delays and ramp rates

Activation Delay (s)	Ramp Rate (pu/s)	$droop_{WT}$(%)
0.1	0.1	10
	0.2	4
	0.3	2
	0.4	2
	0.5	2
0.25	0.1	10
	0.2	4
	0.3	2
	0.4	2
	0.5	2

(*Continued*)

Table 5.2. Recommended droop settings for different activation delays and ramp rates (*Continued*)

Activation Delay (s)	Ramp Rate (pu/s)	$droop_{WT}(\%)$
0.5	0.1	10
	0.2	4
	0.3	2
	0.4	2
	0.5	2
0.75	0.1	8
	0.2	4
	0.3	2
	0.4	2
	0.5	2
1	0.1	6
	0.2	4
	0.3	2
	0.4	2
	0.5	2

However, it should be noted that droop settings of the WTs for emergency overfrequency control should be developed based on a systematic approach. Droop settings also depend on degrees of wind power penetration. Detailed analysis of overfrequency emergency control from WTs integrated to a power system can be found in [45].

5.7 References

1. T. Kåberger, "Progress of renewable electricity replacing fossil fuels," *Global Energy Interconnect*, vol. 1, no. 1, pp. 48–52, Jan. 2018.
2. P. Fairley, "For the first time, offshore wind power will be profitable without subsidies—IEEE Spectrum," 2017. Available online: https://spectrum.ieee.org/green-tech/wind/for-the-first-time-wind-power-will-be-profitable-without-subsidies.
3. Global Wind Energy Council (GWEC), "Global wind energy outlook," 2016. Available online: https://gwec.net/global-wind-energy-outlook-2016/
4. T. Ackermann (ed.), *Wind Power in Power Systems*, John Wiley & Sons, 2nd ed., 2012.
5. R. C. Bansal, and T. S. Bhatti, *Small Signal Analysis of Isolated Hybrid Power Systems: Reactive Power and Frequency Control Analysis*, Alpha Science, UK, 2008.
6. R. C. Bansal (ed.), Handbook of Distributed Generation Electric Power Technologies, Economics and Environmental Impacts, Springer, Switzerland, 2017.
7. A. F. Zobaa, and R. C. Bansal, *Handbook of Renewable Energy Technology*, World Scientific Publishing Co. Pte. Ltd., Singapore, 2011.
8. M. Beck, and M. Scherer, "Overview of ancillary services," April 2010, SwissGrid.
9. Energinet, "Ancillary services to be delivered in Denmark," 2017. Available online: https://en.energinet.dk/Electricity/Rules-and-Regulations#NetworkCodes
10. K. Das, M. Litong-Palima, P. Maule, and P. E. Sorensen, "Adequacy of operating reserves for power systems in future european wind power scenarios" in *IEEE Power & Energy Society General Meeting*, 2015.

11. union for the co-ordination of transmission of electricity (UCTE), "P1-Policy 1: Load-frequency control and performance [C]," 2009. Available online: https://www.entsoe.eu/fileadmin/user_upload/_library/publications/entsoe/Operation_Handbook/Policy_1_final.pdf
12. M. Altın, Ö. Göksu, R. Teodorescu, P. Rodriguez, B.-B. Jensen, and L. Helle, "Overview of recent grid codes for wind power integration" in *IEEE 12th International Conference on Optimization of Electrical and Electronic Equipment*, 2010.
13. A. Causebrook, and B. Fox, "Modernising grid codes to accommodate diverse generation technologies, especially modern windfarms," *Wind Engineering*, vol. 28, no. 1, pp. 75–86, 2004.
14. Energinet, "Technical regulation 3.2.5 for wind power plants above 11 kW," 2016. Available online: https://en2016.energinet.dk/-/media/BD322E7805694462AB125E5B5D0D79BC.PDF?la=en&hash=DB619E3EAD98AD2F2691CEA0B68BE8B438F4B3FD
15. National Grid Electricity Transmission plc, "The grid code," 2017. Available online: https://www.nationalgrid.com/sites/default/files/documents/8589935310-Complete%20Grid%20Code.pdf
16. TenneT, "Grid connection regulations," 2019. Available online: https://www.tennet.eu/fileadmin/user_upload/The_Electricity_Market/German_Market/Grid_customers/tennet-nar-2019_04_EN.pdf
17. EIRGRID, "EirGrid grid code," 2015. Available online: http://www.eirgridgroup.com/site-files/library/EirGrid/GridCodeVersion6.pdf
18. Federal Energy Regulatory Commission (FERC), "Reactive power requirements for non-synchronous generation, Docket No. RM16-1-000; Order No. 827," 2016. Available online: https://www.federalregister.gov/documents/2016/06/23/2016-14764/reactive-power-requirements-for-non-synchronous-generation
19. WECC, "WECC variable generation planning reference book: A guidebook for including variable generation in the planning process," 2013. Available online: https://www.pnnl.gov/main/publications/external/technical_reports/PNNL-22142_Vol2.pdf
20. Australian Government, "Australian Standard AS 4777—2005 Grid Connection of Energy Systems via Inverters—AG—Australian Business Licence and Information Service," 2005. Available online: https://www.saiglobal.com/PDFTemp/Previews/OSH/as/as4000/4700/4777.1-2005.pdf
21. European Commission, "Network code on requirements for grid connection of generators," 2016. Available online: https://eur-lex.europa.eu/legal-content/EN/TXT/?uri=OJ:JOL_2016_112_R_0001
22. European Network of Transmission System Operators for Electricity (ENTSO-E), "Nordic balancing philosophy," 2016. Available online: https://docstore.entsoe.eu/Documents/Publications/SOC/Nordic/Nordic_Balancing_Philosophy_160616_Final_external.pdf
23. K. Das, M. Altin, A. D. Hansen, and P. E. Sørensen, "Inertia dependent droop based frequency containment process," *Energies*, vol. 12, no. 9, p. 1648, Apr. 2019.
24. Hydro-Quebec TransEnergie, "Transmission provider technical requirements for the connection of power plants to the Hydro-Quebec transmission system," 2009. Available online: http://www.hydroquebec.com/transenergie/fr/commerce/pdf/exigence_raccordement_fev_09_en.pdf
25. A. D. Hansen, P. Sørensen, F. Iov, and F. Blaabjerg, "Centralised power control of wind farm with doubly fed induction generators," *Renew Energy*, vol. 31, no. 7, pp. 935–951, Jun. 2006.
26. M. A. Poller, "Doubly-fed induction machine models for stability assessment of wind farms," *IEEE Bologna Power Tech Conference Proceedings,* vol. 3, pp. 653–658, 2003.
27. J. R. Kristoffersen, and P. Christiansen, "Horns rev offshore windfarm: Its main controller and remote control system," *Wind Engineering*, vol. 27, no. 5, pp. 351–359, Sep. 2003.
28. A. D. Hansen, and M. Altin, "Impact of advanced wind power ancillary services on power system," 2015. Available online: https://backend.orbit.dtu.dk/ws/portalfiles/portal/107186047/DTU_Wind_Energy_E_0081.pdf

29. A. D. Hansen, M. Altin, I. D. Margaris, F. Iov, and G. C. Tarnowski, "Analysis of the short-term overproduction capability of variable speed wind turbines," *Renewable Energy*, vol. 68, pp. 326–336, Aug. 2014.
30. I. D. Margaris, A. D. Hansen, P. Sørensen, and n.d. Hatziargyriou, "Illustration of modern wind turbine ancillary services," *Energies*, vol. 3, no. 6, pp. 1290–1302, Jun. 2010.
31. A. D. Hansen, and G. Michalke, "Fault ride-through capability of DFIG wind turbines," *Renew Energy*, vol. 32, no. 9, pp. 1594–1610, Jul. 2007.
32. A. D. Hansen, and G. Michalke, "Modelling and control of variable-speed multi-pole permanent magnet synchronous generator wind turbine," *Wind Energy*, vol. 11, no. 5, pp. 537–554, Sep. 2008.
33. P. (Prabha) Kundur, N. J. Balu, and M. G. Lauby, *Power System Stability and Control*. McGraw-Hill, 1994.
34. A. D. Hansen, G. Michalke, P. Sørensen, T. Lund, and F. Iov, "Co-ordinated voltage control of DFIG wind turbines in uninterrupted operation during grid faults," *Wind Energy*, vol. 10, no. 1, pp. 51–68, Jan. 2007.
35. H. Geng, C. Liu, and G. Yang, "LVRT capability of DFIG-based WECS under asymmetrical grid fault condition," *IEEE Transactions Industrial Electronics*, vol. 60, no. 6, pp. 2495–2509, Jun. 2013.
36. A. D. Hansen, and G. Michalke, "Multi-pole permanent magnet synchronous generator wind turbines' grid support capability in uninterrupted operation during grid faults," *IET Renewable Power Generation*, vol. 3, no. 3, pp. 333–348, 2009.
37. V. Akhmatov, and A. H. Nielsen, "Simulation model of the transmission grid for a large offshore windfarm, used in education and research at the technical university of Denmark," *Wind Engineering*, vol. 30, no. 3, pp. 255–263, 2006.
38. P. Deglaire, S. Eriksson, and A. Solum, "Simulation and control of direct driven permanent magnet synchronous generator (PMSG)," *Uppsala University 2009*. Available online: http://www.elkraft.ntnu.no/smola2005/Topics/2.pdf
39. D. Svechkarenko, "Simulations and control of direct driven permanent magnet synchronous generator," *Nordic PhD course on Wind Power, Smøla, Norway* (2005): 5–11, 2005
40. G. Michalke, A. D. Hansen, and T. Hartkopf, "Variable speed wind turbines—fault ride-through and grid support capabilities" in *European Wind Energy Conference and Exhibition*, European Wind Energy Association (EWEA), 2008.
41. S. De Boeck, D. Van Hertem, K. Das, P. E. Sørensen, V. Trovato, J. Turunen, M. Halat, "Review of defence plans in Europe: Current status, strengths and opportunities," *CIGRE Science Engineering*, vol. 5, pp. 6–16, 2016.
42. K. Das, A. D. Hansen, and P. E. Sørensen, "Aspects of relevance of wind power in power system defense plans" in *12th International Workshop on Large-scale Integration of Wind Power Into Power Systems As Well As on Transmission Networks for Offshore Wind Power (WIW 2013)*, Energynautics GmbH, 2013.
43. Union for the Co-ordination of Transmission of Electricity (UCTE), "Final report system disturbance on 4 November 2006," 2006. Available online: https://www.entsoe.eu/fileadmin/user_upload/_library/publications/ce/otherreports/Final-Report-20070130.pdf
44. Energinet, "TR 3.2.5 Signal list for wind power plants," 2015. Available online: https://en.energinet.dk/-/media/D27EA05070F441E8A2291825834C32D1.PDF?la=en&hash=219947225746E57190E3A78980A096634B4A9AFB
45. K. Das, M. Altin, A. D. Hansen, P. E. Sørensen, D. Flynn, and H. Abildgaard, "Wind power support during overfrequency emergency events," *CIGRE Science Engineering*, vol. 9, pp. 73–83, 2017.

Section II: Solar PV Systems

CHAPTER 6

Solar Thermal Electric Power Plants

Keith E. Holbert[1]

[1]*School of Electrical, Computer and Energy Engineering, Arizona State University, Tempe, AZ 85287-5706. keith.holbert@asu.edu*

Abstract Concentrating solar power appears to be the method of choice for large capacity, utility-scale electric generation in the near term. In particular, distributed trough systems represent a reasonably mature approach, and the power tower configuration is also a viable candidate. Both of these technologies have the possibility of energy storage and auxiliary heat production during sunlight unavailability. Other systems examined in this chapter include dish-Stirling engines, solar chimneys, and solar ponds. Economics inevitably drives the choice of technology.

Keywords: Concentrating solar power, Parabolic trough, Dish-Stirling, Thermal energy storage, Solar chimney.

6.1 Introduction

Heliothermal is the process of conversion of solar energy to thermal (heat) energy. Solar thermal can be used directly for heating purposes (e.g., domestic hot water)—and this was the approach taken by many early residential solar installations. In terms of large-scale electric utility power generation, solar energy can be utilized to heat a working fluid and eventually produce steam to drive a traditional turbine–generator set, which produces AC power (in contrast to solar cells that have DC output). The approach known as concentrating solar power (CSP) is the major subject of this chapter. The solar thermal domestic implementation remains important because some electric utilities provide monetary incentives to customers who install such solar systems, and that subject is addressed in Chapter 12.

6.2 Solar Thermal Systems

The leading contenders for exploiting solar assets for electric power generation are photovoltaic (PV) and CSP facilities. Other schemes for converting sunlight to power include solar salt ponds and solar chimneys. Presently, PV appears to be leading CSP for utility-scale power generation due to its lower cost. In addition, there are two other advantages of solar thermal over PV. First, most CSP units can be constructed with an integral thermal energy storage (TES) system thereby providing the capability of generating electricity into the evening hours. Second, solar thermal plants can be equipped with auxiliary burners in order to generate electricity when sunlight is unavailable (e.g., at nighttime and during inclement weather).

Most CSP plants employ a thermal cycle, as depicted in Fig. 6.1, similar to those of coal and nuclear power plants except that the heat source is from sunlight. The thermal-to-mechanical energy conversion efficiency of a solar thermal electric generating station is Carnot cycle limited, that is, the maximum theoretical thermal efficiency is

$$\eta = 1 - T_L/T_H \tag{6.1}$$

where T_L and T_H are the minimum and maximum absolute temperatures within the cycle. Because T_L is approximately equal to the ambient environmental temperature at the power plant site, this relation motivates the design of a thermal cycle that adds heat at the highest possible temperature (T_H); and hence, it promotes the utilization of sunlight concentrators to achieve high-temperature operation.

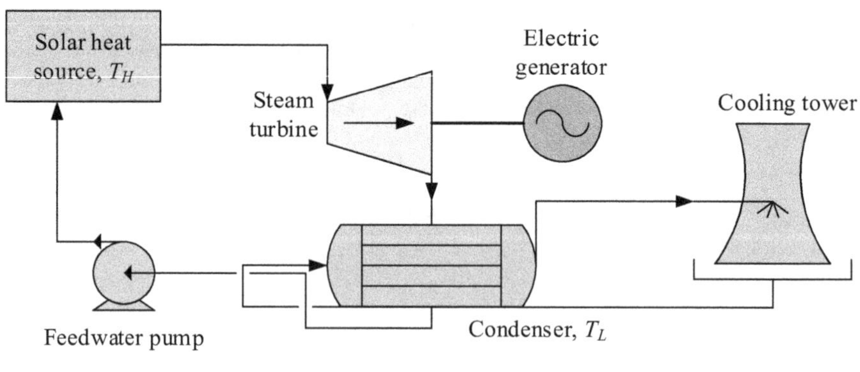

Figure 6.1. Solar thermal plant employing the Rankine cycle.

Solar thermal technologies can be classified in terms of the temperature reached by the working fluid, as exhibited in Fig. 6.2. Flat-plate collectors are employed in domestic hot water heating applications. Because lower thermal efficiency typically results in higher capital costs, electric utilities will tend to adopt concentration methods that achieve higher temperatures and efficiencies. The utility-scale CSP systems may be categorized as

- *Large point focus*: employing heliostats (mirrors) to concentrate light at a central receiver (power tower systems), with plants sizes of 100 kWe to 100 MWe.
- *Small point focus*: using parabolic hemispherical dishes to reflect light to a focal point for each individual dish (5–25 kWe).
- *Line focus systems*: utilizing parabolic-shaped troughs or linear Fresnel reflectors but having lower efficiency than the above, with plant output ranging from a few to hundreds of MWe.

In each of these three systems, a concentrator redirects (reflects) sunlight to a receiver (absorber). As can be noted in Table 6.1, working fluid temperatures for concentrating collectors are noticeably above ambient; hence, receivers may utilize vacuum barriers to prevent absorber heat loss due to convection to the surroundings.

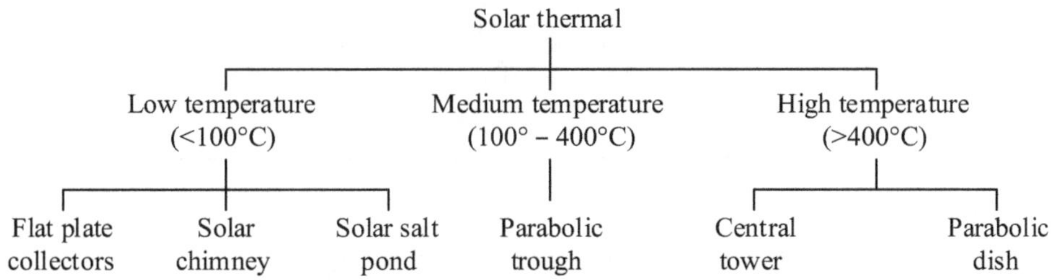

Figure 6.2. Categorization of solar thermal technologies according to temperature.

Table 6.1. Representative solar thermal collector operating parameters [1–4].

Thermal Collector Type	Concentration Ratio	Working Fluid Temperature
Flat plate and solar salt pond	1	40°–100°C
Line focus (parabolic trough)	10–50	150°–350°C
Point focus (parabolic dish and central receiver)	100–1500	250°–1500°C

6.2.1 Solar Thermal Storage

A key advantage of solar thermal compared to PVs is the capability of integrating TES within a CSP plant. Not only can such a TES system allow the plant to continue generating electricity during brief periods of sunlight loss (i.e., cloudiness), more importantly, electricity production can be continued after sunset and into the evening, which generally corresponds to a peak utility load period. Similarly, solar thermal plants can incorporate auxiliary burners to produce heat at night and during inclement weather. In the near term, TES is expected to provide capacity factors approaching 40%, and in the long term up to 70% [5].

Figure 6.3 illustrates how a solar thermal power plant can attain base load generation by diverting thermal energy to storage during the midday hours while also directly supplying electric power to the grid. In the early evening hours, the stored thermal energy is withdrawn and converted to electricity. Once the stored energy is depleted, auxiliary burners utilizing biomass, hydrogen, or fossil fuels could be employed to produce electricity during off-peak times. In this manner, CSP facilities can accomplish grid dispatchable power generation, that is, firm power delivery.

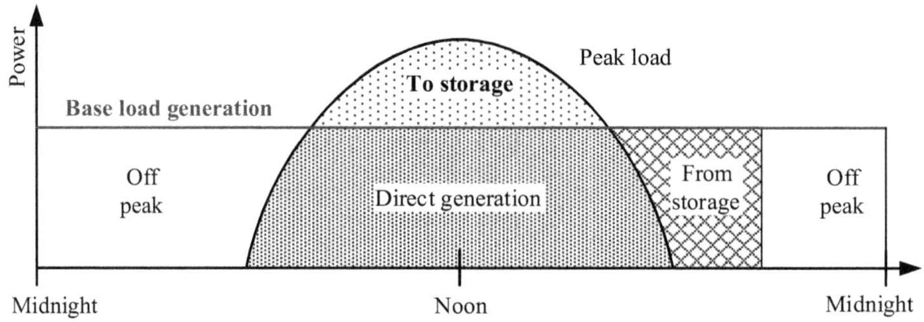

Figure 6.3. Storage and withdrawal of solar thermal energy.

Two types of TES systems have been demonstrated on a significant scale. The first, called a thermocline, utilizes a single storage tank with a thermal gradient of hot fluid on top and cooler fluid at the bottom. The second employs dual storage tanks with one serving to hold the cold fluid, whereas the other contains the hot medium. The latter dual-tank system permits a greater fraction of its stored energy to be extracted.

Because the solar field must be oversized for a CSP plant with thermal storage, the facility may be described in terms of its solar-to-electric capacity ratio (β), which is the ratio of the solar field thermal output capacity to the plant electric power output.

6.2.2 Hybrid Facilities

There are also hybrid solar thermal plants in which fossil or biomass fuels, for instance, can be used as a backup heat source. As depicted in Fig. 6.4, the Solar Electric Generating Systems (SEGS) units built in the 1980s employ natural gas to provide up to 25% of the thermal energy for steam [6]. Hybridization can be accomplished in trough, power tower, and dish engine systems, although implementation in the latter is more difficult [7]. It is noteworthy that present hybrid CSP facilities have only moderate thermal efficiency, leading to greater emissions on a per kWh basis from fossil fuel burning than would occur in a modern fossil power plant; however, designs exist that utilize an integrated solar combined cycle system (ISCCS). An ISCCS could employ a Brayton top cycle using a fossil-fired gas turbine and a Rankine bottom cycle that receives its heat from both the gas turbine waste heat and solar heat supplication, as illustrated in Figure 6.5. Various permutations of the

ISCCS exist, including those designed to save fuel and others intended to boost the power plant electric output [8].

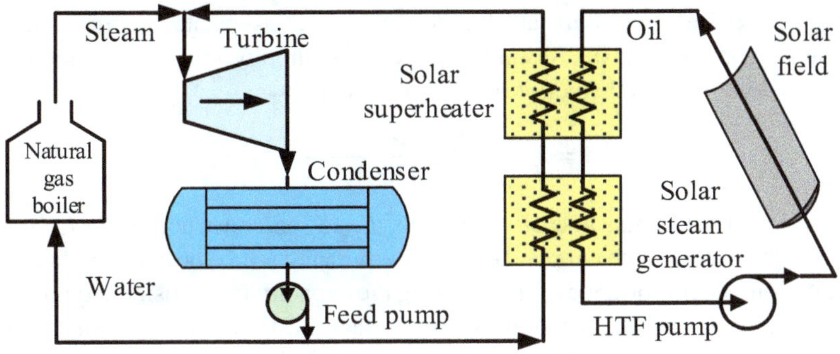

Figure 6.4. Schematic of SEGS hybrid power plant using oil as heat transfer fluid (HTF), after [4,9]. Reheat is implemented in later SEGS units [6].

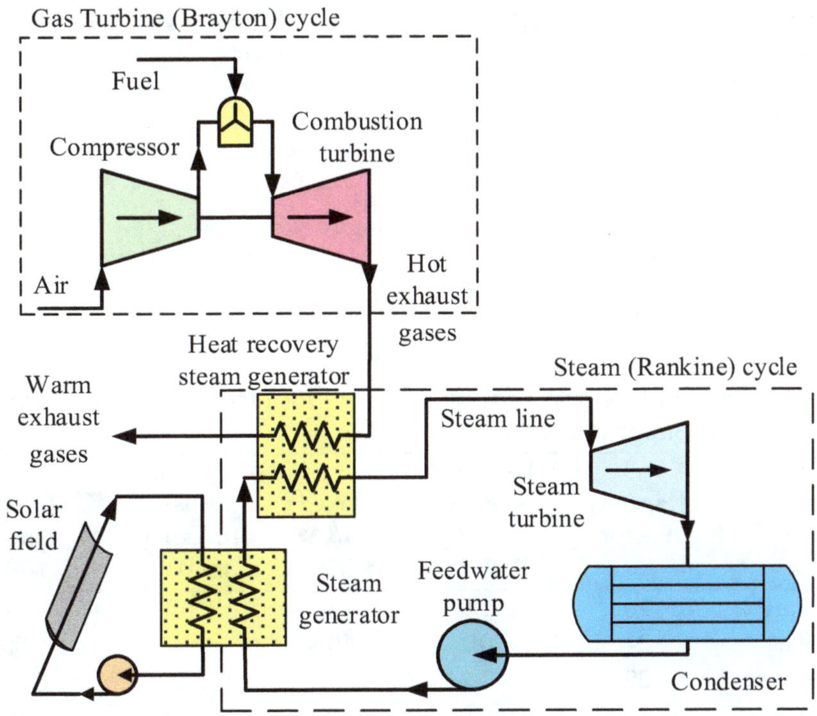

Figure 6.5. An ISCCS.

6.2.3 Economic Considerations

Table 6.2 compares projected costs for central receiver solar thermal and single-axis flat-plate PV power plants versus two other leading green energy sources. The table shows that the capital and the operating and maintenance (O&M) costs for solar thermal plants are high. The high O&M expenses are attributed to the costs associated with servicing and replacing plant components [10]. Assuming a pure (i.e., nonhybrid) solar plant, there are no fuel costs, and the levelized electricity generation cost is

$$e = \frac{IF_B + OM}{P_{e,rate}\, CF\,(8760\,\text{h/yr})}, \tag{6.2}$$

where I is the levelized annual fixed-charge rate; F_B is the construction cost; OM is the annual O&M costs; $P_{e,rate}$ is the net plant electric output; and CF is the capacity factor. From the above equation, the importance of maximizing the capacity factor using techniques such as sensible heat storage becomes obvious. Capacity factor estimates vary from 25% to 29% without storage and from 40% to 56% with storage [11–13].

Because central receiver and trough plants utilize conventional steam turbine–generator and balance of plant equipment, these plants benefit from economy of scale leading to unit sizes on the order of 100 MW. Dish systems, because of their smaller physical size, benefit from the possibility of mass production at a dedicated fabrication facility. In the case of the SEGS plants, the levelized cost of electricity was reduced from 24¢/kWh for the first unit to about 8¢/kWh for the ninth (and last) system [15]. Other estimates indicate overall CSP generation costs from power tower, parabolic trough, and dish-Stirling engines to be 9.0, 13.4 and 16.7 ¢/kWh, respectively [7]. Besides improving marketability, CSP plants with thermal storage decrease capital costs on a per kWh basis because smaller capacity turbine–generator and balance of plant equipment is sufficient.

6.3 CSP Systems

Within this section, three CSP plants are examined:

- Parabolic trough collectors (PTC).
- Central receiver systems (power towers).
- Dish-Stirling engine systems.

Table 6.2. Estimated levelized cost of electricity for plants entering service in the United States in 2022 [14].

Technology	Capacity Factor (%)	Levelized Capital Cost ($/MWh)	Fixed O&M ($/MWh)	Trans-mission Investment ($/MWh)	Total System Cost ($/MWh)
Solar thermal	20	191.9	44.0	6.1	242.0
Photovoltaic	24	70.2	10.5	4.4	85.0
Wind (onshore)	39	47.2	13.7	2.8	63.7
Wind (offshore)	45	133.0	19.6	4.8	157.4

An overall comparison of the three technologies is provided in Table 6.3. The gross solar-to-electric conversion efficiency can be described by

$$\eta_{gross} = \eta_{con}\,\eta_{rec}\,\eta_{cyc}\,\eta_{gen} \tag{6.3}$$

where η_{con} is the concentrator efficiency; η_{rec} is the receiver efficiency; η_{cyc} is the thermal cycle efficiency; and η_{gen} is the electric generator efficiency. The net efficiency would also account for in-house (parasitic) use of electricity by the power plant itself.

Table 6.3. Comparison of CSP technologies [7,16].

Parameter	Parabolic Trough	Power Tower	Dish-Stirling Engine
Peak efficiency	21%–24%	22%–23%	29%–30%
Annual capacity factor without thermal storage	23%–25%	25%–29%	25%
Annual capacity factor with thermal storage for $\beta = 1.8\times$	33% (4 h)	48% (8 h)	n/a
Net annual solar-to-electric efficiency	13%	13%	15%
Acres/MW of collectors	5	8	4

Modern solar concentrators are of two major configurations: (1) heliostats, which are mirrors that focus sunlight at a single point as shown in Fig. 6.6(a), and (2) parabolic reflectors, which redirect sunlight to single or multiple points. The parabolic concentrators include trough geometries that focus light along a line, and dish arrangements that concentrate at a single focal point as depicted in Fig. 6.6(b) and 6.6(c), respectively. Dish systems and heliostats require dual-axis solar tracking, whereas troughs need only use single-axis rotation.

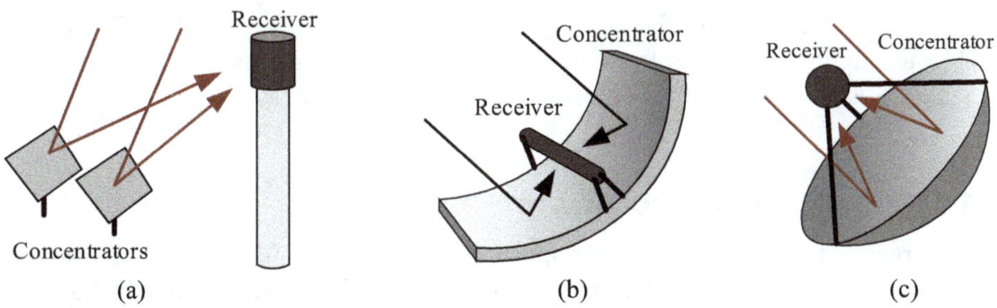

Figure 6.6. Three primary solar concentrating systems: (a) Central receiver using heliostats; (b) parabolic trough; and (c) parabolic dish.

Carnot efficiency considerations of Eq. (6.1) demonstrate the need for heat addition at high temperature, which is accomplished by concentrating sunlight. The concentrating ratio is defined as

$$C = A_{ap}/A_{rec} \tag{6.4}$$

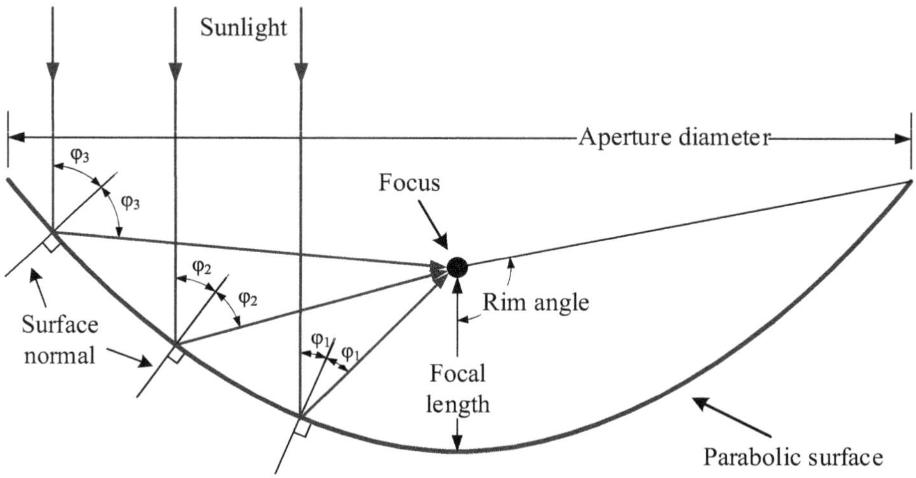

Figure 6.7. Sunlight concentration by a parabolic surface. The focus is a focal point in the case of dish systems and a focal line for trough concentrators.

where A_{ap} is the aperture (projected) area of the concentrator and A_{rec} is the receiver area. Figure 6.7 illustrates the important dimensions associated with a parabolic reflector. The maximum concentrations for linear and dish concentrators are 212 and 45,000, respectively [17].

A listing of major CSP facilities is given in Table 6.4. Luz International Limited constructed the series of SEGS in the Mojave Desert of California (CA) [9]. The SEGS plants benefitted from both favorable power contracts with an electric utility, and federal and state tax incentives [10]. It is noteworthy that Luz went bankrupt in 1991 when tax incentives (which had been as high as 35% in 1984–1986) and mandatory purchase contracts were withdrawn [18]. Presently, parabolic trough plants represent the most mature CSP technology [16], as might be surmised from the number of operating plants in Table 6.4. Specific plants from this table are further discussed later in this chapter.

6.3.1 Parabolic Trough Collectors

As an alternative to the central receiver approach, distributed collectors, such as parabolic trough concentrators (PTCs), may be employed. A PTC is a line focus system in that a long, mirrored concentrator in the shape of a parabolic trough is directed to heat the fluid within a conduit located on the focal line of the trough. The level of concentration within the PTC is less than that of a power tower, resulting in lower peak temperatures. In addition, a power plant requires many PTCs. At the present time, new solar thermal construction appears to favor PTCs, such as that presented in Fig. 6.8, over the power tower configuration. PTCs are typically aligned with the focal line along the north–south axis. A horizontal north–south trough generally collects slightly more energy than an east–west alignment on an annual basis, although an E–W orientation typically provides a more constant annual output [4]. Single-axis tracking is sufficient for PTCs.

Table 6.4. Major concentrating solar power (CSP) plants.

Facility	Unit Size (MW)	Location	Operating Date(s)	Solar Field Area (m^2)
Central Receiver (Power Tower)				
Solar One	10	Barstow, CA	1982–1988	72,650
Solar Two	10	Barstow, CA	1995–1998	82,750
PS10	11	Seville, Spain	2007	74,880
PS20	20	Seville, Spain	2009	150,520
Ivanpah	392	Mojave Desert, CA	2014	2,637,200
Crescent Dunes	110	Tonopah, NV	2015	1,210,000
Line Focus (Parabolic Trough)				
SEGS I	13.8	Daggett, CA	1985	82,960
SEGS II	30	Daggett, CA	1986	190,338
SEGS III	30	Kramer Junction, CA	1987	230,300
SEGS IV	30	Kramer Junction, CA	1987	230,300
SEGS V	30	Kramer Junction, CA	1988	250,500
SEGS VI	30	Kramer Junction, CA	1989	188,000
SEGS VII	30	Kramer Junction, CA	1989	194,280
SEGS VIII	80	Harper Lake, CA	1990	464,340
SEGS IX	80	Harper Lake, CA	1991	483,960
NV Solar One	64	Nevada, USA	2007	357,200
Genesis	250	Blythe, CA	2014	1,900,000
Solana	280	Gila Bend, AZ	2013	2,200,000

The receiver tube in the PTC is constructed using two concentric tubes placed on the focal line of the trough. As shown in Fig. 6.9, the inner metallic tube, which transports the working fluid, is separated from the outer transparent tube by vacuum. The outer glass tube and vacuum permit sunlight entry while preventing the convection heat loss, although there is transmittance loss through the glass.

Various working fluids may be utilized. For low-temperature applications (<200°C), demineralized water with an ethylene–glycol mixture can be used. However, to achieve efficient high-temperature operation, a synthetic oil, such as biphenyl–diphenyl oxide, is employed [21]. More recently, approaches using high-pressure steam and molten salts have been investigated. If water–steam can be used directly, then the need for and cost of the heat

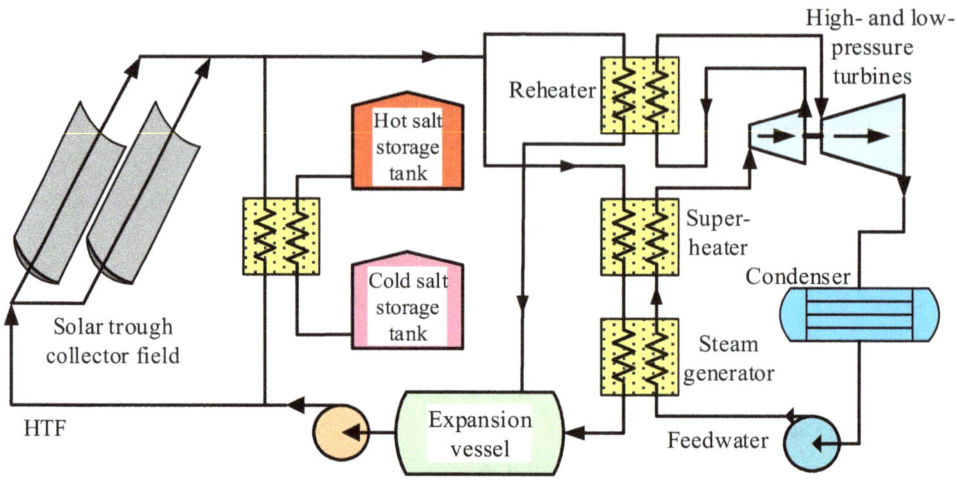

Figure 6.8. Trough type CSP plant with thermal storage, after [19].

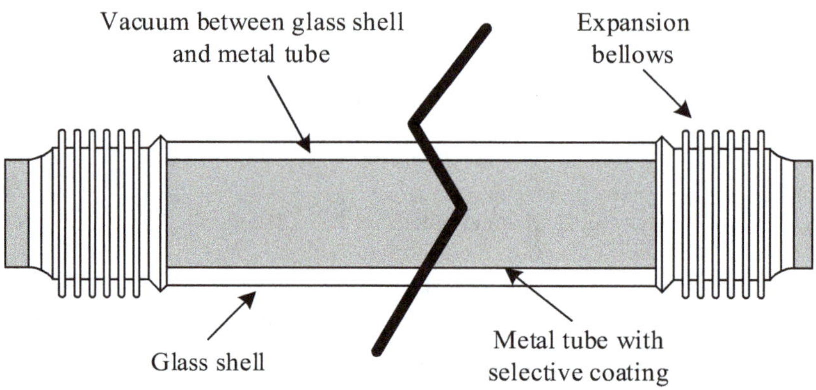

Figure 6.9. Parabolic trough receiver tube, adapted from [20,21].

exchangers and HTF can be avoided; however, heat storage may become more complicated. The DISS (DIrect Solar Steam) Project has successfully demonstrated DISS generation in parabolic trough collectors [22] and results favor the recirculation approach over a once-through method [23]. Multiple trough field collector layouts exist as illustrated in Fig. 6.10. The inverse return layout results in an even pressure drop (Δp) within the collector field, whereas the direct return and center feed require valving to ensure a uniform Δp; however, the central feed requires less piping than the other two options.

The linear Fresnel system, as depicted in Fig. 6.11, is a cross of sorts between the parabolic trough and power tower systems. Each receiver is placed about 10 m above the ground, and multiple reflectors are oriented near ground level to redirect sunlight to the fixed receiver. A secondary reflector above the receiver prevents loss of light. In a variation, known as the compact linear Fresnel reflector (CLFR), the reflectors can be directed

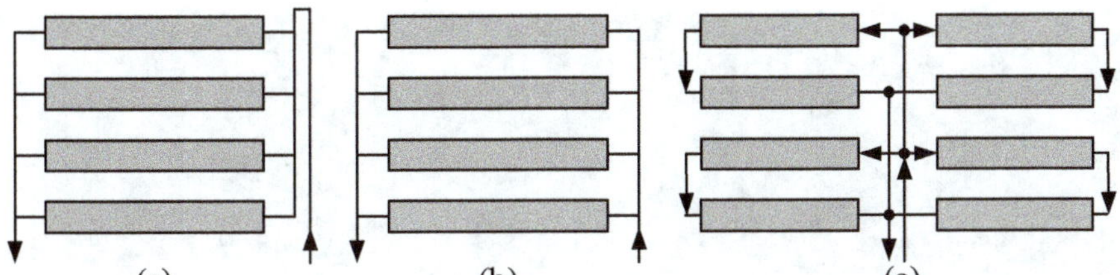

Figure 6.10. Layouts for a trough collector solar field, after [24]: (a) Inverse return; (b) direct return; and (c) central feed.

toward an alternative receiver that would make most effective use of the sunlight depending on the time of day. The interleaving of closely packed reflectors in the CLFR should increase the utilization of the available solar irradiance [25]. A CLFR coupled to a molten salt test loop is shown in Fig. 6.12. A 125-MWe CLFR system began operation in 2014 at Rajasthan, India.

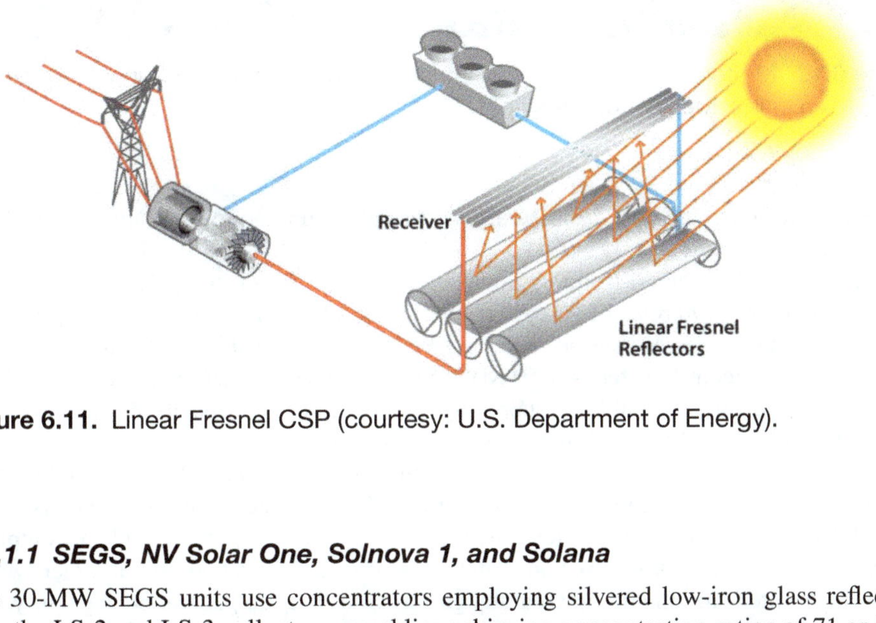

Figure 6.11. Linear Fresnel CSP (courtesy: U.S. Department of Energy).

6.3.1.1 SEGS, NV Solar One, Solnova 1, and Solana

The 30-MW SEGS units use concentrators employing silvered low-iron glass reflectors with the LS-2 and LS-3 collector assemblies achieving concentration ratios of 71 and 82, respectively [26]. The first SEGS plant employed a 3-h thermal storage unit using mineral oil (Therminol VP-1). The SEGS I dual-tank storage system cost \$25/kWt with the oil constituting 42% of the cost [27]. However, in 1999, an accidental fire destroyed the storage unit [7].

The 64-MWe Nevada Solar One was completed in 2007 at a cost of \$266 million. This is the largest CSP trough facility to be built in the world since 1991. Using a concentration ratio of 71, the 760 parabolic troughs heat the working fluid (Therminol) from a field inlet temperature of 300°C to a field outlet of 390°C. Nevada Solar One does not have a TES

Figure 6.12. Molten salt test loop at Sandia National Laboratories in Albuquerque, New Mexico (courtesy of DOE).

unit and is allowed to use only 2% natural gas to steady temperature fluctuations in the HTF and for freeze protection [28].

The Solnova 1, 3, and 4 trough systems heat a synthetic oil to 393°C. Each collector field is composed of 90 rows of four trough modules, which are each 12.5 m long by 5.76 m wide, yielding a mirrored surface area of 3×10^5 m^2. The heat is transferred to produce 390°C, 100-bar steam that feeds a 50-MWe turbine–generator set. The plant includes a natural gas fired boiler capable of supplying 12%–15% of its capacity during low solar radiation conditions. The overall plant efficiency is near 19%.

The 280-MWe gross Solana Generating Station is located about 110 km southwest of Phoenix, Arizona. The temperature of the Therminol VP-1–Xceltherm MK1 HTF is raised from 293°C to 393°C in the solar field, see Fig. 6.13. The 100-bar steam cycle net output is 250 MWe using wet cooling. Six hours of TES is provided via a two-tank indirect system employing molten salt. In addition, natural gas provides backup generation capability.

6.3.2 Central Receiver Systems

Central receiver systems employ a large field of mirrors known as heliostats. The heliostats are individually oriented to redirect sunlight toward the top of the power tower. The heat transfer media of the thermal power cycle may be pumped to the top of the central receiver or a mirror can be located at the top of the tower to reflect the light to a ground-level receiver [29–31]. Configurations using multiple towers have also been proposed [32]. The HTF,

Figure 6.13. Solar trough collectors at Abengoa's Solana Plant in Gila Bend, Arizona (courtesy of DOE).

typically a molten salt mixture, is circulated to transport its received heat to a secondary fluid such as water, as depicted in Fig. 6.14. The fact that the HTF can also be used as a TES medium is an advantage, but the high freezing point of a salt requires deployment of heat tracing and its concomitant parasitic energy use. A representative molten salt is a mixture of 40% potassium nitrate (KNO_3) and 60% sodium nitrate ($NaNO_3$) as was employed in Solar Two. This salt melts at 220°C and is thermally stable to ~600°C [27,33].

The heliostats are a significant component of the plant capital costs. For example, in the case of the Solar One plant, heliostats composed 45% of the total cost [34]. In addition to an individual dual-axis positioning mechanism, a heliostat requires structural support that can withstand high winds. Analysis has determined that excessive shading of heliostats is avoided with reflector-to-land area ratios from 0.1 to 0.4, and averaging around 0.25 [35]. Optimal heliostat field shapes have been determined based on plant power rating, as shown in Fig. 6.15 [36].

6.3.2.1 Solar One and Solar Two

The Solar One central receiver was constructed near Barstow, California, at a cost of $140 million, and operated from 1982 to 1988. The receiver in the 94.5-m tall tower heated the feedwater to generate steam for a direct Rankine cycle. The 1818 heliostats of 39.1 m^2 each were arranged in a surround field configuration [38]. As depicted in Fig. 6.16, the Solar One unit employed a thermocline storage system composed of oil within a gravel and sand mixture. The storage medium had an energy density of 0.05 MWh/m^3 within a temperature range of 220°C–300°C [34].

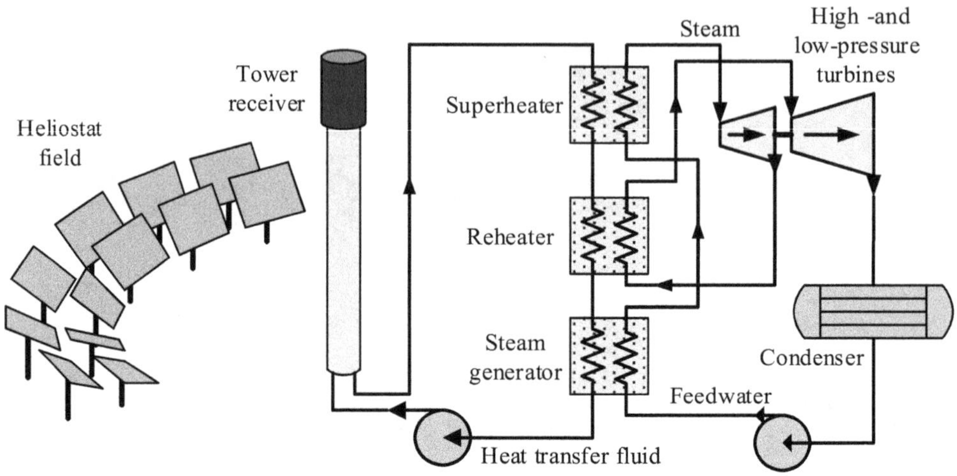

Figure 6.14. Power tower CSP plant without thermal storage.

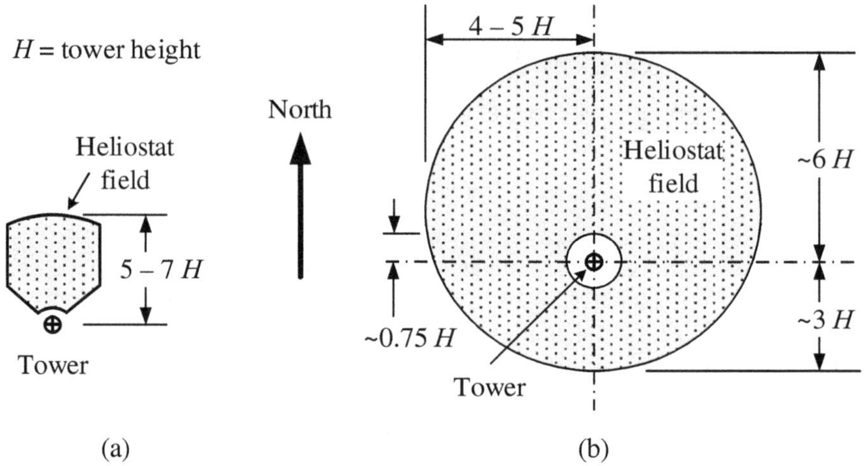

Figure 6.15. Heliostat field shapes in the Northern Hemisphere, after [34,37]. For small thermal output, the heliostats are strictly north of the tower.(a) North field for small plant (<100 MWt) and (b) surrounding field for large plant (>500 MWt).

Solar One was upgraded to Solar Two by utilizing molten salt and increasing the reflector field, and it operated from 1996 to 1999 [39]. With a total reflective surface area of 81,400 m^2, the heliostats provided an average solar flux of 430 kW/m^2 and a peak of 800 kW/m^2 [33]. The molten nitrate salt was used both as a heat transfer medium from the receiver to a steam generator and for 3 h of TES using a dual-tank system, as shown in Fig. 6.17. The molten salt was heated from 290°C to 565°C from its passage through the central tower receiver. Sited in Spain, the Gemasolar Thermosolar Plant (originally named Solar Tres) is a 20-MWe power plant utilizing Solar Two technology.

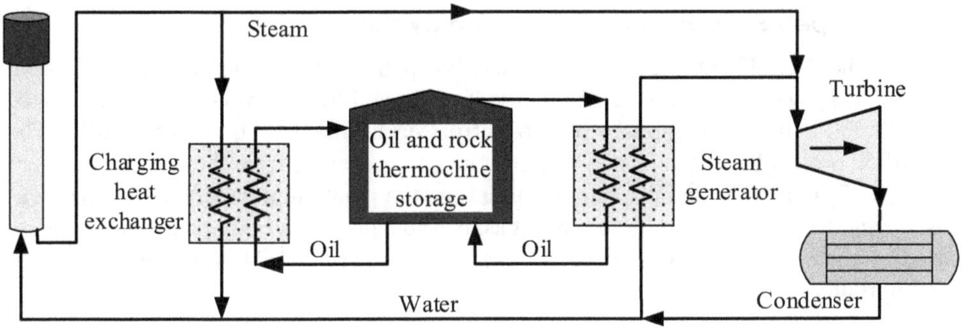

Figure 6.16. Simplified schematic of Solar One process flow; after [37,38].

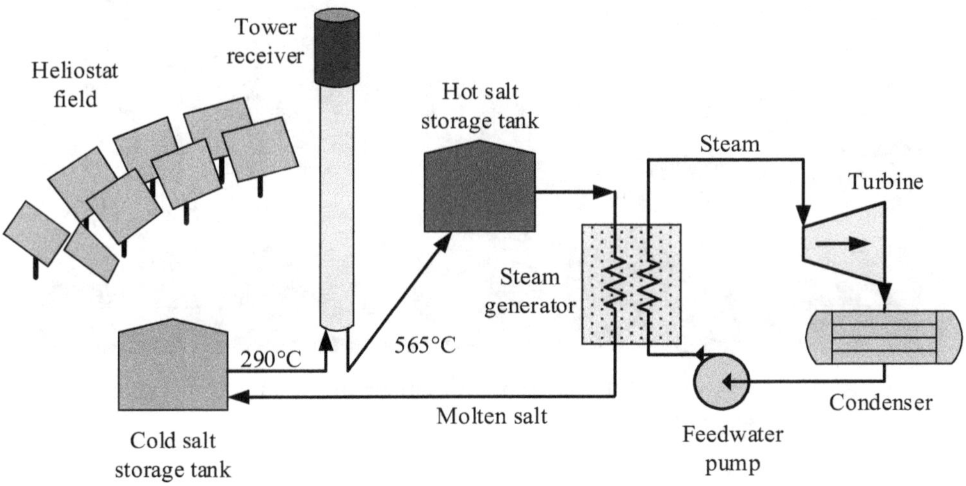

Figure 6.17. Simplified schematic of Solar Two process flow; adapted from [33,39].

6.3.2.2 Planta Solar 10 and 20

The 11-MWe Planta Solar (PS10) solar tower power plant began operation in 2007 on 150 acres near Seville, Spain. The facility utilizes 624 heliostats that each has an area of 120 m^2 in a north field configuration, yielding a reflective area of 7.5×10^4 m^2. This unit differs from Solar One and Two in that 257°C, 40 bar steam is directly made in the 115-m tower cavity receiver. As backup, the plant has the capability of storing an hour's worth of steam, and natural gas firing can supply 12%–15% of its capacity. The steam cycle efficiency is 27%; and the overall solar-to-electric energy conversion efficiency is near 17%. In April 2009, Abengoa Solar began operation of a larger (20 MW) power tower plant, designated PS20, which uses a 160-m tall tower and double the heliostat field size of the PS10.

6.3.2.3 *Ivanpah Solar Electric Generating System*

Presently, the 392 MWe gross (377 MWe net) Ivanpah Solar Electric Generating System (ISEGS) is the world's largest CSP facility. The three 140-m towers receive irradiance from 173,500 heliostats composed of two mirrors each [40], as shown in Fig. 6.18. The $2.2 billion project benefited from a $1.6 billion loan guarantee from the U.S. Department of Energy. To reduce potential water consumption, the plant employs dry, air-cooled condensers. Sited about 60 km south of Las Vegas, Nevada, pilots have complained of the glare caused by the over 2.6 million m^2 of mirrored surface [41]. In addition, concerns due to avian mortality have been expressed [42].

Figure 6.18. The three solar fields of the Ivanpah Solar Electric Generation System (courtesy of DOE).

6.3.2.4 *Crescent Dunes Solar Energy Facility*

Located about 300 km northwest of Las Vegas, Nevada, the 110-MWe Crescent Dunes Solar Energy Facility boasts 10 h of full-load storage (1.1 GWh). Molten salt is heated atop the 195-m tower in a 30.5-m tall receiver composed of nickel alloy steel tubes. The dual-axis tracking of each of the 10,347 heliostats is controlled individually. The TES is accomplished using hot molten salt with the cold tank at 288°C and the hot tank at 566°C with each tank being 12.2 m tall and 42.7 m in diameter, see Fig. 6.19. A hybrid cooling system is employed for the Rankine cycle with most heat dissipated using an air-cooled condenser, but a limited amount of water cooling is utilized during peak demand periods.

6.3.3 Stirling Engine

A Stirling cycle that employs a piston engine can be employed in combination with a dish collector that uses dual-axis tracking. The engine shaft is connected to an electric generator.

Figure 6.19. Crescent Dunes central receiver with molten salt storage tanks in foreground (courtesy of DOE).

Such dish-based Stirling engine systems can achieve high solar-to-electric efficiencies, but an individual unit tends to provide smaller electric outputs—on the order of 25 kW [43]. Assuming an efficiency of 25% and using an irradiance of 1 kW/m^2, one quickly realizes that the dish for a 25-kW system would have a diameter greater than 11 m. To provide large-scale electric generation, the dishes can be installed as arrays of individual dishes. In addition, the modular nature of dish systems makes them appropriate for distributed generation and remote generation (e.g., end of the power line installations). However, dish-Stirling systems are estimated to have capital costs of approximately $10,000 per kWe [44].

The Stirling system consists of four major components: (1) the dish concentrator, (2) a cavity receiver, (3) the Stirling engine, and (4) an electric generator or alternator. The combination of the latter two items forms what is referred to as the power conversion unit (PCU), which is annotated in Fig. 6.20. The dish concentrator was addressed earlier, and the electric generator employed is not novel, therefore, the cavity receiver and Stirling engine and cavity are emphasized herein. However, as shown in Fig. 6.21, one feature of many dish-Stirling systems is the use of multiple surfaces (mirror facets) to concentrate the sunlight.

The cavity receiver, which is located at the focal point of the dish, absorbs the concentrated heat energy and transfers the heat to the cycle working fluid. The working fluid, which is typically helium or hydrogen, is sealed within the Stirling engine. The Stirling cycle uses a four-stage process of heating, expansion, cooling, and contraction of the sealed working gas (fluid) [45], and there are no valves. The alternating heating and cooling of the gas produces more power during the heating–expansion stages than is consumed during

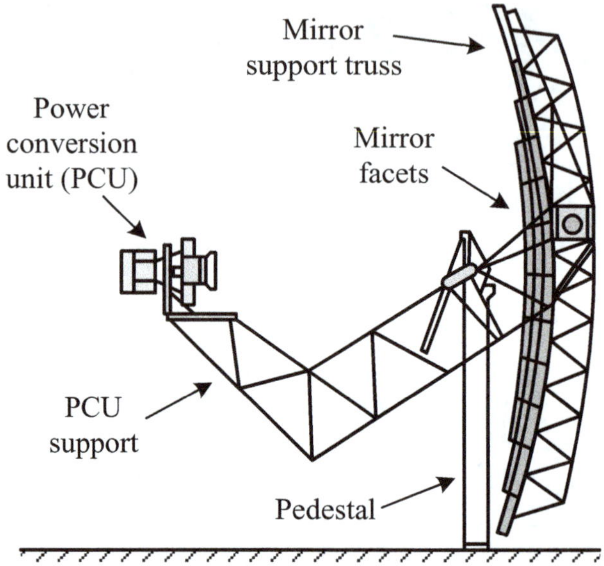

Figure 6.20. Dual-axis tracking dish concentrator with truss support, after [43].

Figure 6.21. Solar dish-Stirling system (courtesy of Stirling Energy Systems, Inc.).

the cooling–contraction phases, thereby yielding a net power output. Although a Stirling engine has the potential to be very efficient in converting heat into mechanical work, dish-Stirling systems are not suitable for TES.

6.4 Low-Temperature Solar Thermal Approaches

In addition to the mainstream CSP technologies, other solar thermal electricity-generation schemes that the reader should be cognizant of include solar chimneys and salt ponds, which are addressed below.

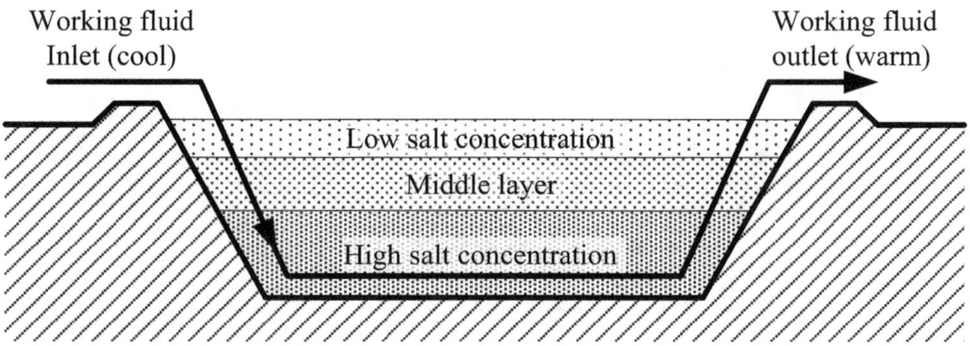

Figure 6.22. Solar salt pond side view.

6.4.1 Solar Salt Pond

In the case of a solar salt pond, a shallow (~2 m deep) pool with salt forms the solar collector, as shown in Fig. 6.22. Unlike a fresh body of water in which warm water rises to the surface and its heat is lost by evaporation, a solar salt pond traps the warm water near the bottom of the pool. Two different implementations of solar salt ponds exist [46]: (1) the saturated salt pond and (2) the nonconvective gradient pond. In either case, piping carrying a low boiling point working fluid (e.g., ammonia) can be placed near the bottom of the pond to extract the heat (~100°C) for use in a thermal power conversion cycle. It is noteworthy that the solar salt pond itself forms an integral TES mechanism.

Solar salt ponds have been demonstrated on a small scale. In particular, the largest constructed was located beside the Dead Sea in Israel and was initially sized for a 5-MWe output using a 1 km^2 pond. The actual unit operated with 1% solar-to-electric efficiency [47] until it was shut down in 1989. The following are several issues with solar salt ponds:

- Water loss due to evaporation
- The need to maintain a critical water–salt chemistry balance to form the required salt concentration gradient
- Inefficiency due to the small temperature difference, which, via Carnot, means that the thermal efficiency is very low
- The surface area of the pond needs to be large for a reasonable power output, which leads to high capital cost

6.4.2 Solar Chimney

A solar chimney plant is a hybrid plant that utilizes sunlight-induced temperature differences to create a natural draft within a tall tower, as shown in Fig. 6.23. A greenhouse of sorts is established within the outer solar collector. The draft created within the tower is then harnessed using wind turbines, thereby making the solar chimney a hybrid solar–wind system. Thermal mass elements can be placed within the collector region of the facility so that heat storage occurs during daylight, and then during nighttime or inclement conditions the heat is released and the draft continues at a reduced level for some limited time period.

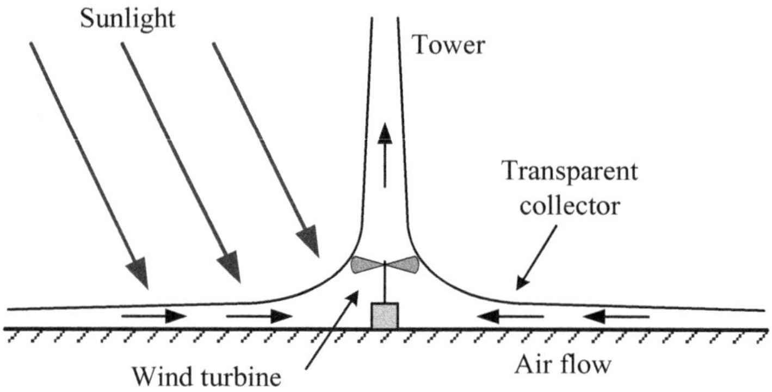

Figure 6.23. Solar chimney utilizing greenhouse effect to create air draft.

A pilot solar chimney plant was constructed at Manzanares, Spain, and operated from 1982 to 1988. The 50-kWe plant had a tower of 194.6 m height and 10.16 m diameter, with collector area covering 46,800 m^2 at an average height of 1.85 m [48]. The pilot plant produced a temperature rise of 20°C, yielding an upward airflow of 9 m/s under load conditions. The plant achieved a solar-to-electricity efficiency of only 0.53% [49].

6.5 Environmental Impact

All electricity-generation schemes have some environmental impact. As a thermal cycle, CSP has the potential to generate thermal pollution. Moreover, the solar thermal approaches that employ a steam turbine cycle require condenser cooling water—such clean water is generally at a premium in the arid regions of the world in which sunlight is correspondingly so plentiful [50]. Straining water resources is not the only impact of solar thermal facilities; for example, McCrary et al. [51] found an avian mortality rate of ~2 birds/week at the Solar One plant. Other possible environmental impacts include land use, accidental chemical releases, aesthetics, and effects on the local ecosystem, including loss of habitat [52–54]. With respect to land use, CSP facilities require about 2 ha (5 acres) per MW of installed capacity. However, use of CSP has the potential to offset emissions, such as CO_2, SO_2, and NO_X, from other electricity-generation technologies.

Conversely, the environment also has an effect on solar plants. Interestingly, the June 1991 eruption of Mount Pinatubo in the Philippines led to a 15%–20% reduction in the direct beam radiation at midlatitudes [55]. The SEGS units recorded a direct normal insolation of 6.32 kWh/m^2/day in 1992 compared to an average of 7.94 kWh/m^2/day for 1988–1990 [6]; the operators attributed the 20% reduction to a combination of Mt. Pinatubo and the weather phenomenon known as El Niño, which brings above normal rainfall to California.

6.6 Conclusion

With concerns over climate change (global warming) due to carbon dioxide emissions from fossil fuel burning, green electricity-generation technologies, such as solar, will see increasing interest and investment. It is also noteworthy that besides electricity generation, CSP facilities have been proposed for other uses including desalination [56,57], coal gasification [58], water splitting to yield hydrogen [59], and destruction of hazardous chemicals [60].

6.7 References

1. J. Weisman, and R. Eckart, *Modern Power Plant Engineering*, Prentice-Hall, 1985.
2. R. L. San Martin, *Solar Energy AccessScience*, McGraw-Hill, 2005.
3. G. W. Crabtree, N. S. Lewis, "Solar energy conversion," *Physics Today*, vol. 60, no. 3, pp. 37–42, 2007.
4. S. A. Kalogirou, "Solar thermal collectors and applications," *Progress in Energy and Combustion Science*, vol. 30, no. 3, pp. 231–295, 2004.
5. M. Mehos, "Concentrating solar power" in *Physics of Sustainable Energy, Using Energy Efficiently and Producing It Renewably*, D. Hafemeister, B. Levi, M. Levine and P. Schwartz. (eds.), vol. 1044, pp. 331–339, AIP Conference Proceedings, American Institute of Physics, 2008.
6. G. J. Kolb, "Evaluation of power production from the solar electric generating systems at Kramer Junction: 1988 to 1993," ASME International Solar Energy Conference, Lahaina, HI, pp. 19–24, Mar. 1995.
7. A. Leitner, *Fuel from the Sky: Solar Power's Potential for Western Energy Supply*, National Renewable Energy Laboratory, NREL/SR-550–32160, 2002.
8. G. J. Kolb, Economic Evaluation of Solar-only and Hybrid Power Towers Using Molten Salt Technology, Sandia National Laboratories, SAND96-2454C, 1996.
9. Luz International Limited, "Solar electric generating stations (SEGS)," *IEEE Power Engineering Review*, vol. 9, no. 8, pp. 4–8, 1989.
10. R. Ramakumar, N. G. Butler, A. P. Rodriguez, and S. S. Venkata, "Economic aspects of advanced energy technologies," *Proceedings of the IEEE*, vol. 81, no. 3, pp. 318–332, 1993.
11. H. W. Price, and S. Carpenter, *The Potential For low-Cost Concentrating Solar Power Systems*, National Renewable Energy Laboratory, NREL/CP-550-26649, 1999.
12. L. Stoddard, J. Abiecunas, and R. O'Connell, *Economic, Energy, and Environmental Benefits of Concentrating Solar Power in California*, National Renewable Energy Laboratory, NREL/SR-550-39291, 2006.
13. H. Price, and D. Kearney, *Reducing the Cost of Energy from Parabolic Trough Solar Power Plants*, National Renewable Energy Laboratory, NREL/CP-550-33208, 2003.
14. Energy Information Administration, Levelized Cost and Levelized Avoided Cost of New Generation Resources in the Annual Energy Outlook 2017, U.S. Department of Energy, p. 8, 2017.
15. M. Lotker, Barriers to Commercialization of Large-Scale Solar Electricity: Lessons Learned from the Luz Experience, Sandia National Laboratories, SAND91-7014, 1991.
16. International Energy Agency, *Renewables for Power Generation 2003: Status and Prospects*, Organization for Economic Co-operation and Development, 2003.
17. S. K. Roy, and H. Liu, "Solar heating" in *Encyclopedia of Electrical and Electronics* Engineering, J. G. Webster (ed.), vol. 2, pp. 649–658, Wiley & Sons, 1999.
18. C. Philibert, "The present and future use of solar thermal energy as a primary source of energy," *The InterAcademy Council*, 2005. Available online: http://philibert.cedric.free.fr/Downloads/solarthermal.pdf
19. U. Herrmann, and D. W. Kearney, "Survey of thermal energy storage for parabolic trough power plants," *Journal of Solar Energy Engineering*, vol. 124, no. 2, pp. 145–152, 2002.
20. E. Zarza, L. Valenzuela, and J. León, "Solar thermal power plants with parabolic-trough collectors," Proceedings of the 4th International Conference on Solar Power from Space, pp. 91–98, Granada, Spain, ESA SP-567, 2004.
21. H. Price, E. Lüpfert, D. Kearney, E. Zarza, G. Cohen, R. Gee, R. Mahoney, "Advances in parabolic trough solar power technology," *Journal of Solar Energy Engineering*, vol. 124, no. 2, pp. 109–125, 2002.
22. E. Zarza, "The DISS Project: Direct steam generation in parabolic trough systems. Operation and maintenance experience and update on project status," *Journal of Solar Energy Engineering*, vol. 124, no. 2, pp. 126–133, 2002.

23. M. Eck, and W.-D. Steinmann, "Direct steam generation in parabolic troughs: First results of the DISS project," *Journal of Solar Energy Engineering*, vol. 124, no. 2, pp. 134–139, 2002.
24. M. Romero-Alvarez, and E. Zarza, "Concentrating solar thermal power" in *Handbook of Energy Efficiency and Renewable Energy*, F. Kreith, and D. Y. Goswami, CRC Press, 2007.
25. D. R. Mills, and G. L. Morrison, "Compact linear Fresnel reflector solar thermal powerplants," *Solar Energy*, vol. 68, no. 3, pp. 263–283, 2000.
26. G. E. Cohen, D. W. Kearney, and G. J. Kolb, Final *Report on the Operation and Maintenance Improvement Program for Concentrating Solar Power Plants*, Sandia National Laboratories, SAND99-1290, 1999.
27. Pilkington Solar International GmbH, *Survey of Thermal Storage for Parabolic Trough Power Plants*, National Renewable Energy Laboratory, NREL/SR-550-27925, 2000.
28. R. Peltier, "Nevada Solar One, Boulder City, Nevada," *Power*, vol. 151, no. 12, pp. 40–43, 2007.
29. M. Romero, R. Buck, and J. E. Pacheco, "An update on solar central receiver systems, projects, and technologies," *Journal of Solar Energy Engineering*, vol. 124, no. 2, pp. 98–108, 2002.
30. A. Yogev, A. Kribus, M. Epstein, and A. Kogan, "Solar 'tower reflector' systems: A new approach for high-temperature solar plants," *International Journal of Hydrogen Energy*, vol. 23, no. 4, pp. 239–245, 1998.
31. A. Rabl, "Tower reflector for solar power plant," *Solar Energy*, vol. 18, no. 3, pp. 269–271, 1976.
32. P. Schramek, and D. R. Mills, "Multi-tower solar array," *Solar Energy*, vol. 75, no. 3, pp. 249–260, 2003.
33. C. Tyner, G. Kolb, M. Prairie, G. Weinrebe, A. Valverde, M. Sanchez, *Solar Power Tower Development: Recent Experiences*, Sandia National Laboratories, SAND96-2662C, 1996.
34. M. M. El-Wakil, *Powerplant Technology,* McGraw-Hill, 1984.
35. A. F. Hildebrandt, L. L. Vant-Hull, "Power with heliostats," *Science*, vol. 197, no. 4039, pp. 1139–1146, 1977.
36. K. W. Battleson, Solar Power Tower Design Guide: Solar Thermal Central Receiver Power Systems, a Source of Electricity and/or Process Heat, Sandia National Laboratories, SAND81-8005, 1981.
37. W. B. Stine, and R. W. Harrigan, *Solar Energy System and Design*, John Wiley, 1985.
38. L. G. Radosevich, and A. C. Skinrood, "The power production operation of Solar One, the 10 MWe solar thermal central receiver pilot plant," *Journal of Solar Energy Engineering*, vol. 111, no. 2, pp. 144–151, 1989.
39. J. E. Pacheco, H. E. Reilly, G. J. Kolb, and C. E. Tyner, "Summary of the Solar Two test and evaluation program," Renewable Energy for the New Millennium Conference, Sydney, 2000.
40. Bright Source Energy, *Ivanpah Solar Electric Generating System*, white paper, 2014.
41. C. K. Ho, C. A. Sims, and J. M. Christian, "Evaluation of glare at the Ivanpah Solar Electric Generating System," *Energy Procedia*, vol. 69, pp. 1296–1305, 2015.
42. C. K. Ho, T. Wendelin, L. Horstman, and J. Yellowhair, "A method to assess flux hazards at CSP plants to reduce avian mortality," *AIP Conference Proceedings*, vol. 1850, p. 030026, 2017.
43. W. B. Stine, and R. B. Diver, *A Compendium of Solar Dish/Stirling Technology*, Sandia National Laboratories, SAND93-7026 UC-236, 1994.
44. T. Mancini, P. Heller, B. Butler, B. Osborn, W. Schiel, V. Goldberg, R. Buck, R. Diver, C. Andraka, J. Moreno, "Dish-Stirling systems: An overview of development and status," *Journal of Solar Energy Engineering*, vol. 125, no. 2, pp. 135–151, 2003.
45. B. Kongtragool, S. Wongwises, "A review of solar-powered Stirling engines and low temperature differential Stirling engines," *Renewable and Sustainable Energy Reviews*, vol. 7, no. 2, pp. 131–154, 2003.
46. H. Tabor, "Solar ponds," *Solar Energy*, vol. 27, no. 3, pp. 181–194, 1981. Also see article correction in *Solar Energy,* vol. 30, no. 1, p. 85, 1983.
47. H. Z. Tabor, and B. Doron, "The Beith Ha'arava 5 MW(e) solar pond power plant (SPPP) — progress report," *Solar Energy*, vol. 45, no. 4, pp. 247–253, 1990.

48. W. Haaf, K. Friedrich, G. Mayr, and J. Schlaich, "Solar chimneys. Part I: Principle and construction of the pilot plant in Manzanares," *International Journal of Solar Energy*, vol. 2, no. 1, pp. 3–20, 1983.
49. D. Mills, "Advances in solar thermal electricity technology," *Solar Energy*, vol. 76, no. 1–3, pp. 19–31, 2004.
50. K. E. Holbert, and J. Haverkamp, "Impact of solar thermal power plants on water resources and electricity costs in the Southwest," Proceedings of the North American Power Symposium, Starkville, Mississippi, Oct. 4–6, 2009.
51. M. D. McCrary, R. L. McKernan, R. W. Schreiber, W. D. Wagner, T. C. Sciarrotta, "Avian mortality at a solar energy power plant," *Journal of Field Ornithology*, vol. 57, no. 2, pp. 135–141, 1986.
52. T. Tsoutsos, N. Frantzeskaki, and V. Gekas, "Environmental impacts from the solar energy technologies," *Energy Policy*, vol. 33, no. 3, pp. 289–296, 2005.
53. S. A. Abbasi, and N. Abbasi, "The likely adverse environmental impacts of renewable energy sources," *Applied Energy*, vol. 65, no. 1–4, pp. 121–144, 2000.
54. D. Pimentel, M. Herz, M. Glickstein, M. Zimmerman, R. Allen, K. Becker, J. Evans, B. Hussain, R. Sarsfeld, A. Grosfeld, T. Seidel, "Renewable energy: Current and potential issues," *BioScience*, vol. 52, no. 12, pp. 1111–1120, 2002.
55. J. J. Michalsky, R. Perez, R. Seals, and P. Ineichen, "Degradation of solar concentrator performance in the aftermath of Mount Pinatubo," *Solar Energy*, vol. 52, no. 2, pp. 205–213, 1994.
56. F. Trieb, J. Nitsch, S. Kronshage, C. Schillings, L.-A. Brischke, G. Knies, G. Czisch, "Combined solar power and desalination plants for the Mediterranean region — sustainable energy supply using large-scale solar thermal power plants," *Desalination*, vol. 153, no. 1–3, pp. 39–46, 2003.
57. F. Trieb, and H. Müller-Steinhagen, "Concentrating solar power for seawater desalination in the Middle East and North Africa," *Desalination*, vol. 220, no. 1–3, pp. 165–183, 2008.
58. P. V. Zedtwitz, and A. Steinfeld, "The solar thermal gasification of coal — energy conversion efficiency and CO_2 mitigation potential," *Energy*, vol. 28, no. 5, 441–456, 2003.
59. A. Kogan, "Direct solar thermal splitting of water and on site separation of the products I. Theoretical evaluation of hydrogen yield," *International Journal of Hydrogen Energy*, vol. 22, no. 5, pp. 481–486, 1997.
60. G. C. Glatzmaier, R. G. Nix, and M. S. Mehos, "Solar destruction of hazardous chemicals," *Journal of Environmental Science and Health*, vol. A25, no. 5, pp. 571–581, 1990.

CHAPTER 7

Solar Energy Calculations

Keith E. Holbert[1] and Devarajan Srinivasan[2]

[1] *School of Electrical, Computer and Energy Engineering, Arizona State University, Tempe, Arizona 85287-5706. keith.holbert@asu.edu*
[2] *Poundra, Tempe, Arizona. srini@poundra.com*

Abstract This chapter presents calculations of the sun's position as a function of location, date, and time. In addition, direct, diffuse, and reflected components of solar insolation on a tilted surface are also determined. This computational approach is applicable to both thermal collection/conversion processes and photovoltaics (PVs). Although sun charts are widely used, there are situations in which charts are inadequate and precise computations are preferred. For example, consider fixed PV modules atop residential or industrial buildings in which the value of the electricity varies with date and time. These calculations can be used to determine angles for positioning one and two axis tracking systems to maximize solar energy utilization.

Keywords: Albedo, Azimuth angle, Collector angle, Extinction coefficient, Solar constant, Solar insolation, Solar spectra, Zenith angle.

7.1 Introduction

Three energy conversion processes are normally associated with the sun [1]:

(i) heliochemical—which is principally the photosynthesis process,
(ii) helioelectrical—which is commonly exploited in solar cells (i.e., photovoltaics), and
(iii) heliothermal—which is the conversion of sunlight into thermal heat as employed within concentrating solar power (CSP) plants.

In the case of the latter two processes, it is oftentimes necessary to calculate the solar energy incident to the energy-collection apparatus. This chapter describes general-purpose computational approaches to determine optimal solar panel positions and expected energy collection. Solar energy is dilute and intermittent. Hence, solar energy-harvesting facilities have to be optimized to improve their economic return. Although solar irradiance is dilute, estimates have shown that the annual extraterrestrial radiation to the contiguous United States is 4.43×10^{16} kWh [2].

The fundamental goal of such calculations is to determine the solar insolation on a given area, so solar collectors can be oriented to maximize collected energy. The total incident solar energy on the panels is a function of several variables—including the day of the year, time of day, latitude, longitude, orientation of the collectors, and atmospheric conditions. In general, one begins by computing angles based on the particular day of the year, and then the time of the day. This is followed by considering the specific collector-panel surface position and orientation with respect to the incident sunlight. The sections of this chapter are arranged with this procedure in mind.

7.2 Earth's Orbit

The earth, as a member of the solar system, circles the sun once a year. The orbital path of the earth around the sun, as shown in Fig. 7.1, is slightly elliptical and not a perfect circle [3–5]. The imaginary plane of Earth's elliptical orbit is called the orbital plane (also known as the ecliptic plane). The direction of Earth's orbit is counterclockwise when the orbital plane is viewed from above the North Pole.

The sun is slightly offset from the center of the ellipse. At perihelion, on January 3, the earth reaches its closest approach to the sun of around 147.09 million km. The earth is farthest from the sun at aphelion, on July 4, at around 152.10 million km. On average, the earth is about 149.60 million km from the sun. The earth travels about 940 million km during a year in its orbit around the sun.

In addition to orbiting the sun, the earth also spins, every 24 h, on an axis that passes through the North and South Geographic Poles. Looking down at the earth from above the North Pole, the earth's rotation is counterclockwise, from the west to the east. Earth's axis is not perpendicular to the orbital plane, but is tilted at 23.45° to the perpendicular. The Northern Hemisphere tilts toward the sun during summer and away from the sun during winter, and vice versa for the Southern Hemisphere. The tilt of the earth's axis and its motion around the sun causes variation:

(i) in the intensity and duration of sunlight received at different places on the earth, which results in the seasons, and
(ii) in the apparent motion of the sun across the sky.

Four associated astronomical events are the vernal and autumnal equinoxes, and the summer and winter solstices. During the March equinox (~ March 21) and September equinox (~ September 21), the sun is directly over the equator and the lengths of day and night are equal (~12 h). During the June solstice, the sun is directly over the Tropic of Cancer (23.45° N latitude), and thus at the equator, the sun appears at its northernmost position (23.45°). In June, it is summer in the Northern Hemisphere and the day is longer than the night; and it is winter in the Southern Hemisphere and the day is shorter than the night. During the December solstice, the sun is directly over the Tropic of Capricorn (23.45° S latitude), and hence, at the equator, the sun appears at its southernmost position (–23.45°).

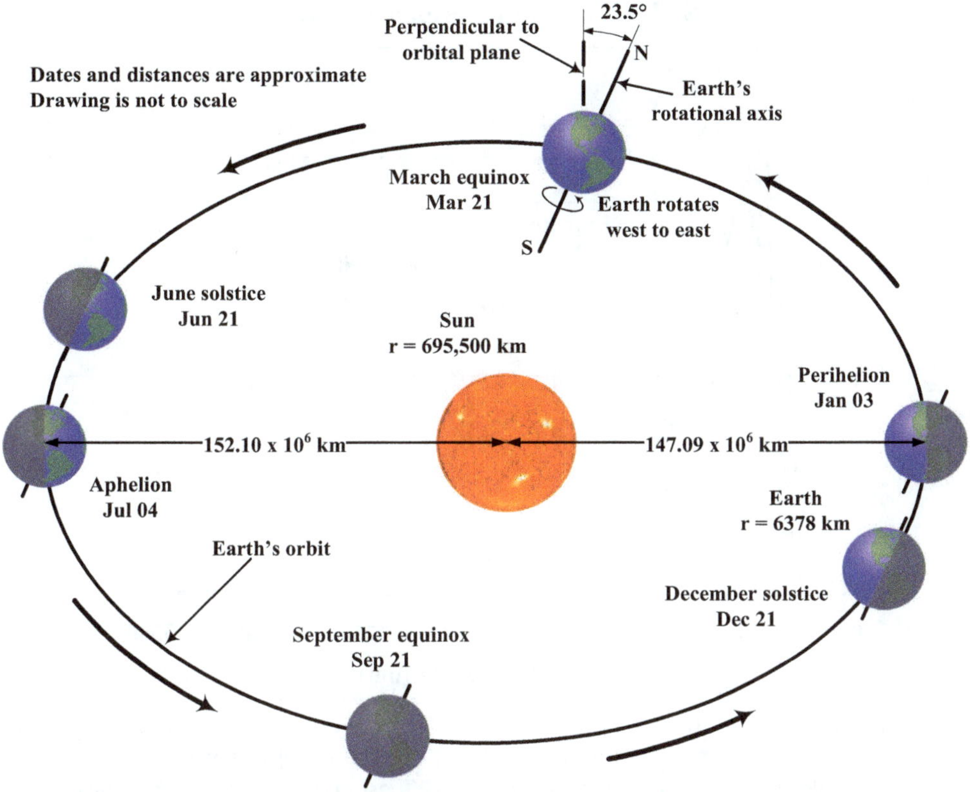

Figure 7.1. Earth's orbit around the sun.

7.3 Solar Constant and Solar Spectra

The orbit of the earth is elliptical; hence, the distance between the earth and sun varies over the year, leading to apparent solar irradiation values throughout the year approximated by [6]

$$I_0 = I_{SC}\left[1 + 0.033\cos\left(\frac{N}{365} \times 360°\right)\right] \tag{7.1}$$

where I_{SC} is the solar constant, and N is the day number of the year, with January 1 equal to 1. The amount of solar radiation, just above the earth's atmosphere, measured in a plane perpendicular to the rays of the sun, is termed the *solar constant*. The solar constant generally accepted by the space community is approximately 1.3661 kW/m^2 [7]. Solar radiation consists of several frequencies, in addition to visible light [8,9]. The extraterrestrial solar spectrum is called the Air Mass Zero (AM0) spectrum and is defined by the ASTM E490 standard [7]. The AM0 spectrum is shown in Fig. 7.2.

The spectrum of sunlight reaching the earth's surface is affected by atmospheric conditions and varies by location and time of day. The photovoltaic (PV) industry accepts a standard spectrum for testing and rating PV devices. This standard (AM1.5) spectrum is

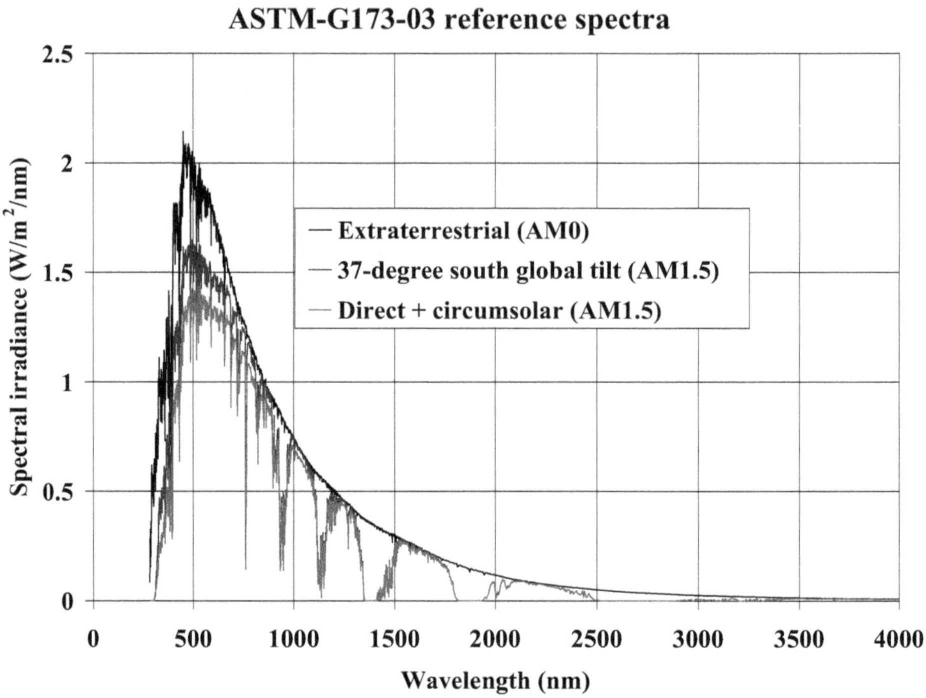

Figure 7.2. Extraterrestrial and photovoltaic reference solar spectra. The direct plus circumsolar spectrum is the direct normal component of the 37° south global tilt (AM1.5) spectrum. The diffuse component (not shown) is the difference between the two AM1.5 spectra. Data are from [8].

defined in the ASTM G173-03 standard [10] for specific atmospheric conditions on a south-facing surface tilted at 37° (i.e., the average latitude for the contiguous United States). The selected conditions are considered an average for the contiguous United States over a one-year period.

7.4 Solar Angles [11,12]

The declination is the angular distance of the sun north or south of the earth's equator. The *declination angle, δ*, for the Northern Hemisphere (reverse the declination angle sign for the Southern Hemisphere) is [13]

$$\delta = 23.45^\circ \sin\left[\frac{N+284}{365} \times 360^\circ\right] \tag{7.2}$$

The earth is divided into latitudes and longitudes. For longitudes, the global community has defined 0° as the prime meridian, which is located at Greenwich, England. The longitudes are described in terms of how many degrees they lie to the east or west of the prime meridian. A 24-h day has 1440 min, which when divided by 360° means that it takes 4 min to move each degree of longitude.

The apparent solar time, *AST* (or local solar time), is calculated from [14,15]

$$AST = LST \pm (4\ \text{min}/\text{deg})\,(LSTM - Long) + ET \tag{7.3}$$

where the sign is positive west of Greenwich and negative east of Greenwich, and
LST = local standard time or clock time for that time zone (may need to adjust for daylight savings time, *DST*, i.e., *LST* = *DST* − 1 h),
Long = local longitude at the position of interest, and
LSTM = local longitude of standard time meridian, which is

$$LSTM = 15^\circ \times \left(\frac{Long}{15^\circ}\right)_{\text{round to integer}} \tag{7.4}$$

The difference between the true solar time and the mean solar time changes continuously day-to-day with an annual cycle. This quantity is known as the *equation of time*. The equation of time, *ET* in minutes, is approximated by [16]

$$ET = 9.87\ \sin(2D) - 7.53\ \cos(D) - 1.5\ \sin(D)$$
$$\text{where}\quad D = 360^\circ\,\frac{(N-81)}{365} \tag{7.5}$$

As illustrated in Fig. 7.3, there are several angles of interest in describing the position of the sun at a particular time instant. The *hour angle, H*, is the azimuth angle of the sun's rays caused by the earth's rotation, and *H* can be computed from [17]

$$H = \frac{AST - 720\ \text{min}}{4\ \text{min}/\text{deg}} \tag{7.6}$$

The hour angle as defined here is negative in the morning and positive in the afternoon ($H = 0^\circ$ at noon).

The solar *altitude angle* (β_1) is the apparent angular height of the sun in the sky if you are facing it. The *zenith angle* (θ_z) and its complement the altitude angle (β_1) are given by

$$\cos(\theta_z) = \sin(\beta_1) = \cos(L)\cos(\delta)\cos(H) + \sin(L)\sin(\delta) \tag{7.7}$$

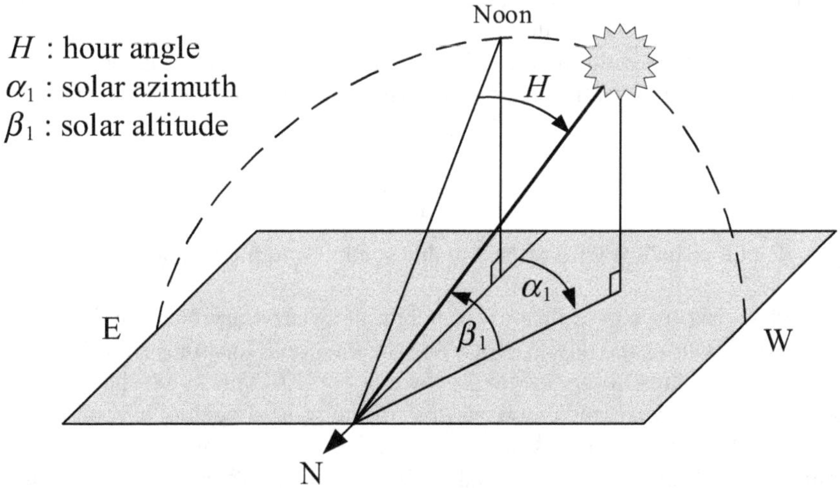

Figure 7.3. Important solar angles.

where L is the latitude (positive in either hemisphere) [0° to +90°]; δ is the declination angle (negative for Southern Hemisphere) [–23.45° to +23.45°]; and H is the hour angle. The noon altitude is $\beta_N = 90° - L + \delta$.

The sun rises and sets when its altitude is 0°, not necessarily when its hour angle is ±90°. The hour angle at sunset or sunrise, H_S, can be found by using Eq. (7.7) when $\beta_1 = 0$

$$\cos(H_S) = -\tan(L)\tan(\delta) \tag{7.8}$$

where H_S is negative for sunrise and positive for sunset. The apparent sunrise (apparent sunset) occurs earlier (later) than the value computed above as atmospheric refraction bends the sunlight. In addition, because the sun is about 0.5° wide as viewed from the earth, whether one considers the first appearance of sunlight as sunrise also plays a role. Absolute values of $\cos(H_S)$ greater than unity can occur in the arctic zones when the sun neither rises nor sets. The length of a day is twice the absolute value of H_S.

The *solar azimuth,* α_1, is the angle away from south in the Northern Hemisphere (referenced to north in the Southern Hemisphere)

$$\cos(\alpha_1) = \frac{\sin(\beta_1)\ \sin(L) - \sin(\delta)}{\cos(\beta_1)\ \cos(L)} \tag{7.9}$$

where α_1 is positive toward the west (afternoon), and negative toward the east (morning), and therefore, the sign of α_1 should match that of the hour angle.

If $\delta > 0$, the sun can be north of the east-west line in the Northern Hemisphere. The time at which the sun is due east and west can be determined from

$$t_{E/W} = 4\,\frac{\text{min}}{\text{deg}}\left\{180° \mp \arccos\left[\frac{\tan(\delta)}{\tan(L)}\right]\right\} \tag{7.10}$$

where these times are given in minutes from midnight AST.

Using the above equations, the apparent route of the sun through the sky can be graphically represented by a sun-path chart as shown in Fig. 7.4. The elevation (or altitude angle) of the sun is plotted against the azimuth for various times during the year. This type of chart is typically used for shading studies. By plotting the elevation versus azimuth of various obstructions such as buildings, trees, and mountains on the same chart, the days and times of potential shading can be identified.

Shading from a zero-width pole can be estimated by considering its height (H_{SH}) and the solar elevation and azimuth angles as shown in Fig. 7.5. The length of the shadow is given by

$$L_{SH} = \frac{H_{SH}}{\tan(\beta_1)} \tag{7.11}$$

The angle of the shadow with respect to due south is complementary (from 180°) to the azimuth angle.

The shadow cast by a pole of a given height at an arbitrary location (origin) can be plotted for every hour of the day. A plot using the data (and location) from Fig. 7.4 and a pole of height 1 m at an arbitrary origin is shown in Fig. 7.6. This graph shows the reach of the shadow during the year, and can be used to locate solar collectors beyond the reach of shade-causing obstructions such as trees, utility poles, or buildings.

Commercial or free three-dimensional (3-D) modelling software is also typically used for shading studies. Detailed 3-D models of surrounding topography and obstructions can

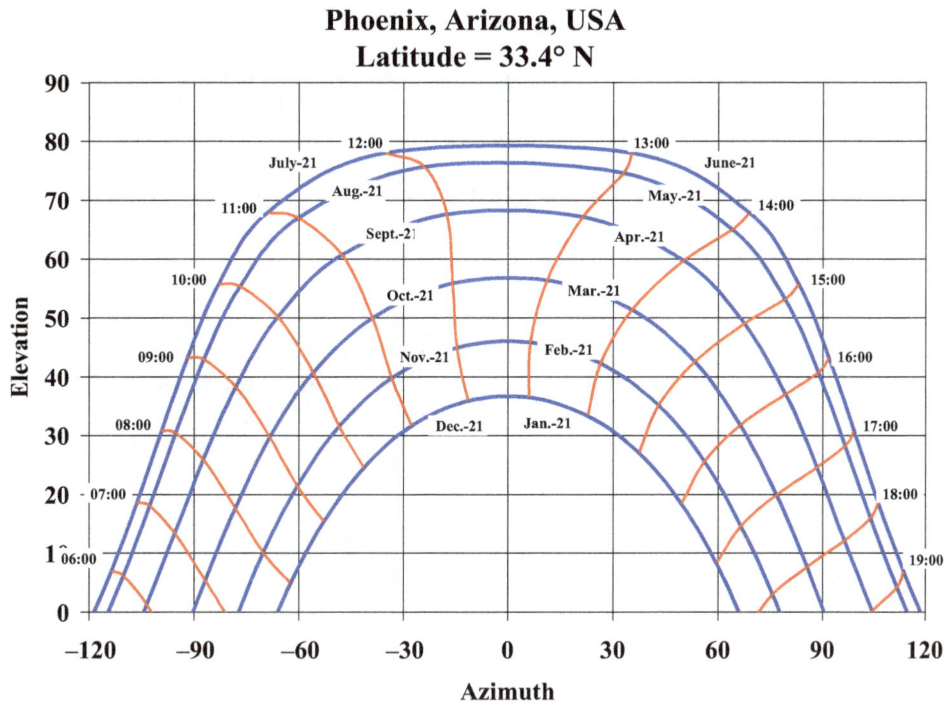

Figure 7.4. Elevation and azimuth angles of the sun at various dates and times during the year for 33.4° N latitude.

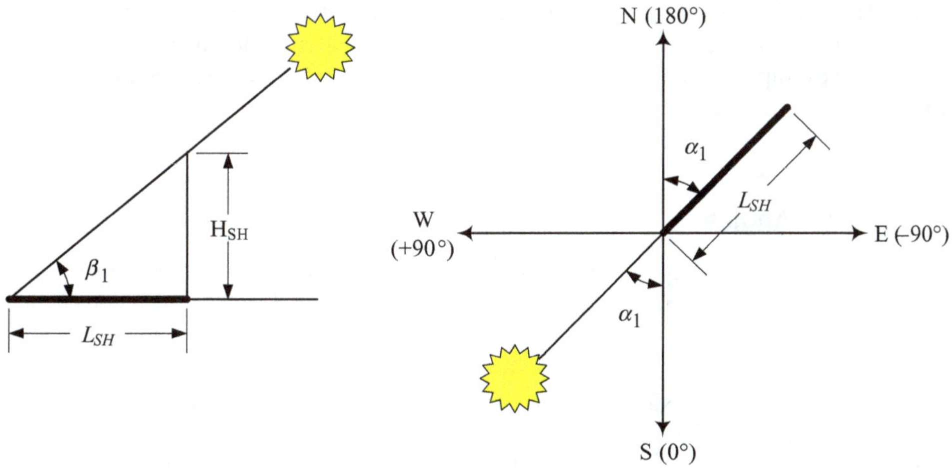

Figure 7.5. Shade cast by a pole of height H_{SH} and zero width.

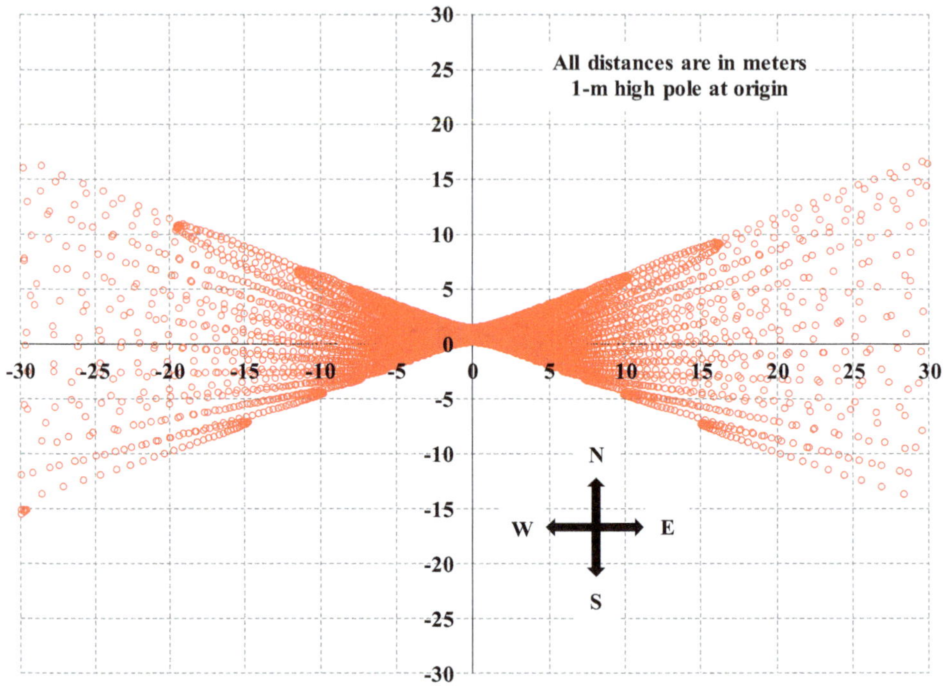

Figure 7.6. Shade profile of a pole of 1-m height at arbitrary origin, using data from Fig. 7.4.

be developed and the shade profile from the sun path across the sky can be plotted. Several modelling software packages include the sun-path calculations or provide plugins. Commercial examples include PVSyst® and AutoCAD Rivet®. Google Sketchup® is a free software that can also be used for shading studies.

7.5 Collector Angles

With knowledge of the position of the sun with respect to a particular location on earth, the solar collector orientation may be determined. The collector angle (θ) between the sun and normal to the surface is

$$\cos(\theta) = \sin(\beta_1)\cos(\beta_2) + \cos(\beta_1)\sin(\beta_2)\cos(\alpha_1 - \alpha_2) \tag{7.12}$$

where α_2 is the azimuth angle normal to the collector surface, and β_2 is the tilt angle from the ground, as shown in Fig. 7.7. If θ is greater than 90°, then the sun is behind the collector. Some collector-panel angles of interest are summarized in Table 7.1.

Given the collector angle, the total daily radiation to a surface is determined from integrating the solar irradiance, $I_0\cos(\theta)$, from sunrise to sunset. For a horizontal surface, the collector and zenith angles are identical ($\theta = \theta_z$). If the extraterrestrial irradiance were to

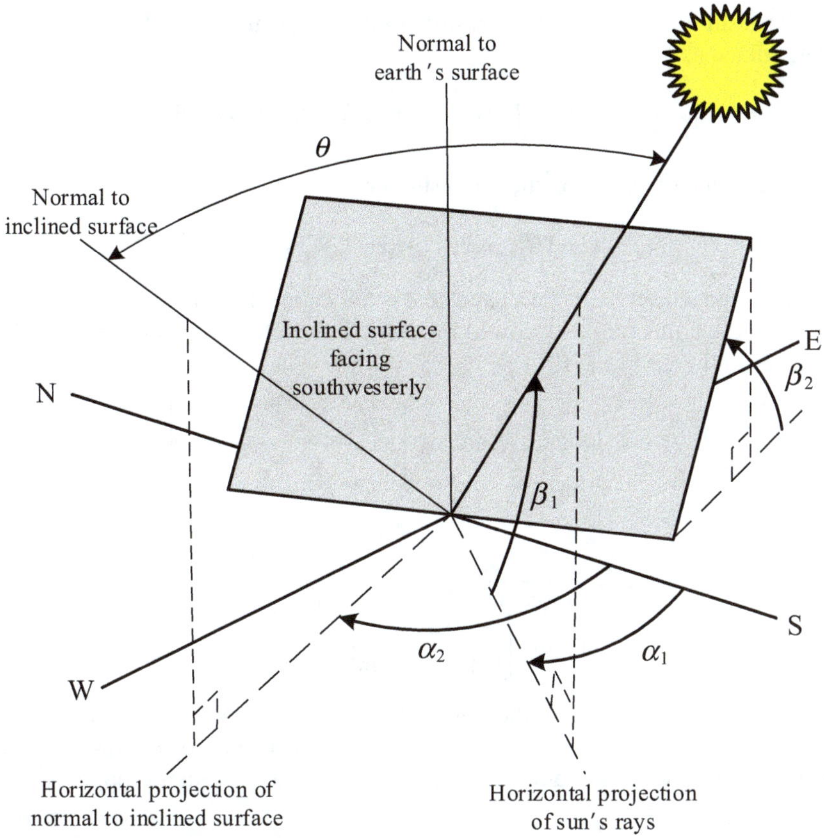

Figure 7.7. Solar angles on a tilted flat surface; diagram derived from [17].

Table 7.1. Solar collector angles of special interest.

Azimuth, α_2	Tilt, β_2	θ	Orientation
n/a	0°	90°–β_1	Horizontal (flat)
–	90°	Varies	Vertical wall
0°	90°	Varies	South facing vertical
–90°	90°	Varies	East facing wall
+90°	90°	Varies	West facing wall
α_1	90°–β_1	0°	Tracking system

reach such a horizontal panel, then the total radiation received is [18]

$$\begin{aligned} H_O &= \int_{-H_S}^{H_S} I_0\ [\cos(L)\cos(\delta)\cos(H) + \sin(L)\sin(\delta)]\ \frac{24}{2\pi}\, dH \\ &= I_0\, \frac{24}{\pi}\ \left[\cos(L)\cos(\delta)\sin(H_S) + H_S\sin(L)\sin(\delta)\right] \end{aligned} \tag{7.13}$$

In similar fashion, the total extraterrestrial radiation to an equator-facing tilted surface can be determined as

$$H_O = I_0 \frac{24}{\pi} \left[\cos(L - \beta_2)\cos(\delta)\sin(H_S) + H_S \sin(L - \beta_2)\sin(\delta) \right] \tag{7.14}$$

where H_S is the time of sunset on the tilted surface

$$H_S = \min\left[H_S, \arccos\left(-\tan(L - \beta_2)\tan(\delta)\right)\right] \tag{7.15}$$

The sunrise and sunset hours on a panel that is not facing directly toward the equator are different when the collector is shadowed by itself. The collector-panel sunrise and sunset hours may be computed from [19]

$$H_{SRS} = \mp \min\left\{ H_S, \arccos\left[\frac{ab \pm \sqrt{a^2 - b^2 + 1}}{a^2 + 1}\right]\right\} \tag{7.16}$$

where

$$\begin{aligned} a &= \frac{\cos(L)}{\sin(\alpha_2)\tan(\beta_2)} + \frac{\sin(L)}{\tan(\alpha_2)} \\ b &= \tan(\delta)\left[\frac{\cos(L)}{\tan(\alpha_2)} - \frac{\sin(L)}{\sin(\alpha_2)\tan(\beta_2)}\right] \end{aligned} \tag{7.17}$$

For sunrise the negative sign of the leading "$\mp$" sign is selected, and for sunset this sign is positive. The choice of sign with respect to the "$\pm$" sign depends on α_2: if sunrise is being calculated, the sign used is the same as that for α_2; for sunset, the sign is opposite to that of α_2.

7.6 Solar Irradiance

As depicted in Fig. 7.8, there are three basic components to the solar irradiance on a surface:

(i) direct (beam),
(ii) diffuse-scattered, and
(iii) reflected short wavelength (solar) radiation from other surfaces.

Solar irradiance is the amount of solar energy incident on a surface per unit time, per unit area. The irradiance is typically expressed in kW/m^2. Solar insolation is the integration of the irradiance with respect to time, and may be quantified in kWh or kWh/m^2.

The direct normal irradiance at the earth's surface is [6]

$$I_{DN} = A \exp\left(-\frac{p}{p_0}\frac{B}{\sin(\beta_1)}\right) \tag{7.18}$$

where A is the apparent extraterrestrial solar intensity, B is the atmospheric extinction coefficient (mainly due to changes in atmospheric moisture), and p/p_0 is the pressure at the location of interest relative to a standard atmosphere, given by

$$p/p_0 = \exp(-0.0001184\, z) \tag{7.19}$$

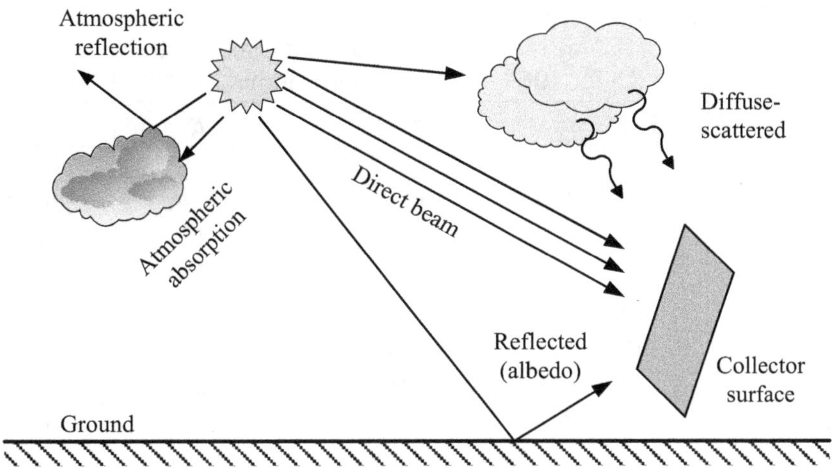

Figure 7.8. Components of solar radiation.

where z is the altitude in meters above sea level. The values for A and B are summarized in Table 7.2. The direct normal radiation at sea level then is

$$I_{DN}(0\,\text{m}) = A\ \exp\left[\frac{-B}{\sin(\beta_1)}\right] \tag{7.20}$$

As noted earlier, calculation of the energy received must account for the position of the panel relative to the sun. The direct radiation flux onto the collector is

$$I_D = I_{DN}\cos(\theta) \tag{7.21}$$

The diffuse-scattered radiation, also referred to as sky radiation, flux onto the collector is

$$I_{DS} = C\,I_{DN}\left[\frac{1+\cos(\beta_2)}{2}\right] \tag{7.22}$$

where C is the ratio of diffuse radiation on a horizontal surface to the direct normal irradiation. The reflected, or albedo, irradiance for a non-horizontal surface may be approximated by

$$I_{DR} = I_{DN}\,\rho\left(C+\sin\left(\beta_1\right)\right)\left[\frac{1-\cos\left(\beta_2\right)}{2}\right] \tag{7.23}$$

where ρ is the foreground reflectivity with some values given below:

ρ	Surroundings Condition
0.2	Ordinary ground or vegetation
0.8	Snow cover
0.15	Gravel roof

Table 7.2. Apparent solar irradiation (*A*), atmospheric extinction coefficient (*B*), and ratio of diffuse radiation on a horizontal surface to the direct normal irradiation (*C*) for 1964 in the Northern Hemisphere [13].

Date	Day of Year	A (W/m^2)	B	C
Jan. 21	21	1230	0.142	0.058
Feb. 21	52	1215	0.144	0.060
Mar. 21	80	1186	0.156	0.071
Apr. 21	111	1136	0.180	0.097
May 21	141	1104	0.196	0.121
June 21	172	1088	0.205	0.134
July 21	202	1085	0.207	0.136
Aug. 21	233	1107	0.201	0.122
Sept. 21	264	1152	0.177	0.092
Oct. 21	294	1193	0.160	0.073
Nov. 21	325	1221	0.149	0.063
Dec. 21	355	1234	0.142	0.057

For the Southern Hemisphere: to determine A, divide the A value by I_0 on the date of interest, and then multiply it by I_0 for half a year later (I_0 is computed using Eq. (7.1)); similarly, for B and C, use the values six months after the date of interest [6].

The total radiation flux, also referred to as the global irradiance, is then

$$I_{Tot} = I_D + I_{DS} + I_{DR} \tag{7.24}$$

However, for CSP, only the direct component is of importance as the mirrored surfaces are positioned to redirect and concentrate the direct beam sunlight; and at the mean sun–earth distance, the sun subtends an angle of 32 min. The intensity of the direct beam, as compared to the diffuse component, is shown in Fig. 7.9 for a collector tilted 33.4° (at latitude) in Phoenix, Arizona, where the direct beam constitutes 83% of the 7.1 kWh/m^2/day of energy received that day. Further illustrated in the graph is the increased energy collection made possible by tracking the sunlight, which in this case increases the total irradiance to 10.6 kWh/m^2/day.

Small variations in tilt from horizontal (±10°) and orientation from south (±20°) do not reduce performance significantly [13]. The overall annual incident energy for tilts ranging from horizontal (0°) to vertical (90°) and collector azimuth angles from south facing to west facing are presented in Fig. 7.10 for a latitude of 33.4°.

In some localities electricity pricing differs during the day, that is, higher rates are charged during peak usage periods. Whereas maximum energy production occurs for panels oriented directly south at a fixed angle near latitude, maximum economic return depends on the electricity rate structure, which may include different charges due to seasonal and time-of-use energy demand. Under such circumstances, fixed PV panels may be more optimally oriented to maximize power production during on-peak times [20]. The total annual return for electricity generation is computed from

$$R = A \sum_{i=1}^{365} \int \eta_i(t)\, I_i(t)\, r_i(t)\, dt \tag{7.25}$$

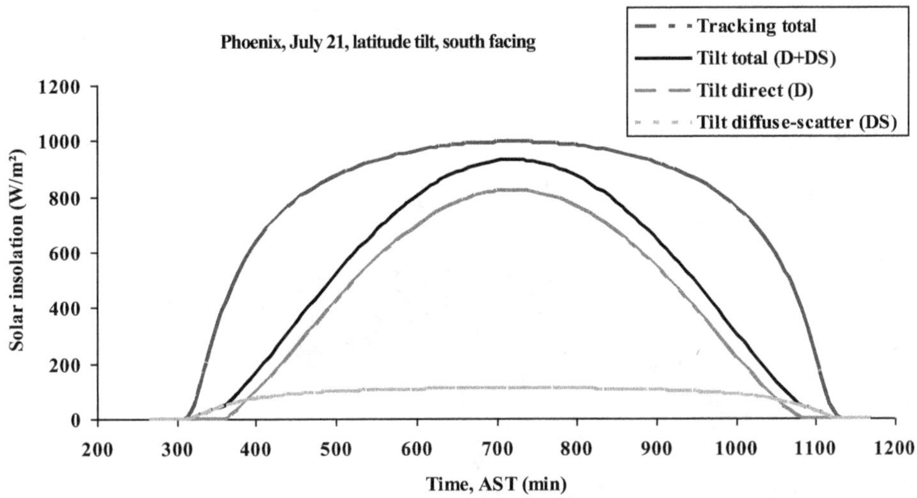

Figure 7.9. Solar irradiance for a summer day for a collector tilted at 33.4° compared to a panel utilizing dual-axis tracking. For the tilted panel, early in the morning and late in the evening, the total irradiance is from the diffuse-scattered component, but during the remainder of the daytime, the direct component provides the majority of the insolation.

where A is the active panel area; η is the overall conversion efficiency of the PV system; $I_i(t)$ and $r_i(t)$ are the solar insolation and value of the electricity, respectively, for the i-th day of the year at time t; and the integration is carried out over the hours in which the panel is illuminated.

Not all the incident solar insolation is absorbed by a particular body. The emissivity (ε) quantifies the energy absorption

$$E_{abs} = \varepsilon_{surf} I_{Tot} \tag{7.26}$$

Representative surface ε at solar energies [17] include

- asphalt pavement: $\varepsilon = 0.82 - 0.88$
- white (flat) paint: $\varepsilon = 0.12 - 0.25$
- red brick, tile, concrete: $\varepsilon = 0.65 - 0.77$

7.7 Comparison to Measured Data

The solar energy received is also affected by varying atmospheric conditions. For example, weather conditions—such as rain, wind, pressure, temperature, humidity, and pollution—inhibit the solar to energy conversion processes. Therefore, the availability of long-term measured data can also be useful; however, such data are typically acquired for a limited number of collector orientations.

For instance, direct beam, diffused, and total radiation on a horizontal surface, along with meteorological data for various locations in the United States and its territories, are available from the National Solar Radiation Database (NSRDB) maintained by the National Renewable Energy Laboratory. Using these data, the incident radiation for flat-plate collectors at different orientations can be calculated using Eqs. (7.12) and (7.21)–(7.23). Those

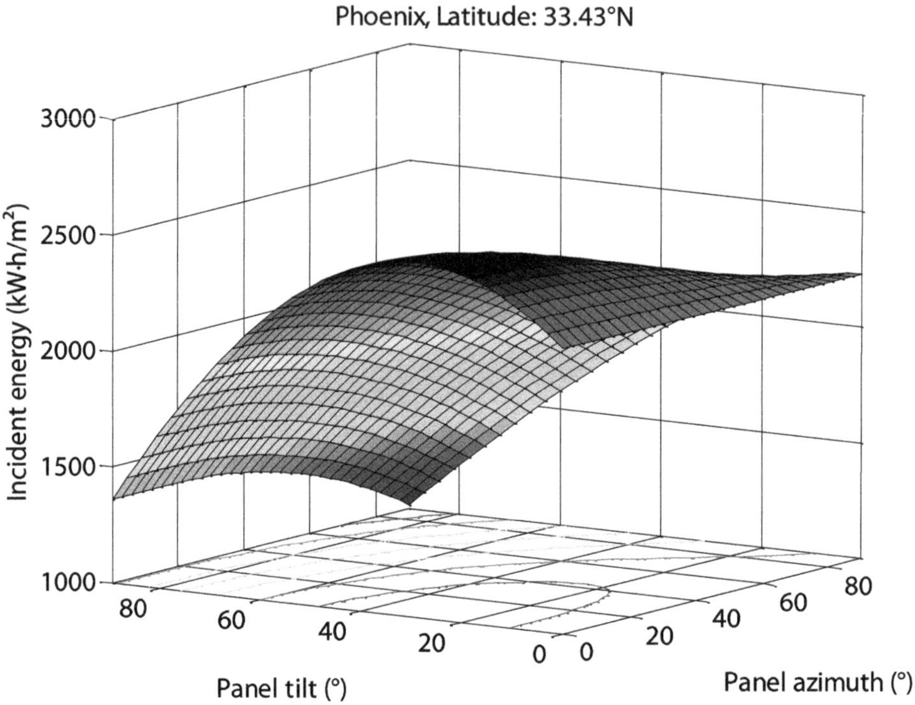

Figure 7.10. Annual incident energy as a function of collector position. The maximum of 2528 kWh/m^2 occurs at a tilt of 29° and azimuth of 0°.

NREL data for Phoenix, Arizona, from a TMY2 (Typical Meteorological Year) model are plotted in Fig. 7.11 for a fixed tilt of 0° (horizontal) and 90° (vertical). The annual average solar radiation for horizontal and south-facing vertical flat-plate collectors is 5.8 and 4.0 kWh/m^2/day, respectively. The graph also shows the solar radiation for the more optimal configuration of a south-facing tilt at latitude (33.43°), which gathers an annual average of 6.6 kWh/m^2/day. Figure 7.11 also demonstrates that by using either single or dual-axis tracking, further increases in the total annual energy collection are achieved. In particular, non-tilted single-axis and dual-axis trackers received 7.5 and 9.1 kWh/m^2/day, respectively. Such measured data may be used to refine analytical predictions in situations where measured data are unavailable for unique collector geometries.

To enhance the data quality for modeled ground sites, and for locations without available ground measured or modeled radiation data, satellite imagery-based irradiance models can prove helpful. Satellites provide multispectral information about the earth's atmosphere and cloud cover, typically for a large area, with high geographic and time resolution [22]. Data for ground-irradiance modelling are typically derived from geosynchronous satellites like the U.S. Geostationary Operational Environmental Satellite (GOES) series, the European METEOSAT series, and the Japanese Multifunctional Transport Satellite (MTSAT) [23]. Validation and comparison of satellite irradiance models show that these data sets are

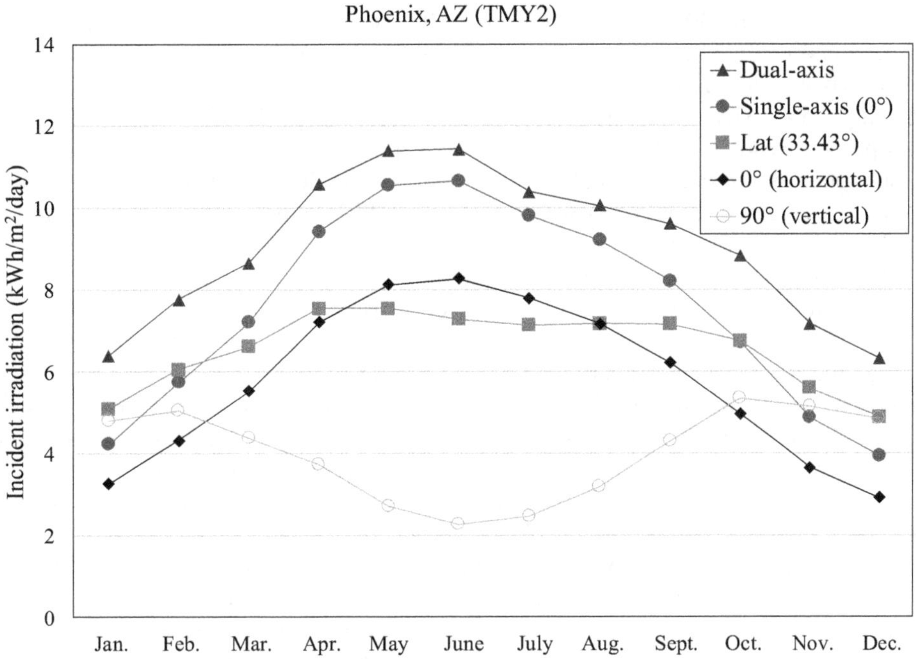

Figure 7.11. Solar radiation for flat-plate collectors facing south at various tilts. Data are from [21].

becoming more accurate and that differences between models are becoming smaller [22]. Two broad methods are mainly used to derive ground-irradiance (direct and diffused) data from satellite images: empirical and physical [23].

The traditional, simpler empirical models use relationships between site-specific ground measurements and satellite measurements. Most of these empirical models depend on some form of an unitless clearness index, defined as the ratio of global irradiance (I_{Tot}, Eq. (7.24)) on a horizontal surface to the apparent solar radiation (I_0, Eq. (7.1)) on an extraterrestrial horizontal surface

$$k_t = \frac{I_{Tot}}{I_0 \cos(\theta)} \tag{7.27}$$

Satellite imagery is used to calculate a cloud index, which quantifies the radiation reflected back from the atmosphere. A high cloud index indicates that the ground is receiving a smaller portion of the solar radiation. Site-specific, empirical, or look-up table relations between the cloud index and clearness index are used to determine k_t. The global horizontal irradiance is then estimated from Eq. (7.27).

The amount of solar radiation reaching the earth is reduced due to absorption, scattering, and reflection from clouds and other aerosols in the atmosphere such as ozone, oxygen, water vapor, and carbon dioxide. This phenomenon described by physics-based, radiative-transfer models relate the radiation reaching the earth's surface to atmospheric properties retrieved from satellite data. Physical models are gaining popularity due to availability of fast computing hardware and software, and improved uncertainties in modelling ground irradiance. NREL's NSRDB (National Solar Radiation Database) [21] has a 4-km × 4-km gridded data set with 30-min resolution from 1998 to 2014 for North and South America and from 2000 to 2014 for India and Southeast Asia. The annual estimated uncertainty of global horizontal data is around 5%. This database is derived from the GOES measurements and using the Physical Solar Model (PSM) developed by NREL [23]. The PSM retrieves cloud properties such as cloud-mask, cloud-type, particle size, snow and surface albedo, aerosol optical depth and precipitable water vapor from multispectral (visible and infrared) satellite measurements, and other ancillary sources. PSM then calculates the global horizontal and direct normal irradiance at a given ground site using a radiative-transfer model called FARMS (Fast All-sky Radiation Model for Solar applications).

Solar radiation data, historical or predicted, measured or modeled, are characterized by atmospheric conditions as explained above. The irradiance can change rapidly due to passing clouds. Statistical properties of clearness index defined in Eq. (7.27) can be used to characterize sky conditions [24]. The daily mean of instantaneous clearness index is a measure of how much irradiance reaches the surface for that day. High values indicate sunny days, and low values indicate cloudy days. The daily standard deviation of the instantaneous clearness index is a measure of irradiance variations due to cloud cover. A high standard deviation is indicative of fast moving clouds. Hence, these and other statistical metrics can be used to classify sky categories as cloudy, partly cloudy, sunny, and clear [25]. NREL [24] uses 1-min clearness index data mean and standard deviation to group irradiance variability into five classes: I to V. Classes I, II, and III have low variability (clearer skies), whereas Classes IV and V have ramps in clearness index that indicate cloudy days.

Solar radiation measurements are believed to be amongst the most uncertain in any measurement discipline [23]. Models (empirical or physical), in addition to model inaccuracies, are limited by measurement uncertainties. Measurement uncertainty analysis is typically based on GUM (Guide to Uncertainty in Measurements) [26]. Measurements approximate physical quantities; and uncertainty analysis, which considers every element (and step) of the measurement chain, provides a quantitative evaluation of measurement accuracy. In GUM, uncertainty sources are classified into random and bias errors. The total uncertainty is the square-root of sum of squares value of the random and bias errors. Sources of uncertainty in irradiance measurements include calibration, sensor spectral response, sensor bandwidth, sensor cosine response, sensor accuracy, sensor linearity, sensor stability, sensor orientation, solar zenith angle, data acquisition precision, bias, and accuracy, external interference (such as electromagnetic interference), and operation/maintenance (such as regular sensor cleaning) of the measurement system. A high-quality, well-maintained, ground-based irradiance measurement system can achieve uncertainties below 5% [26]. NREL's NSRDB satellite model has about 10% uncertainty in the annual global horizontal irradiance.

7.8 PV Energy Conversion

The electrical output of PV modules depends on the electrical, thermal, solar spectral, and optical properties of the module (or array) as well as the angle and amount of incident radiation [27]. The amount of input radiation has a linear (direct) effect on the current output of PV modules. The effect of variation in the solar spectrum on module output is relatively small for air mass values between 1 and 2 for crystalline PV modules. Empirical methods are available to estimate the effect of changing solar spectrum on PV module currents.

For surfaces not normal to the incident sunlight, the density of incident radiation is reduced, due to geometry, by a factor equal to the cosine of the angle of incidence as shown in Fig. 7.12 (see Eq. (7.21)).

In addition to this typical cosine loss of input radiation, large collector angles lead to optical losses due to reflectance from the surface of PV modules. These optical losses result in a lower incident radiation on the PV cell under the glass, than what is suggested in Eq. (7.21). For crystalline silicon PV modules, the effect of optical losses is significant for collector angles greater than 55° [27]. The effect of collector angle on PV module current is given by a fifth-order curve fit:

$$f_2(\theta) = b_0 + b_1\theta + b_2\theta^2 + b_3\theta^3 + b_4\theta^4 + b_5\theta^5 \tag{7.28}$$

where the coefficients b_0 to b_5 are determined using regression analysis on field data [28]. For a monocrystalline silicon module, the response of collector angle on module current is shown in Fig. 7.13.

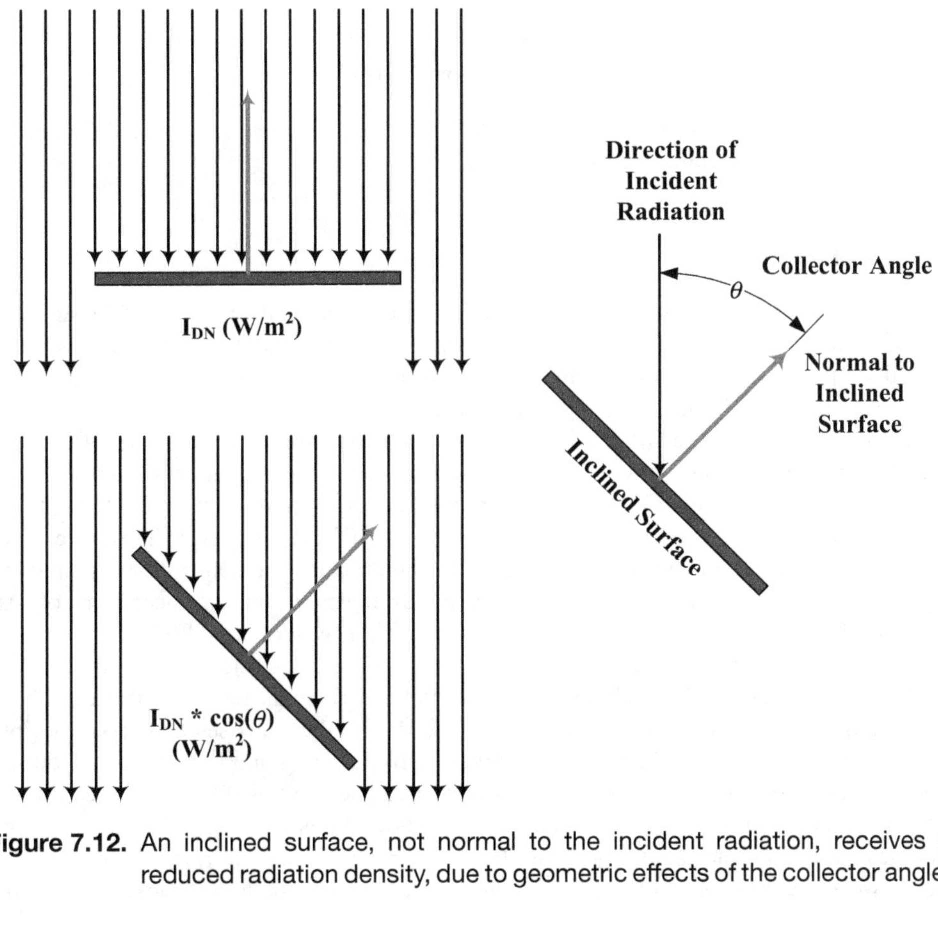

Figure 7.12. An inclined surface, not normal to the incident radiation, receives a reduced radiation density, due to geometric effects of the collector angle.

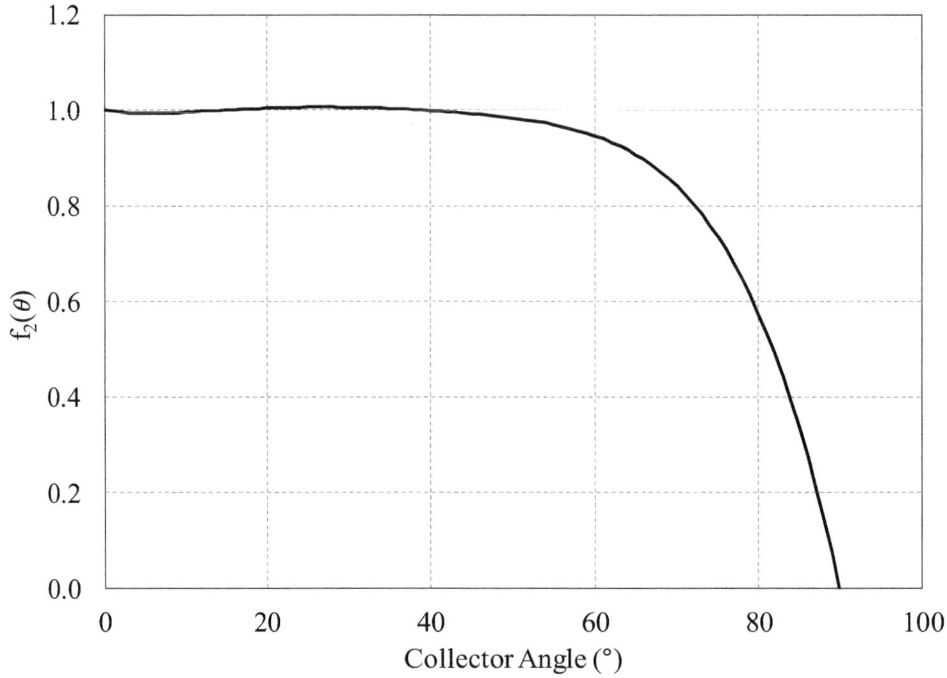

Figure 7.13. Collector angle response $f_2(\theta)$ for a monocrystalline PV module. Regression coefficients are from [28].

7.9 Conclusion

This chapter has provided straightforward methods to determine the position of the sun with respect to an earth-based solar collector. Using measured data that incorporate atmospheric conditions provides coefficients to the defining formulae. The calculations and measured data can be used for design, shading, and tracking.

Such algorithms can be utilized for studies of solar collector performance and for active solar-tracking equipment. For concentrating solar systems, tracking accuracies better than 0.5° are typically required. Flat-plate, single-axis tracking systems can follow the optimal solar position within 5° with less than 1% loss in annual output energy. There are more precise, and correspondingly more complex, computational approaches. For example, an algorithm to compute the solar zenith and azimuth angles within an uncertainty of just ±0.0003° for Julian calendar years from −2000 to 6000 is described in [29].

Another application of the solar position and shading calculations is backtracking in single-axis trackers — PV modules rotated around a horizontal, north–south axis. These modules are rotated from east to west during the day as the sun rises and sets. Multiple north–south rows are placed next to each other. Early morning and late afternoon, during low sun elevations, the modules can be shaded from adjacent rows if they are tilted toward the sun. In order to avoid row–row shading, backtracking is implemented. At dawn, the modules start in the horizontal position. Early morning, as the sun rises in the east and moves west, the modules move east from horizontal. Hence, the term "backtracking." As the sun's elevation increases, the modules then move west with the sun. Late afternoon, as the sun sets in the west, the modules "backtrack" again toward the east and end up in the horizontal position.

7.10 References

1. G. Zhao, H. Kozuka, H. Lin, and T. Yoko, "Solar energy conversion" in *Wiley Encyclopedia of Electrical and Electronics* Engineering, J. Webster (ed.), pp. 638–649; John Wiley & Sons, New York, 1999.
2. M. M. El-Wakil, *Powerplant Technology,* McGraw-Hill, New York, 1984.
3. M. Pidwirny, *Fundamentals of Physical Geography,* 2nd ed. Ch. 6, University of British Columbia, Okanagan, 2006. Available online: http://www.physicalgeography.net/fundamentals/contents.html.
4. OK-FIRST Project, "Seasons," *Oklahoma Climatological Survey*, Norman, OK, 1997. Available online: http://okfirst.mesonet.org/train/meteorology/Seasons.html.
5. D. R. Williams, "Earth Fact Sheet," National Space Science Data Center, NASA Goddard Space Flight Center, Greenbelt, MD, 2001. Available online: http://nssdc.gsfc.nasa.gov/planetary/fact sheet/earthfact.html.
6. P. J. Lunde, *Solar Thermal Engineering: Space Heating and Hot Water Systems*, John Wiley & Sons, New York, pp. 62–100, 1980.
7. ASTM E490-00a, *Standard Solar Constant and Zero Air Mass Solar Spectral Irradiance Tables*, ASTM International, West Conshohocken, PA, 2014.
8. NREL, "Solar Spectra," Renewable Resource Data Center, National Renewable Energy Laboratory (NREL), Golden, CO, 2020. Available online: http://rredc.nrel.gov/solar/spectra/.
9. C. Gueymard, "Simple Model of the Radiative Transfer of Sunshine," National Renewable Energy Laboratory (NREL), Golden, CO, 2001. Available online: http://www.nrel.gov/rredc/smarts/.
10. ASTM G173-03, *Standard Tables for Reference Solar Spectral Irradiances: Direct Normal and Hemispherical on 37° Tilted Surface*, ASTM International, West Conshohocken, PA, 2012.
11. I. R and A. Andreas, *Solar Position Algorithm for Solar Radiation Applications,* National Renewable Energy Laboratory, Technical Report, NREL/TP-560-34302, 2005.
12. M. Rymes, "Solar Position and Intensity," Renewable Resource Data Center, National Renewable Energy Laboratory (NREL), Golden, CO, 2001 Available online: http://rredc.nrel.gov/solar/codesandalgorithms/solpos/.
13. "Solar energy use," ASHRAE Handbook – HVAC Applications (SI), Ch. 33, American Society of Heating, Refrigeration, and Air-Conditioning Engineers (ASHRAE), Atlanta, GA, 2007.
14. A. Rabi, *Active Solar Collectors and Their Applications,* Oxford University Press, New York, p. 30, 1985.
15. S. A. Kalogirou, *Solar Energy Engineering: Processes and Systems*, 2nd ed. Elsevier, Amsterdam, Netherlands, 2014.
16. D. S. Chauhan, and S. K. Srivastava, *Non-Conventional Energy Resources*, 2nd ed. New Age International, New Delhi, p. 47, 2006.
17. A. W. Culp, *Principles of Energy Conversion*, 2nd ed. McGraw-Hill, New York, pp. 97–107, 1991.
18. B. Y. H. Liu, and R. C. Jordan, "Daily insolation on surfaces tilted toward the equator," *ASHRAE Journal*, vol. 3, no. 10, pp. 53–59, 1961.
19. S. A. Klein, "Calculation of monthly average insolation on tilted surfaces," *Solar Energy*, vol. 19, pp. 325–329, 1977.
20. K. E. Holbert, "An analysis of utility incentives for residential photovoltaic installations in Phoenix, Arizona," Proceeding Thirty-ninth North American Power Symposium, Las Cruces, NM, pp. 189–196, 2007.

21. NREL, National Solar Radiation Database (NSRDB), National Renewable Energy Laboratory, 2020. Available online: https://nsrdb.nrel.gov/index.php.
22. J. S. Stein, "PV performance modeling methods and practices, results from the 4th PV performance modeling collaborative workshop," International Energy Agency-Photovoltaics Power Systems Programme, Report IEA-PVPS T13-06, St. Ursen, Switzerland, 2017. Available online: https://iea-pvps.org/wp-content/uploads/2020/01/T13_Report_PV_Performance_Modeling_Methods_and_Practices_FINAL_March_2017.pdf
23. M. Sengupta, A. Habte, S. Kurtz, A. Dobos, S. Wilbert, E. Lorenz, T. Stoffel, D. Renné, C. Gueymard, D. Myers, S. Wilcox, P. Blanc, and R. Perez, *Best Practices Handbook for the Collection and Use of Solar Resource Data for Solar Energy Applications,* National Renewable Energy Laboratory, Technical Report, NREL/TP-5D00-63112, 2015.
24. M. Hummon, E. Ibanez, G. Brinkman, and D. Lew, "Sub-Hour Solar Data for Power System Modeling From Static Spatial Variability Analysis," National Renewable Energy Laboratory, NREL/CP-6A20-56204, Golden, Colorado, 2012.
25. A. Colli, D. Pavanello, W. J. Zaaiman, J. Heiser, and S. Smith, "Statistical analysis of weather conditions based on the clearness index and correlation with meteorological variables," *International Journal of Sustainable Energy*, vol. 35, no. 6, 523–536, 2016.
26. Working Group 1 of the Joint Committee for Guides in Metrology, "Evaluation of measurement data — Guide to the expression of uncertainty in measurement," BIPM. Bureau International des Poids et Mesures International Bureau of Weights and Measures; France, Report JCGM 100: 2008, First Edition, Sept. 2008.
27. D. L. King, W. E. Boyson, and J. A. Kratochvil, *Photovoltaic Array Performance Model,* Sandia National Laboratories, Technical Report, SAND2004-3535, 2004.
28. A. H. Fanney, M. W. Davis, B. P. Dougherty, D. L. King, W. E. Boyson, and J. A. Kratochvil, "Comparison of photovoltaic module measurements," International Solar Energy Conference ISEC2005, Orlando, Florida, Aug. 2005.
29. I. Reda, and A. Andreas, "Solar position algorithm for solar radiation applications," *Solar Energy*, vol. 76, pp. 577–589, 2004; Corrigendum in vol. 81, p. 838, 2007.

CHAPTER 8

Parameter Estimation of Solar Photovoltaic and Impact of Environmental Conditions on Its Performance

Abhishek Chauhan and Surya Prakash

Department of Electrical and Instrumentation Engineering, Thapar Institute of Engineering and Technology, Patiala, Punjab, India. abhishek.chauhan@thapar.edu, surya.prakash@thapar.edu

Abstract This chapter elaborates the importance of parameter estimation of solar photovoltaic (PV) cell and its impact on the performance characteristics of solar PV array. Moreover, various equivalent circuits are rigorously analyzed on the basis of their complexities and the number of unknown parameters that are to be estimated. The literature on parameter estimation is also critically reviewed and classified on the basis of techniques used, that is, analytical, metaheuristic, and hybrid objective function considered and equivalent circuits. Furthermore, performance of 200-W KC200GT solar PV module under different environmental conditions at two different locations, that is, L_1 and L_2, are critically studied, and on the basis of their *I-V* and *P-V* characteristics, it is perceived that the performance of solar PV is significantly dependent on the environmental conditions, where the module under low-temperature condition perform well with appreciable margins.

Keywords: Parameter estimation, solar PV, optimisation techniques, renewable energy sources.

8.1 Introduction

The electrical power is one of the most emergent forms of energy that have been studied since antiquity. At the most rudimentary level, electrical power was used for lighting purpose only, but industrialization in the eighteenth century brands this magical force as a vital part of human life. Moreover, socioeconomic fabric of a country is also stitched with the thread of its per capita consumption of power. Hence, rapid expansion in limitless set of applications, which include heating, lighting, communication, transport, computation, and to keep economic competition among countries at its swing, leads the countries toward the identification of various modes of harnessing electrical power. Where, exploitation of hydrocarbon fuels is most prominently done. As a consequence, hydrocarbon-based fuels exhibit their toxic behavior on environment and on human beings, which stimulate communities to identify alternate sources for harnessing power, that is, renewable sources of power generation. Where, direct conversion of solar energy to usable electrical power, zero white space technology, and customized range keep the solar photovoltaic (PV) systems at top in the hierarchy among all renewable sources of power generation.

As analysis of PV system requires simulation models prior to practical implementation [1–4], where simulation model results are highly dependent on the parameter of PV model that are given in manufacturer's datasheets. Solar PV is articulated in electrical circuits with various electrical parameters whose precise estimation leads toward the defined assessment of maximum power point tracking (MPPT), efficiency monitoring, and performance characteristic evaluation [3–5]. The prediction of the electric circuit parameters of solar PV, such as light-induced current, diode dark saturation current, diode ideality factor, series resistance, and shunt resistance, is labelled as parameter estimation of solar PV [5–12]. In this chapter, the basic building block of solar PV has been discussed and impact of environmental conditions on power–voltage (*P-V*) and voltage–current (*V-I*) characteristics are analyzed under different sets of temperatures and irradiance at different altitude locations. Moreover, various methods of parameter estimation using different models of solar PV in literature are also analyzed.

8.2 Solar PV Cell Models and its Physics

Solar PV cells behavior has been explained as of simple *p–n* junction when it is exposed to sunlight [1,3].

On the basis of quantum theory, it is elucidated that the electrons in the conduction band are accountable for the current flow, where in semiconductor materials the conduction band is empty at absolute zero temperature, moreover only one out of 10^{10} electrons in silicon exist in conduction band at room temperature [1,3].

When *p–n* junction is exposed to the source of photons, that is, sunlight, they are absorbed and hole–electron pair may be formed (as photons have sufficient energy to strike electron from an atom). When mobile charges reach in the vicinity of the junction, the holes in the depletion region will push the holes into *p*-side and push the electrons into *n*-side due to the electric field of the depletion region as shown in Fig. 8.1. The *p*-side accumulates holes and the *n*-side accumulates electron, which creates a voltage that can be used to deliver current to a load. If electrical contacts made to the top and bottom of the cell, the electrons flow out of the *n*-side into the connecting wire, through the load and back to the *p*-side [1]. When the electrons reach the *p*-side, they recombine with holes and completing the circuit as shown in Fig. 8.2.

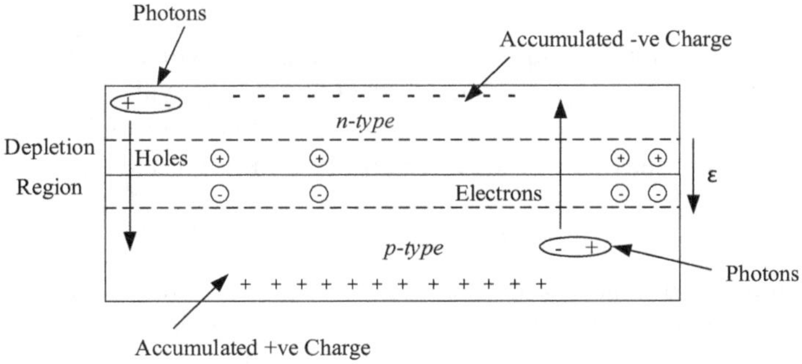

Figure 8.1. Charge accumuation in *p–n* junction diode [1].

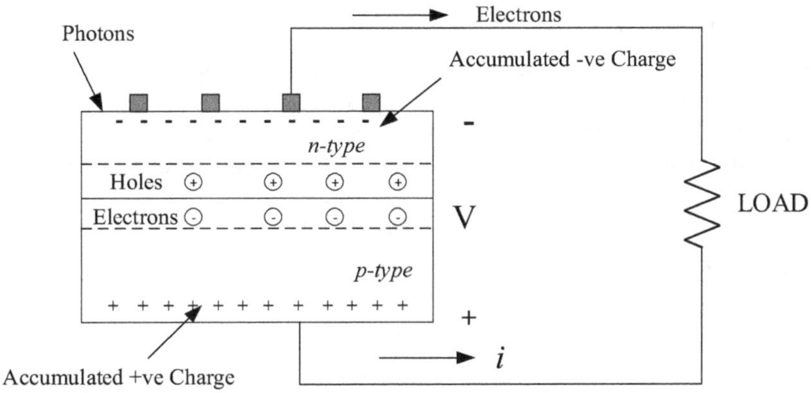

Figure 8.2. Flow of electrons and conventional current.

8.2.1 Ideal PV Cell Model

PV cell constitutes two layers of differently doped semiconductor, whose p–n junction is exposed to photons [1–12] as shown in Fig. 8.1. In the absence of solar irradiation, the PV cell behaves as a simple *p–n* junction diode whose *I-V* curve is expressed as Shockley equation [1–12]. The incident photons create potential difference and that results in the flowing of charge carriers toward external circuit resulting in a current, that is, denoted by "*i*." Ideal PV cell model is depicted in Fig. 8.3.

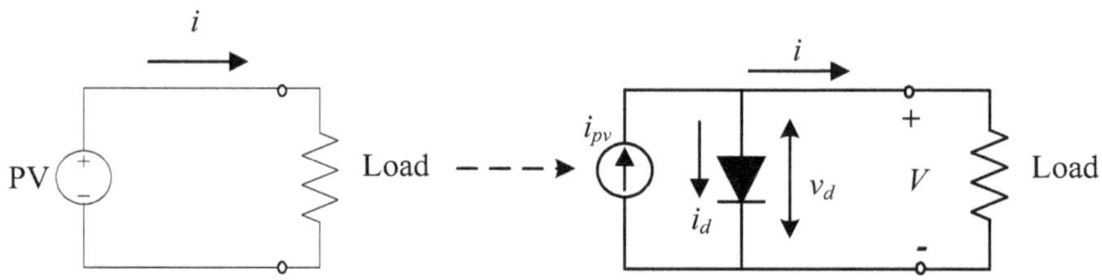

Figure 8.3. Ideal PV cell model.

$$i_d = i_{sat}\left(e^{qv_d AkT} - 1\right) \tag{8.1}$$

The expression for diode current is observed in Eq. 8.1. Where v_d and i_d are the diode voltage and current, respectively, saturation current of diode i_{sat} with an ideality factor of *A*, *q* is the electric charge of an electron, *T* is the temperature in Kelvin, and *K* is the Boltzmann constant. The *I-V* characteristic of PV cell can be drawn using Eq. 8.2.

$$i = i_{pv} - i_d = i_{pv} - i_{sat}\left(e^{qv_d AkT} - 1\right) \tag{8.2}$$

At short-circuit condition, $v_d = 0$, which results $i = i_{pv} = i_{sc}$, whereas during open-circuit condition, $v_d = V_{oc}$ and $i = 0$ and on solving Eq. 8.2 for open circuit, V_{oc} explains as

$$V_{oc} = \frac{AkT}{q}\text{In}\left(\frac{i_{sc}}{i_{sat}} + 1\right) \tag{8.3}$$

8.2.2 Single-Diode R_s Model

The R_s model of solar PV is more demonstrative if it is compared with ideal PV cell, as R_s model encompasses an important fact as R_s, that is, the contact resistance of silicon and connection lead [1,3] that holds the ability to contribute substantial changes in PV characteristic if remained unaddressed. A single-diode R_s model is represented in Fig. 8.4.

$$V = v_d + iR_s \tag{8.4}$$

Current *i* for R_s PV model is expresses as

$$i = i_{pv} - i_{sat}\left(e^{\left(\frac{V + R_s i}{AkT}\right)} - 1\right). \tag{8.5}$$

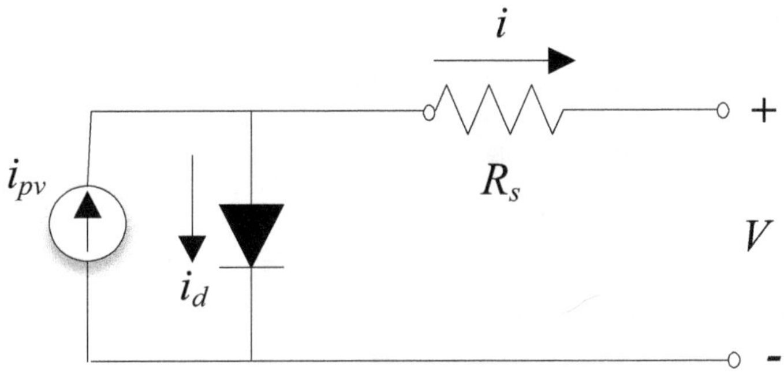

Figure 8.4. Single-diode R_s PV model.

8.2.3 Single-Diode R_p Model

It is important to understand that the generated photocurrent is significantly very sensitive to the presence of light, where under shading condition, the current through the current source becomes zero and the diode is reverse biased. During this only reverse saturation current passes, hence a simple equivalent circuit gives no power output to the load if any one of the cells of a module is shaded [1], as shown in Fig. 8.5. Therefore, to counter this, equivalent circuit is modified by using a parallel leakage resistance R_p, as shown in Fig. 8.6, and this showcases that current delivers to the load through diode and the R_p [1–12].

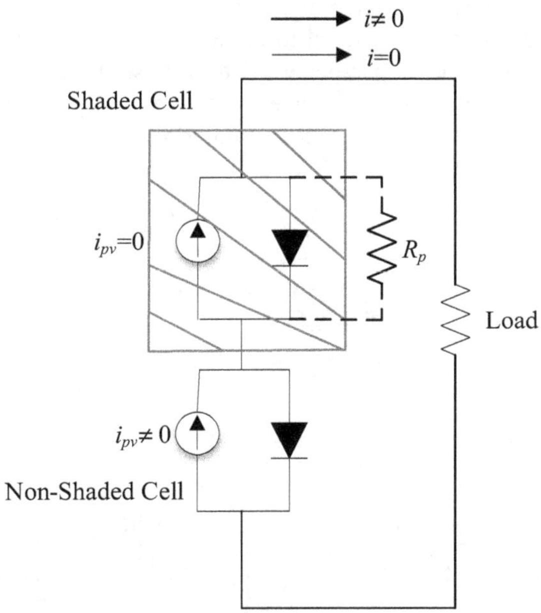

Figure 8.5. Shaded cell of a module.

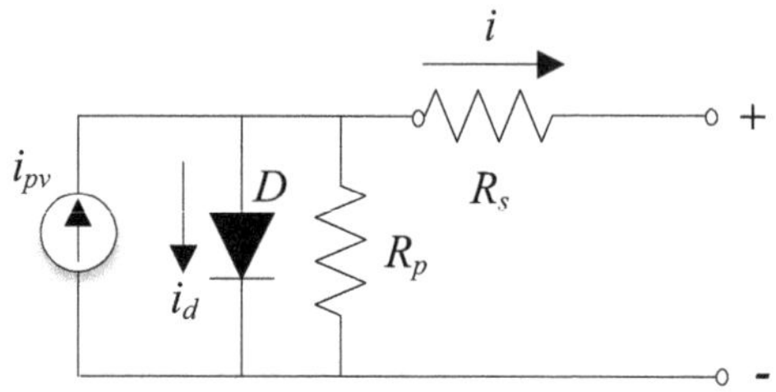

Figure 8.6. Single-diode R_s and R_p PV model.

Suitable transaction among precision and simplicity results in the R_p being considered the most commonly used PV model. The five unknown parameters that are to be estimated prior simulation by using single-diode R_p model are i_{pv}, i_{sat}, R_s, R_p, and A [1–16]. Whereas R_s model has four unknown parameters that excludes R_p. The current to the load can be expressed as

$$i = i_{pv} - i_{sat}\left(e^{\left(\frac{q(v_d + R_s i)}{AkT}\right)} - 1\right) - \frac{V + R_s i}{R_p}. \tag{8.6}$$

8.2.4 Two-Diode PV Model

In order to consider the effect of recombination current loss in the depletion region [1–16], another diode D_2 is connected in parallel to the R_p model. As the number of diode sets increase, the number of unknown parameters also increase. Hence, this two-diode model has seven unknown parameters, which include five single-diode R_p model parameters along with two additional parameters, that is, diode saturation current (i_{sat2}) and ideality factor (A_2) for D_2 [2–16] (Fig. 8.7).

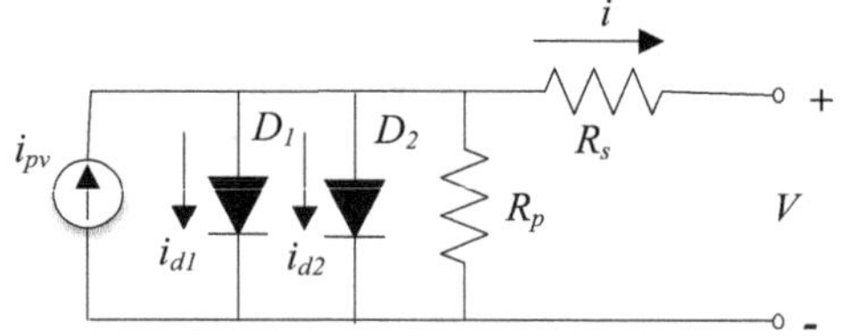

Figure 8.7. Two-diode PV cell model.

$$i = i_{pv} - i_{sat1}\left(e^{\left(\frac{q(V + iR_s)}{A_1 kT}\right)} - 1\right) - i_{sat2}\left(e^{\left(\frac{q(V + iR_s)}{A_2 kT}\right)} - 1\right) - \frac{V + iR_s}{R_p} \tag{8.7}$$

8.2.5 Junction Capacitance Single-Diode R_p Model

This model is the least considered model for the analysis and simulation of solar-based power systems. This model includes a capacitance that considers the *p–n* junction capacitance, that is, capacitance of the depletion region [17].

This model is shown in Fig. 8.8. Junction capacitance model has six unknown parameters, five as of R_p model and C_{jc}. It is illustrated that the capacitance can be estimated by using

$$C_{jc} = \frac{\Delta t_i \times \overline{i_{jc}}}{\Delta V_c}. \tag{8.8}$$

where Δt_i is the time period in μsec when load is instantaneously connected and for the fraction of time the voltage remains unchanged, $\overline{i_{jc}}$ is the mean current flown over the capacitor throughout transient, and ΔV_c is the change in voltage from steady-state charge and discharge *p–n* junction capacitance C_{jc}.

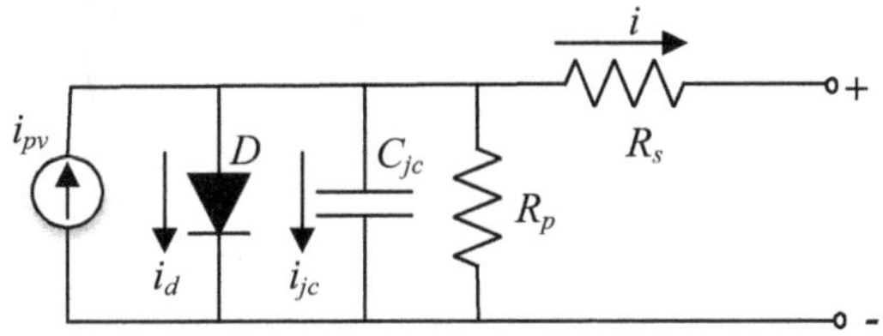

Figure 8.8. Single-diode junction capacitance, R_s and R_p PV model.

8.2.6 Selection of R_p and R_s

On considering the equivalent circuit with R_p, Eq. 8.9 showcase that for a particular voltage, the R_p results in a drop in load current by a fraction V/R_p, also illustrated in Fig. 8.9.

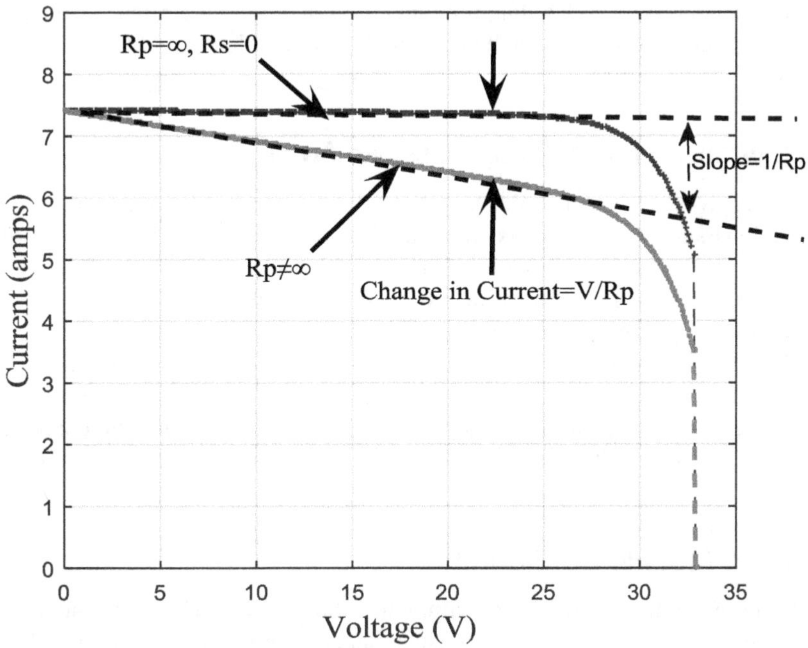

Figure 8.9. Drop in load current by a function V/R_p.

$$i = \left(i_{pv} - i_d\right) - \frac{V}{R_p}. \tag{8.9}$$

If a cell has a loss of only 1% due to R_p, then $R_p > 100\ V_{oc}/I_{sc}$, for cells having $I_{sc} = 7$ A and $V_{oc} = 0.6$ V then the R_p should be greater than 9Ω.

Whereas, on considering R_s, load current expresses as

$$i = \left(i_{pv} - i_d\right) = i_{pv} - i_{sat}\left(e^{\left(\frac{qv_d}{AkT}\right)} - 1\right) \tag{8.10}$$

$$v_d = V - iR_s \tag{8.11}$$

$$i = i_{pv} - i_{sat}\left(e^{\left(\frac{q(V + R_s i)}{AkT}\right)} - 1\right). \tag{8.12}$$

Hence, from Eq. 8.12, it is evident that in order to keep "i" near to its rated the series resistance should be as small as possible. For instance, for a cell that has 1% of loss due to R_s, then, $R_s < 0.01V_{oc}/I_{sc}$, it means if $I_{sc} = 9$ A, $V_{oc} = 0.6$ V, then the R_s should be less than 0.0009Ω. So, it can be concluded that, for better current output at particular voltage, the R_p should have higher values, whereas the R_s should be as small as possible [1].

8.3 Configuration of Solar PV Modules and Array

As a solar cell gives an output of 0.5 V only, and this voltage level feeds only some limited applications such as calculators, watches. Whereas more feasible applications need customization of PV cells in various configurations. In view of this, solar PV modules act as the building block for PV applications. PV modules consist of prewired cells in series, cased in robust and weatherproof structure [1].

A module of 36 series cells is designated as 12-V module, whereas 72-cell modules are the most commonly used. A series combination of 72-cell configuration results in an output of 24 V, where two parallel strings of 36 cells result in 12-V output. The combination of modules for a particular voltage and current output is known as array as shown in Fig. 8.10.

8.3.1 Cell to Module Configuration

When PV cells are connected in series combination they carry same current and the voltages adds. As shown in Fig. 8.11, there are 36 cells that are connected in series and the voltage of the modules becomes 36 × 0.7 = 21.6 V, whereas the current remains almost same.

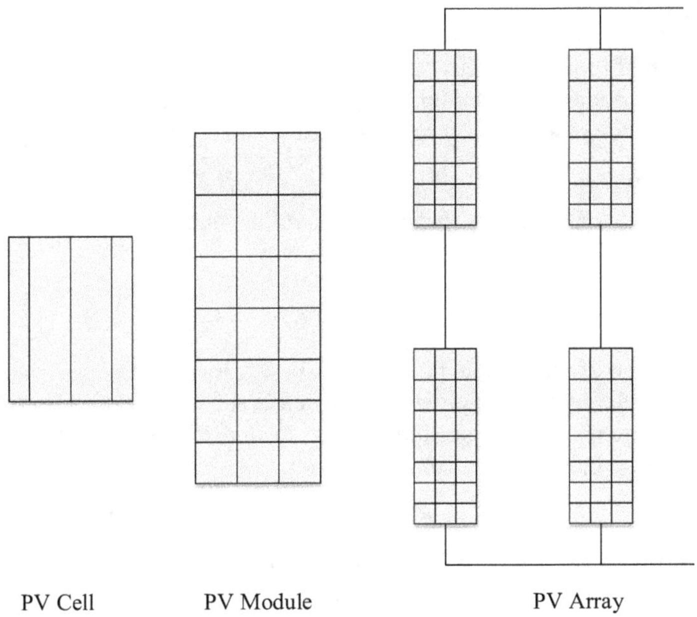

Figure 8.10. Solar PV cell, module, and array.

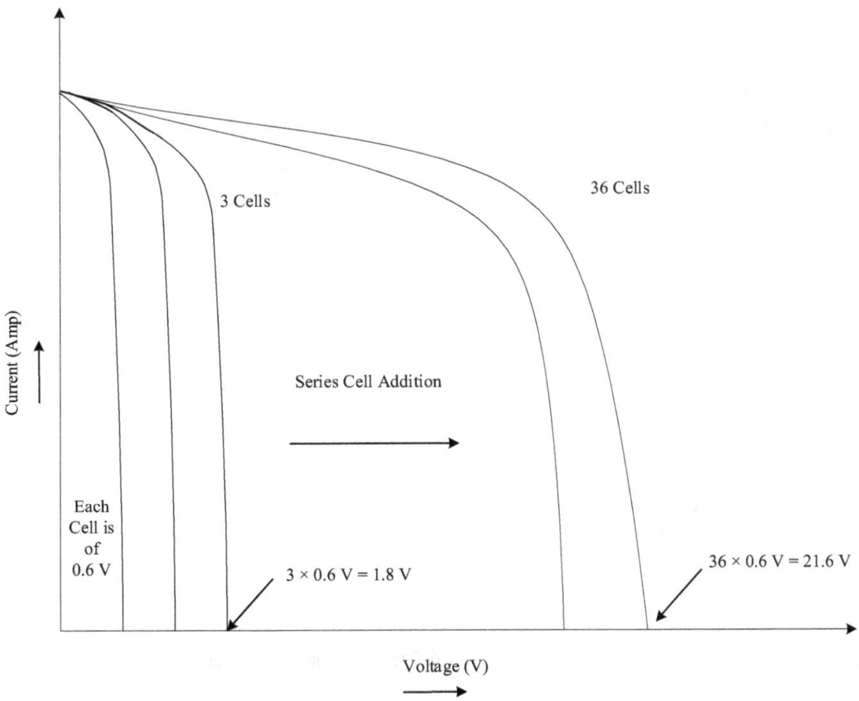

Figure 8.11. Addition of cell voltage for a 36-cell module [1].

Example 8.1

In the condition of full sun and at a current of 2.3 A, if 1 cell out of a 36-cell module having $R_p = 6\Omega$ each is shaded, what would be the new output voltage and power of the module? Also determine the voltage drop across the shaded cell if voltage of the module is 19.7 V.

Solution: By using

$\Delta V = \frac{V}{n} + IR_p$, where ΔV is the change in voltage and n is the number of cells of PV module.

$$\Delta V = \frac{19.7}{36} + 2.3 \times 6 = 14.34 \text{ V}$$

So, new voltage of the module is 19.7 – 14.34 = 5.36 V

Power developed by the module when one cell is shaded = $VI = 5.36 \times 2.3 = 12.32$ W.

As all the current 2.3 A passes through the R_s and R_p, hence voltage drop across the shaded cell, if $R_s = 0.008\Omega$, is calculated as

$$V_{cd} = I\left(R_s + R_p\right) = 2.3\,(0.008 + 6) = 13.81 \text{ V}.$$

Power dissipated across cell = $V_{cd} \times I = 31.763$ W

So, it is observed that shading of a single cell leads to a drop of 31.763 W of power across the cell that dissipated as heat across the cell.

Example 8.2

Consider a PV module having 72 identical cells wired in series. The sun insolation is 1000 W/m^2, each cell has $i_{pv} = 7$ A, and at STC $i_{sat} = 6 \times 10^{-10}$ A. If $R_p = 7.8\Omega$, $R_s = 0.008\Omega$, and $v_d = 0.5$ V then calculate the current, voltage, and power delivered by the module.

Solution: By using Eq. 8.6

$$i = i_{pv} - i_{sat}\left(e^{\left(\frac{q\left(V + R_s i\right)}{AkT}\right)} - 1\right) - \frac{V + R_s i}{R_p}.$$

At STC Eq. 20.6 becomes

$$i = i_{pv} - i_{sat}\left(e^{(38.9 v_d)} - 1\right) - \frac{v_d}{R_p} = 7 - 10^{-10}\left(e^{38.9 \times 0.50} - 1\right) - \frac{0.50}{7.8} = 6.76 \text{ A},$$

as $V + iR_s = v_d$,

so, $V_{module} = n\left(v_d - iR_s\right) = 72\,(0.50 - 6.76 \times 0.008) = 32.10$ V

$$P = V \times i = 32.10 \times 6.76 = 216.9 \text{ W}.$$

8.3.2 Module to Array Configuration

When the modules are connected in series/parallel combination, then the formation is known as array as shown in Fig. 8.12. It is also important to know that in order to increase the voltage the modules are connected in series, whereas for increasing current, the modules are to be connected in parallel. Moreover, for obtaining large powers from an array, the modules are connected with some series as well as parallel connections.

The *I-V* characteristics of series combination, added along the voltage axis and the total voltage is simple, the sum of the voltage of each module individually is as shown in Fig. 8.12.

Furthermore, the *I-V* characteristic for parallel combination of module move along the current axis and the total current of the array is the algebraic of individual current of each module as shown in Fig. 8.13.

The high-power configuration needs both series and parallel combination as explained above and can be classified into two arrangements:

(i) Modules are connected in series to form strings, and the strings are connected in parallel (Fig. 8.14 (a)).
(ii) Modules are connected in parallel to form strings, and the strings connected in series as elaborated (Fig. 8.14 (b)).

Configuration (i) of parallel strings are preferred, and if a string is removed for service, then the system still supply the same voltage with reduced current.

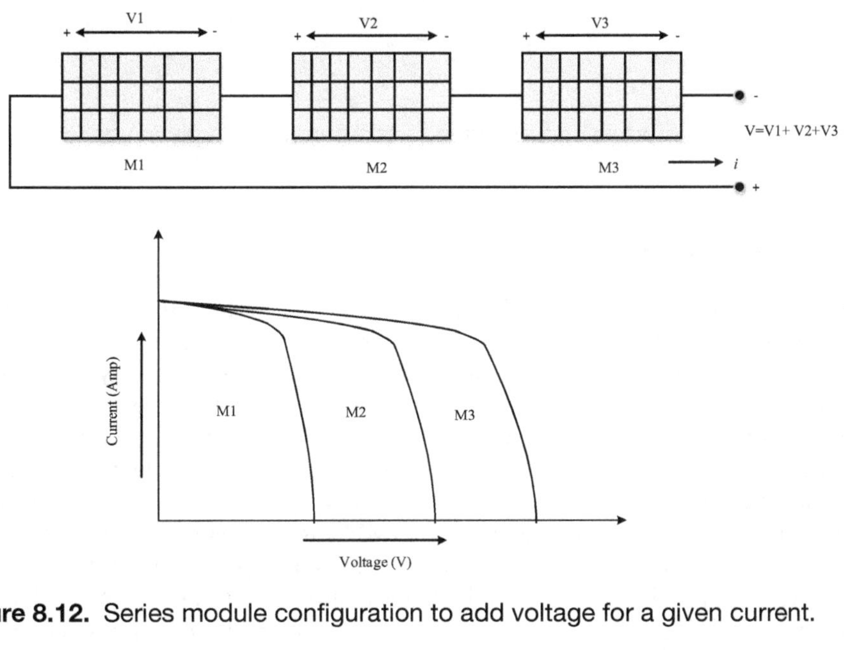

Figure 8.12. Series module configuration to add voltage for a given current.

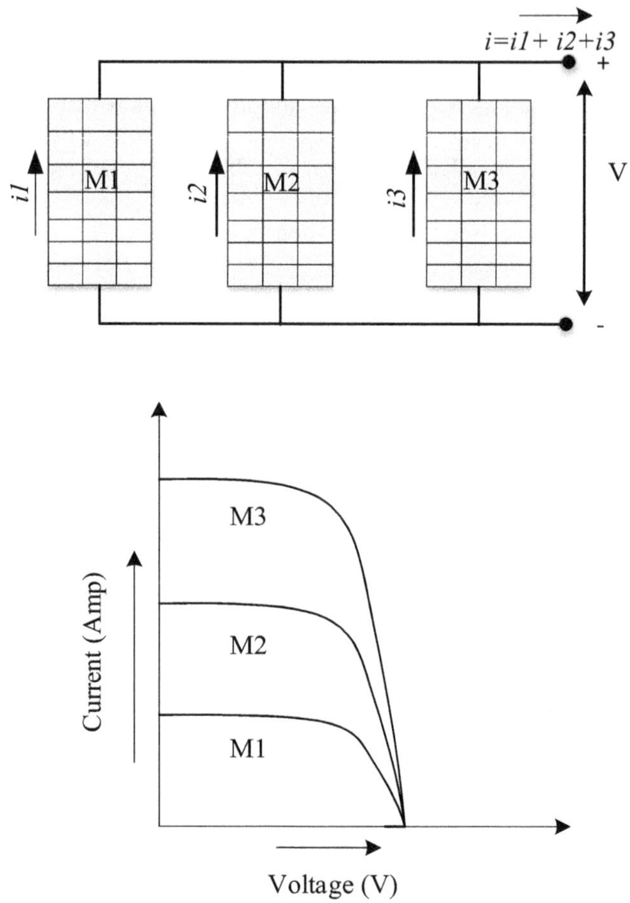

Figure 8.13. Parallel module configuration to add current for a given voltage.

8.4 The *I-V* and *P-V* Characteristic of Solar PV

Let a PV module be exposed to sunlight and no load is connected to the output terminals, then the voltage across the terminal is the open-circuit voltage, that is, V_{oc} only and the current remains zero. Whereas if the terminals are short circuited, then only i_{sc} flows and V_{oc} becomes zero, but as the power is the product of V and i, in both the cases the power remains zero or in other words the power delivered to the load remains zero as shown in Fig. 8.15. When the actual load is connected to the PV module, it starts extracting the current from the module for a particular voltage and results in a power delivered to the load. As shown in Fig. 8.15, the maximum power point (MPP) is the point where the product of voltage and current is maximum, the current and voltage corresponding to MPP also known as V_{mp} and I_{mp}, whereas in ideal test condition it is known as V_{rated} and I_{rated}.

In order to locate the MPP, the biggest possible rectangle fit under *I-V* curve can be considered. As shown in Fig. 8.16, the sides of the rectangle have V and I, hence the area under the rectangle illustrates the power [1].

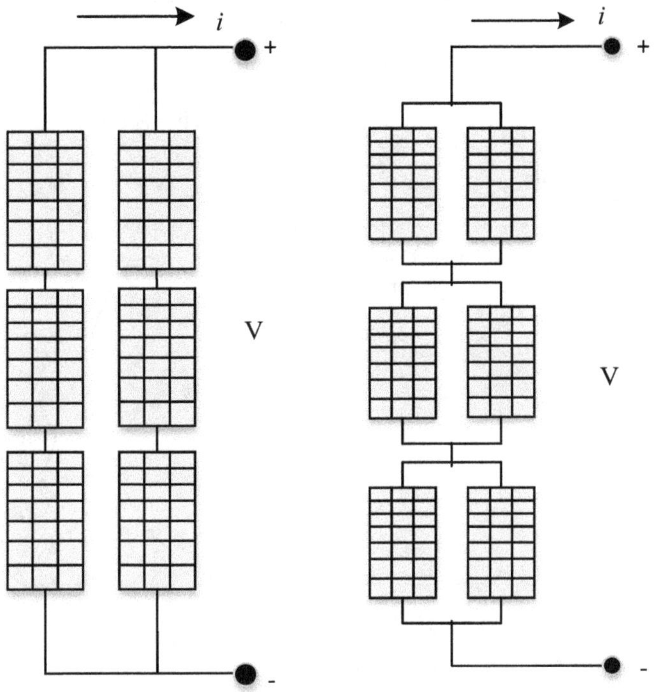

(a) (b)

Figure 8.14. Solar PV array configuration: (a) series module strings connected in parallel; (b) parallel module string connected inseries.

Moreover, fill factor is also used for module performance analysis where

$$\text{Fill factor} = \frac{V_{rated} \times I_{rated}}{V_{oc} \times I_{sc}} \tag{8.13}$$

8.5 Impact of Environmental Conditions on Performance Characteristics of PV

It is important to understand that PVs are very sensitive to the environmental changes. Temperature (T) and irradiance (G) are the environmental parameters that make substantial impact on the performance of solar PV. In this section, the impact of temperature and irradiance individually on the i_{pv} and V_{oc} is discussed. Moreover, the combined impact of T and G on i_{pv} and V_{oc} is also discussed. In Fig. 8.17, it is observed that with the increase in T the V_{oc} decreases, where a slight increase in i_{pv} is recorded. Hence, astonishingly it is observed that the modules/array perform better at low temperatures [1,18].

Whereas Fig. 8.18 illustrates that, with the increase in G the i_{pv} increases, and the slight change in V_{oc} is observed.

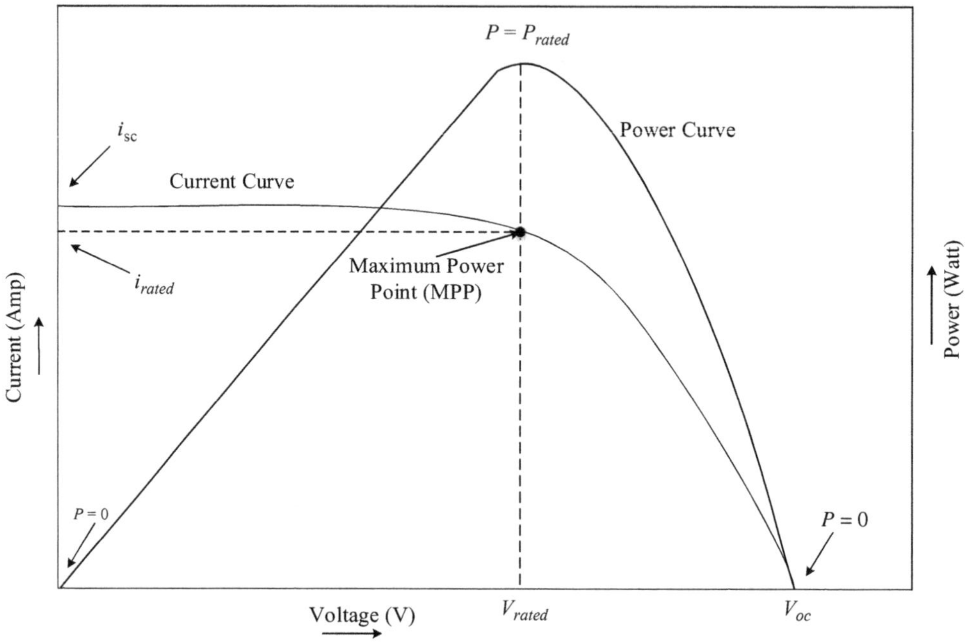

Figure 8.15. *I-V* and *P-V* curve for a PV module.

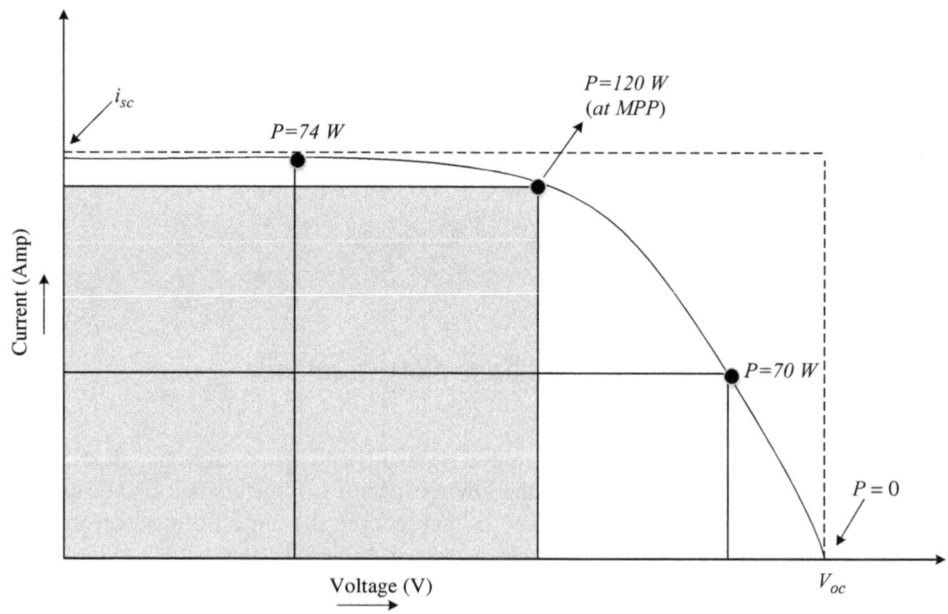

Figure 8.16. Fill factor.

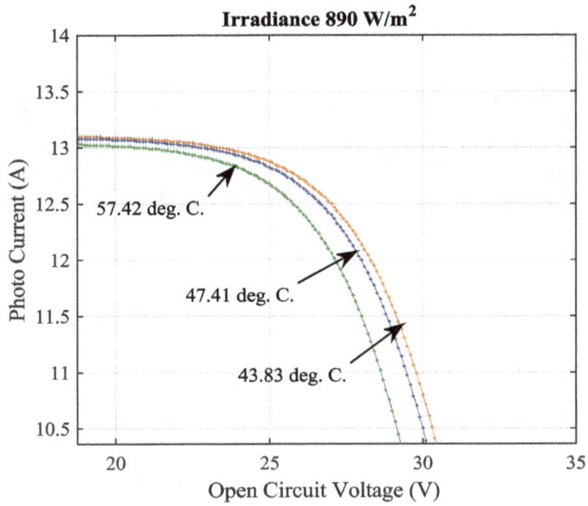

Figure 8.17. *I-V* Impact of change in temperature at fixed irradiance for KC200GT module.

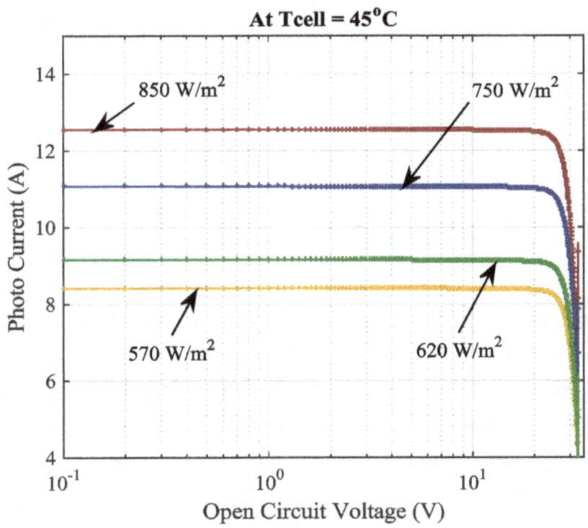

Figure 8.18. *I-V* Impact of change in irradiance at fixed temperature for KC200GT module.

The impact of environmental changes on i_{pv} and V_{oc} can be depicted in Eqs. 8.14 and 8.15, respectively [3–6].

$$i_{pv}(T, G) = (i_{pv,STC} + k_i(T - T_{STC}))\frac{G}{G_{STC}} \tag{8.14}$$

where $i_{pv,\ STC}$ T_{STC}, and G_{STC} represent photocurrent, temperature, and irradiance at standard test conditions (STCs). The STCs is defined as the condition having the temperature of 25°C with an irradiance of 1000 W/m^2 and airmass 1.5. The temperature coefficient of voltage and current are represented by k_v and k_i.

$$V_{oc}(T, G) = V_{oc,STC} + k_v(T - T_{STC}) + V_T \ln\left(\frac{G}{G_{STC}}\right) \quad (8.15)$$

Not only the i_{pv} and V_{oc} but also the parameters such as V_{mp}, I_{mp}, i_{sat}, R_s, R_p, and A are subtle to the environmental changes, as expressed in Eq. 8.16–8.21.

$$V_{mp}(T, G) = V_{mp,STC} + k_v(T - T_{STC}) + V_T \ln\left(\frac{G}{G_{STC}}\right) \quad (8.16)$$

$$I_{mp}(T, G) = (I_{mp,STC} + k_i(T - T_{STC}))\frac{G}{G_{STC}} \quad (8.17)$$

$$i_{sat}(T, G) = i_{sat,STC}\left(\frac{T}{T_{STC}}\right)^3 e^{\left[\frac{E_g}{K}\left(\frac{1}{T_{STC}} - \frac{1}{T}\right)\right]} \quad (8.18)$$

$$R_s(T, G) = R_{s,STC}\frac{T}{T_{STC}}\left(1 - 0.217 In\frac{G}{G_{STC}}\right) \quad (8.19)$$

$$R_p(G) = R_{p,STC}\left(\frac{G_{STC}}{G}\right) \quad (8.20)$$

$$A(T) = A_{STC}\left(\frac{T}{T_{STC}}\right) \quad (8.21)$$

The V_{mp} and I_{mp} are the voltage and current at maximum power, respectively, with V_T as thermal voltage of the diode. The $i_{sat,STC}$, $R_{s,STC}$, $R_{p,STC}$, G_{STC}, and A_{STC} are the diode's saturation current, series resistance, parallel resistance, irradiance, and diode's ideality factor at STC and E_g represents the energy gap for material used. It is important to know that, by each degree increase in cell temperature, V_{oc} of a crystalline silicon cell drops by 0.37%, and new V_{oc} can be calculated by using Eq. (8.22), whereas i_{pv} increases by 0.05%.

$$V_{oc(new)} = V_{oc(rated)}\left[1 - 0.0037(T_{cell} - 25)\right] \quad (8.22)$$

On the basis of ambient temperature (T_a), temperature of cell, that is, T_{cell}, is calculated by using,

$$T_{cell} = T_a + \left(\frac{NOCT - 20^\circ}{0.8}\right) \times G \quad (8.23)$$

whereas power is reduced by 0.5%/°C and calculated as

$$P_{Max.(new)} = P_{Max.(rated)}\left[1 - 0.005\left(T_{cell} - 25\right)\right] \quad (8.24)$$

T_{cell} at T_a of 20°C, insolation of 0.8 kW/m^2 at wind speed of 1 m/s, is known as nominal operating cell temperature ($NOCT$).

Example 8.3

At ambient temperature of 25°C and 40°C, irradiance of 1000 W/m^2, calculate the T_{cell}, change in V_{oc}, and maximum power output of 150W-BP2150S. If the NOCT is 47°C.

Solution: By using

$$T_{cell} = T_a + \left(\frac{NOCT - 20^{\circ}}{0.8}\right) \times G$$

at 25°C

$$T_{cell} = 25 + \left(\frac{47 - 20^{\circ}}{0.8}\right) \times 1 = 58.75°\text{C}$$

at 40°C

$$T_{cell} = 40 + \left(\frac{47 - 20^{\circ}}{0.8}\right) \times 1 = 73.75°\text{C},$$

for T_{cell} of 58.75°C the drop in V_{oc} is

$$V_{oc(new)} = 42.7\,[1 - 0.0037(58.75 - 25)] = 37.36\text{ V},$$

whereas for T_{cell} 73.75 °C

$$V_{oc(new)} = 42.7\,[1 - 0.0037(73.75 - 25)] = 35\text{ V}.$$

The power can be calculated for new V_{oc} as

$$P_{Max.(new)} = P_{Max.(rated)}\left[1 - 0.005\left(T_{cell} - 25\right)\right]$$

$$P_{Max.(new)} = 150\,[1 - 0.005\,(58.75 - 25)] = 124\text{ W}$$

$$P_{Max.(new)} = 150\,[1 - 0.005\,(73.75 - 25)] = 113\text{ W}$$

8.6 Importance of Modelling and Parameter Estimation of Solar PV

For the analysis of a PV module or array primarily the apposite equivalent circuit of the PV cell is selected, and on the basis of the set of equation the PV system is modelled and simulated. It is important to highlight that practical implementation requires prior simulation analysis of PV system, where accuracy of simulation analysis is highly dependent on the parameters of selected PV model that are not mentioned in manufacturer's datasheet [2–16]. As in Section 8.1, various equivalent circuits along with their unknown parameters that are not mentioned in manufacturer's datasheet are described [2–16]. Where, single-diode R_p model is the most widely used model of solar PV cell for the modelling and analysis of solar PV module or array. In this model, i_{pv}, i_{sat}, R_s, R_p,andA are the unknown parameters that are to be estimated precisely.

8.6.1 Existing Parameter Estimation Techniques

Effective estimation of parameters tends toward more exact overlapping of PV characteristic. As precise values of PV cell model parameters affect efficiency and MPPT computations, finding model parameters that provide high accuracy in simulation is of very prominence and significance. As a result, it is a highly identified problem among researchers. In literature, there are many different approaches of parameter estimation have been explained. Broadly these methods are classified into three main categories: analytical approach, metaheuristic techniques, and Hybrid methods.

8.6.1.1 Analytical Approach

Analytical methods use mathematical equations to solve problems and are rapid. These approaches deliver comparatively correct solutions, even if failing to solve nonlinear equations with increased number of unknowns [3]. Single-diode R_s model has four unknown parameters—i_{pv}, A, i_{sat}, and R_s—and these unknown parameters estimation is possible with four equations as prescribed in [3,4]. Whereas manufacturers only expose the value of V_{oc}, i_{sc}, V_{mp}, I_{mp}, K_i, and K_v. Diverse investigators consider different sets of conditions to select these four equations for single-diode R_s model. As in [19], four equations are formulated by choosing short-circuit points, open-circuit point, slope of I-V curve at open-circuit point, where ideality factor of diode is subjectively chosen as 1. In [4,20], short-circuit point, open-circuit point, and the derivative of power with respect to voltage at maximum power is set to zero for the extraction of equation for single-diode R_s model. In single-diode R_p model five parameters—i_{pv}, A, i_{sat}, R_s, and R_p—are to be estimated [3]. For these parameters, five equations are required [2,3]. Table 8.1 summarizes the information used for the preparation of equations in single R_p model in literature. R_{so} and R_{po} is the slope of I-V at short-circuit and open-circuit points, respectively.

In [3], parameter estimation is analytically conferred by using Newton–Raphson Method and is validated by experimented characteristics. Most of the engineering problems are nonlinear in nature, such as the implicit transcendental I-V equations for R_p model of PV modules that consists of five unknown parameters. These problems cannot be solved explicitly for its output current and voltage using common basic methods, rather an iterative technique such as Newton–Raphson method can be employed for obtaining roots of transcendental equations.

Table 8.1. Analytical approaches.

Ref.	V_{oc}	i_{sc}	V_{mp}	I_{mp}	R_{so}	R_{po}	$dp/dv = 0$
[5–19,21]	✓	✓	✓	✓	✓	✓	–
[10]	✓	✓	–	–	✓	✓	✓
[11]	✓	✓	✓	✓	–	✓	✓
[12]	✓	✓	✓	✓	✓	–	–
[13]	✓	✓	✓	✓	–	–	✓

*[12] adopt the value of ideality factor as 1, 2

For two-diode model seven parameters—i_{pv}, R_s, R_p, A_1, A_2, i_{sat1}, and i_{sat2}—require seven equations, most of the approaches handle the problem by using Eq. (8.20). Three equations can be identified by using three critical points of open-circuit voltage, short-circuit condition, and MPP. Whereas, the slope of *I-V* characteristics at open-circuit and short-circuit points leads to two more equations. To extract sixth equation derivative of power with respect to voltage at MPP set to zero. In [14], it is underlined that seventh equations can be mined by setting the sum of two ideality factor to 3.

8.6.1.2 Metaheuristic Approach

After critical analysis, it is observed by many researchers that analytical approaches have their own limitations. Analytical approaches face convergence problem as well as in some cases it is important to estimate initial parameters. Furthermore, the solution of resultant transcendent equation is problematic to be solved. Hence, some researches look forward for the translation of parameter estimation problem toward optimization problem. Metaheuristic optimization algorithm is probabilistic, and so also are population-based optimization techniques. These are inspired by nature and efficient in resolving challenging engineering optimization glitches, hence these metaheuristic optimization techniques are popular nowadays for parameter estimation of solar PV. Various researches in literature using metaheuristic approaches are integrated in Table 8.2. Root mean square error (RMSE) is considered as objective function in most of the researches [18].

Where $I_{i,exp}$ is the *i*th experiment current of PV cell, $I_{i,est}$ represents the *i*th estimated current of PV cell, and *n* is the number of *I-V* points.

$$RMSE = \sqrt{\frac{\sum_{i=1}^{n}\left(I_{i,\exp} - I_{i,est}\right)^2}{n}} \tag{8.25}$$

8.6.1.3 Hybrid Analytical-Metaheuristic Approach

As the name suggests in these approaches, both analytical and metaheuristic techniques are used to solve parameter estimation problem. Several parameters are estimated by using analytical methods, whereas the rest of the parameters are optimally estimated through metaheuristic optimization techniques. It is imperative to distinguish that, in literature, photocurrent and short-circuit current are popularly computed analytically. Literature based on hybrid techniques is unified in Table 8.3. In literature, most of researchers consider mean absolute error (*MAE*) as objective function [18].

$$MAE = \frac{\sum_{i=1}^{n}\left(I_{i,\exp} - I_{i,est}\right)}{n} \tag{8.26}$$

$$\gamma_1 = \left|\frac{i_{sc,e}}{i_{sc,d}} - 1\right|, \gamma_2 = \left|\frac{V_{oc,e}}{V_{oc,d}} - 1\right|, \gamma_3 = \left|\frac{P_{mp,e}}{P_{mp,d}} - 1\right|, \mathrm{OF} = \left|\gamma_1 + \gamma_2 + \gamma_3\right|,$$

where OF denotes objective function, $i_{sc,e}$, $V_{oc,e}$, and $P_{mp,e}$ are the estimated short-circuit current, estimated open-circuit voltage, and estimated maximum power, respectively. The $i_{sc,d}$, $V_{oc,d}$, and $P_{mp,d}$ are the datasheet values of short-circuit current, open-circuit voltage, and maximum power, respectively.

Table 8.2. Literature for metaheuristic approaches [18].

Ref.	Model Analyzed	Data Used	Optimization Algorithm Used	Objective Function	Conclusion Underlined
[15]	Single–diode R_p model	*I-V* data experimental	Artificial Bee Colony	RMSE	✓ By comparative analysis it is concluded that ABC is more superior to GA, PSO, HS, and BFOA in terms of robustness and accuracy
[16]	Single- and two-diode model	*I-V* data experimental	SA	RMSE	✓ SA outperforms the result analysis at [22, 23]
[3]	Single-diode R_p model	Data available in manufacturer's datasheet	SA and Newton–Raphson	RMSE	✓ *I-V* characteristic has been drawn for different irradiance ✓ Relative error plots have been drawn among module current and experimental data ✓ Drawback of single-diode model over other PV models has been highlighted, as it deviates from actual value at low irradiance values
[24]	Single- and two-diode model	*I-V* data experimental	Modified artificial Bee Colony explained in [25]	RMSE	✓ Modified ABC identified its importance over PS, HS, SA and chaotic PSO ✓ Results showcases the suitability of single-diode model over two-diode model, as the difference is only of 0.79%.

[26]	Two-diode model	Data from two-diode model (synthetic)	Penalty-based DE	RMSE	✓ Different types of solar modules used for analysis (mono-crystalline, multicrystalline, and thin filmed) ✓ Penalty based DE give improved results over SA, PSO, and GA ✓ More robust, accurate, consistency with good convergence speed
[27]	Single-diode R_p model	Data available in manufacturer's datasheet	Adaptively controlled scaling factor and crossover rate	RMSE	✓ Separate extraction of parameters for different irradiance values ✓ It is underlined that it is more superior than penalty based explained in [23]
[28]	Single-diode R_p model	Data not available in datasheets	DE and adaptive strategy for scaling factor and crossover rate	RMSE	✓ Identify the dominance of proposed adaptive DE over GA, DE, SA, PSO, and analytical approaches ✓ Proposed adaptive DE does not require the tuning of any control parameter
[13]	Single-diode R_p model	Data available in datasheets	Lambert W function for *I-V* characteristic.	RMSE	✓ An inclusive modelling and parameter estimation method has been proposed ✓ Improved model of solar PV has been classified and optimized approach by using iterative method and approximate method is presented ✓ It is underlined that the proposed model is more realistic when subjected to ideality factor variations

(*Continued*)

Table 8.2. Literature for metaheuristic approaches [18]. (*Continued*)

Ref.	Model Analyzed	Data Used	Optimization Algorithm Used	Objective Function	Conclusion Underlined
[29]	Single-diode R_p model	Data available in datasheets	Curve fitting algorithm	-	✓ Proposed method estimate minimum absolute error between *P-V* curves generated by the proposed electrical model and the *P-V* curves provided by the manufacturers' datasheets under temperature and irradiance. ✓ Parameter estimation has been done by using *P-V* curves instead of *I-V* curves, as in some applications use electrical model with accurately extracted power. ✓ A new error metric MAEP was defined.
[30]	Single- and double-diode model	Available in datasheet	Symmetrized shiftedGompertz function (SSGF)	RMSE	✓ Number of fitting parameters, computational cost and convergence problems has been reduced if compared with previous modelling

Table 8.3. Literature for hybrid approaches [18]

Ref.	Model Analyzed	Data Used	Optimisation Algorithm Used	Objective Function	Conclusion Underlined
[31]	Single-diode R_p model	Data available in datasheet	DE/best/1/bin	OF	✓ Validation through three different modules, that is, mono-crystalline, multicrystalline and thin film. Photocurrent and short-circuit current are analytically estimated rest through DE/best/1/bin
[32]	Single-diode R_p model	Data available in datasheet	Inertia weight PSO	$\gamma_1, \gamma_2, \gamma_3$	✓ Illustrates the difference of maximum power with estimated parameters and the maximum power by experiment
[33]	Single- and double-diode model	Data unavailable in datasheet	GA	MAE	✓ Results are analyzed for different atmospheric conditions where photocurrent and short-circuit current are analytically estimated
[34]	Single-diode model	Data available in datasheet	Hybrid GA-IP	–	✓ Exemplify the outperformance of GA-IP over PSO
[35]	Single-diode model	Data available in datasheet	PS	Eq. (8.27)	✓ Parameters are estimated for crystalline PV array, photocurrent, and short-circuit current are analytically estimated, rest estimated through PS
[36]	Single-diode R_p model	Data available in datasheet	BFOA	–	✓ BFOA showcase its superiority over GA and AIS, in terms of accuracy and robustness

$$\text{Objective Function for [35] is } (I_{mp} - I_{mp,est})^2 + \frac{dI}{dV}(mpp) + \frac{I_{mp}}{V_{mp}}, \quad (8.27)$$

where $I_{mp,est}$ is the estimated current at MPP.

8.7 Analysis of 200 W Module

To understand the impact of environmental conditions on the performance of PV module more aptly, analysis of a 200-W module, that is, KC200GT has been carried out. The KC200GT module is modelled by using single-diode R_p model, where parameters are derived from [3] using Newton–Raphson method. By using 22-year insolation and temperature data [37] of two different altitude locations, designated as L_1 and L_2, the *I-V* and *P-V* characteristics of KC200GT is drawn, analyzed, and compare. Five types of plots have been analyzed for each location, that is: (1) *I-V* with different insolation and T_{cell}, (2) *I-V* with fix insolation and variable T_{cell}, (3) *I-V* with fix T_{cell} and variable insolation, (4) change in power throughout year with change in T_{cell}, and (5) *P-V* at different T_{cell} and insolation. By using Eqs. (8.23) and (8.24), the T_{cell} and P_{max} are calculated analytically for each location by using respective T_a and G throughout the year.

8.7.1 *I-V* and *P-V* Characteristics of 200-W KC200GT at L_1

The location L_1 having coordinates of 29.58° N 79.64° E is considered, this is, the highest altitude location among the two locations considered for analysis, that is, 1,638 m above sea level.

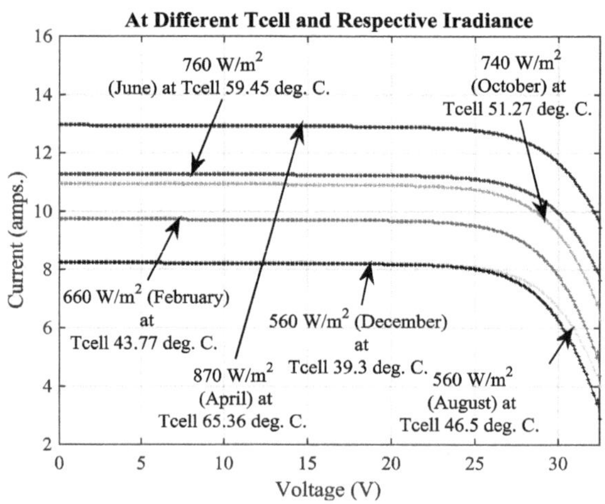

Figure 8.19. *I-V* characteristic for KC200GT at different T_{cell} and respective irradiance during every alternate month at L_1.

Throughout year, KC200GT delivers an average voltage of 29.74 V and an average power output of 174.12 W, that is, a drop of 9.60% and 12.9% in rated voltage and power,

Table 8.4. Monthwise Variation Chart of L_1 for Solar *PV* Module KC200GT.

S. No	Month	Irradiance (W/m^2)	T_{cell} (C)	Power (W)	Voltage (V)
1.	Jan.	560	37.3	187.7	31.40
2.	Dec.	560	39.3	185.7	31.15
3.	Feb.	660	43.77	181.23	30.61
4.	Nov.	660	46.17	178.83	30.32
5.	Aug.	560	46.5	178.5	30.28
6.	Sept.	660	49.37	175.63	29.93
7.	Jul.	740	49.68	175.32	29.89
8.	Oct.	740	51.27	173.73	29.70
9.	Mar.	780	55.62	169.38	29.17
10.	Jun.	760	59.45	165.55	28.70
11.	Apr.	870	65.36	159.64	27.98
12.	May	900	66.77	158.23	27.81
Average		704.16	50.88	174.12	29.74

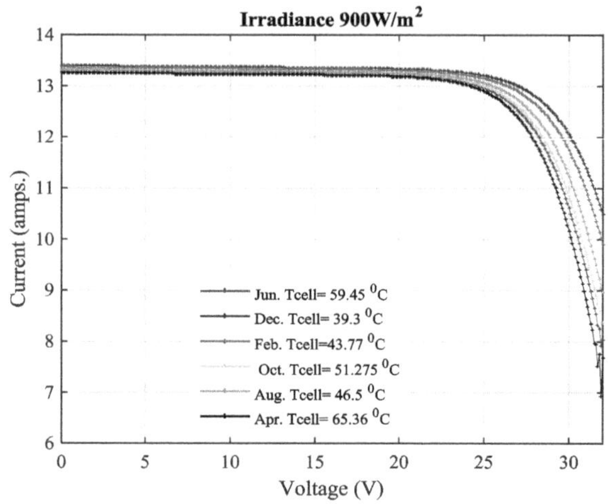

Figure 8.20. *I-V* characteristic for KC200GT at fix irradiance and different T_{cell} during every alternate month at L_1.

respectively. Moreover, Table 8.4 represents that January is the best month in terms of harnessing power at highest voltage, whereas in May, voltage drops by approximately 11% and a drop of 15.7% in output power. Figures 8.19–8.23 illustrate the *I-V* and *P-V* characteristics under different set of condition of T_{cell} and irradiance at L_1. Where, from Fig. 8.21, it is observed that the power output seems deteriorating as the cell temperature increases, which concludes that the solar power systems are more efficient at low temperatures.

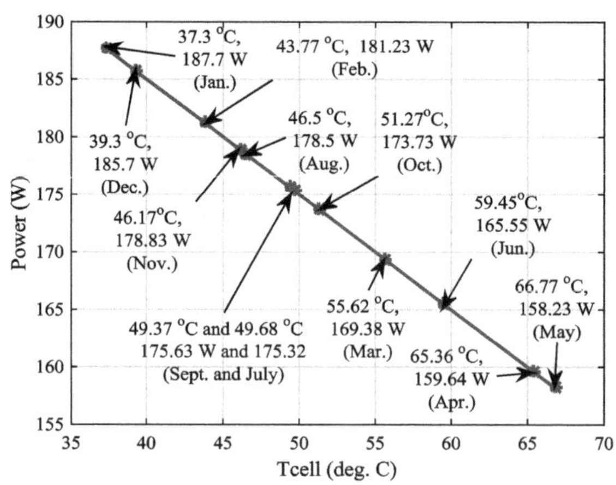

Figure 8.21. Variation of KC200GT power output with respective T_{cell} throughout year at L_1.

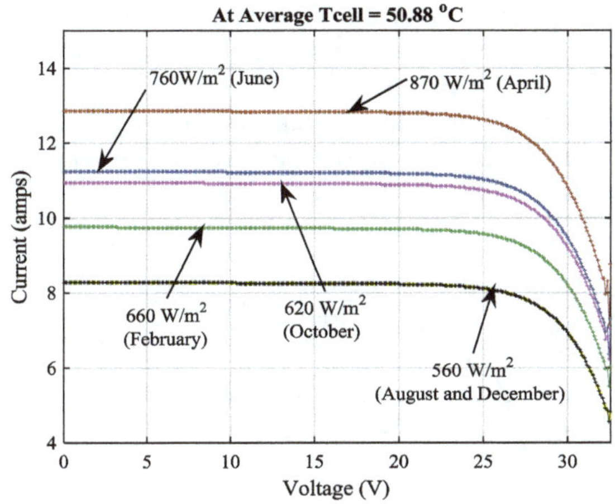

Figure 8.22. *I-V* characteristic for KC200GT at fix T_{cell} and different irradiance during every alternate month at L_1.

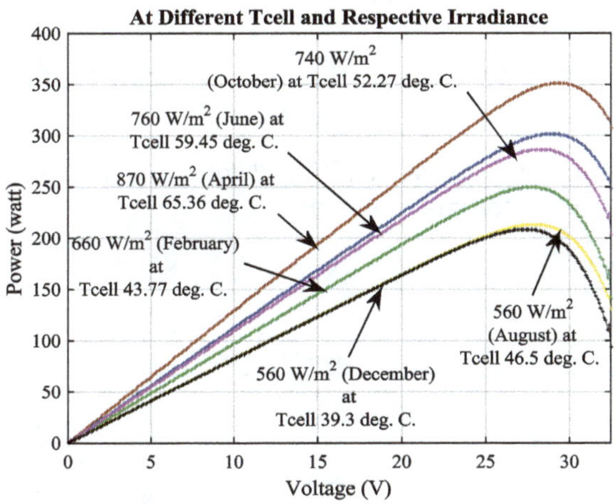

Figure 8.23. *P-V* characteristic of KC200GT with variable T_{cell} and respective irradiance during every alternate month at L_1.

8.7.2 *I-V* and *P-V* Characteristics of 200-W KC200GT at L_2

The L_2 is the location with coordinates 9.4097° N, 78.3643° E (coastal region), that is, 38 m above sea level.

Analytical analysis result of KC200GT at location L_2 summarized in Table 8.5 reveals that 26.40% T_{cell} variation and 32.09% irradiance variation lead toward an average drop of 11.08% in voltage and 15.47% in power, respectively, from rated levels during year.

Whereas, November is the most efficient month with 30.07 V and 176.77 W. Furthermore, KC200GT harnesses power of 159.47 W with voltage of 27.96 V during March, that is, minimum throughout year. On the other hand, average power and voltage is 169.05 W and 29.00 V, respectively, as summarized in Table 8.5, whereas, Figs. 8.24–8.28 show the performance chart of KC200GT solar PV if installed at L_2. The basic advantage of installing the module at L_2 is minimum module voltage variation if compared with L_1, that is, only 3.20% from minimum to maximum output voltage level throughout year or in other words a steady voltage output. Hence, L_2 hosts world's largest solar power plant at single location, with a capacity of 648 MW.

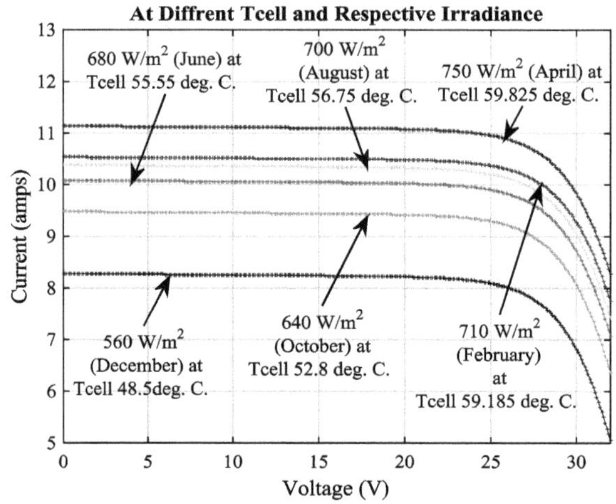

Figure 8.24. *I-V* characteristic for KC200GT at different T_{cell} and respective irradiance during every alternate month at L_2.

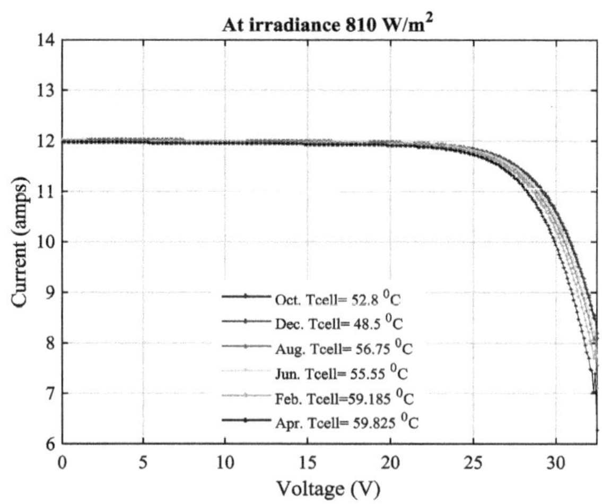

Figure 8.25. *I-V* characteristic for KC200GT at fix irradiance and different T_{cell} during every alternate month at L_2.

Table 8.5. Monthwise variation chart of L_4 for solar PV module KC200GT.

Sl. No	Month	Irradiance (W/m^2)	T_{cell} (°C)	Power (W)	Voltage (V)
1.	Nov.	550	48.22	176.77	30.07
2.	Dec.	560	48.50	176.50	30.03
3.	Oct.	640	52.80	172.20	29.51
4.	Jan.	630	52.96	172.04	29.46
5.	Jul.	670	55.51	169.49	29.186
6.	Jun.	680	55.55	169.45	29.181
7.	Aug.	700	56.75	168.25	29.03
8.	May	750	58.11	166.88	28.86
9.	Sept.	750	58.41	166.59	28.83
10.	Feb.	710	59.18	165.81	28.73
11.	Apr.	750	59.82	165.18	28.66
12.	Mar.	810	65.53	159.47	27.96
Average		683.33	55.94	169.05	29.001

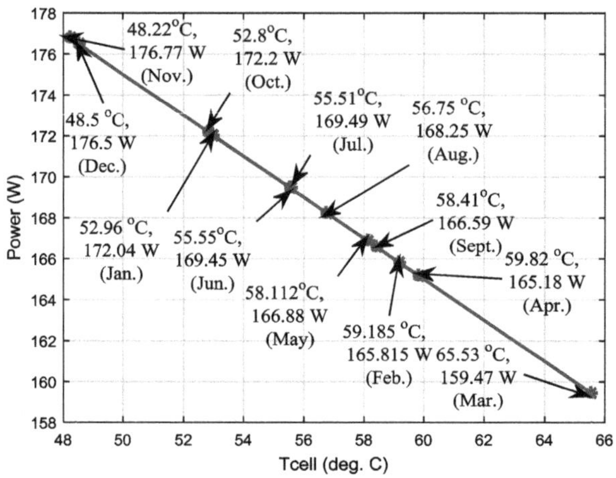

Figure 8.26. Variation of KC200GT power output with respective T_{cell} throughout year at L_2.

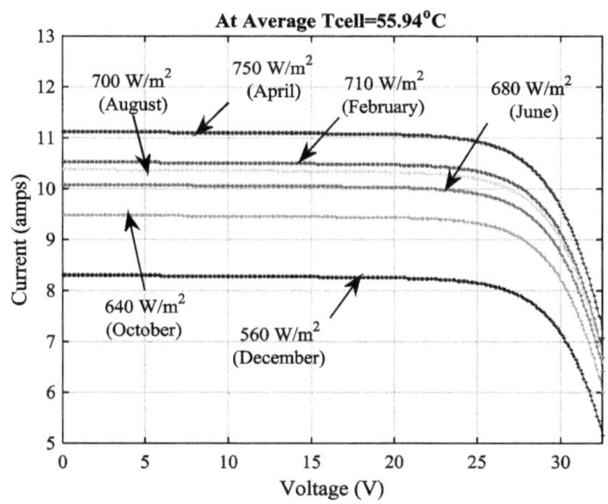

Figure 8.27. *I-V* characteristic for KC200GT at different T_{cell} and variable irradiances during every alternate month at L_2.

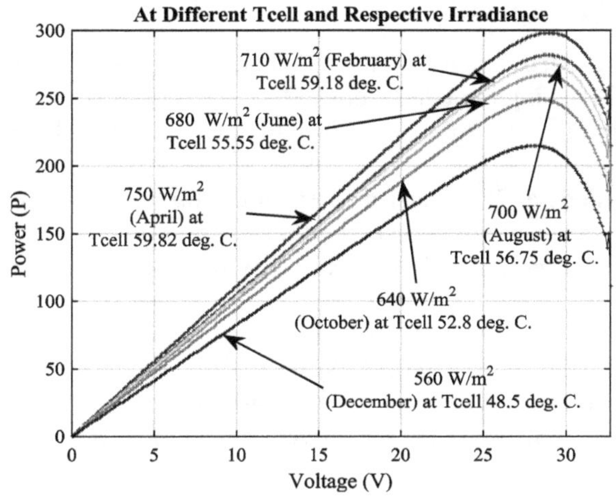

Figure 8.28. *P-V* characteristic of KC200GT with variable T_{cell} and respective irradiance during every alternate month at L_2.

8.8 Conclusion

This chapter concluded that parameter estimation is one of the most emerging nonlinear problem in recent times, as due to the nonlinearity of the problem some sophisticated approaches can be employed, where hybrid techniques gain acceptance widely by using single-diode R_p model.

Moreover, Table 8.6 substantiates the impact of environmental condition on the PV module with the help of analysis carried in Section 8.7. As the L_1 is the location with higher altitude and have low average temperature, hence, it is observed that the power output and the voltage output throughout the year are superior to the L_2, where the consistency in voltage output is observed at L_2.

Table 8.6 Conclusion of analysis

Parameters	L_1	L_2
Avg. power (yearly)	179.72 W	169.05 W
% Drop of power output	(10.14%)	(15.47%)
% Drop of power output (among min and max levels)	16.8%	9.7%
Avg. voltage (yearly)	30.40 V	29.001 V
% Drop of voltage output	(7.57 %)	(11.8 %)
% Drop of voltage output (among min and max levels)	11.7%	7.01%
Avg. light hours (yearly)	12.11 h	12.08 h
T_{cell} variation (yearly)	51.52%	26.40%
Irradiance variation (yearly)	38.20%	32.09%

8.8.1 Unsolved Examples

1. Explain the impact of temperature on the *I-V* characteristics of a solar module?
2. List the advantages of single-diode R_p model?
3. Importance of parameter estimation of solar PV and list the techniques available for parameter estimation?
4. If a *p–n* junction diode at 25°C carries a current of 200 mA when the diode voltage is 0.67 V, what is the reverse saturation current?
5. Calculate the T_{cell} and power delivered by 200 W module, if it is assumed that power loss is 0.5%/°C and,
   (a) $NOCT = 55°C$, $T_a = 25°C$, $G = 1000$ W/m^2
   (b) $NOCT = 45°C$, $T_a = 25°C$, $G = 500$ W/m^2
6. A PV module of 36 cell has an idealized *I-V* curve with $I_{sc} = 3$ A and $V_{oc} = 24$ *V*. If single cell has an $R_p = 6\Omega$, draw the *I-V* curve if one of the cell is shaded.

8.9 References

1. G. M. Masters, "Photovoltaic materials and electrical characteristics" in *Renewable and Efficient Electric Power Systems*, 2nd ed. John Wiley & Sons, Hoboken, New Jersey, 2013.
2. V. J. Chin, Z. Salam, and K. Ishaque, "Cell modelling and model parameters estimation techniques for photovoltaic simulator application: A review," *Applied Energy*, vol. 154, pp. 500–519, Sept. 2015.
3. T. T. Yetayew, and T. R. Jyothsna, "Parameter extraction of photovoltaic modules using newton raphson and simulated annealing techniques", Proceedings. IEEE International Conference on Power, Communication, Information and Technology, Bhubaneswar, India, pp. 239–244, Mar. 2015.
4. N. N. B Ulapane, C. H. Dhanapala, Shyama M. Wickramasinghe, S. G. Abeyratne, Nimal Rathnayake, and P. J. Binduhewa, "Extraction of parameters for simulating photovoltaic panels," Proceedings. 6th IEEE International Conference on Industrial and Information Systems (ICIIS), Kandy, Srilanka, pp. 539–544, Oct. 2011.
5. M. Ade Blas. L Torres E Prieto and A. García, " Selecting a suitable model for characterizing photovoltaic devices," *Renewable Energy*, vol. 25, no. 3, pp. 371–380, Mar. 2002.
6. A. Orioli, and A. D. Gangi, "A procedure to calculate the five-parameter model of crystalline silicon photovoltaic modules on the basis of the tabular performance data," *Applied Energy*, vol. 102, pp. 1160–1177, 2013.
7. Aissa Chouder, Santiago Silvestre, Nawel Sadaoui, and Lazhar Rahmani, " Modelling and simulation of a grid connected PV system based on the evaluation of main PV module parameters," *Simulation Model Practice and Theory*, vol. 20, no. 1, pp. 46–58, Jan. 2012.
8. Lo Brano, Valerio and Orioli, Aldo and Giuseppina, Ciulla and Di Gangi, Alessandra, "An improved five-parameter model for photovoltaic modules," *Solar Energy Materials and Solar Cells,* vol. 94, no. 8, pp. 1358–1370, Aug. 2010.
9. O. Mares, M. Paulescu, and V. Badescu, "A simple but accurate procedure for solving the five-parameter model," *Energy Conversion and Management*, vol. 105, pp. 139–148, Nov. 2015.
10. Jianbo Bai, Sheng Liu, Yuzhe Hao, Zhen Zhang, Meng Jiang and Yu Zhang, "Development of a new compound method to extract the five parameters of PV modules," *Energy Conversion and Management,* vol. 79, pp. 294–303, Mar. 2014.
11. D. Sera, R. Teodorescu, and P. Rodriguez, "PV panel model based on datasheet values", Proceedings. IEEE International Symposium on Industrial Electronics (ISIE), Vigo, Spain, pp. 2392–2396, 2007.

12. F. Adamo, F. Attivissimo, A. Di Nisio, A. M. L. Lanzolla, M. Spadavecchia, "Parameters estimation for a model of photovoltaic panels," Proceedings. 11th IMEKO World Congress Fundamental Applied Metrology, Lisbon, Portugal, pp. 6–11, Sept. 2009.
13. L. Peng, Y. Sun, and Z. Meng, "An improved model and parameters extraction for photovoltaic cells using only three state points at standard test condition," *Journal of Power Sources*, vol. 248, pp. 621–631, Feb. 2014.
14. A. A. Elbaset, H. Ali, and M. Abd-El Sattar, "Novel seven-parameter model for photovoltaic modules," *Solar Energy Materials and Solar Cells*, vol. 130, pp. 442–455, Nov. 2014.
15. D. Oliva, E. Cuevas, and G. Pajares, "Parameter identification of solar cells using artificial bee colony optimization," *Elsevier: Energy*, vol. 72, pp. 93–102, Aug. 2014.
16. K. M. El-Naggar, M. R. AlRashidi, M. F. AlHajri, A. K. Al-Othman, "Simulated annealing algorithm for photovoltaic parameters identification," *Solar Energy*, vol. 86, pp. 266–274, 2012.
17. P. Suskis, and I. Galkin, "Enhanced photovoltaic panel model for MATLAB-simulink environment considering solar cell junction capacitance," Proceedings IEEE Industrial Electronics Society, IECON, Vienna, Austria, pp. 1613–1618, 2013.
18. A. R. Jordehi, "Parameter estimation of solar photovoltaic (PV) cells: A review," *Renewable and Sustainable Energy Reviews*, vol. 61(c) pp. 354–371, 2016.
19. G. R. Walker, "Evaluating MPPT converter topologies using a MATLAB PV model," *Journal of Electrical and Electronics Australia*, vol. 21, no.1, pp. 49–55, 2001.
20. K. Ding, X. Bian, H. Liu, and T. Peng, "A MATLAB simulink based PV module model and its application under conditions of non-uniform irradiance," *IEEE Transactions of Energy Conversion*, vol. 27, no. 4, pp. 864–872, Dec. 2012.
21. A. H. Arab, F. Chenlo, and M. Benghanem, "Loss-of-load probability of photovoltaic water pumping systems," *Solar Energy*, vol. 76, no. 6, pp. 713–723, 2004.
22. T. Easwarakhanthan, J. Bottin, I. Bouhouch, and C. Boutrit, "Nonlinear minimization algorithm for determining the solar cell parameters with microcomputers," *International Journal of Solar Energy*, vol. 4, pp. 1–12, 1986.
23. K. Bouzidi, M. Chegaar, and N. Nehaoua, "New method to extract the parameters of solar cells from their illuminated *I-V* curve," Proceedings 4th IEEE International Conference on Computer Integrated Manufacturing, Setif, Algeria, pp. 1–4, 2007.
24. A. Askarzadeh, and A. Rezazadeh, "Artificial bee swarm optimization algorithm for parameters identification of solar cell models," *Applied Energy*, vol. 102, pp. 943–949, Feb. 2013.
25. A. Askarzadeh, and A. Rezazadeh, "A new artificial bee swarm algorithm for optimization of proton exchange membrane fuel cell model parameters," *Journal of Zhejiang University Science C*, vol. 12, pp. 638–646, 2011.
26. Kashif Ishaque, Zainal Salam, Saad Mekhile, Amir Shamsudin, "Parameter extraction of solar photovoltaic modules using penalty-based differential evolution," *Applied Energy*, vol. 99, pp. 297–308, Nov. 2012.
27. D. H. Muhsen, A. B. Ghazali, T. Khatib, and I. A. Abed, "Extraction of photovoltaic module model's parameters using an improved hybrid differential evolution/electromagnetism-like algorithm," *Solar Energy*, vol. 119, pp. 286–297, 2015.
28. L. L. Jiang, D. L. Maskell, and J. C. Patra, "Parameter estimation of solar cells and modules using an improved adaptive differential evolution algorithm," *Applied Energy*, vol. 112, pp. 185–193, 2013.
29. E. A. Silva, F. Bradaschia, C. Marcelo, and A. J. Nascimento, "Parameter estimation method to improve the accuracy of photovoltaic electrical model," *IEEE Transactions on Journal of Photovoltaics,* vol. 6, no. 1, pp. 278–285, 2016.
30. A. Molina-Garcia, J. Guerrero-Pérez, M. C. Bueso, M. Kessler and E. Gómez-Lázaro, "A New Solar Module Modeling for PV Applications Based on a Symmetrized and Shifted Gompertz

Model," in *IEEE Transactions on Energy Conversion*, vol. 30, no. 1, pp. 51–59, March 2015, doi: 10.1109/TEC.2014.2330741.

31. K. Ishaque, and Z. Salam, "An improved modelling method to determine the model parameters of photovoltaic (PV) modules using differential evolution (DE)," *Solar Energy*, vol. 85, pp. 2349–2359, 2011.
32. J. J. Soon, and K. S. Low, "Photovoltaic model identification using particle swarm optimization with inverse barrier constraint," *IEEE Transactions on Power Electronics*, vol. 27, no. 9, pp. 3975–3983, Sept. 2012.
33. M. Ismail, M. Moghavvemi, and T. Mahlia, "Characterization of PV panel and global optimization of its model parameters using genetic algorithm," *Energy Conversion and Management,* vol. 73, pp. 10–25, 2013.
34. A. M. Dizqah, A. Maheri, and K. Busawon, "An accurate method for the PV model identification based on a genetic algorithm and the interior-point method," *Renewable Energy*, vol. 72, pp. 212–222, 2014.
35. J. Y. Park, and S. J. Choi, "Datasheet-based circuit parameter extraction method for maximum power point simulation of photovoltaic array," Proceedings. 9th IEEE International Conference on Power Electronics and ECCE Asia, Seoul, South Korea, pp. 1413–1419, 2015.
36. N. Rajasekar, N. K. Kumar, and R. Venugopalan, "Bacterial foraging algorithm based solar PV parameter estimation," *Solar Energy*, vol. 97, pp. 255–265, Nov. 2013.
37. Atmospheric science data centre, "NASA Surface Meteorology and Solar Energy." 2017. Available online: https://eosweb.larc.nasa.gov/cgi

CHAPTER 9

Power Electronics for Solar Photovoltaic System: Configuration, Topologies, and Control

Rishabh Dev Shukla[1], Navdeep Singh[2], and Subhajit Roy[1]

[1]*Department of Electrical Engineering, Budge Budge Institute of Technology, Kolkata, West Bengal 700137, India. shukla.rishabhdev@gmail.com; subhajitroy111@gmail.com*
[2]*Department of Electrical Engineering, Madan Mohan Malaviya University of Technology, Gorakhpur 273016, India. navdeepsingh26nov@gmail.com*

Abstract Renewable energy sources are a replacement of conventional sources to overcome the environmental pollution issues, rising prices, and limited stock. Solar energy is the best alternative to provide abundant and clean source of energy. This chapter presents a brief introduction of the solar photovoltaic (PV) energy systems, dependence on different parameter, possible operating modes,that is, on-grid and off-grid, different converter topologies, and their control with all the aspects covering standards and specifications. Moreover, to integrate the solar PV DC–AC converter with grid/load, there are two methods such as single-stage operation and two-stage operation. This chapter is focused on the application of the different types of DC–DC converters and inverters employed for

DC–AC power conversion. In this conversion, single-stage operation and two-stage converter, that is, DC–DC converter for DC load system and DC–AC converter for AC load system to make the efficient output of PV as per the requirement of grid/load.

Keywords: Renewable energy, DC–DC converter, DC–AC converter, Grid-connected solar PV, PV panels/module, Sand-alone solar PV, Solar invertertopologies, Structure-based topologies, Transformer-less topologies.

9.1 Introduction

Renewable energy generation systems produce viable energy by converting natural phenomena into practical energy forms. These systems use the energy that comes from the sun and its direct and indirect effects on the earth (radiation and temperature, various plants, i.e., biomass, wind, and falling water), forces of gravitation (tides on ocean), and the earth's core heat (geothermal) as the basics from which heat/energy/power can be produced [1]. These renewable energy resources have huge energy potential, but they are generally dispersed and not completely reachable: mostly they are intermittent and have different regional uncertainty. These aspects give rise to certain technical and economical challenges. Now, considerable advancement has been made by improving the conversion and collection efficiencies, lowering the capital and maintenance costs, and increasing the applicability and reliability.

Renewable energy conversion systems (RECSs) are the appropriate solution to provide green energy demand in dealing with the worldwide energy issues such as global warming. Moreover, the use of these systems (i.e., RECSs) has increased during the last decade due to development in the grid integration technologies [2,3]. Certainly, solar energy is a renewable energy source, which is available freely, sustainable, and completely inexhaustible, nothing like fossil fuels, which are limited [4,5]. Also, it is the most plentiful energy resource on earth, which has the potential to respond to the global energy demand [6]. Hence, solar power has significant future for numerous countries including both developing and developed.

In the case of passive solar energy technology, solar energy is only collected but will not be converted to light or heat forms and includes, for instance, maximizing the utilization of day light/heat by means of building design [7]. On the contrary, in the case of active solar energy technology, the solar ray's occlusion takes place to store or convert for different applications. The active solar energy technology can be largely categorized into two groups: (1) solar thermal systems and (2) photovoltaic (PV) systems. In PV technology, the solar ray falling on the semiconductor material, the radiant energy, which is contained in solar light quanta, is directly converted to electrical energy due to electron excitation, and as a result conductivity of semiconductor material has increased greatly [8]. As PV technology has the capability of generating clean and quiet electrical energy and is also reliable enough, so in this current situation, this is considered a promising RECS. Currently, PV is the third most significant RECS after hydro and wind in reference to the globally installed capacity [9,10].

9.1.1 Solar PV Energy Generation Systems

Solar energy can be converted into electrical energy using PV systems. This converted electrical energy can be used directly to a load connected with the system (i.e., off-grid configuration) or it can supply power to the utility grid connected with on-grid configuration as well [11–18].

In off-grid or standalone configuration, shown in Figure 9.1(b), the PV system operates autonomously from the grid or power system. Conversely, the on-grid configurations work directly connected with the grid/power system and PV system feed the main grid. A usual grid-connected PV system is shown in Figure 9.1(a), which consists of PV panels along with some other auxiliary devices such as transformers, certain filters, and other energy conversion devices.

Here, some power electronics-based devices are used to convert DC/DC and DC/AC electrical energy and are known as power converters. On-grid system directly supplies power and work in a parallel connected system with the utility grid. In this case, no energy storage device is required. When the amount of power converted is more than the load demand, the PV system is able to supply to the grid.

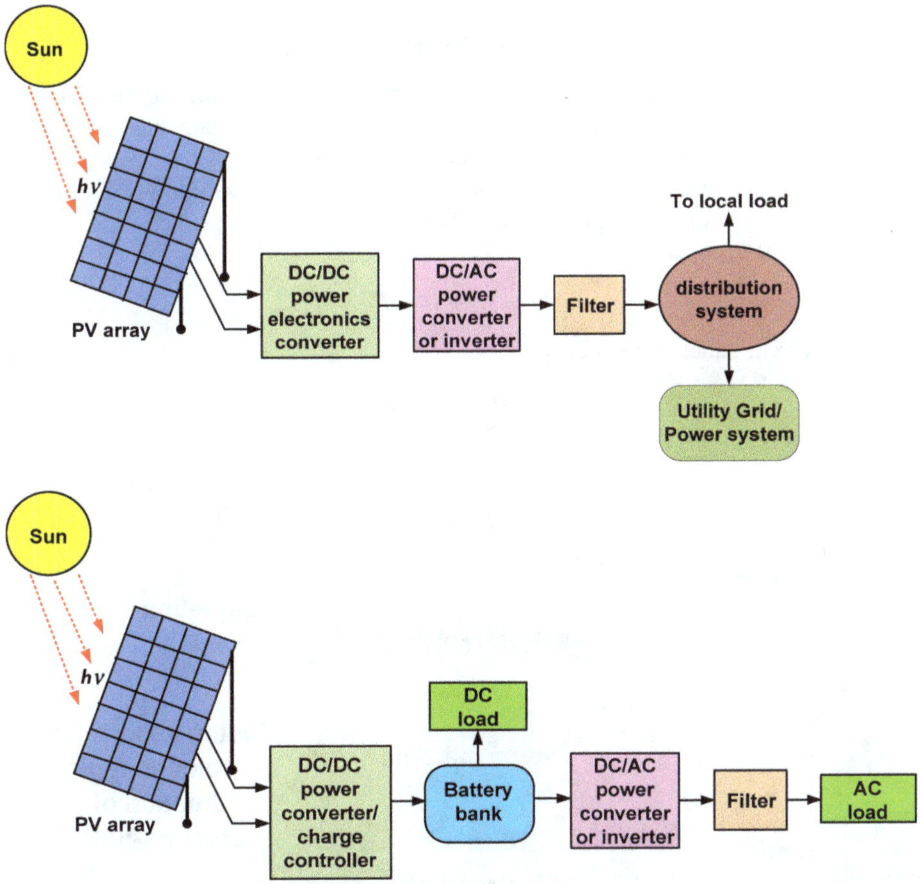

Figure 9.1 Typical configurations of the solar PV energy generation systems. Schematic view of (a) on-grid and (b) off-grid PV systems.

9.1.2 Solar PV Cell

Solar cell used to convert solar energy into electrical energy has a similar structural characteristic of a general *p–n* junction diode. The only difference is that its working capability depends on solar radiation and surface temperature [15]. The solar cell is having a suitable surface area of *p–n* junction to allow light, because *n*-type material is generally thin. Electricity is generated in the depletion zone when photon is incident on the *p–n* junction and if an external load is connected with the help of wire to the *p–n* junction. Thus, electron and hole are attracted to the positive and negative charges of the *p*- and *n*-type materials, respectively. The current flows through the external load as shown in Figure 9.2.

To build a PV module, often the solar cells are connected in either series or parallel configuration, as the output of a single solar cell is very small to serve a load. Many PV modules are used to build complete PV panels and some PV panels are connected. Finally, we build PV array to obtain any desired current–voltage characteristics as per requirement.

9.1.3 Equivalent Electrical Circuit for Solar PV Cell

If we try to understand the internal electrical circuit of a solar PV cell, it is current source connected in parallel with a rectifying diode. For practical operation, solar cell has two resistances, which are connected as series resistances R_S and shunt resistances R_P as shown in Figure 9.3. Series resistance, called intrinsic resistance, has a very low value. Parallel resistance, called equivalent shunt resistance, has a very high value [8, 9].

From Figure 9.3, applying Kirchhoff's law of current

$$I_L = I_D + I_P + I \tag{9.1}$$

The PV current equation [9–12] is

$$I = I_L - I_P - I_D \tag{9.2}$$

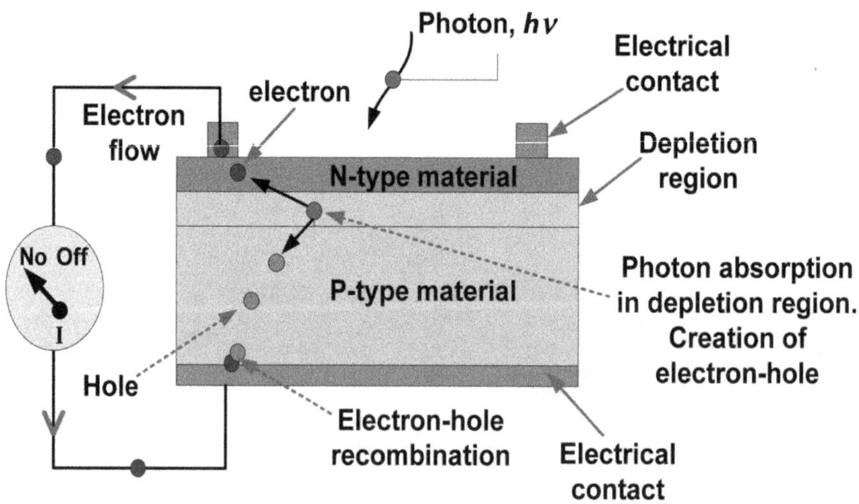

Figure 9.2. Electricity generation by solar PV cell.

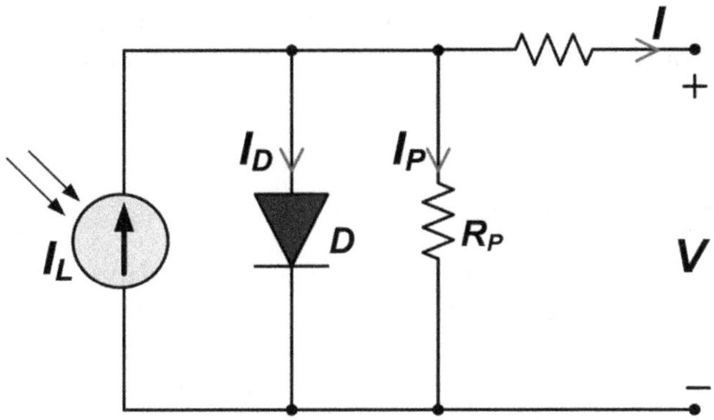

Figure 9.3. Equivalent circuit model of PV cell.

$$I = I_L - I_o[e^{\left\{V+\frac{IR_s}{V_T}\right\}} - 1] - [V + \frac{IR_s}{R_P}], \tag{9.3}$$

where I_L is PV current, I_o is diode saturation current, I is the cell current, V_T is the thermal voltage, and V is the cell voltage.

9.1.4 Status and Scenario of Solar PV System [19–20]

Solar PV has emerged as one of the key technologies to reduce emissions as per the Paris Agreement. For the last couple of years, solar power generation has been a result of public enthusiasm, but currently, it is proving to be an economical way of electricity generation. China overtook Germany in terms of total PV energy generation in 2015 and Japan and United States by 2017. The European Union (EU) was among the leading PV energy-generating countries with 102 GW having around 32% of the total worldwide PV energy generation by the end of 2017. The global PV grid power installed capacity has reached 580 GW by the end of 2019, which is almost 3% of the total worldwide power demand. As per 2019, top 5 nations with PV installed capacity is being China 205 GW, USA 76 GW, Japan 63 GW, Germany 49 GW, and India 38 GW [19]. In recent years, the yearly PV solar system installation has also increased drastically from 29.5 GW in 2012 to 125 GW in 2020 [19].

China targets to build over 1,200 GW of solar and wind power by 2030, and aim to achieve 25% non-fossil in its primary energy by 2030. US expects to have 325 GW of PV installed capacity by 2030. Japan and Germany target to have 150 GW and 100 GW of PV installed capacity by 2030, respectively. Solar PV installations is expected to grow over fourfold in Australia from 18 GW in 2020, to 80 GW in 2030. India recently reached the 5th place in the global deployment of solar power by outperforming Italy. The solar output climbed from 2.6 GW in March 2014 to 30 GW in July 2019 by more than 11 times over the last five years. India has set one of the most ambitious targets to have over 280 GW of PV capacity out of installed of 450 GW renewable energy target by 2030. Solar tariffs in India are currently highly competitive and grid parity has been reached due to which several GW capacity PV plants have come to picture which includes Bhadla Solar Park 2.25 GW in Jodhpur district in Rajasthan, Shakti Sthala solar power (Karnataka) 2.05 GW, 1.0 GW PV Plant Kurnool district in Andhra Pradesh, 0.75 GW Rewa solar park (Madhya Pradesh) and 0.65 GW Kamuthi solar park in Ramanathapuram district in Tamilnadu [20].

9.2 Types of Solar PV Cell [23]

Solar cells possess the characteristics of semiconducting material, as such absorbing property of sunlight. Sunlight has been absorbed as such directly from the earth's surface or indirectly from the space surface. Solar cells are designed as single-layer light-absorbing or multiple-layer material. Its working is based on charge separation mechanisms. Solar cell is categorized into the following three types:

(a) First-generation cells (conventional solar cell): It is manufactured by crystalline silicon, which includes poly- and monocrystalline silicon.
(b) Second-generation cells (thin-film solar unit): It is designed with amorphous silicon, CdTe, and CIGS cells for stand-alone solar generation.
(c) Third-generation cells (multiple number of thin film): It is not commercially applied and is still in the research phase [24].

9.2.1 Monocrystalline Silicon PV Panels

Solar cells are manufactured by monocrystalline silicon, also known as single-crystalline silicon. It is easily recognizable by high-purity silicon coloring and uniform look, as shown in Figure 9.4.

Monocrystalline solar cells have a cylindrical shape for getting optimized performance at lower costs. The difference between mono- and polycrystalline cells is that polycrystalline cell has a rectangular shape with no rounded edges.

Figure 9.4. Monocrystalline silicon PV panels.

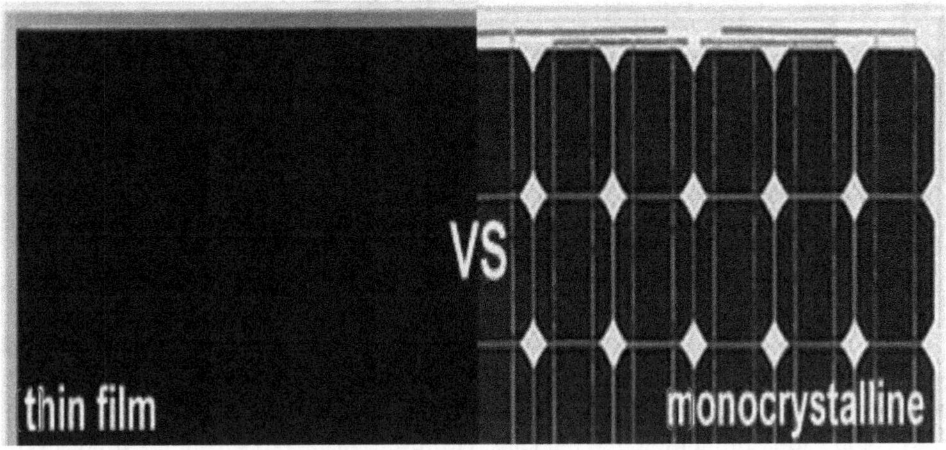

Figure 9.5. Monocrystalline silicon PV panels.

Advantages of monocrystalline solar cells are (1) highest efficiency rates due to high-grade silicon, (2) space-efficient: high power requirement need least amount of space compared, (3) long life, and (4) low-light conditions.

Disadvantages of monocrystalline solar cells are (1) expensive, (2) shading issues, and (3) temperature dependent.

9.2.2 Amorphous/Thin-Film Solar Panels

The low output of power is generated in amorphous silicon-based solar panel. The difference between monocrystalline and thin film is shown in Figure 9.5. It has small power application like a pocket calculator. Several amorphous silicon layers produce higher efficiency rates of 6%–8% due to stacking. In amorphous silicon solar panel, 1% of the silicon is used in amorphous silicon cells for reducing the cost.

9.2.3 Hybrid Silicon Solar Panels

Hybrid silicon solar panels are designed with organic and inorganic semiconductors. Organic materials of hybrid panel absorb the sunlight and follow the mechanism of the donor and transport holes. On the other hand, inorganic materials have acceptor and electron transporter for absorbing the sunlight. The hybrid panels are low cost for scalable solar system. Heterojunction photoactive layer is utilized due to its prominent power conversion efficiency in comparison to single layer. In hybrid panels, it is designed by the two materials. One of the materials, the photon, works as absorber and exciton donor, whereas the other works as exciton dissociation at the heterojunction. The problem of short exciton diffusion length is resolved by heterojunction structure in place of phase-separated bilayer.

9.3 Different Configurations of Solar PV Energy System

Different configurations of solar PV energy system have been discussed in Section 9.3.1, 9.3.2, 9.3.3 and 9.3.4.

9.3.1 Single-Stage DC–DC Converter for DC Load

A single-stage DC–DC converter configuration supplying the DC load is shown in Figure 9.6. The unregulated variable voltage obtained from solar PV array has been boost up by DC–DC converter for getting the constant voltage at the load side. Stepping up the low-level input DC voltage obtained from solar PV cell to high-level output DC voltage will also step down the high-level input current obtained from solar PV cell to low-level output current and maintain the power close to the conserved state, for the desired efficient system operation of power electronics DC–DC converter.

An important design criterion for DC–DC converters for PV applications is the ability to produce a high-voltage output from a low-voltage input. The operating voltage for commercially available PV modules is low-voltage DC, typically around 20 volts direct current (VDC) [21]. However, in the case of DC–AC inverter, a higher voltage input is quite necessary for an efficient conversion to AC. One option is to increase this voltage through series combination of multiple PV modules. However, this approach suffers from several disadvantages. First, in the case of series combinations reliability reduces much; if any one module connected in a series system gets faulty, then the whole system fails to work. As the current through series-connected components must always be equal, the lowest performing module always control or dictate the string current of the system. This is especially problematic in partial-shading situations, where shading of a single module will diminish the power output of the entire string. So, there is the conclusion that we should look more to use a parallel-connected DC–DC converter system instead of DC–AC inverters. The output voltage of this DC–DC converter must be in the range of 400 VDC for a typical DC-link input voltage for a three-phase 208/120 V or single-phase 220- or 240-V inverter.

9.3.2 Single-Stage DC–AC Converter for AC Load

The DC–DC converter is like a transformer used to step-up and/or step-down the level of the input voltage. Power electronics DC–DC chopper with controller is used to buck and boost input voltage from the unregulated DC solar power. This regulated DC power has been converted to AC supply by DC–AC voltage source inverter shown in Figure 9.7.

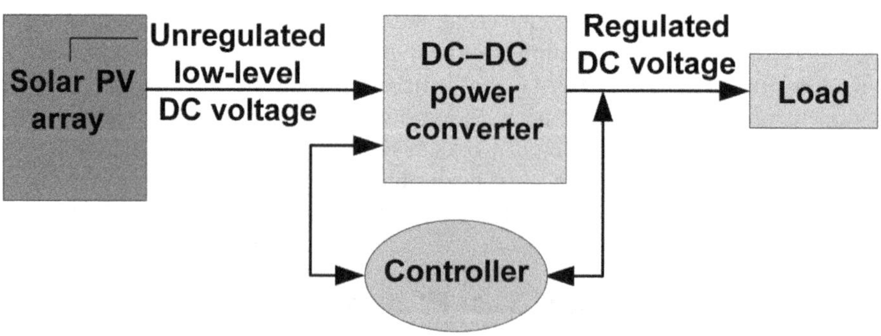

Figure 9.6. Block diagram of closed loop system [23].

9.3.3 Two-Stage DC–DC Converter for DC Load

High-frequency link (HFL) DC–DC converter is rectifier-type HFL as shown in Figure 9.8. It consists of high-frequency DC/AC step-up and step-down inverter output to feed high-frequency transformer, then followed by high-frequency AC output voltage to AC/DC converter for getting high-gain DC output voltage. HFL-isolated DC–DC converter has AC–DC conversion at load side (on the secondary side of transformer), that is, from AC/DC converter and DC/AC inverter [25].

9.3.4 Two-Stage DC–AC Converter for AC Load

HFL inverter is rectifier-type HFL inverter (RHFL) shown in Figure 9.9. RHFL consists of high-frequency DC/AC inverter output to feed high-frequency transformer and a set of AC/DC converters along with a pulsating DC/AC inverter. RHFL technique has two stages of conversion at load side (on the secondary side of transformer), that is, from AC/DC converter and DC/AC inverter.

9.4 Grid Connection Topologies (Structure-Based) for Solar PV System

Centralized technology is shown in Figure 9.10. In the case of centralized inverter system, a huge number of PV modules fed the power to the grid [27–30]. Here, we see PV modules

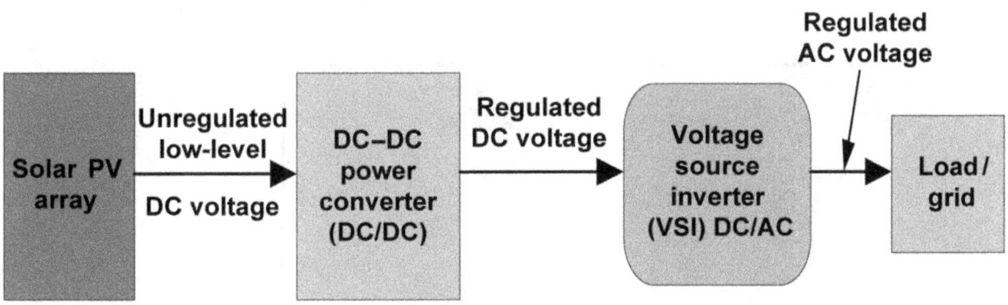

Figure 9.7. Block diagram of solar energy-based DC–AC system [21].

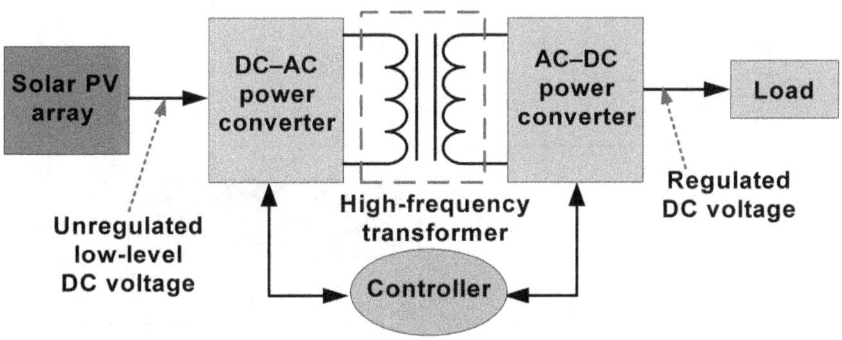

Figure 9.8. Block diagram of HFL two-stage DC–DC converter for DC load.

are connected into herded or string formation, that is, series connections where each of the module generates an adequately high voltage to avoid further magnification. These strings are then connected using a special arrangement called string diodes in parallel, to achieve a high-power output [30].

This centralized inverter topology offers some strict limitations, for instance, high-voltage DC cables between the inverter and PV modules, losses due to mismatching between the PV modules, centralized maximum power point tracking (MPPT) power losses, string diode losses, and a stiff design where the reimbursement of mass production could

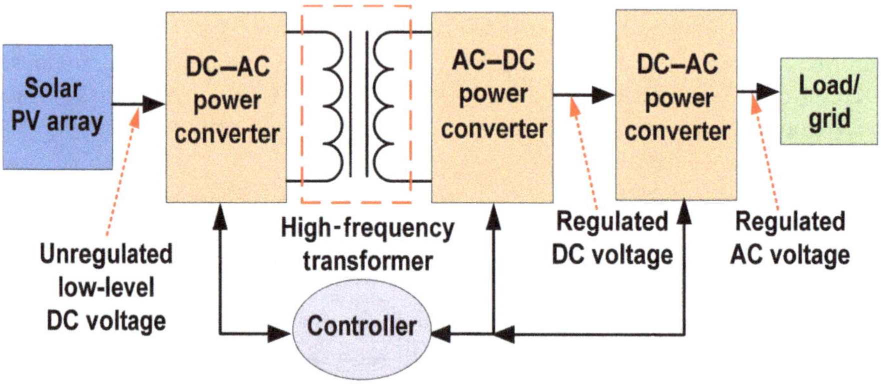

Figure 9.9. Block diagram of RHFL inverter topology [26].

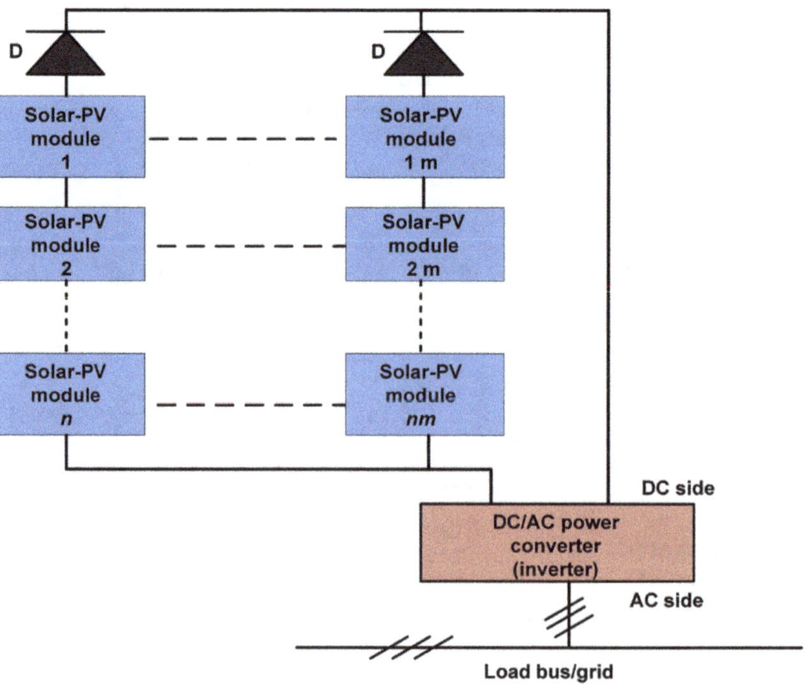

Figure 9.10. Centralized inverter technology.

not be reached. The failure of the central inverter results in that the whole PV plant fails to operate [31–34].

The string inverter topology, shown in Figure 9.11, is the reduced edition of the centralized inverter topology. In the string inverter topology, each of the PV module strings is connected to an inverter [28]. The input DC voltage may be sufficiently higher to avoid voltage magnification. The losses related to the string diodes are eliminated and separate MPPT techniques can be applied to each of the string inverter. This increases the energy yield and enhances the supply reliability through decrease in the losses due to mismatching and partial shading. Furthermore, increases the overall efficiency as compared to the centralized inverter topology, and the mass production reduces the price.

The block diagram representation of the multi-string inverter topology is shown in Figure 9.12. Here, in Figure 9.12, we are able to understand the proper working of this new extended string-type technology, where the strings are actually interfaced to their individual DC/DC converter and finally to the common DC/AC inverter. Separate DC/DC converter for each of the strings allows implementing specific techniques to the particular strings. As a result, a compact and economical solution is obtained, which is having the benefits of central and string topologies. This topology allows for the incorporation of PV strings of different technologies and of various orientations. The application area of it includes the PV plants of 3–10 kW [34,35].

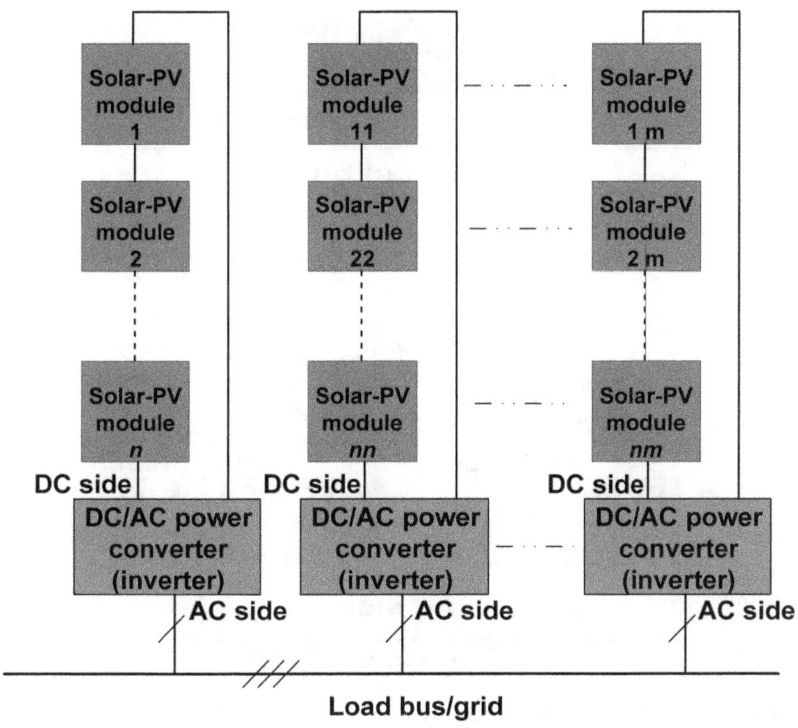

Figure 9.11. String inverter topology (separate MPPT systems).

The AC module illustrated in Figure 9.13 is the incorporation of the inverter and PV module into one electrical device. As here we are working with only one PV module, there is no issue of differences of losses between different PV modules and also it allows the proper adjustment of the PV module and the inverter and also the individual MPPT. Due to the modular structure, there is the possibility of a facilitated extension of the system (Table 9.1).

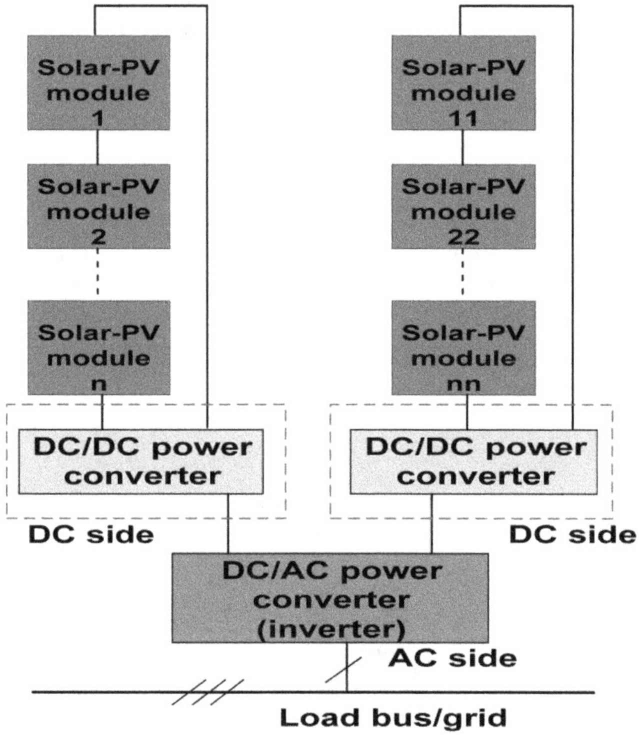

Figure 9.12. Multi-string topology.

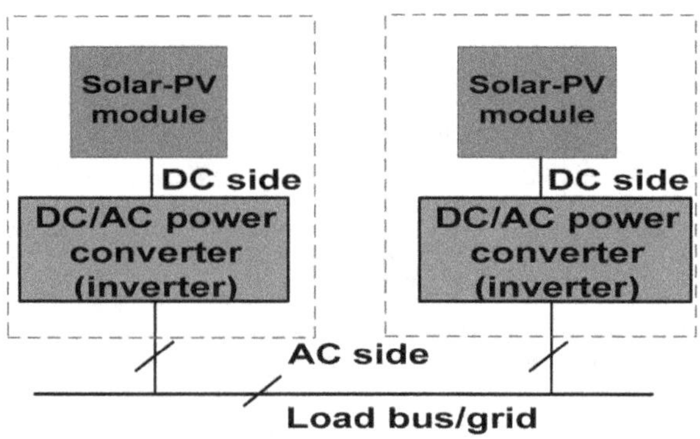

Figure 9.13. AC module-based topology.

Table 9.1. Merits and demerits of topologies (structure-based) of grid-connected solar PV systems [35]

Topology	Merits	Demerits
Centralized inverter	For HVHP application (higher voltage and power)	Out of couple of major demerits, these are due to common MPPT very higher power loss, mis-match of different module losses due to string diodes, pretty much expensive, even for minor fault may the whole plant get shut down.
String inverter	Work better than the centralized inverter system as there are no diode losses; it is much cheaper than the other configuration, for individual string separate MPPT technique can be applied.	Lower efficiency can be achieved than the other existing techniques, and also generation cost get higher for this process.
Multi-string inverter	In this technique, a number of PV modules can be controlled individually as string, so we can have different strings having different MPPT technologies and also of different orientation; here small DC-link capacitors are used, which in result gives a high energy level and better monitoring can be achieved.	Cost per kilowatt-generated solar PV energy is pretty high due to application of DC–DC and DC–AC converters.
AC module inverter	One of the most economic designs of PV module string, but it also ensures higher power extraction from the module, but the best feature is easier to expand than the modular-structured system.	Although higher power extraction is achieved by this technique, but overall efficiency gets decreased, but the main issue for the designers is to make an inverter that can amplify a pretty much small voltage to a usable voltage.

9.4.1 MPPT in Solar PV Systems

The concept of MPPT is to continuously monitor the terminal voltage and current and update the control signal accordingly to achieve maximum power point (MPP). The main thinking behind the MPPT is to always vary the voltage and current level due to pulsation of instantaneous power; after examining the individual instantaneous power levels the system is able to track the closest point to achieve the maximum power output [36].

The maximum power given by the solar PV array is represented by the following relation:

$$P_{\max} = V_{\mathrm{mpp}} \cdot I_{\mathrm{mpp}} \tag{9.4}$$

where V_{mpp} and I_{mpp} are the optimal operating voltage and current of PV array at the condition of maximum power output, respectively. Solar radiation, load impedance, and module temperature are the three factors that affect the maximum power extraction from solar PV module. *I-V* curve of PV module is a function of insulation and temperature that affects output current and voltage. The increased temperature decreases the open circuit voltage (V_{oc}), whereas the increased intensity of solar radiation increases short-circuit current (I_{sc}). Therefore, *I–V* and *P–V* curves change according to the operating conditions, which alter MPP accordingly. Basically, the VI characteristic curve of a PV cell shows a nonlinearity. So an MPPT tracker must need to achieve maximum power with having all environmental conditions in consideration [37]. The MPPT can be achieved with means of Chopper and DC–AC power inverters along with PV arrays. To maintain the electrical quantity values nearer to the MPP, the MPPT controllers try to maintain a certain Duty Ratio (D), which is figured out using the measured voltage and current values [38]. There are a couple of MPPT techniques that are already in use, but the most extensively used one involves incremental conductance method, perturbation and observation (P & O) method, Hill climbing technique, three-point weighted comparison method, open circuit voltage method, short-circuit current method, Fuzzy logic, and neural network-based method [39,40].

9.5 DC–DC Converter for DC Load Used in Solar PV Systems

The DC–DC converter is also known as a Chopper, which is a power electronics circuit and actually converts a DC quantity to another DC quantity. Here, in the case of PV system, this circuit operates to change the voltage level to usable and essential voltage level. The MPPT in solar PV systems can be achieved easily via DC–DC power converters. Different DC–DC converter topologies have been described in the literature, which can be broadly categorized into isolated and nonisolated topologies. The nonisolated-type topologies have transformer-less connection and used in the application where voltage gain factor is low. Electrical isolation is preferred for safety reasons in grid-tied PV applications [41]. Therefore, isolated-type converter topologies are used, which use transformer for isolation purposes. A couple of topologies are used to build a DC–DC converter for use in PV system. These are for stepping up purpose boost converter, for stepping down purpose Buck converter, and also for hybrid configuration buck–boost converter or Cuk converter [42]. There are two distinct modes of operations in these DC–DC converters: (1) continuous conduction mode (CCM) and (2) discontinuous conduction mode (DCM). In the case of solar PV systems, CCM is preferred in general [43]. The performance of DC–DC converters can be reported by the following parameters [44]: (1) voltage gain; (2) current gain; (3) input impedance; (4) boundary filter inductance; and (5) minimum filter capacitance.

In Figure 9.14(a), we can see the configuration of boost converter or boost chopper, which is actually used to step up the DC–DC voltage and extensively used for low voltage to grid setup purpose. In addition, it acts as an MPPT device/controller under the nominal utility situations [45].

The output voltage is stepped up by varying the duty cycle of solid-state switch SSW (such as BJT, MosFET, and IGBT) in boost converter. This converter has some demerits such as high-voltage stress across the switching device, more conduction losses in the switch due to higher on-state resistance, reverse recovery problem of output diode, leakage inductance losses, and parasitic capacitance [46].

The output voltage level is always lesser than the input voltage level in the case of buck converter. The buck converter is also used in many applications for solar PV system such as step down of voltage for front-end application. For battery charging, it is pretty much necessary, and it is also used for MPPT application. The fundamental circuit diagram of the buck converter (or step-down chopper) is illustrated in Figure 9.14(b). Although it is extensively used in many applications, its demerits come in account as it works with pulsating source current. It generates large amount of electromagnetic interference noise, which has negative impact on the system [42].

In buck–boost DC–DC converter (or step-up/step-down chopper), the output voltage level may be lesser or higher than the input voltage level. By cascading the buck-type and boost-type topology, buck–boost-type DC–DC converter can be achieved. The fundamental circuit diagram of the buck–boost converter for solar PV applications is represented in Figure 9.15(a). The input current is discontinuous for all the time in the case of buck–boost converter [43]. High-input voltage ripple and high electrical stresses to the switch are major disadvantages of buck–boost topology. Cuk converter can work like a hybrid one similar as buck–boost converter having capabilities of both step-up as well as step-down voltage with reversing of polarity with the means of common terminal of input voltage. One of the differences of Cuk converter with general buck–boost converter is that the Cuk converter has a capacitor and an inductor fitted with its circuit [46,47]. We can understand the circuit of a Cuk converter from Figure 9.15(b).

The flyback topology (isolated-type converter) [47], shown in Figure 9.16, is one of the most frequently used switch mode power supply (SMPS) for low power applications. A flyback converter can be used for dual output (DC and DC) from single DC input source. It consists of a transformer that offers isolation between the supply and load output voltages. The flyback-type SMPS circuit's output power may fluctuate from few watts to maximum up to 100 W.

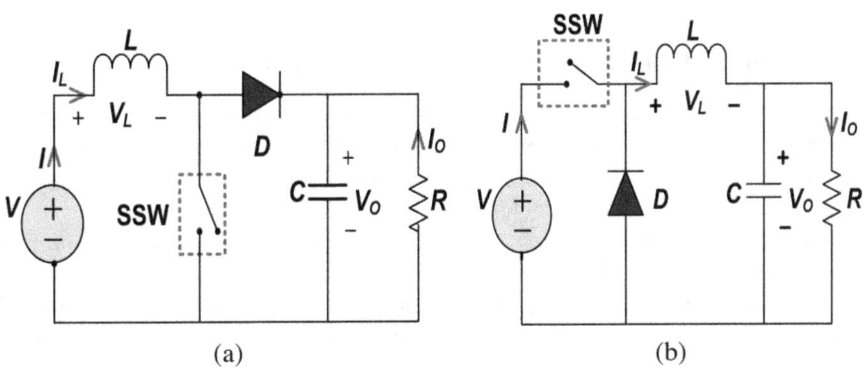

Figure 9.14. DC–DC converter topology: (a) boost type and (b) buck type.

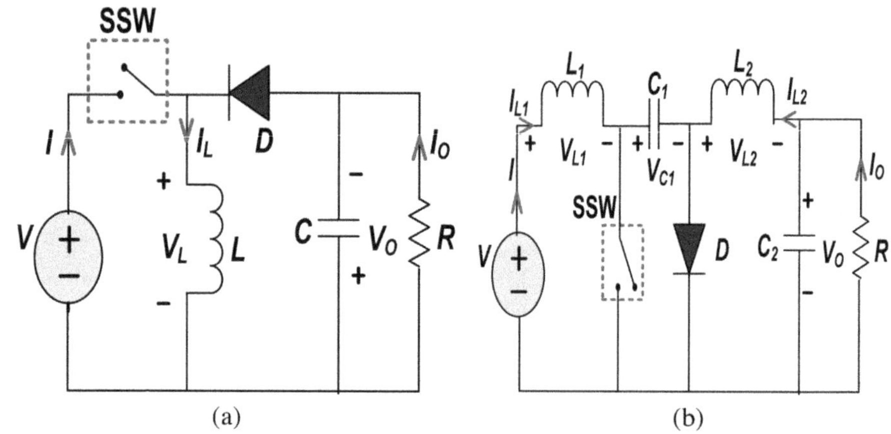

Figure 9.15. DC–DC converter topology. (a) Buck–boost type and (b) Cuk type.

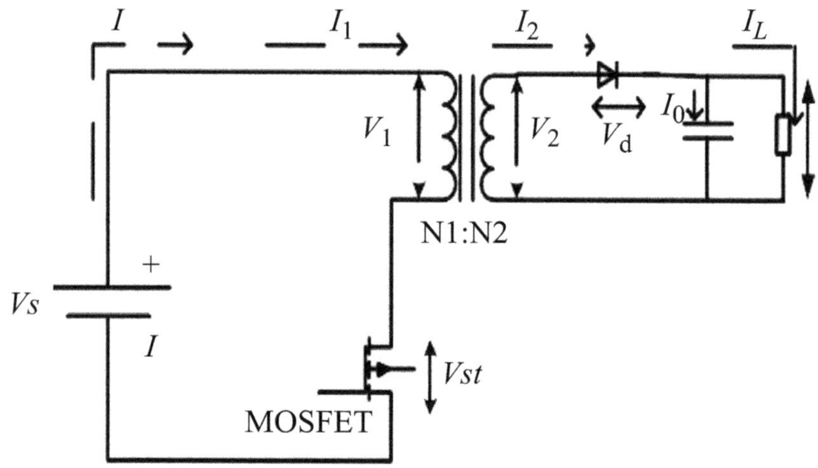

Figure 9.16. Circuit diagram of the flyback converter topology.

9.6 Single-Phase Inverter (Transformer-Less) Topologies

There are three basic requirements to the energy/power from PV arrays to utility grids: (1) conversion of the DC voltage into AC voltage; (2) voltage boosting, if the PV array voltage is lower than the grid voltage; (3) to insure maximum power exploitation of the PV modular.

The two typically used converter system configurations are illustrated in Figure 9.17. A transformer at line frequency is employed to step up the voltage subsequent to the DC–AC converter (i.e., inverter) in the system configuration shown in Figure 9.17(a). Typically, a line frequency transformer is related with loud acoustic noise, huge size, and cost. Also, the inverter has to be bulky to deal with the wide change in PV array voltage. The inverter KVA rating is doubled if the PV voltage changes at a range of 1:2. Thus, to remove the transformer and to minimize the requisite inverter, KVA rating as in numerous applications, a high-frequency DC–DC converter (i.e., chopper circuit) is employed to step up the voltage to a constant value as illustrated in Figure 9.17(b). But, the solid-state switches used in the DC–DC converter increases the overall cost and degrade the efficiency of the system [48,49]. Here, in following subsection, widely used transformer less single-phase inverter topologies are described and discussed.

In Figure 9.18, we can see a half H-bridge inverter topology for the PV application, where it works with two switches fitted with a capacitive divider within them.

There is a chance of leakage current through the parasitic capacitance of PV module, which can decrease or fluctuate the common mode voltage level. So, to get a constant common mode voltage the capacitive divider middle point is grounded [50]. This inverter configuration has some benefit along with some drawbacks as it is cheaper than the other

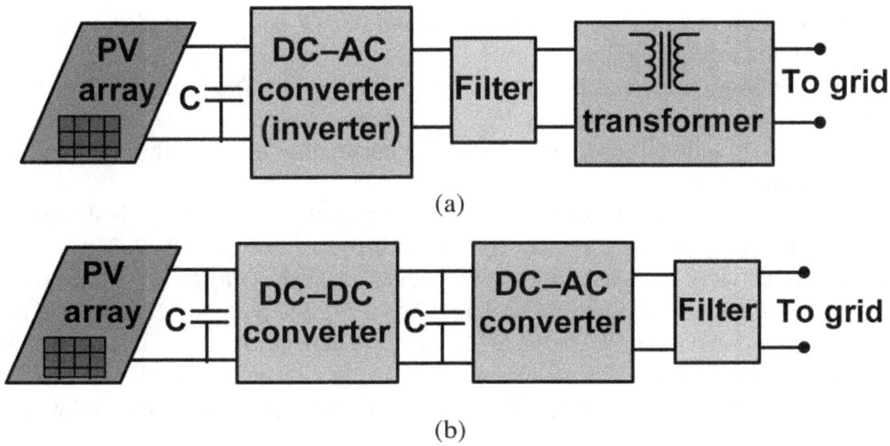

Figure 9.17. Traditional PV systems: (a) DC–AC with step-up transformer and (b) DC–AC with DC–DC boost.

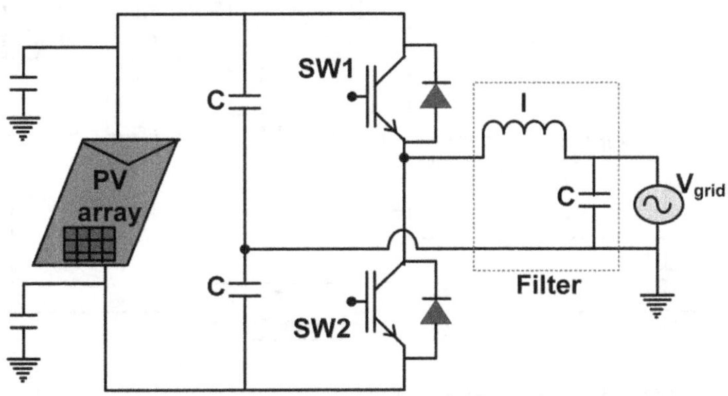

Figure 9.18. Half H-bridge inverter topology for the PV application.

configuration, but the switches have to work with two levels of waveform of output voltages. So, switches have to hold on high voltage and highly distorted current causes high electromagnetic interference [51]. Therefore, the multilevel half H-bridge inverter topologies for single-phase grid-connected PV systems is proposed in [52,53].

In Figure 9.19, we can see the full H-bridge inverter topology, where four switches are fitted unlike two switches of half H-bridge inverter for single-phase PV system. Unipolar pulse width modulation is generally used for PV systems and here a high-frequency common mode voltage of $\underline{V}_{dc}/2$ is given to the PV panels. In Figure 9.20, we can see a highly efficient and reliable inverter concept (HERIC) [54], where it will have a full-bridge inverter along with AC bypass topology. In HERIC inverter configuration, a few of the branches are coupled in parallel with the filter at the output side and have the benefits of condensed leakage current, higher efficiency, and three-level voltage at the output side by employing unipolar pulse width modulation with reduced ripple across the filter [55].

The H-5 inverter topology, illustrated in Figure 9.21, uses one extra DC-bypass switch compared to the full-bridge inverter topology [56]. Switches SW1 and SW2 are functioning at grid-frequency, whereas switches SW3, SW4, and SW5 are operating at high frequency. In the case of current freewheeling period, the switch SW5 is open to decouple the PV modules from the inverter [57].

This H5 topology (patented by SMA Solar Technology AG) of solar inverter has the advantages of approximately constant V_{cm} (i.e., common mode voltage), high efficiency mainly under the fractional loading, and reduced conduction losses with proper selection of semiconductor [58].

The H6 topology of PV inverter [59], derivative of the H5 topology, is designed by adding an extra semiconductor switch (SW6) as illustrated in Figure 9.22. The clamping branch creates a divider of capacitance and two diodes that can clamp the V_{cm} (i.e., common mode voltage) to half of the input DC voltage. The SW1 and SW4 are commutated at the line frequency and in antiparallel to the SW2 and SW3 depending on positive and negative half periods of the grid voltage cycle. During the time of freewheeling, diodes FD1 and FD2 can conduct (i.e., short circuited) depending on the amount of potential of the free-wheeling

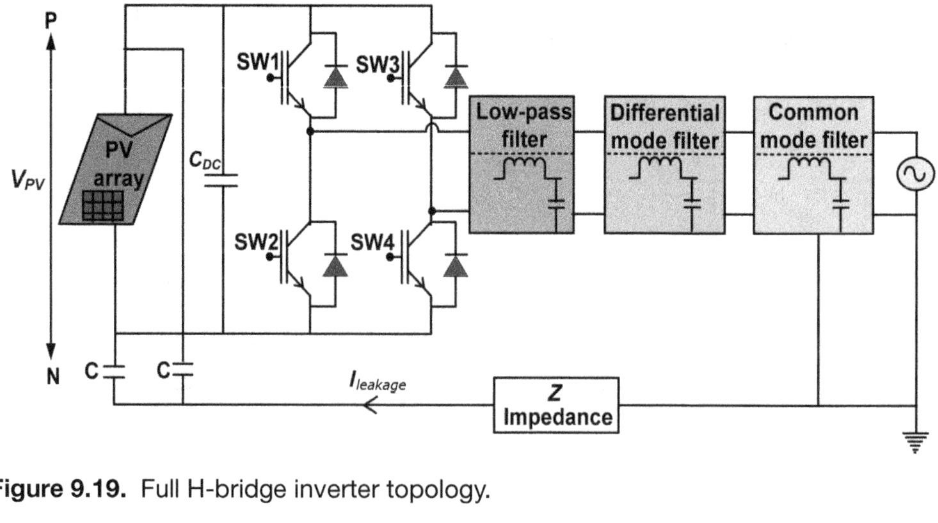

Figure 9.19. Full H-bridge inverter topology.

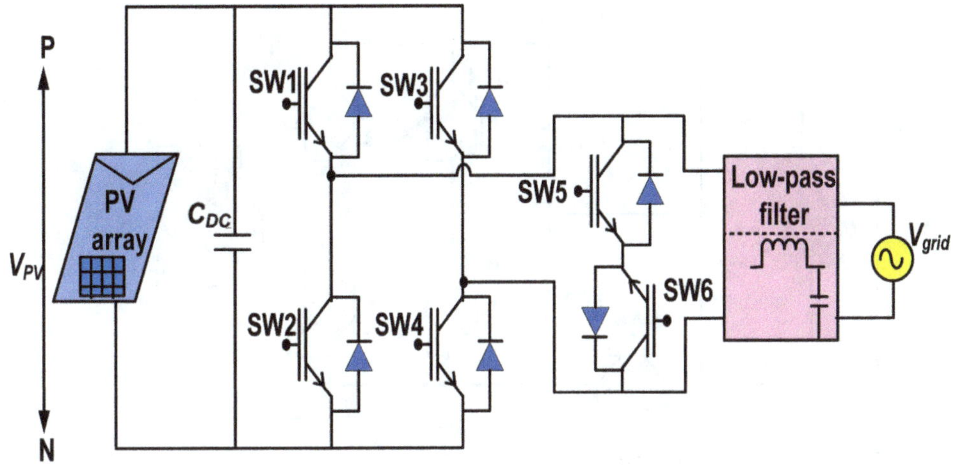

Figure 9.20. Highly efficient and reliable inverter concept (HERIC) topology.

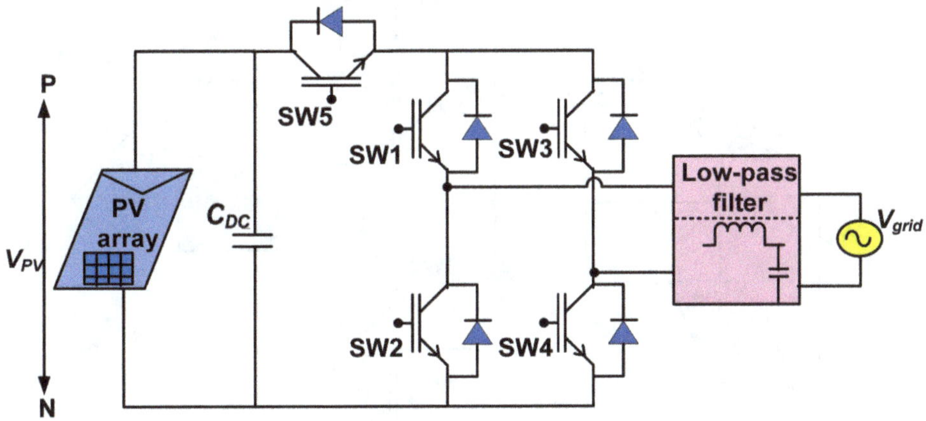

Figure 9.21. H5 inverter topology.

path in reference to the DC link voltage, that is, above or below. In H6 topology, unipolar PWM and double frequency PWM control methods are commonly used and the leakage current elimination depends on the clamping diodes' turn on speed [60]. A high conduction loss is key disadvantage of the H6 topology owing to the inductor current circulating in four active switches [61].

In neutral point clamped (NPC) topology, multilevel inverter technology is implemented for the single phase. A three-level half bridge voltage source inverter [62] based on NPC topology is illustrated in Figure 9.23.

The NPC topology has only one branch consisting of four switches (SW1, SW2, SW3, and SW4) and two diodes FD1 and FD2 for clamping the middle point voltage. In the freewheeling time, there is a setup of two clamping diodes, which actually gives the freewheeling path to the outgoing current and zero output voltage state [63]. This inverter topology

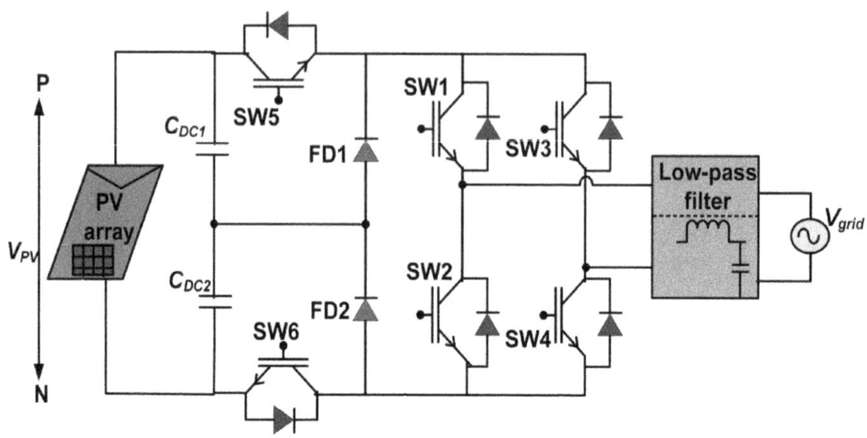

Figure 9.22. H6 inverter topology.

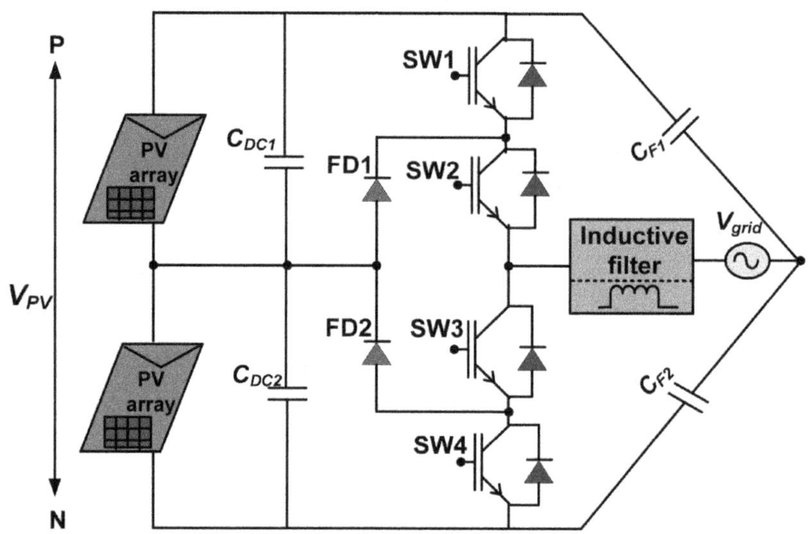

Figure 9.23. NPC inverter topology.

has some advantages of gain of high efficiency, but at the same time it requires high input voltage compared to the full H-bridge topology and also requires an extra capacitor bank to work properly [64].

The active NPC topology, shown in Figure 9.24, is a modified version of NPC topology where the clamping diodes are replaced by controlled power semiconductor switches (SW5 and SW6). NPC topology of the inverter is modified by replacing the clamping as compared to NPC topology [65].

Several different ways are provided due to the use of controlled power semiconductor switches (SW5 and SW6) for clamping the midpoint voltage. Turn on of the SW2 and SW5 provides the upper clamping path, whereas the actuation of the SW3 and SW6 provides the lower clamping path [66]. The active NPC topology is having inner voltage boosting capability and has the merits high efficiency and low power loss in reference to the solar PV applications.

The flying capacitor (FC) topology, illustrated in Figure 9.25, is fabricated by replacing clamping diodes (FD1 and FD2) from NPC topology with a capacitor known as flying capacitor (C_{FC}). As the name implies, the flying capacitor (C_{FC}) is floating in respect to the DC source reference and provides the additional levels of clamping voltage [67,68].

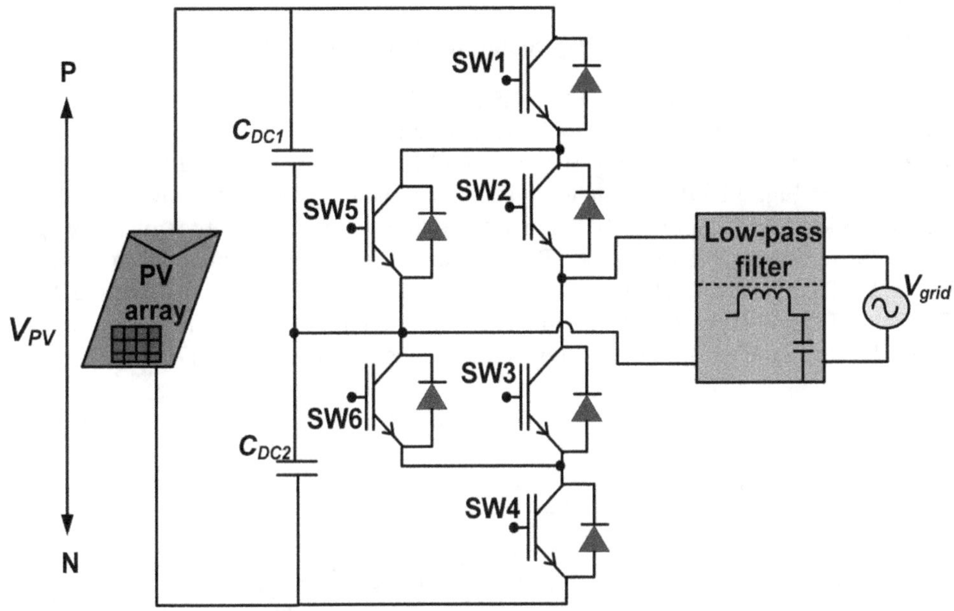

Figure 9.24. Active neutral point clamped (ANPC) inverter topology.

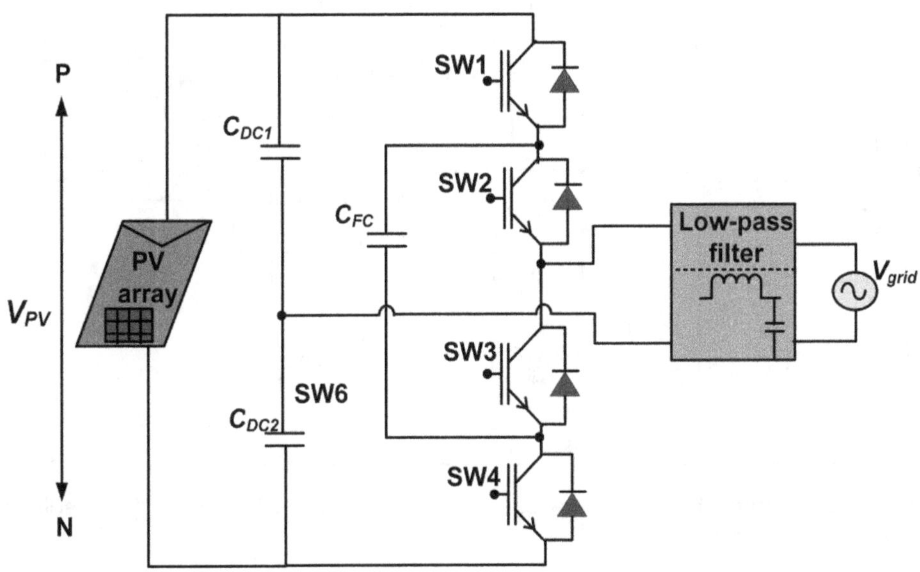

Figure 9.25. FC NPC inverter topology.

The beauty of the FC configuration is that it has competence of fault-tolerant operation due to higher number of levels. The FC topology has the demerits of high voltage during imbalance, which may demolish inverter system [69].

9.7 Three-Phase Inverter (Transformer-Less) Topologies

As similar to preceding section, the three-phase solar PV inverter topologies are obtained from the single-phase inverter topologies (described earlier) by same way in all the three phases. Three-phase three-level inverter based on NPC topology for solar PV energy system is illustrated in Fig. 9.26. The complete system consists of a DC-to-DC converter, solar PV array, three-level NPC topology-based inverter, output LC filter, and three-phase AC grid. Each of the three branches of the solar PV inverter consists of four semiconductor switches (SW1a, SW2a, SW3a, and SW4a) with four free-wheeling diodes and two clamping diodes (FD1a and FD2a). Three-phase three-level NPC topology-based inverter can create five voltage levels on the AC grid depending on the frequency of switching [70].

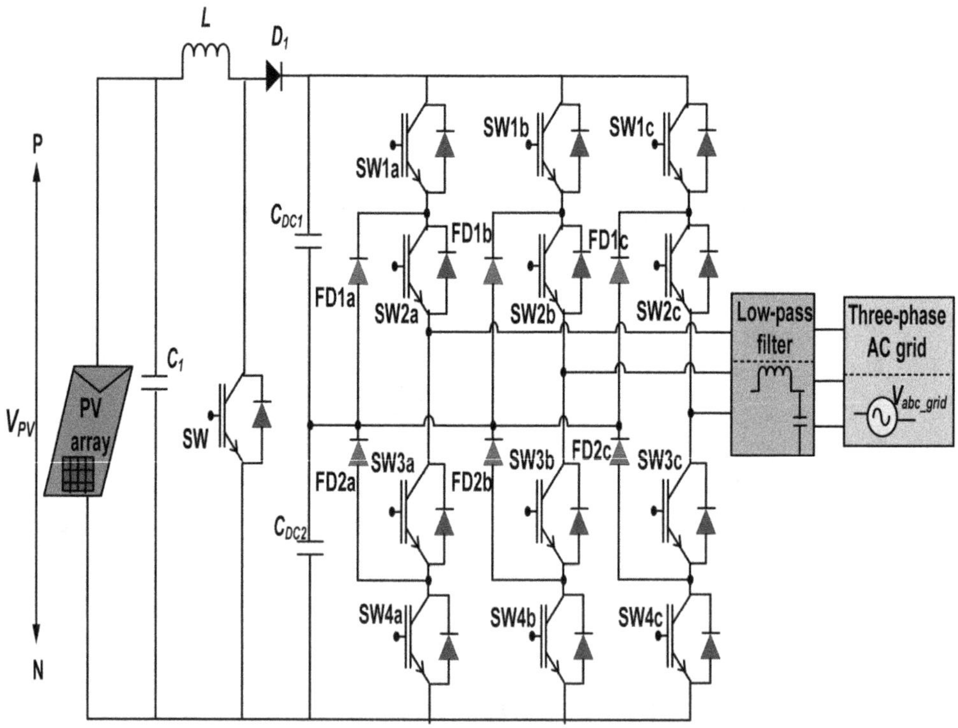

Figure 9.26. Three-phase three-level inverter based on NPC topology for solar PV energy system.

9.8 Different Control Schemes for Grid-Tied PV Inverter System

The ABC control of the grid-connected PV inverter system is shown in Fig. 9.27. In the case of ABC control, a dedicated controller for each phase of grid current is given as illustrated in the figure. Some of the controllers reported in literature are hysteresis current controller, dead-beat controller, PI and PI-R controller [31,71–74].

In the case of $\alpha\beta$-control technique, the transformation of three-phase grid currents into $\alpha\beta$ or stationary reference frame from ABC-reference frame is done. The advantages of this controller are higher current gain around the resonant frequency, increased dynamic characteristics, and lower steady-state error between the controlled signal and its reference [75].

The configuration of d-D-Q control technique is shown in Fig. 9.28. The D-Q control method for three-phase grid-tied PV inverters is implemented using PI controller in synchronous reference frame [76], and combination of fuzzy logic with PID controller [77] has been reported in literature. In D-Q control technique, the three-phase grid currents are transformed into D-Q or synchronous reference frame, that is, three-phase grid currents (I_{g_abc}) are converted into direct axis component (I_{d}) and quadrature axis component (I_{q}) voltage and currents are transformed into a rotating reference system with the grid voltage, using the ABC–DQ transformation in the synchronous reference frame. This transformation is used to decouple the three AC currents into active (I_d, direct component) and reactive (I_q, quadrature component) power components. The I_d and I_q current components are synchronized to reduce the error between the measured and reference values of active and reactive powers [78].

A complete integrated control of grid-connected solar PV system is illustrated in Fig. 9.29. Basically, the integrated control system is having two subsections: DC–DC/MPPT controller and grid side converter (GSC) or grid side inverter (GSI) controller. The grid side DC-to-AC inverter assists in regulating the power factor through supplying the sinusoidal current to the AC grid. The direct current power/energy produced by the solar PV is transformed into the alternating current power/energy than it is adequately supplied to the AC grid via grid side converter/inverter. The appropriate function of the GSI is established through designing fast and accurate control scheme (described in Figs. 9.27 and 9.28). Therefore, the GSI controller is one of the most important part of the grid-connected

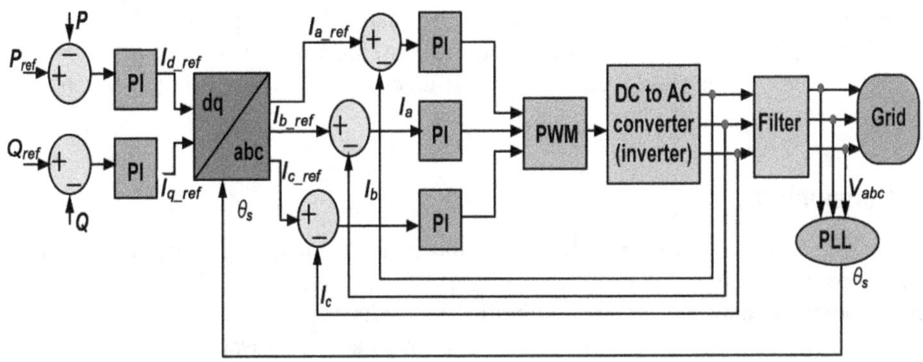

Figure 9.27. Three-phase or ABC control of solar PV DC/AC converter or inverter.

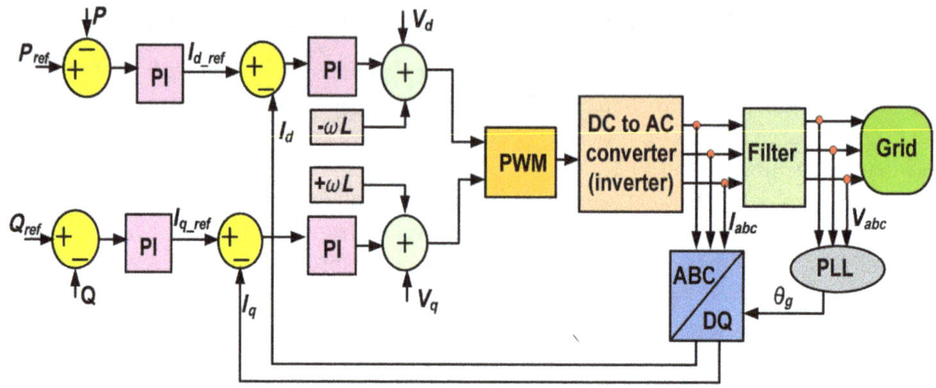

Figure 9.28. D-Q control of solar PV DC/AC converter or inverter.

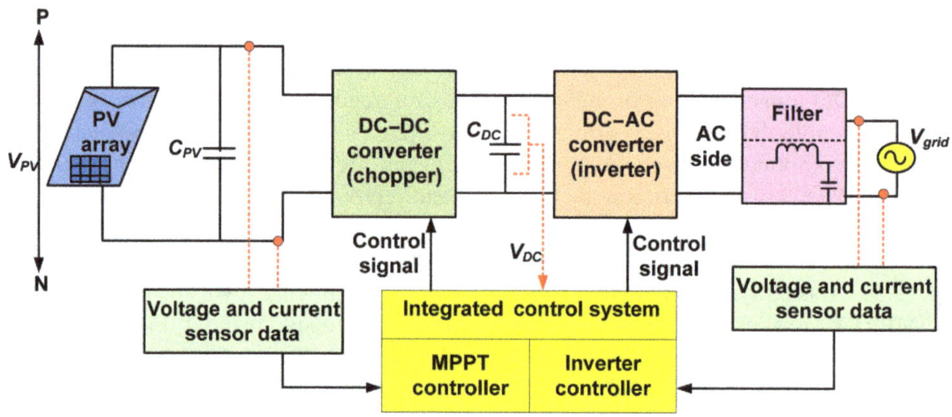

Figure 9.29. Complete integrated control of solarPV system.

solar PV system. The main function of DC–DC converter is to track the MPP in solar PV applications according to the used MPPT control algorithm. The MPPT controller is the heart of the solar PV integrated control system. Basically, the voltage output from PV string is very low regardless of the use of MPPT controller. This needs the utilization of front-end DC–DC converter to be employed with a boost-up competency for grid connection.

9.9 Conclusion

This chapter presents a brief introduction of the solar PV energy systems, dependence on different parameter, possible operating modes, that is, on-grid and off-grid, different converter topologies, and their control with all the aspects covering standards and specifications. In addition, integration of solar PV with grid has been discussed via DC–DC and DC–AC converters. Two configurations have been demonstrated for these PV systems such as single-stage and two-stage operations. This chapter has been focused on the application of the different types of DC–DC converters and inverters employed for DC–AC power conversion. In the conversion, single-stage operation and two-stage converter, that is, DC–DC converter for DC load system and DC–AC converter for AC load system to make the efficient output of PV as per the requirement of grid/load.

9.10 References

1. S. Kalogirou, "Solar thermal collectors and applications," *Progress in Energy and Combustion Science*, vol. 30, no. 3, pp. 231–295, 2004.
2. G. Tsengenes, and G. Adamidis, "Investigation of the behavior of a three-phase grid connected photovoltaic system to control active and reactive power." *Electric Power System Research*, vol. 81, no. 1, pp. 177–184, 2011.
3. K. Zeb, W. Uddin, M. Adil Khan, M. Umair Ali, N. Christofides, and H. Kim, "A comprehensive review on inverter topologies and control strategies for grid connected photovoltaic system." *Renewable and Sustainable Energy Reviews*, vol. 94, pp. 1120–1141, 2018.
4. P. Garg, and T. C. Kandpa, "Renewable energy education in developing countries," Proceedings of 2nd World Renewable Energy Congress, ScienceDirect, 1992.
5. T. Muneer, M. Asif, and S. Munawwar. "Sustainable production of solar electricity with particular reference to the Indian economy," *Renewable and Sustainable Energy Reviews*, vol. 9, pp. 444–473, 2005.
6. National Petroleum Council, "Is an Oil and Natural Gas Advisory Committee to the Secretary of Energy." 2007. Available online: http://www.npc.org/
7. D. D. Chiras, *The Solar House: Passive Solar Heating and Cooling*. Chelsea Green Publishing Company, 2002.
8. B. Sorensen, Renewable Energy: Its Physics, Engineering, use, Environmental Impacts Economy and Planning Aspects, 2nd ed. Academic Press, Sand Diego, CA, 2000.
9. K. Zeb, W. Uddin, M. A. Khan, Z. Ali, M. U. Ali, N. Christofides, and H. J. Kim, "A comprehensive review on inverter topologies and control strategies for grid connected photovoltaic system." *Renewable and Sustainable Energy Reviews*, vol. 94, pp. 1120–1141, 2018.
10. P. T. Manditereza, and R. C. Bansal, "Review of technical issues influencing the decoupling DG converter design from the distribution system protection strategy. *IET-Renewable Power Generation*, vol. 12, no. 10, pp. 1091–1100, 2018.
11. R. C. Bansal, *Handbook of Distributed Generation: Electric Power Technologies, Economics and Environmental Impacts*, Springer, Switzerland, 2017.
12. U. Jadli, P. Thakur, and R. D. Shukla, "A new parameter estimation method of solar photovoltaic." *IEEE Journal of Photovoltaics*, vol. 8, no. 1, pp. 239–247, Jan. 2018.
13. W. Libo, Z. Zhengming, and L. Jianzheng, "A single-stage three-phase grid-connected photovoltaic system with modified MPPT method and reactive power compensation." *IEEE Trans Energy Convers,* vol. 22, no. 4, pp. 881–886, 2007.
14. O. P. Mahela, and A. G. Shaik, "Comprehensive overview of grid interfaced solar photovoltaic systems," *Renewable and Sustainable Energy Reviews*, vol. 68, no. P1, pp. 316–332, 2017.
15. A. Zobaa, and R. Bansal, *Handbook of Renewable Energy Technology,* World Scientific Pub. Co., Singapore, 2011
16. M. ElNozahy, and M. Salama, "Technical impacts of grid-connected photovoltaic systems on electrical networks—A review." *Journal of Renewable and Sustainable Energy*, vol. 5, no. 3, p. 032702, 2013.
17. R. Dubey, R. Joshi, and R. C. Bansal, "Optimization of solar photovoltaic plant and economic analysis," *Electric Power Components and Systems*, vol. 44, no. 18, pp. 2525–2535, 2016.

18. S. Manju, and N. Sagar, "Progressing towards the development of sustainable energy: A critical review on the current status, applications, developmental barriers and prospects of solar photovoltaic systems in India." *Renewable and Sustainable Energy Reviews,* vol. 70, pp. 298–313, 2017.
19. The International Renewable Energy Agency (IRENA) renewable power generation capacity statistics: https://www.irena.org/publications/2021/March/Renewable-Capacity-Statistics-2021
20. NS Energy, Available online: https://www.nsenergybusiness.com/industry_taxonomy/power/
21. V. Rustagi, (Managing Director, BRIDGE-TO-INDIA), "Detailed scientific Report: INDIA SOLAR COMPASS 2017Q2" Available online: https://www.bridgetoindia.com/wp-content/uploads/2018/12/BRIDGE-TO-INDIA-India-Solar-Compass-Q2-2017-Full-report.pdf
22. M. Das, and V. Agarwal, "Design and analysis of a high-efficiency DC–DC converter with soft switching capability for renewable energy applications requiring high voltage gain," *IEEE Transactions on Industrial Electronics*, vol. 63, no. 5, pp. 2936–2944, 2016.
23. A. M. Bagher, M. M. A. Vahid, and M. Mohsen, "Types of solar cells and application," *American Journal of Optics and Photonics*, vol. 3, no. 5, pp. 94–113, 2015.
24. S. Kumar, R. Kumar, and N. Singh, "Performance of closed loop SEPIC converter with DC-DC converter for solar energy system," 4th International Conference on Power, Control & Embedded Systems (ICPCES), IEEE, Allahabad, pp. 1–6, 2017.
25. S. B. Kjaer, J. K. Pedersen, and F. Blaabjerg, "A review of single-phase grid-connected inverters for photovoltaic modules," *IEEE Transactions on Industry Applications*, vol. 41, no. 5, pp. 1292–1306, 2005.
26. R. Reshma Gopi, and S. Sreejith, "Converter topologies in photovoltaic applications—A review," *Renewable and Sustainable Energy Reviews*, vol. 94, pp. 1–14, 2018.
27. L. Zhang, K. Sun, L. Feng, H. Wu, and Y. Xing, "A family of neutral point clamped full-bridge topologies for transformer less photovoltaic grid-tied inverters," *IEEE Transactions on Power Electronics*, vol. 28, no. 2, pp. 730–739, 2013.
28. S. B. Kjaer, J. K. Pedersen, and F. Blaabjerg, "A review of single-phase grid-connected inverters for photovoltaic modules," *IEEE Transactions on Industry Applications*, vol. 41, no. 5, pp. 1135–1141, 2005.
29. J. Jana, H. Saha, and K. D. Bhattacharya, "A review of inverter topologies for single-phase grid-connected photovoltaic systems," *Renewable and Sustainable Energy Reviews*, vol. 72, pp. 1256–1270, 2017.
30. R. Teodorescu, and F. Blaabjerg, "Overview of renewable energy system," ECPE Seminar Renewable Energy, ISET, Kassel, Germany, 2006.
31. R. Teodorescu, F. Blaabjerg, M. Liserre, and P. C. Loh, "Proportional-resonant controllers and filters for grid-connected voltage-source converters," *IEE Electric Power Applications*, vol. 153, no. 5, pp. 750–762, 2006.
32. H. Haeberlin, "Evolution of inverters for grid connected PV-systems from1989 to 2000," Proceedings of the 17th European Photovoltaic Solar Energy Conference, Munich, Germany, pp. 426–430, Oct. 22–26, 2001.
33. S. B. Kjaer, J. K. Pedersen, and F. Blaabjerg, "Power inverter topologies for photovoltaic modules—A review," *Conference Recreational IEEE-IAS Annual Meeting*, vol. 2, pp. 782–788, 2002.
34. T. Shimizu, M. Hirakata, T. Kamezawa, and H. Watanabe, "Generation control circuit for photovoltaic modules," *IEEE Transactions on Power Electron*, vol. 16, no. 3, pp. 293–300, 2001.
35. M. Calais, J. Myrzik, T. Spooner, and V. G. Agelidis, "Inverters for single-phase grid connected photovoltaic systems—An overview," *Proceedings of the IEEE PESC'02*, vol. 2, pp. 1995–2000, 2002.

36. D. Casadei, G. Grandi, and C. Rossi, "Single-phase single-stage photovoltaic generation system based on a ripple correlation control maximum power point tracking," *IEEE Transactions Energy Conversion*, vol. 21, no. 2, pp. 562–568, 2005. doi:10.1109/TEC.2005.853784
37. C. L. Kuo, C. H. Lin, H. T. Yau, and J. L. Chen, "Using self-synchronization error dynamics formulation-based controller for maximum photovoltaic power tracking in microgrid systems." *IEEE Journal Emerging Selected Topics Circuits Systems*, vol. 3, no. 3, pp. 459–467, 2013. doi:10.1109/JETCAS.2013.2272839
38. B. Bendib, H. Belmili, and F. Krim, "A survey of the most used {MPPT} methods: Conventional and advanced algorithms applied for photovoltaic systems," *Renewable and Sustainable Energy Reviews,* vol. 45, pp. 637–648, 2015.
39. S. Jain, and V. Agarwal, "Comparison of the performance of maximum power point tracking schemes applied to single-stage grid-connected photovoltaic systems," *IET Electric Power Applications*, vol. 1, no. 5, pp. 753–62, 2007.
40. B. Subudhi, and R. Pradhan, "A comparative study on maximum power point tracking techniques for photovoltaic power systems," *IEEE Transactions Sustainable Energy*, vol. 4, no. 1, pp. 89–98, 2013.
41. J. Zeng, W. Qiao, and L. Qu, "A single-switch isolated dc-dc converter for photovoltaic systems," Proceedings of the IEEE Energy Conversion Congress and Exposition (ECCE), Raleigh, NC, USA, pp. 3446–3452, 2012.
42. M. Veerachary, "Fourth-order buck converter for maximum power point tracking applications," *IEEE Transactions Aerospace Electronic Systems,* vol. 47, no. 2, pp. 896–911, 2011. doi:10.1109/TAES.2011.5751233
43. E. Durn, J. Andjar, F. Segura, and A. Barragn, "A high-flexibility (DC) load for fuel cell and solar arrays power sources based on DC–DC converters," *Applied Energy*, vol. 88, no. 5, pp. 1690–702, 2011. doi:10.1016/j.apenergy.2010.11.002
44. M. Farahat, H. Metwally, and A. A. E. Mohamed, "Optimal choice and design of different topologies of DC–DC converter used in (PV) systems, at different climatic conditions in Egypt," *Renew Energy,* vol. 43, pp. 393–402, 2012.
45. H. L. Jou, Y. H. Chang, J. C. Wu, and K. D. Wu, "Operation Strategy for a Lab-Scale Grid Connected Photovoltaic Generation System Integrated with Battery Energy Storage," *Energy Conversion and Management*, vol. 89, pp. 197–204, 2015.
46. M. Das, and V. Agarwal, "A novel high efficiency, high gain, front end dc-dc converter for low input voltage solar photovoltaic applications," Proceedings of the 38th Annual Conference on IEEE Industrial Electronics Society, Montreal, QC, Canada, pp. 5744–5749, 2012.
47. M. Shayestegan, M. Shakeri, H. Abunima, S. M. Salim Reza,M. Akhtaruzzaman, B. Bais, S. Mat, K. Sopian, and N. Amin, "An overview on prospects of new generation single-phase transformer less inverters for grid-connected photovoltaic (PV) systems," *Renewable and Sustainable Energy Reviews*, vol. 82, pp. 515–530, 2018.
48. M. Calais, J. Myrzik, T. Spooner, and V. C. Agelidis, "Inverters for single-phase grid connected photovoltaic systems-an overview," *Proceedings IEEE 33rd Annual Power Electronics Specialists Conference*, vol. 4, no. 23–27, pp. 1995–2000, Jun. 2002.
49. T. F. Wu, C. H. Chang, and Y. K. Chen, "A multi-function photovoltaic power supply system with grid-connection and power factor correction features," *Proceedings IEEE 31st Annual Power Electronics Specialists Conference*, vol. 3, no. 18–23, pp. 1185–1190, Jun. 2000.
50. C. L. Shen, and S. T. Peng, "A half-bridge PV system with bi-direction power flow controlling and power quality improvement," Proceedings of the 7th International Conference on Power Electronics and Drive Systems, PEDS '07, pp. 725–731, 2007.
51. I. Patrao, E. Figueres, F. Gonzlez-Espn, and G. Garcer, "Transformer less topologies for grid-connected single-phase photovoltaic inverters," *Renewable and Sustainable Energy Reviews,* vol. 15, no. 7, pp. 3423–3431, 2011.

52. M. Calais, and V. Agelidis, "Multilevel converters for single-phase grid connected photovoltaic systems-an overview," *Proceedings of the IEEE International Symposium on Industrial Electronics, ISIE '98*, vol. 1, pp. 224–229, 1998.
53. Y. Xue, and M. Manjrekar, "A new class of single-phase multilevel inverters," Proceedings of the 2nd IEEE International Symposium on Power Electronics for Distributed Generation Systems (PEDG), Hefei, China, pp. 565–571, 2010.
54. Y. Yang, F. Blaabjerg, and H. Wang, "Low-voltage ride-through of single-phase transformer less photovoltaic inverters," *IEEE Transactions Industry Applications*, vol. 50, no. 3, pp. 1942–52, 2014. doi:10.1109/TIA.2013.2282966
55. P. Q. Dzung, D. n.d.at, N. B. Anh, L. C. Hiep, and H. H. Lee, "Design of HERIC inverter for PV systems by using hardware in the loop (HIL) concept," IEEE 9th Conference on Industrial Electronics and Applications (ICIEA), Hangzhou, China, pp. 2035–2040, 2014.
56. L. Zhang, K. Sun, Y. Xing, and J. Zhao, "Parallel operation of modular single-phase transformer-less grid-tied PV inverters with common DC bus and AC bus," *IEEE Journal Emerging Selected Topics Power Electronics*, vol. 3, no. 4, pp. 858–869, 2015.
57. W. Li, Y. Gu, H. Luo, W. Cui, X. He, and C. Xia, "Topology review and derivation methodology of single-phase transformer-less photovoltaic inverters for leakage current suppression," *IEEE Transactions Industrial Electronics*, vol. 62, no. 7, pp. 4537–4551, 2015.
58. H. Xiao, S. Xie, Y. Chen, and R. Huang, "An optimized transformer-less photovoltaic grid connected inverter," *IEEE Transactions Industrial Electronics*, vol. 58, no. 5, pp. 1887–1895, 2011.
59. R. Gonzalez, J. Lopez, P. Sanchis, and L. Marroyo, "Transformer-less inverter for single phase photovoltaic systems," *IEEE Transactions Power Electronics*, vol. 22, no, 2, pp. 693–697, 2007.
60. L. Zhang, K. Sun, Y. Xing, and M. Xing, "H6 transformer-less full-bridge PV grid-tied inverters," *IEEE Trans Power Electron*, vol. 29, no. 3, pp. 1229–1238, 2014.
61. M. Islam, and S. Mekhilef, "H6-type transformer-less single-phase inverter for grid-tied photovoltaic system," *IET Power Electronics*, vol. 8, no. 4, pp. 636–644, 2015.
62. I. Roasto, D. Vinnikov, and V. Vodovozov, "Simulation and evaluation of control methods for the rolling stock static auxiliary converter based on three-level NPC inverter topology," Proceedings of the International Conference on Power Engineering, Energy and Electrical Drives, POWERENG '09, pp. 593–598, 2009.
63. K. Oliveira, M. Cavalcanti, and J. Afonso, A. Farias, and F. Neves, "Transformer-less photovoltaic systems using neutral point clamped multilevel inverters," Proceedings of the IEEE International Symposium on Industrial Electronics (ISIE), pp. 1131–1136, 2010.
64. R. Latha, C. Bharatiraja, R. Palanisamy, S. S. Sudeepbanerji, Dash, "Hysteresis current controller based transformerless split inductor-NPC for grid connected PV system," *Procedia Engineering, International Conference on Design and Manufacturing (IConDM2013)*, vol. 64, pp. 224–233, 2013.
65. N. Achilladelis, E. Koutroulis, and F. Blaabjerg, "Optimized pulse width modulation for transformer-less active-NPC inverters," Proceedings of the 16th European Conference on Power Electronics and Applications (EPE'14-ECCE Europe), Lappeenranta, Finland, pp. 1–10, 2014.
66. Y. Wang, and F. Wang, "Novel three-phase three-level-stacked neutral point clamped grid-tied solar inverter with a split phase controller," *IEEE Transactions Power Electronics*, vol. 28, no. 6, pp. 2856–2866, 2013.
67. S. Daher, J. Schmid, and F. Antunes, "Multilevel inverter topologies for stand-alone PV systems," *IEEE Transactions Industrial Electronics*, vol. 55, no. 7, pp. 2703–2712, 2008.
68. M. Gao, M. Chen, C. Zhang, and Z. Qian, "Analysis and implementation of an improved flyback inverter for photovoltaic ac module applications," *IEEE Trans Power Electron*, vol. 29, no. 7, pp. 3428–3444.

69. M. B. Latran, and A. Teke, "Investigation of multilevel multifunctional grid connected inverter topologies and control strategies used in photovoltaic systems," *Renewable and Sustainable Energy Reviews*, vol. 42, pp. 361–376, 2015.
70. Y. Kim, H. Cha, B. M. Song, K. Lee, "Design and control of a grid-connected three phase 3-level NPC inverter for building integrated photovoltaic systems," Proceedings of the IEEE PES Innovative Smart Grid Technologies (ISGT), pp. 1–7, 2012.
71. D. Zmood, and D. Holmes, "Stationary frame current regulation of PWM inverters with zero steady-state error," *IEEE Transactions Power Electronics*, vol. 18, no. 3, pp. 814–22, 2003.
72. Q. Chunqing, Y. Yong, and S. Ji, "Deadbeat decoupling control of three-phase photovoltaic grid-connected inverters," Proceedings of the International Conference on Mechatronics and Automation, ICMA, pp. 3843–3848, 2009.
73. H. Su Fen, Y. Ling-Zhi, L. Ju Cheng, Y. Zhe Zhi, and P. Han Mei, "Design of three-phase photovoltaic grid connected inverter based on RBF neural network," Proceedings of the International Conference on Sustainable Power Generation and Supply, SUPERGEN '09, Nanjing, China, pp. 1–5, 2009.
74. L. Hassaine, E. OLias, J. Quintero, and V. Salas, "Overview of power inverter topologies and control structures for grid connected photovoltaic systems," *Renewable and Sustainable Energy Reviews*, vol. 30, pp. 796–807, 2014.
75. A. B. Rey-Bou, R. Garca-Valverde, F. de, A. Ruz-Vila, and J. M. Torrelo-Ponce, "An integrative approach to the design methodology for 3-phase power conditioners in photovoltaic grid-connected systems," *Energy Conversion Management*, vol. 56, pp. 80–95, 2012.
76. M. Liang, and T. Zheng, "Synchronous PI control for three-phase grid-connected photovoltaic inverter," Proceedings of the Chinese Control and Decision Conference (CCDC), Xuzhou, China, pp. 2302–2307, 2010.
77. M. Louzazni, and E. Aroudam, "Control and stabilization of three-phase grid connected photovoltaics using PID-Fuzzy logic," In Proceedings of the IEEE International Conference on Intelligent Energy and Power Systems (IEPS), Kiev, Ukraine, pp. 279–284, 2014.
78. M. Schonardie, A. Ruseler, R. Coelho, and D. Martins, "Three-phase grid-connected pv system with active and reactive power control using dq0 transformation," Proceedings of the 9th IEEE/IAS International Conference on Industry Applications (INDUSCON), Sao Paulo, Brazil, pp. 1–6, 2010.

CHAPTER 10

Design Aspects of Solar PV Modules, Selection of Battery, and Protection Schemes

Saji Chacko[1], R. N. Patel[2], and Ramesh C. Bansal[3]

[1]Department of Electrical, Government Polytechnic Durg, Durg, Chhattisgarh, India. chackosaji68@gmail.com
[2]Department of Electrical Engineering, National Institute of Technology Raipur, Raipur, Chhattisgarh, India. rnpatel.ee@nitrr.ac.in
[3]Department of Electrical Engineering, University of Sharjah, United Arab Emirates, and Department of Electrical, Electronic and Computer Engineering, University of Pretoria, Pretoria, South Africa, email: rcbansal@ieee.org

Abstract The chapter focuses on the various design aspects of solar PV (photovoltaic) system based on consumer requirements and load pattern. Sizing of PV panel for application with and without batteries, selection of batteries as storage system, choice of battery type, sizing of cables, and most importantly the design of protection system against overcurrent, overvoltage, and lightning surges, have been elaborated upon.

Section 10.1 deals with the type of consumer and the electrical appliances to be operated. Load size and energy demand, time of operation, and calculation of solar energy irradiation of the locality have been discussed. Section 10.2 deals with the study of PV panel specification and sizing of panels for application without battery with consideration of maximum demand. Section 10.3 deals with the types of battery, battery parameters, and selection of battery capacity considering days of autonomy. Sizing of PV for different load configurations with battery, need of battery equalization circuit, and discussions on active balancing techniques for removal of charges from higher capacity battery and delivering it to lower capacity battery have been discussed. Subsequently, Section 10.4 deals with the sizing of cables and factors to be considered, protection of PV modules, and location of protection equipments. Effect of shading or mismatch of PV cell characteristics on the series/parallel connection of PV cells to form arrays/modules have also been discussed in the same context.

Keywords: Design of solar PV systems, Load estimation, Series-parallel operation, Solar pumping system, Energy irradiation, Battery sizing, Power/energy density, Battery selection, Battery equalization, PV arrays, Protection schemes.

10.1 Load Estimation

The design of a reliable and efficient solar photovoltaic (PV) system commences with the calculation of the daily energy consumption pattern of the installed load. The system design based on the erroneous calculation of the load may lead to frequent breakdown and finally system failure [1, 2]. Therefore, the first step involved in the design of a reliable PV system is to have accurate information of the load. The designer needs to know the ratings of the appliances/equipment used and also their usage pattern. The load size determines the rating as well as the size of PV arrays/modules and batteries. If the information accessed is inaccurate, the calculated size of the PV system could be too high leading to an increase in its initial cost or too low leading to system failure.

Conventionally, for an existing load, the rating of the transformer is considered as the size of the connected load. Very often the transformer with a higher rating is selected for power supply to a given load keeping in view the future load growth demand. Selection of a PV system based on the above load calculation will lead to increased initial cost; therefore, it becomes pertinent to estimate the load profile throughout the year if the said load is to be retrofitted with a PV system.

The second step involved is to determine the type of load such as light or fan load whose load profile throughout the year will remain almost constant for either domestic or commercial use. On the other hand, loads such as refrigerators or air conditioners are of different types where the load demand depends on the season, that is, summer or winter; similarly, the irrigation pumps have the load profile that depends on the season as well as the time of the day. Finally, loads that are critical in nature such as the loads in an operation theater or loads related to surveillance or communication systems, where not only the hours of operation, but also the mode of operation, that is, transmitting or receiving as well as the hours of standby operation, should be included in the load calculations.

The third step involved is the use of more efficient load. For example, an existing fluorescent or incandescent lamp can be replaced by a more efficient CFL or LED lamps. In the same way, an existing conventional drive for air conditioning or refrigeration system may be replaced by energy-efficient inverter drive compressor and fan motors. It should also be observed while selection of load whether the load is adaptable for AC and DC supply or for both. The emphasis should be on DC loads so that the need of DC–AC converter can be eliminated resulting in improved efficiency. The insolation on a flat surface varies greatly with season as well, with the insolation level minimum during the winter season and maximum during the summer season. Depending on the load demand, an insolation profile with the best fit to that of the load profile is achieved by tilting the PV array at suitable angle [3, 4]. Accordingly, a different tilt angle may be chosen if the load is relatively constant every month due to change in orientation of the sun.

The load (or the energy consumed) is commonly expressed in watt hours (Wh). For example, the Wh calculation of a 40-watt incandescent lamp turned on for 8 h and 60 watt LED television turned on for 5 h is as follows:

(i) Wh of incandescent lamp = 40 W × 8 h = 320 Wh
(ii) Wh of LED television = 60 W × 6 h = 360 Wh

Then the total load in Wh is 680 Wh.

For PV system with battery backup, the total load is calculated in ampere hour (Ah), where $\text{Ah} = \frac{\text{Wh}}{V}$, where V is the nominal system voltage.

10.2 Load Calculation

The preparation of the load table is based on the type of load. The chapter focuses on two types of load:

(i) Domestic/commercial load
(ii) Water pumping system

10.2.1 Domestic/Commercial Load

The first step in designing a PV system is to prepare the load table. The load table enables to calculate the per day Wh consumption of all the connected loads. The following columns should be considered while preparing the load table:

(i) Type of load whether AC or DC loads. It is better to prepare separate load tables for AC and DC loads.
(ii) Rating of each load in wattage.
(iii) Load usage is uniform throughout the year or it is season dependant.
(iv) Total time in hours of the load operation per week obtained by multiplying the per day operation of the load with the number of days of such operation per week.

A typical load table of a middle class house as discussed on the above points is summarized in Table 10.1.

Table 10.1. Load calculation.

Sl. No.	Load Description	Type of Load (AC/DC) (1)	Load Rating (Watts) (2)	Number of Devices (Nos) (3)	Operating hours/day (4)	Load Usage in Months (5)	Remarks
1	Tube lights	AC	40	5	5	Uniform	
2	LED lamps	AC	20	6	8	Uniform	
3	Refrigerator	AC		1		Summer/ Winter	Refer Note 1
4	Water purifier	AC	60	1	3	Uniform	
5	Microwave oven	AC	900	1	0.5	Uniform	
6	Iron press	AC	1000	1	0.25	Uniform	
7	Water Geyser	AC	2000	1	1	Winter	Refer Note 2
8	Air conditioner	AC	1600	2	6	Summer	Refer Note 3

Notes:

1. Average load consumption throughout the year is 766,500 Wh. The assumption is that the compressor rating of the refrigerator is 300 W with an average operation of 7 h/day.

2. Geyser operation is assumed for only 3 months in winter.

3. Air conditioner operation is assumed for only 4 months in summer.

Using the above load table, the total Wh consumption of uniform loads.

$$W_1 = 40 \times 5 \times 5 + 20 \times 6 \times 8 + 60 \times 1 \times 3 + 900 \times 1 \times 0.5 + 1000 \times 1 \times 0.25 + \frac{766500}{365} = 4940\,\text{Wh}$$

The Wh consumption of loads during winter season

$$W_2 = 2000 \times 1 \times 1 = 2000\,\text{Wh}$$

The Wh consumption of loads during summer season

$$W_3 = 1600 \times 2 \times 6 = 1920\,\text{Wh}$$

Hence, the average total Wh/day during winter season is

$$Wh_{av(winter)} = W_1 + W_2 = 4940 + 2000 = \frac{6940}{24} = 289$$

Similarly, the average total Wh/day during summer season is

$$Wh_{av(summer)} = W_1 + W_3 = 4940 + 19200 = \frac{24140}{24} = 1006$$

10.2.2 Solar PV Water Pumping System

The use of solar PV modules finds a prominent place in water pumping applications due to several advantages offered by it. A typical solar water pumping system consists of PV modules, motor, pump, and a storage tank. Figure 10.1 shows the block diagram of a typical solar PV water pumping system. The two broad categories of pump used for the PV water pumping system are the centrifugal pumps and the volumetric/displacement pumps. Table 10.2 summarizes the comparative features of these two pumps. In the case of centrifugal pumps, a DC motor can be directly coupled with solar PV panel without the requirement of power conditioning circuit. However, the presence of brush limits its use in submersible application. AC motor can also be used with an inverter, which converts DC power of a PV module into AC power with variable voltage and frequency. Apart from this, brushless DC motors are also used with appropriate electronic control and conversion system. In addition, a maximum power point tracking (MPPT) device is used to match the PV module output impedance with that of motor so as to extract maximum power throughout the day by the designed PV-based water pumping system. Depending upon the water requirement (whether normal drinking water or for irrigation purpose), PV pumping system can be designed.

The design of solar PV pumping system requires knowledge about how much water needs to be pumped, from what depth water should be pumped, how many solar panels will be required for given water requirement, what should be the ratings of the motor used with PV panels, and so on. The important steps to design solar PV pumping system are as follows:

(i) Determine the amount of water required per day.
(ii) Determine the total dynamic head (TDH) for water pumping.
(iii) Determine the hydraulic energy required per day (Wh/day).

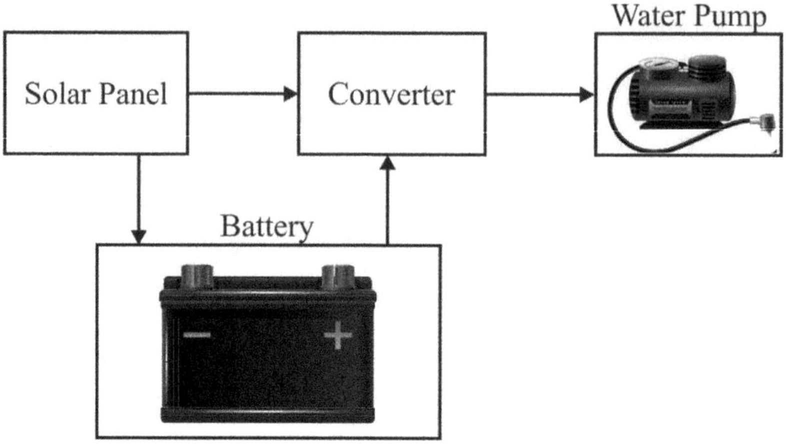

Figure 10.1. Block diagram of solar pumping module.

Table 10.2. Salient features of centrifugal and volumetric pumps.

Centrifugal Pumps	Volumetric Pumps
a. Centrifugal pumps find their application for low pressure/low head conditions	**a.** Volumetric or displacement pumps are used for applications requiring high head with low pump rates
b. The characteristic of a Centrifugal pump is that it produces torque proportional to the square of the speed	**b.** Volumetric pumps are characterized by flat torque versus speed characteristics
c. Direct connection with the solar panels is possible	**c.** Direct connection with the solar panels is not possible and requires power conditioning circuitry
d. High efficiency at high speeds	**d.** High efficiency for large heads
e. It is highly reliability, low cost, and easy maintenance	**e.** Generally requires gear transmission system to match the motor speed and therefore increases the system complexity
f. Used as submersible pumps due to its inability of self-priming	**f.** It has self-priming capability

(A) Data requirement

(i) Amount of water to be pumped per day = 25,000 L = 25 m^3

(ii) Total vertical lift = 12 m (5 m elevation, 5 m standing water level, 2 m drawdown)

(iii) Water density = 1000 kg/m^3

(iv) Acceleration due to gravity, g = 9.8 m/s^2

(v) Pump efficiency = 30% (can be between 0.25 and 0.4)

(B) Water pumping load

The steps involved for calculation of load in Wh/day

Step 1: The total daily water requirement = 25 m^3/day

Step 2: Total vertical lift = 12 m

Frictional losses = 5% of the total vertical lift = 12 × 0.05 = 0.6 m

Therefore, TDH = 12 + 0.6 = 12.6 m.

Step 3: Hydraulic energy required to raise water level

$$\text{Mass} \times \text{g} \times \text{TDH} = \text{Density} \times \text{volume} \times \text{g} \times \text{TDH}$$
$$= 1000 \times 25 \times 9.8 \times 12.6 = 3.087\,\text{MJ} = \frac{3.087}{3.6} = 0.8575\,\text{kWh}$$

where 1kWh = 3.6 MJ

Therefore, the rating of the pump must be equal to the potential energy of the water, which is to be raised during pumping and for the above example it is equal to 857.5 Wh/day.

10.2.3 Sizing of PV Array

The proper sizing of PV array plays a vital role in ensuring maximum reliability. The following basic information is required for array sizing.

Location information: For proper array sizing, the most important requirement is insolation at a given location. Once the insolation level at a given location is obtained, the daily energy incident on a horizontal flat surface is calculated with knowledge of the latitude of the location [5–7]. The other important factors to be considered are the height above mean sea level, volume of air mass, and local atmospheric condition such as ozone layer, water vapour content in the air column, suspended particles in air. With the above information, the average solar energy incident (KWh/m^2/day) on a horizontal/tilted surface of a PV array for the given location is calculated. For this, an algorithm is developed with the following input parameters using any of the software tools such as MATLAB/Octave. The average daily maximum temperature of the location is also important for selection of array voltage, which can impact the choice of the modules.

Load information: Total Wh and Ah consumed per day as calculated in load N_s calculation table.

With the above information, the selection of the number of modules to be connected in series or parallel can be found by using the following formula.

The total number of modules to be connected in series to obtain the desired voltage is found by the following formula:

$$N_s = \text{Nominal desired system voltage/Nominal module voltage}$$

where N_s is the number of modules connected in series to form the array.

Following similar lines, the number of modules to be connected in parallel to obtain the total Ah is as follows:

$$I_{array} = \frac{\text{Total average daily load in Wh}}{\text{Energy irradiance (kWh/m}^2\text{/day)} \times \text{derating factor}}$$

Based on the maximum current rating I_{amp} of the module chosen, the number of modules that has to be connected in parallel to produce the required level of array current can be found by using the following formula:

$$N_p = I_{array}/I_{mp} \tag{10.1}$$

where N_p is the number of arrays connected in parallel. It is seen from the above formula that higher the maximum value of module current, lesser will be the number of modules required to be connected in parallel. Moreover, the average solar insolation on a particular location has also an impact on the fulfilment of load demand. It is therefore the duty of the designer to select the optimum number of modules required without compromising the system reliability and significant increase in cost.

The next step is to determine the total number of modules required to deliver the daily average energy and it is given as

$$N = N_s \times N_p \tag{10.2}$$

10.2.4 Energy Irradiation Calculation

The energy incident on a horizontal flat PV surface H_o is given as

$$H_o = 24 * \mathrm{k} * \frac{L_{sc}}{pi} * \cos\phi * \cos\delta * \sin\omega_{sr} + \omega_{sr} * \sin\phi * \sin\delta \tag{10.3}$$

where $\phi, \delta, \omega_{sr}$ are the latitude of the given location, the declination, and the sun rise angle in degrees, respectively. Once the energy incident on a flat plate is calculated, the energy incident on a tilted plane H_{ot} is calculated. With tilted angle β taken into consideration, the above equation modifies to

$$H_{ot} = 24 * \mathrm{k} * \frac{L_{sc}}{pi} * \cos(\phi - \beta) * \cos\delta * \sin\omega_{sr} + \omega_{sr} * \sin(\phi - \beta) * \sin\delta \tag{10.4}$$

The main aim of a PV system designer is to extract maximum energy irradiance from the PV array surface. To achieve this, the panel surface has to be tilted by a certain angle, denoted by tilt angle β. With atmospheric effect of a given location taken into consideration the energy irradiation is given as

$$H_{at} = \mathrm{K}_T * H_o * R_D - \mathrm{H}_d * R_D + H_d * \frac{1+\cos\beta}{2} + \mathrm{H}_a\rho\frac{1-\cos\beta}{2} \tag{10.5}$$

where K_T is known as the clearness index and ρ is the reflective coefficient depending on the topography location on which the PV array is installed.

For example, let us assume the total load of a consumer as

$$Wh_{load} = \left(Wh_{day} + Wh_{night}\right) \text{ KWh/day}$$

where Wh_{day} is the energy required by the load say from 6 a.m. to 6 p.m. k kWh/day and Wh_{night} is the energy required say from 6 p.m. to 6 a.m.. For the study purpose, the Wh_{day} is assumed as 160 and 40 kWh/day, respectively, and a power tariff (C_{load}) of Indian Rupees (INR) 4 per kWhof energy consumption. The tariff considered is based on the average prevailing energy charges in India fixed by the respective state power regulatory commission authority.

10.3 Case Studies and Examples

To understand the PV sizing, the following four Cases I–IV are considered as examples (Figs. 10.2–10.5).

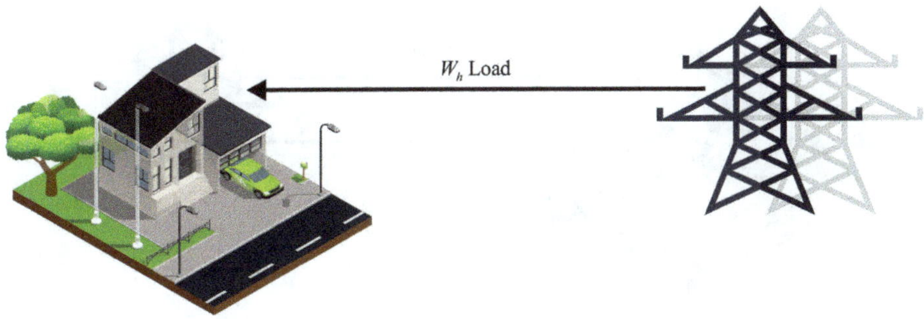

Figure 10.2. Load demand from utility grid.

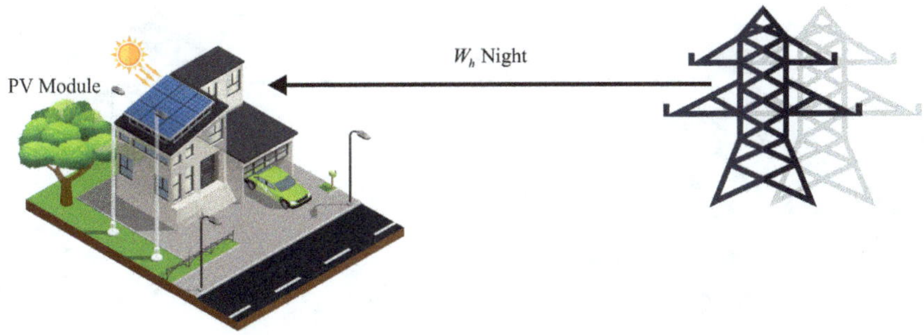

Figure 10.3. Day load demand from PV system.

Case I: Load energy demand fulfilment through utility grid

$$\text{Energy bill} = Wh_{day} \times C_{load} + Wh_{night} \times C_{load}$$
$$= (160 \times 4 + 40 \times 4) = \text{Rs. } 800/\text{day}$$

Case II: Use of PV power to meet the day load energy demand

Based on the selection criteria as specified in section I for PV array sizing, let us assume a PV panel rating, which meets the day load energy requirement, that is, $Wh_{PV} = Wh_{day}$. Under this condition, the energy demand from the utility grid only for the night load is given as $Wh_{load} - Wh_{PV} = Wh_{day} + Wh_{night} - Wh_{day}$. The corresponding bill then comes out to be $Wh_{night} \times C_{load}$ = Rs. 160/day.

Case III: Use of PV power to meet the day load energy demand and with zero energy bill

Let as assume that the generating energy from PV modules equals Wh^*_{PV} kWh/day and the total load energy demand is Wh_{load}, same as above. The tariff for energy supplied from PV system is equal to C_{PV} per kWh. For net energy cost to be zero, the revenue generated by the PV system through energy supply to the grid should be equal to the cost incurred for drawing energy from the grid by the load, that is,

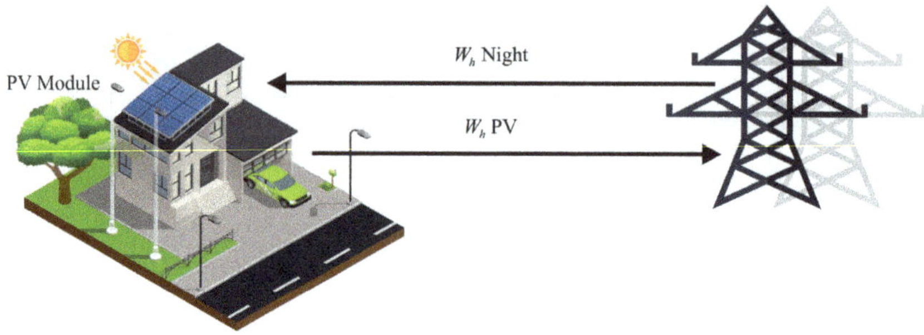

Figure 10.4. Load demand requirement from PV system without battery.

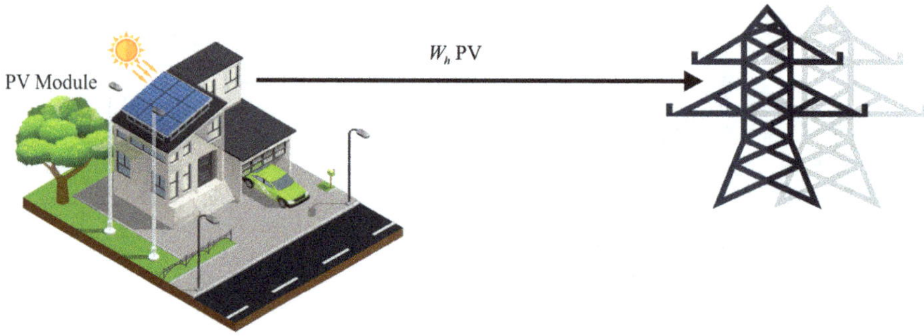

Figure 10.5. Load demand requirement from PV system with battery.

$$Wh^*_{PV} \times C_{PV} = Wh_{load} \times C_{load}$$

$$\text{OR } Wh^*_{PV} = \frac{Wh_{load} \times \text{C}_{load}}{C_{PV}} \tag{10.6}$$

From the equation, it is seen for net zero energy bill the sizing of the PV system has to be increased as the PV system considered is without battery and has to fulfil the night load energy demand.

Case IV: Use of PV power to meet the day load energy demand with battery

To understand the PV sizing requirement consider the load profile of a consumer with the load table summarized in Table 10.3.

The total Wh consumption of the load is calculated as follows:

Load during daytime where the daytime is considered from 7.00 a.m. to 5.00 p.m. and the remaining time period is treated as nighttime.

$$Wh_{dayload} = 1000 \times 1 + 5 \times 72 \times 10/60 + 900 \times 0.5 = 1510\,\text{Wh}.$$

$$Wh_{nightload} = 40 \times 5 \times 5 + 20 \times 6 \times 8 + 1000 \times 1 \times 2 + 7 \times 72 \times 10/60 = 4044\,\text{Wh}.$$

Table 10.3. Load demand table.

Sl. No.	Load Description	Type of Load (AC/DC) (1)	Load Rating (Watts) (2)	Number of Devices (Nos) (3)	Operating hr/day (4)	Load Usage in Months (5)	Remarks
1.	Tube lights	AC	40	5	5	Uniform	Refer Note: 1
2.	LED lamps	AC	20	6	8	Uniform	Refer Note: 1
3.	Submersible water pump	AC	1000	1	3	Uniform	Refer Note: 2
4.	Water puri-fier	AC	72	1	2	Uniform	Refer Note:3
5.	Microwave oven	AC	900	1	0.5	Uniform	Refer Note:4

Notes:

1. Time of operation from 6 p.m. to 11 p.m.
2. Time of operation three times per day of 1-h duration: One before sunrise, once at noon, and once after sunset.
3. Ten-minute operation during every 2 h of the day.
4. Time of operation only during daytime 10.00 a.m. to 2.00 p.m.

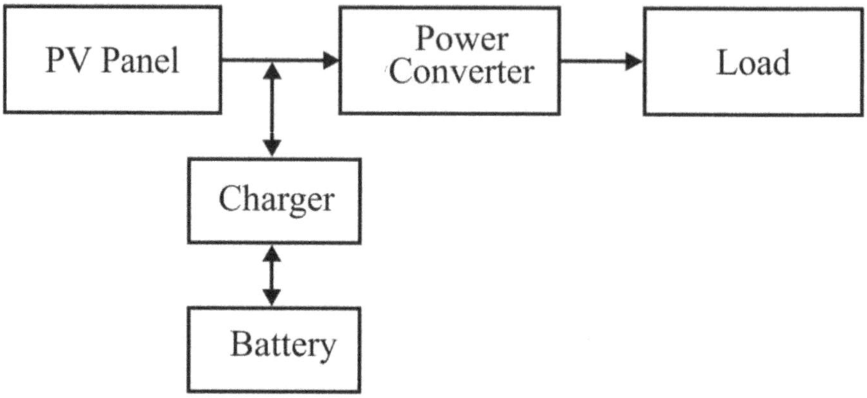

Figure 10.6. Block diagram of PV system with energy storage.

When a PV system with battery is considered, the purpose is to fulfil the energy requirement of the load by the load during the nighttime. Then the Wh energy of the load to be supported by the battery is

$$Wh_{load} = \frac{Wh_{nightload}}{\eta_B} \tag{10.7}$$

Other than this, there will be some days during a year when the PV system could not meet the day load energy requirement due to cloudy weather condition. The days of such operation are known as autonomy days (η_a). During this period, the battery has to support the day load and the Wh energy of the load to be supported by the battery is given as

$$Wh_{load} = \frac{Wh_{nightload}}{\eta_B} + \frac{Wh_{dayload} + Wh_{nightload}}{\eta_B} \times \eta_a \tag{10.8}$$

As seen in Section 10.2, it is noticed that when load is directly connected using power converter through PV panel, batteries are not required. But sometimes load demands steady-state power and power from PV panel is highly fluctuating due to fluctuating insolation. Therefore, we need to have an energy buffer, so that PV panel can charge it and also buffer would supplement with energy when PV panel output is weak. So the primary functions of battery as an energy storage device is to first store the surplus energy generated by PV array during low load hours. Second, to supply energy to load during peak load hours and during night hours and cloudy days, and the third and most important function is to provide energy at stable voltage and current. Figure 10.6 shows the block diagram of a PV system with battery.

10.4 Battery for PV Systems

Before selection of battery for a particular application, understanding the various battery terminology is important. The following paragraph briefly explains the battery terminologies one should know for its use in PV applications:

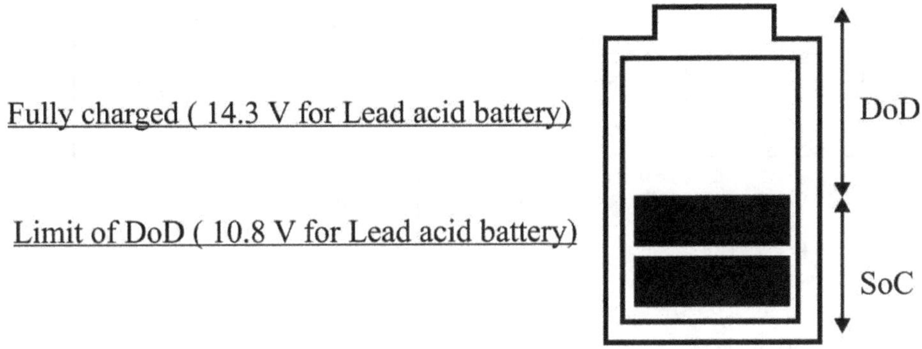

Figure 10.7. Battery charge level.

Capacity: Capacity of battery means how much amount of energy a battery can discharge to load. The term commercially used for indicating capacity of battery is Ah. The capacity of a battery is given by the following equation:

$\int V_b I_b dt$, if the battery voltage V_b is constant then the equation modifies to $\int I_b dt = \text{C Ah}$. The battery current I_b is considered positive during charging and vice versa during discharge.

State of Charge: The state of charge (SoC) refers to the amount of energy existing in a battery expressed as a percentage of total energy stored in a fully charged battery. For example, a battery that has been discharged to 60% is said to be at 40% SoC.

Depth of Discharge: The depth of discharge (DoD) refers to the amount of energy that has been extracted from a battery compared to its total fully charged capacity given as $\text{SoC} = (1 - (\text{DoD}/100)) \times 100$

As the limit of DoD increases, the life of battery decreases. A typical lead acid battery level showing SoC and DoD is shown in Figure 10.7.

C-Rate: It is the rate at which the capacity of battery is built up (charged) or given up (discharged) to load. The C-rate is symbolically expressed as C/N or CN, where C indicates capacity of battery and N indicates the number of hours for full discharge. For example, C/10 or C10 indicates battery has capacity to supply an average current for 10 h. Alternatively, it can be expressed as, for example, C/0.5 or C0.5 or 2C, where it indicates the battery can supply an average current for 2 h.

Dynamic Resistance: Practically, the voltage across terminals of battery does not remain constant always. The drop in voltage is due to the dynamic resistance R_b. It is nonlinear and is a function of many parameters such as chemistry of battery and DoD.

Efficiency: This is a very important parameter of battery. In the case of battery, efficiency is measured in terms of energy.

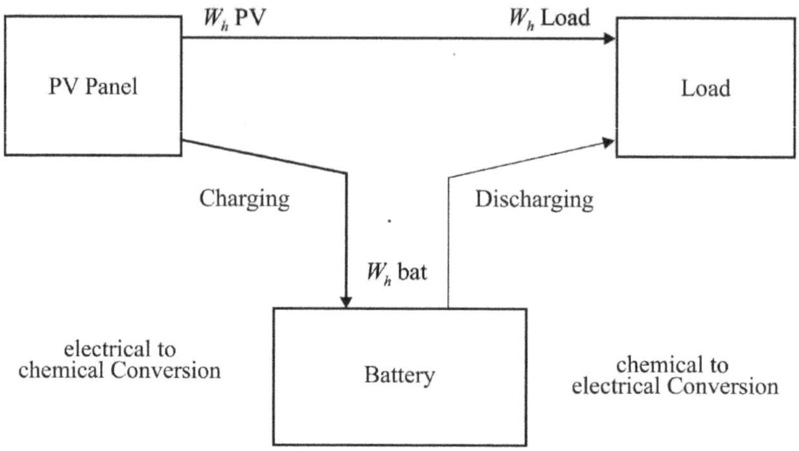

Figure 10.8. Block diagram of power flow of a PV system with battery.

As shown in Figure 10.8, there is a conversion loss during both charging and discharging process. The efficiency for most of the batteries is around 70%.

10.4.1 Classification of Battery

Batteries are classified into two categories. The first is Primary Battery used in low-power household devices and can be used for only one discharge and the second is Secondary Battery for small- to medium-power applications [8]. This battery can be charged and discharged several times depending on the type of battery. Some of the commercially available batteries, which come under this classification, are PbO_2 (lead acid), NiMH (nickel metal hydride), $LiSO_2$ (lithium sulphur dioxide), and LiPo (lithium ion polymer). For PV application, the lead acid battery is used because of its low initial cost and ease of maintenance and repair.

10.4.2 Energy and Power Density

This is one of the important parameters for sizing of battery. Energy and power density together play an important role in determining the volume, sizing, and space requirement for a given application in housing the battery. The energy density can be categorized as either Gravimetric or Specific Energy (Wh/Kg) or Volumetric (Wh/L). Similarly, the power density can be categorized as either Gravimetric or Specific Power (W/Kg) or Volumetric (W/L). For a given battery either energy density or power density will be high and both cannot be high in a given battery due to battery chemistry. So, these two offset each other. Table 10.4 summarizes the comparison of some of the commercially available batteries in market.

Table 10.4. Comparison of batteries.

Sl. No.	Battery Type	Wh/kg	W/kg	Wh/L	Operating Tempera-ture (°C)	Wh Efficiency (%)	Life (no. of cycle)
1.	Lead Acid	35	100	80	Ambient	70	1000
2.	Ni Metal Hydride	55	100	150	Ambient	70	2000+
3.	Sodium Sulphate	100	150	300	350	75	1000
4.	Lithium Polymer	150+	200+	450	Ambient	80	10000

10.4.3 Battery Selection

The following steps are observed for selection of a battery:

First, identify and select battery chemistry. The selection is based on the type of application, availability, and capital investment.

Second, find out the allowable DoD for battery type. For example, for the lead acid battery used for starting, lighting, ignition (SLI) operation, which require high-power density but less energy density, the DoD is only 20%, whereas for a tubular lead acid battery the DoD is 80%.

Third, find out the Wh requirement of load $\left(Wh_{load}\right)$ as discussed above to decide the capacity of battery. The battery capacity in Ah then is given by the following equation:

$$Ah_{batt} = \frac{\frac{Wh_{load}}{DoD}}{V_b} \tag{10.9}$$

where V_b is the nominal battery voltage.

Fourth, select the C-rate from the battery data sheet to find out the average battery discharge current (I_b) for the given application and approximately choose C-rate from the C curve given as $I_b^n t = \text{C}$, where $n = 1$, for low-discharge current, and $n = 2$, for high-discharge current.

10.4.4 PV System Design with Battery Sizing

Considering the load profile as summarized in load Table 10.3, let us determine the battery rating for the said application. For ease of calculation, the following assumptions are made. First, the loads are considered as DC loads with load supply voltage equals 48 V DC, and second, the days of autonomy N_a and therefore days of charge replenish N_r equal to zero.

Now, we will see how to size PV panel is decided when used along with battery.

To determine the battery rating, find the peak load current that may occur in the worst condition and thereafter the average load current. The peak load current is calculated based on the simultaneous operation of all the connected loads on a particular time duration of the day. For example, the peak load current I_{loadp} calculated as power output per device multiplied by the number of devices upon voltage rating of the device for the load table profile as stated above is given as

$$I_{loadp} = (40 \times 5/48) + (20 \times 6/48) + (1000/48) + (900/48) + (72/48) = 47.74\ \text{Amp}$$

Whereas, the average load current, considering the time duration of device operation, is given as

$$\begin{aligned} I_{load(av)} = {} & (200/48 \times 5/24) + (120/48 \times 8/24) + (1000/24 \times 3/24) \\ & + (900/24 \times 0.5/24) = 7.69\ \text{Amp} \end{aligned}$$

This is average discharge current of battery for whole day.

With the second assumption and using Eq. (10.7)

$$Wh_{load} = \frac{4044}{0.7} = 5780\ \text{Wh}$$

and the $Ah_{batt} = \dfrac{Wh_{load}}{\text{DoD} \times V_b} = \dfrac{5780}{0.8 \times 48} = 150\ \text{Ah}$

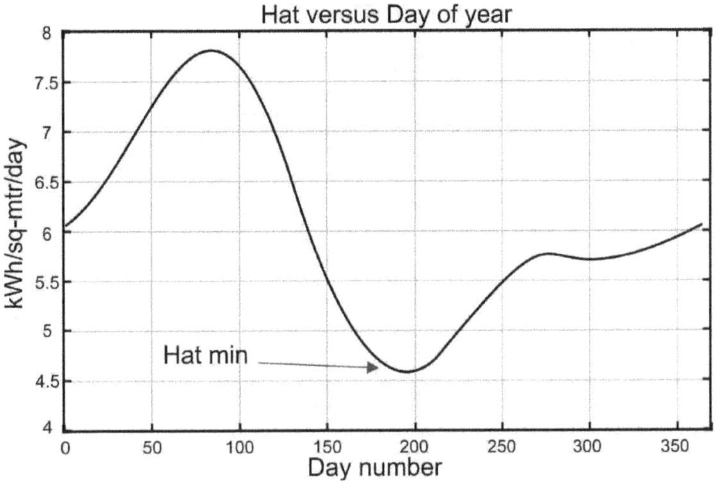

Figure 10.9. Plot of the energy incident versus the number of days in a year.

Based on the above results, the battery size chosen is 200 Ah with C-rate chosen as C10 considering average load current.

Thereafter the PV array sizing is given by equating Eq. (10.8) as

$$Wh_{PV/day} = Wh_{dayload} + \frac{Wh_{nightload}}{\eta_B} + \frac{Wh_{dayload} + Wh_{nightload}}{\eta_B} * \eta_a \tag{10.10}$$

Then the array peak P_m in watts is given by

$$P_m = \frac{Wh_{PV/day}}{H_{at\,\min}} \tag{10.11}$$

$$Wh_{PV/day} = 1510 + \frac{4044}{0.7} + \frac{(1510 + 4044)}{0.7} \times 0 = 7290.$$

Figure 10.9 shows the plot of the H_{at} for a whole year.

From the plot of Figure 10.9, the $H_{at\,\min} =\sim 4.6\,\text{kWh/m}^2\text{/day}$. Applying the value of $H_{at\,\min}$ in Eq. 10.11, we get the peak PV array rating equals 1585 W.

Considering the same assumptions as discussed above the PV array rating required for water pumping system is calculated as follows:

The input rating of the pump motor is given by the ratio of pump power output to the pump efficiency. For the example considered earlier in the load calculation table, the pump power input = $857.5/0.3 = 2858.33\,\text{Wh/day}$.

Applying the value of $H_{at\,min}$ in Eq. 10.11 and considering the water pumping will be done only during daytime, we get the peak PV array rating equal to 621 W.

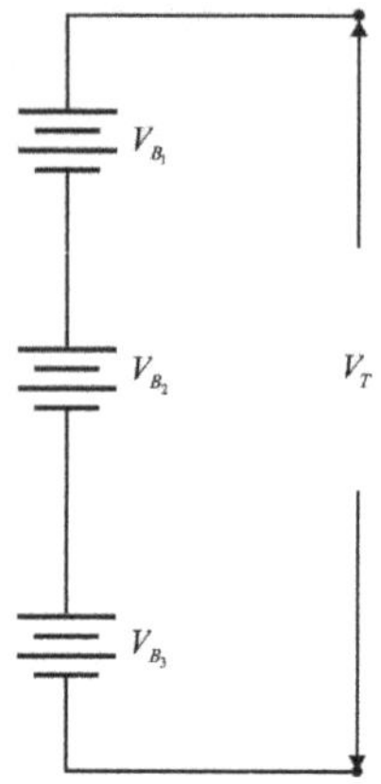

Figure 10.10. Series connection.

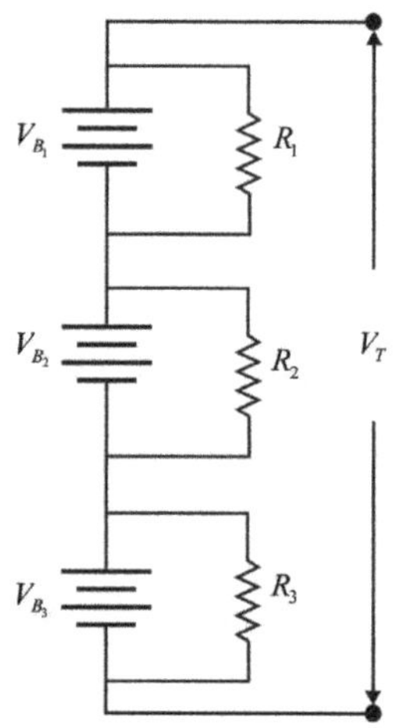

Figure 10.11. Equalization using shunt resistors.

10.5 Battery Equalization

Batteries are connected in series combination, where the negative terminal of first battery is connected with the positive terminal of the second battery and so on till the required terminal voltage levels are obtained. The total Ah of the battery will be same as the Ah of a single battery if all the batteries connected in the series string have the same capacity, otherwise, the capacity will be limited to that of the battery with the lowest capacity [9]. Figure 10.10 illustrates the series connection of batteries.

In Figure 10.10, if all the batteries have the same specifications, then $V_{B1}, V_{B2}, V_{B3} = V_B$ or $V_T = 3V_B$. In practice, however, each battery will charge and discharge at different rates. Therefore, charge balancing is required. The charge equalization methods can be classified

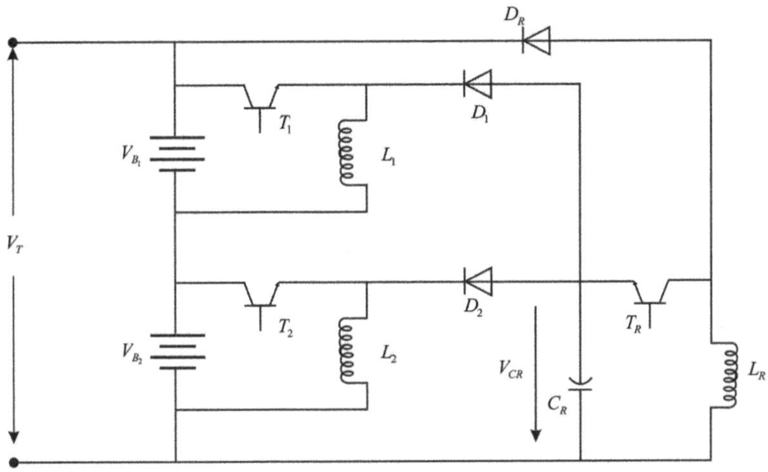

Figure 10.12. Charge pump-based charge equalization circuit.

as active or passive. Passive methods are used where the batteries can be overcharged safely and also where the number of batteries in the series string is less. Many active methods are available for charge equalization and one of the low-cost methods is the use of shunt resistors as shown in Figure 10.11. The method is generally employed in low-power circuit due to power dissipation in the shunt resistors. A low-cost active charge equalization circuit known as charge pump for two series-connected batteries with similar specification is shown in Figure 10.12.

The charge equalization of batteries will work as follows. Assume that the voltage across battery V_{B1} is overcharged to a voltage greater than its rated voltage V_B. From the control circuit as shown in Figure 10.13, it can be seen that the comparator 1 (COMP 1) output becomes positive as V_{B1} at noninverting terminal becomes greater than the input at the inverting terminal, which is $\frac{V_T}{N}$, where V_T is the rated terminal voltage and N is the number of batteries connected in series. The switch T_1 turns ON and the excess charge in V_{B1} will be stored in inductor L_1. Once T_1 turns OFF the excess charge from L_1 will be transferred to battery V_{B1} through V_{CR} and D_1. The switch T_1, inductor L_1, and diode D_1 act as a boost converter. Similarly, switch T_2 turns ON if battery V_{B2} voltage becomes greater than its rated value. The excess charge is stored in V_{CR}, which acts as a reservoir capacitor. The switch T_R turns ON when the output of COMP 3 becomes positive and the charge stored in V_{CR} is fed to both the batteries through inductor L_R and D_R.

Batteries connected in parallel offer more than one electrical path for the current to flow, which is governed by the number of battery strings connected in parallel. The voltage across the circuit terminal will be same as the voltage across each parallel branch and the total capacity will be sum of the parallel branch capacities. The drawback is that a minor voltage difference between the parallel strings will result in circulating current to flow and thus damage the batteries. This problem can be overcome by connecting diodes in series as shown in Figure 10.14.

The other issue associated with parallel operation of batteries is the unequal current sharing by the parallel strings. The unequal current sharing would be the difference in either

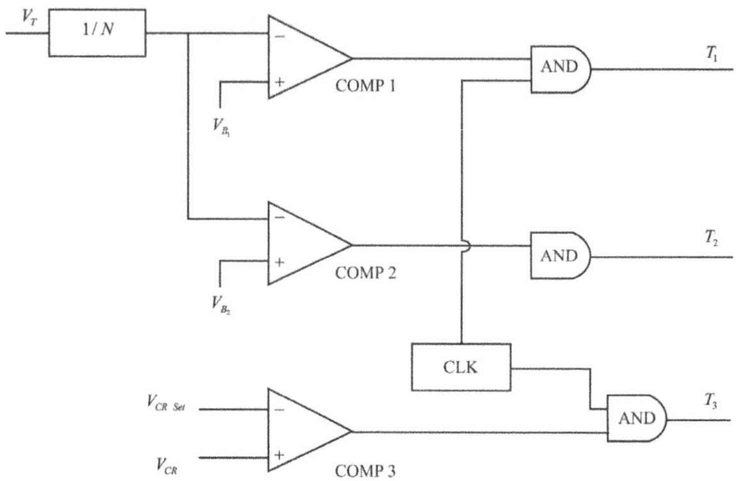

Figure 10.13. Control circuit diagram for charge equalization.

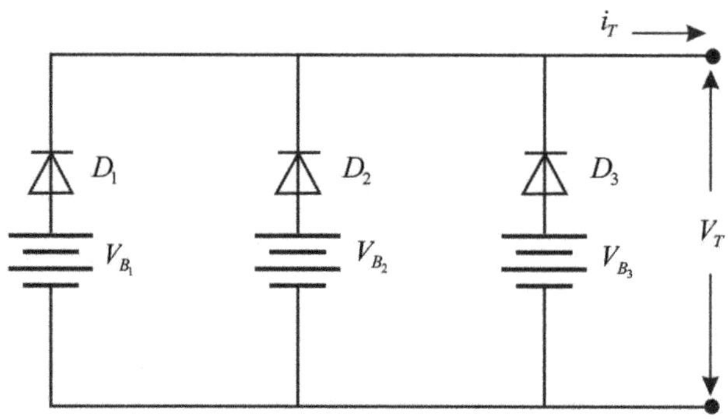

Figure 10.14. Parallel connection of batteries.

the internal resistance of the battery/batteries of each parallel string or the circuit resistance due to the varying length of the connecting wire of each parallel path. In the case of lead acid batteries, an undercharge will result in the battery to sulphate prematurely and an overcharge will cause corrosion, thus reducing the battery life. Figure 10.15 shows a typical charge equalization circuit for batteries connected in parallel.

10.6 Protection Schemes

The larger PV system requires more attention during installation as small design faults may lead to system malfunctioning, resulting in complete collapse of the system. The first important installation guideline is with respect to the array sizing. The first step could be

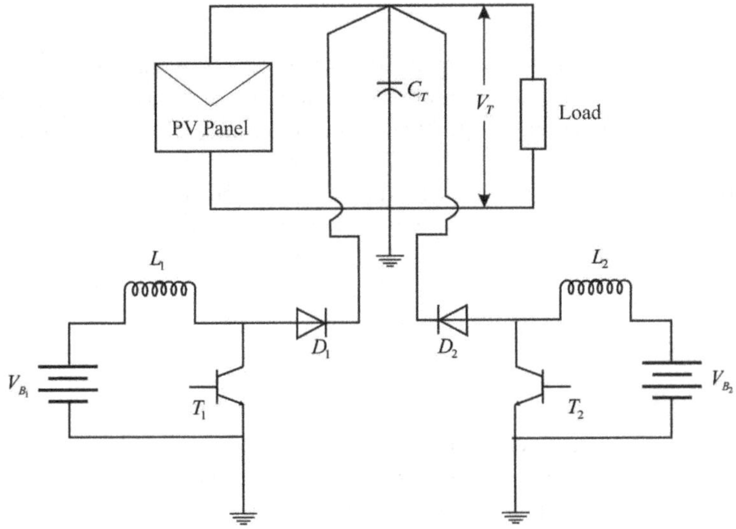

Figure 10.15. Charge equalization circuit for batteries connected in parallel.

splitting the arrays into smaller subarrays to produce the required DC system voltage [10]. Once the numbers of subarrays are finalized, we can decide the number of charge controllers required for each subarray unit. It is easier to maintain a PV system with large number of subarrays and charge controllers. The major advantage is the ease of location of the fault point and isolation of it. This will ensure system with higher reliability as the entire system is not required to shut down. Second, one can choose a lesser rating of charge controller instead of choosing a single controller required to handle the full array current. Third, the cable sizing reduces, reducing its cost with increased ease of handling and installation of the cable.

The number of subarrays $N_{sub\ array}$ chosen for a large PV system is the current handling capacity of the charge controller. It can be calculated as

$$N_{\text{subarray}} = \frac{N_p \times I_{mp}}{I_{ccp}}$$

where N_{subarray} is the number of subarrays required, N_p the total number of modules connected in parallel to form the array, I_{ccp} the peak current of charge controller, and I_{mp} the peak current of the module.

The second most important requirement for reliable operation is the installation of protective devices of appropriate ratings. In a PV system, there are four locations where circuit breakers are needed. They are (1) between the array and the charge controller, (2) between the controller and the battery, (3) between the battery and the load centre and last, and (4) between the battery and the inverter. In all these locations, DC circuit breakers are used and if not available AC circuit breakers of twice the rating of DC circuit breakers are used.

10.6.1 Battery Protection

The output conductors of the battery bank shall be protected against overcurrent, by high rupturing capacity (HRC) fuses or suitable DC-rated circuit breakers. A battery bank in a short-circuit condition can discharge hundreds or thousands of amps causing damage to equipment and human fatalities. The battery bank, which is electrically floating, requires necessary protection to be provided for both positive and negative battery leads. The protective device should be installed as close to the battery terminals so that no fault flashover results due to ignition of the hydrogen emitted if any from the batteries, especially during charging.

10.6.2 Array Protection

When multiple strings are connected in parallel, there is a risk of feeding high current from combined strings to a faulty string due to short circuit or shading condition. For fault isolation, either DC circuit breakers of appropriate rating or HRC fuses are to be used. Other than this when the series and parallel combination of subarrays are connected to the DC junction box, blocking and bypass diodes are also to be used. The following section will focus on the issues related to series and parallel connection of PV cells to form subarrays or modules.

(A) Series and parallel operation of PV cells/modules

Loads with high-power rating require PV cells with high open-circuit voltage and short-circuit current ratings. Conventionally, a single PV cell cannot fulfil the high-power requirement, and for this many PV cells must be connected in series and parallel, collectively called PV modules to get higher power. The important issue now is: Can the PV cells be connected directly in series or parallel for high-power requirement? Some of the gray areas are (1) even if the PV cells are manufactured in the same batch by a manufacturer, they will not have identical Volt/current characteristics as there will be variation in the rated open-circuit voltage V_{OC} and current I_{SC} and (2) the effect of partial shading in one of the modules will vary its voltage and current outputs. The effect of nonidentical VI characteristics of PV cells on the operation of series and parallel connected PV cells/modules will be discussed in detail below.

(B) PV cells in series

Figure 10.16 shows two PV cells connected in series. For simplicity of understanding, the PV cells are assumed to be ideal and have nonidentical VI characteristics for reasons as mentioned above. Let us consider PV cell 2 is under partial shading condition. As the open-circuit voltage is in logarithmic proportion and the short-circuit current is directly proportional to the solar insolation, the VI characteristics of the two cells is drawn assuming the open-circuit voltages V_{OC1} and V_{OC2} to be same and I_{SC1} and I_{SC2} to be different as shown in Figure 10.16.

Under open-circuit condition, that is, with $R_O = \infty$, the load line will lie on the horizontal axis with operating point shown as A. At this operating point, the terminal voltage $V_T = V_{OC} = V_{T1} + V_{T2} = V_{OC1} + V_{OC2}$ and $I_T = I_{SC1} + I_{SC2} = 0$. Now for finite value of load resistance R_{01}, the load line $\frac{1}{R_{01}}$ is denoted by 1. For this particular load line, the terminal current is I_{T1} and the terminal voltage $V_T = V_{T1} + V_{T2}$, where V_{T1} and

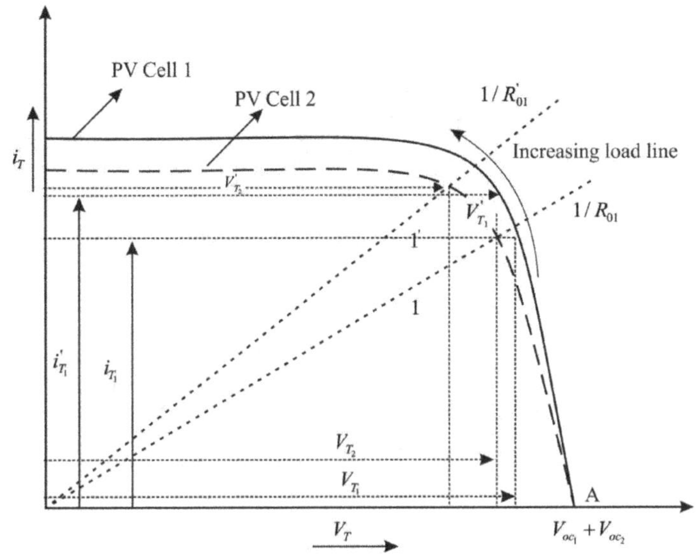

Figure 10.16. *I-V* characteristic of series-connected PV cells.

V_{T2} are the terminal voltages of PV cells 1 and 2, respectively. With further decrease in load resistance R_0, the load line slope increases as shown by $\frac{1}{R_{01'}}$ with the terminal voltage of cell $V_{T2'}$ less than $V_{T1'}$. Further decrease in load resistance will lead to the load line intersecting with the combined PV VI characteristics at point B as shown in Figure 10.17.

From Figure 10.17 it is seen that the terminal voltage V_{T2} of PV cell 2 becomes zero and the power delivered by PV cell 2 becomes zero. The load terminal voltage will now be equal to the terminal voltage of PV cell 1. Increase in load line beyond point B will lead to cell 2 acting as sink and starts consuming power as shown. This results in cell 2 becoming hot, deteriorating the cell 2 condition and also leads to decrease in overall power generation by the combined cell in series.

(C) Protecting series-connected PV cells

As seen from Figure 10.17, PV cell 1 act as a source and PV cell 2 as a sink beyond point B. To protect PV cell 2, one can use a bypass diode across cell 2 as shown in Figure 10.18. As observed when PV cell 2 acts as a sink, the polarity across cell 2 gets reversed and the bypass diode starts conducting. The bypass diode thus does two things: first, it prevents the shaded PV cell, that is, cell 2 to act as a sink, and second, it improves the overall efficiency of the series-connected system.

(D) PV cells in parallel

The issues as discussed in series-connected PV cells due to difference in VI characteristics can also be observed when the PV cells are connected in parallel. Consider two cells PV cell 1 and PV cell 2 with *I-V* characteristics as shown in Figure 10.19. PV cells 1 and 2 have different short-circuit current, that is, I_{SC1} and I_{SC2} and open-circuit voltage V_{OC1} and V_{OC2}, respectively. Due to partial shading of PV cell 2, the cell short-circuit current and open-circuit voltage are less.

For $R_o = 0$, the load line is vertical as shown with operating point A, where the terminal voltage $V_T = 0$ and $I_T = I_{SC1} + I_{SC2}$. The load resistance R_o is now increased with load line intersecting at point B. The terminal voltage and the corresponding load

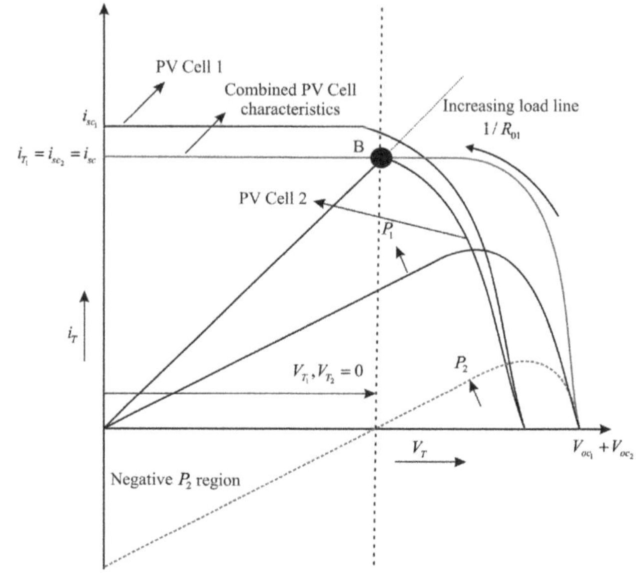

Figure 10.17. Effect of changing load on nonidentical series-connected PV cells.

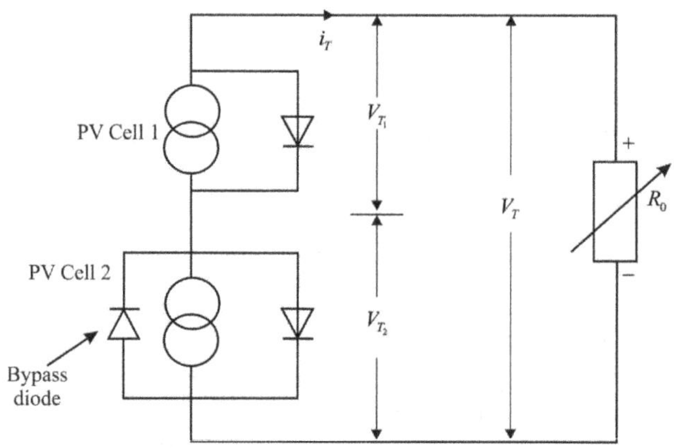

Figure 10.18. Protection of series-connected PV cells.

current shared by the two cells are shown in Figure 10.20. The load resistance is further increased, that is, the load line slope will decrease and we get operating point C with load line as $\frac{1}{R_{O1'}}$. From Figure 10.19, it is observed that the cell terminal voltage has become equal to its open-circuit voltage V_{Oc2} and the current shared by the two cells $I_{T1'}$ and $I_{T2'}$ of cell 2 equal to zero. With further increase in load resistance, that is, load line equal to $\frac{1}{R_{O1''}}$, it is seen that PV cell 2 curve goes in the fourth quadrant of the I-V characteristics, hence $I_{T2''}$ becomes negative and $I_{T1''}$ and $I_{T2''}$ cancel out each other.

(E) Protecting cells in parallel

Due to sinking/sourcing issue, the PV cell dissipates power during sinking and the same issues as observed in series operation occur here as well. To avoid the negative

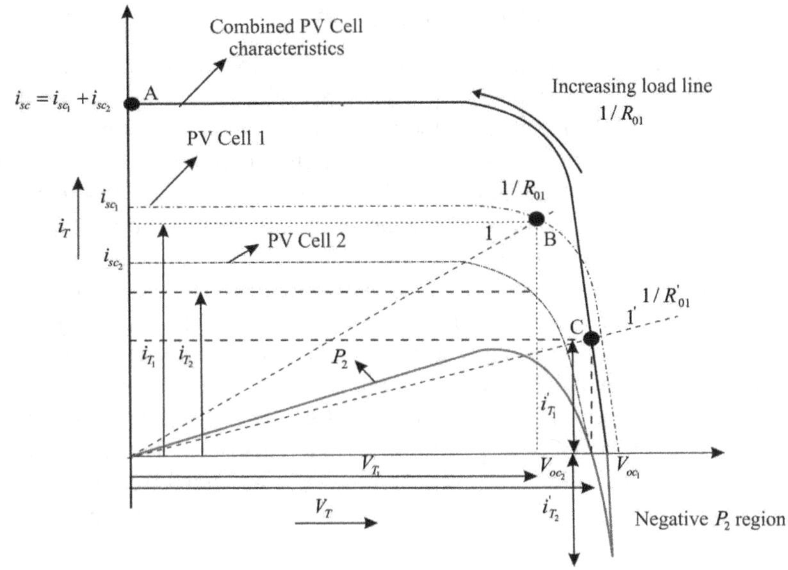

Figure 10.19. Effect of changing load on nonidentical parallel-connected PV cells.

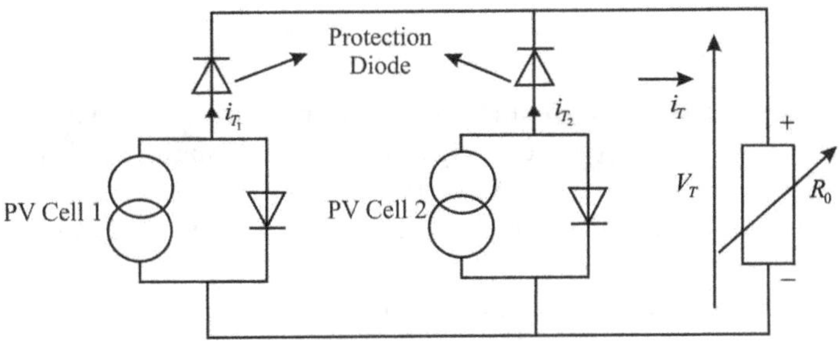

Figure 10.20. Protection of parallel-connected PV cells.

power region, we can put diode in series with PV cells, which will avoid negative flow of current as shown in Figure 10.20.

10.6.3 Cabling

The proper wiring will result in efficient and reliable PV systems. The important considerations while selection of cables is as follows. The cable insulation should withstand and should not deteriorate under environmental conditions. Reinforced or double-insulated cables should be used when cables are laid in metallic tray or conduit so as to protect it from inadvertent contact and mechanical damage. As per IEC 63248 standards, the total voltage drop in all DC cables should be less than 3% and in all AC cables it should be less than 2%.

10.6.4 Earthing and Lightning Protection

System earthing in electrical circuits has the primary function of safety. A heavy fault current flows when an electrical conductor makes contact with any metallic component. This results in human fatalities if one touches the live metallic part. An earthing wire provides a highly conductive path to the energy source, through which this heavy current is made to bypass the human body. The large current flow through the earth wire will result in the protective overcurrent devices to operate and disconnect the healthy circuit from the faulty circuit. It is a general practice to earth all exposed metal surfaces like the module frames, array mounting structures, and panel housings of all electrical load distribution centres.

The primary purpose of a lightning protection system is to attract lightning strikes away from the installation structures to an air terminal and direct it to the ground through ground wires without causing harm to installation structures or people. The elements of a lightning protection system are an air terminal, a down conductor of adequate size and conductance to carry the current to earth, and a proper earthing system that can safely dissipate the sudden current surge to earth. In addition, the PV system is to be protected from induced voltages and the subsequent side flashes due to it using Metal Oxide Varistors (MOV).

10.7 Future Issues and Challenges

The power electronic interface for PV-grid systems has the following tasks:

- **(i)** Amplification of generated DC power and its inversion into AC current, which is compatible to the grid.
- **(ii)** Design of maximum power point (MPP) tracking system for maximizing the energy capture from solar insolation according to latitude and longitude of the place and orientation of the sun.
- **(iii)** Effective energy-storage device that can make the system work during islanding operation during grid failure.
- **(iv)** Protection of PV system from various disturbances in the grid.

All the above tasks should be achieved with the highest possible efficiency, over a wide power range, regardless of seasonal variations. The design challenges should involve production of a sinusoidal voltage and current free from harmonics [11–13]. The current injected into the grid must obey the regulations, such as the EN61000-3-2 and the IEEE std. 1547, which state the maximum allowable amount of injected current harmonics. Design challenge must also involve a suitable mechanism to detect the islanding operation, which may result due to either grid disturbance or high penetration of renewable energy sources in the grid.

10.8 Conclusion

This chapter elaborates upon various load parameters to be considered while deciding the size of PV modules. It also takes into account the impact of atmospheric condition on solar energy irradiation. The chapter also focuses on the salient parameters to be considered for battery selection and its sizing under days of autonomy. It also briefs on the need of battery equalization for increased battery life used for PV system along with various protection features for enhancing the reliability of the PV system. Finally, a brief overview is given on the issues pertaining to grid integration without compromising on the standards and regulations.

10.9 References

1. C. Singh Solanki, *Solar Photovoltaics: Fundamental Technologies and Applications*, Prentice Hall of India Pvt. Ltd., 2009
2. C. Lupangu, and R. C. Bansal, "A review of technical issues on the development of solar photovoltaic systems," *Renewable and Sustainable Energy Reviews*, vol. 73, pp. 950–965, 2017.
3. H. A. M. Maghraby, M. H. Shwehdi, and G. K. Al-Bassam, "Probabilistic assessment of photovoltaic (PV) generation systems," *IEEE Translational Power Systems*, vol. 17, no.1, pp. 205–208, 2002.
4. J. Weniger, T. Tjaden, and V. Quaschning, "Sizing of residential PV battery systems," *Energy Procedia*, vol. 46, pp. 78–87, 2014.
5. R. Kumar, and L. Umanand, "Estimation of global radiation using clearness index model for sizing photovoltaic system," *Journal of Renewable Energy*, vol. 30, no.15, pp. 2221–2233, 2005
6. M. Kohle, "Techno-economic optimum sizing of a stand-alone solar photovoltaic system", *IEEE Translational Energy Conversion*, vol. 24, no. 2, pp. 511–519, Jun. 2009.
7. T. Markvart, "PV system sizing using observed time series of solar radiation," *Solar Energy*, vol. 8, no. 1, pp. 46–50, 2006.
8. S. M. Lukic, J. Cao, R. C. Bansal, F. Rodriguez, and A. Emadi, "Energy storage systems for automotive applications," *IEEE Translational Industrial Electronics*, vol. 55, no. 6, pp. 2258–2267, 2008.
9. Y. V. Pavan Kumar, and R. Bhimasingu, "Renewable energy based microgrid system sizing and energy management for green buildings," *Journal of Modern Power Systems and Clean Energy*, Vol. 3, no. 1, pp. 1–13, Mar. 2015
10. P. T. Manditerezaa, and R. C. Bansal, "Renewable distributed generation: The hidden challenges—A review from the protection perspective," *Renewable and Sustainable Energy Reviews*, vol. 58, pp. 1457–1465, 2016
11. G. Nottona, V. Lazarovb, and L. Stoyanovab, "Optimal sizing of a grid-connected PV system for various PV module technologies and inclinations, inverter efficiency," *Renewable Energy*, vol. 35, no. 2, pp. 541–554, Feb. 2010.
12. A. Kumar, A. R Singh, Y. Deng, X. He, P. Kumar, and R. C. Bansal, "A novel methodological framework for the design of sustainable rural microgrid for developing nations", *IEEE Access*, vol. 8, pp. 24925–24951, 2018.
13. V. Hengsritawata, T. Tayjasananta, and N. Nimpitiwanb, "Optimal sizing of photovoltaic distributed generators in a distribution system with consideration of solar radiation and harmonic distortion," *International Journal of Electrical Power & Energy Systems*, vol. 39, no. 1, pp. 36–47, Jul. 2012.

CHAPTER 11

Solar PV Single-Phase Supply Systems

Nhamo Dhlamini, Maxwell Sibanyoni, and S. P. Daniel Chowdhury

Department of Electrical Engineering, Tshwane University of Technology (TUT), Pretoria, Republic of South Africa. nhamodhlamini@hotmail.com; peace.sibanyoni@gmail.com; spchowdhury2010@gmail.com

Abstract This chapter provides a detailed analysis on the performance of a photovoltaic (PV) generator operating under varying irradiance levels. The PV generator is connected to a single-phase grid supply and the nonlinear behavior of the system is analyzed. The PV generator produces total harmonic distortion (THD) of current fed to the grid; improvements are evaluated to stabilize the quality of supply injected into the grid. Increasing the size of the inductor is proposed to stabilize and maintain the PV generator current fed to the grid. The response of the maximum power point tracking (MPPT) controller is analyzed by simulating the PV generator system with and without an MPPT controller. It is noted that the introduction of MPPT can maintain the maximum PV generator power output irrespective of changes in irradiation.

Keywords: Solar PV, Single-phase grid, Inverter, Maximum power point, Maximum power point tracking, Time step.

11.1 Introduction

Variations in irradiance and temperature on the PV cell have an effect on the output voltage and current. Cell temperature is the parameter that affects the output power of a solar cell.

The cell temperature and the voltage have an inverse relationship. This relationship shows that as the cell temperature increases, the voltage decreases. At a particular solar irradiation G, a large power decrease can be caused as a result of increasing solar temperature. The overall efficiency of a solar cell is adversely affected by temperature, which affects the power output.

The maximum power from the PV array decreases as the temperature increases [1]. The current generated by the solar cell is a function of the irradiance. The current increases as the irradiance increases and a decrease in irradiation causes a corresponding reduction in the current output. This relationship shows that the irradiation is directly proportional to the current output.

In this chapter, a PV model is developed in Matlab/Simulink, which is suitable for the integration with grid-connected power systems [2,3]. In the following sections, a description of a single-phase PV system connected to the grid is presented and modeled. Photovoltaic (PV) generator sources connected to the grid present a new technological advancement in power systems and a better knowledge through analytical studies of the electrical behavior of these systems when connected to the grid can assist in improving grid stabilization [3,4].

In the following sections, descriptions of the various elements that make up a complete single-phase PV generator system suitable for connecting to the grid system are presented and their suitability to modeling are evaluated. The well-known models in Matlab/Simulink are interfaced with the PV generator to complete the single-phase-connected grid system. To prove how valid the chosen approach is, the simulation results of the voltage and current waveforms and the harmonic content are compared with results from field measurement. The results of the PV generator with maximum power point tracking (MPPT) are compared with the generator without an MPPT system incorporated. The MPPT enables the PV generator to produce continuous maximum power most of the time. The MPPT continuously finds the voltage at maximum point or the current to enable maximum power to be produced.

11.2 Single-Phase Grid-Connected System

A simplified single-phase-connected grid system is built in Matlab/Simulink and is made of building blocks consisting of the following equipment:

- PV generator
- Inverter
- Galvanic transformer
- Grid

The PV generator systems convert solar energy into electricity through the deployment of PV cells to capture solar irradiation. The galvanic transformer is represented by an equivalent linear inductance L_{tr}/a_{tr}^2 connected to the DC/AC inverter and the grid, where L_{tr} is the transformer inductance and a_{tr}^2 is the transformer winding number of turns. The linear inductance term L_{tr}/a_{tr}^2 is for both the primary and the secondary sides of the transformer and is referred to the PV generator side. A single-phase DC/AC inverter is used consisting of a simplified full-bridge inverter.

11.2.1 DC/AC Single-Phase Inverter

A single-phase bridge inverter of the line-commutated type is considered for this study, with all switches modeled as ideal. An ideal switch is one in which the resistance is considered to be zero when closed and infinite when it is in the open position. By using the bipolar switching scheme, the two switching states of the full-bridge single-phase grid-connected inverter are summarized in Table 11.1.

The inverter switches are tracked at every time step of operation. The current of each switch in the closed position and the voltage of open switches are monitored. When the current of each switch pair nears zero, it is considered as open in preparation for the next time step. When the firing angle is close to the set value due to the voltage turning positive, the corresponding switches are closed and will start conducting. The diagram of a line-commutated switching scheme is shown in Fig. 11.1.

The control of the inverter switching is done through sinusoidal pulse width modulation (SPWM) gating signals, which drive the gate of the line-commutated inverter [5]. The bipolar switching scheme is employed to generate the gating signals (V_g), which require two inputs for the comparator, which are the sinusoidal modulating signal (V_m) and the triangular carrier signal (V_{cr}). At the higher amplitude signal of V_m compared to V_{cr}, the comparator will be high (ON) and otherwise it will be low (OFF). The gating signal waveform is illustrated in Fig. 11.2.

The line-commutated inverter and the main components that make up a grid-connected system are shown in Fig. 11.3. In Fig. 11.3, the connection of a single-phase grid to the PV system is presented. The PV system is linked to the grid through a galvanic transformer; the inverter is used to convert the DC generated by the PV system to AC. The AC system is then transformed to the grid voltage by the galvanic transformer.

Table 11.1. Switching state of full bridge line-commutated inverter—single-phase.

Switching State	Switch On	Switch Off	Voltage Pulse Output
1	S1, S2	S3, S4	$+V_{DC}$
2	S3, S4	S1, S2	$-V_{DC}$

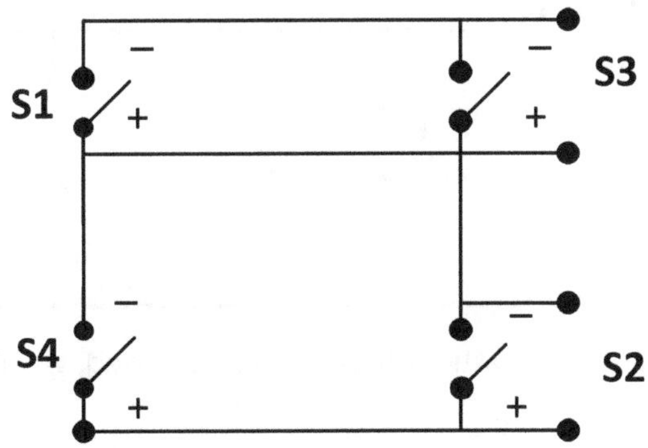

Figure 11.1. Single-phase full bridge inverter.

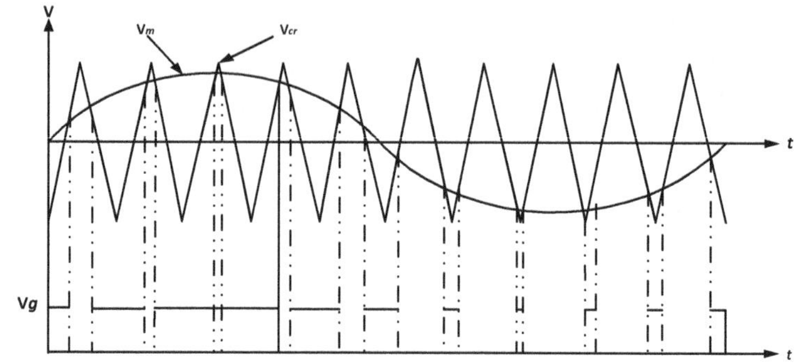

Figure 11.2. Gating signal generation of V_g.

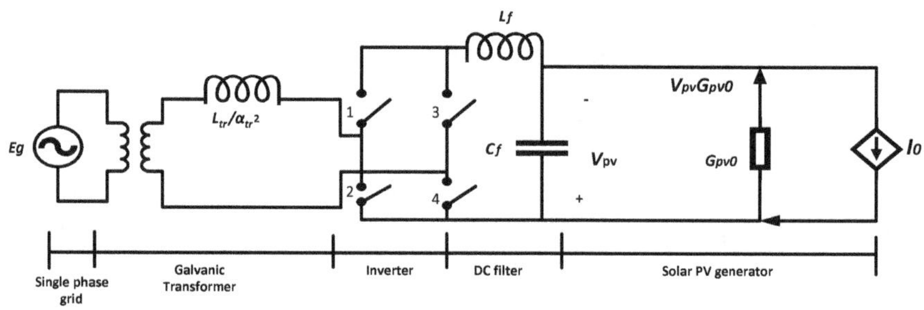

Figure 11.3. PV model of a single-phase grid-connected system.

Figure 11.4 shows the configuration diagram, which represents the topology of single-phase grid with switches 1 and 2 in the ON position, whereas switches 3 and 4 are in the OFF position.

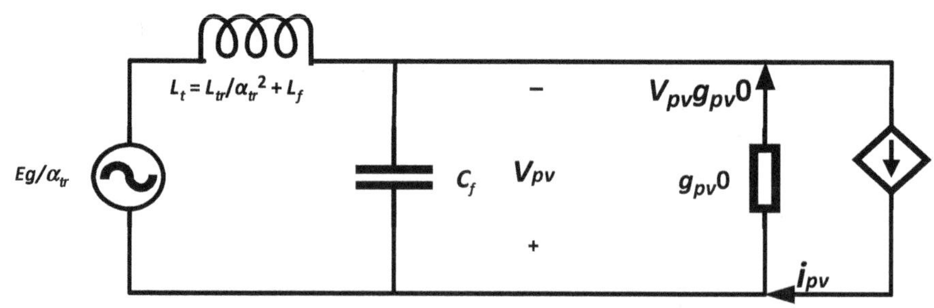

Figure 11.4. PV configuration with switches 1 and 2 closed, whereas switches 3 and 4 open.

11.3 PV-Connected Single-Phase Grid in Matlab/Simulink

The block elements used for building the PV grid-connected single-phase test system are shown in Fig. 11.5. The building blocks of the single-phase grid presented in Fig. 11.5 are the solar cells, the DC filter, the inverter, the isolation transformer, and the grid, and their functions are explained in detail in Section 11.4.

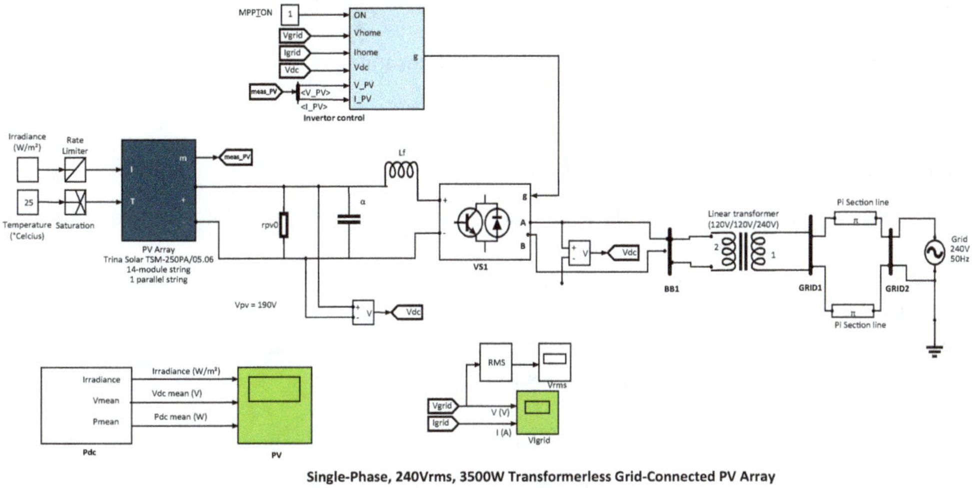

Figure 11.5. Single-phase grid-connected system.

11.4 Model of the PV Generator Model

The power generated by a single cell is considerably small. In order to generate power at a commercial scale, series connection of solar cells is used to generate higher voltage and parallel connection of solar cells is done for higher current output. Solar cells connected in series make up a module and modules connected together are referred to as an array. An array consists of modules that are connected in series-parallel type of arrangement. The output power generated from the solar PV generator can be connected directly to DC-connected loads or can be connected via inverters. The PV generator in this model is connected to the DC filter and inverter. The building blocks of the single-phase generator are explained in the sections that follow.

11.4.1 DC–LC Filter

The DC filter is connected between the PV generator and the inverter. The basic filter consisting of an inductor and capacitor is used. The main purpose of the filter is to smooth the DC harmonics and limit the rate of change of DC current.

11.4.2 Inverter Model

The inverter consists of two pairs of thyristors and the line-commutated conduction method is adopted for this study. The control strategy used for the inverter is that switches 1 and 2 are triggered at angle α, whereas switches 3 and 4 are triggered by a firing angle that is delayed by π radians per each cycle.

11.4.3 Isolation Transformer

The transformer provides galvanic isolation for the grid and the PV generator. The transformer used for this study has unity ratio and is represented for both primary and secondary sides as an equivalent inductor.

11.4.4 Utility Grid

The utility grid is represented as an energy sink by an ideal alternating voltage source.

11.5 Simulation and Analysis of Results

The system is tested on two insolation levels of 350 and 450 W/m^2. The data used for the test system are summarized in Table 11.2 [6].

The test data in Table 11.2 do not provide all the required parameters to provide inputs to the PV generator model. The inputs required are V_{oc}, I_{sc}, I_{mpp}, and V_{mpp} [7]. To obtain the input values for the stated parameters, equations characterizing the single diode have been used. The equation for the PV current I_{pv} is then given as

$$I_{pv} = I_{ph} - I_D \left[e^{W(V_{pv} + i_{pv} r_s)} - 1 \right] \tag{11.1}$$

where I_{ph} is photo current and I_D is the diode saturation current [8]. W is then found using

$$W = \frac{q}{wKT_{\alpha}} \tag{11.2}$$

where q is the electron charge, W is the curve fitting constant, and the Boltzmann's constant is represented by K [9]. After trial simulations, the relationship between v_{pv} and i_{pv} can be derived. The required input data for the PV generator is obtained with assistance from data in Table 11.2. Table 11.3 summarize data obtained from the process above under different irradiation levels.

Suitable firing angles of the inverter had to be obtained for the two irradiation levels to have the PV voltage settle at approximately 190 V. The firing angle for irradiance at 450

Table 11.2. Test data single-phase grid system.

AC grid Frequency = 50 Hz
$V_{\text{rms}} = 220\text{V}$
Transformer $L_{\text{tr}} = 320\ \mu\text{H}$
$\alpha_{\text{tr}} = 1$
DC filter $L_f = 30$ mH
$C_f = 3.3$ mF
PV generator $V_{pv} = 210$ V
$r_s = 1.95\ \text{m}\Omega$
$I_{ph} = 21.423$ A, G = 450 W/m^2
$I_{ph} = 15.325$ A, G = 350 W/m^2
$\text{I}_D = 339.5\ \mu\text{A}$
$\text{W} = 0.54932\ \text{V}^{-1}$

Table 11.3. PV generator input data for varying irradiance.

PV Generator Input Data	
G = 350 W/m^2	V_{OC} = 202 V
	I_{sc} = 18 A
	V_{mpp} = 156 V
	I_{mpp} = 15 A
G = 450 W/m^2	V_{OC} = 208 V
	I_{sc} = 24 A
	V_{mpp} = 160 V
	I_{mpp} = 21 A

W/m^2 is set at $\alpha = 147.5°$ and at 350 W/m^2 is set at $\alpha = 172°$. The simulation time step is set at $\Delta t = 10^{-0}$ s. Figure 11.6 presents the settling voltage at the two irradiance levels of 450 and 350 W/m^2. When the insolation is reduced from 450 to 350 W/m^2, the inverter goes into nonconduction mode as shown in Fig. 11.6 and verified in Fig. 11.7. This is because of the decrease in the current of the PV generator. The change in the current injected into the grid and the current i_{pv} generated from the PV generator are shown in Fig. 11.8. It is therefore of interest to note that as the irradiation level changes from 450 to 350 W/m^2, the current supplied by the PV generator also decreases in response to the fall in the irradiation level. The lowest value of i_{pv} is seen at point t_E for the two irradiation levels in Fig. 11.8.

The operating points trajectory on the $i_{pv} - v_{pv}$ plane are shown in Fig. 11.9 for period (0, t_E). At time t_E, the minimum value i_{pv} is obtained. Point B is equal to t_B = 0 s and point t_E equals point E. These are the points where the generator currents i_{pv} are shown at their minimum values, which are the same as in Fig. 11.8. The simulation results of the PV generator results are compared with the $i_{pv} - v_{pv}$ curves calculated from Eq. (11.1) obtained from the manufacturer's data sheet, and a good correlation of the two curves is shown in Fig. 11.9. The small mismatch is as a result of the approximations used in Eq. (11.1) to determine values to match the manufacturer's curves. Despite the mismatch, the conclusion derived shows that the PV generator model matches with the nonlinear model in Eq. (11.1). Figures 11.6–11.9 are further validated by the model presented in Theocharis [6] and a perfect match is observed. This gives credibility to the Matlab/Simulink model presented in this section.

Harmonics in a power system are ripples of the fundamental waves and have a great impact on the quality of supply if not properly controlled. Harmonics cause distortions in the fundamental wave. The harmonic content of the fundamental wave is measured for this study. The measured values of the harmonic AC current fed to the utility grid are compared with predicted values as shown in Fig. 11.10(a) and (b). Measured values were retrieved from a similar system used by McNeil [10] and also used by Theocharis [6]. The values of the harmonic content for both the measured and predicted values are extremely high. It is observed that the third harmonic order is extremely high in both irradiance levels with the 350 W/m^2 irradiance experiencing a higher value compared to the 450 W/m^2 irradiance level. The seventh harmonic shows a ripple and is greater than the third harmonic in the 350 W/m^2 irradiance level. The total harmonic distortion (THD) of the predicted

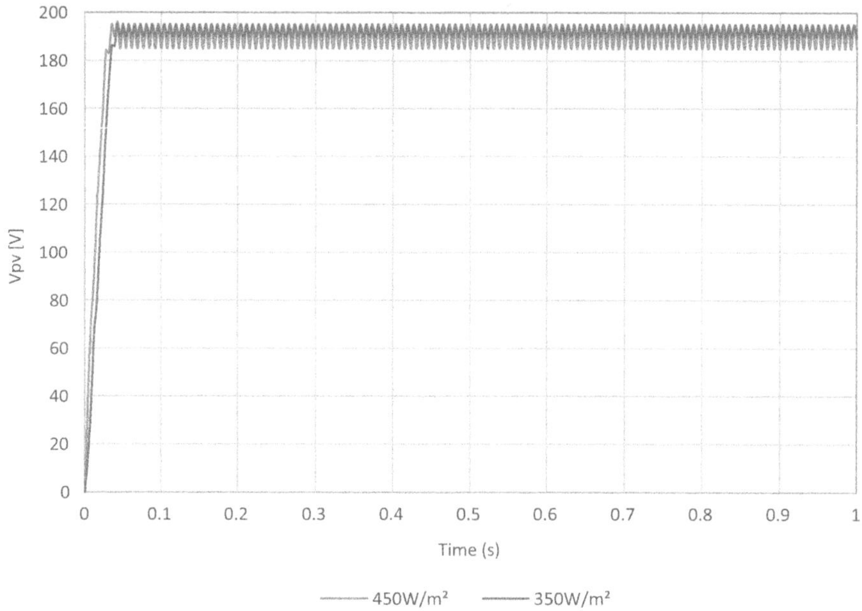

Figure 11.6. PV generator voltage v_{pv} at 450 and 350 W/m^2 irradiance levels.

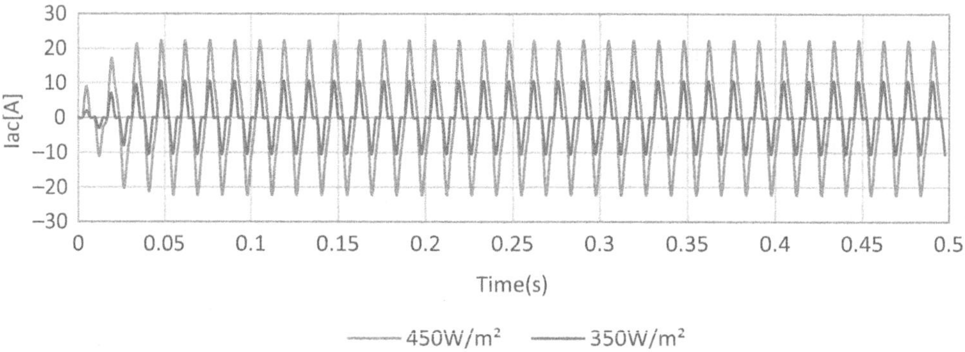

Figure 11.7. Current i_{ac} injected into the grid at 450 and 350 W/m^2 irradiance levels.

current fed into the utility network at the 350 W/m^2 irradiance level is 27.5%, whereas at 450 W/m^2 irradiance level, it is 14.9%. The graphs of the predicated values are shown in Fig. 11.10. The values are inconsistent with the required limit of 5% of the THD of AC current to be injected into the utility grid [11]. The inverter is turned into noncontinuous mode, which then causes the inconsistency in the values obtained. The dominant third harmonic is compared with the predicted and measured values. An error margin of 20.8% for the 450 W/m^2 and 14.8% for the 350 W/m^2 irradiance levels is observed. The values obtained by simulation in Matlab/Simulink are in agreement with values obtained in Theocharis [6].

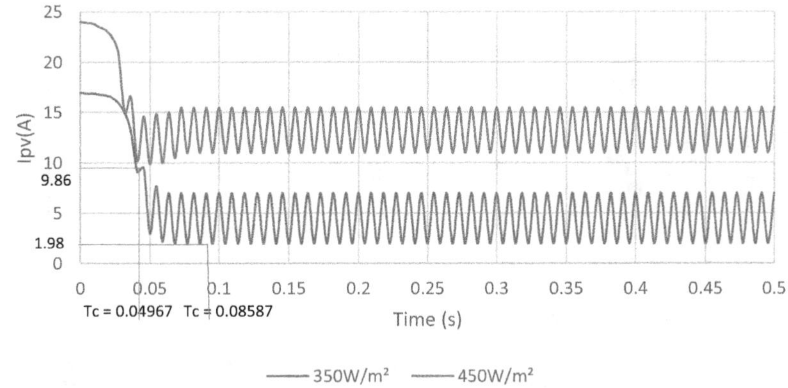

Figure 11.8. PV-generated current i_{pv} at 450 and 350 W/m^2 irradiance levels.

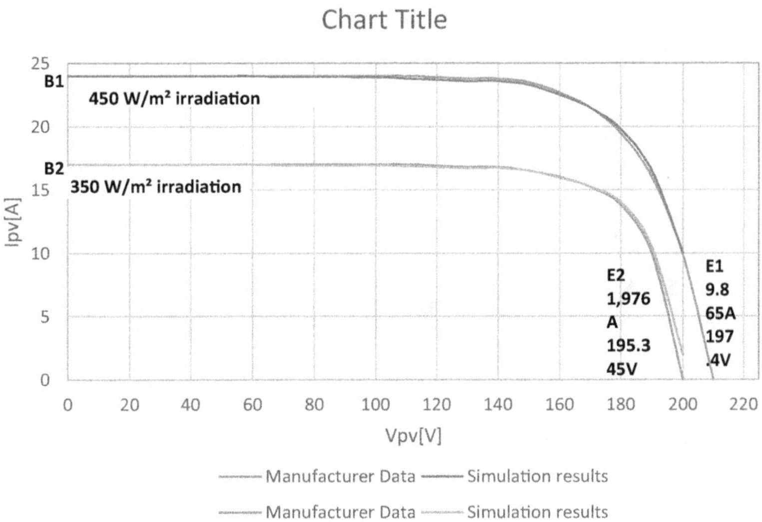

Figure 11.9. Operating point trajectory for period (0,t_E) for 450 and 350 W/m2 irradiance levels.

The discrepancies observed for the THD for the measured and predicted values are mainly because the inverter and transformer are considered ideal for simplicity as explained earlier. In general, accurate model results can only be achieved if the correct input parameters of the transformer and the inverter are implemented.

It is critical to consider that the model presented and designed with parameters in Table 11.2 is supported in McNeil [12] and Theocharis [6]; however, it does not represent the requirements of a commercially integrated grid system. This is as a result of the THD of current fed to the utility grid, which is extremely high for the two irradiance levels and

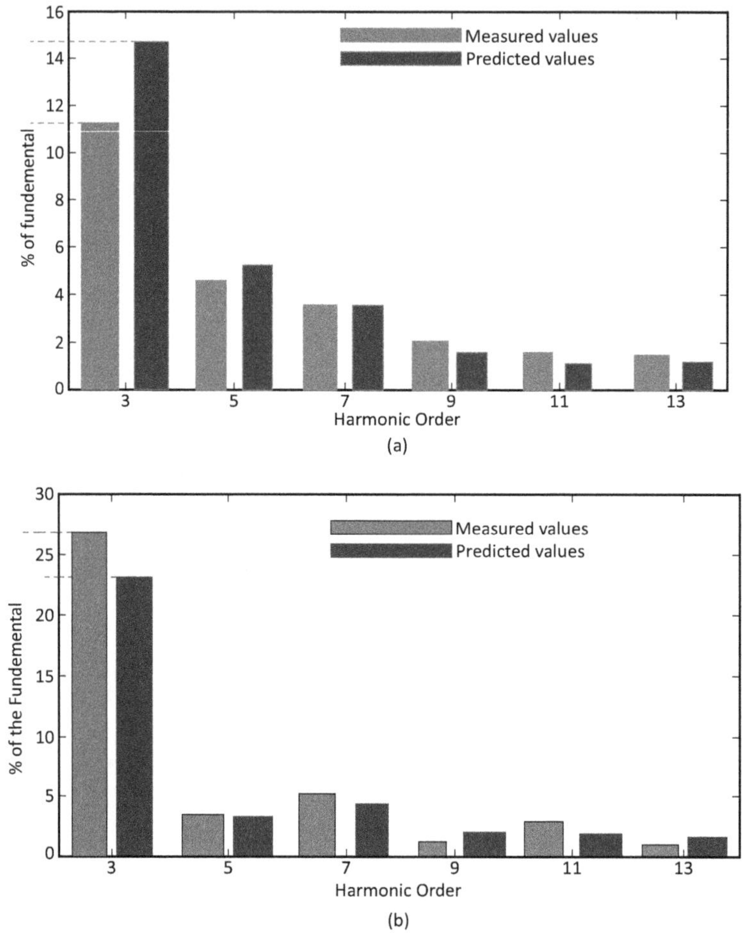

Figure 11.10. Harmonic distortion of current fed to the grid: Harmonic content at (a) 450 W/m^2 and (b) 350 W/m^2.

moreover a large variation of PV-generated current, is witnessed in Fig. 11.7. To improve, a study carried out in Theocharis [6] with regard to the performance of the system for different sizes of the inductor L_f, capacitor C_f, and the PV generator voltage v_{pv} is analyzed. A sensitivity analysis carried out in Theocharis [6] on the THD of the current fed to the grid and the variation of the PV-generated current, with different values of the DC filter, was performed with irradiation level of 350 W/m^2. It is noted in Theocharis [12] that a capacitor has no effect on the THD of current supplied to the grid because a capacitor affects the voltage waveforms only, whereas an inductor's size influence on the THD of current supplied to the grid is enormous. It is noted in Theocharis [13] that the inductor can significantly reduce the magnitude of the THD of the injected current by significantly increasing the size of the inductor L_f. As the state of the inverter switching changes from below to above continuous conduction mode, the current waveform becomes more sinusoidal than being a square wave [12]. When the inductor size is increased, this has an effect of increasing the power factor. The tendency to increase the inductor size will keep the inverter closer to its continuous conduction mode as much as possible. The variation of the output current i_{pv} from the PV generator is less impacted by the current from the inductor but is mainly influenced by the PV generator voltage v_{pv}. The reason being that it is the same in magnitude as the capacitor voltage. Simulation results presented in Theocharis [6] show that if the inductor size is varied between 12 and 320 mH with a capacitor size of 3.3 μF and a voltage v_{pv} = 195 V, large variations of i_{pv} can be eliminated. The variation of the PV current i_{pv} can be more specifically reduced by the reduction of the variation of the PV generator voltage v_{pv} by selecting the correct size of the capacitor C_f. Simulation trials were carried out in Matlab/Simulink, and it has been proven that above 55 mF, the voltage fluctuations v_{pv} become insignificant by varying the capacitor between 2 and 340 mF. Simulation results are presented in Fig. 11.11 for three capacitor sizes of 3.3, 10, and 55 mF. The fluctuation of i_{pv} can be seen for the different capacitor sizes. In Fig. 11.12, the results from simulating the system with a capacitor size of 55 mF, inductance of 25 mH, and PV generator voltage v_{pv} = 190 V, at an appropriate firing angle is shown. Interpretation of the simulation results show that when the voltage v_{pv} from the PV generator is maintained constant, the variation of the current i_{pv} is negligible.

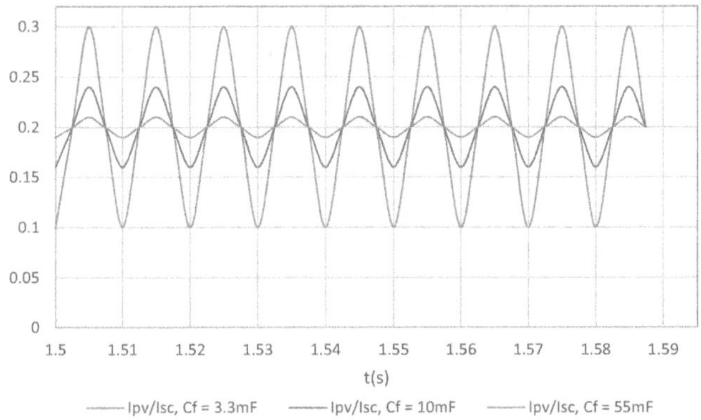

Figure 11.11. PV generator current i_{pv} for different capacitor size C_f.

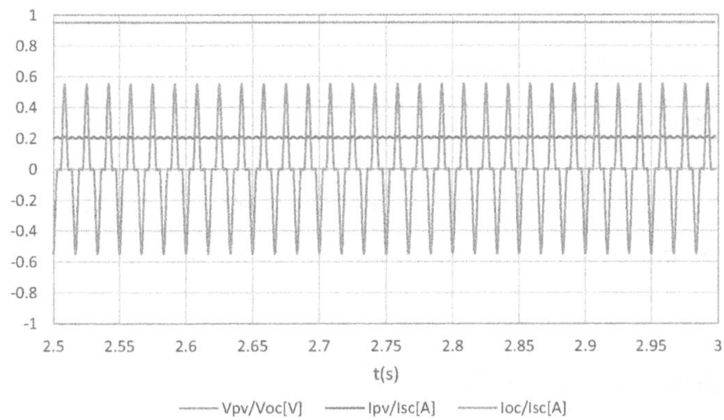

Figure 11.12. Simulation graph for capacitor size 55 mF.

11.6 MPPT Controller

The PV array operates electrically at a certain voltage V_{out}, which corresponds to the maximum power point (MPP) at the given operating conditions, that is, the temperature and irradiation [14,15]. The MPP of a PV cell is never constant over time and is always varying due to irradiation and cell temperature variations. The addition of a MPPT controller enables the tracking of the MPP [16]. The controller incorporates control algorithm for extracting and monitoring the tracking of maximum power from the PV generator system at all times. The MPPT controller will always track the system to always operate at MPP, thereby improving the PV generator efficiency. The previously studied single-phase system will further be studied with an MPPT controller incorporated. The results are compared with the PV generator system without MPPT controller (Fig. 11.5). A single-phase grid system is shown in Fig. 11.13 with an MPPT controller incorporated.

The MPPT controller uses control algorithm for monitoring output power from the PV generator system as shown. The MPPT operation principle is to vary the duty cycle of the DC/DC converter using pulse width modulation (PWM) in response to varying input current and voltage and hence the output power, placed on the MPP. Several algorithms are presented in the literature. The two mostly used MPPT controllers are the Perturb and Observe algorithm (P&O) and the Incremental Conductance algorithm (IncCond).

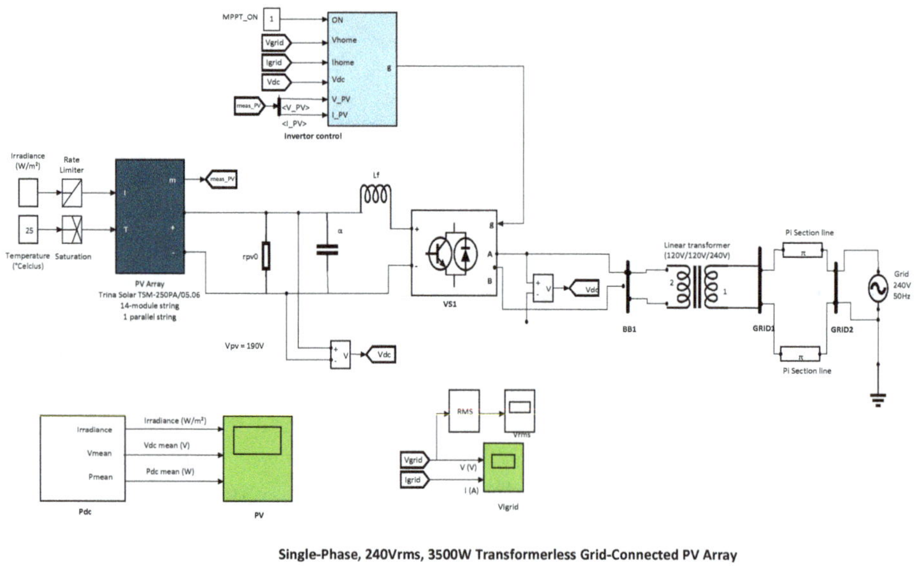

Figure 11.13. Single-phase grid system with MPPT incorporated.

11.7 Simulation and Results Analysis

The data used for this study are the same as in Tables 11.2 and 11.3; however, a capacitor value $C_f = 55$ mF has been used. The MPPT controller has been tested for irradiance level of 350 W/m^2. The simulation time step of $\Delta t = 10^{-0}$ s is chosen. The MPPT algorithm is updated after every five cycles and is not in the ON state at $t > 0$ s to allow the system to attain some stability. A comparison of the results with the system without MPPT at maximum power will be done. Simulations were run without MPPT to determine the appropriate firing angle, which gives the maximum PV generator power P_{max} [17]. The firing angle, which gives P_{max}, is found at $\alpha = 137.4°$ and the system is run at 350 W/m^2 irradiation level. The PV generator voltage v_{pv} is presented in Fig. 11.14. The maximum voltage from the simulation is $V_{max} = 156.7$ V. The PV generator current i_{pv} is given in Fig. 11.15. The maximum current $I_{max} = 15.24$ A. The PV generator power corresponding to a firing angle of $\alpha = 137.4°$ is given in Fig. 11.16. The maximum value of the power is about $P_{max} =$ 2388 W. The results obtained with MPPT are a perfect match with results obtained without MPPT.

Simulations in the following sections are conducted for initial values lower and higher of the firing angle of the MPP to test the effectiveness of the MPPT. Figure 11.13 shows the PV generator with an MPPT controller incorporated that is used for the simulation.

The MPPT is first tested for lower firing angle $\alpha_0 = 125°$ in Fig. 11.17 and its response to reach MPP is analyzed. The tracking shows that the MPPT is responding according to the algorithm set and settles at $\alpha = 137.4°$, which is the same as the previous simulations and calculations. The voltage v_{pv} is shown in Fig. 11.18. The MPPT controller reaches steady state after $t = 4.4$ s at a maximum voltage $V_{mpp} = 158$ V. This value is in agreement with the simulation result obtained without MPPT at optimal firing angle. Figure 11.19 shows the PV generator current i_{pv}. The PV generator current reaches its steady state after about $t = 4.4$ s. It is also evident from Fig. 11.19 that the PV generator current decreases as the PV generator voltage v_{pv} increases. The final steady-state maximum current settles at $I_{mpp}=$ 15.07 A, the value is in agreement with the simulated results without MPPT at optimal firing angle. The maximum power curve is shown in Fig. 11.20 and the PV generator power settles at $P_{max} = 2384$ W, which is in agreement without MPPT controller at optimal firing angle. Figure 11.21 is the simulated results for the RMS current injected into the utility grid

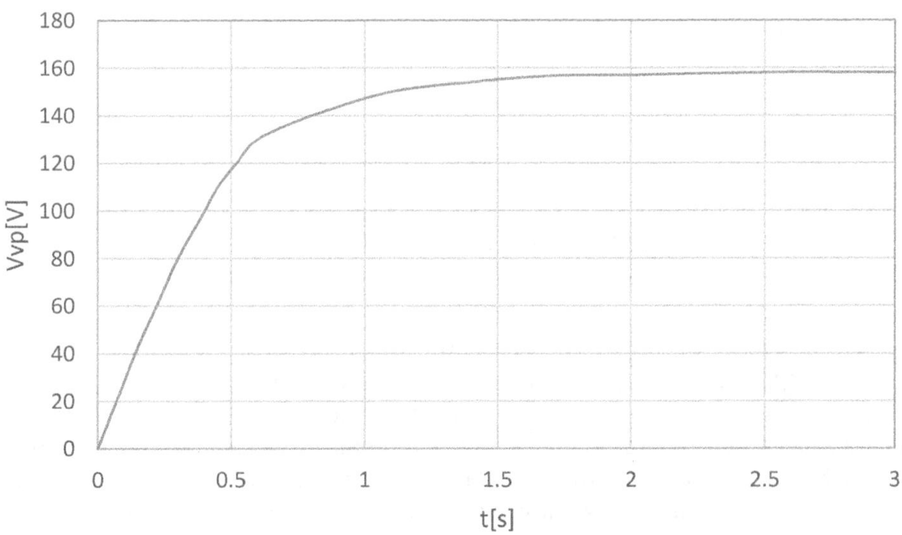

Figure 11.14. V_{pv} for $\alpha = 137.4°$ at irradiance 350 W/m^2.

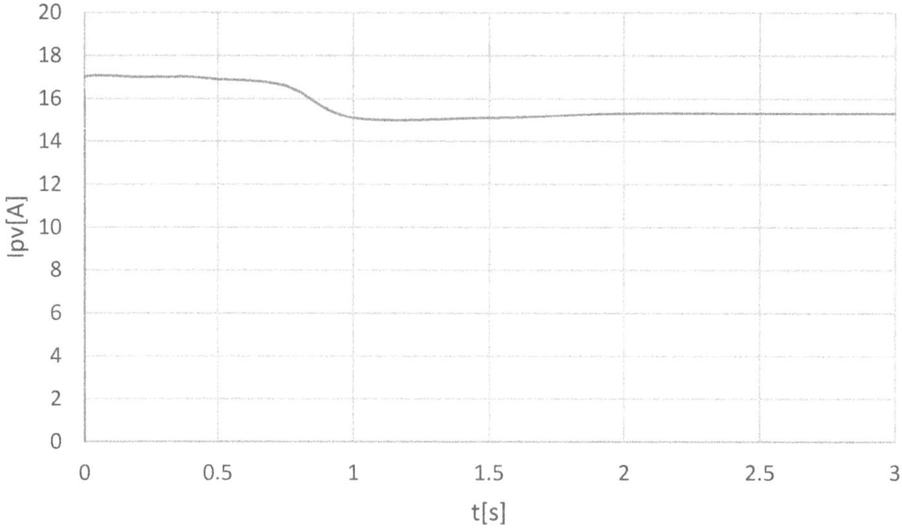

Figure 11.15. I_{pv} for α = 137.4° at irradiance 350 W/m^2.

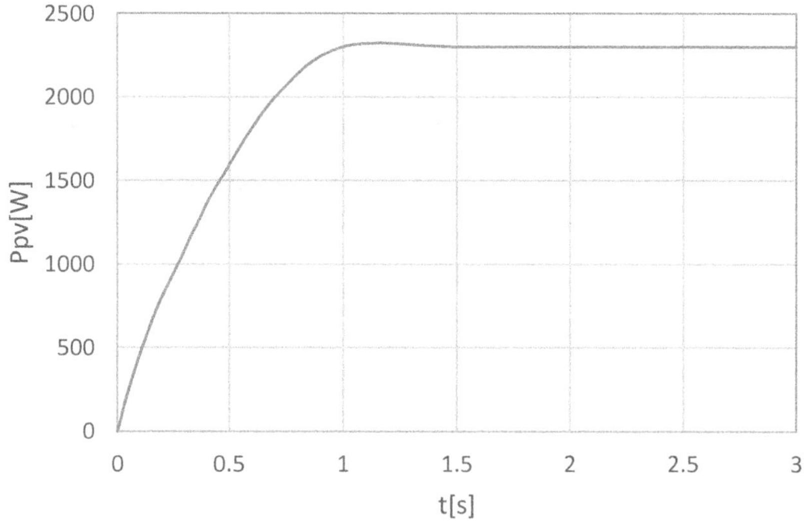

Figure 11.16. P_{pv} for α = 137.4° at irradiance 350 W/m^2.

r_{acrms}. The fluctuations at $t > 2$ s are due to MPPT, this is because the initial current is not close to the MPP before it settles at its maximum steady value.

The second simulation analysis is carried out with a firing angle of α_0 = 145°, which is higher than expected. Figure 11.22 presents the optimal tracking angle with respect to time. The MPPT algorithm is activated and is monitored to track the firing angle that gives P_{max}, which is found at α = 137.4°, which is the expected value. The PV generator voltage is shown in Fig. 11.23 and settles at V_{mpp} = 158 V at about t = 3.6 s, which is in perfect

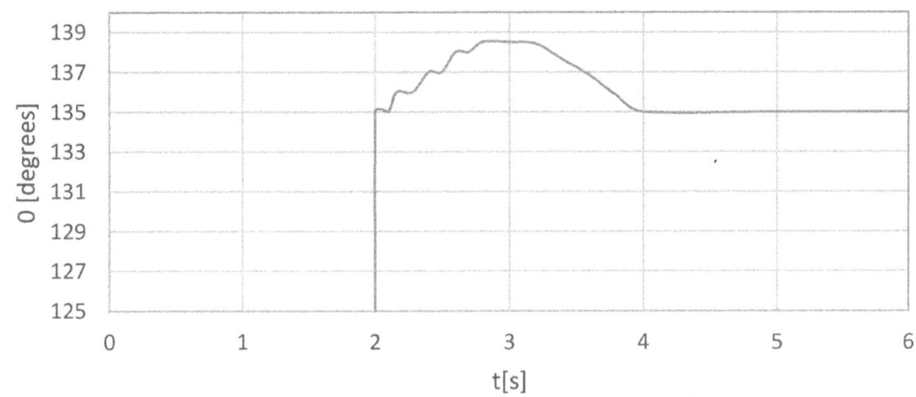

Figure 11.17. Firing angle tracking for α_0 = 125° at 350 W/m² irradiance level.

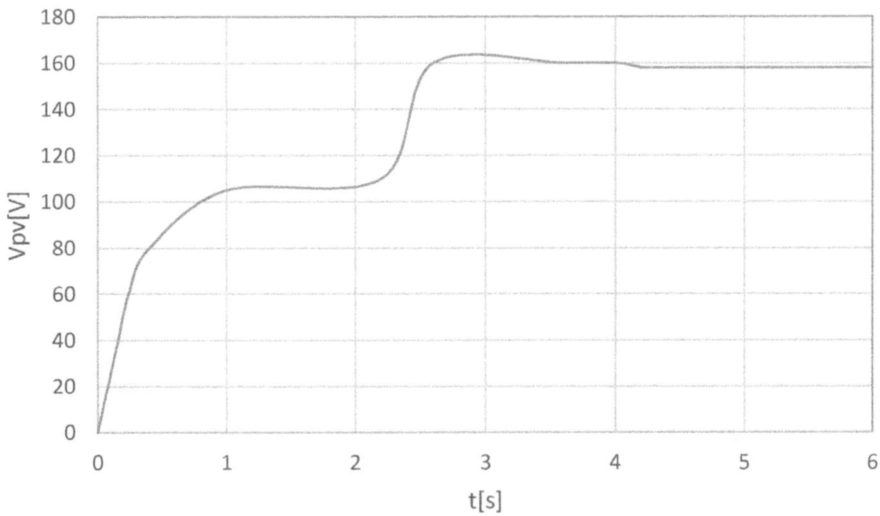

Figure 11.18. V_{pv} for α_0 = 125° at 350 W/m² irradiance level.

agreement with the expected results from previous simulation. The PV generator current is shown in Fig. 11.24 and settles at I_{mpp} = 15.07 A, which also agrees with simulation without MPPT at optimal firing angle. The power curve of the PV generator is shown in Fig. 11.25 at $P_{\max}$ = 2383.7 W, which agrees well with the simulation without MPPT at optimal tracking point.

Figure 11.26 is the simulation results for the RMS current injected into the utility grid r_{acrms}. The fluctuations at $t > 2$ s due to MPPT are because the initial current is not close to the MPP before it settles at its maximum steady value.

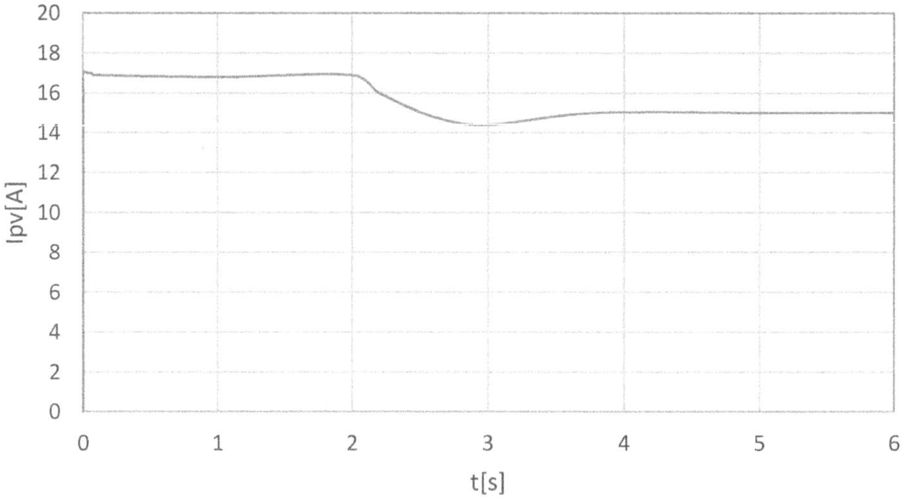

Figure 11.19. i_{pv} for α_0 = 125° at 350 W/m^2 irradiance level.

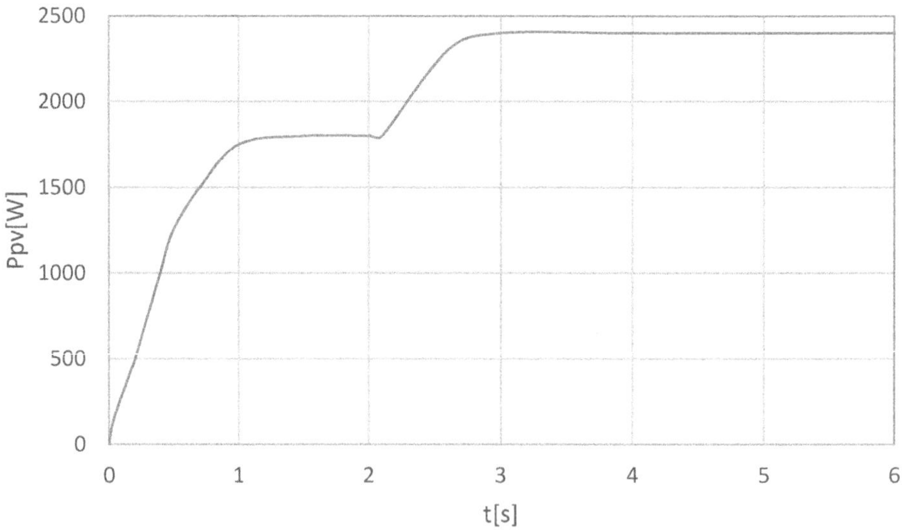

Figure 11.20. P_{pv} for α_0 = 125° at 350 W/m^2 irradiance level.

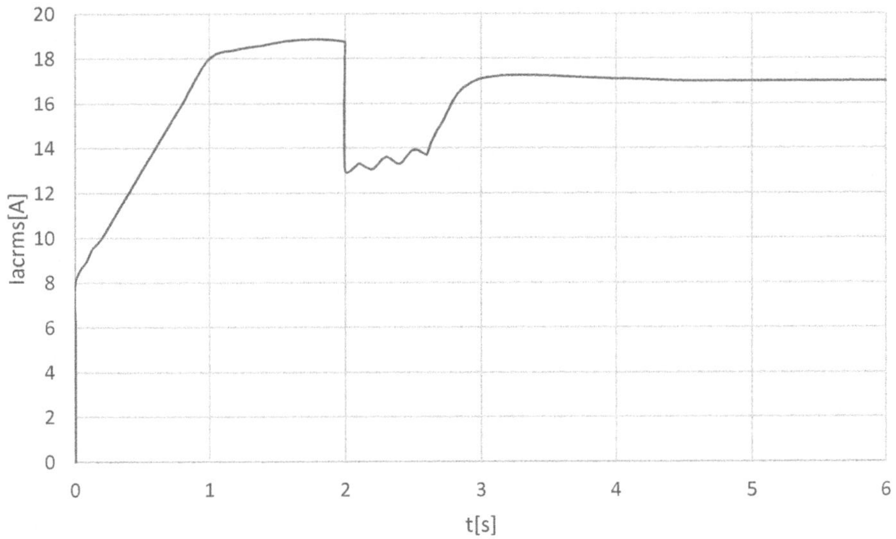

Figure 11.21. Current injected into the utility grid for $\alpha_0 = 125°$ at 350 W/m^2 irradiance level.

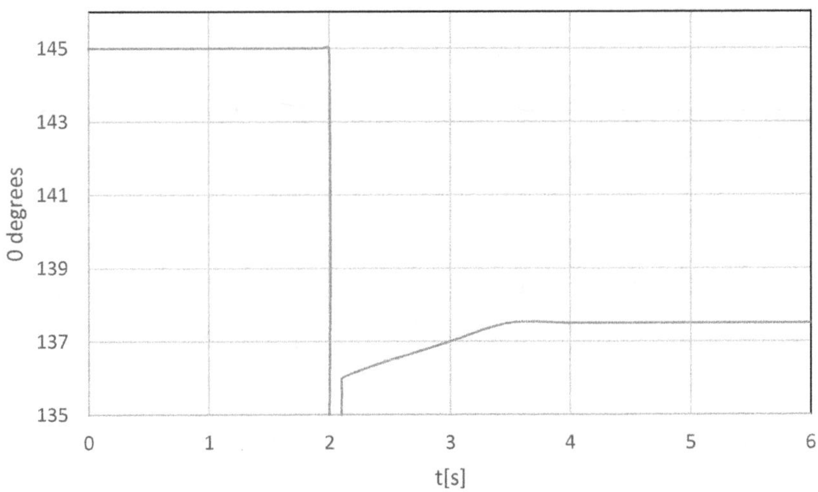

Figure 11.22. Firing tracking angle for $\alpha_0 = 145°$ at 350 W/m^2 irradiance level.

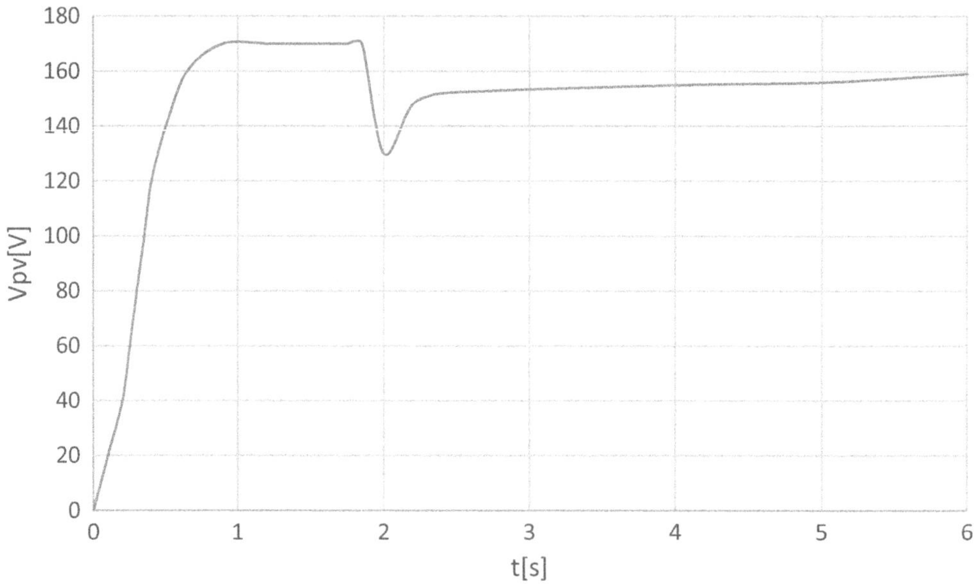

Figure 11.23. V_{pv} for α_0 = 145° at 350 W/m² irradiance level.

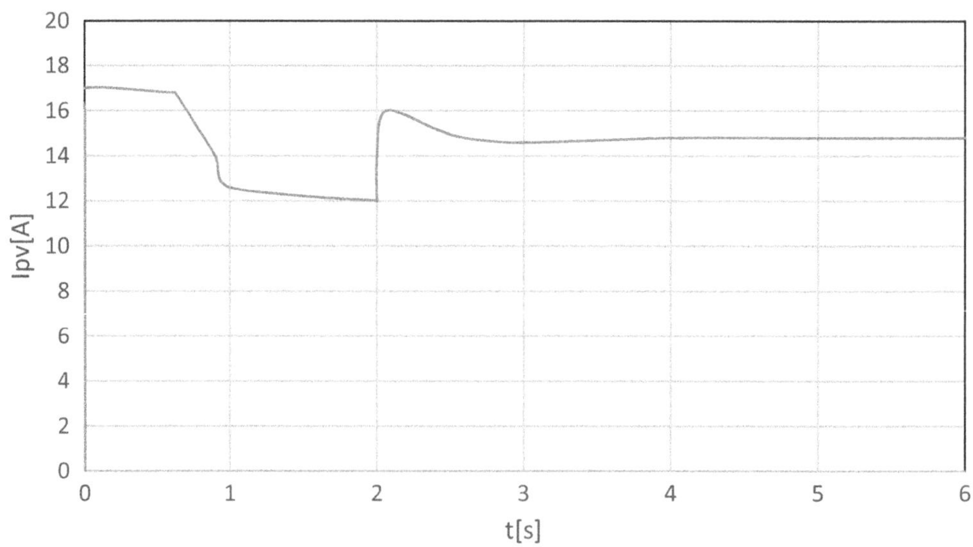

Figure 11.24. i_{pv} for α_0 = 145° at 350 W/m² irradiance level.

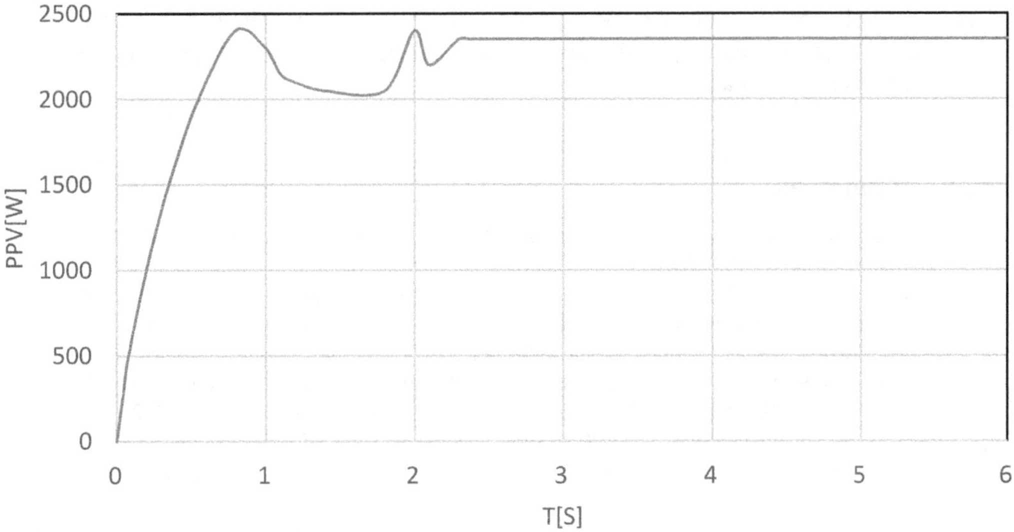

Figure 11.25. P_{pv} for α_0 = 145° at 350 W/m² irradiance level.

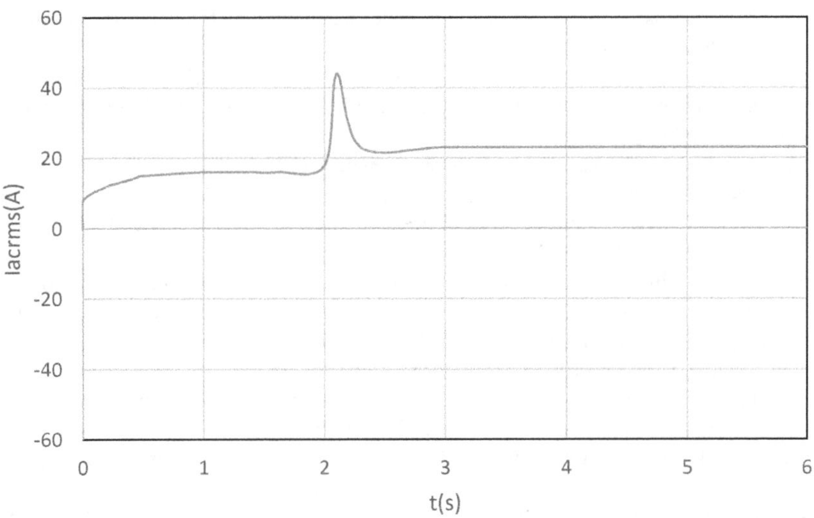

Figure 11.26. Current injected into the grid i_{ac} for α_0 = 145° at 350kW/m² irradiance level.

11.8 Conclusion

In conclusion, the PV generator in Matlab/Simulink is integrated with grid-connected single-phase PV generator system. The results confirm that the THD of current fed to the grid by the PV generator is improved by increasing the size of the inductor. Increasing the size of the inductor maintains the PV generator current stable when the PV generator voltage is also stable. It is noted that the MPPT maintains the maximum PV generator power despite changes in irradiance.

11.9 References

1. *Planning and Installing Photovoltaic Systems; A Guide for Installers, Architects, and Engineers, DGS*, 2nd ed, pp. 85–93, 2010.
2. J. Widen, E. Wäckelgård, J. Paatero, and P. Lund, "Impacts of distributed photovoltaics on network voltages Stochastic simulations of three Swedish low-voltage distribution grids," *Electric Power Systems Research*, vol. 80, pp. 1562–1571, 2010.
3. J. Solanki, S. K. Solanki, and V. Ramachandran, "Steady state analysis of high penetration PV on utility distribution feeder," IEEE Conference, Orlando, FLA USA, pp. 1–6, 2012.
4. R. Shah, N. Mithulananthan, and R. C. Bansal, "Oscillatory stability analysis with penetrations of large scale photovoltaic generation," *Energy Conversion and Management*, vol. 65, pp. 420–429, 2013.
5. M. J. Hosain, T. K. Saha, N. Mathlananthan, and H. R. Pota, "Robust control strategy for PV system integration in distribution systems," *Applied Energy*, vol. 99, pp. 355–362, 2012.
6. H. Patel, and V. Agarwal, "MATLAB-based modelling to study the effects of partial shading on PV array characteristics," *IEEE Transactions on Energy Conversion*, vol. 23, pp. 302–310, 2008.
7. J. Hernandez, G. Gordillo, and W. Vallejo. "Predicting the behavior of a grid-connected photovoltaic system from measurements of solar radiation and ambient temperature," *Applied Energy*, vol. 104, pp. 527–537, 2013.
8. S. K. Kim, J. H. Jeon, C. H. Cho, E. S. Kim, and J. B. Ahn. "Modeling and simulation of grid-connected PV generation system for electromagnetic transient analysis," *Solar Energy*, vol. 83, pp. 664–678, 2009.
9. J. Milias-Argitis, and T. Zacharias, "Transient phenomena of RLC circuits—The photovoltaic input," *Solar Energy materials and Solar Cells*, vol. 55, pp. 363–378, 1998.
10. P. Jose, and P. R. Jose. "Grid connected photovoltaic system with fuzzy logic control based MPPT," *IJEIT*, Vol. 3, Nov. 8, 2014.
11. "MATLAB Simulink Simulation and Model Based Design User's Guide." Version6.3. Available Online: http//www.mathworks.com/access/helpdesk/help/pdfdoc/simulink/slusing.pdf. The Mathworks—MATLAB & SIMULINK, Vol. 6, 2006.
12. Siemens solar module sm110/sm100. Solar Energy, Vol. 57, pp. IV, 1996.
13. G. Bayrak, "A remote islanding detection and control strategy for photovoltaic-based distributed generation," *Applied Energy*, vol. 104, pp. 527–537, 2013.
14. T. Zacharias, J. Milias-Argitis, and V. Makios, "First-order circuits driven by a photovoltaic generator," *Solar cells*, vol. 31, no. 1, pp. 57–75, 1991.
15. G. Notton, V. Lazarov, and L. Stoyanov. "Optimal sizing of a grid-connected PV system for various PV module technologies and inclinations, inverter efficiency characteristics and locations," *Renewable Energy*, vol. 35, pp. 541–554, 2010.
16. J. H. R. Enslin, M. S. Wolf, D. B. Snyman, and W. Swiegrs, "Integrated photovoltaic maximum power point tracking converter," *IEEE Transactions on Industrial Electronics*, Vol. 44, No.6, pp, 769–773, 1997.
17. C. Riodan, and R. Hulstrom, "What is air mass 1.5 spectrum," *Solar Energy Research Institute*, vol. 2, pp. 1085–1088, 1990.

CHAPTER 12

Techno-Commercial Analysis of Microinverters as a Future Technology in Solar PV Power Generation

Ruchira[1], S. K. Sinha[1], and R. N. Patel[2]

[1]*Department of Electrical and Electronics Engineering, Amity University Noida, Uttar Pradesh, India. rsingla@amity.edu; sksinha6@amity.edu*
[2]*Department of Electrical Engineering, National Institute of Technology Raipur, Raipur, Chhattisgarh, India. rnpatel.ee@nitrr.ac.in*

Abstract The demand for energy is increasing at a high pace with the development of new technologies. Also, the fast rate of depleting of fossil fuels has compelled the user to switch to some alternative sustainable sources of energy. One such alternative is solar cells, which convert the sunlight into electrical power. The output of the solar-based generated power is DC and the existing infrastructure requires the AC voltage to interconnect with the grid. Therefore, the use of inverter in solar power generation system is important to convert the power into more usable form, which is also compatible with the electrical appliances. This chapter discusses the three major inverter configurations: central inverter (CI), string inverter, and microinverter. The various standards for the photovoltaic (PV) installation, the grid and the PV module requirements have been presented. The discussion further continues with the effects of shading on the performance of CI and microinverter (MI) connected to low-power solar-based domestic or rural applications. A case study has been included wherein the comparison of MI and CI has been made on the basis of experimental results. At the end the techno-economical analysis of MIs has been discussed wherein the energy saving with the use of MI technology has been highlighted.

Keywords: Solar PV systems, Solar inverters, Microinverter, Central inverter, String inverter, Partial shading, Dusting, PV economics.

12.1 Introduction

The demand for energy is increasing at a high pace with the development of technology. Also, the fast rate of depletion of fossil fuels has compelled the user to switch to alternative sources of energy. One such alternative is solar photovoltaic (PV) systems, which convert the sunlight into electrical power. The output solar power generated is DC and the existing infrastructure requires AC voltage to connect with the grid. Therefore, the inverter is an important component in solar power generation system to convert the power into more usable form that is compatible with the electrical appliances.

A grid interactive system is a solar PV system, which is connected to the grid but can also work stand alone in the absence of the grid. If the grid experiences a problem, grid tied systems are required by safety standards to be disconnected from the grid. Grid interactive systems can be designed with or without battery storage. The major advantage of this system is that power can be fed into the grid or can be drawn from the grid as required. The components of grid interactive PV system are PV array, inverter or power conditioning unit, breaker box, connectors, and meters. Inverter is the backbone of grid interactive system. The output AC power from the inverter is synchronized with the grid's voltage levels depending on the size of the PV system installed at a particular place and the load requirement. The function of the power electronic interface for PV-grid systems is to amplify and invert the generated DC power into AC current suitable for the grid and to control the PV module so as to track the maximum power point (MPP) for maximizing the energy capture. The functions of the inverter are

(i) It converts solar-generated DC power into AC power.
(ii) It can also work as a battery charger and store energy in a battery designed for residential and commercial applications.
(iii) During outage, it can convert the DC output of battery into usable AC power and charge the battery bank during the day from the PV modules or generator. It can also turn a generator on as per the need.

Broadly, inverters can be classified into different ways based on different criterion. For example, based on output waveform, it can be classified as square wave inverter, sine wave inverter, or modified sine wave inverter. Another way of classification is based on input used. Accordingly, it is known as current source inverter or voltage source inverter. Similarly, we can classify inverter based on load or conversion technique used. These are some of the general classifications of inverter. But when we use inverter integrated with PV system, there are three major inverter configurations: central (CI), string, and microinverter (MI) discussed in Section 12.2.

12.2 Evolution of the Grid-Connected Inverter

In this section, various inverter configurations are discussed.

12.2.1 The Past

The old technique is based on the centralized inverters, where huge number of PV modules are interfaced by the centralized inverters to the grid as depicted in Fig. 12.1(a). The PV modules are connected in series, producing reasonably high voltage eliminating the need for further amplification. The string of such PV modules is further connected in parallel through diodes to achieve high-power output. But this type of system has several drawbacks, for example, power losses due to a centralized maximum power point tracker (MPPT), high-voltage DC cables between the PV modules and the inverter, losses in the string diodes, mismatch losses between the PV modules, and nonflexible design. Therefore, advantage of large-scale production could not be obtained. The grid-connection stage was line commutated through thyristors having large current harmonics and low quality of power. This drawback of large harmonics led to the new inverter topologies/system layouts [1]. Many benefits of using the CI are as follows:

(i) Low capital price per watt.
(ii) High efficiency.
(iii) Comparative ease of installation.

The CI has also certain limitations such as

(i) High-voltage DC cable is required between the PV modules and the inverter, which increases the cost and losses.
(ii) Mismatch losses between the modules occur in case of different ratings.
(iii) Losses in the string diodes.
(iv) Problem of hotspot formation.
(v) Due to a nonflexible design the benefits of mass production cannot be reached.
(vi) System size becomes large.
(vii) More noise and harmonic generation.
(viii) The entire system fails because of single potential point.

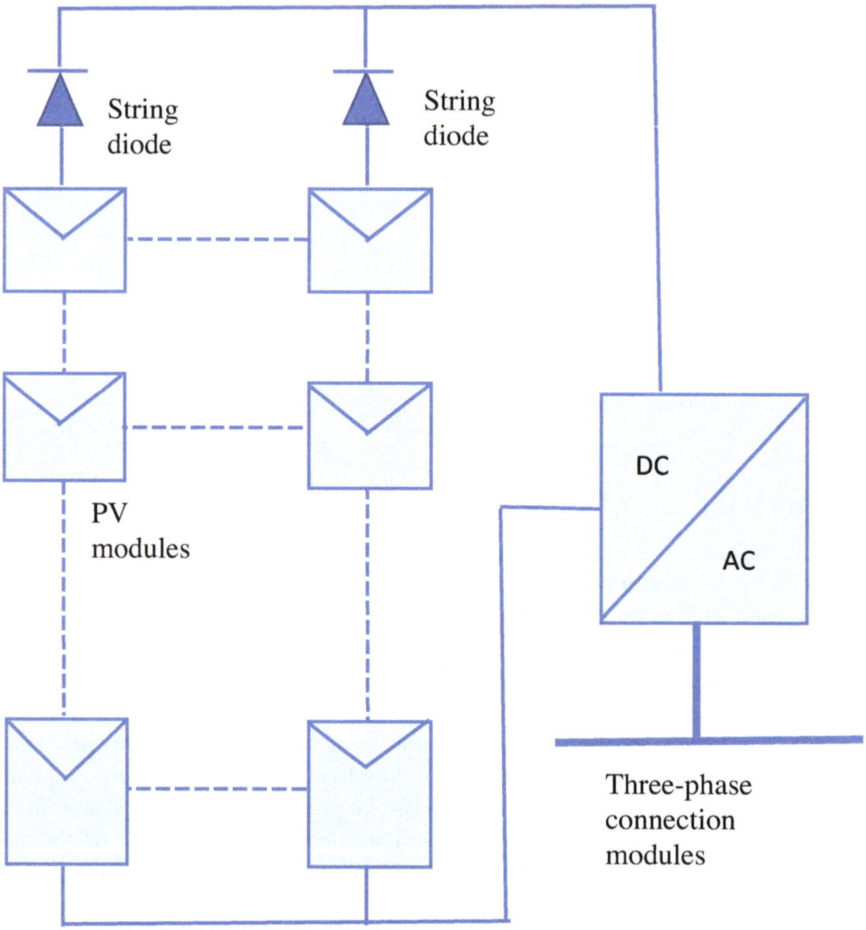

Figure 12.1(a). Central inverter.

12.2.2 The Present

String inverter is a simplified version of the CI where a single string of PV modules is connected to the inverter as shown in Fig. 12.1(b) [1]. In this system, voltage amplification is avoided due to high-input voltage and this requires roughly 16 PV modules in series. The total open-circuit voltage for 16 PV modules reaches around 720 V, which requires 1000 V metal oxide semiconductor field-effect transistor (MOSFET)/insulated-gate bipolar transistor (IGBT) to allow 75% voltage derating of the semiconductors. The normal operation voltage is approximately 450–510 V. If DC–DC converter or line-frequency transformer (LFT) is used for voltage amplification, then PV modules are required to be connected in series. Individual MPPTs can be applied to each string and no losses occur with string diode. The overall efficiency of string inverter is better than the centralized inverter. String inverters are low in cost also due to large-scale production [1]. The various advantages of string inverter are as follows:

(i) This is relatively more flexible and robust.
(ii) High efficiency.
(iii) Cost is low.

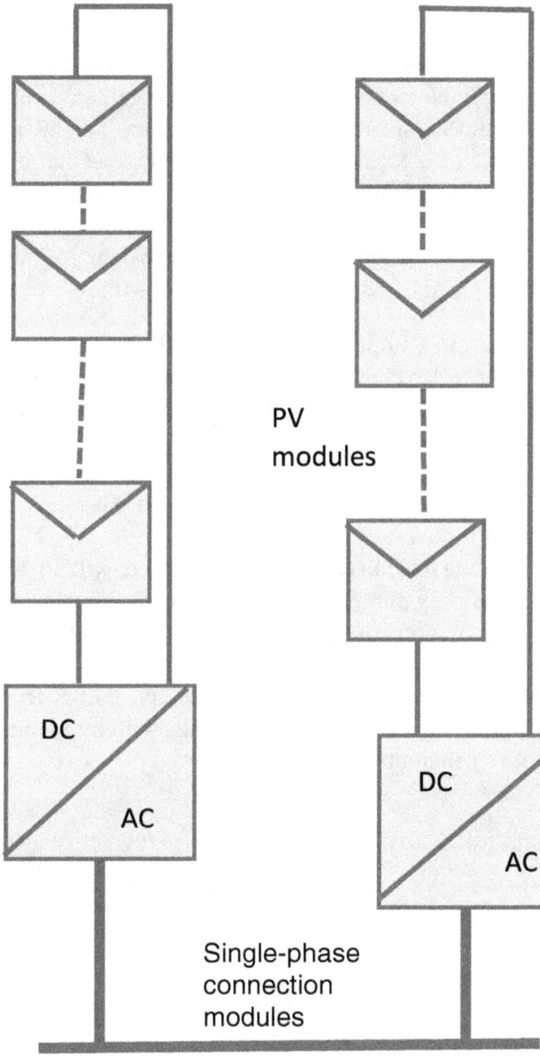

Figure 12.1(b). String inverter technology.

(iv) Three-phase variations available.
(v) Well supported (with all trusted brands).
(vi) Remote system-monitoring capabilities.

The limitations of string inverter are as follows:

(i) Module level MPPT is required.
(ii) Panel level monitoring is not possible.
(iii) High-voltage levels develop a potential safety hazard.

12.2.3 The Future

Figure 12.1(c) shows the multistring AC module, which is the combination of the PV module and the inverter in a single electrical unit [2]. This arrangement is known as MI. Due to single PV module, it mitigates the mismatch losses. One important advantage of the above-mentioned device is the ease of extension. Another advantage is that it acts as a "plug and play" device, which can be used by a layman also. The AC modules have low manufacturing cost/retail price due to intended large production. Self-commutated DC–AC inverters through IGBTs/MOSFETs having high-power quality and when used in line with the standards give a better solution [1]. The various advantages of MI are:

(i) **Shade tolerant:** The most important benefit of MI is that it performs quite better than the centralized inverter in case of shading. The partial shade on a part of a solar array does not affect the overall output of the array. This will be discussed in detail in the later sections of the chapter.
(ii) **No air cooling:** Air cooling is not required in case of MIs, because less heat loss is there. This is because the size of MI is very small. However, separate cooling is required in case of CIs.
(iii) **Minimum maintenance loss:** In case of any fault, there will not be a complete breakdown of the system as only part of it stops working. So this is another major benefit of MI. This makes the maintenance task also easier. Moreover, the replacement cost will also be lesser here.
(iv) **Remote monitored:** Typically in an MI architecture, output from each solar panel and the MI is monitored via a communication bus, thereby identifying and isolating a problem much faster than other inverters.

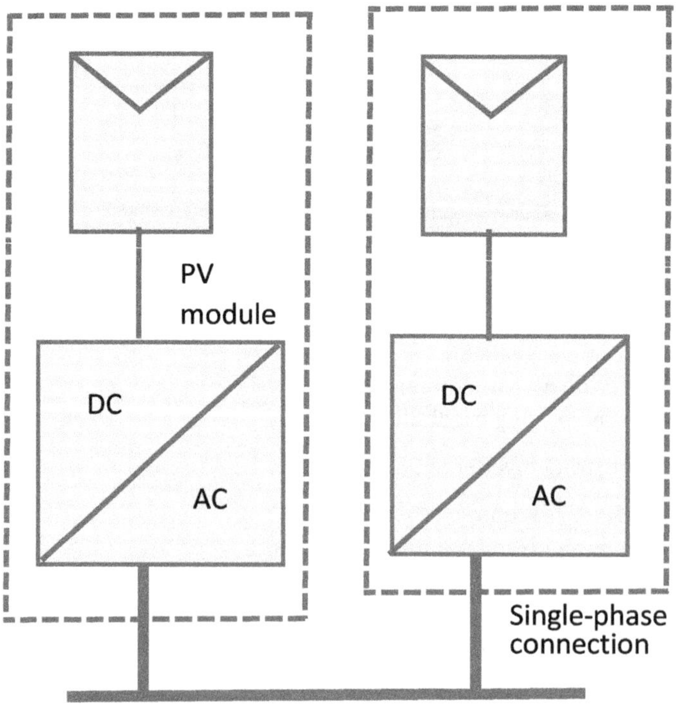

Figure 12.1(c). AC module technology.

(v) **Smaller cable size:** As output is converted from DC to AC (230 V) at the back of the panel, the cable required to carry the current can be of lower diameter.
(vi) **Easily expandable:** As all the panels work independently, system expansion is easier in this case without affecting the already installed infrastructure.
(vii) **Suitable for residential buildings:** Due to its modularity and its size, it's very suitable for residential buildings where each panel can be differently oriented or have different shading patterns during the day.
(viii) **Easy installation:** As there are no specific restrictions on the orientation of solar panels, anyone can install a MI-based solar on grid system with relative ease.

However, there are certain disadvantages of MI. These are:

(i) **Higher initial cost:** The major drawback of MI is its high initial cost. But this cost gets covered with time, as in case of shading the entire system performs equally well, which is not the case with other configurations. But the payback period is large.
(ii) **Extra monitoring device:** As each inverter is independently located below a solar panel, a communication bus and a common monitoring system are required as there is onboard display system like in string inverters.
(iii) **Higher cost of replacing:** As the inverter is below the solar panel, it might be very difficult to replace the MI due to existing site conditions and might require more than one technician on site to replace.

12.3 Standards and Requirements for the Future Inverters

12.3.1 Standards of PV Systems

Because of fast growth in demand of PV applications, international codes and standards are evolving. IEEE, International Electro-technical Commission (IEC), the European Committee for Electrotechnical Standardization (CENELEC), and the Association for Electrical, Electronic, and Information Technologies (VDE) are standards organizations. United States follows the National Electrical Code (NEC). For grid-connected systems, these are the standards that handle detection of anti-islanding, galvanic isolation presence, total harmonic distortion (THD), and injection of DC current [2].

(A) Galvanic isolation

Galvanic isolation protects PV systems as separation of circuit is ensured from stray currents. In the absence of galvanic isolation, there is leakage of ground current as capacitances develop between PV cells and metallic frame, which are grounded. The inclusion of transformers can reduce these leakage currents as they provide galvanic isolation of the PV modules from the grid. But, additional cost and reduced efficiency come into effect due to inclusion of transformers. As per NEC 690.35 and 690.41 standards, ground-fault protection device can be used for the protection of PV sources and output circuits. Alternatively, protection can also be provided by solidly grounding one of the conductors of a PV system when the output voltage is more than 50 V [3].

(B) Anti-islanding detection

When the grid trips, PV inverter has to be interrupted from providing power. This is done by anti-islanding detection. According to IEEE 1547 and IEC61727 standards, the islanding conditions of distributed generation (DG) should be identified and separated within 2 s. Different methods of islanding detection of DG have been discussed in [4–8].

(C) Total harmonic distortion

As per IEEE1547 and IEC61727 standards, the THD should not be more than 5%. These standards limit DC current injection up to a maximum of 1%, whereas as per the standard, VDEV 0126-1-1, 1 A is the maximum allowable DC current. The reduction in THD of the output current can be done by connecting a power conditioner in parallel with the PV plant. Soft computing methods are employed for harmonics elimination by improving switching modulation technique [9–12].

(D) Reactive power control

In grid-integrated PV systems, injection of reactive power increases voltage in the grid. Hence, reactive power and phase shifts are required to be regularly monitored. As per the German Federal Association of the Energy and Water Industry guidelines, operators should ensure that reactive power (both inductive and capacitive) be fed into the grid with a shift factor of 0.95. It is recommended to maintain a shift factor of 0.90 for the low-voltage grid connected to a PV system. To accomplish this, different schemes of reactive power control have been proposed in [13–15].

12.3.2 Performance Requirements of PV Converters

MPP tracker accomplishes the maximum power operation of the PV module. This maximum power operation must be ensured by MI. Its efficiency should be maintained for a wide range of voltage and power variations for continuously changing parameters such as ambient temperature and solar radiation. Additionally, it should be compact and reliable.

(A) Efficiency

Efficiency has always been the most important factor for the design engineers. Therefore, the components are selected keeping the related losses in mind while designing the MI. As discussed in [16], the maximum efficiency reported for single-stage isolated topology is 94%, whereas that of multistage topology is 96.2% [17,18]. The peak efficiency of 99% is reported for a single-phase nonisolated transformer-less inverter [19]. Generally MOSFET is used for switching in most of the inverter topologies. Use of GaN high-electron-mobility transistors or SiC MOSFETs reduces the switching losses as they are very fast switching devices. Optimum core design and the winding turns ratio reduce core losses. Current stress and losses are also reduced by the use of multicore transformers.

(B) Installation cost

PV module and the converter system are the major components that account for cost of a grid-connected PV system. In the last 10 years, there has been a drastic reduction in the cost of PV modules. Simultaneously, other ancillary costs, such as labor and construction materials, have also reduced. These two costs have reduced by 68% and 56%, respectively, over this period [19]. Hence, it is essential to reduce the cost of MI to limit the grid-connected

PV system expenditure. As reported in [20], the price of MI varied in the range of 116–542 USD for a unit of 215–300 W. The overall system cost also varies depending on the difference in land, labor, and other charges from one region to another.

(C) Life span

The capacitance rating of the power-balancing decoupling capacitor decides the reliability and hence life span of a MI. The value of this capacitor can be determined by the following equation:

$$C_d = \frac{P_{in}}{w_o V_{dc} \Delta V} \tag{12.1}$$

where P_{in} is the PV panel rated power, V_{dc} is the DC voltage of the decoupling capacitor C_d, ω_o is the line frequency, and ΔV is the peak-to-peak voltage ripple. By placing the capacitor on a high-voltage DC link or by including a decoupling circuit on the PV side, the reliability of the MI can be enhanced [21]. Alternatively, the use of a Li-ion ultra-capacitor on the PV side also increases the life span of the MI.

(D) Power density

As MI is connected to each PV module, it should be very compact in design, which can be achieved by using high switching frequency, which also helps in greatly reducing the sizes of transformer and inductors. The highest power density of 4.86 kW/L at a switching frequency of 80 kHz for nonisolated inverter has been achieved [22].

12.4 MI Design Objective

MI is module-integrated device whose major function is to convert the DC output coming from the PV into AC form. This conversion is important to make the power into usable form because most of the appliances demand AC supply and we have AC infrastructure adopted in our country. The important feature of MI is that it is connected at the backside of the panel. It is a plug and play device, which is easy to install. While designing the MI for any application, it should be ensured that it provides a high rate of efficiency. This is required to harness the maximum energy from the module. The design should be such that it consists of a minimum number of electronic components. When inverter is interfaced with the PV module(s) and the grid, it should ensure the following: (1) PV module operates at the MPP and (2) sinusoidal current is injected into the grid. The MPPT is also done at the module level. The design should be such that it should be able to convert even small amount of power in case of any cloudy condition.

Thermal modelling is one important factor that can increase the life cycle of MI up to 20 years. Therefore, a reliable MI depends on the reliability of the components used. While connecting to the interface, the inverter must comply with the requirements of the utility. The voltage, frequency, harmonics, power factor, and DC injection are measured at the interface. The MI should not inject a DC current into the AC supply in excess of 0.5% of its rated AC output current under any operating conditions. Galvanic isolation (GI) is one way to keep DC from being coupled to the grid. The MIs must also be designed to operate well under islanding conditions.

12.5 Topologies of MI

Based on the number of processing stages, it can be categorized as single-stage MI and multistage MI.

12.5.1 Single-Stage MI

Figure 12.2 shows the basic block diagram of a single-stage MI, which performs mainly two important functions: Firstly, the MPP tracking and, secondly, the conversion from DC into AC. In single stage, the PV module is connected with the inverter, which is further connected to the grid. It is the responsibility of the inverter to control the essential parameters such as voltage amplification, MPPT, and grid current control. In this case, the peak power is two times the nominal power. An effort is made to raise the system efficiency by using minimum number of components. The experimentally investigated proposed single-stage topologies are compared in Table 12.1. The comparison is in terms of input voltage range, power rate, THD, efficiency, and power factor.

In [23], a new topology was proposed wherein a single-stage, power-decoupling, and DC–AC conversion could be done. The conventional flyback was modified by including a parallel switch on the input side. By the combination of predictive control and pulse width modulation (PWM) strategy, switching signals were acquired. A film capacitor was used, which reduced decoupling capacitance value. The developed model was of 100 W having maximum efficiency around 91% and harmonics under 1.7%.

In another work [24], a single-stage zeta MI was demonstrated. The key function of the design was its capacity to function in continuous conducting mode for different range of load. The 220-W MI design gave 93% as maximum efficiency and a THD value of 3.05%. In [25], a double-boost inductor-coupled inverter was realized as a single-stage MI. The developed prototype showed THD less than 3% with power factor of 0.99 and efficiency of 97.5%. In [26], a dual buck-boost converter-based MI was proposed. The THD was 4.7% and power factor was 0.98. The design exhibited a novel control and modulation technique with efficiency of 93%.

In [27], the authors have discussed issues related to controller design of interleaved flyback converter with fully controlled MI. The investigation was done experimentally where peak efficiency of 93.9 with THD within 5% was achieved. In another work [28], a mathematical model was developed for a flyback converter combined with inverter connected

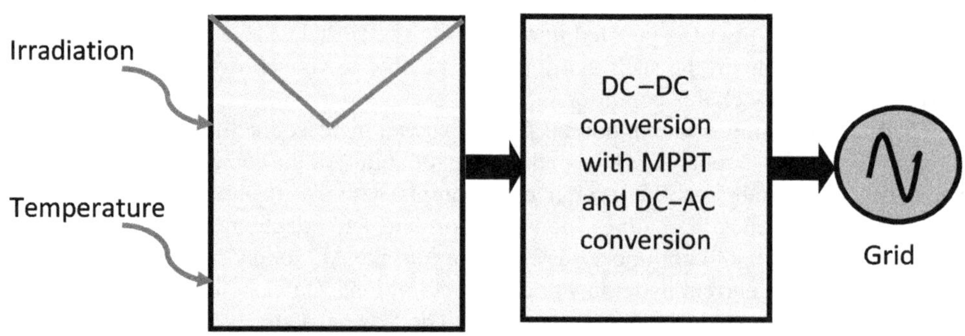

Figure 12.2. Single-stage microinverter.

Table 12.1. Summary of comparison of single-stage MI.

Reference	Power Rating (W)	Input Voltage (V)	THD (%)	Power Factor	Efficiency (%)
[23]	100	60	<1.7	–	90.6
[24]	220	30–48	3.05	–	93
[25]	260W	34.8 (V_{mpp})	<3	>0.99	>97.5
[26]	150	100	4.7	>0.96	93
[27]	200	36–60	<5	High	90.5–93.9
[28]	200	36	2.45	–	>93
[29]	250	22.7–33.3	–	–	>95
[30]	120	48.2	<5	~1	92.4
[31]	100	60	<1.7	–	91–92
[32]	100	60	<1.9	–	>90
[33]	230	29.67	<5	–	93
[34]	250	30–48	–	–	>92
[10]	205	31–40	3.77	–	>92
[35]	100	25	–	–	–
[36]	120	25–50	<5.2	–	83–94
[37]	100	35	<5	–	70
[38]	300	100	3.8		–
[39]	300	85 V	<4	–	–
[25]	–	–	<3	0.99	97.5

to grid. In this work, a control scheme was proposed to reduce the THD. To validate the model, a prototype of the rating 200 W was developed where maximum efficiency around 94% and harmonics of the output current of 2.46% were obtained. In [29], a model was proposed for synchronous rectifier-based interleaved flyback inverter. In one of the researches, a 120-W prototype of single-stage forward MI was fabricated in which an efficiency of 92.4% and a THD of the output current of less than 5% were achieved. This model also had the capability of MPPT [30]. In [31], a novel work was done where it was demonstrated that a three-port flyback PV MI topology provided the capability to decouple power. The 100-W prototype showed an efficiency of around 94% with current THD below 1.7%. In another work, which was an extension of the previous effort, an extra switch was employed in MI-based flyback in converter. The switch facilitated easy separation of the decoupling capacitor from the converter [32]. In another effort [33], a flyback converter with single-stage MI having MPP control was developed. The 250-W prototype gave an efficiency of 93% with THD of output current less than 5%. The highlight of this prototype was great reduction in components. In a novel research work [34], a PWM cycloconverter-based MI was developed for PV applications. In [10], a model was proposed for AC PV applications where a single-stage flyback inverter was proposed and high efficiency was claimed. In another practical design work, two topologies were introduced for single-stage flyback PV MI. These topologies had the capability to achieve power decoupling without the use of electrolytic capacitors. The scheme also mitigates the problem of leakage current interruption [35].

In a research work done on optimal control of PV grid-connected flyback MI, one-phase Discontinuous Conduction Mode (DCM) and two-phase DCM were combined. A model of 200 W was fabricated and efficiency between 90.5% and 93.9% was achieved [36]. In [37], Toshihisa Shimizu came up with a design for a 100-W utility interactive inverter circuit topology for AC module that provided solution to problems such as short lifetime and large size of the system. The topology is of flyback type and gives stable AC current injection into the grid.

In 2007, Chen et al. presented an integrated single-stage grid-connected inverter that works on the principle of Z-source [38]. Jain proposed low-cost solution that not only increases the usually low PV array voltage, but also provides high-quality AC power with the use of MPPT. Thus, in this design, the THD is less than 4% as per the IEEE-519 standard [39]. The coupled-inductor double-boost inverter (CIDBI) topology proposed by Yu Fang generates an AC output voltage more than the DC input, which is dependent on factors such as instantaneous duty cycle and turns ratio of the coupled inductor. The system shows THD less than 3% with power factor of 0.99% and 97.5% efficiency [25].

12.5.2 Multistage MI

Unlike single-stage, multistage MI topology performs the task of DC–DC conversion followed by DC–AC conversion in different stages as depicted in Fig. 12.3. The MPPT operation is done in DC–DC stage and the second stage where conversion of DC into AC is used to control current and hence power to the grid. Additionally, this DC–AC stage also supports reactive power and grid connection compatibility. In contrast to single stage, the multistage control structure is not complex as the tasks are performed in different stages.

In [40], a novel three-stage PV MI is proposed in which a double parallel boost converter is applied in the first stage, which performs MPPT and raises the level of input voltage. In the second stage, rectification of current is done by the flyback converter. Full bridge converter is the third stage where rectified current at low switching frequencies is unfolded. The efficiency obtained is 93% and THD is 4.38%.

Kusakawa proposed a transformer-less topology having wide voltage-matching ability with efficiency around 85% over the whole power range. However, the system shows 91% efficiency measured at rated capacity [41]. In 1995, Anderson proposed a 200-W module-integrated converter with low cost. It is a two-stage MI with a high-frequency soft switching PWM Power Conversion interface, analog and logic control, and protection circuits and an analog MPPT [42].

Ho et al. [43] has designed a dual-stage MI using a buck/boost converter and single-phase inverter. It is a 30-W MI having overall efficiency of 90.2%. Rodriguez et al. presented a three-stage current source inverter (CSI) design whose efficiency is 89% by fulfilling the goal of long life of the inverter. The design delivers a maximum power of 150 W [44]. Gonzalez proposed a multilevel and transformer-less topology as shown in Fig. 12.4. The system ensures that DC is not injected into the grid. The results are obtained for a 5-kW prototype and the maximum efficiency obtained is 98.16% [45]. In 2009, Choi et al. presented DC–DC converter for MI applications. The system provides better conversion ratios and consists of two transformers. The author proposed a dual-stage scheme and tested for a 260-W MI and attained 97% efficiency. The system applied the zero-current switching (ZCS) scheme [46].

In 2011, Krishnaswami presented a high-frequency link using ZVS and high-frequency transformer. He proposed a double-stage power conversion scheme as shown in Fig. 12.5, in which the high-frequency link is created by the first stage. The second stage inverts the high-frequency component into grid compatible form. The losses are reduced with the use of soft switching [47].

Chen et al. [48] proposed a 250-W MOSFET transformer-less inverter with MOSFET-based phase leg to solve the problem of conduction losses and low magnetic utilization. Figure 12.6 shows the proposed scheme, which shows around 98% magnetic utilization with no dead time requirement and 99.01% peak efficiency at full load condition.

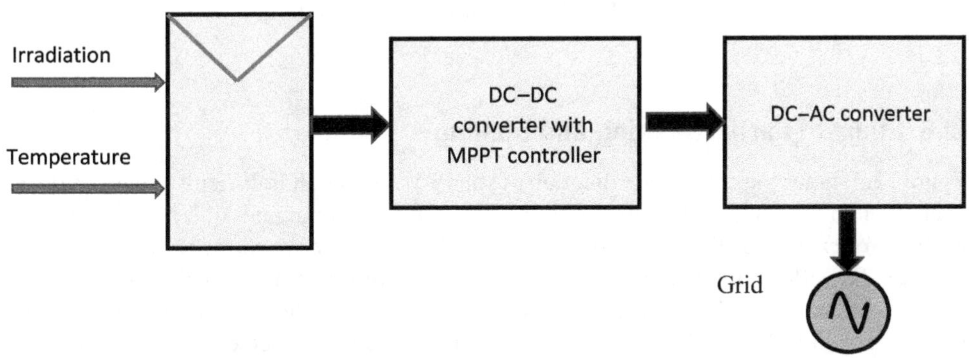

Figure 12.3. Multistage MI.

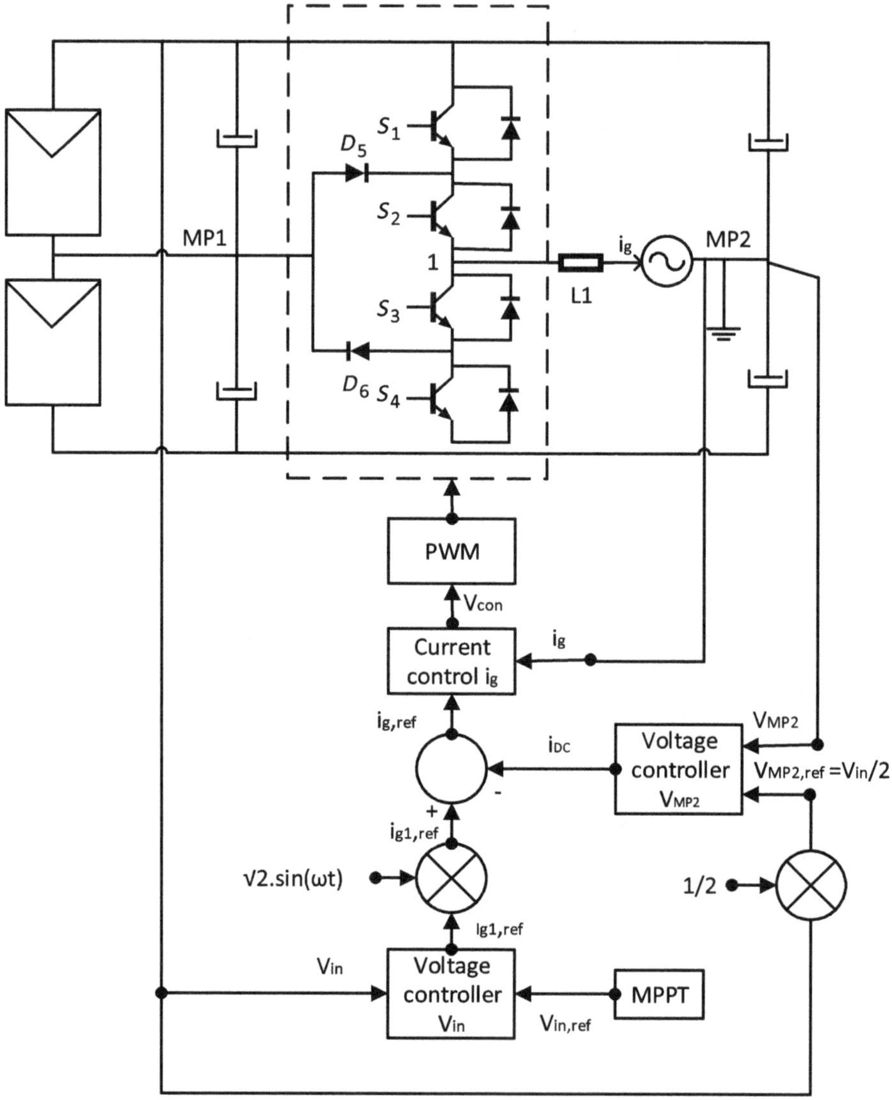

Figure 12.4. Multilevel design of MI [45].

12.6 Effect of Partial Shading and Dusting

Figure 12.7 shows the connection diagram of solar PV module. It is difficult to have uniform irradiation on the PV panels throughout the day. There are many possible reasons to have shading, for example, cloud formation, building shadow or tree shadow, or daily sun angle changes. The effect of shading is the power loss, current mismatch within a PV string, and voltage mismatch between parallel strings. The bypass diode is used to reduce the effect of shading. When there is no shading, this diode gets reverse biased and restricts the flow of current due to high impedance. These diodes are connected in parallel with the

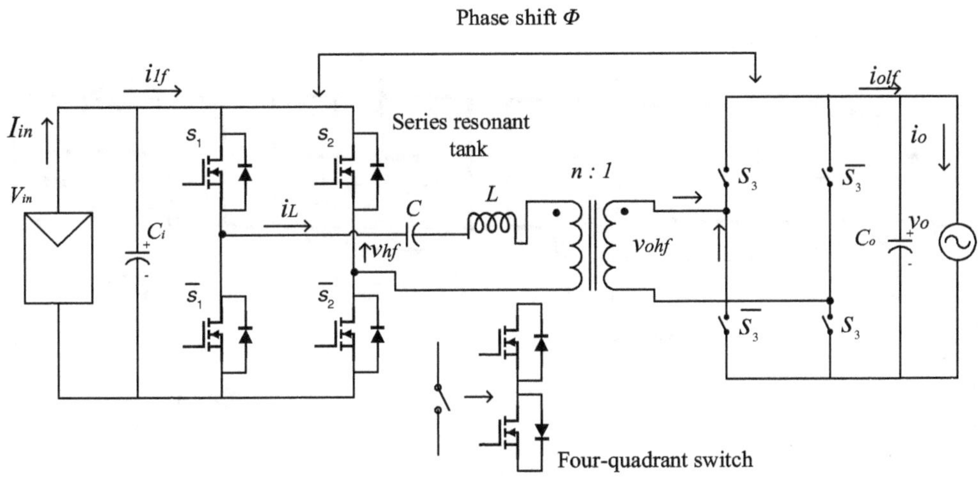

Figure 12.5. Krishnaswami's topology [47].

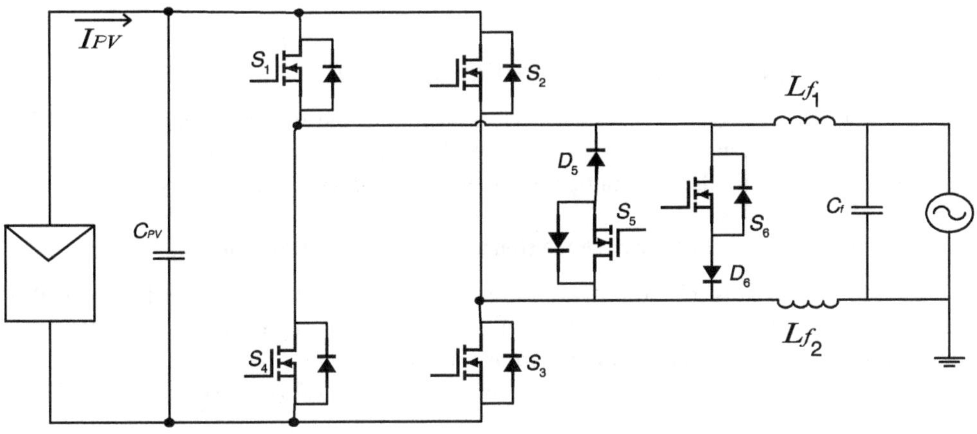

Figure 12.6. Baifeng Chen's topology [48].

panel to shunt the current around it. The voltage of the neighboring cell forward biases the diode, which lowers its impedance as compared to the shaded cell. Hence, the current passes through the diode instead of the cell. This prevents the formation of hot spots. These are formed when the power generated by series-parallel connected modules is not same for the same specifications of the panel. The low-power module will behave as a sink absorbing the power from the high-power generating module. This raises the cell temperature and sometimes develops crack in the glass shield. This problem is called hot spot formation. This can be resolved by the use of bypass diodes.

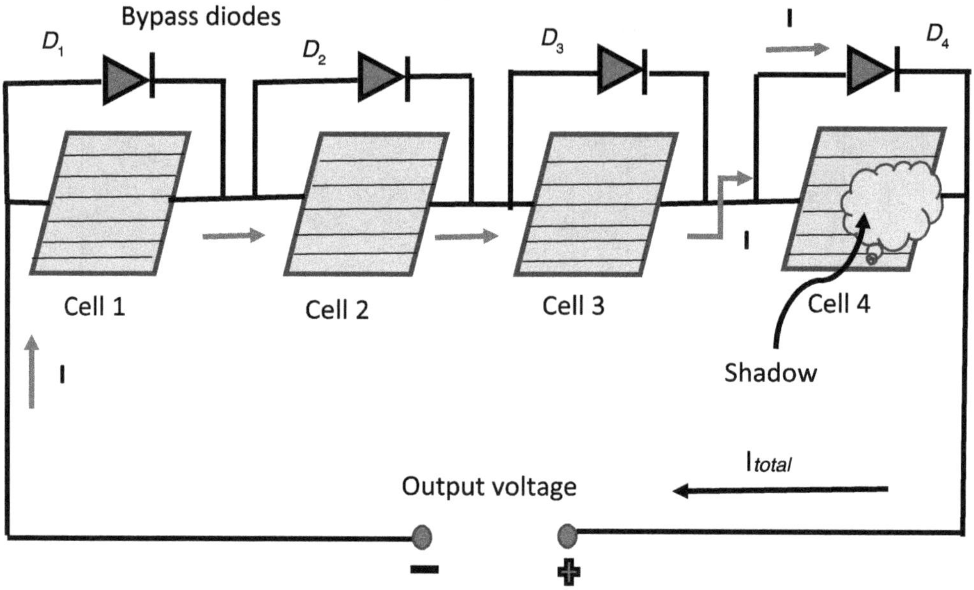

Figure 12.7. Solar PV Module showing partial shading.

12.6.1 Strategies to Deal with Partial Shading

The shading has a major effect on the performance and the return on investment (ROI) from a solar power system. This effect is more if the shading due to the object is regular. Following are the methods that can be implemented to reduce the effect of shading:

(A) Site your solar panel array where there will be no regular shading

The first and foremost is to find out the sources of shading at the time of installation. The solar system installer should be very vigilant while deciding the place of panels. Because sometimes as per the current situation there is no object to cause regular shading. But it is possible to have a tree nearby, which will grow in the near future and will cause shading.

(B) Use a string inverter that has MPP tracking capability

Inverters with MPP tracking facility extract maximum usable energy from solar panels even in the condition of shading. This is achieved by adjusting the voltage to suit the inverter's preferred input range. In summary, an MPP tracker minimizes the output losses due to partial shading and other panel output conditions. Inverters without MPPT capability have low output from the weaker string once it passes below the required output threshold.

(C) Use of MIs

The use of MIs provides an optimal solution to the problem of shading. This happens because all the modules are not affected in case we use MI. Each module works independently and reduces the problem of multiple maxima that are developed. Hence, the system operates at MPP and thereby gives the maximum output.

12.6.2 Experimental Investigation for 1 kW PV System

An experimental investigation has been carried out to study the comparison of CI and MI connected with 1 kW PV system. The data for the study were collected from a 1-kW roof-top grid-tied solar PV panel installation at a house in Bhilai, Chhattisgarh, India. The setup consists of a 1-kW system where four panels of 250 W each, four MIs, and one CI form the major part of the test system. The PV system was connected to either of the CI and MI through Double Pole Single Throw (DPST) switches. The use of DPST switch gives us more constructive results as the panel location and input conditions remain the same. The MI MI-250 model of make Hoymiles has been used for the current study. The specifications of the MIs used for study are summarized in Table 12.2.

Apart from the system's equipment, some measuring devices such as lux meter, digital voltmeter, and ammeter have been used to perform the study. The results are shown in Figs. 12.8 and 12.9, which display how the power output varies during a typical sunny day with CI and MI connections under normal and shading conditions [49].

The presented graphs show observations taken only at discrete intervals of every half an hour of the day. In normal condition, the observations have been taken for both MI and CI under same weather conditions. In normal condition, it has been assumed that there is no continuous shading due to any tree or any controllable obstacle. However, the shading due to clouds is possible, which is not in our control. The second condition is the shade condition where a partial shading condition was created manually to consider the effect of shading, which is usually through tree and neighboring wall, and so on. The study of effect of shading at different places and times has been done.

The presented graph in Fig. 12.9 only shows a typical condition of partial shading (of the order of less than 20% module being shaded by a tree shadow) to demonstrate the effect of shading on the power output of a PV system in case both types of inverters are considered. Thus, we observe that the power output of the CI significantly reduces under shading condition compared to a normal condition. The higher the percentage shading, the reduction in power would be even more prominent. In contrast, the MI can overcome this effect to a great degree. Thus, it is found that MIs have performed better than CIs in case of shading condition.

Table 12.2. Specifications of Hoymiles micro PV inverter utility interactive MI-250.

Input Ratings	Output Ratings
Maximum input voltage = 60 V	Nominal output voltage = 230 V AC
Maximum input current = 10 A DC	Nominal output frequency = 50 Hz
Maximum input short-circuit current = 15 A	Output power factor = 0.99
	Maximum output current = 1.08 A AC
	Maximum continuous output power = 250 W
	Maximum output overcurrent protection = 20 A

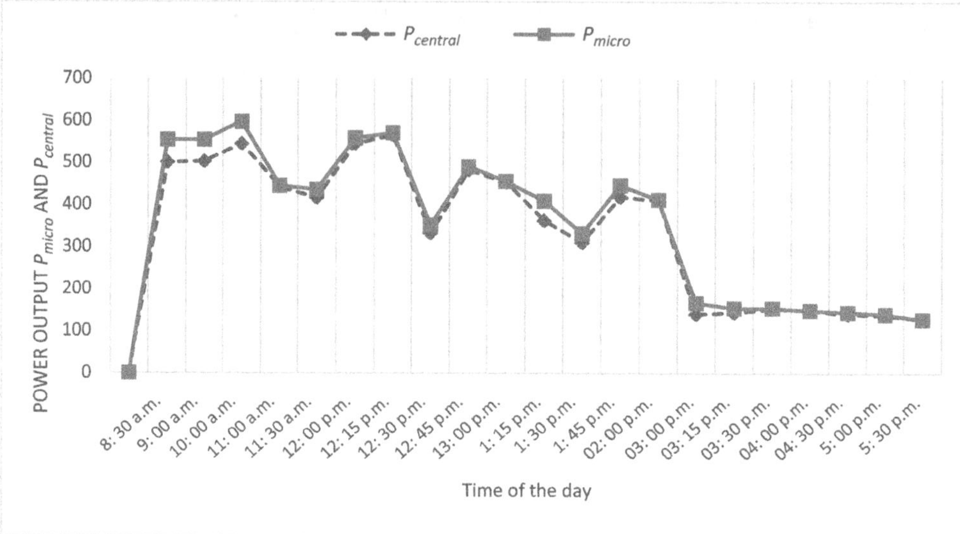

Figure 12.8. Comparison of power developed with CI and MI under normal condition.

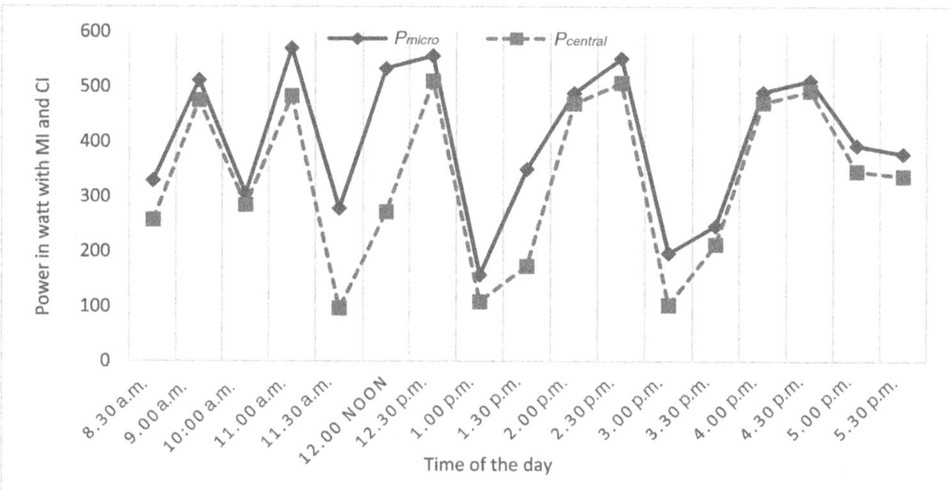

Figure 12.9. Power output comparison of standard PV system connected with central and MI under shading condition.

12.6.3 Analysis of Experimental PV System Under Dusting Condition

The deposition of dust on the PV module cells lowers the efficiency of the solar cell day by day especially in the locality where dust concentration in atmosphere is high. The deposited dust reduces the intensity of sun rays falling over PV module, which in turn reduces the output.

(A) Dust settlement and energy generation

This study was conducted in two parts. In the first part, we examined the dust collection cycle and then we conducted comparative analysis of the system with CI and MI. Bhilai is an industrial town and host to one of the most important steel plants of India. Thus, it is very prone to dust with dust accumulation a very common occurrence in the city. The data were collected for a period of 2 weeks with readings taken every hour during the day from sunrise (around 7 a.m.) to sunset (around 6 p.m.). The first cycle was during Feb. 15–20, 2018 and the second cycle was during Feb. 25–Mar. 1, 2018. Figure 12.10 presents the daily energy generation for the first cycle (with 12 readings for each day). We can note from Fig. 12.10 that energy generation reduces day by day as dust accumulates in the first cycle. Figure 12.11 presents the power generated by the panel during the day for the current cycle. Under the specified conditions the total energy generation of the PV module goes down. The total energy reduction per day was about 3.18% for this particular system. Such reduction is likely to be higher for large installation. The total reduction in energy generation during a week cycle was 15.47%.

(B) Comparison of power outputs

Figure 12.12 shows the comparison of output powers in case of dusting condition with CI and MI for a typical dusty module. The comparison gives a rough estimate of the reduction in power output of a solar module in case of both the inverter configurations. Thus, we can understand that the MI is capable of generating a higher power even in case of dust accumulation on the solar PV modules. In conclusion, the difference in power generation capacity becomes more prominent for MI and CI, under dust accumulation and under partial shading conditions. And hence, the MI technology could be a more promising alternative under such conditions.

12.7 Economics of PV

There are mainly two types of costs, capital cost and recurring cost, which are to be borne during the installation of PV systems. The capital cost includes the cost of all the major components of solar PV system, that is, solar PV module/array, MPPT, charge controller, battery, and inverter. The price of solar modules is common for all types of configurations. Here, we are interested in studying the difference in costing of different inverter configurations. It is found that the cost of a single centralized inverter or multiple string inverters may be slightly less than MI modules put together. The installation or skilled labor cost in case of MI gets reduced as there is no need of installing high-voltage DC switch boxes and protection circuits, which requires a skilled and trained person. The maintenance and repair cost factor comes under the recurring cost. This cost is higher in case of solar PV systems

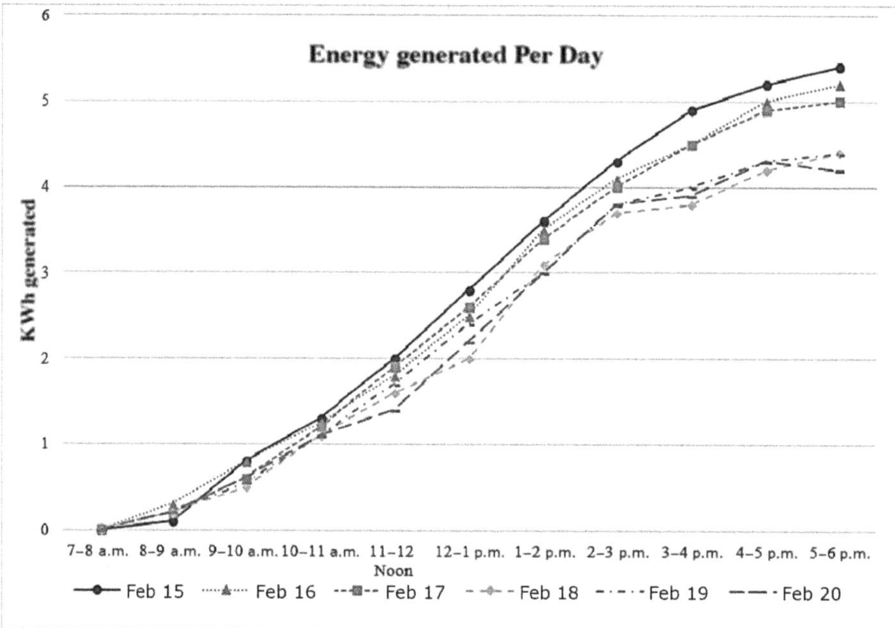

Figure 12.10. The comparison of per day energy generation.

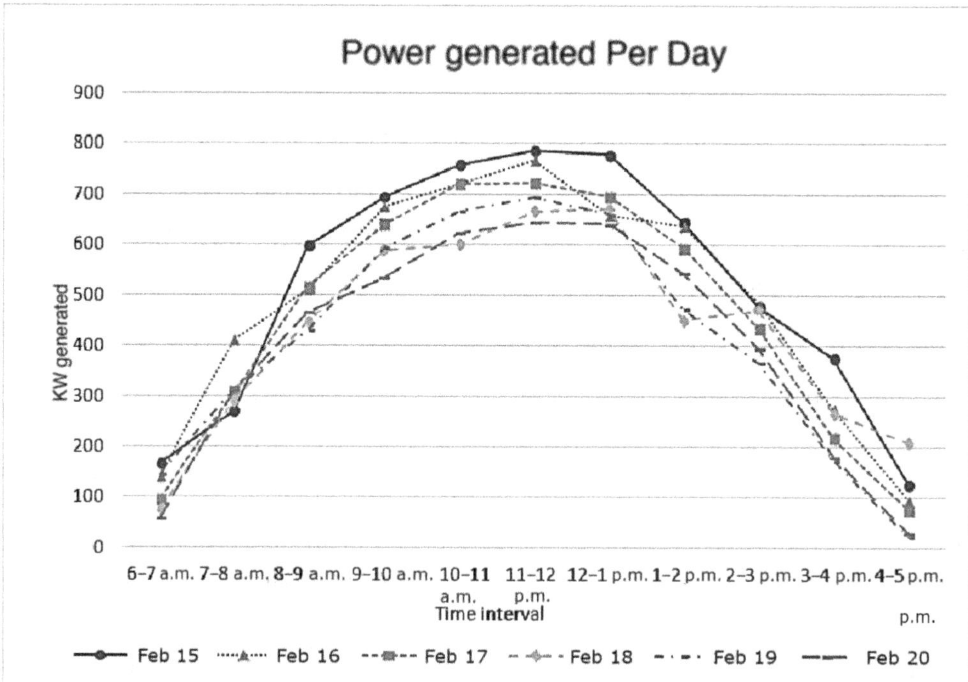

Figure 12.11. Power generated during the day from the panel for the study.

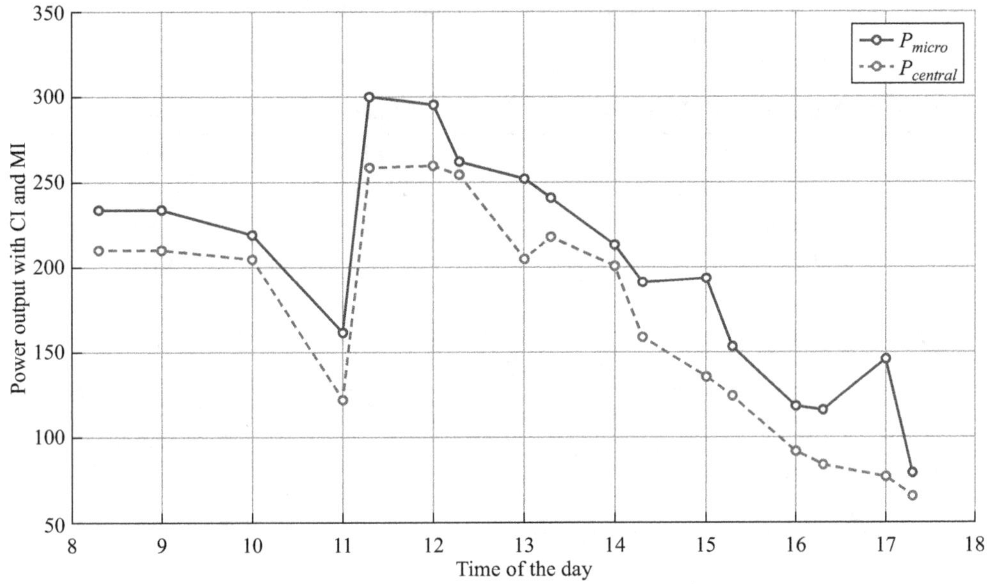

Figure 12.12. Comparison of power developed with MI and CI under dusting condition.

with central or string inverters than the MIs. This is because fault detection and repair both are easy in case of AC modules, which is very difficult in other case as monitoring can only be done at a system level. In contrast, AC units can be monitored individually even online through a Web-enabled interface thereby facilitating the exact location of faults. The PV modules have a life span of around 25 years. However, the string inverters have a limited warranty period of perhaps 5–10 years. It is expected that MIs in the near future will probably achieve a lifetime of 15–20 years. It is important for the consumer to understand the statistics while deciding the inverter configuration. In simple words, the lifetime is inversely proportional to cost of replacement of component. A field experience for a 7-kWp solar PV system is presented in [50], according to which the cost of MI is less than the CI. The price comparison is summarized in Table 12.3. Here the module cost, installation cost, and maintenance cost are all included. In another study [50], which was done for a comparatively higher rating (25 kWp), it is again seen that the pricing of CIs is more than the MI employed for the same system. The comparison is summarized in Table 12.4. A 25-kW system would require about 10 CIs versus over 200 MIs. However, this comparison holds good only at lower volumes. With larger rating PV systems (of several MWp), advantages of AC module inverters begin to reduce and centralized inverters prove to be a better option [50]. System optimization also plays a major role in deciding the economy of the PV systems [51].

12.8 Conclusion

This chapter discusses the three major inverter configurations namely CI, string inverter, and MI. Various standards for the MI, grid, and PV module requirements have been discussed.

Table 12.3. Price comparison of string inverter with Enecsys MIs for 7 kW system [50].

Sl. No.	Name of the Head	MI ($)	String Inverter ($)
1	Electronic module cost	3500	5250
2	Installation cost	12,500	16,500
3	Maintenance cost	739	5192
	Total	16,739	26,942

Table 12.4. Price of central and MI for 25 kWp BIPV Solarex MST50 MV thin-film 0 MV T50 MV Sola [50].

Sl. No.	Name of the Head	Roof Integrated (Laminated Module) $/Wp DC		Roof Stand-off (Framed Module) $/Wp DC	
		CI	MI	CI	MI
1	Modules	3	3	3.6	3.6
2	Balance of system (BOS) and installation	2.86	2.57	2.13	1.92
3	Inverter accessories	1.71	1.43	1.71	1.43
	Total	7.57	7.00	7.44	6.94

The discussion further continues with the effects of shading and dust on the performance of CI and MI connected to a low-power solar-based domestic or rural application. The study contributes to the technical and commercial viability of MI based on experimental results and it showed that MI is better suited under shading and dusting conditions. The following are some of the key learning outcomes of this chapter:

1. It will help the user to understand the need of inverter and the different configurations of the inverter.
2. It will help the user to know the reasons affecting the performance of PV module.
3. It will help the user to decide the MI topology depending on the size of the system.
4. It will be helpful for the user to know about the pros and cons of the different topologies of MIs.

12.9 References

1. S. B. Kjaer, J. K. Pedersen and F. Blaabjerg, "A review of single phase grid connected inverters for photovoltaic modules," *IEEE Transactions on Industry Applications*, vol. 41, no. 5, pp. 1292–1306, 2005.
2. H. Rasedul, S. Mekhilef, M. Seyedmahmoudian, and B. Horan, "Grid connected isolated PV micro-inverters: A review," *Renewable and Sustainable Energy Reviews*, vol. 67, pp. 1065–1080, 2017.
3. L. Schwartfeger, and D. Santos-Martin, "Review of distributed generation inter-connection standards," EEA Conference and Exhibition, Auckland, 2014.

4. J. Munoz-Cruzado-Alba, J. Villegas-Nunez, J. A. Vite-Frias, J. M. Carrasco-Solis, and E. Galvan-Diez, "New low distortion qf droop plus correlation anti-islanding detection method for power converters in distributed generation systems," *IEEE Transaction Industrial Electronics*, vol. 62, no. 8, pp. 5072–5081, 2015.
5. D. Velasco dela Fuente, C. L. Trujillo Rodriguez, and E. A. Narvaez, "Review of anti-islanding methods: Analysis by figures of merit tools for controllers re-configuration in micro grids," *IEEE Latin America Transaction*, vol. 13, pp. 679–686, 2015.
6. S. Raza, H. Mokhlis, H. Arof, J. Laghari, and L. Wang, "Application of signal processing techniques for islanding detection of distributed generation in distribution network:are view," *Energy Conversion Management*, vol. 96, pp. 613–624, 2015.
7. E. J. Estébanez, V. M. Moreno, A. Pigazo, and M. Liserre, "Performance evaluation of active islanding-detection algorithms in distributed-generation photovoltaic systems: Two inverters case," *IEEE Transactions on Industrial Electronics*, vol. 58, pp. 1185–1193, 2011.
8. X. Guo, D. Xu, and B. Wu, "Overview of anti-islanding US patents for grid-connected inverters," *Renewable and Sustainable Energy Reviews*, vol. 40, pp. 311–317, 2014.
9. L. Chen, C. Hu, Q. Zhang, K. Zhang, and I. Batarseh, "Modeling and triple-loop control of ZVS Grid-Connected DC/AC converters for three-phase balanced micro-inverter application," *IEEE Transactions on Power Electronics*, vol. 30, pp. 2010–2023, 2015.
10. A. C. Nanakos, G. C. Christidis, and E. C. Tatakis, "Weighted efficiency optimization of flyback micro-inverter under improved boundary conduction mode (i-BCM)," *IEEE Transactions on Industrial Electronics*, vol. 30, pp. 5548–5564, 2015.
11. M. I. Hamid, and A. Jusoh, "Reduction of wave form distortion in grid-injection current from single-phase utility interactive PV-inverter," *Energy Conversion and Management*, vol. 85, pp. 212–226, 2014.
12. A. M. Amjad, and Z. Salam, "A review of soft computing methods for harmonics elimination PWM for inverters in renewable energy conversion systems," *Renewable and Sustainable Energy Reviews*, vol. 33, pp. 141–153, 2014.
13. Y. Hoseynpoor, and T. P. Ashraf, "Maximum power point tracking and reactive power control of single stage grid connected photovoltaic system," *Research Journal of Applied Sciences, Engineering and Technology*, vol. 3, pp. 1430–1436, 2011.
14. A. Maknouninejad, N. Kutkut, I. Batarseh, and Z. Qu, "Analysis and control of PV inverters operating in VAR mode at night," IEEE PES Innovative Smart Grid Technologies (ISGT), pp. 1–5, Anaheim, 2011.
15. R. Kabiri, D. Holmes, and B. P. Mc Grath, "The influence of PV inverter reactive power injection on grid voltage regulation," IEEE 5th International Symposium on Power Electronics for Distributed Generation Systems (PEDG), 2014, Galway, pp. 1–8.
16. N. Sukesh, M. Pahlevaninezhad, and P. K. Jain, "Analysis and implementation of a single-stage flyback PV micro-inverter with soft switching," *IEEE Transactions on Industrial Electronics*, vol. 61, pp. 1819–1833, 2014.
17. W. J. Cha, Y. W. Cho, J. M. Kwon, and B. H. Kwon, "Highly efficient micro-inverter with soft-switching step-up converter and single-switch-modulation inverter," *IEEE Transactions on Industrial Electronics*, vol. 62, pp. 3516–3523, 2015.
18. B. Chen, B. Gu, L. Zhang, et al., "A high-efficiency MOSFET transformerless inverter for non isolated micro-inverter applications," *IEEE Transactions on Power Electronics*, vol. 30, pp. 3610–3622, 2015.
19. S. Kouro, J. Leon, D. Vinnikov, and L. G. Franquelo, Grid-connected photovoltaic systems: an overview of recent research and emerging PV converter technology," *IEEE Industrial Electronics Magazine*, vol. 9, pp. 47–61, 2015.
20. S. Krauter, and J. Bendfeld, "Cost, performance, and yield comparison of eight different micro-inverters," Proceedings of the IEEE 42nd Photovoltaic Specialist Conference (PVSC), pp. 1–4, New Orleans, 2015.

21. H. B. Hu, S. Harb, N. Kutkut, I. Batarseh, and Z. J. Shen, "A review of power decoupling techniques for micro-inverters with three different decoupling capacitor locations in PV systems," *IEEE Transactions on Power Electronics*, vol. 28, pp. 2711–2726, 2013.
22. Y. Ohnuma, K. Orikawa, and J.-i. Itoh, "A single-phase current-source PV inverter with power decoupling capability using an active buffer," *IEEE Transactions on Industry Applications*, vol. 51, pp. 531–538, 2015.
23. H. Hu, S. Harb, X. Fang, D. Zhang, Q. Zhang, Z. J. Shen, and I. Batarseh, "A three-port fl thre for PV micro-inverter applications with power pulsation decoupling capability," *IEEE Transactions on Power Electronics*, vol. 27, no. 9, pp. 3953–3964, 2012.
24. R. K. Surapaneni, and A. K. Rathore, "A single-stage CCM zeta Micro-inverter for solar photovoltaic AC module," *IEEE Journal of Emerging and Selected Topics in Power Electronics*, vol. 3, no. 4, pp. 892–900, 2015.
25. Y. Fang, and X. Ma, "A novel PV micro-inverter with coupled inductors and double-boost topology," *IEEE Transactions on Power Electronics*, vol. 25, no. 12, pp. 3139–3147, 2010.
26. D. Petreuş, S. Daraban, I. Ciocan, T. Patarau, C. Morel, and M. Machmoum, "Low cost single stage micro-inverter with MPPT for grid connected applications," *Solar Energy*, vol. 92, pp. 241–255, 2013.
27. Z. Zhang, X. F. He, and Y. F. Liu, "An optimal control method for photovoltaic grid-tied interleaved flyback micro-inverters to achieve high efficiency in wide load range," *IEEE Transactions on Power Electronics*, vol. 28, no. 11, pp. 5074–5087, 2013.
28. M. Gao, M. Chen, C. Zhang, and Z. Qian, "Analysis and implementation of an improved flyback inverter for photovoltaic AC module applications," *IEEE Transactions on Power Electronics*, vol. 29, no. 7, pp. 3428–3444, 2014.
29. Y. H. Kim, J. W. Jang, S. C. Shin, and C. Y. Won, "Weighted-efficiency enhancement control for a photovoltaic AC module interleaved flyback inverter using a synchronous rectifier," *IEEE Transactions on Power Electronics*, vol. 29, no. 12, pp. 6481–6493, 2014.
30. D. Meneses, O. Garcia, P. Alou, J. A. Oliver, and J. A. Cobos, "Grid-connected forward micro-inverter with primary-parallel secondary-series transformer," *IEEE Transactions on Power Electronics*, vol. 30, no. 9, pp. 4819–4830, 2015.
31. H. Hu, Q. Zhang, F. Fang, Z. J. Shen, and I. Batarseh, "A single stage micro-inverter based on a three-port flyback with power decoupling capability," Proc. IEEE Energy Conversion Congress and Exposition: Energy Conversion Innovation for a Clean Energy Future, ECCE 2011, pp. 1411–1416, Phoenix, 2011.
32. H. Hu, S. Harb, N. H. Kutkut, Z. J. Shen, and I. Batarseh, "A single-stage micro-inverter without using eletrolytic capacitors," *IEEE Transactions on Power Electronics*, vol. 28, no. 6, pp. 2677–2687, 2013.
33. C. M. Lai, "A single-stage grid-connected PV micro-inverter based on interleaved flyback converter topology," Proc. International Symposium Computer Consumer, and Control, IS3C 2014, p. 187–190, Taichung, 2014.
34. R. K. Surapaneni, "A novel single-stage isolated PWM half-bridge micro-inverter for solar photovoltaic modules," Energy Conversion Congress and Exposition (ECCE), pp. 4550–4556, Montreal, 2015.
35. S. Chakraborty, and S. Chattopadhyay, "Topology variations and design improvements of a single-stage flopolo PV micro-inverter," IEEE Applied Power Electronics Conference and Exposition (APEC), pp. 3026–3033, Fort Worth, TX, USA, 2014.
36. Z. Zhang, M. Chen, M. Gao, Q. Mo, and Z. Qian, "An optimal control method for grid connected photovoltaic micro-inverter to improve the efficiency at light-load condition," IEEE Energy Conversion Congress and Exposition: Energy Conversion Innovation for a Clean Energy Future, ECCE, Proceedings, pp. 219–224, Phoenix, 2011.
37. T. Shimizu, K. Wada, and N. Nakamura, "Flyback-type single-phase utility interactive inverter with power pulsation decoupling on the dc input for an AC photovoltaic module system," *IEEE Transactions on Power Electronics*, vol. 21, pp. 1264–1272, 2006.

38. Z. Chen, X. Zhang, and J. Pan, "An integrated inverter for a single-phase single-stage grid-connected PV system based on Z-source," *Bulletin of the Polish Academy of Sciences: Technical*, vol. 55, pp. 263–272, 2007.
39. S. Jain, and V. Agarwal, "A single-stage grid connected inverter topology for solar PV systems with maximum power point tracking," *IEEE Transactions on Power Electronics*, vol. 22, pp. 1928–1940, 2007.
40. G. Zhu, L. Mu, and J. Yan, "A novel multi-function PV micro-inverter with an optimized harmonic compensation strategy," *Journal of Power Electronics*, vol. 16, no. 6, pp. 2284–2293, 2016.
41. M. Kusakawa, H. Nagayoshi, K. Kamisako, and K. Kurokawa, "A new type of module integrated converter with wide voltage matching ability," 2nd World Conference and Exhibition Photovoltaic Solar Energy Conversion, Vienna, 1998.
42. M. Anderson, and B. Alvsten, "200W low cost module integrated utility interface for modular photovoltaic energy systems," Proceedings of IECON '95—21st Annual Conference on IEEE Industrial Electronics, pp. 572–577, 1995.
43. B. M. T. Ho, and H. S. H. Chung, "An integrated inverter with maximum power tracking for grid-connected PV systems," *IEEE Transactions on Power Electronics*, vol. 20, pp. 953–962, 2005.
44. C. Rodriguez, and G. A. J. Amaratunga. "Long-lifetime power inverter for photovoltaic AC modules," *IEEE Transactions on Industrial Electronics*, vol. 55, pp. 2593–2601, 2008.
45. R. González, E. Gubía, J. López, and L. Marroyo, "Transformerless single-phase multilevel-based photovoltaic inverter," *IEEE Transactions on Industrial Electronics*, vol. 55, pp. 2694–2702, 2008.
46. W. Y. Choi, S. M. Kim, S. J. Park, K. H. Kim, and Y. C. Lim, "High step-up DC–DC converter with high efficiency for photovoltaic module integrated converter systems," INTELEC 2009—31st International Telecommunications Energy Conference, pp. 1–4, Incheon, 2009.
47. H. Krishnaswami, "Photovoltaic micro-inverter using single-stage isolated high-frequency link series resonant topology," IEEE Energy Conversion Congress and Exposition, pp. 495–500, Phoenix, 2011.
48. B. Chen, B. Gu, L. Zhang, Z. U. Zahid, J. Sheng (Jason) Lai, Z. Liao, and R. Hao, "A high-efficiency MOSFET transformerless inverter for nonisolated micro-inverter applications," *IEEE Transactions on Power Electronics*, vol. 30, no. 7, pp. 3610–3622, 2015.
49. Ruchira, A. Sao, R. N. Patel, and S. K. Sinha, "GUI supported experimental investigation of central inverter and microinverter integrated with 1kW PV system," 3rd International Conference and Workshops on Recent Advances and Innovations in Engineering (ICRAIE), pp. 1–6, Jaipur, India, 2018.
50. H. P. Ikkurtin, and S. Saha, "A comprehensive techno-economic review of micro-inverters for Building Integrated Photovoltaics (BIPV)," *Renewable and Sustainable Energy Reviews*, vol. 47, pp. 997–1006, 2015.
51. R. Dubey, D. Joshi, and R. C. Bansal, "Optimization of solar photovoltaic plant and economic analysis," *Electric Power Components and Systems*, vol. 44, no. 18, pp. 2525–2535, 2016.

CHAPTER 13

Protection of Feeders with High Rooftop PV Penetration

Hadi Hosseinian Yengejeh[1], Farhad Shahnia[2], and Syed M. Islam[3]

[1]*School of Electrical Engineering, Computing and Mathematical Sciences, Curtin University, Perth, Australia*
[2]*Discipline of Engineering and Energy, Murdoch University, Perth, Australia. F.Shahnia@Murdoch.edu.au*
[3]*School of Science, Engineering and Information Technology, Federation University, Ballarat, Australia*

Abstract This chapter focuses on short-circuit fault on the feeders supplying residential customers with rooftop photovoltaic (PV) systems. The chapter discusses the sensitivity of the short-circuit fault level and the corresponding current drawn from the distribution transformer, in addition to the node voltages along the low-voltage feeder, following a short-circuit fault. The chapter also demonstrates that various factors such as the PVs' installation point and rating, as well as the location and type of the fault, impact the study parameters. Furthermore, the chapter illustrates that the time and sequence of the PVs' isolation following the fault will vary depending on the impedance and location of fault, the PV generation–demand ratio, as well as the network's earthing type.

Keywords: Rooftop photovoltaic systems, Earthing, Short-circuit fault, PV disconnection, Sensitivity analysis, Stochastic analysis.

13.1 Introduction

The past decade has observed a substantial effort toward the proliferation of distributed generation units within power distribution networks [1, 2]. One of the most common types of distributed resources of energy embedded within the low-voltage feeders in many countries is the residential rooftop single-phase photovoltaic units, which are denoted by "PVs" in the subsequent sections.

Australia has installed more than one million PVs over the past 6 years [3] due to the state and federal subsidies [4, 5]. In [6], the electricity tariffs and bills across various zones of Australian power distribution companies have been compared, based on which in [7], an efficient seasonal time of use feed-in tariff for residential PVs has been proposed, whereas the economic benefits and the payback periods of PVs are evaluated in [8]. Furthermore, in [9, 10], a dynamic feed-in tariff to further encourage the households to install PVs while benefiting the distribution network operators have been proposed. Likewise, the concept of solar leasing has been discussed in [11–13] suggesting new ways for private investors to get involved in the PV business at residential levels.

Certain frameworks are available for the integration of large centralized PV farms into distribution networks to guarantee their system-level performances, matching the existing and expected standards [14, 15]. However, such frameworks do not exist for household-owned PVs. One key issue of several distributed household-owned PVs for the utilities is the total amount of power generated by these systems. One possible solution is having smart meters installed in each house that can measure and transmit the generated power by each PV to the utility. Another approach is using a single pyranometer to measure the solar irradiance and then using some estimation techniques to estimate the overall generated power in the neighborhood [16, 17].

The planning and operational control and management of distribution networks with PV penetrations are discussed in [18, 19]. The impact of large PV penetration in low-voltage feeders has been evaluated from voltage and power quality perspectives in [20–24] and it is concluded that a large PV penetration may cause bidirectional flow of power in the low-voltage feeder [25–27], voltage unbalance [28, 29], and variation in the network's fault current level and direction. Additionally, the unexpected change of voltage within the feeders that are dominated by PVs following the cloud movements in the sky has been studied in [30] followed up by some enhancing methods that are proposed in [31] to overcome the fast voltage variation issues. Likewise, employing a different control technique for the inverters of the PVs to inject reactive power and regulate the voltage at their connection point [32, 33], adding battery energy storages and siting them such that the unbalances are minimized [34], dynamically switching the connection of one house from one phase to another [35–37], coordinated control of the inverters of adjacent PVs in the network [38–41] or under a droop control [42] or in collaboration with demand-side management [43–47], and employing various power electronics-based converters with different controls [48–55] are some of the many different approaches to minimize the technical issues of PVs in the networks.

Along with the problems of voltage rise and fluctuation, as well as power quality, many utilities around the world are currently concerned with the impacts of large PV penetration on the mis-coordination of protection systems in their networks [56–60]. For example, in [61], the protection system issues in distribution networks with a large PV penetration have been discussed. In [62], a technique to determine the appropriate settings of the protection systems for medium-voltage networks that accommodate large PV penetrations in their low-voltage feeders has been proposed. Moreover, in [63], a technique that can determine the impact of PVs on the network's short-circuit faults has been proposed, which is focused on current phase comparison at various nodes within the medium-voltage feeder.

All the above protection-related studies have focused on the impact of PVs on medium-voltage feeders, but have not discussed the effects of PVs on the low-voltage feeders to which they are connected. Furthermore, they do not focus on the nominal ratings of the PVs nor their distribution along the low-voltage feeder. Such factors need to be well-thought-out in protection system studies for low-voltage networks with a large PV penetration and are discussed in this chapter.

On the other hand, the utilities around the world try to minimize the possible negative impacts of PVs by limiting their penetration in the networks. As an example, majority of electrical utilities in Australia have developed a 20%–30% maximum penetration limit for the PVs in each low-voltage feeder [64, 65] because of voltage rise, maximum harmonic levels, reverse power flow, and protection mis-coordination. Such a limit stops other households to proceed with installing PVs. The utilities are concerned on the protection side because a large PV penetration may lead to a possibility in which the PVs do not permit the voltage in the feeder to reduce when short-circuit faults occur. This will result in PVs the constantly injecting power to the fault, even after the proper operation of the upstream circuit breakers. In [66], it has been stated that investigating the possibility of PVs continuously injecting power into the faulty feeder even after the opening of the upstream side of the network is a very important issue.

The magnitude of the voltage in a low-voltage feeder's point of coupling to a PV plays a crucial role in the PVs' isolation (disconnection) following short-circuit faults on that feeder. Likewise, the current flowing through a distribution transformer's secondary is another important parameter for the operation (disconnection) of the switch fuse in case of short-circuit faults on the low-voltage feeder.

This chapter initially presents a sensitivity analysis based on which the PVs' impact on the fault level and the current observed at the transformer secondary, in addition to the node voltages along the low-voltage feeder are assessed in the course of short-circuit faults. PVs' location and rating, as well as the location and type of short-circuit faults, are the focused factors for this analysis. Then, using a stochastic analysis, the chapter considers the effect of multiple PVs, which might have dissimilar ratings and distributed unevenly among the network's phases. Moreover, the time and sequence of the PVs' disconnection are also evaluated considering various factors such as the impedance and location of fault and PV generation–demand ratio. This chapter also discusses the impact of network earthing type on the disconnection time of PVs and compares the voltage profile along between two feeders that have non-effectively grounded (NEG) and multiple earthed neutral (MEN) systems [66, 67] following a short-circuit fault.

13.2 The Considered Network

Consider Figure 13.1(a) that portrays the schematic of a typical low-voltage distribution network delivering power to urban residential customers in Australia. Figure 13.1(b) shows the equivalent circuit diagram of Figure 13.1(a).

Figure 13.1(b) shows a three-phase, three-wire medium-voltage line feeding a three-phase, four-wire low-voltage line via a three-phase Dyn transformer [68]. The transformers are frequently 25–630 kVA, whereas the low-voltage lines are usually 400 m long that are distributed by the help of 10–12 poles, located 30–40 m from each other. The switch fuses located at the distribution transformer's secondary operate in case of a short-circuit fault

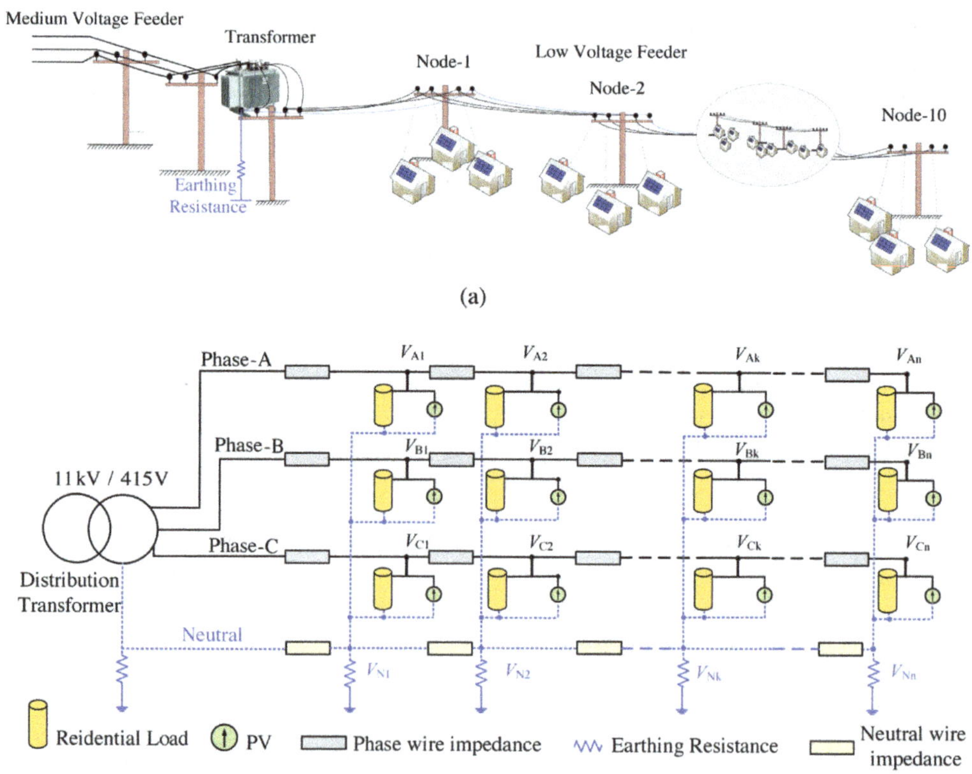

Figure 13.1. (a) Assumed low-voltage feeder with large PV penetration, supplied from a medium-voltage feeder. (b) Equivalent single-line representation of the considered low-voltage feeder.

in the low-voltage feeder to protect them. The residential houses are mostly single-phase customers. A similar type of network is used in the majority of the European and Asian countries; however, it is different from those at North America [69].

The properly designed low-voltage feeders have MEN earthing system in which the neutral conductor is earthed at each load (through an earthing resistance) and at the transformer secondary [70], as shown in Figure 13.2(a). However, old low-voltage feeders or those low-voltage feeders developed without proper engineering supervisions may have an NEG earthing system in which the neutral wire is grounded only at the transformer, as seen from Figure 13.2(b).

The PVs are thought as single-phase constant power-type sources that operate at unity power factor, as recommended by IEEE for interfacing PVs with a utility feeder [71]. PVs in Australia usually are 1–5 kW and the number of PVs connected to each of the phases of the low-voltage feeder is uncontrolled (i.e., random), because the utilities do not aim to balance their numbers per phase. All PVs have an under-/over-voltage islanding protection feature according to the Australian Standard on inverter-based grid connection of energy systems [72], as can be seen from the datasheet of available PVs on the market [73–75] to disconnect after a short-circuit fault in the low-voltage feeder [66]. The PVs also have a current limiting technology to prevent their output current exceed by 150% of their nominal value [73].

The considered low-voltage feeder in Figure 13.1(b) consists of 30 dwellings that are energized through a 150-kVA distribution transformer. Three dwellings are thought to be energized by service cables originated at each pole, where these radially connected poles are assumed to be 40 m apart from each other. Each house is represented as a single-phase, constant-impedance load of 4.4 kVA (i.e., the after diversity maximum demand of townhouses of Perth, Australia [76]). Each load is assumed to have a power factor of 0.95 lagging. The loads are distributed equally among the phases. Each PV is presumed to be 4.4 kW (i.e., almost the median of the PVs at Perth [77]). It is to be noted that the loads can be also assumed operating under a demand-side management program [78] without affecting the studies in this chapter. Furthermore, it is assumed that the PV penetration (denoted by the proportion of the PVs' output power over the network's peak demand [68]) is 100% to yield a worst case study. In [79], a fast load flow technique to evaluate such systems has been discussed, whereas Table 13.1 provides the detailed parameters used in modelling this system.

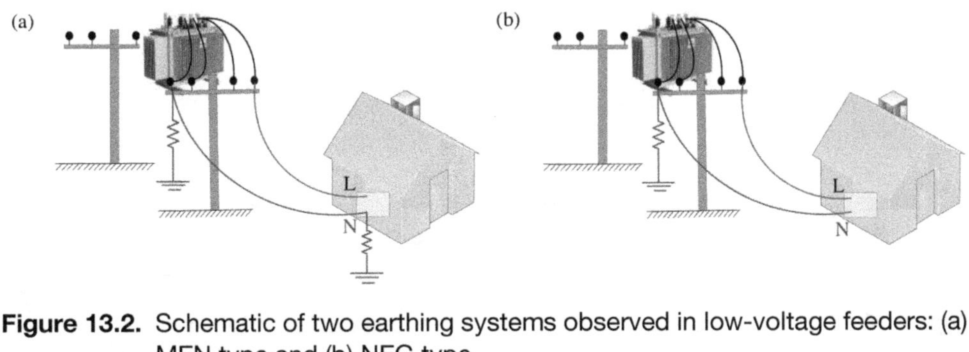

Figure 13.2. Schematic of two earthing systems observed in low-voltage feeders: (a) MEN type and (b) NEG type.

Table 13.1. Assumed technical parameters of the considered network of Figure 13.1 (b)

Transformer: 150 kVA, 11 kV/415 V, 50 Hz, Dyn-type, $Z = 5\%$
Medium-voltage line: 11 kV L-L rms, 2 km, ACSR 50 mm^2 bare conductor, three-phase three-wire system, $R = 2.16\ \Omega/\text{km}$, $X = 2.85\ \Omega/\text{km}$ [80]
Low-voltage line: 415 V, 400 m, AAC 75 mm^2 bare conductor, three-phase four-wire system with ABCN horizontal configuration on 120-cm crossarms [68] and a total of 400-m length, 10 nodes with a distance of 40 m from each other, $R = 0.452\ \Omega/\text{km}$, $X = 0.27\ \Omega/\text{km}$ [80]
PVs: PF = 1, $\eta = 100\%$, $I_{\text{max at Fault}} = 150\%\ I_{\text{rated}}$
Residential demand: Single-phase constant-impedance loads, $S = 4.4$ kVA, PF = 0.95 lagging

13.3 System Modelling and Analysis

Let us consider Figure 13.3(a) and (b), which shows, respectively, a PV at the down-stream and upstream of a short-circuit fault on a low-voltage feeder. In each of these cases, the fault current consists of the PVs' output current and the current drawn from the transformer. Hence, the fault level rises in the network of Figure 13.3(a) and (b) versus the network of Figure 13.3(c), which doesn't have a PV. A larger rise is experienced as the PV's rating increases. This chapter aims to illustrate the impact of the PV on the low-voltage feeder's protection, including the fault level, current drawn through the transformer, and the node voltages within the low-voltage feeder.

Assume a short-circuit fault occurring at PV upstream (Figure 13.3(a). In this case, with the PV being situated nearer to the low-voltage feeder's far end nodes, a larger voltage will be observed in the node to which the PV is connected. This is because the PV's output current flows a bigger length of the low-voltage feeder; and therefore, all nodes between the PV and fault observe a larger voltage. The voltage lessens from the PV location towards the fault point. In case of a bolted (i.e., a zero impedance) short-circuit fault, the PV does

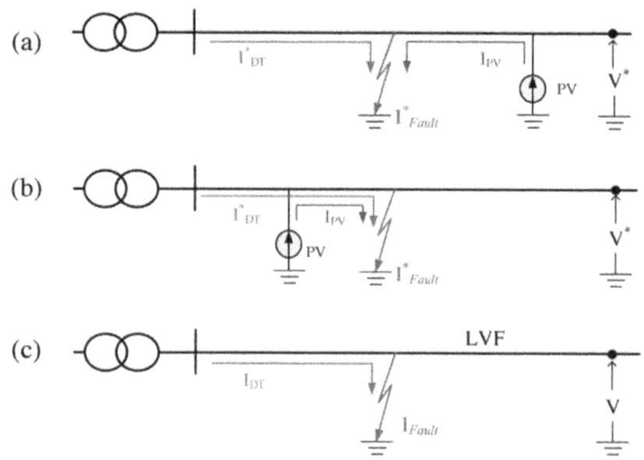

Figure 13.3. Various situations of PV and short-circuit fault location: (a) no-PV, (b) fault occurring at PV upstream, and (c) fault occurring at PV downstream.

not have an impact on the voltages of the nodes between the fault and the transformer. Furthermore, the current drawn through the transformer is not impacted. On the other hand, for a realistic short-circuit fault (with a small impedance), the PV's current will increase the node voltage of the fault location that will slightly lessen the current drawn from the transformer. The impact is stronger for a PV with a higher rating.

Now, assume a short-circuit fault occurring at PV downstream (Figure 13.3(b)). In this case, the node voltage of PV location increases versus that in Figure 13.3(c). Hence, the current drawn from the transformer reduces. This impact becomes stronger when the PV is situated nearer to the low-voltage feeder's beginning. Thus, the fault level marginally decreases but is still higher than the fault level observed in Figure 13.3(c). The impact is greater when the PV has a larger rating.

To calculate I_{Fault}, I_{DT}, and V, the network of Figure 13.1(b) is used. The network in this figure resembles a traditional three-phase power system if the PVs are ignored. However, because of unequal and random number and rating of the PVs in every phase and node, the network of Figure 13.1(b) is studied using the sequence component method [81]. The key difference between this unbalanced network with other unbalanced systems is the presence of voltage's negative and zero sequences, as discussed below.

Let us assume the same distance (impedance) among adjacent poles (nodes) in the low-voltage feeder of Figure 13.1b. Now, the low-voltage feeder's impedance matrix can be expressed by

$$[Z^{lv}]_{\text{ABC}} = z_f.I_m \tag{13.1}$$

for phases A, B, and C, in which Z_f denotes the impedance between two neighboring nodes in each phase, whereas I_m represents the order-3 identity matrix. Likewise, the impedance matrix for the medium-voltage feeder and transformer can be expressed by

$$[Z^{\text{mv}}]_{\text{ABC}} = z'_{\text{f}}.I_m \tag{13.2}$$

$$[Z^{\text{dt}}]_{\text{ABC}} = z_{\text{dt}}.I_m \tag{13.3}$$

where z'_f and z_{dt}, respectively, denote the impedance of the medium-voltage feeder between every two neighboring nodes and the distribution transformer in each phase.

Now, the low-voltage feeder's impedance can be expressed in sequence components from (13.1) as

$$[Z^{\text{lv}}]_{012} = A^{-1}[Z^{\text{lv}}]_{\text{ABC}}A \tag{13.4}$$

where

$$\mathbf{A} = \begin{bmatrix} 1 & 1 & 1 \\ 1 & a^2 & a \\ 1 & a & a^2 \end{bmatrix}$$

$$a = 1\ \angle 120°$$

Likewise, the sequence components for the impedance can be computed for the medium-voltage feeder and the transformer.

Detailed modelling of all network equipment including PVs is essential in dynamic studies; however, for short-circuit fault studies, the network components can be presented using their simpler forms. As such, the PVs can be modelled as current sources, but with considerations like maximum output current or disconnection regarding a voltage variation in their coupling point during short-circuit faults [82]. Similarly, although the short-circuit fault impedance has dynamic characteristics during the short-circuit fault [83], these dynamics are ignored. Thus, I^{PV} denotes the PV's output current, whereas Z_{Fault} is the fault impedance, both in a steady-state condition. Therefore, the matrix denoting the PV's output current, when installed at node-i (where $1 \leq i \leq N$) and in phases A, B, and C, can be expressed by

$$[I_i^{\text{pv}}]_{\text{ABC}} = [\, I_A^{\text{pv}} \; I_B^{\text{pv}} \; I_C^{\text{pv}} \,]_i^T \tag{13.5}$$

in which T denotes the transpose function. Now, the PV's output current at each node can be expressed in sequence components from (13.5) by

$$[I_i^{\text{pv}}]_{012} = A^{-1}[I_i^{\text{pv}}]_{\text{ABC}}A \tag{13.6}$$

Now, assuming an asymmetrical short-circuit fault occurring at node-k, the equivalent sequence networks can be defined from (13.4) and (13.6), as shown in Figure 13.4(a). This figure can be streamlined by deriving the Thevenin and Norton equivalent circuit from the fault point.

The output currents of the PVs located at the fault downstream can be aggregated using the Norton equivalent circuit by

$$\begin{aligned} I_{no}^1 &= \sum\nolimits_{i=k}^{N} I_i^{\text{pv}^1} \\ I_{no}^2 &= \sum\nolimits_{i=k}^{N} I_i^{\text{pv}^2} \\ I_{no}^0 &= \sum\nolimits_{i=k}^{N} I_i^{\text{pv}^0} \end{aligned} \tag{13.7}$$

in which the subscript $_{no}$ represents Norton equivalent, as shown in Figure 13.4(b).

The network components and PVs located at the fault upstream can be depicted by its Thevenin equivalent and can be expressed by

$$\begin{aligned} V'^{\,1}_{th} &= V_{S_{th}}^{\;1} + \sum\nolimits_{i=1}^{k-1} I_i^{\text{pv}^1}\left(Z^{\text{mv}^1} + Z^{\text{dt}^1} + i.Z^{\text{lv}^1}\right) \\ V'^{\,2}_{th} &= \sum\nolimits_{i=1}^{k-1} I_i^{\text{pv}^2}\left(Z^{\text{mv}^2} + Z^{\text{dt}^2} + i.Z^{\text{lv}^2}\right) \\ V'^{\,0}_{th} &= \sum\nolimits_{i=1}^{k-1} I_i^{\text{pv}^0}\left(Z^{\text{dt}^0} + i.Z^{\text{lv}^0}\right) \end{aligned} \tag{13.8}$$

where

$$\begin{aligned} Z_{\text{th}}^{\;1} &= Z^{\text{mv}^1} + Z^{\text{dt}^1} + kZ^{\text{lv}^1} \\ Z_{\text{th}}^{\;2} &= Z_{\text{th}}^{\;1} \\ Z_{\text{th}}^{\;0} &= Z^{\text{dt}^0} + kZ^{\text{lv}^0} \end{aligned}$$

whereas $_{\text{th}}$ represents Thevenin equivalent, as shown in Figure 13.4(b).

Equations (13.7) and (13.8) can be further simplified as Figure 13.4(c) in which

$$\begin{aligned} V_{\text{th}}^1 &= V'^{\,1}_{\text{th}} + I_{\text{no}}^1 Z_{\text{th}}^1 \\ V_{\text{th}}^2 &= V'^{\,2}_{\text{th}} + I_{\text{no}}^2 Z_{\text{th}}^2 \\ V_{\text{th}}^0 &= V'^{\,0}_{\text{th}} + I_{\text{no}}^0 Z_{\text{th}}^0 \end{aligned} \tag{13.9}$$

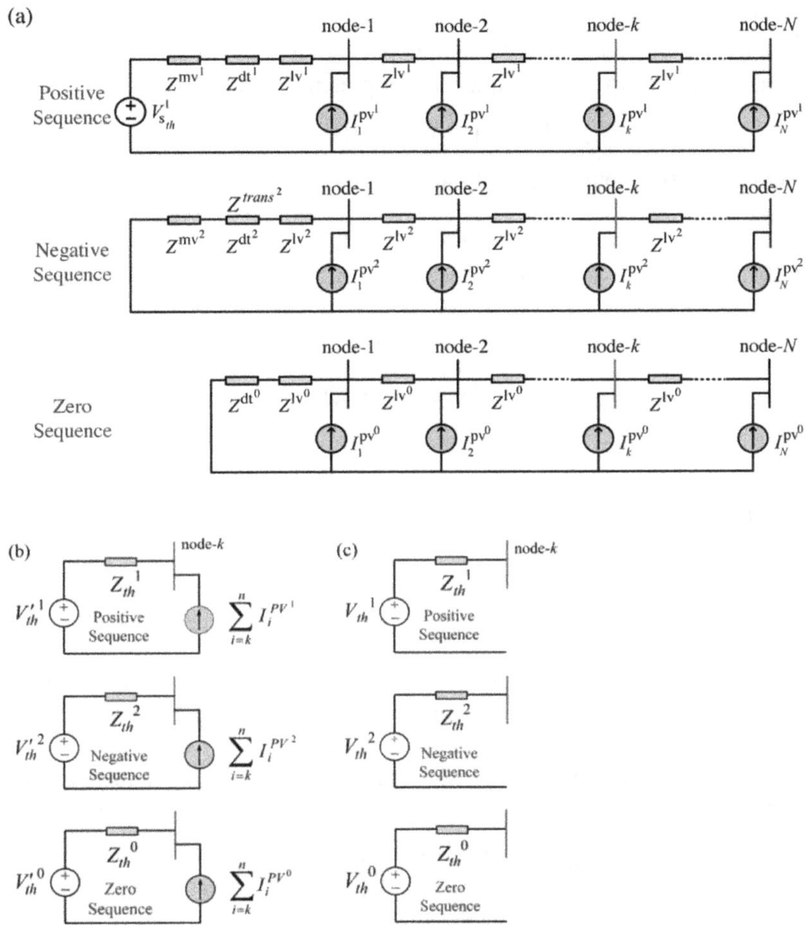

Figure 13.4. (a) Schematic illustration of the sequence networks of Figure 13.1(b) for a short-circuit fault occurring at node-*k*. (b) Equivalent sequence networks from fault point. (c) Streamlined Thevenin equivalent from fault point.

or

$$[V_{th}]_{012} = [V'_{th}]_{012} + [I_{no}]_{012}\ [Z_{th}]_{012} \tag{13.10}$$

which shows that zero and negative voltage sequence also emerge because of unequal distribution of PVs among the phases of the network on top of the positive voltage and zero and negative impedance sequences [59].

In case of an LG fault, the fault level can be computed from the current sequence components. The fault level is calculated as

$$[I^F]_{\mathrm{ABC}} = 3I\,[\,\lambda_A\ \ \lambda_B a^2\ \ \lambda_C a\,]^T \tag{13.11}$$

in which

$$I = \frac{V_{\mathrm{th}}{}^1 + V_{\mathrm{th}}{}^2 + V_{\mathrm{th}}{}^0}{Z_{\mathrm{th}}{}^1 + Z_{\mathrm{th}}{}^2 + Z_{\mathrm{th}}{}^0 + 3Z_{\mathrm{Fault}}}$$

and $\lambda_j = 1$ when the LG fault is in phase j (where $j = A, B$, or C). If not, $\lambda_j = 0$.

After calculating the fault level, the fault point's voltage can be calculated by employing the Kirchhoff's voltage law in the sequence components from

$$[V_F]_{012} = [V_{th}]_{012} - [I]_{012}\,[Z_{th}]_{012} \tag{13.12}$$

and later converted into the ABC phase domain by

$$[V_F]_{\mathrm{ABC}} = A\ [V_F]_{012} \tag{13.13}$$

to calculate the voltage at the fault point. Then, the node voltage within the low-voltage feeder of Figure 13.2(a) is determined.

Likewise, the fault level is calculated for an LL fault in the ABC domain from

$$[I^F]_{\mathrm{ABC}} = A\,[I_F]_{012} \tag{13.14}$$

in which

$$I^1 = \frac{V_{\mathrm{th}}{}^1 - V_{\mathrm{th}}{}^2}{Z_{\mathrm{th}}{}^1 + Z_{\mathrm{th}}{}^2 + Z_F}$$

$$I^2 = -I^1$$

$$I^0 = 0$$

when the fault is between phase-B and C. At this stage, first the fault level and then the voltage magnitude in every node along the low-voltage feeder can be determined by the current sequence components, as was discussed above.

Similarly, for an LLG fault between phase-B and C, the current sequence components are

$$\begin{aligned} I^1 &= \frac{(1+a)V_{\mathrm{th}}{}^1 + V_{\mathrm{th}}{}^2 - Z_{\mathrm{th}}{}^2 I^2}{(1+a)\,Z_{\mathrm{th}}{}^1} \\ I^2 &= \frac{\beta_1 - \beta_3\beta_4}{\beta_2 - \beta_3\beta_5} \\ I^0 &= -aI^1 - a^2 I^2 \end{aligned} \tag{13.15}$$

where

$$\beta_1 = V_{\mathrm{th}}{}^0 + a^2 V_{\mathrm{th}}{}^1 + aV_{\mathrm{th}}{}^2$$

$$\beta_2 = (1 + a - 2a^2)Z_{\mathrm{Fault}} + aZ_{\mathrm{th}}{}^2 - a^2 Z_{\mathrm{th}}{}^0$$

$$\beta_3 = (1 + a^2 - 2a)Z_{\mathrm{Fault}} + a^2 Z_{\mathrm{th}}{}^1 - aZ_{\mathrm{th}}{}^0$$

$$\beta_4 = \frac{(1+a)V_{\mathrm{th}}{}^1 + V_{\mathrm{th}}{}^2}{(1+a)Z_{\mathrm{th}}{}^1}$$

$$\beta_5 = \frac{Z_{\mathrm{th}}{}^2}{(1+a)Z_{\mathrm{th}}{}^1}$$

At this stage, first the fault level and then the voltage magnitude in every node along the low-voltage feeder can be determined as was discussed above. Interested readers can refer to [61] to see how to calculate the fault current for faults among other phases.

13.4 Impact of One PV

In this section, a sensitivity analysis [84, 85] is used to define and evaluate a PV's impact on the low-voltage feeder's fault level and the drawn current through the transformer, as well as the node voltages within the low-voltage feeder during an LG fault on phase-*A*, in addition to an LL and LLG fault between phases *A* and *B*, for a bolted fault. To this end, the fault location and PV installation point are varied discretely from the beginning until the end of the feeder (i.e., from nodes 1 to 10) in Figure 13.1(b).

13.4.1 Voltage Profile

One factor that is very important in PVs' disconnection in the course of a short-circuit fault in the low-voltage feeder is the voltage magnitude observed at the PV's installation point. The changes of the node voltages within the low-voltage feeder in the presence of a PV are discussed here.

First, the "no-PV" case (i.e., the network of Figure 13.1(b) without any PVs) is considered. Then we have considered a "PV-available" case in which a 2-kW PV (that has an output current of 8.3 A at 240 V and unity power factor) is assumed, which has an output current limited 12.5 A (i.e., 150% of its nominal rating) during the short-circuit faults.

Suppose an LG fault on phase-*A* at node-1, that is, the beginning of the feeder, whereas the PV's location varies from nodes 1 to 13. The node voltages within the low-voltage feeder in the PV-available case are greater than the no-PV case. Figure 13.5(a) shows this voltage difference. From this figure, it can be seen that the end nodes of the low-voltage feeder observe a larger voltage difference versus the beginning nodes. This is because as the PV is situated closer to node-10 (i.e., the end of feeder), its output current flows through a greater length of the low-voltage feeder; hence, each node among the fault and the PV locations observes a larger voltage. Similarly, it is observed that voltage difference rises from the feeder's beginning toward the PV. Nevertheless, it remains unchanged for the nodes at PV downstream. Correspondingly, the largest voltage difference is observed in the case that the PV is situated at feeder's far end nodes.

Now, consider a fault occurring at the end of the feeder. Under such a condition, the voltage difference becomes very little versus the situation in which the short-circuit fault occurs at node-1, as seen in Figure 13.5(b). This voltage difference rises from the beginning of the feeder on the way to the PV location, but starts to reduce toward the fault location. This is due to the fact that the PV and the transformer currents flow from the PV location toward the fault point. Hence, a greater voltage is observed at PV downstream. The largest voltage difference is detected when the PV is situated at node-5 (i.e., in the middle of the feeder).

Now consider a short-circuit fault occurring at the middle of the feeder. As can be seen from Figure 13.5(c), if the PV is at fault upstream, the voltage difference resembles the characteristic of Figure 13.5(a), whereas its pattern is like Figure 13.5(b) in the case the PV is situated at fault downstream, which is acceptable based on the reasons discussed earlier.

Now, to assess the impact of PV rating, consider an LG fault occurring at node-8, whereas the PV situated at node-4 has a rating that varies from 0.7 to 3.5 kW. In this case, a larger voltage difference is seen as the PV's output power increases, as seen in Figure 13.5(d). This is due to the fact that a PV with a larger rating has a larger output current. Therefore, a larger voltage is experienced.

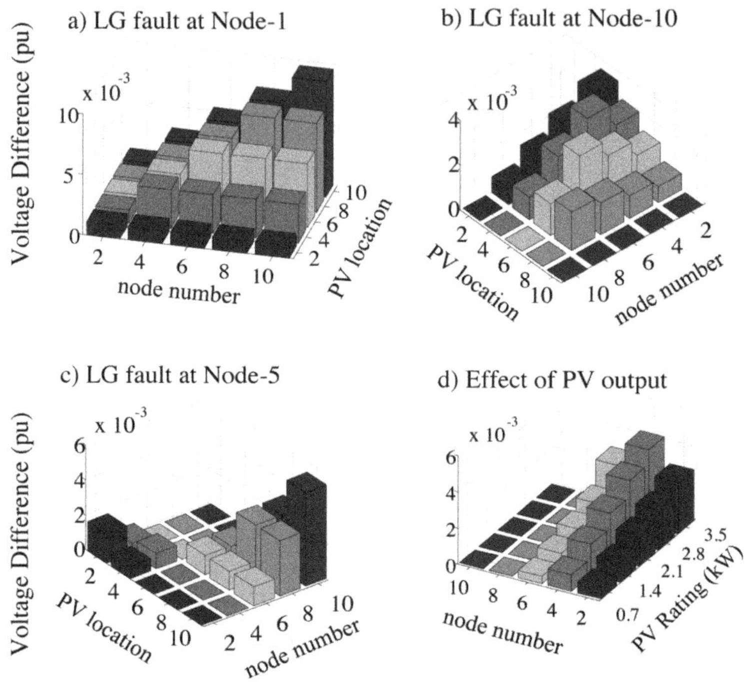

Figure 13.5. Voltage difference along the low-voltage feeder during an LG fault.

Now, consider an LL fault at the beginning of feeder. In such a condition, the voltage difference pattern is like the results of LG fault; however, it is nearly half of that, as seen from Figure 13.6(a). For the fault at the end of the feeder, the voltage difference deviation is larger at PV upstream, whereas it is less at PV downstream, as can be seen from Figure 13.6(b). Also, the voltage difference rises by PV being situated toward the feeder's end.

Now, consider an LL fault at the middle of the feeder. Two trends are seen for the voltage difference in this case. If the PV is situated at fault downstream, the voltage difference rises toward the end of the feeder; however, if the PV is at fault upstream, the voltage difference is large at PV upstream but small for PV downstream, as can be seen from Figure 13.6(c).

Now, consider a PV with a rating of 0.7 to 3.5 kW, connected to node-4, whereas a short-circuit fault arises at node-8. Under such a situation, the change in the PV rating does not affect the voltage difference on the nodes at fault downstream, as can be seen from Figure 13.6(d). However, the voltage difference increases for the nodes at fault upstream as the PV rating rises.

The analyses for LLG faults have similar results to those of LL faults for voltage difference and are not provided here.

13.4.2 Transformer Current and Short-Circuit Fault Level

The short-circuit fault and the current sensed by the switch fuses at the transformer secondary is the other crucial parameter in protecting the low-voltage feeders. Any alteration in

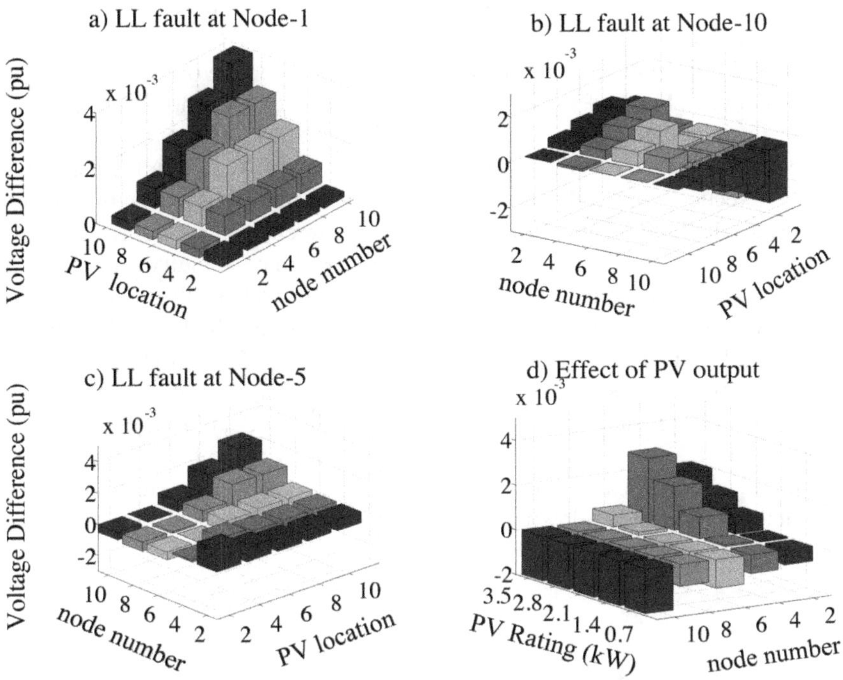

Figure 13.6. Voltage difference along the low-voltage feeder during an LL fault.

the fault current or the sensed current at the transformer secondary will deviate the expected operation time of the switch fuse. Let us denote these deviations as the fault current difference and transformer current difference. This section evaluates these factors.

Suppose a 2-kW PV, in which both PV and fault are relocated from the beginning toward the end of the feeder.

For the short-circuit faults at PV downstream, by shifting the fault toward the end of the feeder, the fault current difference decreases, as seen from Figure 13.7(a). This is because of the variation of the node voltages due to the short-circuit fault location that accordingly affects the current drawn from the transformer. When the short-circuit fault is at PV upstream, by shifting the fault from the beginning of the feeder toward the PV location, the fault current difference increases, as can be seen from Figure 13.7(a). The highest fault current difference is detected when the short-circuit fault is at PV's installation point.

In the case that PV is situated at fault upstream, the current drawn from the transformer is lower than the no-PV mode, as seen by negative numbers in Figure 13.7(b). This is due to the fact that the voltage at PV's installation node results in a lower drawn current from the upstream side. When the PV is at fault downstream, the transformer's current is equal to the no-PV mode, as seen from the zero transformer current difference in Figure 13.7(b). This is because the voltages at fault upstream are not affected. The transformer secondary current is much lower from the no-PV mode when PV is situated toward the feeder's beginning nodes.

Now, consider an LG fault occurring at node-8, whereas a PV with a rating of 0.7 to 3.5 kW is located at node-4. As can be seen from Figure 13.7(c), the fault current difference becomes larger as the PV rating increases, whereas Figure 13.7(d) shows that the transformer current difference reduces more.

In case of an LL fault at feeder's beginning nodes, irrespective of the installation point of the PV, the fault level in the PV-available mode is greater than the no-PV mode, as can be seen from the positive fault current difference in Figure 13.8(a). However, the fault current difference becomes negative as the fault occurs at the far end nodes. In the case of a fault occurring at the feeder's end nodes, with the PV being shifted from the beginning nodes of the feeder toward the fault location, the fault current difference increases. However, it does not change when the PV is at fault downstream. In the case that the fault occurs at beginning nodes, the fault current difference reduces by PV being shifted from the beginning nodes of the feeder toward the fault. Nevertheless, it is zero for a PV installed at fault downstream.

In case of an LL fault, the transformer current in the PV-available mode is smaller versus the no-PV mode, as can be seen from the negative values in Figure 13.8(b). This is because the node voltage of PV location is higher; thus leading to a reduction in the current drawn from the transformer. For the LL faults, as the PV moves from the beginning of the feeder toward the fault location, the transformer current difference decreases. However, it does not change when the PV is at fault downstream, as seen from Figure 13.8(b).

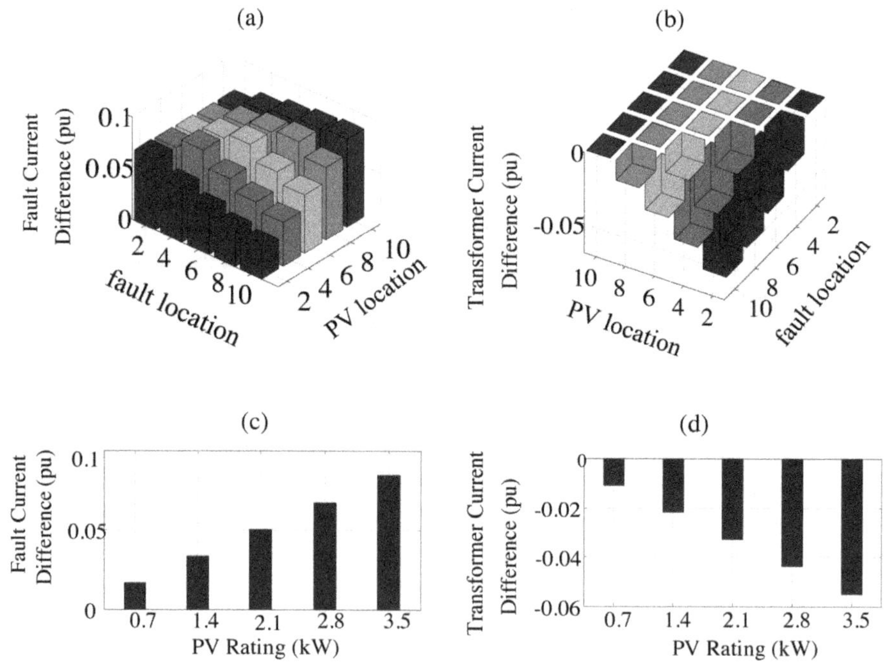

Figure 13.7. Fault current difference and transformer current difference during an LG fault.

Now, consider an LL fault at node-8, whereas a PV with a rating of 0.7 to 3.5 kW is located at node-4. As seen in Figure 13.8(d), for PVs smaller than 2.8 kW, the fault current difference is negative. However, it is positive for larger PVs. With the rise of the PV rating,

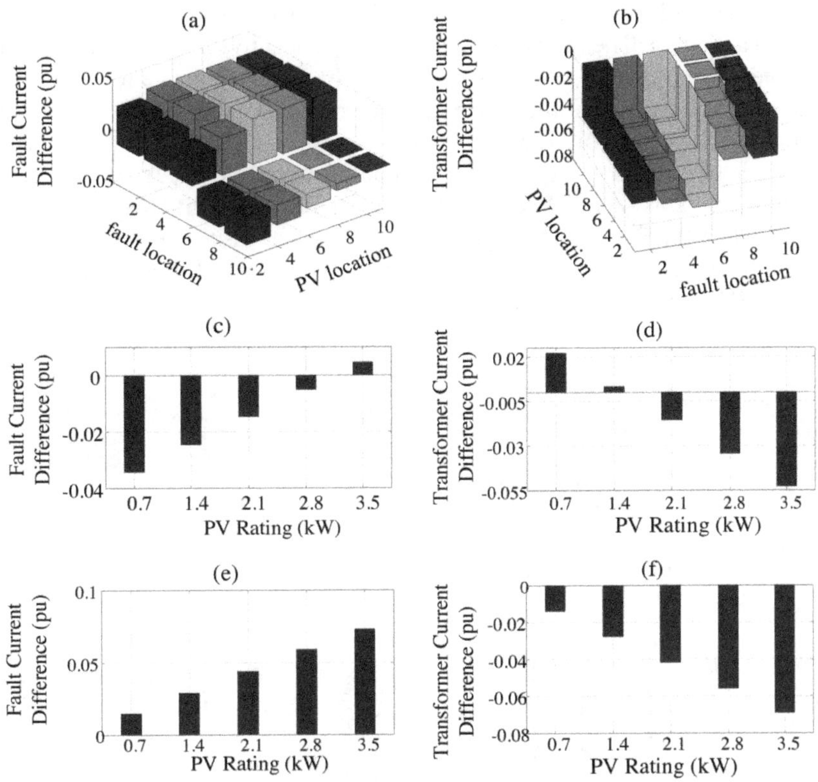

Figure 13.8. Fault current difference and transformer current difference during an LL fault.

the fault current difference decreases, as seen in Figure 13.8(c), which shows that for PVs smaller than 1.4 kW, the transformer current difference is positive, whereas negative for PV ratings more than that.

Now, consider an LL fault occurring at node-4 in which the PV is situated at node-8. As seen in Figure 13.8(e), in this condition, the fault current difference is positive and rises for larger PVs. Even though the transformer current difference increases for larger PVs, it is negative, as seen from Figure 13.8(f). This is because, for a PV situated at fault downstream, the node voltage at the fault point is greater than that of the no-PV mode. A greater difference is seen for larger PVs. Thus, a smaller current is drawn from the transformer.

For an LLG fault, the fault current difference rises as the PV is situated closer to the fault location; but it does not change when the PV is at fault downstream, as can be seen from Figure 13.9(a). In case of an LLG fault, when the PV is at fault downstream, the transformer current difference is zero. Nevertheless, when the PV is situated at fault upstream, the transformer current difference reduces as the PV moves from the beginning of the feeder toward the fault. Presuming that the PV is situated at one node, the transformer current difference rises as the fault is relocated from the PV location toward the end of the feeder, as can be seen from Figure 13.9(b).

Now, consider an LLG fault occurring at node-8, where a PV with a rating of 0.7–3.5 kW is located at node-4. The fault current difference is negative and increases for large PV ratings, as seen from Figure 13.9(c), whereas it nearly does not change for PVs larger than 2.1 kW. In contrary, the transformer current difference is negative and observes a rise for large PV ratings, as seen from Figure 13.9(d).

Now, consider an LLG fault occurring at node-4 when the PV is situated at node-8. Under this condition, as seen in Figure 13.9(e), the fault current difference is negative and nearly equal when the PV's rating is below 2.8 kW, whereas it is positive for more than that. In addition, it can be seen from Figure 13.9(f) that the transformer current difference is zero for all PV ratings in this case.

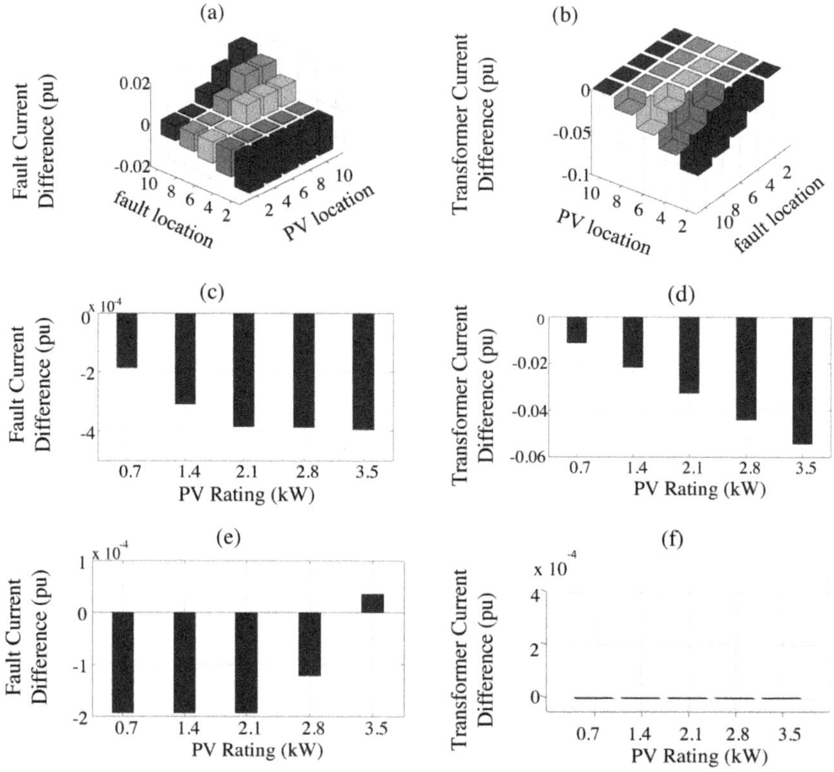

Figure 13.9. Fault current difference and transformer current difference during an LLG fault.

13.4.3 Important Notes

The sensitivity analysis results of Sections 4.1 and 4.2 illustrate the deviations of the voltages of nodes, the fault level, and the transformer secondary current with respect to the variations in fault and PV location along the low-voltage feeder and the PV rating. The studies show that a PV at fault upstream has dissimilar influences on above factors versus a PV situated at fault downstream. In addition, the studies demonstrate that these properties are different for different types of faults.

The studies show that the fault level is always larger if a PV exists in the feeder in the course of an LG fault, whereas it can be less or larger in case of LL and LLG faults, subject to the PV and fault location. The studies also show the PV rating has a completely dissimilar effect on fault level and current drawn from the transformer even for the same PV and fault location and fault type. The studies also show that in case of LG and LLG faults, the current at the transformer secondary is less when the network hosts a PV, and a greater difference is observed as the fault is relocated to the end of the feeder. Nevertheless, a greater change can be observed for LL faults, if the fault occurs at the middle of the feeder. As the current drawn from the transformer reduces, the switch fuse's operating time will observe a slight delay.

13.5 Impact of Multiple PVs

Because of the randomness in PV rating, its connection to a phase and a node along the low-voltage feeder, along with the fault location and impedance, a Monte Carlo-based stochastic analysis [86–89] is used instead of a deterministic approach to investigate the fault level and the node voltages within the low-voltage feeder. The uncertainties that are considered within the Monte Carlo analysis include [61].

- PV penetration in the low-voltage feeder
- PVs' connection points to phases and nodes within the feeder
- PVs' ratings
- Location of the fault
- Impedance of the fault

and the main objective is to analyze

- the node voltages within the feeder
- the current observed at the transformer secondary.

Figure 13.10 illustrates the flowchart of the stochastic analysis. As seen from this figure, the short-circuit fault-type and PV penetration levels or PV ratings are selected and separate Monte Carlo analysis is conducted for each set of selections. The PV and fault locations are thought with an equal distribution probability over 10 poles with equal distances of 40 m over a 400-m long feeder. PV ratings are presumed with discrete values with a 2.5-kW average and a 0.8-kW variance under a normal probability density function (PDF) [90]. The impedance of the LG fault is presumed as a uniform distribution between 0 and 0.3 Ω, whereas that of LL and LLG faults is thought like the historical data-based histogram of Figure 13.11 [91].

A *Time* parameter is considered in the flowchart of Figure 13.10, which denotes the time of the study over 24 hours. It is then normalized between 0 and 1, as seen from Figure 13.12, to remove the undesired input combinations for the Monte Carlo analysis. The availability of sunlight is supposed from 6:00 to 19:00 and it is assumed that the PVs experience the highest generation at 12:00 (based on the solar irradiance data at Perth, Australia [92]). The *Time* factor is used to only choose correlated random numbers for PVs' output currents, whereas the other inputs of the Monte Carlo analysis are assumed *Time*-independent [93]. Otherwise, one may use the methods of [94, 95] to define the PVs' output power according to the available solar irradiance.

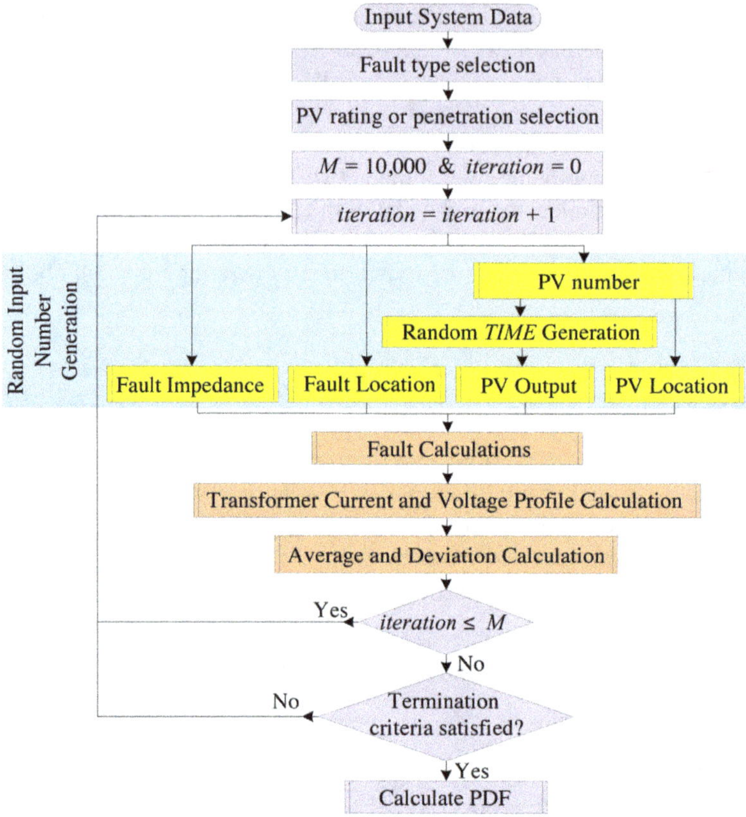

Figure 13.10. Monte Carlo analysis flowchart.

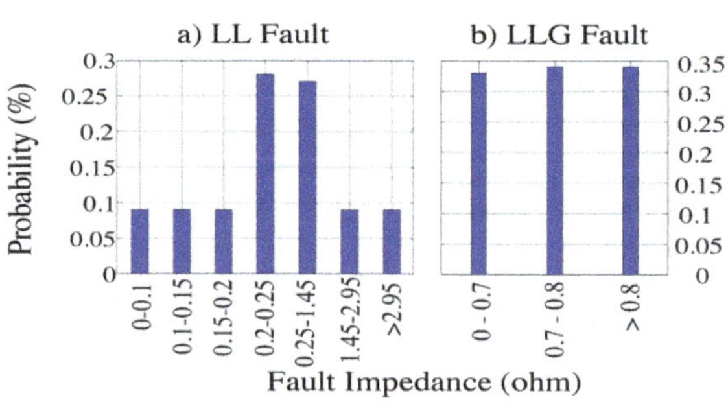

Figure 13.11. Assumed fault impedance histograms for (a) LL faults and (b) LLG faults.

The termination (stopping) criterion of the Monte Carlo analysis is selected with respect to attaining a satisfactory convergence for the average and variance of the fault level, and node voltages along the low-voltage feeder, with a confidence degree of 95%, whereas a minimum of 10,000 iterations is used to prevent early termination. When the Monte Carlo analysis converges, the PDF of all outputs are determined and expressed in terms of the distribution function, expected average, and standard deviation.

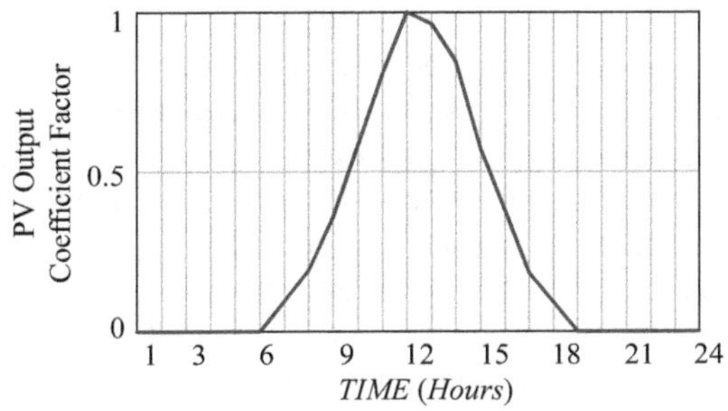

Figure 13.12. Assumed variation of the output power of PVs versus *TIME*.

The stochastic analysis aims to investigate the highest allowed PV penetration level in the considered low-voltage feeder of Figure 13.1(b) from the short-circuit fault point of view only. It is to be noted that other factors including voltage rise and harmonic distortion also need to be evaluated when selecting the overall limit.

This study aims to define the highest PV penetration level considering the successful probability of the operation of switch fuse and PV disconnection. The earlier shows the probability of switch fuse's operation after a short-circuit fault, whereas the latter depicts the PVs' disconnection probability from the low-voltage feeder after a short-circuit fault following the voltage deviation at the PV connection point, according to IEEE Standard 929-2000 [71]. Using the Monte Carlo analysis-derived PDF for the current sensed at the transformer secondary, the probability of observing a current higher than the threshold for the operation of the switch fuse (e.g., 1.25 pu) is calculated, and referred to as the switch fuse operation probability. Likewise, using the calculated PDF for node voltages, the probability of observing a voltage exceeding the thresholds for the disconnection of a PV (e.g., higher than 1.1 or below 0.88 pu [71]) is calculated and referred to as the PV disconnection probability. From the short-circuit fault point of view, the highest allowed PV penetration level is then defined as the matching PV penetration level for the minimum probability of the successful PVs' disconnection and switch fuse operation.

13.5.1 Voltage Profile

First, the Monte Carlo analysis is performed for each fault type, assuming a PV penetration level of 50%. Based on *Time*, different output powers are chosen for the PVs in the Monte Carlo analysis iterations. The results of this analysis at the beginning of feeder, middle of feeder, and end of the feeder of the faulty phase are shown in Table 13.2. As seen from this table, the expected average voltage of each node has a slight increase for large PV ratings. It is to be noted that no substantial variation is observed in the standard deviation for the variation of PV ratings. For LL and LLG faults, the data of one of the faulty phases have been captured. The results for LL fault show a decrease in the expected average voltages of each node for large PV ratings. The standard deviation of the node voltages also reduces with an increase in the PV rating. Moreover, this table reveals that the expected average and standard deviation of the voltages at the beginning of feeder increase for large PV ratings; however, they decrease for the middle and end of the feeder.

Now, assume various PV penetrations although all PVs have an output power of 2 kW. Table 13.3 lists the results of this analysis and shows that in the case of LG faults, the expected average voltage of each node increases for large PV penetration levels; however, it is not considerable. Also, for large PV penetration levels, the standard deviation for node voltages slightly increases. For LL fault, it is seen from the table that the expected average node voltage and their standard deviation also increase with an increase in PV penetration level and rating. In the case of LLG faults, the expected average voltage at the beginning of feeder increases for large PV penetration levels; however, it decreases at the middle of the feeder and end of the feeder. The standard deviation of voltages in all nodes increases for large PV penetration levels.

The expected average voltage at each node of the low-voltage feeder is listed in Table 13.4 for various PV penetration levels under different fault types. This table shows that, irrespective of the PV penetration level, the voltages on the faulty phase drop below 0.88 pu in the case of LG and LLG faults. Nevertheless, the expected average node voltages at the beginning of the feeder are above the threshold.

Using the calculated PDFs, the PV disconnection probability is determined for every node along the feeder, as provided in Table 13.5. This study shows that with a probability of over 89%, the PVs within the faulty phase will disconnect according to [71], in the course of LG and LLG faults. Nevertheless, this probability is 52%–63% in the case of LL fault. Thus, the impact of the PV penetration level is more tangible in the case of LL faults.

13.5.2 Transformer Current

The above analysis is replicated to determine the expected PDF of the current drawn from the transformer in the course of short-circuit faults. Table 13.6 lists the results of this analysis. As from this table, the ratings and penetration levels of PVs do not affect the expected average current at transformer secondary for an LG fault. Furthermore, the increase in the standard deviation with respect to the increase of PV rating and penetration level is not substantial. On the other hand, it can be seen that the average and standard deviation of transformer current increases for large PV ratings or penetration levels in case of LL faults, whereas they decrease for LLG faults.

Table 13.7 also illustrates the calculated switch fuse operation probability to each short-circuit fault. It can be seen from this table that the switch fuse licated at the transformer secondary will see the short-circuit fault in the low-voltage feeder with a probability of 89%, for all PV penetration levels in every short-circuit type.

13.5.3 Important Notes

The ratio of the standard deviation to the expected average voltage or current signifies the data diversity in each of the Monte Carlo analysis studies. It can be seen from Table 13.2 that this ratio is less than 18% for LL and LLG faults, whereas it is 23%–52% for LG faults. Furthermore, this ratio is 7%–24% when the fault is at the beginning of feeder, whereas it is 12%–40% and 15%–52% when the fault is, respectively, at the middle and end of the

Table 13.2. Node voltages (pu) of the faulty phase in different PV ratings under various short-circuit fault types

Fault	PV (kW)	Beginning of Feeder		Middle of Feeder		End of Feeder	
		Expected Average	Standard Deviation	Expected Average	Standard Deviation	Expected Average	Standard Deviation
LG	0.7	0.7482	0.1742	0.5578	0.2236	0.4865	0.2513
	1.4	0.7492	0.1739	0.5605	0.2239	0.4904	0.2525
	2.1	0.7514	0.1736	0.5632	0.2247	0.4932	0.2537
	2.8	0.7547	0.1746	0.5675	0.2255	0.4962	0.2543
	3.5	0.7568	0.1751	0.5704	0.2247	0.4991	0.2539
LL	0.7	0.9552	0.0693	0.8639	0.1193	0.8354	0.1400
	1.4	0.9475	0.0687	0.8537	0.1191	0.8238	0.1401
	2.1	0.9395	0.0677	0.8425	0.1187	0.8115	0.1394
	2.8	0.9316	0.0669	0.8319	0.1175	0.7999	0.1383
	3.5	0.9262	0.0651	0.8250	0.1165	0.7923	0.1373
LLG	0.7	0.5354	0.0917	0.4625	0.0583	0.4469	0.0700
	1.4	0.5393	0.0930	0.4604	0.0568	0.4413	0.0683
	2.1	0.5418	0.0942	0.4577	0.0565	0.4355	0.0679
	2.8	0.5461	0.0965	0.4552	0.0561	0.4294	0.0667
	3.5	0.5487	0.0973	0.4534	0.0556	0.4251	0.0662

Table 13.3. Node voltages (pu) of the faulty phase in different PV penetration levels under various short-circuit fault types

Fault	PV Penetration Level (%)	Beginning of Feeder		Middle of Feeder		End of Feeder	
		Expected Average	Standard Deviation	Expected Average	Expected Average	Standard Deviation	Expected Average
LG	20	0.7481	0.1733	0.5583	0.2229	0.4882	0.2516
	40	0.7492	0.1745	0.5615	0.2242	0.4893	0.2518
	60	0.7502	0.1760	0.5635	0.2253	0.4902	0.2536
	80	0.7551	0.1756	0.5697	0.2258	0.4977	0.2549
	100	0.7580	0.1763	0.5715	0.2265	0.5018	0.2552
LL	20	0.9551	0.0702	0.8641	0.1202	0.8360	0.1405
	40	0.9482	0.0697	0.8551	0.1192	0.8255	0.1399
	60	0.9425	0.0668	0.8462	0.1183	0.8157	0.1392
	80	0.9331	0.0658	0.8340	0.1172	0.8020	0.1386
	100	0.9246	0.0650	0.8230	0.1167	0.7900	0.1378
LLG	20	0.5357	0.0913	0.4626	0.0536	0.4467	0.0687
	40	0.5395	0.0926	0.4599	0.0539	0.4405	0.0689
	60	0.5423	0.0938	0.4568	0.0546	0.4347	0.0694
	80	0.5468	0.0963	0.4554	0.0550	0.4294	0.0695
	100	0.5489	0.0971	0.4533	0.0552	0.4244	0.0699

Table 13.4. Node voltages (pu) of the low-voltage feeder in various PV penetration levels and faults

Fault	PV Penetration Level (%)	Expected Average (%) at Node-									
		1	2	3	4	5	6	7	8	9	10
LG	20	74.8	68.3	63.1	58.8	55.8	55.4	52.6	50.6	49.2	74.8
	40	74.9	68.4	63.3	59.0	56.1	55.5	52.8	50.8	49.4	74.9
	60	75.0	68.8	63.7	59.5	56.3	56.0	53.4	51.4	50.0	75.0
	80	75.5	68.9	63.8	59.6	56.9	56.0	53.6	51.6	50.2	75.5
	100	75.8	69.4	64.4	62.0	57.1	56.8	54.1	52.1	50.7	75.8
LL	20	95.5	92.4	89.8	88.2	86.4	85.1	84.3	83.8	83.5	83.6
	40	94.8	91.6	89.0	87.3	85.5	84.2	83.3	82.8	82.5	82.5
	60	94.2	97.0	88.0	85.9	84.6	83.0	82.2	81.6	81.3	81.5
	80	93.3	89.9	87.2	85.0	83.4	82.0	81.1	80.5	80.2	80.2
	100	92.4	89.1	86.3	84.1	82.3	81.1	80.2	79.6	79.2	79.0
LLG	20	53.5	59	48.9	47.3	46.2	45.4	45.0	44.7	44.6	44.6
	40	53.9	51.1	48.9	47.2	45.9	45.0	44.5	44.2	44.1	44.0
	60	54.2	51.3	48.9	47.1	45.6	44.8	44.1	43.8	43.6	43.4
	80	54.6	51.5	49.0	47.0	45.5	44.4	43.7	43.2	43.0	42.9
	100	54.8	51.7	49.0	46.9	45.3	44.0	43.2	42.7	42.5	42.4

Table 13.5. The corresponding PV disconnection probability in various PV penetration levels and faults

Fault	PV Disconnection Probability (%) at Node-										PV Disconnection Probability (%)
	1	2	3	4	5	6	7	8	9	10	
LG	77.7	84.9	88.9	91.4	92.6	92.1	93.1	93.7	94.0	94.1	90.3
	77.4	84.5	88.7	91.2	92.3	92.0	92.9	93.5	93.9	94.0	90.1
	77.0	84.1	88.2	90.7	92.1	91.5	92.5	93.1	93.5	93.8	89.7
	76.2	83.8	87.9	90.4	91.6	91.2	92.2	92.9	93.2	93.4	89.3
	75.6	83.3	87.5	90.1	91.4	90.9	92.0	92.6	93.0	93.2	89.0
LL	14.3	48.0	49.3	49.9	55.3	59	61.0	62.1	62.7	62.4	52.4
	16.5	48.3	49.6	52.6	58.3	61.8	63.8	64.8	65.4	65.3	54.7
	17.7	48.8	50.0	57.7	61.3	65.4	67.0	68.1	68.6	68.0	57.3
	21.1	49.1	50.4	61.1	65.2	68.4	70.2	71.2	71.6	71.5	60.0
	25.0	49.5	50.8	64.4	68.7	71.0	72.7	73.5	74.1	74.5	62.5
LLG	99.9	100	100	100	100	100	100	100	100	100	99.9
	99.9	100	100	100	100	100	100	100	100	100	99.9
	99.9	100	100	100	100	100	100	100	100	100	99.9
	99.9	100	100	100	100	100	100	100	100	100	99.9
	99.9	100	100	100	100	100	100	100	100	100	99.9

Table 13.6. Current at the transformer secondary (pu) of different PV ratings under various short-circuit fault types

Fault	PV (kW)	Transformer Current	
		Expected Average	Standard Deviation
	0.7	6.9221	2.9889
	1.4	6.9364	2.9944
LG	2.1	6.9421	2.9819
	2.8	6.9227	3.0054
	3.5	6.9328	3.0312
	0.7	4.8637	2.9879
	1.4	4.8810	2.9948
LL	2.1	4.8927	3.0241
	2.8	4.9465	3.0275
	3.5	4.9534	3.0289
	0.7	8.7255	2.8067
	1.4	8.7156	2.8037
LLG	2.1	8.7041	2.7928
	2.8	8.6715	2.7885
	3.5	8.6608	2.7770

Table 13.7. Current at the transformer secondary (pu) of different PV penetration levels under various short-circuit faults and the expected probability of switch fuse operation (%)

Fault	PV Penetration Level (%)	Transformer Current		Successful Operation Probability (%)
		Expected Average	Standard Deviation	
	20	6.9381	2.9884	97
	40	6.9336	2.9904	
LG	60	6.9490	2.9949	
	80	6.9040	3.0230	
	100	6.9050	3.0412	
	20	4.8413	2.9862	89
	40	4.8655	3.0103	
LL	60	4.9017	3.0168	
	80	4.9224	3.0206	
	100	4.9327	3.0293	
	20	8.7538	2.8227	99
	40	8.7405	2.8111	
LLG	60	8.6751	2.7859	
	80	8.6642	2.7756	
	100	8.6464	2.7676	

feeder. Similarly, Table 13.4 shows that this ratio is 32%–62%. Thus, the short-circuit fault impedance and location (with the assumed PDFs) are more dominating factors versus other factors such as the PV rating and penetration level. Furthermore, as the fault occurs closer to the feeder's end nodes, the diversity of data rises because the other factors are becoming dominant, too.

Table 13.7 shows that the switch fuse operation probability is above 89% for all faults regardless of the PV penetration level. Likewise, Table 13.5 shows that the PV disconnection probability is above 89% for LG and LLG faults, whereas it is 52%–63% for LL faults. Therefore, only considering the switch fuse operation probability, the switch fuse at the transformer secondary will detect the short-circuit fault in the low-voltage feeder successfully with a probability of over 89% even at a PV penetration level of 100%. Nevertheless, considering the PV disconnection probability for all faults, the disconnection of PVs connected to the faulty phase has a probability of 52%–57% when the PV penetration level is below 80%, and 60%–63% for a PV penetration level more than that.

13.6 PV Disconnection

PVs are required to disconnect from the low-voltage feeders when a short-circuit fault arises in the network. There is a possibility that the feeder is kept energized by the PVs if they are not disconnected, even if the upstream switch fuse or circuit breaker has operated. In such conditions, there is a risk of islanding in the case the PVs' output power is larger than the network demand. Such an issue can lead to hazards for the utility personnel or even damage of the electrical equipment. Although no records are available based on [60] and it can be a rare situation, but appropriate protection concepts are essential to be used to avoid that. IEEE Recommended Practice for Utility Interface of PVs [71] has determined the normal and abnormal voltage limits for PVs of less than 10 kW. In [71], the PVs need to disconnect from the feeder if their connection point voltage reduces below 0.88 pu or rises above 1.1 pu. On the other hand, the Australian standard on inverter-based grid connection of energy systems [72] gives a comparable recommendation for the PVs' disconnection when experiencing abnormal voltages in the network. These references also recommend the largest tripping period for PVs' disconnection subject to the level of the observed abnormal voltage, as summarized in Table 13.8. The levels defined in [71] are considered in the rest of this chapter.

PVs are required to have protection schemes to avoid islanding in the low-voltage feeders such as passive anti-islanding protection functions [96]. The under-/overvoltage and under-/overfrequency are the basic protection functions for the available PVs on the Australian market [73–75]. In addition, PVs usually have low-voltage ride-through (LVRT) capability [95, 97], which enables them to continue injecting power in the case the feeder voltage falls below the normal level. This is a very useful capability specifically when the network experiences temporary short-circuit faults. In [97], it has been reported that a small delay (e.g., 0.2 or 0.5 s) can be utilized to prevent PVs' needless disconnection under such conditions. In the case the fault was not temporary and is not cleared, the feeder's voltage drop will not improve either, thus the PV will disconnect from the feeder.

In general, the voltage will drop along the feeder in the faulty-phase after an LG fault in the low-voltage feeder (depending on the fault impedance), whereas it will rise in the healthy phases. Nowadays, the utilities are concerned that a large generation by PVs and network earthing in addition to the impedance and location of fault may prevent the voltage from dropping below 0.88 pu, as a result of which the PVs will not disconnect and will keep feeding the fault. In this condition, the feeder voltage of healthy phases may not increase beyond 1.1 pu; thus, their PVs may keep supplying the fault through the transformer. This may go on until the operation of an upstream switch fuse or overcurrent-protected circuit breaker. It is only at this stage that the voltages in both faulty and healthy phases will reduce considerably, leading to PVs' disconnection.

To understand the network situation during a single- or three-phase fault, this section discusses the node voltages within the feeder and the PVs' disconnection time considering

- PV generation–demand ratio
- Fault impedance
- Fault location along the feeder
- Network earthing system

13.6.1 Single-Phase Faults

In this section, the PVs' disconnection is discussed after an LG fault in the low-voltage feeder.

Assume the network of Figure 13.1. The upstream circuit breaker is assumed to have an extremely inverse time–current curve with a time multiplier setting of 0.02. The low-voltage feeder is assumed to have a PV generation–demand ratio of unity. An LG fault with an impedance of 2 Ω is considered on phase-*A* at the middle of the feeder at $t = 0$ (i.e., the network steady-state condition). The voltages of the faulty phase drop below 0.88 pu straightaway; hence, all PVs within phase-*A* disconnect concurrently at 4.8 ms. Likewise, the voltages of the healthy phases increase by the fault-occurrence. As an example, the voltages of node-1 in phase-*B* and nodes 1–4 in phase-*C* rise above 1.1 pu at 6 and 7.5 ms. As a result, the PVs that are connected to these nodes disconnect. The remainder of the PVs that are connected to phases *B* and *C* disconnect in a similar pattern by 13.7 ms. Hence, within a cycle after fault-occurrence, all PVs disconnect. LVRT was not considered in the above study; nevertheless, in the case the PVs have an enabled LVRT feature, they will disconnect in one cycle after the predefined delay time of LVRT. The upstream circuit

Table 13.8. Disconnection time of PV in the case of experiencing abnormal voltages

IEEE Recommendation (IEEE Standard 929-2000) [71]		Australian Standard (AS4777-2013) [72]	
Observed Abnormal Voltage	**Largest Tripping Period (Cycle)**	**Observed Abnormal Voltage**	**Largest Tripping Period (Cycle)**
$0.5 < V < 0.88$ or $1.1 < V < 1.37$ pu	120	$V < 0.78$ or $V > 1.13$ pu	100
$V < 0.5$ pu	6		
$V > 1.37$ pu	2		

breaker opens at 1.306 s. Figure 13.13 illustrates the schematic of the fault-occurrence, disconnection time of the PVs, and upstream circuit breaker for the considered study case. This network is now analyzed in detail considering different impedances and locations of fault and PV generation–demand ratios, as discussed below:

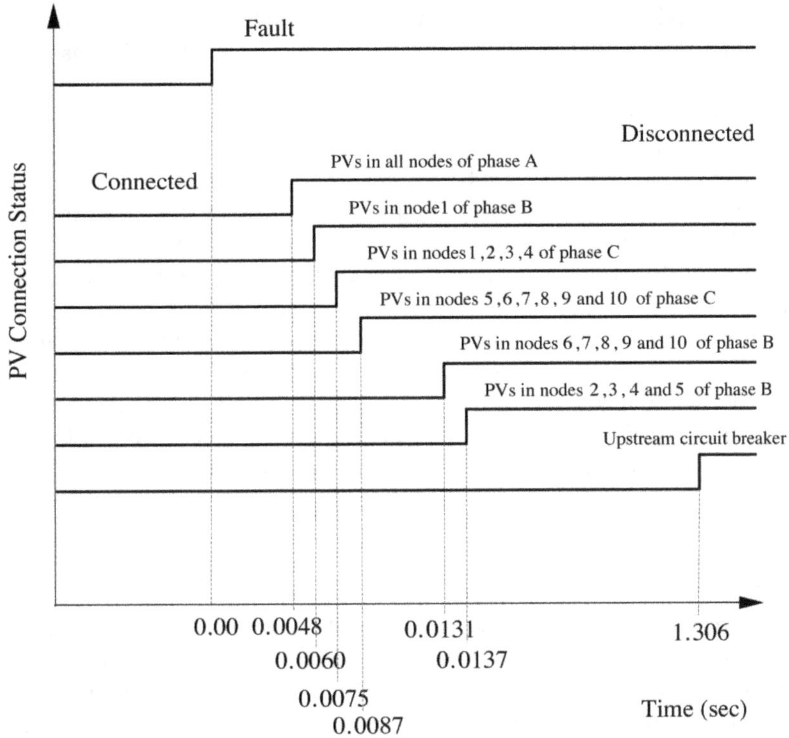

Figure 13.13. PVs' disconnection time and sequence along with the operation of the upstream circuit breaker following an LG fault.

13.6.1.1 Fault Impedance

Assume the network of Figure 13.1(b) with the PV generation–demand ratio of unity. An LG fault occurs on phase-*A* at the middle of the feeder. In this study, the impedance of fault varies from 2 m Ω to 20 Ω. Figure 13.14 illustrates the node voltages within the feeder between fault-occurrence and the first disconnection of PVs or the opening of the circuit breaker, for the NEG and MEN systems, separately. It can be seen from this figure that, all nodes' voltages within the faulty phase reduce below 0.88 pu for all fault impedance except 20 Ω, in both NEG and MEN systems. Hence, all PVs in the faulty phase disconnect after fault-occurrence. For high-impedance faults (such as the impedance of 20 Ω considered above), the PVs in both healthy and faulty phases do not see the fault and remain connected to the feeder until the opening of the upstream circuit breaker. This figure also shows that for fault impedance of 20 Ω in the MEN system and larger than 1 Ω in the NEG system, the voltage of the healthy phases *B* and *C* stays within the normal operation bandwidth of 0.88

and 1.1 pu. Thus, the PVs connected to the healthy phases do not see the fault and keep feeding the fault.

The voltages shown in Figure 13.14 are recorded between fault-occurrence and the earlier of upstream circuit breaker and the first PV's disconnection. Thus, this figure does not illustrate the voltage variations after the first PV's disconnection. Therefore, even if the voltages of some nodes are within the normal limits of 0.88 to 1.1 pu in Figure 13.14, they may observe abnormal voltages after the disconnection of the other PVs, based on which they will disconnect.

PVs' disconnection time for various fault impedances is presented in the radar charts of Figure 13.15 in which a circle's radius represents the time, whereas numbers 1 to 10 listed around the chart denote the node numbers in the low-voltage feeder. The considered low-voltage feeders are radial (as seen from Figure 13.1(b)); hence, the circular alignment of the nodes in the radar charts should not be misinterpreted as a low-voltage network with loop configuration. It can be seen from Figure 13.15 that, when the fault impedance is higher than 2 Ω, the PVs connected to both healthy and faulty phases do not disconnect for both NEG and MEN systems. It is also observed that for each fault impedance, all PVs within the same phase disconnect approximately simultaneously (i.e., within half a cycle difference). PVs' disconnection time increases as the fault impedance increases. It can also be seen that for the MEN system, PVs' disconnection time is nearly equal for both faulty and healthy phases, whereas this time is more in the low-voltage feeder with NEG-type earthing compared to the MEN-type.

13.6.1.2 Fault Location

Assume the network of Figure 13.1(b), where the PV generation–demand ratio is unity in which a short-circuit fault with an impedance of 2 mΩ occurs on phase-*A* at $t = 0$. The fault location varies from the beginning toward the end of the feeder. Figure 13.16 shows the node voltages within the feeder in this case. It can be seen from this figure that for all fault locations, the node voltages in the faulty phase are approximately equal and all are lower than 0.2 pu in the MEN and less than 0.4 pu in the NEG system. Hence, all PVs connected to the faulty phase disconnect irrespective of the fault location. After fault-occurrence, the node voltages within the healthy phases rise above 1.1 pu, and thus, their PVs also disconnect. However, the results are different for the NEG system. As seen in Figure 13.16, some nodes voltage in phase-*B* of the NEG system do not increase beyond 1.1 pu even though the fault impedance is very small. Hence, the PVs situated in the middle of this phase and its end do not disconnect. PVs' disconnection time is shown in Figure 13.17. It can be seen from this figure that all PVs situated in both healthy and faulty phases disconnect within few millisecond differences from each other and their disconnection time does not depend on the fault location, which is also true for both NEG and MEN systems.

13.6.1.3 PV Generation–Demand Ratio

Assume the network of Figure 13.1b in which a short-circuit fault with an impedance of 2 mΩ occurs on phase-*A* at the middle of the feeder at $t = 0$. The PV generation–demand ratio varies between 50% and 200%. Figure 13.18 shows the node voltage within the feeder in this case. It can be seen from this figure that for all PV generation–demand ratios, the voltages of all nodes in the faulty phase are almost equal and all are below, respectively,

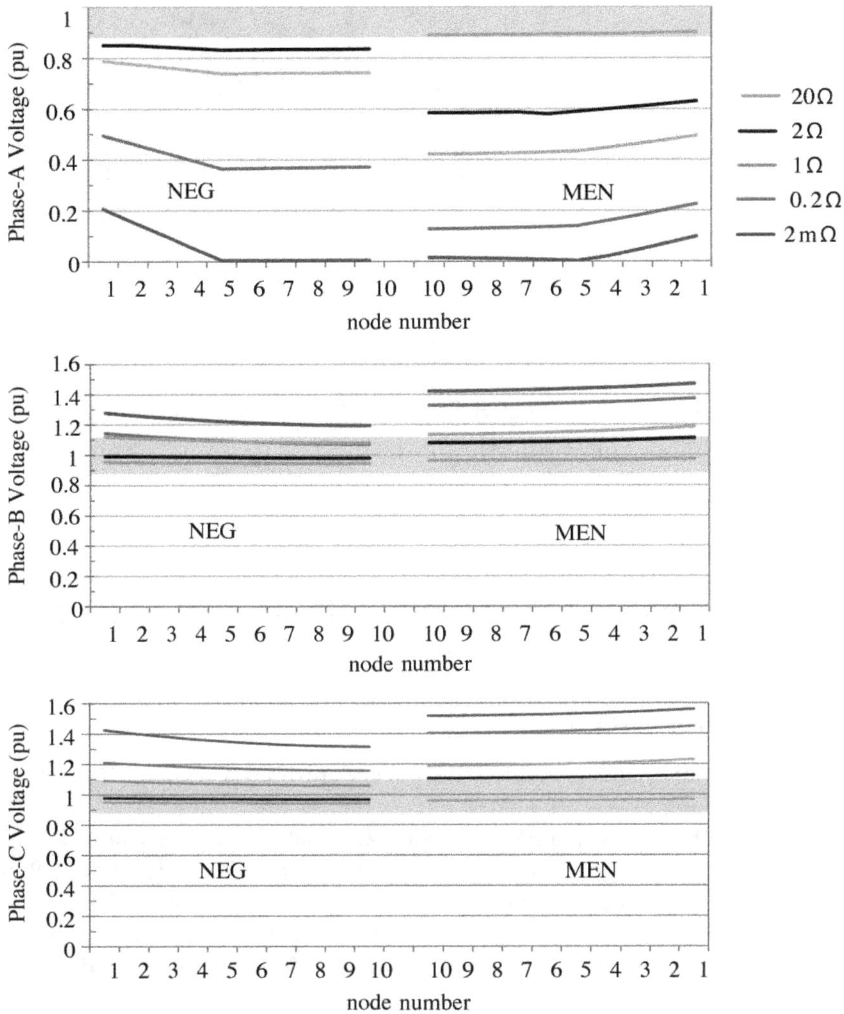

Figure 13.14. Node voltages within the feeder during an LG fault between fault-occurrence and the first tripping of a PV/upstream circuit breaker for various fault impedances.

0.1 and 0.25 pu in the NEG and MEN systems. Hence, irrespective of the PV generation–demand ratio, the PVs within the faulty phase disconnect. After fault-occurrence, the node voltages within the healthy phases increase beyond 1.1 pu, and therefore, all PVs in these phases also disconnect. PVs' disconnection time is shown in Figure 13.19 for this case. This figure shows that as the PV generation–demand ratio increases, the disconnection time of those PVs connected to the healthy phases reduces, which is true for both NEG and MEN systems. All PVs connected to the faulty phase disconnect within a cycle after fault-occurrence.

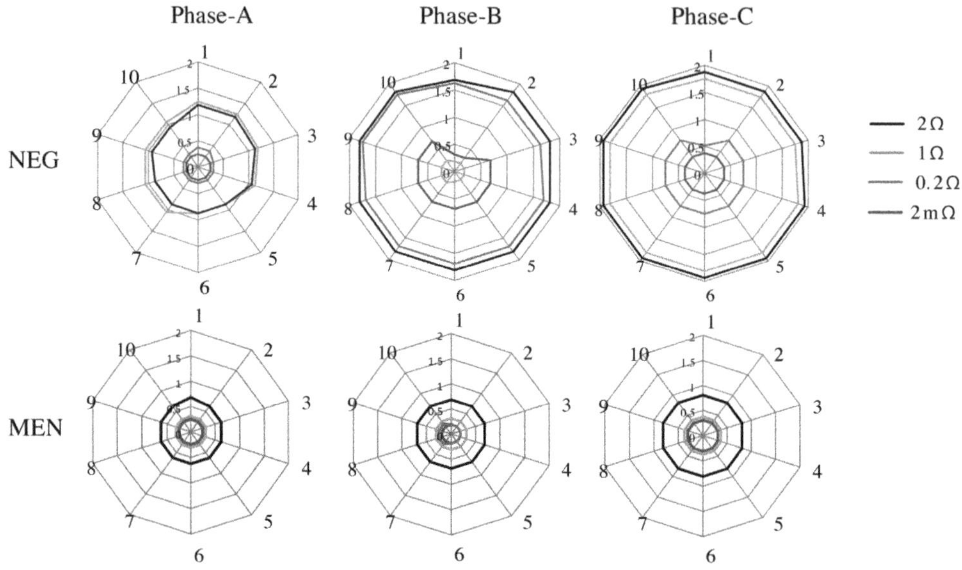

Figure 13.15. PVs' disconnection time after an LG fault for different fault impedances.

13.6.2 Three-Phase Faults

In this section, PVs' disconnection time and sequence are discussed after a three-phase fault in the low-voltage feeder.

Assume Figure 13.1(b) with a PV generation–demand ratio of unity in which a three-phase fault with an impedance of 2 mΩ occurs at the middle of the feeder at $t = 0$. Promptly after fault-occurrence, all node voltages of all phases reduce below 0.88 pu; hence, all PVs disconnect nearly concurrently within a cycle. This is followed by the opening of the upstream circuit breaker within two cycles. Figure 13.20 shows the PVs' disconnection time in this case.

13.6.2.1 Fault Impedance

Assume the network of Figure 13.1(b) with the PV generation–demand ratio of unity in which a three-phase fault occurs at the middle of the feeder at $t = 0$, where the impedance of fault varies from 2 to 20 Ω. Figure 13.21(a) shows the node voltages within the feeder between fault-occurrence and the earlier of the opening of the upstream circuit breaker and the first PV's disconnection. It is seen from this figure that all node voltages in all phases reduce below 0.88 pu for all fault impedance except 20 Ω. Thus, all PVs in the feeder disconnect after fault-occurrence except the very high-impedance condition, which is true for NEG and MEN systems. Figure 13.22(a) shows PVs' disconnection time for different fault impedances and illustrates that none of the PVs disconnect in the case the fault impedance is above 2 Ω. It is also seen that all PVs situated in one phase disconnect nearly simultaneously (i.e., within half a cycle), for each fault impedance. The PVs situated at the very beginning of the low-voltage feeder may not see the fault, and hence, may not

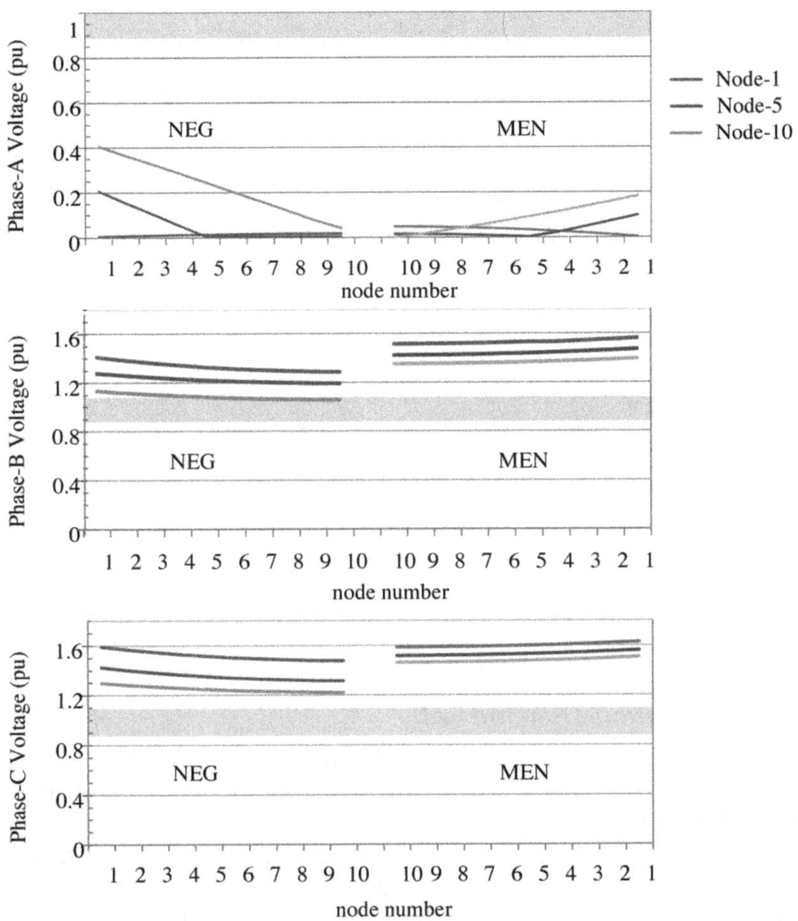

Figure 13.16. Node voltages within the feeder during an LG fault between fault-occurrence and the first tripping of a PV/upstream circuit breaker for various fault locations.

disconnect even for a fault impedance of 2 Ω since their coupling point voltage does not drop below 0.88 pu, as shown in Figure 13.21(a).

13.6.2.2 Fault Location

Assume the network of Figure 13.1(b) with the PV generation–demand ratio of unity in which a three-phase fault with an impedance of 2 mΩ occurs at $t = 0$, where the fault location varies from the beginning toward the end of the feeder. Figure 13.21(b) shows the node voltages within the feeder in this case and illustrates that all node voltages fall below 0.88 pu for all fault locations. Therefore, all PVs in the low-voltage feeders of NEG and MEN disconnect within half a cycle following the fault-occurrence, as shown in Figure 13.22(b).

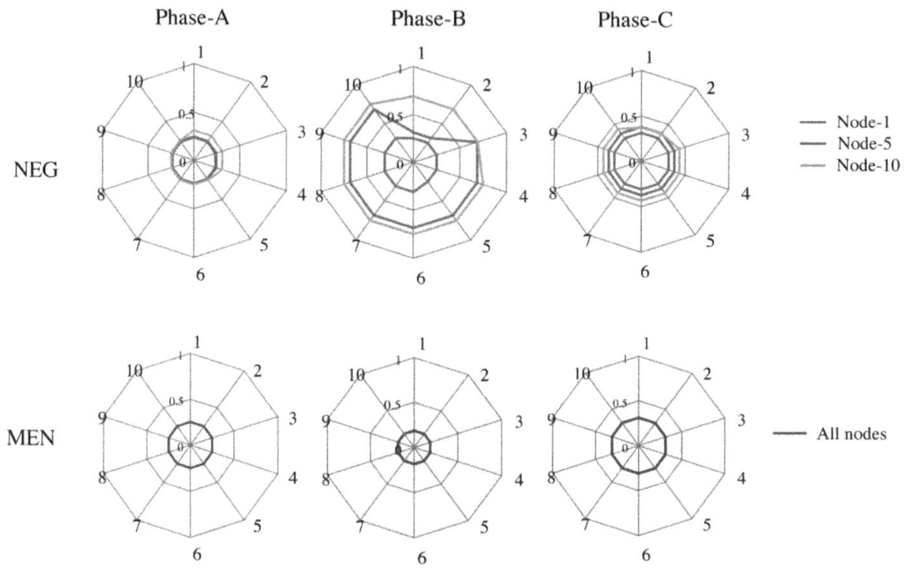

Figure 13.17. PVs' disconnection time after an LG fault for different fault locations.

13.6.2.3 PV Generation–Demand Ratio

Assume the network of Figure 13.1(b) in which a three-phase fault with an impedance of 2 mΩ occurs at the middle of the feeder at $t = 0$, where the PV generation–demand ratio is varied from 50% to 200% in steps of 50. Figure 13.21(c) shows the voltage profile along the feeder for this case. It can be seen from this figure that for all different considered PV generation–demand ratios, the voltage of all feeder nodes drop below 88%, and thus, all PVs disconnect within half a cycle following the fault-occurrence, as shown in Figure 13.22(c). Although in this case, the node voltages are marginally higher for a low-voltage feeder with NEG-type earthing versus the MEN-type, the PVs disconnection time is not affected.

13.6.3 Important Notes

The above study demonstrates that following an LG fault, all PVs connected to the faulty phase disconnect in an MEN system because their voltages drop below 0.88 pu immediately, except for large fault impedances (e.g., 20 Ω). However, they may not disconnect before the operation of the upstream circuit breaker for fault impedances of larger than 2 Ω in the case of high PV generation–demand ratio (e.g., 2) or an end of feeder fault.

The studies also show that the amount of voltage drop in the MEN system is higher than that of the NEG system. The only exception is when an LG fault occurs at the beginning of the feeder. The studies also show fault location and the PV generation–demand ratio does not impact PVs' disconnection in MEN systems substantially; nevertheless, they may lead to unsuccessful disconnection of PVs in NEG systems.

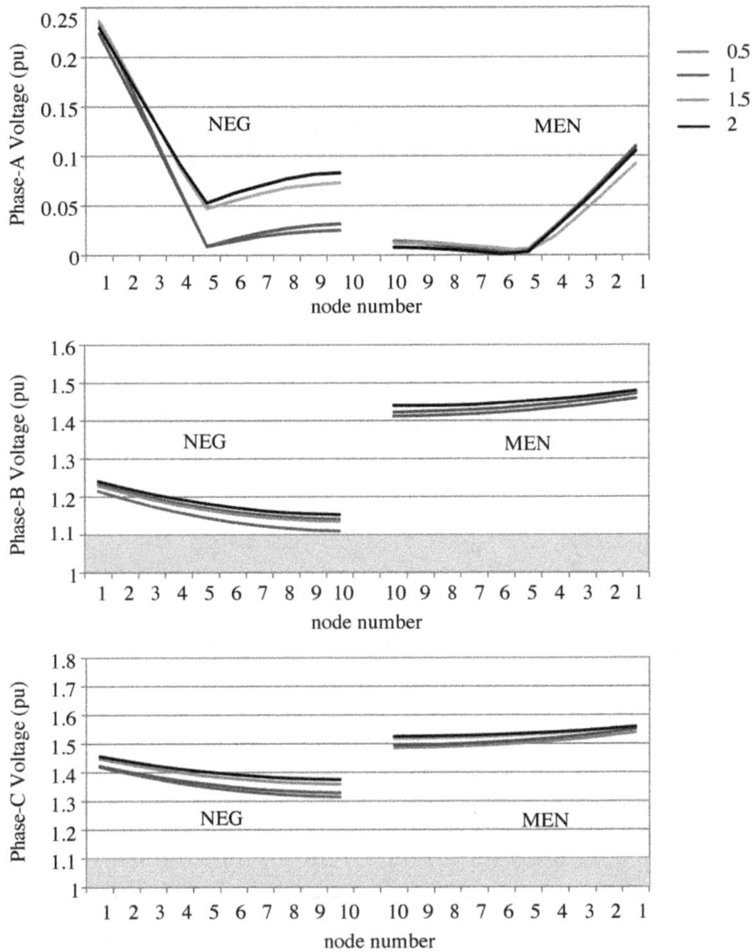

Figure 13.18. Node voltages within the feeder during an LG fault between fault-occurrence and the first tripping of a PV/upstream circuit breaker for various PV generation–demand ratios.

13.7 Summary

This chapter has evaluated the contribution of residential single-phase rooftop PVs on the short-circuit faults in the supplying low-voltage feeders. The typical Australian low-voltage three-phase, four-wire, MEN system, supplied by three-phase distribution transformers, were focused and the effect of PVs on the voltage profile along the feeder and the current sensed at the secondary of the distribution transformer is studied during various types of faults. Three parameters of short-circuit fault level, transformer current, and voltages of the nodes along the low-voltage feeder were studied. Through the sensitivity analyses, the chapter shows that the short-circuit fault level increases by 10% when a PV is present in the feeder, whereas the transformer secondary current may decrease by 7%. The results of the stochastic analysis show that the expected average current at transformer secondary and

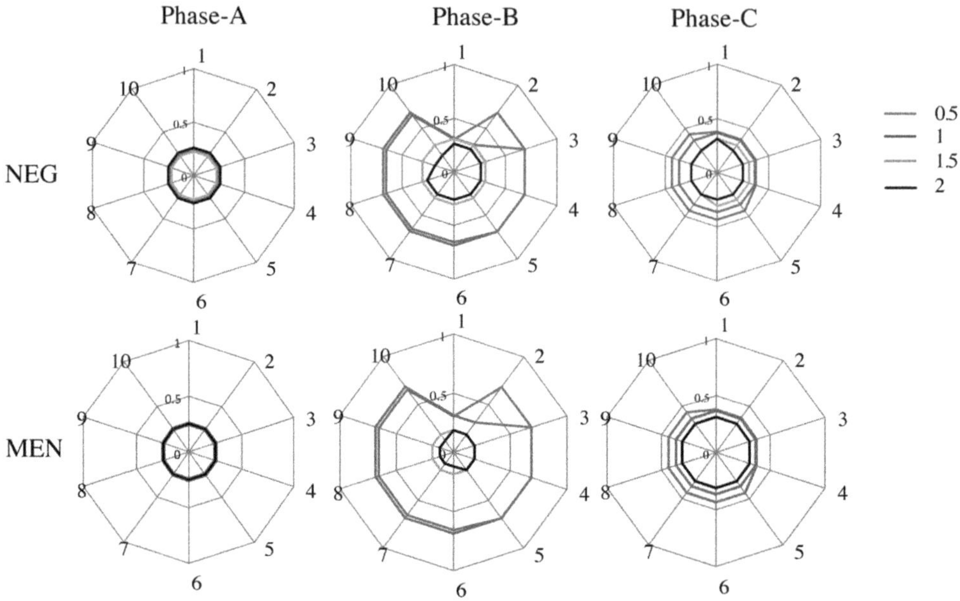

Figure 13.19. PVs' disconnection time after an LG fault for different PV generation–demand ratios.

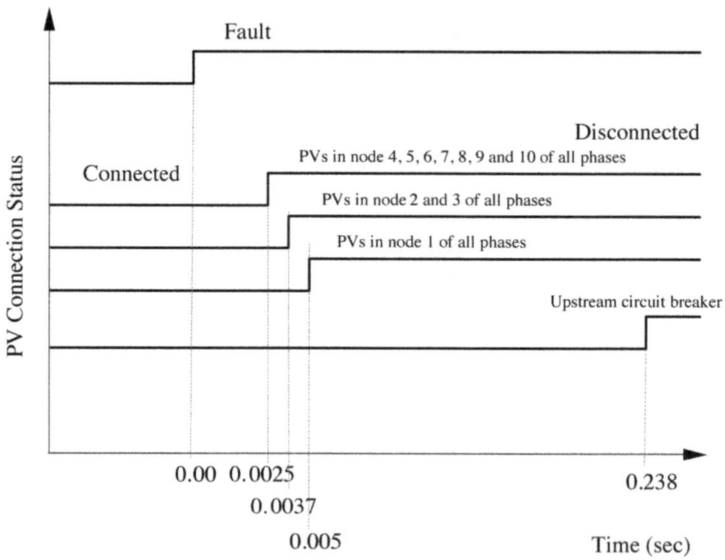

Figure 13.20. PVs' disconnection time and sequence along with the opening of the upstream circuit breaker following a three-phase fault.

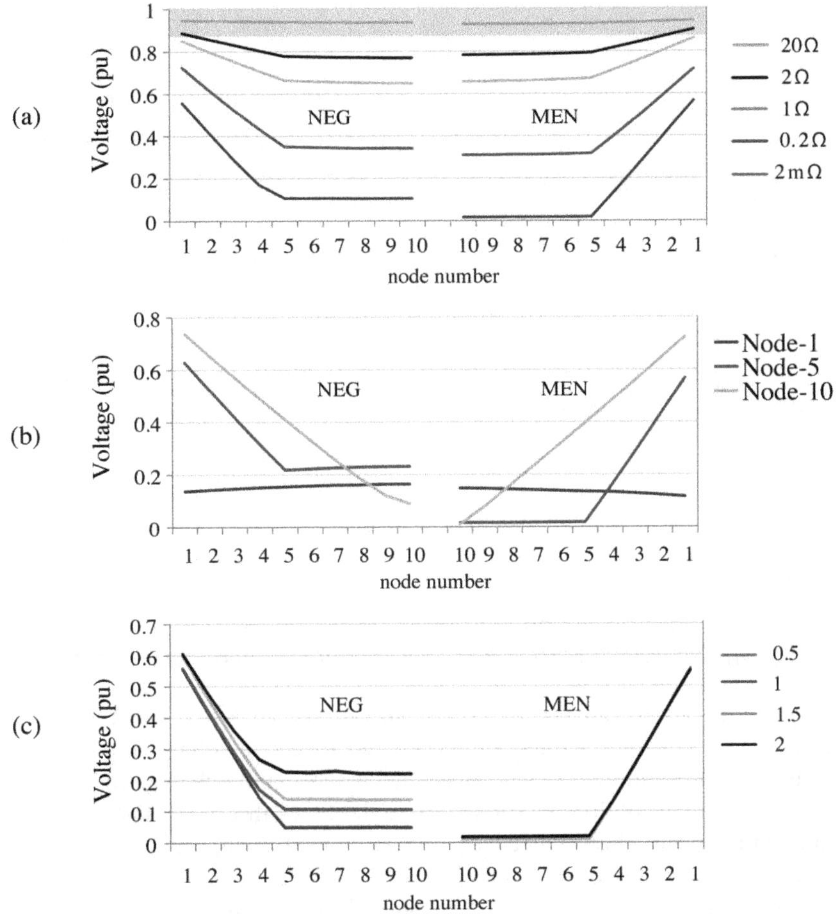

Figure 13.21. Node voltages within the feeder during a three-phase fault between fault-occurrence and the first tripping of a PV/upstream circuit breaker for various: (a) fault impedances, (b) fault locations, and (c) PV generation–demand ratios.

expected average voltage of the nodes along the feeder are not influenced strongly by the PV ratings and penetration levels, and it is seen that the fault location and impedance are the dominant factors to be considered.

The chapter also showed that with a probability of over 89%, the switch fuse will detect the faults on typical Australian low-voltage feeders. It also shows that with a minimum probability of 52%–57%, the disconnectors will isolate the PVs when the penetration level of the PVs is below 80%. However, at penetration levels of higher than 80%, the PV disconnection probability increases to 60%–63%.

This chapter also discussed the disconnection time of single-phase rooftop PVs following a single-phase or a three-phase short-circuit fault on the low-voltage feeders. The chapter reveals that for an MEN system, during a single-phase short-circuit fault, the PVs in

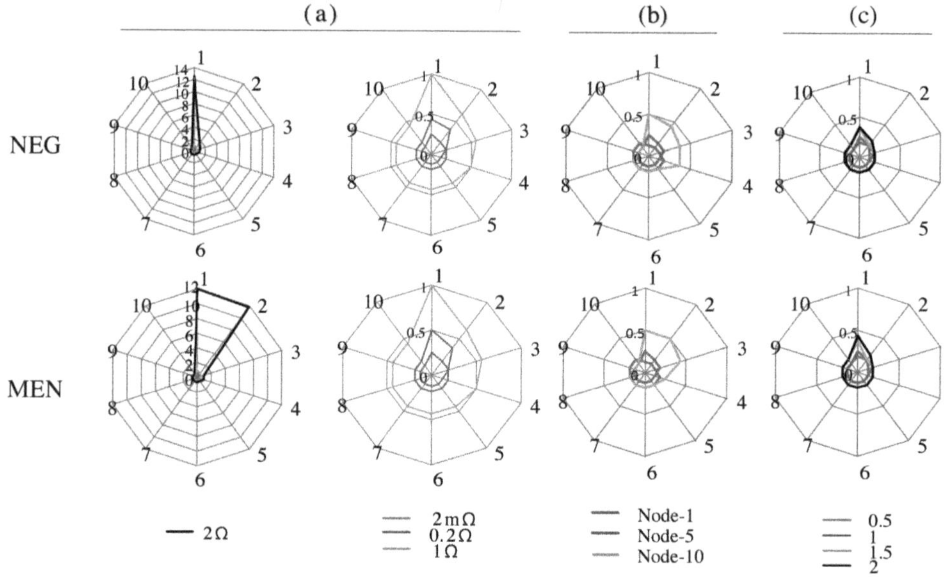

Figure 13.22. PVs' disconnection time after a three-phase fault for different: (a) fault impedances, (b) fault locations, and (c) PV generation–demand ratios.

the faulty phase sense the fault and disconnect immediately in less than a cycle after fault-occurrence or after the required delay time for LVRT since the voltage of their PCC drops below the limit of 88%. Similarly, the PVs located in the healthy phases sense the fault and disconnect as the voltage of their PCC rises above the 110% threshold. The exception is for faults with relatively large impedances (e.g. 20 Ω or more). Also, if the LV feeder is NEG, there are situations in which the PVs may not disconnect.

The chapter also discussed that in case of three-phase and three-phase-to-ground faults, depending on the system earthing type, the location and impedance of the fault and the ratio of the PVs' generation capacity versus the load demand, there are situations in which the PVs will not disconnect following the fault. Such PVs will remain connected until the upstream circuit breaker operates, after which they disconnect successfully. These situations are more probable in NEG systems or when the fault impedance is high in MEN systems. Therefore, only utilizing the under-/overvoltage protection function will not guarantee the successful disconnection of PVs in such situations.

13.8 References

1. A. F. Zobaa, and R. C. Bansal, *Handbook of Renewable Energy Technology*, World Scientific Publishers, Singapore, 2011.
2. R. C. Bansal, *Handbook of Distributed Generation: Electric Power Technologies, Economics and Environmental Impacts*, Springer, Switzerland, 2017.
3. "The Critical Decade: Australia's Future—Solar Energy," 2013. http://www.climatecouncil.org.au/uploads/497bcd1f058be45028e3df9d020ed561.pdf
4. Australian Energy Statistics. "Energy Update." 2011. http://data.daff.gov.au/data/warehouse/pe_abares99010610/EnergyUpdate_2011_REPORT.pdf

5. L. Romanach, Z. Contreras, and P. Ashworth, "Australian householders' interest in the distributed energy market—National survey results," CSIRO report prepared for the Australian Photovoltaic Association, 2013. http://www.csiro.au/Organisation-Structure/Flagships/Energy-Flagship/Australian-householders-interest-in-the-distributed-energy-market.aspx
6. M. A. Hayat, F. Shahnia, and A. Arefi, "Comparison of the electricity tariffs and bills across the zones of Australian power distribution companies," 26th Australasian Universities Power Engineering Conference (AUPEC), pp. 1–6, Brisbane, 2016.
7. M. A. Hayat, F. Shahnia, and A. Arefi, "Efficient seasonal time of use feed-in tariff for residential rooftop solar panels in Australian electricity market," in *Transition Towards 100% Renewable Energy*, A. Sayigh (ed.), pp. 171–175, Springer, 2018.
8. M. Hayat, F. Shahnia, A. Arefi, et al., "Comparison of the economic benefits and the payback periods of rooftop solar panels in Australia," 3rd International Conference on Power Generation Systems and Renewable, pp. 1–5, Johor Bahru, 2017.
9. M. A. Hayat, F. Shahnia, and G. Shafiullah, "A novel dynamic feed-in tariff for residential rooftop PVs," 45th IEEE Photovoltaic Specialists Conference (PVSC) and 7th IEEE World Conference on Photovoltaic Energy Conversion (WCPEC), pp. 1–6, Hawaii, 2018.
10. M. A. Hayat, F. Shahnia, and G. Shafiullah, "Replacing flat rate feed-in tariffs for rooftop photovoltaic systems with a dynamic one to consider technical, environmental, social and geographical factors," *IEEE Trans. Industrial Informatics*, vol. 15, pp. 3831–3844, 2018. doi:10.1109/TII.2018.2887281
11. M. A. Hayat, F. Shahnia, and G. Shafiullah, "Economic viability of roof leasing for rooftop photovoltaic systems from a leasing company's perspective," 45th IEEE Photovoltaic Specialists Conference (PVSC) and 7th IEEE World Conference on Photovoltaic Energy Conversion (WCPEC), pp. 1–6, Hawaii, 2018.
12. K. N. Amarawardhana, S. D. G. Jayasinghe, and F. Shahnia, "Grid-interactive rooftop photovoltaic clusters with third-party ownership," International Conference on Smart Power and Internet Energy Systems (SPIES), Melbourne, 2019.
13. M. A. Hayat, F. Shahnia, and G. Shafiullah, "Technical analysis of the charge/discharge rate and the capacity of batteries in meeting the residential peak demand within a community solar," 9th IEEE PES Innovative Smart Grid Technologies Asian Conference (ISGT-Asia), pp. 1–6, Chengdu, 2019.
14. J. Susanto, F. Shahnia, and D. Ludwig, "A framework to technically evaluate integration of utility-scale photovoltaic plants to weak power distribution systems," *Applied Energy*, vol. 231, pp. 207–221, 2018.
15. R. Shah, N. Mithulananthan, R. C. Bansal, and V. K. Ramachandaramurthy, "A review of key power system stability challenges for large-scale PV integration," *Renewable and Sustainable Energy Reviews*, vol. 41, pp. 1423–1436, 2015.
16. H. A. H. Al-Hilfi, F. Shahnia, and A. Abu-Siada, "Comparison of different correlational techniques in estimating the total generated power of neighboring photovoltaic systems," 26th Australasian Universities Power Engineering Conference (AUPEC), pp. 1–6, Brisbane, 2016.
17. H. A. H. Al-Hilfi, F. Shahnia, A. Abu-Siada, "Enhancing the estimation of the overall produced power by several adjacent photovoltaic systems using existing correlational factors," IEEE 27th International Symposium on Industrial Electronics (ISIE), pp. 43–47, Cairns, 2018.
18. F. Shahnia, A. Arefi, and G. Ledwich, *Electric Distribution Network Planning*, Springer, 2018.
19. A. Arefi, F. Shahnia, and G. Ledwich, *Electric Distribution Network Management and Control*, Springer, 2018.
20. M. J. E. Alam, K. M. Muttaqi, and D. Sutanto, "An approach for online assessment of rooftop solar PV impacts on low-voltage distribution networks," *IEEE Trans. Sustainable Energy*, vol. 5, no. 2, pp. 663–672, 2014.
21. G. Mokhtari, G. Nourbakhsh, F. Zare, and A. Ghosh, "Improving the penetration level of PVs using dc link for residential buildings," *Energy and Buildings*, vol. 72, pp. 80–86, 2014.

22. F. Shahnia, R. Majumder, A. Ghosh, et al., "Voltage imbalance analysis in residential low voltage distribution networks with rooftop PVs," *Electric Power Systems Research*, vol. 81, no. 9, pp. 1805–1814, 2011.
23. F. Shahnia, A. Ghosh, G. Ledwich, and F. Zare, "Voltage unbalance improvement in low voltage residential feeders with rooftop PVs using custom power devices," *International Journal of Electrical Power & Energy Systems*, vol. 55, pp. 362–377, 2014.
24. G. Mokhtari, G. Nourbakhsh, F. Zare, and A. Ghosh, "Overvoltage prevention in LV smart grid using customer resources coordination," *Energy and Buildings*, vol. 61, pp. 387–395, 2013.
25. F. Shahnia, A. Ghosh, G. Ledwich, and F. Zare, "An approach for current balancing in distribution networks with rooftop PVs," IEEE Power and Energy Society General Meeting, pp. 1–6, San Diego, 2012.
26. S. Mazumder, A. Ghosh, F. Shahnia, et al., "Excess power circulation in distribution networks containing distributed energy resources," IEEE Power and Energy Society General Meeting, pp. 1–8, San Diego, 2012.
27. F. Shahnia, S. M. Ami, and A. Ghosh, "Circulating the reverse flowing surplus power generated by single-phase DERs among the three phases of the distribution lines," *International Journal of Electrical Power & Energy Systems*, vol. 76, pp. 90–106, 2016.
28. F. Shahnia, A. Ghosh, G. Ledwich, and F. Zare, "Voltage unbalance reduction in low voltage distribution networks with rooftop PVs," 20th Australasian Universities Power Engineering Conference (AUPEC), pp. 1–5, Christchurch, 2010.
29. F. Shahnia, and A. Ghosh, "High penetration of rooftop photovoltaic cells in low voltage distribution networks: Voltage imbalance and improvement," in *Renewable Energy Integration, Challenges and Solutions*, J. Hossain and A. Mahmud (eds.), pp. 69–95, Springer, 2014.
30. N. Jabalameli, M. Masoum, T. H. Mehr, and F. Shahnia, "Impact of battery rating on performance of rooftop PV supporting household loads, regulating PCC voltage and providing constant output power to grid," 23rd Australasian Universities Power Engineering Conference (AUPEC), pp. 1–6, 2013.
31. N. Jabalameli, n.d.eilami, M. A. S. Masoum, and F. Shahnia, "Rooftop PV with battery storage for constant output power production considering load characteristics," 8th International Conference on Electrical and Electronics Engineering (ELECO), Bursa, 2013.
32. F. Shahnia, R. Majumder, A. Ghosh, et al., "Sensitivity analysis of voltage imbalance in distribution networks with rooftop PVs," IEEE Power and Energy Society General Meeting, pp. 1–8, Minneapolis, 2010.
33. X. Su, J. Liu, Y. Fu, et al., "An efficient joint coordination of volt-var support in medium and low voltage distribution feeders," 20th IEEE International Conference on Industrial Technology (ICIT), pp. 1–6, Melbourne, 2019.
34. X. Su, J. Liu, Y. Fu, et al., "A chronological sensitivity-based approach for BESSs placement in unbalanced distribution feeders," IEEE Power and Energy Society General Meeting, pp. 1–6, Atlanta, 2019.
35. F. Shahnia, P. Wolfs, and A. Ghosh, "Smart inter-phase switching of residential loads in low voltage distribution feeders," in *Computational Intelligence for Decision Support in Cyber-Physical Systems*, Z. H. Khan, S. Ali, and Z. Riaz (eds.), pp. 329–354, Springer, 2014.
36. F. Shahnia, P. J. Wolfs, and A. Ghosh, "Voltage unbalance reduction in low voltage feeders by dynamic switching of residential customers among three phases," *IEEE Transactions on Smart Grid*, vol. 5, no. 3, pp. 1318–1327, 2014.
37. F. Shahnia, P. J. Wolfs, and A. Ghosh, "Voltage unbalance reduction in low voltage feeders by dynamic switching of residential customers among three phases," IEEE Power and Energy Society General Meeting, pp. 1–5, Vancouver, 2013.
38. N. Safitri, F. Shahnia, and M. A. S. Masoum, "Coordination of single-phase rooftop PVs in unbalanced three-phase residential feeders for voltage profiles improvement," *Australian Journal of Electrical and Electronics Engineering*, vol. 13, no. 2, pp. 77–90, 2016.

39. N. Safitri, F. Shahnia, and M. A. S. Masoum, "Different techniques for simultaneouly increasing the penetration level of rooftop PVs in residential LV networks and improving voltage profile," IEEE PES Asia-Pacific Power and Energy Engineering Conference (APPEEC), pp. 1–5, Hong Kong, 2014.
40. N. Safitri, F. Shahnia, and M. A. S. Masoum, "Coordination of single-phase rooftop PVs to regulate voltage profiles of unbalanced residential feeders," 24th Australasian Universities Power Engineering Conference (AUPEC), pp. 1–5, Perth, 2014.
41. N. Safitri, F. Shahnia, and M. A. S. Masoum, "Monte Carlo-based stochastic analysis results for coordination of single-phase rooftop PVs in unbalanced residential feeders," *Intelligent Industrial Systems*, vol. 1, no. 4, pp. 359–371, 2015.
42. F. Shahnia, and A. Ghosh, "Decentralized voltage support in a Low Voltage feeder by droop based voltage controlled PVs," 23rd Australasian Universities Power Engineering Conference (AUPEC), pp. 1–6, Hobart, 2013.
43. F. Shahnia, M. T. Wishart, and A. Ghosh, "On-line demand management of low voltage residential distribution networks in smart grids," in *Computational Intelligence for Decision Support in Cyber-Physical Systems*, Z. H. Khan, S. Ali, and Z. Riaz (eds.), pp. 293–328, Springer, 2014.
44. F. Shahnia, M. T. Wishart, A. Ghosh, et al., "Smart demand side management of low-voltage distribution networks using multi-objective decision making," *IET Generation, Transmission & Distribution*, vol. 6, no. 10, pp. 986–1000, 2012.
45. M. T. Wishart, F. Shahnia, A. Ghosh, and G. Ledwich, "Multi objective decision making method for demand side management of LV residential distribution networks with plug-in electric vehicles," IEEE Power and Energy Society General Meeting, pp. 1–8, 2011.
46. F. Shahnia, M. T. Wishart, and A. Ghosh, "Voltage regulation, power balancing and battery storage discharge control by smart demand side management and multi-objective decision making," 23rd Australasian Universities Power Engineering Conference (AUPEC), pp. 1–5, Hobart, 2013.
47. B. Behi, A. Arefi, H. Pezeshki, and F. Shahnia, "Distribution transformer lifetime analysis in the presence of demand response and rooftop PV integration," *Renewable Energy and Environmental Sustainability*, vol. 2, pp. 1–6, 2017.
48. A. Ghosh, and F. Shahnia, "Applications of power electronic devices in distribution systems" in *Transient Analysis of Power Systems: Solution Techniques, Tools and Applications*, J. A. Martinez-Velasco (ed.), pp. 248–279, Wiley-IEEE, USA, 2014.
49. H. Hatami, F. Shahnia, A. Pashaei, and S. H. Hosseini, "Investigation on D-STATCOM and DVR operation for voltage control in distribution networks with a new control strategy," IEEE Power Tech, pp. 2207–2212, 2007.
50. F. Shahnia, A. Ghosh, G. Ledwich, and F. Zare, "Voltage correction in low voltage distribution networks with rooftop PVs using custom power devices," 37th Annual Conference of the IEEE Industrial Electronics Society (IECON), Melbourne, 2011.
51. F. Shahnia, R. P. S. Chandrasena, A. Ghosh, and S. Rajakaruna, "Application of DSTATCOM for surplus power circulation in MV and LV distribution networks with single-phase distributed energy resources," *Electric Power Systems Research*, vol. 117, pp. 104–114, 2014.
52. F. Shahnia, and A. Ghosh, "Coupling of neighbouring low voltage residential distribution feeders for voltage profile improvement using power electronics converters," *IET Generation, Transmission & Distribution*, vol. 10, no. 2, pp. 535–547, 2016.
53. F. Shahnia, "Application and control of a DSTATCOM to couple neighboring low voltage feeders," 18th IEEE International Conference on Industrial Technology (ICIT), pp. 476–481, Taipei, 2016.
54. F. Shahnia, "Protection considerations of using a DSTATCOM when coupling neighboring low voltage feeders," 18th IEEE International Conference on Industrial Technology (ICIT), pp. 529–534, Taipei, 2016.
55. F. Shahnia, S. Rajakaruna, and A. Ghosh, *Static Compensators (STATCOMs) in Power Systems*, Springer, 2014.

56. M. E. Baran, H. Hooshyar, H. Shen, et al., "Impact of high penetration residential PV systems on distribution systems," IEEE Power and Energy Society General Meeting, Detroit, 2011.
57. H. Hooshyar, M. Baran, and L. Vanfretti, "Coordination assessment of overcurrent relays in distribution feeders with high penetration of PV systems," IEEE PowerTech. Conference, pp. 1–6, 2013.
58. V. Calderaro, S. Corsi, V. Galdi, and A. Piccolo, "Optimal setting of the protection systems in distribution networks in presence of distributed generation," 40th International Universities Power Engineering Conference (UPEC), Cork, 2005.
59. H. H. Yengejeh, F. Shahnia, and S. Islam, "Contributions of single-phase rooftop PVs on short circuits faults in residential feeders," 24th Australasian Universities Power Engineering Conference (AUPEC), Australia, 2014.
60. A. Bruce, S. Heslop, I. MacGill, and M. Watt, " Magnetic Island and Townsville solar city: A case study of increasing PV penetration in electricity networks," Technical report, Australian PV Association, 2013.
61. H. H. Yengejeh, F. Shahnia, and S. Islam, "Impact of distributed rooftop photovoltaic systems on short-circuit faults in the supplying low voltage networks," *Electric Power Component & Systems*, vol. 45, no. 20, pp. 2257–2274, 2017.
62. H. Hooshyar, and M. E. Baran, "Fault analysis on distribution feeders with high penetration of PV systems," *IEEE Trans. Power Systems*, vol. 28, no. 3, pp. 2890–2896, 2013.
63. T. G. Alvin, I. Z. Abidin, H. Hashim, and A. Z. Abidin, "Phase comparison protection for distribution networks with high PV penetration," IEEE PES Innovative Smart Grid Technologies-Asia Conference (ISGT Asia), pp. 216–221, 2014.
64. B. Noone, "PV integration on Australian distribution networks: Literature review," The Australian PV Association, 2013.
65. Energex/Ergon Energy connection standard on Small Scale Parallel Inverter Energy Systems up to 30 kVA, 2014.
66. D. Dong, Z. Qi, and Y. H. Yang, "Non-effectively grounded system line selection device based on ARM," China International Conference on Electricity Distribution (CICED), pp. 1–6, 2008.
67. WesternPower, "Western Australian Distribution Connections Manual." 2015. http://www.westernpower.com.au/documents/WA_Distribution_Connections_Manual.pdf
68. A. Hoke, R. Butler, J. Hambrick, and B. Kroposki, "Steady-state analysis of maximum photovoltaic penetration levels on typical distribution feeders," *IEEE Transaction Sustainable Energy*, vol. 4, no. 2, pp. 350–357, 2013.
69. T. A. Short, *Electric Power Distribution Handbook*, CRC Press, 2004.
70. WesternPower, "Distribution Construction Standards Handbook-Part 6." 2007. http://www.westernpower.com.au/documents/low-voltage-overhead-distribution-construction-standards-han.PDF
71. IEEE Recommended Practice for Utility Interface of Photovoltaic (PV) Systems, IEEE Std 929-2000.
72. Australian Standard on Grid Connection of Energy Systems via Inverters, AS 4777, 2013.
73. "Photovoltaic System Protection Application Guide." 2012. http://www1.cooperbussmann.com/pdf/1b416a65-f5ac-4730-ab77-9e2faa147945.pdf
74. "Clenergy Product Catalogue." http://www.supplypartners.com.au/media/pdf/Clenergy%20Product%20Catalogue.pdf
75. "ABB Solar Inverters Catalogue." http://new.abb.com/docs/librariesprovider22/technical-documentation/solar-inverter-pvs300-flyer.pdf?sfvrsn=2
76. WesternPower. "Residential Design After Diversity Maximum Demand Calculation." http://www.westernpower.com.au/land-developers-designers-residential-dadmd-calculator.html
77. "Residential Solar PV System Prices, Website." http://www.solarchoice.net.au/blog/residential-solar-pv-system-prices-may-2015

78. A. Ashourpouri, A. Ghosh, and F. Shahnia, "Droop-based demand dispatch schedule in smart grid," 25th Australasian Universities Power Engineering Conference (AUPEC), pp. 1–6, Wollongong, 2015.
79. X. Su, J. Liu, Y. Fu, et al., "An improved direct load flow approach for unbalanced active distribution networks with multiple voltage levels and configurations," 20th IEEE International Conference on Industrial Technology (ICIT), pp. 1–6, Melbourne, 2019.
80. "Olex Aerial Conductors Catalogue." 2012. http://www.olex.com.au/australasia/2012/olc12641_aerialcat.pdf
81. J. G. Grainger, and W. D. Stevenson Jr., *Power System Analysis*, McGraw-Hill, 1994.
82. H. H. Yengejeh, F. Shahnia, S. Islam, "Disconnection of single-phase rooftop PVs after short-circuit faults in residential feeders," *Australian Journal of Electrical & Electronic Engineering*, vol. 13, pp. 1–15, 2016.
83. M. Dewadasa, *Protection of Distributed Generation Interfaced Networks*, PhD Thesis, Queensland University, Brisbane, 2010.
84. E. Pashajavid, and F. Shahnia, "Sensitivity of electric vehicles demand profile to the batteries departure state-of-charge," 24th Australasian Universities Power Engineering Conference (AUPEC), pp. 1–6, Perth, 2014.
85. F. Shahnia, A. Ghosh, G. Ledwich, and F. Zare, "Voltage unbalance sensitivity analysis of plug-in electric vehicles in distribution networks," 21st Australasian Universities Power Engineering Conference (AUPEC), pp. 1–6, Brisbane, 2011.
86. M. Batool, S. M. Islam, and F. Shahnia, "Stochastic modeling of the output power of photovoltaic generators in various weather conditions," 26th Australasian Universities Power Engineering Conference (AUPEC), pp. 1–5, 2016.
87. N. Safitri, F. Shahnia, and M. A. S. Masoum, "Stochastic analysis results for coordination of single-phase rooftop PVs in unbalanced residential feeders," 7th IEEE PES Asia-Pacific Power and Energy Engineering Conference (APPEEC), pp. 1–6, Brisbane, 2015.
88. F. Shahnia, A. Ghosh, G. Ledwich, and F. Zare, "Predicting voltage unbalance impacts of plug-in electric vehicles penetration in residential low voltage distribution networks," *Electric Power Components & Systems*, vol. 41, no. 16, pp. 1594–1616, 2013.
89. C. Robert, and G. Casella, *Monte Carlo Statistical Methods*, Springer, 2004.
90. "Perth Solar City Annual Report." 2011. http://www.westernpower.com.au/documents/psc_-_2012_annual_report_-_final_for_distribution_(lo-res_18.pdf
91. V. Garcia, J. C. Cebrian, and N. Kagan, "Evaluation of probability functions related to short circuit random variables using power quality meters," IEEE/PES Transmission & Distribution Conference Latin America, Sao Paulo, 2010.
92. Bureau of Metrology. "Average Annual and Monthly Sunshine Duration." http://www.bom.gov.au/jsp/ncc/climate_averages/sunshine-hours/index.jsp
93. H. H. Yengejeh, F. Shahnia, and S. Islam, "Disconnection time and sequence of rooftop PVs under short-circuit faults in low voltage networks," North American Power Symposium (NAPS), pp. 1–6, Charlotte, 2015.
94. "Clipsal Solar Product Overview Catalogue." 2013. http://updates.clipsal.com/ClipsalOnline/Files/Brochures/I0000114.pdf
95. N. Hoang, and A. Yokoyamaa, "Influence of penetration levels and fault ride through characteristics of photovoltaic generation on voltage stability," *Journal of International Council on Electrical Engineering*, vol. 3, pp. 283–289, 2014.
96. W. Bower, and M. Ropp, "Evaluation of islanding detection methods for utility-interactive inverters in photovoltaic systems," Technical report, Sandia National Laboratories, 2002.
97. C. Y. Tang, Y. T. Chen, and Y. M. Chen, "PV power system with multi-mode operation and low-voltage ride-through capability," *IEEE Transaction Industrial Electronics*, vol. 62, no. 12, pp. 7524–7533, 2015.

Section III: Hybrid Powers/Micro Grid/Distributed Generation

CHAPTER 14

Design and Evaluation of Microgrids Based on Renewable Energy Technologies with a Perspective of Sustainable Development

Abhishek Kumar[1], Arvind R. Singh[2], Nand K. Meena[3], Yan Deng[1], Xiangning He[1], Praveen Kumar[4], and Ramesh C. Bansal[5]

[1]College of Electrical Engineering, Zhejiang University, Hangzhou, P. R. China. abhi@zju.edu.cn; dengyan@zju.edu.cn; hxn@zju.edu.cn
[2]School of Electrical Engineering, Shandong University, Jinan, P. R. China. arvindsinghwce@sdu.edu.cn
[3]School of Engineering and Applied Science, Aston University, Birmingham, United Kingdom. n.meena@aston.ac.uk
[4]Department of Electronics and Electrical Engineering, Indian Institute of Technology Guwahati, Guwahati, India. praveen_kumar@iitg.ac.in
[5]Department of Electrical Engineering, University of Sharjah, Sharjah, United Arab Emirates, and Department of Electrical, Electronic and Computer Engineering, University of Pretoria, Pretoria, South Africa. rcbansal@ieee.org

Abstract Global warming, energy poverty issues along with prolonged dependency on conventional sources of energy such as fossil fuels to meet the energy demand have inspired the majority of nations to use nonconventional clean sources of energy. Moreover, the affordability to energy access (cooking/heating and electricity) even for the household for the fast-developing nations such as Brazil, India, and China have many significant challenges. Majority of the populace suffering from energy poverty issue is confined mainly in the rural and remote areas of developing nations such as India and sub-Saharan Africa. Nevertheless, even with technological advancement, providing energy access to such areas is far from reality. Also, to achieve the inclusion of renewable energy technologies in the present power system, thoughtful consideration should be given covering not only technical and economic aspects but also social, environmental, and institutional issues. Microgrid design utilizing the locally available nonconventional energy sources, such as solar, wind, and small hydro, could be a key to providing economical and reliable energy access to such areas with a sustainability perspective.

Keywords: Renewable energy; Microgrid; Rural electrification; Energy access; Sustainable development.

The design process has become onerous and convoluted due to the involvement of several key metrics, eventualities and shareholders. A correct approach is therefore required which can consider simultaneously the evaluation metrics and differences of view of the various actors involved in multiple scenarios. The chapter first establishes a general relationship between energy access with human development and linkage with other sustainable development goals (SDGs) along with its impact on them. Also, it illustrates clearly the current energy scenario with specific discussion toward the role of microgrid-based renewables and hurdles in their deployment to overcome the energy access problem. Finally, the chapter iterates a generalized comprehensive trilevel framework combining the synergies of system modeling, multiobjective optimization, and multicriteria decision analysis (MCDA) models to design microgrids based on renewables with a sustainability perspective. The trilevel framework is designed to accommodate not only technical and economic factors but also social, environmental, and institutional/political traits.

14.1 Introduction

Energy is one of the crucial factors that affects the economic, technical, and social development of any nation and its population [1,2]. For any community to march on the path of development and thrive, access to energy is essential, which has profound implications in the various dimensions (social, economic, and environmental) of human development [1–4]. Better energy access contributes to the social well-being of humans by meeting their primary energy needs for cooking food, education, and health [1–3]. Modern energy services also contribute toward economic growth, which has been shifted from agro-based economy to an industry-based economic system requiring ample amount of energy [1,2]. These recent infrastructural developments in the economic system have entirely changed the energy consumption patterns and levels, which have resulted in a significant rise in energy demand [1,2,5–7]. Efficient energy solutions can also provide a better environmental condition by reducing pollution and deforestation to a great extent [1,2,7]. Reliable access to modern energy services (clean cooking and electricity facilities) can be proved to be a critical enabler in the development at disintegrated levels, that is, at household and community levels [1,2,6,8].

14.1.1 Energy and Its Relevance to Human Development Index

International Energy Agency (IEA) defines energy access for a household as, "the availability of reliable and economical services both for clean cooking (access to modern fuels such as LPG, biogas, or efficient biomass cook stoves) and electricity (lighting, phone charging, and radio) to meet their present basic demands and then an increase in the electricity levels in time to reach respective regional average" [1,2].

Lack of modern energy services results not only in the slow economic progress of a nation but also to the low Human Development Index (HDI) values. According to Human Development report [8], "HDI is the measure of crucial dimensions of human development which includes a longer and healthier lifespan, being well-informed & educated and should have a decent living standard." Economy, energy consumption patterns, and HDI are very closely linked to each other [2,7]. The higher HDI-ranked countries also have higher energy consumption (primary energy and electricity uses per capita) patterns, which are very closely related to the higher gross domestic product (GDP) per capita [7]. World economies with high GDP/capita (developed economies), such as United States, United Kingdom, and Germany, have higher electricity consumption patterns as well as higher HDI values as illustrated in Fig. 14.1(a, b). Developed countries have a better infrastructure providing reliable and economical modern energy services to its people, thus leading to a higher HDI compared to the nations, which lack these services [4,7,8].

As nations progress, their energy consumption levels also increases over a period as illustrated in Fig. 14.1(a). An increase in the electricity consumption levels can be seen in developing economies such as China and India. However, a spectacular increase can be observed in the case of China in a decade from 992.34 kWh/capita in 2000 to 3927.04 kWh/capita in 2014 [9,10]. In sub-Saharan African countries, such as Tanzania, Ethiopia, and Central African Republic, where little or no economic progress has been witnessed, the energy consumption levels have negligibly changed as shown in Fig. 14.1(a). Moreover, the HDI levels in these sub-Saharan African nations are also inferior as illustrated in Fig. 14.1(b).

These inequalities in development that exist in the world today can also be witnessed similarly in the case of energy consumption levels as well as HDI [7,10]. Thus, a strong link exists between the energy consumption, development, and HDI that are linked to the modern energy access situations based on geographical context [3–5,7,8,10].

14.1.2 Sustainable Development Goals and Its Relationship with Energy

Highlighting the gravity of energy snag and how modern energy access is critical for overall development of any nation, United Nation (UN) general assembly unanimously declared the decade 2014–2024 as the "*Decade of Sustainable Energy for All,*" namely to "*ensure access to affordable, reliable, sustainable and modern energy for all* [11]." Furthermore, the general assembly in September 2015 marked a very important day in history when 193 countries (developed and developing alike) adopted the sustainable development goals

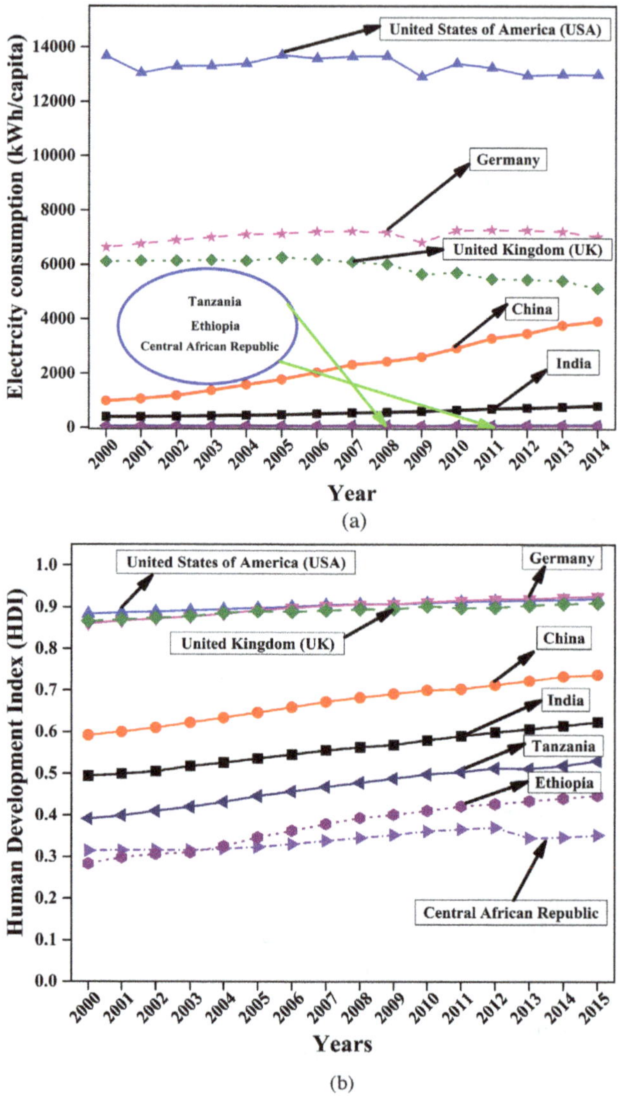

Figure 14.1. Electricity consumption and HDI pattern of various nations: (a) Electricity consumption (kWh/capita) [9,10] and (b) HDI values from 2000–2015 [8,10] (Data Source: UNDP [8] and World Bank [10]).

Figure 14.2. 17 Sustainable development goals (Source: United Nations [3, 2]).

(SDGs), which comprises 17 new SDGs as shown in Fig. 14.2 [3,12]. Moreover, an exceptional recognition and particular focus is given toward modern energy access as one of the SDGs, which can certainly help in achieving the other development goals [3].

The SDGs are integrated into nature and have direct or indirect linkage with each other, especially with access to modern energy services (SDG 7.1). A poor or lack of energy access services can make it very challenging for any nation, especially the least developing economies such as sub-Saharan African nations, to progress facing numerous issues. These hurdles are very closely related to SDGs such as poverty (SDG 1), air pollution and healthcare services (SDG 3), education (SDG 4), climate and environmental changes (SDG 11), food security (SDG2), economic progress and jobs (SDG 8), gender equality (SDG 5), and sustainable industrial development (SDG 9) [1–3,10,11]. Energy access has explicit or implicit contributions in achieving the other SDGs. Few of the explicit linkages are explained in detail below.

14.1.2.1 Energy Access and Its Impact on Health (SDG 3)

The usages of traditional energy sources, such as cow dung cakes, wood, coal, for cooking and other polluting fuels such as kerosene for lightning has severe implications on the health [1,2,13]. The inefficient usages of such traditional sources is a significant cause for the estimated premature death of 3.8 million per year [1,13]. All these premature deaths are due to indoor household pollution, which is a significant cause for diseases such as pneumonia (27%), heart disease (27%), chronic obstructive pulmonary disease (20%), stroke (18%), and lung cancer (8%) [13]. Reliable modern access to energy for clean cooking and lighting can greatly increase the air quality of households and provide women and children with better health, resulting in significant reductions in mortalities [1,2,5,8,13].

Apart from direct health benefits energy access also has implications for healthcare facilities, which depends on electricity to operate and function [1,2]. In the case of least developing nations, such as sub-Sharan Africa, approximately 60% of healthcare facility have no electricity access [14], whereas only 34% hospitals and 28% of healthcare equipment have reliable electricity access in the region [1,15]. Approximately 60% of the cold storage units for medical supplies in health clinics in Africa lose more than half of essential health vaccines due to unreliable electricity access, and 70% electrical medical equipment fail in developing nations due to the poor quality of power supply [1,15–17]. The energy requirements are sure to grow independently for better-equipped healthcare institutions and storage facilities for vaccines, particularly in developing economies in future years [1,15–18].

14.1.2.2 Energy Access and Its Impact on Gender Equality (SDG 5)

A well-defined and essential linkage exists between the energy access and gender equality [1,2]. Absence of up-to-date energy facilities has an influence on both male and female. But, owing to the disparities that are presently predominant in social statuses, monetary autonomy and sex-demarcated responsibilities are responsible majorly for a misappropriate effect on females when compared to male [1,2].

Women in developing countries bear most of the responsibility for collecting and preparing fuel, such as firewood, dung cakes, and so on, for cooking, on average spending more than 3 h/day [19]. In Africa, on a daily basis, women carry 20–25 kg of cooking fuel, which have an adverse effect on their health [17]. In South Asian nations women spend an average of 4 h/day for cooking using the traditional cooking equipments [20]. Moreover, women also take care of other household responsibility such as cleaning and washing clothes, which takes long durations due to the lack of modern energy services [1,2,8,17]. The access to affordable, economic and reliable modern energy facilities will certainly enhance the well-being of women and provides them with additional revenue streams [1,2,18]. Moreover, the time saved can be better utilized for education, social, and family activities. Education plays a vital role in reducing poverty and income opportunities and also in improving gender equality in developing nations. In rural areas, girls with access to electricity will be more successful in completing their schooling by the age of 18 and self-employed women with better access to modern energy services are likely to increase their income when compared to the self-employed women without energy access [1,2,21]. Reliable and improved energy access have raised female employment opportunities in countries such as South Africa by more than 10% in rural communities [22] and Nicaragua by 23% [23].

14.1.2.3 Energy Access for Productive Uses (SDGs 1, 8, and 9)

As per IEA special report on world energy outlook [2], productive usage is defined as "*the activities that create goods or services which in turn enhances the income potential or value which includes the income generation in agriculture, industry, mining and commercial activity as well as education and health services which also contribute to make economic gains.*"

For developing nations, to achieve their development goals and to support its expanding economy, surplus energy is the major key. When any country progresses or undergoes transformation and shift toward more modern and diverse economic system, which is based on technological innovations with a greater role being played by industry and services, energy consumption levels rise [1,2]. For economic development in case of most of the developing nations such as India and sub-Saharan Africa, access to energy for agricultural purposes is of utmost significance, which accounts for more than 18% of their GDP. Most of the people in developing nations are confined toward rural areas where agriculture plays a vital role in eradicating poverty and providing employment opportunities (SDGs 8 and 9).

In a rural development area, where humans and animal resources are often primary source of energy in agriculture activities, a decent access to energy services can dramatically increase yields and productivity [1,2]. Energy inputs in the current scenario are insufficient to agro-food-based chain in developing and least developed nations due to which food production, handling and distribution, and storage become very challenging. India loses almost 14 billion USD worth of food per year due to lack of better storage facilities in the country [24]. Accessibility for energy could therefore help to prevent food waste at significant levels through improved processing and storage units, and therefore mitigate to a certain extent hunger and poverty (SGD 1, SDG 2) [25]. Affordable and reliable electricity is a significant input to small business enterprises, hospitals, schools, and shops. Most of the small business in developing economies find unreliable electricity access as a major obstacle in their economic development [26]. Reliable and cost-efficient access to electricity can thus help companies to start and grow businesses that provide services to sectors and communities that strengthen the economy and prosperity (SDG 3) [1,2,26].

Hence, from above, it can be seen how energy access is in the center of achieving all the other SDGs. Better economical and reliable modern energy services can not only improve health by providing clean cooking and lightning facilities but can also provide a safer indoor environment for women and children (SDG 4 and 11). It also helps in achieving lower greenhouse gas (GHG) emission levels as modern and efficient energy technologies emit less CO_2 (SDG 13 and 14).

14.1.3 Global Energy Access Overview and Challenges in Improving Energy Access

Despite many intensified efforts by many countries, 2792 million people (38%) of World's population do not have access to clean cooking facility in 2015 out of which 2500 million still depend on the traditional sources to meet their basic needs [1,2]. Approximately 80% of the population suffering from energy poverty are concentrated in rural areas of developing countries such as India and sub-Saharan Africa [1,2].

A collective populace corresponding of the European Union and the United States still do not have electricity access in India, with the majority of its residents depending on biomass to fulfil their day-to-day domestic energy need [27]. Figure 14.3(a, b) illustrates the regional summary of the population without access to the clean cooking facility and depend on biomass as per IEA data till 2015. As depicted in Fig. 14.3(a) the majority of the population without access to clean cooking facility belongs to sub-saharan Africa (846 million) and India (836 million) followed by China (457 million), other developing Asian nations (312 million), and other Southeast Asian nations (188 million). The people lacking these facilities and mostly concentrated toward the rural and remote areas are dependent on the traditional biomass sources as illustrated in Fig. 14.3(b). With regard to electricity access as per 2016 data, China has achieved 100% electricity access but is still struggling to provide clean cooking facility to the majority of its population [1,2, 28].

A majority of people without electricity access globally as illustrated with Fig. 14.4 are confined to sub-Saharan Africa (588 million), India (239 million), and other developing Asian (135 million) and Southeast Asian (42 million) nations [1].

Even with the existence of new policy scenarios by the governments of various nations and UN's SDGs, the availability of energy (clean cooking and electricity) to all is full of challenges. To summarize the following significant problems exist, which have to be adequately addressed to achieve the goal for "*Energy access to all*":

(i) Due to enormous population growth energy demand is going to rise furthermore in developing nations by the year 2040 [29]. This case is prevalent mostly in sub-saharan Africa, India, and other developing Asian nations (Bangladesh, Nepal, Pakistan, DPR Korea, Mongolia, and Sri Lanka) [30–33].

(ii) In fast-developing economies, mainly China and India, the energy demand is anticipated to grow faster mainly owing to their national development goals such as roads and infrastructure expansions, industrial development for the manufacturing sector, low-cost transportation, commercialization of various agricultural, technical produce, and so on. Sufficient energy supply is the crucial factor that can help the fast-developing economies (China and India) to succeed in their national development objectives [27,29]. By the year 2040, it is projected that 48% rise will be there in energy demand globally with major contributions from non-OCED (Organization for Economic Cooperation and Development) countries mainly due to their national and long-term national economic growth [33,34].

(iii) Lack of proper infrastructure, tougher geographical terrain leading to poor road connectivity and higher cost of connection, an adverse climatic condition affecting the electricity supply from the grid, and so on, are few more reasons adding up to this challenge in case of remote communities [35–37].

(iv) Also, when the economic condition of any country or region tends to get better, people will migrate from traditional sources of energy to modern sources of energy such as electrical appliances for heating, cooking, lighting purposes, and thus, the energy demand is going to increase [38–43].

(v) Even this transition from traditional energy sources to modern sources is a challenging one due to technical suitability and availability, cost, adequacy and quality, social and cultural acceptability, and reliability of energy supply [1,44–46].

(vi) Apart from these, the other prolonged issues such as depleting reserves of fossil fuels, greenhouse gas emission, and environmental pollution caused due to thermal-based power plants, rapid industrial and infrastructure development, and so on, are also few factors that can undoubtedly have direct inferences on the energy demand and proper availability of energy services to all [28,43,47–51].

From above, it is clearly understood how energy access has direct implications not only for the development of a country as a whole but also at disintegrated level on the livelihood of its people. Modern energy access has explicit effects on their health, education, gender equality, poverty, employment, economic progress, and so on. Even providing energy with

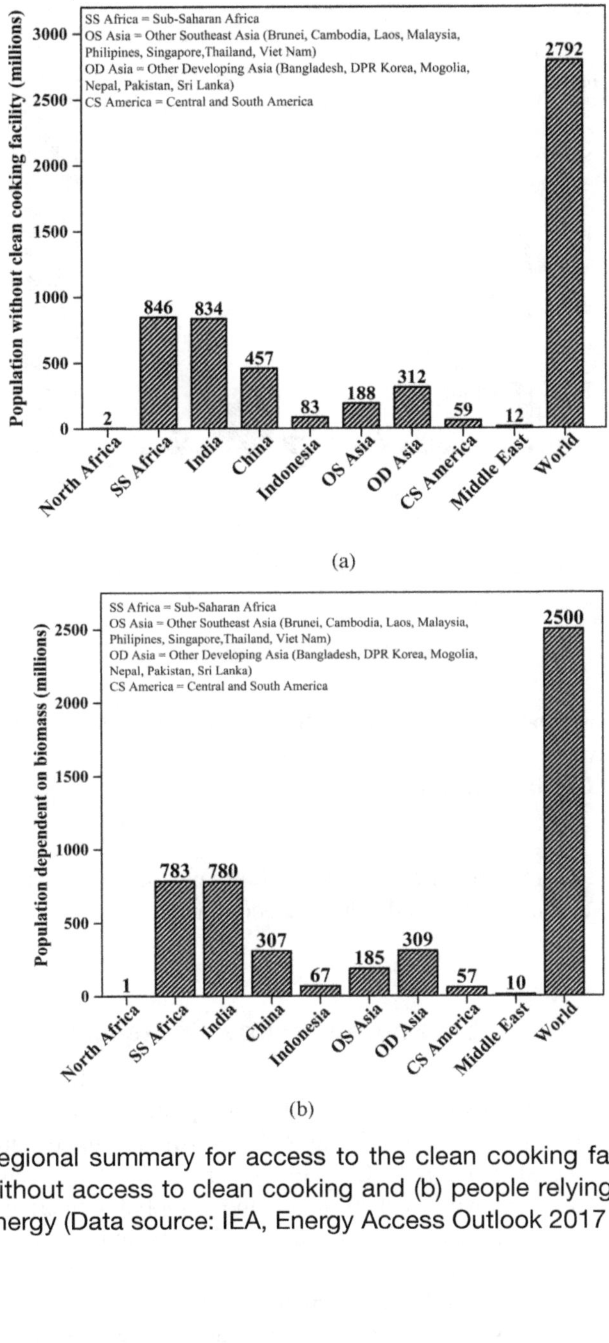

Figure 14.3. Regional summary for access to the clean cooking facility: (a) People without access to clean cooking and (b) people relying on biomass for energy (Data source: IEA, Energy Access Outlook 2017 [1]).

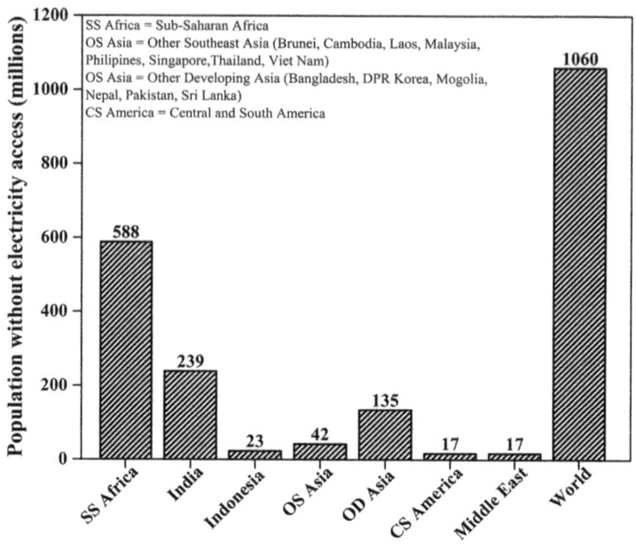

Figure 14.4 Regional summary for electricity access (Data source: IEA, Energy Access Outlook 2017 [1]).

a perspective of sustainability is full of challenges and obstacles specific to the remote locations in the developing nations that have the majority suffering from energy poverty. So, one significant question arises here,

> "How to achieve energy access with a sustainability perspective for rural and remote locations of developing and least developing economies?"

14.1.4 Role of Renewable Energy Sources in Realizing Energy for All

To combat the threats of repeated climatic variations and to establish a step forward for sustainable development accommodating least carbon footprint, the global leaders signed an agreement at the 21st Conference of the Parties (COP 21) of the United Nations Framework Convention on Climate Change (UNFCCC) held in Paris [52]. The agreement is primarily focused to take certain steps worldwide to maintain the global temperature growth under 2°C for this era and to create the essential framework to bound the temperature rise not exceeding 1.5°C [52,53]. Power generation technologies that are environmental friendly and have the ability to create an additional carbon sink can certainly play a key role in reducing the greenhouse gas emission. Renewable energy-based power schemes have ascertained to be a viable alternative in meeting the growing energy demand as well as decreasing the emission without obstructing economic progress.

Renewable energy sources (RESs) such as wind, solar, and hydro are one of the core solutions that can provide energy access to the majority of people globally. Recent growing global support for renewables due to the signing of the COP 21 agreement, technological advancement in harnessing the RESs, and declining cost of the technologies can provide the edge to meet obstacles in the way of achieving the universal energy access targets effectively and promptly [54].

14.2 Microgrids Based on Renewables: A New Paradigm for Energy Access

As mentioned Section 14.1, significant populace (80%) suffering from energy poverty are located in rural and remote locations where technical, economic, social, as well as environmental hurdles exist for providing reliable and affordable energy solutions. Providing electricity access with provisions of high-efficient appliances can play a significant role in extending access to lighting, cooking, and heating in the isolated and rural communities [1,2,27,55]. However, electricity access provided by government utilities in developing nations are often unreliable, and their quality of supply is also questionable [1,2,6,56]. For providing access to modern energy needs to these masses, microgrids based on a combination of locally available energy sources such as solar, wind, small hydro, and biomass could be the key to success [1,2,10,19,27,54,55,57–60]. Many classifications and definitions of microgrids currently exist in literature given by many academicians and expert engineers [55,61,62]. The concept of microgrids was introduced and pioneered first by Thomas Edison as a concept of small-scale isolated system in the latter half of the nineteenth century [55]. In 2002, R. H. Lasseter in Power Society General Meeting provided an overview of the microgrid concept including a basic architecture with control and power strategy [63]. A general definition potently cited in the literature based on United States Department of Energy is as follows:

> *"A microgrid can be defined as a group of interconnected loads and distributed energy resources within clearly defined electrical boundaries that acts as a single controllable entity with respect to the grid. A microgrid can connect and disconnect from the grid to enable it to operate in both grid-connected or island mode* [55,63,64]."

Another popular definition based on European Union Research Project is as follows:

> *"Microgrids comprise LV distribution systems with Distributed Energy Resources (DER) (microturbines, fuel cells, PV, etc.) together with storage devices (flywheels, energy capacitors, and batteries) and flexible loads. Such systems can be operated in a non-autonomous way if interconnected to the grid, or in an autonomous way, if disconnected from the main grid. The operation of microsources in the network can provide distinct benefits to the overall system performance, if managed and coordinated efficiently* [65,66]."

However still the term "microgrid" could not be universally defined and in *literature* often terms such as mini-grids, pico-grids, small hybrid renewable energy systems, and so on are also being used [23,55–57,67–72]. To summarize, the microgrids are formed using interconnected available generating sources such as photovoltaics (PVs), small wind turbine systems, microhydro, and biomass-based generation system, which can transfer power using a low-voltage distribution system to relative small group of customers [71,72]. In context to the rural and isolated community, the microgrid can be defined as a small local power system that is capable of generating power based on the locally available energy sources, which could meet their load demand efficiently, reliably, and economically [70–72]. These rural microgrids can have a different installed capacity ranging from few kilowatts (kW) to larger values close to hundreds of kWs depending upon the generation technologies, load type, and number of consumers [71,72].

Rural microgrids do not use one-size or one-type-fit-all approach for design purpose, hence classifying rural microgrids is a bit complicated [55,73]. Also, in *literature*, many types of classifications in the context of rural areas exist. Most of the authors have considered an isolated or off-grid type of microgrid for rural electrification with no consideration toward the possible grid interactive-type microgrids [57,74,75]. Mandelli et al. [57] have classified the rural microgrids into two dominant types, which are as follows:

(i) Decentralized-type microgrids are mostly the small energy systems based on a single generating unit with proper storage for single or many users with no interaction between the conversion and distribution network with other units having all the necessary power electronics converters and control strategy.

(ii) Distributed type can be based on multiple generating units with storages for single or multiple users with a distribution network of its own [57]. The energy-generating source, storage, and load are interfaced with essential power electronic devices with a centralized energy management or control strategy as outlined in [57]. These type of microgrids are termed as hybrid microgrids by Mandelli, et al.

IEEE Power Energy Society (PES) subcommittee task force on renewable energy technologies led by Nehir [76] have categorized the microgrids based on the integration schemes of RESs and based on power frequencies. The microgrid can be either isolated with provisions for grid connections to supply power to AC or DC loads, which are as follows [77–81]:

(i) DC-coupled microgrid system
In a DC-coupled system, the energy-generating sources whether AC sources such as wind, microturbines, or DC sources such as PVs along with storage system are connected to a common DC bus with necessary power electronics converters. In the presence of DC loads, which can be connected directly to DC bus or, if required, through DC/DC converter, the microgrid system could be efficiently used to supply power to AC loads as well as have provision for connection to the utility grid (50/60 Hz) using a bidirectional converter.

(ii) AC-coupled microgrid system
These can be further subclassified into two types as follows:

a. High-frequency AC (HFAC) System [81]: In this type, the energy sources (AC or DC) along with storage systems with basic converter interfaces are coupled to HFAC bus along with HFAC loads. The HFAC loads have very high frequency requirements as 400 Hz such as submarines, space applications, and airplanes [81–83].

b. Power-frequency AC (PFAC) System: In the PFAC system, all the generating sources along with storage systems are integrated directly to the PFAC bus (50/60 Hz) with their converter interfaces.

Both PFAC and HFAC have necessary provisions to supply DC-type loads using a rectifier. Moreover, HFAC system also has provisions to include a separate PFAC bus for connection to PFAC loads and utility grid.

(iii) Hybrid-coupled microgrid system [77–80]
In this scheme, the energy sources (AC or DC type) could be connected to AC or DC bus of the hybrid system. The DC energy system can be coupled directly to the DC bus and AC energy system to the PFAC bus with storage system on the DC bus. Similarly, the DC loads are supplied from DC bus and AC loads and utility grid coupled to PFAC bus. The DC and PFAC buses are coupled using a bidirectional converter to manage the power flow from DC to PFAC bus and vice versa.

As per National Renewable Energy Laboratory (NREL), USA [84], the renewable-based rural microgrids can be of two major types with provisions of isolated and grid interaction:

(i) Distributed microgrid system is home-based energy system comprising decentralized generators such as rooftop PVs and small wind turbine with storages and provisions for connection to the utility grid with individual power management system with no transmission and distributions losses with single or multiple consumers.
(ii) Central microgrid systems are a community serving-type local power system having its own transmission and distribution network having single or combination of energy-generating sources with storage system along with interactive grid type having all the necessary converters and centralized control to manage the power flow.

However, there is no generalized conception and classification of microgrids in context to rural electrification, and depending upon one's suitability different categorizations exist in the literature [56,57,62,70–72,84]. In this work, the categorization of rural microgrids is done following broader and application aspects currently being practiced in developing nations such as in India for rural electrification purposes, which will be used in this chapter. Fig. 14.5 illustrates the various schemes for classifying the rural microgrids, which are mainly of two types:

(i) Centralized system
- ➢ The centralized microgrid system is a local power system constituted up by combining single or multiple locally available energy-generating sources and storage system having a transmission and distribution network to supply a community-based load having a number of consumers with centralized control to manage and monitor the power flow.
- ➢ The microgrid can be of either the isolated or grid-interactive types in different architectures depending upon the integration of energy generation sources.
- ➢ In the AC-type scheme, all the energy-generating sources irrespective of the nature of their output will be connected to the AC bus using the necessary power electronic interfaces. Similarly, in DC type all the sources are connected to the DC bus.
- ➢ In the hybrid type, the AC output sources such as wind will be coupled to the AC bus and DC output sources such as PVs on DC bus, respectively.
- ➢ The storage system with DC output nature such as battery storage will be connected to the DC bus in all the schemes whether AC, DC, or hybrid type. However, the storage system with AC output nature such as pumped hydrostorage system will be connected to the AC bus only.
- ➢ These types are beneficial for providing electrification for a community or a village with more number of consumers generally having more than 100 households [85–88].

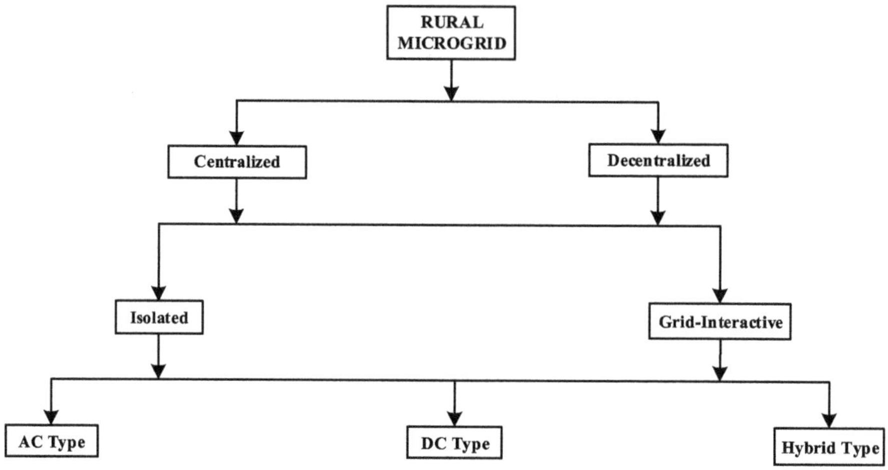

Figure 14.5. Classification schemes of rural microgrid.

(ii) Decentralized system

- ➢ The decentralized microgrid system is a small individual system based on a combination of single or multiple locally available generating sources such as rooftop PVs and small wind turbines with a battery-based storage system to supply a single or a small group of home-based load in the close vicinity having a local power management system.
- ➢ This type can also be of isolated or grid-interactive nature in different architecture schemes such as AC, DC, or hybrid type.
- ➢ The one major point is to be noted that, even if the microgrid is the off-grid type, it should have provisions of connection to the utility grid in future. It can help the households or community people to generate economic opportunities by selling back the excess generation to the grid through smart metering schemes being regulated in the new policy scenario [1,2,27,35,89–91].
- ➢ These types of system are generally suitable for locations having distributed type loads with a small number of consumers having less than 10 households [85–88].

Rural microgrids if deployed and appropriately designed can undoubtedly deliver significant benefits by providing better energy access by replacing the low-quality traditional energy sources. It will definitely help the communities with improved health, education, safety, social benefits along with new economic and income generation opportunities [2,5,8,15,19,35,45,54,71,72,89,91,92]. However, deployment of microgrids with a perspective of sustainability in developing nations is full of challenges. The barriers in the deployment of rural microgrids not only involve the technical and economic issues but also the sociocultural and environmental aspects [45 46,89,91,93], which is detailed in Section 14.2.1.

14.2.1 Problems Associated with the Deployment of Microgrids Based on Renewable Sources

In recent decades the RES-based microgrid power system has been fully made into a more dynamic one by integrating a variety of benchmarks, stakeholders and competing goals, especially for rural electrification [45,91,94]. To achieve extensive inclusion of RESs in the current power system structure, a thoughtful contemplation must be specified in planning prospects bearing in mind economic, institutional, technical, social, and environmental barriers [95]. The complexity with regard to rural electrification power projects has become more prominent due to the presence of numerous models for energy planning, varying policies/regulations, and also existence of multiple participants [45,96–98]. Hence, in contemporary energy planning situation, it would be adolescent to advocate or execute a project without making an allowance for the concern and the predilections of the numerous partcipants/elements [95,97]. To sum up, in the developing countries, the following significant obstacles and vulnerabilities exists for implementation of sustainable energy microgrids based on renewables:

(i) Socio-cultural barriers

Social influences are very prominent on electrification programs in rural and developing countries. Acknowledgment by the community is one of the biggest hurdles to the implementation of the energy related infrastructure projects. The involvement and assessment of local communities can help in deciding energy alternatives, which would not make them feel isolated or threatened by it and thus avoiding energy project failures [28,37,38,45,50]. Lack of public recognition in the majority of cases has real ramifications for rural renewables-based microgrid projects. This often leads to increased upfront investment, delays caused by repeated demonstrations against the projects and even a halt to the energy development projects [54,99].

Other factors such as jobs, health, and economic assistance are closely related to community acceptance [1,2,89,93,99,100]. Due to a possible lack of knowledge and education in the area of primary electricity, the community is not always involved in using modern energy services in developing nations. They often take these as something of a financial strain on their livelihoods and would therefore are against electrifications schemes [89,93,99–101].

(ii) Economic barriers

Rural renewable energy microgrid projects require higher initial capital costs (CC), contributing to increased output costs relative to energy projects based on traditional energy sources [54,102]. One other strongly associated variable in renewable energy projects associated with the higher CC is the clear identification and development of microgrid systems based on low-cost energy resources. This factor is often neglected by the renewable energy planners [37,38,50,54,102,103]. For example, in India, solar power is quite expensive in the *North Eastern States* (NES) compared to rest of its territories. But, NES gets more investment opportunities for solar-based system rather than locally available energy sources such as small microhydropower [27,54,102,104–106]. Moreover, the RETs are also subjected to high operation and maintenance cost.

Moreover, the RETs are also subjected to high operation and maintenance cost. Other financial hurdles include the absence of ample prospects of long-term financing to finance clean energy products [54]. There is often a case of a disparity between the term of supporting funds and the cash flow of the renewable energy projects. Moreover, the financial security and return on investments in energy projects based on the renewables are also questionable. Market constraints such as inconsistency and uneven pricing of renewable products, information irregularities, misrepresentation in power market, indigenous underdeveloped renewable industrial chain, and trade barriers such as imposing heavy import duty taxes, which makes the renewable products quite expensive adds up to the economic barriers also [54].

(iii) Environmental barriers

Zoning regulations and environmental features (harsh climatic condition and tougher geographical terrain) can hinder the deployment of energy projects in specific areas [54]. Land requirements and its acquisition are also one of the major obstacles for microgrids based on the wind, hydropower, solar PVs, and so on. RESs such as hydropower have adverse effects on the river life and occasionally a genuine risk to damaging the biodiversity predicament of several scarce species as outlined in [107]. In addition, other closely related issues are risk of flash flood in case of hydropower, noise pollution in case of wind power, water contagion, erosion, and deforestation for site expansion for the installation and construction phase of electrification projects [37,51,108].

(iv) Institutional barriers

Energy sector's structural and governmental arrangement in developing economies are government-owned. Many times, the lack of cooperation, including lack of authority, causes a turbulent regulatory environment, which considerably leads to a higher risk and even at times further lowers the investments opportunities. This then leads to major obstacles for executing energy development projects [54,102 and 109].

Few other factors include [54,102]:

- Lack of established institutions, policies, and guidelines related to sustainable RES development.
- The absence of exact and well-defined job roles, unclear licensing procedure.
- Difficulty with a nuanced, sluggish, long, and ambiguous project screening process in the acquisition of required lands

(v) Technical barriers

Insufficient technologies and lack of infrastructure are the technical difficulties that are required to support the implementation of renewable energy-based micro grids [54]. The failure to incorporate renewable energy technologies in existing systems and the lack of adequate physical infrastructure for regional transmission and distribution systems, equipment and services are some of the significant obstacles to implementation of renewable energy projects in developing countries [102,110]. For example, in India, underdeveloped states of North Eastern Region (NER) have a great amount of capacity for generation from hydro, which can meet its current as well as future energy requirements and also capable of overcoming the power scarcities for other parts of India [38,51,104,111]. But due to inadequate road infrastructure and a poor power Transmission & Distribution (T&D) network the hydropower potential could not be tapped efficiently due to which a large number of unelectrified communities exist in NER in comparison to other parts of India [112]. Controlled grid connection is another apprehension, predominantly for distributed generation technologies and also in case of a vertically incorporated power sector [54]. Lack of good quality metrological data also remains a big problem due to which it is quite cumbersome and difficult to perform the correct feasibility study to determine the real potential of RESs at a particular location [102]. Further problems in developing countries are the shortage of qualified local human resource forces for implementing and sustaining energy projects [54,108,113].

Thus, for the successful deployment of the rural microgrids based on renewables the barriers related to technology, awareness and capacity, cost, finance, infrastructure, and public acceptance, in addition to policy, regulatory, institutional, political, and administration, need to be properly addressed.

So few questions arise here:

(i) How to achieve the sustainable alloy and address the barriers related to rural microgrid projects?
(ii) What kind of solutions are being offered, practiced, and available in academic community which can address the issues related to microgrids based on the renewables?
(iii) *Is the available solution in the scientific literature enough to cater to all the associated problems in context to the rural microgrids and provide a generalized solution?*

14.3 Proposed Generalized Framework for Sustainable Rural Microgrid Design

This chapter puts forward a comprehensive methodological framework, which is based on coaction of optimization models and decision analysis for designing a robust, reliable, and economically efficient microgrid system considering local available energy resources for rural/remote locations. The proposed design framework has the ability to simultaneously take the different criteria and various views of the multiple players involved in the design process in varying scenarios.

Figure 14.6 shows the proposed generalized framework for a sustainable rural microgrid design for developing nations. The design framework contains three significant levels with a number of subdesign levels. The **first level of design** employs the decision-making technique to find the best promising alternative for the design of microgrid incorporating the numerous performance indices to articulate diverse scenarios. The output of this level will act as input to the next (second) level of framework design.

After the initial solution set of energy alternatives are obtained, the load growth forecasts will be carried out in **Level 2.** Furthermore, a multiobjective optimization tool is applied in the **second-level design** to calculate the feasibility analysis of preeminent alternative to find the optimum size with least cost considering a multiyear load progression. Various optimal solutions based on different microgrid architectures will be obtained after the second level of the design process, having diverse component sizes and cost.

After Level 2, numerous optimal microgrid solutions will be obtained having diverse cost and size. Finally, using hybrid decision analysis methods, the microgrid solution ranking will be carried out taking into account a set of performance indices, which will be validated in the **third level** of the design process. Level 3 will provide a final set of solutions, which are judgmental in nature. Based on the final relative weight scores of each obtained solution, the ranking and selection of the best sustainable microgrid system will be done. The trilevel framework proposed is user-friendly and it iterates a real-world design process at disintegrated levels.

The detailed description of the design methodology is shown in Fig. 14.6, which is explained in the following subsections.

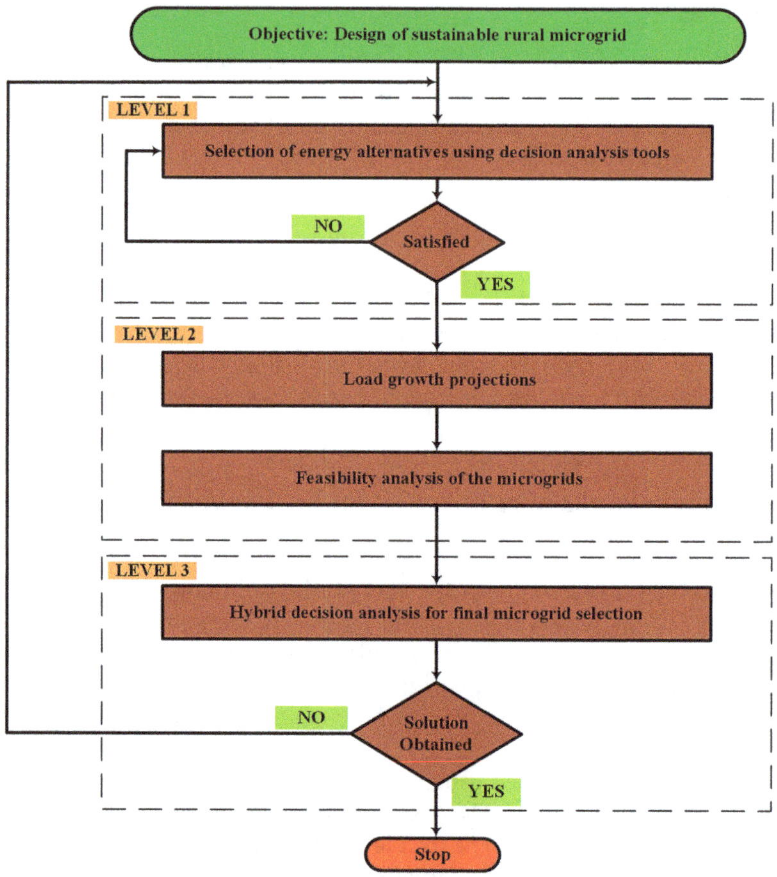

Figure 14.6. The proposed framework for sustainable rural microgrid design.

14.3.1 Level 1—Selection of Energy Alternatives Using Decision Analysis

The first level of the design process illustrates the preliminary evaluation of energy alternatives with a sustainability perspective. Fig. 14.7 shows the various sub design steps involved in the Level 1 of the proposed framework. Figure 14.7

MCDA technique have demonstrated to deliver a superior solution to such a complex project having several criteria, circumstances, and different views of participants in the identical framework [50,98]. As demarcated in literature, none of MCDA techniques can be categorized as greatest or poorest [50,98,114,115]. All decision-making techniques have their individual merits and demerits in significance of applicability to the particular problem [28,48,50,98]. Hence, based on the accessible information, complexity of the problem, flexibility (simultaneously handling of qualitative and quantitative indices), actual location contemplations, and so on, suitable MCDA method is selected for assessment of alternatives on the definite set of factors [48,98,114]. By means of the suitable MCDA method, the relative weights of criteria based on diverse scenarios are initially determined. Then, with respect to the set of criteria the weights of alternatives on each scenario are calculated, and on the basis of the final weight score, the ranking of the alternatives is done. The energy alternative having maximum score is given first rank, the alternatives with second-top score are marked as second rank, and so on for specified scenarios.

For formulated scenarios, a set of judgemental solutions will be obtained with different ranking and final relative score. As the solutions are judgmental, experts (the government entities, engineers, project planners, etc.) advice who are familiar with the selected locality, which should be taken into account to choose either single or multiple solutions. In the

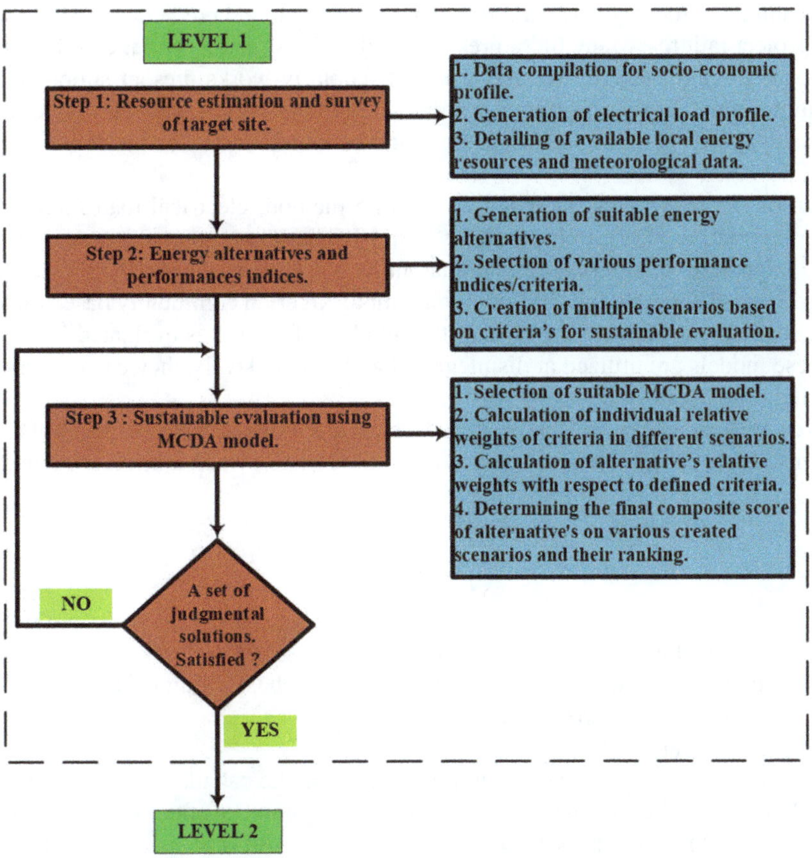

Figure 14.7. Illustration of design process steps involved in Level 1 of the proposed framework.

next level design process, the final selected energy alternative will be input for further evaluation. In some instances, if the solution obtained is not realistic or satisfactory, the whole process can be repeated, which will yield a different set of solutions with different weights depending upon the selected criteria.

14.3.2 Level 2—Load Growth Projections and Feasibility Analysis of Microgrid Alternatives

Level 2 deals with load demand projections and detailed assessment of feasibility of alternatives for design of microgrid obtained as input from Level 1 utilizing multiobjective optimization tool. Majorly of it contain four steps, and each step the design process is divided into smaller tasks, which is shown in Fig. 14.8.

As outlined in Section 14.1, majority of available literature reported have used the recent load demand to perform the analysis and design of energy system for complete project lifetime [116–133]. In such kinds of scenarios, the designed system may not be able to cater to the increasing load demand due to techinical as well as financial issues [134]. Hence, energy planning without considering future increasing demand predictions will lead to successive project failures. Many literatures reported various models and methods to forecast yearly load demand. Very few of the reported scholarly works present simple solutions, for example, trend analysis, growth rate indicators, and so on, and few works reported complex advanced solutions such as engineering-economic model and econometric model [135,136].

Based on complex model and soft computing method, electrical load forecast (ELF) has been reported in [137]. In [138], long-term ELF based on quite a few models such as stochastic, qualitative, time-series, and quarantine has been outlined. The PV-based small power grid with 1% yearly load growth for a remote Nigerian community based on increasing user load demand due to increase in the number of houses is evaluated in [134]. But when these models are utilized at disintegrated and remote levels, they can complicate the forecasting of load demand.

However, Bhattacharyya [135] proposed a simple approach, which can be used for ELF to design suitable energy systems for remote communities, which is expressed in equation (14.1)

$$D_t = D_0(1 + g)^t \tag{14.1}$$

where D_t = Electrical load demand in year t.
D_0 = Electrical load demand of base year, that is, the existing load.
g = Assumed load demand growth rate.
t = time in years.

The assumed electrical load demand growth rate can be calculated using heuristic load data available or can be calculated using the increasing number of consumers per year and economical growth of the targeted community because both these parameters are directly related to yearly load demand growth [38, 51, 134, 135, 139]. On many occasions, date quality historical load demand data availability is also an issue, however, in this scenario, expert advice from a relevant central/local institution for load demand increase can be taken into consideration.

Then, detailed mathematical modeling of individual components for constituting the microgrid based on the final alternatives, i.e., power generation sources and storage sys-

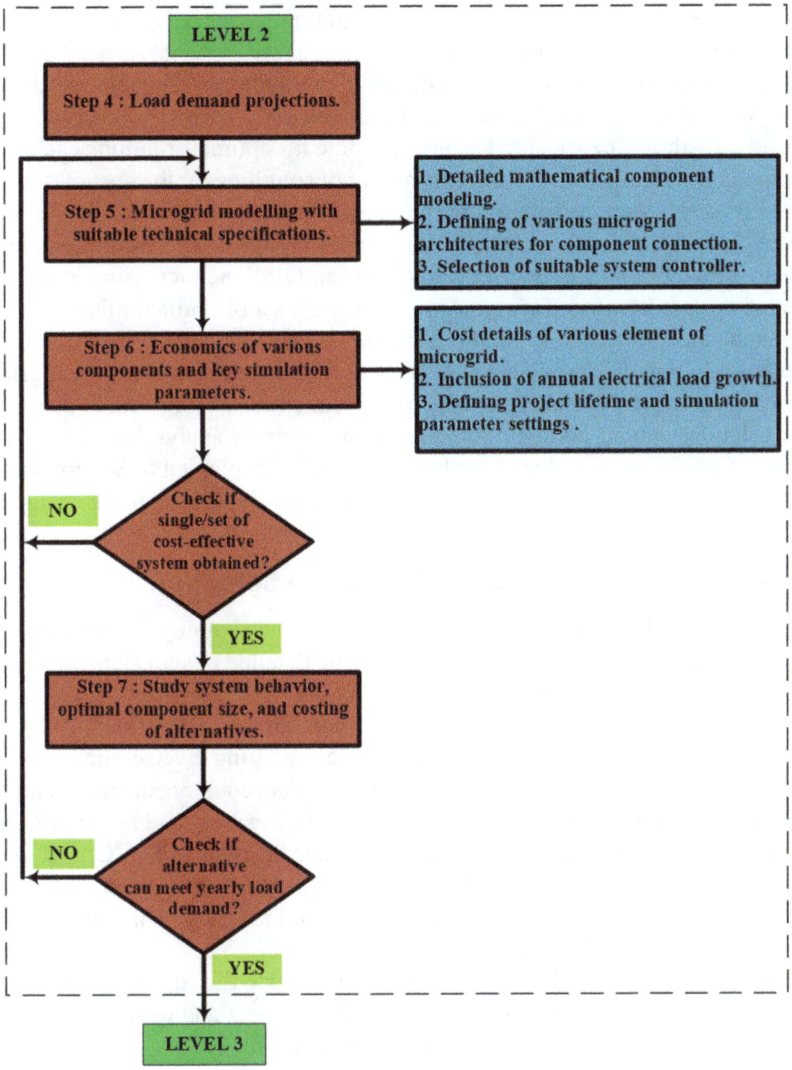

Figure 14.8. Design process steps involved in Level 2 of the proposed framework.

tem in different architectures will be done. Furthermore, to manage the power flow balance between generation and load economically, a suitable power flow controller can be modelled if required. Therefore, the standard cost parameters such as capital investment cost, operation and management cost (O&M) for each component such as PV system and Hydro Kinetic Run of River (HKRoR) turbine can be used to determine overall cost of the system, which can be obtained from regulated and standard government organization for the current market cost. Separately from these major cost details other vital parameters as inflation rates, existing discount rates, project estimated lifetime, yearly load growth, and so on should be included in the simulation. The constraints for various objective functions such as minimum net present cost (NPC), emission, maximum renewable fraction, and maximum reliability should be defined in this process.

When system component design and detailed simulation parameters are defined, the simulation is carried out for each year on an hourly basis (8760 h/year) considering yearly load demand growth for the specified project lifetime till optimal solution is achieved. The results from the simulation provide the optimal set of solutions for the cost of various components and sizes for microgrid alternative having different microgrid architectures. A comprehensive system behavioral analysis is done based on outcomes of countless profiles to verify whether the optimal solution for microgrid can fulfill the increasing load growth till the estimated project lifetime. This level will provide a set of optimal microgrid solutions depending upon the different microgrid architectures. The obtained microgrid solutions will have different component size and cost. At this stage, selection of the best microgrid system cannot be done merely by detecting technical and economical results. So, to decide the best microgrid solution without being unfair, once again decision analysis evalution should be carried out to find the improved set of solutions, which is explained in Section 14.3.3. The solutions found from Level 2 process will act as input for final level of the design process.

14.3.3 Level 3—Final Microgrid Alternative Selection

Level 3 deals with the final stage of the design process, which employs a combined approach based on two different decision-making models. The detailed design steps involved in the final level is illustrated in Fig. 14.9.

At first, the various data and their characterization obtained after the thorough techno-financial investigation of the microgrid alternatives considering diverse microgrid architectures will be done. The technical data involve the individual component sizes of microgrids, renewable fraction, storage depletion and autonomy, power loss, and so on. Similarly, the economic data associated with microgrid design include the total NPC, CC, cost of electricity (COE), and so on. The environmental aspects may include the total emissions, ecological footprint, required land, and so on depending upon the specifics inputs and objectives formulated during the Level 2 process.

Once the data collection is done, it should be presented to the energy project experts and stakeholders for their views. Depending on the available data and opinions of experts, the selection and classification of essential performance indices (EPIs)/criteria will be done. The classification of various cratiria in this stage is done on the basis of technological, economic, performance, and environmental aspects of the mircogrid alternatives. The technological criteria involves factors such as power generation capacity, total electrical production, annual capacity shortage, and load ratio. The economic criteria involve system cost, annualized cost, component cost, operating cost, salvage cost, return on investments, net present worth, grid energy purchases and sells, and so on. The performance criterion is directly related to the performance of the system such as power losses, excess electricity production, unmet electric load, maximum renewable penetration, and total load served. The environmental aspects generally involve the quantities of major pollutants emitted from the generators, boilers, or grid in kg/year.

Majority of times, the data found after techno-financial assessment are quantities with disproportionate units. For these types of issues, hybrid multicriteria decision-making (MCDM) models can be applied for final microgrid project selection. Using the MCDM

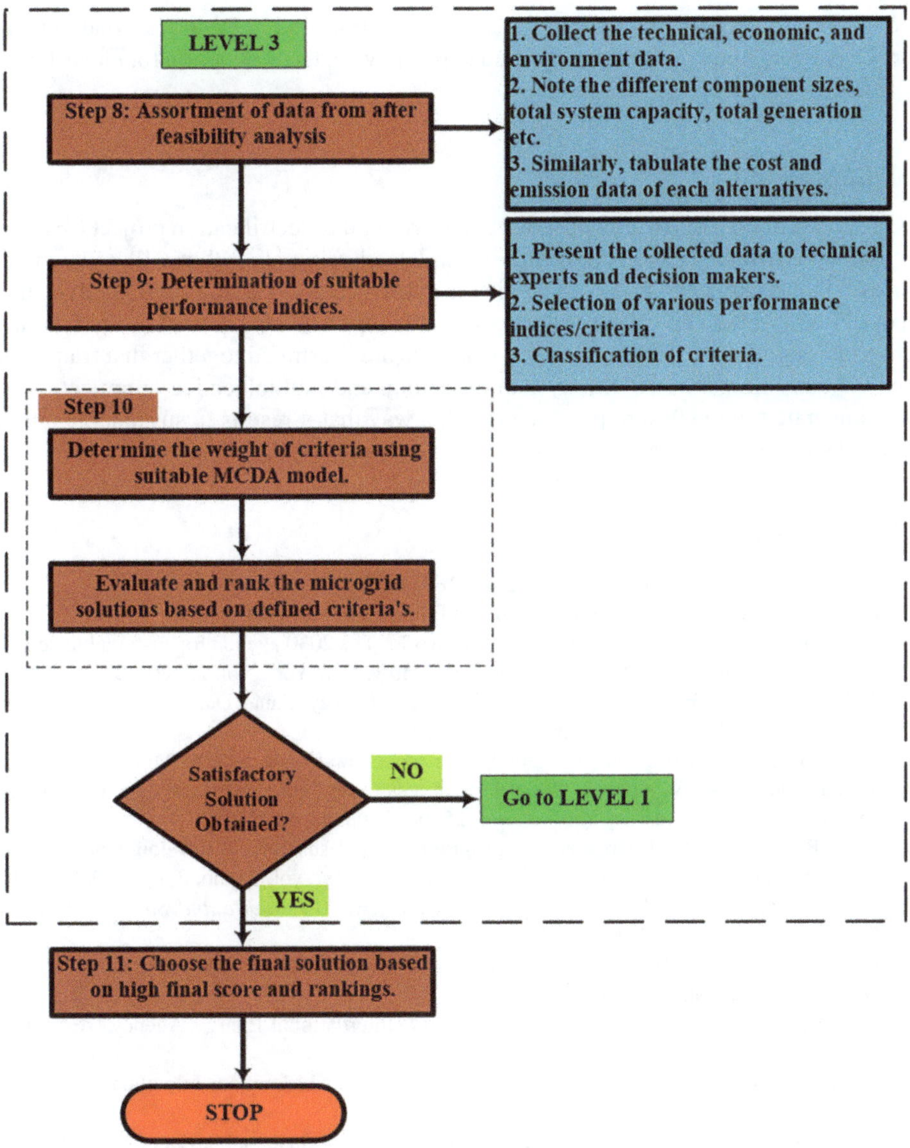

Figure 14.9. Final level of the design framework.

model such as analytic hierarchy process (AHP) or sum weighted method (SWM) or fuzzy logic, the relative weights of the EPIs relevant to the microgrid solutions are determined. Then using other MCDM models such as TOPSIS, which has the ability to handle the incommensurate quantities, the final weight score of each microgrid solutions with respect to the defined criteria will be determined. Based on the final weighted score, the final ranking of the microgrid solutions will be done. The microgrid alternative with the highest weigh score will be assigned the first ranking, followed by the second highest as second rank, and so on. Depending upon the best ranking and expert advice, the best option for electrification is selected.

14.4 Conclusion

This chapter proposes the overall design framework for the electrification project for remote locations. The trilevel framework design is explained using algorithmic illustrations and each level is also explained briefly to obtain the exact idea of designing steps involved in the proposed framework. This chapter addressed many important characteristics of sustainable microgrid design for rural locations. These aspects are illustrated together in a framework, which validates how the MCDM, system modeling, and multiobjective optimization tool can be integrated in the design process in such a way that it results in sustainable optimal microgrid solutions for rural locations.

14.5 References

1. I. E. Agency, *Energy Access Outlook 2017*, IEA, Paris, 2017.
2. I. E. Agency, *World Energy Outlook*, OECD Publishing, Paris, 2017.
3. A. Resolution, "RES/70/1. Transforming our world: The 2030 agenda for sustainable development," Seventieth United Nations General Assembly, New York, vol. 25, 2015.
4. C. Pîrlogea, "The Human Development Relies on Energy. Panel Data Evidence," *Procedia Economics and Finance*, vol. 3, pp. 496–501, 2012.
5. M. Bazilian, P. Nussbaumer, C. Eibs-Singer, A. Brew-Hammond, V. Modi, B. Sovacool, V. Ramana, and P. -K. Aqrawi, "Improving access to modern energy services: Insights from case studies," *The Electricity Journal*, vol. 25, no. 1, pp. 93–114, 2012.
6. S. C. Bhattacharyya, "Energy access programmes and sustainable development: A critical review and analysis," *Energy for Sustainable Development*, vol. 16, no. 3, pp. 260–271, 2012.
7. M. González-Eguino, "Energy poverty: An overview," *Renewable and Sustainable Energy Reviews*, vol. 47, pp. 377–385, 2015.
8. U. N. D. P. (UNDP), *Human Development Report: Human Development Index and its Components*, United Nations, New York, 2016.
9. I. E. Agency, *IEA Statistics © OECD/IEA 2014,* International Energy Agency (IEA), Paris, 2014.
10. W. Bank. "World Development Indicators." 2014. Available online: https://data.worldbank.org/indicator/EG.USE.ELEC.KH.PC
11. United Nation General Assembly. "2014–2024 Decade of Sustainable Energy for All." 2012. Available online: https://www.un.org/press/en/2012/ga11333.doc.htm
12. United Nations, "2030 Agenda for Sustainable Development: 17 Sustainable Development Goals (SDGs)." 2015. Available online: https://www.un.org/sustainabledevelopment/sustainable-development-goals/
13. W. H. Organization, *Household Air Pollution and Health-Fact Sheets*, World Health Organization, Geneva, Switzerland, 2018.
14. Sustainable Energy for All (SE4ALL) Hub, "Powering Affordable, Reliable, and Sustainable Energy," African Development Bank Nairobi. 2014. Avilable Online: https://www.afdb.org/fileadmin/uploads/afdb/Documents/Generic-Documents/SE4ALL_Africa_Hub_-_Powering_Affordable__Reliable_And_Sustainable_Energy.pdf

15. W. H. Organization, *Access to Modern Energy Services for Health Facilities in Resource-Constrained Settings: A Review of Status, Significance, Challenges and Measurement*, World Health Organization, Geneva, 2014.
16. W. H. Organization, *Medical Devices: Managing the Mismatch: An Outcome of the Priority Medical Devices Project*, World Health Organization, Geneva, 2010.
17. U. N. E. P. (UNEP), *Atlas of Africa Energy Resources,* United Nations Environment Programme, Nairobi, 2017.
18. J. Porcaro, S. Mehta, M. Shupler, S. Kissel, M. Pfeiffer, C. F. C. Dora, and H. Adair-Rohani, "Modern Energy Access and Health," World Bank, 2017.
19. P. Action, *Poor People's Energy Outlook 2016: National Energy Access Planning from the Bottom Up*, Practical Action Publishing, Warwickshire, 2016.
20. P. Action, *Gender and Livelihoods Impacts of Clean Cookstoves in South Asia*, Global Alliance for Clean Cookstoves, Washington DC, 2014. Available online: https://www.cleancookingalliance.org/binary-data/RESOURCE/file/000/000/363-1.pdf
21. K. O'Dell, S. Peters, and K. Wharton, Women, *Energy and Economic Empowerment: Applying a Gender Lens to Amplify the Impact of Energy Access*, Deloitte University Press, Washington DC, 2014.
22. T. Dinkelman, "The effects of rural electrification on employment: New evidence from South Africa," *American Economic Review*, vol. 101, no. 7, pp. 3078–3108, 2011.
23. L. Grogan, and A. Sadanand, "Rural electrification and employment in poor countries: Evidence from Nicaragua," *World Development*, vol. 43, pp. 252–265, 2013.
24. Ministry of Food Processing Industries, Annual Report 2016–17, Government of India, New Delhi, 2017.
25. R. Sims, A. Flammini, M. Puri, and S. Bracco, *Opportunities for Agri-food Chains to Become Energy-smart*, pp. 978–92, UN Food and Agricultural Organisation, Rome and GIZ, 2015.
26. J. N. Arlet, D. Diane, T. Parvanyan, J. Srinivasan, E. Tjong, *Doing Business 2017: Getting Electricity: Factors Affecting the Reliability of Electricity Supply,* 122197, vol. 1, World Bank Group, Washington DC, 2017.
27. I. E. A. (IEA), *India Energy Outlook 2015*, International Energy Agency (IEA), Paris, 2015.
28. A. Kumar, A. R. Singh, Y. Deng, X. He, P. Kumar, and R. C. Bansal, "A novel methodological framework for the design of sustainable rural microgrid for developing nations," *IEEE Access*, vol. 6, pp. 24925–24951, 2018.
29. I. E. Agency, *World Energy Outlook 2016-Factsheet*, International Energy Agency (IEA), Paris, 2016.
30. K. Narula, and B. S. Reddy, "A SES (sustainable energy security) index for developing countries," *Energy*, vol. 94, no. Supplement C, pp. 326–343, 2016.
31. S. K. Sahoo, P. Varma, K. P. Lall, and C. K. Talwar, "Energy efficiency in India: Achievements, challenges and legality," *Energy Policy*, vol. 88, pp. 495–503, 2016.
32. C. Arndt, and D. Arent, "Special section of applied energy: Energy and climate change in sub-Saharan Africa," *Applied Energy*, vol. 161, pp. 553–555, 2016.
33. N. S. Ouedraogo, "Modeling sustainable long-term electricity supply-demand in Africa," *Applied Energy*, vol. 190, pp. 1047–1067, 2017.
34. U. S. E. I. A. (EIA), *International Energy Outlook 2016 With Projections to 2040*, Energy Information Administration (US), and Government Publications Office, Washington, DC, 2016.
35. D. Palit, and K. R. Bandyopadhyay, "Rural electricity access in India in retrospect: A critical rumination," *Energy Policy*, vol. 109, pp. 109–120, 2017.
36. A. Kumar, Y. Deng, X. He, P. Kumar, and R. C. Bansal, "Energy management system controller for a rural microgrid," *The Journal of Engineering*, vol. 2017, no. 13, pp. 834–839, 2017.

37. A. Kumar, B. Sah, Y. Deng, X. He, P. Kumar, and R. C. Bansal, "Application of multi-criteria decision analysis tool for design of a sustainable micro-grid for a remote village in the Himalayas," *The Journal of Engineering*, vol. 2017, no. 13, pp. 2108–2113, 2017.
38. A. Kumar, Y. Deng, X. He, and P. Kumar, "A multi criteria decision based rural electrification system," 42nd Annual Conference of the IEEE Industrial Electronics Society (IECON 2016), pp. 4025–4030, Florance, Italy, 2016.
39. A. S. Al-Sumaiti, M. M. A. Salama, and M. El-Moursi, "Enabling electricity access in developing countries: A probabilistic weather driven house based approach," *Applied Energy*, vol. 191, pp. 531–548, 2017.
40. V. Chaturvedi, J. Eom, L. E. Clarke, and P. R. Shukla, "Long term building energy demand for India: Disaggregating end use energy services in an integrated assessment modeling framework," *Energy Policy*, vol. 64, pp. 226–242, 2014.
41. A. Das, and S. K. Paul, "Changes in energy requirements of the residential sector in India between 1993–94 and 2006–07," *Energy Policy*, vol. 53, pp. 27–40, 2013.
42. B. Kumar Singh, "South Asia energy security: Challenges and opportunities," *Energy Policy*, vol. 63, pp. 458–468, 2013.
43. R. Bansal, *Handbook of Distributed Generation: Electric Power Technologies, Economics and Environmental Impacts*, Springer, Switzerland, 2017.
44. K. S. Benjamin, B. Morgan, and T. Michael, "Paradigms and poverty in global energy policy: research needs for achieving universal energy access," *Environmental Research Letters*, vol. 11, no. 6, pp. 064014, 2016.
45. B. K. Sovacool, "Design principles for renewable energy programs in developing countries," *Energy & Environmental Science*, vol. 5, no. 11, pp. 9157–9162, 2012.
46. B. K. Sovacool, "Energy studies need social science," *Nature*, vol. 511, no. 7511, pp. 529, 2014.
47. Q. Wang, and K. Zhou, "A framework for evaluating global national energy security," *Applied Energy*, vol. 188, pp. 19–31, 2017.
48. R. D. Prasad, R. C. Bansal, and A. Raturi, "Multi-faceted energy planning: A review," *Renewable and Sustainable Energy Reviews*, vol. 38, pp. 686–699, 2014.
49. R. D. Prasad, R. Bansal, and A. Raturi, "A review of Fiji's energy situation: Challenges and strategies as a small island developing state," *Renewable and Sustainable Energy Reviews*, vol. 75, pp. 278–292, 2017.
50. A. Kumar, B. Sah, A. R. Singh, Y. Deng, X. He, P. Kumar, and R. C. Bansal, "A review of multi criteria decision making (MCDM) towards sustainable renewable energy development," *Renewable and Sustainable Energy Reviews*, vol. 69, pp. 596–609, 2017.
51. A. Kumar, G. Bam, B. Sah, and P. Kumar, "A hybrid micro grid for remote village in himalayas," 3rd IET Renewable Power Generation Conference (RPG), Naples, Italy, pp. 1–6, 2014.
52. United Nations Framework Convention on Climate Change (UNFCCC), "COP 21, Paris Agreement." Available online: https://unfccc.int/process-and-meetings/conferences/past-conferences/paris-climate-change-conference-november-2015/cop-21
53. United Nations Environment Programme (UNEP), *Climate Commitments of Subnational Actors and Business: A Quantitative Assessment of Their Emission Reduction Impact*, United Nations, New York, 2015.
54. International Renewable Energy Agency (IRENA), *Renewable Energy Policies in a Time of Transition'. IRENA, OECD/IEA and REN21*, International Energy Agency (IEA), Paris, 2018.
55. A. Hirsch, Y. Parag, and J. Guerrero, "Microgrids: A review of technologies, key drivers, and outstanding issues," *Renewable and Sustainable Energy Reviews*, vol. 90, pp. 402–411, 2018.
56. S. Mahapatra, and S. Dasappa, "Rural electrification: Optimising the choice between decentralised renewable energy sources and grid extension," *Energy for Sustainable Development*, vol. 16, no. 2, pp. 146–154, 2012.

57. S. Mandelli, J. Barbieri, R. Mereu, and E. Colombo, "Off-grid systems for rural electrification in developing countries: Definitions, classification and a comprehensive literature review," *Renewable and Sustainable Energy Reviews*, vol. 58, pp. 1621–1646, 2016.
58. B. Johansson, "Security aspects of future renewable energy systems–A short overview," *Energy*, vol. 61, pp. 598–605, 2013.
59. B. W. Ang, W. L. Choong, and T. S. Ng, "Energy security: Definitions, dimensions and indexes," *Renewable and Sustainable Energy Reviews*, vol. 42, pp. 1077–1093, 2015.
60. H. Winkler, A. F. Simões, E. L. l. Rovere, M. Alam, A. Rahman, and S. Mwakasonda, "Access and affordability of electricity in developing countries," *World Development*, vol. 39, no. 6, pp. 1037–1050, 2011.
61. L. Meng, E. R. Sanseverino, A. Luna, T. Dragicevic, J. C. Vasquez, and J. M. Guerrero, "Microgrid supervisory controllers and energy management systems: A literature review," *Renewable and Sustainable Energy Reviews*, vol. 60, pp. 1263–1273, 2016.
62. L. Mariam, M. Basu, and M. F. Conlon, "Microgrid: Architecture, policy and future trends," *Renewable and Sustainable Energy Reviews*, vol. 64, pp. 477–489, 2016.
63. R. H. Lasseter, "MicroGrids," IEEE Power Engineering Society Winter Meeting, vol. 1, pp. 305–308, 2002.
64. D. T. Ton, and M. A. Smith, "The U.S. Department of Energy's Microgrid Initiative," *The Electricity Journal*, vol. 25, no. 8, pp. 84–94, 2012.
65. N. Hatziargyriou, "Microgrids–large scale integration of microgeneration to low voltage grids," CIGRE C6-309, Paris, 2006.
66. N. Hatziargyriou, *Microgrids: Architectures and Control*, John Wiley & Sons, West Sussex 2014.
67. N. W. A. Lidula, and A. D. Rajapakse, "Microgrids research: A review of experimental microgrids and test systems," *Renewable and Sustainable Energy Reviews*, vol. 15, no. 1, pp. 186–202, 2011.
68. O. Hafez, and K. Bhattacharya, "Optimal planning and design of a renewable energy based supply system for microgrids," *Renewable Energy*, vol. 45, pp. 7–15, 2012.
69. R. Siddaiah, and R. P. Saini, "A review on planning, configurations, modeling and optimization techniques of hybrid renewable energy systems for off grid applications," *Renewable and Sustainable Energy Reviews*, vol. 58, pp. 376–396, 2016.
70. P. A. Madduri, J. Poon, J. Rosa, M. Podolsky, E. A. Brewer, and S. R. Sanders, "Scalable DC Microgrids for Rural Electrification in Emerging Regions," *IEEE Journal of Emerging and Selected Topics in Power Electronics*, vol. 4, no. 4, pp. 1195–1205, 2016.
71. D. Schnitzer, D. S. Lounsbury, J. P. Carvallo, R. Deshmukh, J. Apt, and D. M. Kammen, *Microgrids for Rural Electrification: A Critical Review of Best Practices Based on Seven Case Studies*," United Nations Foundation, New York, 2014.
72. D. Schnitzer, D. S. Lounsbury, J. P. Carvallo, R. Deshmukh, J. Apt, and D. M. Kammen, *Microgrids for Rural Electrification*, IEEE Smart Grid, United Nations Foundation, New York, 2014.
73. B. Domenech, L. Ferrer-Martí, P. Lillo, R. Pastor, and J. Chiroque, "A community electrification project: Combination of microgrids and household systems fed by wind, PV or microhydro energies according to micro-scale resource evaluation and social constraints," *Energy for Sustainable Development*, vol. 23, pp. 275–285, 2014.
74. A. Chauhan, and R. P. Saini, "A review on Integrated renewable energy system based power generation for stand-alone applications: Configurations, storage options, sizing methodologies and control," *Renewable and Sustainable Energy Reviews*, vol. 38, pp. 99–120, 2014.
75. A. Chauhan, and R. P. Saini, "Renewable energy based off-grid rural electrification in Uttarakhand state of India: Technology options, modelling method, barriers and recommendations," *Renewable and Sustainable Energy Reviews*, vol. 51, pp. 662–681, 2015.

76. M. H. Nehrir, C. Wang, K. Strunz, H. Aki, R. Ramakumar, J. Bing, Z. Miao, and Z. Salameh, "A review of hybrid renewable/alternative energy systems for electric power generation: Configurations, control, and applications," *IEEE Transactions on Sustainable Energy*, vol. 2, no. 4, pp. 392–403, 2011.
77. K. Agbossou, M. Kolhe, J. Hamelin, and T. K. Bose, "Performance of a stand-alone renewable energy system based on energy storage as hydrogen," *IEEE Transactions on Energy Conversion*, vol. 19, no. 3, pp. 633–640, 2004.
78. K. Agbossou, R. Chahine, J. Hamelin, F. Laurencelle, A. Anouar, J. M. St-Arnaud, and T. K. Bose, "Renewable energy systems based on hydrogen for remote applications," *Journal of Power Sources*, vol. 96, no. 1, pp. 168–172, 2001.
79. K. Sung-Hun, S. R. Lee, H. Dehbonei, and C. V. Nayar, "Application of voltage- and current-controlled voltage source inverters for distributed generation systems," *IEEE Transactions on Energy Conversion*, vol. 21, no. 3, pp. 782–792, 2006.
80. F. A. Farret, and M. G. Simoes, *Integration of Alternative Sources of Energy*, John Wiley & Sons, Hoboken, NJ, 2006.
81. M. H. Nehrir, and C. Wang, *Modeling and Control of Fuel Cells: Distributed Generation Applications*, John Wiley & Sons, Hoboken, NJ, 2009.
82. P. K. Sood, T. A. Lipo, and I. G. Hansen, "A versatile power converter for high-frequency link systems," *IEEE Transactions on Power Electronics*, vol. 3, no. 4, pp. 383–390, 1988.
83. H. J. Cha, and P. N. Enjeti, "A three-phase AC/AC high-frequency link matrix converter for VSCF applications," 34th Annual IEEE Power Electronics Specialist Conference, vol. 4, pp. 1971–1976, 2003.
84. N. Carlisle, J. Elling, and T. Penney, *A Renewable Energy Community: Key Elements*, National Renewable Energy Laboratory, Golden, CO, 2008.
85. Ministry of Power, Government of India, "Power Sector at a Glance All India." 2016. Available online: https://powermin.nic.in/en/content/power-sector-glance-all-india
86. N. S. S. Office, *Household Consumption of Various Goods and Services in India*, Ministry of Statistics and Programme Implementation, Government of India, New Delhi NSS Report No. 558, (July 2011 – June 2012) June, 2014.
87. D. Palit, B. K. Sovacool, C. Cooper, D. Zoppo, J. Eidsness, M. Crafton, K. Johnson, and S. Clarke, "The trials and tribulations of the Village Energy Security Programme (VESP) in India," *Energy Policy*, vol. 57, pp. 407–417, 2013.
88. Rural Electrification Corporation (REC), Ministry of Power, Governement of India, "Garv Grameen Vidyutikaran." 2018. Available online: http://garv.gov.in/dashboard/ue
89. E. Dugoua, R. Liu, and J. Urpelainen, "Geographic and socio-economic barriers to rural electrification: New evidence from Indian villages," *Energy Policy*, vol. 106, pp. 278–287, 2017.
90. D. Silva Herran, and T. Nakata, "Design of decentralized energy systems for rural electrification in developing countries considering regional disparity," *Applied Energy*, vol. 91, no. 1, pp. 130–145, 2012.
91. J. Urpelainen, "Grid and off-grid electrification: An integrated model with applications to India," *Energy for Sustainable Development*, vol. 19, pp. 66–71, 2014.
92. J. Hazelton, A. Bruce, and I. MacGill, "A review of the potential benefits and risks of photovoltaic hybrid mini-grid systems," *Renewable Energy*, vol. 67, pp. 222–229, 2014.
93. T. Urmee, and A. Md, "Social, cultural and political dimensions of off-grid renewable energy programs in developing countries," *Renewable Energy*, vol. 93, pp. 159–167, 2016.
94. J. R. S. C. Mateo, *Multi Criteria Analysis in the Renewable Energy Industry*, Springer, London, 2012.
95. H. Polatidis, D. A. Haralambopoulos, G. Munda, and R. Vreeker, "Selecting an appropriate multi-criteria decision analysis technique for renewable energy planning," *Energy Sources, Part B: Economics, Planning, and Policy*, vol. 1, no. 2, pp. 181–193, 2006.

96. J. C. Rojas-Zerpa, and J. M. Yusta, "Application of multicriteria decision methods for electric supply planning in rural and remote areas," *Renewable and Sustainable Energy Reviews*, vol. 52, pp. 557–571, 2015.
97. E. Strantzali, and K. Aravossis, "Decision making in renewable energy investments: A review," *Renewable and Sustainable Energy Reviews*, vol. 55, pp. 885–898, 2016.
98. S. D. Pohekar, and M. Ramachandran, "Application of multi-criteria decision making to sustainable energy planning—A review," *Renewable and Sustainable Energy Reviews*, vol. 8, no. 4, pp. 365–381, 2004.
99. A. M. González, H. Sandoval, P. Acosta, and F. Henao, "On the acceptance and sustainability of renewable energy projects—A systems thinking perspective," *Sustainability*, vol. 8, no. 11, pp. 1171, 2016.
100. B. Tilt, Y. Braun, and D. He, "Social impacts of large dam projects: A comparison of international case studies and implications for best practice," *Journal of Environmental Management*, vol. 90, pp. S249–S257, 2009.
101. T. Mohn, "It takes a village: Rural electrification in East Africa," *IEEE Power and Energy Magazine*, vol. 11, no. 4, pp. 46–51, 2013.
102. G. Sargsyan, M. Bhatia, S. G. Banerjee, K. Raghunathan, and R. Soni, *Unleashing the Potential of Renewable Energy in India*, The World Bank, 2011.
103. A. Kumar, B. Sah, Y. Deng, X. He, R. Bansal, and P. Kumar, "Autonomous hybrid renewable energy system optimization for minimum cost," 4th International Conference on Renewable Power Generation (RPG), pp. 1–6, Beijing, 2015.
104. Ministry of New and Renewable Energy, Government of India, *Annual Report 2016–17: Chapter 13—Renewable Energy in North-eastern States*, Government of India, New Delhi, 2016–17.
105. Ministry of Development of North Eastern Region, Government of India, "Union Budget (2017–18)." 2017. Available online: https://openbudgetsindia.org/dataset/ministry-of-development-of-north-eastern-region-2017-18
106. Ministry of Development of North Eastern Region, Governement of India "Hydro Power Projects." 2019. Available online: http://www.mdoner.gov.in/node/1307
107. P. Barah, and K. Bhuyan, "India: Endangered species damned by dams," *Nature*, vol. 515, no. 7525, pp. 37–37, 2014.
108. D. Kariuki, "Barriers to Renewable Energy Technologies Development." 2018. Available online: https://www.energytoday.net/economics-policy/barriers-renewable-energy-technologies-development
109. A. Fashina, M. Mundu, O. Akiyode, L. Abdullah, D. Sanni, and L. Ounyesiga, "The drivers and barriers of renewable energy applications and development in Uganda: A review," *Clean Technologies*, vol. 1, no. 1, pp. 3, 2018.
110. P. Cook, "Infrastructure, rural electrification and development," *Energy for Sustainable Development*, vol. 15, no. 3, pp. 304–313, 2011.
111. S. K. Jain, V. Kumar, and M. Saharia, "Analysis of rainfall and temperature trends in northeast India," *International Journal of Climatology*, vol. 33, no. 4, pp. 968–978, 2013.
112. N. A. India Energy Portal, *Transmission and Distribution Losses NITI Aayog*, NITI Aayog/ Planning Commission, Government of India, New Delhi, 2015.
113. T. Urmee, D. Harries, and A. Schlapfer, "Issues related to rural electrification using renewable energy in developing countries of Asia and Pacific," *Renewable Energy*, vol. 34, no. 2, pp. 354–357, 2009.
114. H. Al Moussawi, F. Fardoun, and H. Louahlia-Gualous, "Review of tri-generation technologies: Design evaluation, optimization, decision-making, and selection approach," *Energy Conversion and Management*, vol. 120, no. Supplement C, pp. 157–196, 2016.

115. W. L. Theo, J. S. Lim, W. S. Ho, H. Hashim, and C. T. Lee, "Review of distributed generation (DG) system planning and optimisation techniques: Comparison of numerical and mathematical modelling methods," *Renewable and Sustainable Energy Reviews*, vol. 67, pp. 531–573, 2017.
116. C. L. Azimoh, P. Klintenberg, C. Mbohwa, and F. Wallin, "Replicability and scalability of mini-grid solution to rural electrification programs in sub-Saharan Africa," *Renewable Energy*, vol. 106, pp. 222–231, 2017.
117. A. López-González, B. Domenech, D. Gómez-Hernández, and L. Ferrer-Martí, "Renewable microgrid projects for autonomous small-scale electrification in Andean countries," *Renewable and Sustainable Energy Reviews*, vol. 79, pp. 1255–1265, 2017.
118. M. Qolipour, A. Mostafaeipour, and O. M. Tousi, "Techno-economic feasibility of a photovoltaic-wind power plant construction for electric and hydrogen production: A case study," *Renewable and Sustainable Energy Reviews*, vol. 78, pp. 113–123, 2017.
119. H. S. Das, C. W. Tan, A. H. M. Yatim, and K. Y. Lau, "Feasibility analysis of hybrid photovoltaic/battery/fuel cell energy system for an indigenous residence in East Malaysia," *Renewable and Sustainable Energy Reviews*, vol. 76, pp. 1332–1347, 2017.
120. E. Park, "Potentiality of renewable resources: Economic feasibility perspectives in South Korea," *Renewable and Sustainable Energy Reviews*, vol. 79, pp. 61–70, 2017.
121. A. Ajlan, C. W. Tan, and A. M. Abdilahi, "Assessment of environmental and economic perspectives for renewable-based hybrid power system in Yemen," *Renewable and Sustainable Energy Reviews*, vol. 75, pp. 559–570, 2017.
122. A. Jaiswal, "Lithium-ion battery based renewable energy solution for off-grid electricity: A techno-economic analysis," *Renewable and Sustainable Energy Reviews*, vol. 72, pp. 922–934, 2017.
123. M. J. Khan, A. K. Yadav, and L. Mathew, "Techno economic feasibility analysis of different combinations of PV-Wind-Diesel-Battery hybrid system for telecommunication applications in different cities of Punjab, India," *Renewable and Sustainable Energy Reviews*, vol. 76, pp. 577–607, 2017.
124. V. Tomar, and G. N. Tiwari, "Techno-economic evaluation of grid connected PV system for households with feed in tariff and time of day tariff regulation in New Delhi – A sustainable approach," *Renewable and Sustainable Energy Reviews*, vol. 70, pp. 822–835, 2017.
125. A. Bhatt, M. P. Sharma, and R. P. Saini, "Feasibility and sensitivity analysis of an off-grid micro hydro-photovoltaic-biomass and biogas-diesel-battery hybrid energy system for a remote area in Uttarakhand state, India," *Renewable and Sustainable Energy Reviews*, vol. 61, pp. 53–69, 2016.
126. A. Chauhan, and R. P. Saini, "Techno-economic feasibility study on Integrated Renewable Energy System for an isolated community of India," *Renewable and Sustainable Energy Reviews*, vol. 59, pp. 388–405, 2016.
127. W. M. Amutha, and V. Rajini, "Cost benefit and technical analysis of rural electrification alternatives in southern India using HOMER," *Renewable and Sustainable Energy Reviews*, vol. 62, pp. 236–246, 2016.
128. N. Ramchandran, R. Pai, and A. K. S. Parihar, "Feasibility assessment of Anchor-Business-Community model for off-grid rural electrification in India," *Renewable Energy*, vol. 97, pp. 197–209, 2016.
129. H. S. Das, A. H. M. Yatim, C. W. Tan, and K. Y. Lau, "Proposition of a PV/tidal powered micro-hydro and diesel hybrid system: A southern Bangladesh focus," *Renewable and Sustainable Energy Reviews*, vol. 53, pp. 1137–1148, 2016.
130. H. Rezk, and G. M. Dousoky, "Technical and economic analysis of different configurations of stand-alone hybrid renewable power systems – A case study," *Renewable and Sustainable Energy Reviews*, vol. 62, pp. 941–953, 2016.

131. S. Upadhyay, and M. P. Sharma, "Development of hybrid energy system with cycle charging strategy using particle swarm optimization for a remote area in India," *Renewable Energy*, vol. 77, pp. 586–598, 2015.
132. J. G. Castellanos, M. Walker, D. Poggio, M. Pourkashanian, and W. Nimmo, "Modelling an off-grid integrated renewable energy system for rural electrification in India using photovoltaics and anaerobic digestion," *Renewable Energy*, vol. 74, pp. 390–398, 2015.
133. S. Yilmaz, and F. Dincer, "Optimal design of hybrid PV-Diesel-Battery systems for isolated lands: A case study for Kilis, Turkey," *Renewable and Sustainable Energy Reviews*, vol. 77, pp. 344–352, 2017.
134. D. O. Akinyele, and R. K. Rayudu, "Comprehensive techno-economic and environmental impact study of a localised photovoltaic power system (PPS) for off-grid communities," *Energy Conversion and Management*, vol. 124, pp. 266–279, 2016.
135. S. C. Bhattacharyya, *Energy Economics: Concepts, Issues, Markets and Governance*, Springer Science & Business Media, London, 2011.
136. P. K. Adom, and W. Bekoe, "Conditional dynamic forecast of electrical energy consumption requirements in Ghana by 2020: A comparison of ARDL and PAM," *Energy*, vol. 44, no. 1, pp. 367–380, 2012.
137. R. C. Bansal, and J. C. Pandey, "Load forecasting using artificial intelligence techniques: a literature survey," *International Journal of Computer Applications in Technology*, vol. 22, no. 2/3, pp. 109–119, 2005.
138. M. S. Kandil, S. M. El-Debeiky, and N. E. Hasanien, "Overview and comparison of long-term forecasting techniques for a fast developing utility: Part I," *Electric Power Systems Research*, vol. 58, no. 1, pp. 11–17, 2001.
139. S. C. Bhattacharyya, "Energy access problem of the poor in India: Is rural electrification a remedy?," *Energy Policy*, vol. 34, no. 18, pp. 3387–3397, 2006.

CHAPTER 15

Optimal Sizing of Hybrid Energy Systems

Ranjay Singh[1], Ramesh C. Bansal[2,1], Arvind R. Singh[3], and R.M. Naidoo[1]

[1]Department of Electrical, Electronic and Computer Engineering, University of Pretoria, South Africa. ranjaysingh.c@gmail.com, raj@up.ac.za
[2]Department of Electrical Engineering, University of Sharjah, United Arab Emirates. rcbansal@ieee.org
[3]School of Electrical Engineering, Shandong University, Jinan P.R. China. arvindsinghwce@gmail.com

Abstract This chapter presents a holistic overview of hybrid energy systems consisting of multiple energy generation and energy storage sources. The chapter initially presents the mathematical modelling of the various system components necessary for analysis of optimal system sizing. The various criterions based on which the system is optimized is also articulated with relevant formulations. The chapter presents an all-inclusive review of various optimization techniques and its application for optimal sizing of hybrid renewable energy system. It also articulates the various drawbacks of each optimization technique hence providing the reader with relevant knowledge to make an informed decision with respect to choosing an optimization technique.

Keywords: Hybrid system, Optimization techniques, Renewable energy, Evolutionary algorithm, Bioinspired algorithm, Optimization criterions.

15.1 Introduction

The ever-increasing awareness toward the harmful impact of conventional energy-generating sources has brought attention toward renewable energy sources. The Kyoto protocol in collaboration with the United Nations has committed its parties to reduce carbon emissions and a viable option to achieve this is through renewable energy generation. There are several renewable energy sources such as photovoltaic (PV), wind, geothermal, micro-hydro, tidal, and biomass. By end of 2019, the World grid installed capacity of renewable energy sources was 2537 GW (hydro 1190 GW, Solar 586 GW, wind 623 GW, bioenergy 124 GW, and geothermal 13.9 GW [1]. Wind and PV are two most popular fast growing renewable energy sources [2].

Renewable energy sources not only have the advantage of emission-free (or low emission) energy generation, but also the fuel cost for these generating systems is free. The current economic policies and intermittent nature of energy generation from these sources makes the cost of the renewable system very high. One way of reducing the unpredictability of energy generation is by using multiple sources for energy generation along with energy storage system (ESS), which works in synergy for load agreement. This combination using multiple energy-generating units and ESS is termed as hybrid energy system (HES), a block diagram of a typical HES is shown in Fig. 15.1. Addition of storage units not only aids in increasing the reliability of HES, but also performs the following objectives: power quality improvement, axillary power supply, and peak load shaving for grid-connected system [3]. To further increase the reliability of supply, conventional energy-generating units can be used with a trade-off with emission and an additional cost of fuel. It is a viable option for systems that are off-grid and reliability of power is the prime requirement. A prudent way of reducing the system cost is by using an optimal size of generating units based on the load requirement and environmental factors [4].

In HES, multiple energy-generating sources and ESS units can be used, thus making the system reliability high, but the optimal sizing of the generation and storage units is of prime importance. The HES can be optimized by recognizing the criteria that the system needs to be furnished; these criterions are broadly classified as technical, economic, social–political, and environmental. Based on the chosen criteria(s), several optimization constraints are satisfied to obtain the optimal system configuration. These optimization constraints are loss of power supply probability (LPSP), loss of load (LOL), and so on, for technical criteria; total

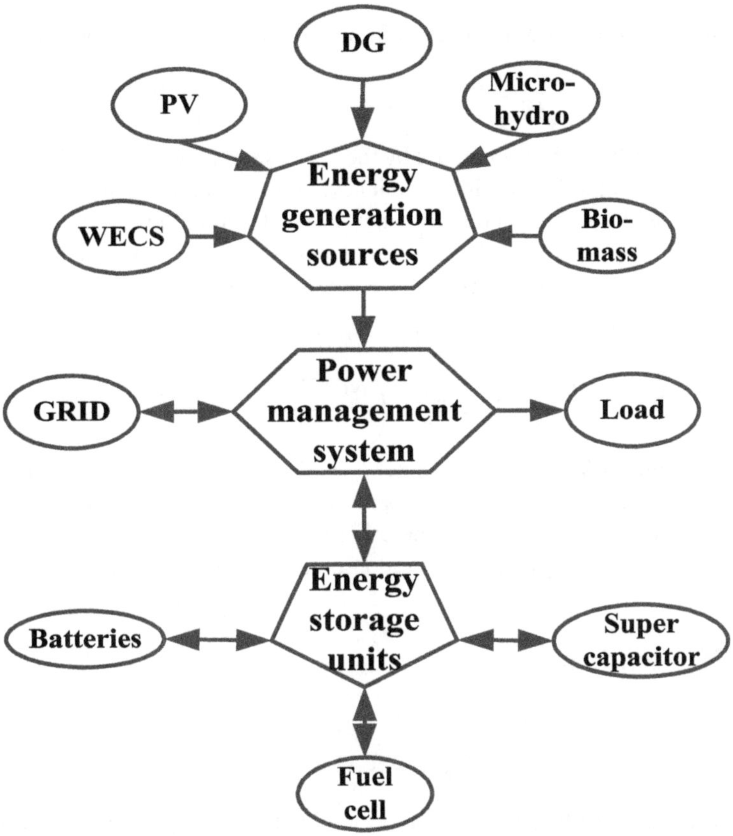

Figure 15.1. Block diagram of a HES. WECS—Wind Energy Conversion System, DG—Diesel Generators

system cost (TSC), cost of energy (COE), and so on, for economic criteria [5]; life cycle emissions (LCEs), fuel consumption (FC), and so on, for environmental criteria; and social cost of carbon (SCC) for social criteria. Once the objective criteria(s) and constraint(s) of the HES are decided, the optimization of the system components can be performed by several techniques. These techniques are broadly classified as global search and optimization algorithm (GSOA), artificial neural network (ANN), fuzzy logic (FL), conditional approximate reasoning (CAR), mathematical iterative algorithms (MIA), and software tools.

15.2 Modelling of HRES

The various components of the HES need to be modelled for the optimization technique to be implemented. The modelling of main components is explained in brief as follows:

15.2.1 Modelling of PV Systems

The hourly energy output for a mono-crystalline PV module can be expressed as shown in (15.1) [6]:

$$E_{PV}(t) = N_{PV} \times A_{PV} \times \eta_{PV} \times I(t) \tag{15.1}$$

where E_{PV} is the hourly energy generated by the PV system in watt hour, N_{PV} is the number of PV modules, η_{PV} is the PV module efficiency, and $I(t)$ is the net solar radiation reaching the PV modules in Wh/m^2. The efficiency of the PV modules is very low and varies with operating cell temperature, which is also dependent on the ambient temperature. The variation of PV efficiency with temperature is modelled using (15.2) and (15.3) [6]:

$$\eta_{PV} = \eta_r \left[1 - \beta_T \left(T_c - T_r\right)\right] \tag{15.2}$$

$$T_c = T_a + \left(\frac{NOCT - T_{a,NOCT}}{I_{NOCT}}\right) \times I(t) \tag{15.3}$$

where, η_r is the rated PV panel efficiency, β_T is the PV temperature coefficient of efficiency, T_c is the computed ambient temperature in °C, T_r is the temperature at rated efficiency in °C, NOCT is the normal operating cell temperature in °C, T_a is the ambient temperature in °C, and I_{NOCT} is the solar radiation at NOCT in Wh/m^2.

15.2.2 WECS Modelling

The hourly energy generated by the WECS is expressed in (15.4) [7]:

$$E_W = \frac{1}{2} \times \rho \times A_w \times \eta_w \times v^3 \tag{15.4}$$

where A_w is the area swept by the wind turbine blades in m^2, η_w is the power conversion efficiency, ρ is the air density in kg/m^3, and v is the wind speed in m/s at hub height. The air density is not a constant parameter and varies with ambient temperature and height, therefore the variability of the air density is modelled in (15.5) and (15.6) [7]:

$$\rho = \frac{P(h) \times MW \times 10^{-3}}{R \times \left(273 + T_a\right)} \tag{15.5}$$

$$\text{where } P(h) = P_o e^{-(1.185 \times 10^{-4} \times H)} \tag{15.6}$$

where MW is the molecular weight of gas in g/mol; usually considered 28.97 g/mol, R is the ideal gas constant, which is 8.2056×10^{-3} m^3.atm.K^{-1}.mol^{-1}, and $P(h)$ is the atmospheric pressure in atm, which is a function of height (H) and varies as expressed in (15.6). The wind speed is usually recorded at a height of 10 m, but the wind speed varies with the increase in height. The wind speed at the hub height of the wind turbine is expressed in (15.7):

$$\left(\frac{v}{v_0}\right) = \left(\frac{H}{H_0}\right)^{\alpha} \tag{15.7}$$

where, v_0 is the wind speed in m/s recorded at height H_0, v is the wind speed in m/s at height H (hub height), and α is the friction coefficient. The friction coefficient for different terrains can vary from 0.1 for a smooth ground to 0.40 for cities with tall building [7].

15.2.3 Modelling of Energy Storage Systems

The charge stored in the BESS is dependent on the net energy generated by energy-generating units (PV, WECS, etc.) and the primary load requirement. The charge of BESS for a particular time interval is governed by (15.8) and (15.9) [8]:

$$E_B(t+1) = E_B(t)(1-\acute{o}) + (surplus\ energy) \times \eta_{BC};\ \text{Charging mode} \tag{15.8}$$

$$E_B(t+1) = E_B(t)(1-\ \acute{o}) - (deficit\ energy) \times \eta_{BD};\ \text{Discharging mode} \tag{15.9}$$

where E_B is the energy in the BESS, η_{BC} and η_{BD} is the charge and discharge efficiency of BESS, respectively, and σ is the self-discharge rate of the BESS. There are certain prerequisites that the BESS should satisfy to prevent failure of the system, that is,

$$E_{B,\min} \le E_B(t) \le E_{B,\max} \tag{15.10}$$

$$E_{B,\min} = DOD \times E_{BR} \tag{15.11}$$

where $E_{B,\min}$ and $E_{B,\max}$ are the minimum and maximum allowable charge storage capacity of BESS, DOD is the depth of discharge of the battery, and E_{BR} is the rated capacity of the battery.

15.2.4 Modelling of Diesel Generator System

Diesel generator (DG) is one of the most common system components used to increase the reliability of the HES. DG has proven to be the most viable option for systems working off-grid. The fuel consumption of a DG is estimated in (15.12) [9]:

$$D_f(t) = \acute{a}_D P_{DG}(t) + \acute{a}_D P_{DGr} \tag{15.12}$$

where $D_f(t)$ is the hourly fuel requirement of the DG in L/h. P_{DG} is the average power generated, P_{DGr} is the rated power of the DG in kW, and α_D and β_D are the fuel consumption coefficients (usually considered as 0.25 and 0.08, respectively). The overall efficiency of the DG is given by (13):

$$\eta_{overall} = \eta_{generator} \times \eta_{thermal} \tag{15.13}$$

It is observed that the DG when running at 70%–90% of its full load capacity gives the most economical fuel consumption. The choice of rating of the DG is governed by what is the primary role of DG in the HES. If the DG is directly connected to the load, then the rating of the DG should be slightly greater than the maximum load requirement. If DG is used to increase the autonomy of the system by charging the BESS, then the charge generated by the DG should not be greater than $C_{\text{Ah}}/5$ A, where C_{Ah} is the ampere-hour rating of the battery.

15.3 Need of Multi-Criterion for HRES Optimization

The primary concern for system planners initially was to reduce the cost of the renewable energy systems, but a single criteria decision-making approach is not sustainable in the current energy scenario. There are numerous stakeholders involved in the process of planning, execution, and operation of HRES, for example, administrative authorities, potential

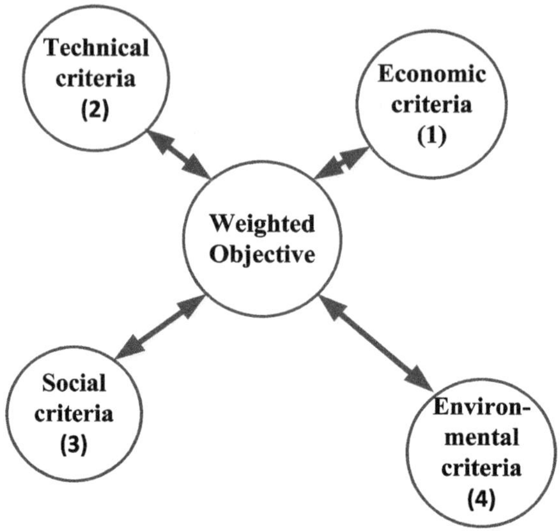

Figure 15.2. Weighted preferences of optimization criteria.

investors, academic institutions, and environmental groups. These authorities have direct or indirect influence in the decision-making process [10]. Therefore, only multiple criteria consideration can lead to the success of the renewable energy project. In the literature, these criteria are broadly classified into four groups: technical criteria, economic criteria, environmental criteria, and social criteria. Technical criteria are mainly concerned with the performance of the HES. The objective of technical criteria is to evaluate feasibility, performance, reliability, efficiency, and so on, of the renewable energy project. The objective of economic criteria is to evaluate the system cost, cost of energy, the return of investment, and availability of funds. The objective of environmental criteria is to evaluate the project performance based on factors such as emissions, fuel consumption, noise, waste management, land use, and topography. The social aspects of the renewable energy project are evaluated based on job creation, land acquisition, and social acceptance of project security. The priority of choosing the constraint for multi-criteria optimization depends on the weighted average of the various criteria as shown in Fig. 15.2. The various parameters for evaluation of optimization criteria are shown in Table 15.1.

15.3.1 Total System Cost (TSC)

This criterion gives the net cost of the system, which includes the cost of components, installation cost, and maintenance cost, through the lifetime of the HES. It is defined by (15.14) [11].

$$TSC = \sum_{n=1}^{n=k} C_C(n) + \sum_{n=1}^{n=k} C_{inst}(n) + \sum_{n=1}^{n=k} C_{M\&O}(n) + C_{BOS} \tag{15.14}$$

where C_C is the capital cost of each component, C_{inst} is the installation cost of the component, $C_{\text{M\&O}}$ is the maintenance and operation cost of each component, and C_{BOS} is the balance of system cost.

15.3.2 Annualized System Cost (ASC)

This criterion gives the annual contribution of the total system cost. This parameter is evaluated for the projects, which have a line of credit. It is calculated by (15.15) and (15.16) [11]:

$$ASC = TSC \times CRF \tag{15.15}$$

$$\text{where } CRF = \frac{r \times (1+r)^L}{(1+r)^L - 1} \tag{15.16}$$

where CRF is known as capital recovery factor, r is the rate of interest, and L is the lifetime of the system. The limitation of this parameter is that change in interest rate and inflation is not considered for the cost estimation.

Table 15.1. Optimization criteria, objectives and evaluation [12,13].

Technical Objectives	Description	Evaluation and Parameters
Feasibility	Assessment of weather and topology of the site for availability of renewable energy.	Solar patterns, wind data, temperature data, rainfall index, topographical data of the site.
Sustainability	Review of the technological requirement of the project in comparison to the available technology.	Evaluation of upcoming and current technologies.
Risk	Review of installation standards and regulation.	Infrastructure constraints such as ventilation level, leakage constraints, distance of nearby buildings, explosion proof barriers.
Reliability	Analysis of the probability of failure of system components. Study of total hours that the system is unable to supply load. Types of load profile to which the system can/cannot cater.	Evaluation of parameters such as LPSP, DPSP, LOLE, LOL, EENS.
Performance	Analysis of component characteristics specific to location.	Evaluation of fuel consumption, maintenance and installation constraints, storage capacity.

(*Continued*)

Table 15.1. Optimization criteria, objectives and evaluation [12,13]. (*Continued*)

Technical Objectives	Description	Evaluation and Parameters
Efficiency	Analysis of systems energy path efficiency.	Evaluation of the cumulative efficiency of all components from the energy source to load.
System integration	Analysis of the types of system components and mode of operation.	Evaluation of possibility of grid connection, available system components.
Economic Objectives	Description	Evaluation and Parameters
System cost	Analysis of economic burden of the project.	Evaluation of TSC, ASC, NPC.
Generation cost	Analysis of the cost of energy generated by the system.	Evaluation of COE, LCE.
Return on investment	Analysis of economic aspects of the system based on the cost of a system versus benefits or savings.	IRR, payback period, cost of energy savings.
Availability of funds	Analysis of available funding from government and international organizations.	Subsidies and rebates applicable for the project.
Environmental Objectives	Description	Evaluation and Parameters
Emission	Analysis of CO_2 equivalent emission reduction.	AEC
Waste management	Analysis of requirement of waste disposal.	Volume of waste generated, type of waste generated, reusability of waste.
Fuel consumption	Minimization of fuel consumption or energy imported from grid.	FC
Lifecycle	Analysis of life cycle of emission generated during commissioning, running and decommissioning of the system.	Life cycle of emission
Social Objectives	Description	Evaluation and Parameters
Political Acceptance	Analysis of policies and subsidies of energy market affecting the energy sector.	Small- and large-scale renewable targets of the government.
Social acceptance	Review of unanimity of public for energy initiatives.	Public participation indicator index.

(*Continued*)

Table 15.1. Optimization criteria, objectives and evaluation [12,13]. (*Continued*)

Technical Objectives	Description	Evaluation and Parameters
Land acquisition	Visual impact, ecosystem altercation, acoustic noise generated, and electromagnetic interference.	Social acceptance indicator index.
Job creation	Analysis of employment opportunities.	Direct and indirect job opportunities in the renewable energy sector.
Security of supply	Analysis of reliability of supply in extreme weather condition and availability of back up generation.	Supply risk management probability.

IRR—Internal Rate of Return, AEC—Annual Emission Cost, FC—Fuel Consumption

15.3.3 Net Present Cost (NPC)

Net present cost is the net cost of the system after it is past its due lifespan. It includes the capital cost, maintenance and operation cost, replacement cost of components, and the salvage cost by selling the system components after its due lifespan. It is calculated by (15.17) [11]:

$$NPC = \sum_{n=1}^{n=k} C_C(n) + \sum_{n=1}^{n=k} C_{inst}(n) + \sum_{n=1}^{n=k} C_{M\&O}(n) + \sum_{n=1}^{n=k} C_{replace}(n) + C_{BOS} - \sum_{n=1}^{n=k} C_{sale}(n) - \sum_{n=1}^{n=k} C_{end}(n) \tag{15.17}$$

where C_{sale} is the cost recovered by selling the components, which need to be replaced, C_{end} is the cost salvage by selling the system components after the HES is past its due lifespan. The limitation of this parameter is that change in fuel cost (in case DG is used), uncertainties in the lifespan of components, and sudden failure of the system component is not incorporated.

15.3.4 Cost of Energy (COE)/Levelized Cost of Energy (LCE)

These parameters are used interchangeably and signify the cost of each energy unit generated by HES. It is calculated by (15.18) [11]:

$$LCE/COE = \frac{C_{TASC}}{\sum_{t=0}^{24} L(t) \times 8760 \times L} \tag{15.18}$$

where L is the load requirement for each day in kW. The limitation of this parameter is that the cost recovered by salvaging the system components is not included.

15.3.5 Lifelong Cost of Battery (LLCB)

It evaluates the cost of BESS inclusive of its initial capital cost and cost of replacement. It is calculated by (15.19) [11]:

$$LLCB = C_{Bat,C} \times \left(1 + \frac{r}{(1+r)^n} + \frac{r}{(1+r)^{2\times n}} + \ldots + \frac{r}{(1+r)^L}\right) + C_{Bat,I} \tag{15.19}$$

where $C_{Bat,\,C}$ is the capital cost of each battery, $C_{Bat,I}$ is the installation cost of each battery, and n is the lifetime of each battery. The limitation of this parameter is that derating of the battery performance during its lifetime is not included.

15.3.6 Loss of Power Supply Probability (LPSP)

It gives us the probability of the load, which cannot be catered by HES. LPSP of 1 signifies that the entire load will be rejected, whereas LPSP of 0 signifies that the entire load is catered by the HES. It is calculated by (15.20) [14]:

$$LPSP = \frac{\sum_{t=1}^{t=T} DE\,(t)}{\sum_{t=1}^{t=T} L\,(t)} \tag{15.20}$$

where $DE\,(t)$ is the hourly deficiency in the energy of the HES and T is the time period of analysis. The limitation of this parameter is that variation in load profile cannot be accessed.

15.3.7 Expected Energy Not Supplied (EENS)

This parameter gives the actual energy not supplied by HES to the load over the entire lifespan. It is calculated by (15.21) [11]:

$$EENS\left(L, P_h\right(= \begin{cases} L - \int\limits_{P_{h\_min}}^{P_{h\_max}} P_h \bullet f_{P_h}\left(P_h\right) dP_h & L > P_{h\_max} \\ \int\limits_{P_{h\_min}}^{L} \left(L - P_h\right) \bullet f_{P_h}\left(P_h\right) dP_h & P_{h\_min} \le L \le P_{h\_max} \\ 0 & L \le P_{h\_min} \end{cases} \tag{15.21}$$

where P_h is the power generated by the HES and f_ph is the probability function of the load not being catered. The limitation of this parameter is the lack of precise estimation of the probability of load not supplied.

15.3.8 Loss of Load (LOL)

This parameter evaluates the annual load not catered by the HES. It is calculated by (15.22) [11]:

$$LOL = \sum_{t=1}^{t=8760} DE\,(t) \tag{15.22}$$

The limitation of this parameter is that the total energy deficient is normalized to the annual deficit.

15.3.9 Deficiency of Power Supply Probability (DPSP)

It evaluates the probability of excess power generation by the HES and is calculated by (15.23) [11]:

$$DPSP = \frac{\sum_{t=1}^{t=T} EPG(t)}{\sum_{t=1}^{t=T} L(t)} \tag{15.23}$$

where EPG is the expected power generated by the HES.

15.4 Optimal Sizing Techniques

The optimization techniques can be broadly classified into six categories. The subclassification of various categories is depicted in Fig. 15.3.

15.4.1 Global Search and Optimization Algorithm (GSOA)

15.4.1.1 Evolutionary Algorithms (EA)

Based on Darwin's theory of natural selection, evolutionary algorithms simulate the process of natural selection by probabilistic search methods. These algorithms arrive at the fittest solution from a population of generated solution through iterative search methods. There are broadly three types of evolutionary algorithms as explained in Section 15.4.1.1.1.

15.4.1.1.1 Genetic Algorithm

Genetic algorithm (GA) was initially introduced by John H. Holland in the 1960s as genetic reproductive plans [15]. There were several modifications that this method has gone through. In its current form, the GA evaluated the fitness of all its population. Once the fitness function is evaluated, new population is generated by performing crossover, reproduction, and mutation operations. GA resembles the process of sexual reproduction in the process of crossover and mutation. The pseudo code of basic operation of GA is illustrated in the following algorithm:

```
Generate population
Evaluate population
While fittest function not found do
    For each member of the population
        t = t + 1
        Regenerate population
        Re-evaluate population
        Selection
        Recombination
        Mutation
```

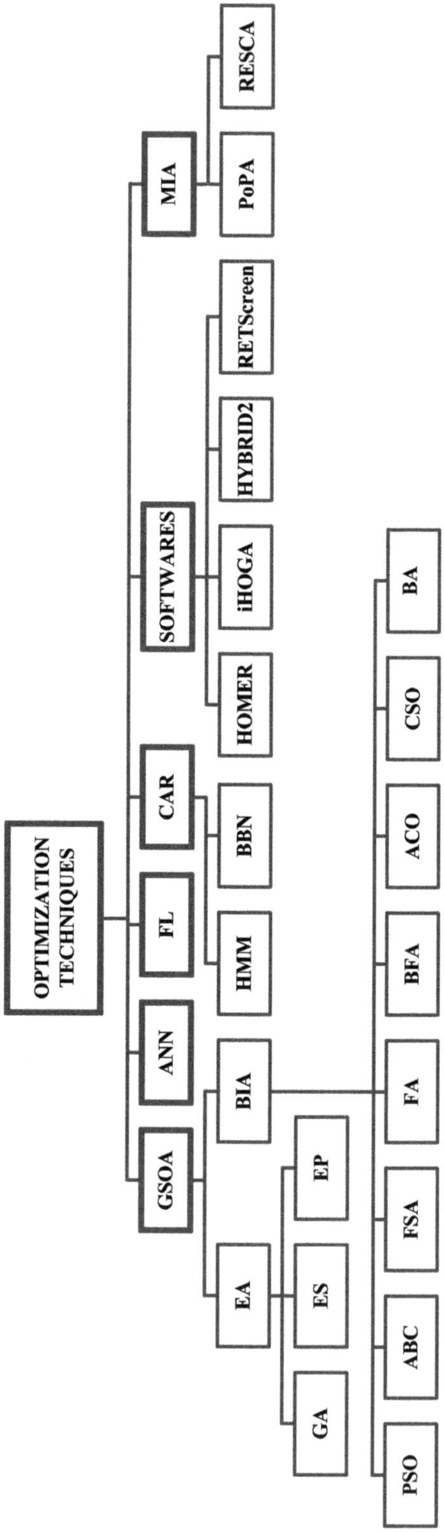

Figure 15.3. Classification of various optimization techniques.

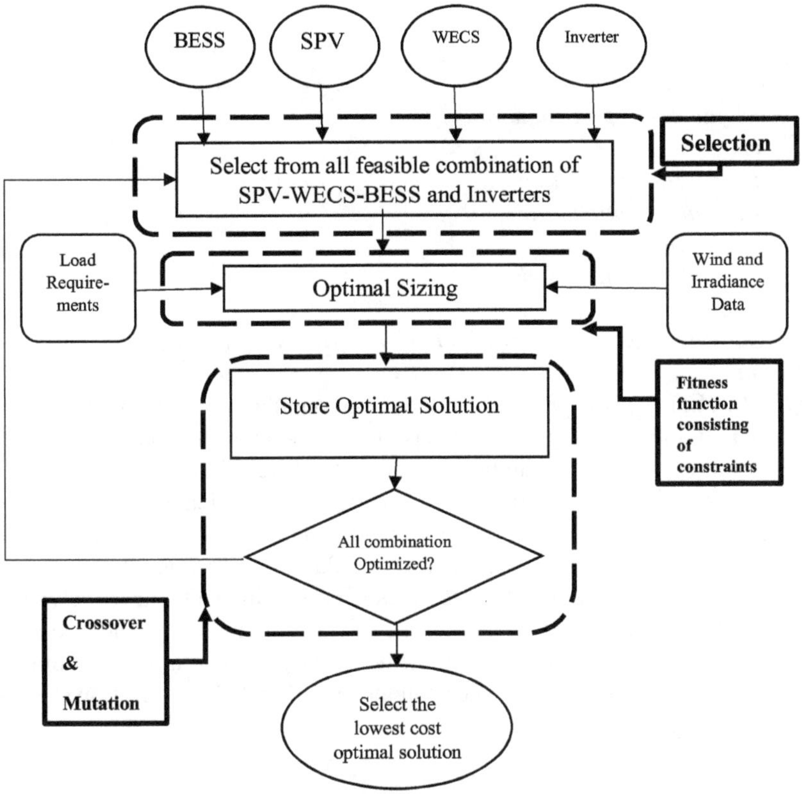

Figure 15.4. Adaptation of GA for HES optimization.

Create new offspring of the population
End For
End While

Initially, the population is generated randomly with all the possible solutions. On the population generated, a fitness function is applied to obtain the fittest solutions, which are used to generate the next population. Then, the recombination process takes place where the fittest solutions are swapped completely or partially to generate the next population. In the mutation process, the nonfit solutions are modified to satisfy the fitness function, which are then used to generate the new population. The fitness of the new population generated is again tested and this process is repeated until a predefined iteration number or until the solution fitness reaches a satisfactory tolerance (optimization constraint). A graphical adaptation of GA for HES optimization is shown in Fig. 15.4 [16].

15.4.1.1.2 Evolutionary Strategies

This technique was introduced by Ingo Rechenberg in 1964 to optimize hydrodynamics process [17]. The modification in this approach is that only a single parent is selected from the generated population, to which mutation is applied to generate the next population. The pseudo code of the evolutionary strategy (ES) is as follows:

Generate population
Evaluate population
While** fittest function not found **do
  ***For** each member of the population*
    t = t + 1
    Regenerate population
    Re-evaluate population
    Selection
    Recombination/crossover
    Mutation
    Substitute worst parents with best offspring's (from t = 2)
  End For
End While The choice of the parent from the initial population is based on a deterministic process. Each individual in the population has three parameters attached to it: a real number signifying its value, a deviation number signifying the level of mutation it can go through, and a vector sign signifying the direction in which the modification can take place. The deviation number and vector sign are used for the mutation process of the parent selected from the initial population to generate the next population. Thereafter, the offspring is generated and the parent competes to generate the next population. Whichever between them is the fittest goes through the mutation and crossover process to generate the next population.

The advantage of this technique is that it does not get stuck at local minima and the solution is immune to the initial estimates. On the other hand, the realization of the optimal solution takes a longer time.

15.4.1.1.3 Evolutionary Programming

This technique was introduced by Lawrence J. Fogel and is very similar to the ES approach [18]. The difference here is that no recombination or crossover takes place. The new population is generated based on the mutation process and the best half of the fittest solution is kept, whereas the rest of the solutions are rejected. Evolutionary programing (EP) has the following pseudo code:
Generate population
Evaluate population
While** fittest function not found **do
  ***For** each member of the population*
    t = t + 1
    Regenerate population
    Re-evaluate population
    Selection
    Mutation
    Keep the best half of the population and delete the rest
  End For
End While

15.4.1.2 Bioinspired Algorithms (BIA)

Interspecies and intraspecies interaction is the outcome of bioinspired algorithm (BIA). Initially, there were only two types of BIA: particle swarm optimization (PSO) and ant colony optimization (ACO). Later, several other optimization methods came into existence such as artificial bee colony (ABC), firefly algorithm (FA), fish swarm algorithm (FSA), and so on, which are explained as follows:

15.4.1.2.1 Particle Swarm Optimization

PSO was developed by Eberhart and Hennedy in 1995. It is inspired by bird flocking, which is addressed as particles in the algorithm [19]. Unlike GA, PSO does not have any evolution, mutation, or crossover operations. In PSO, the particles (birds) fly through the entire search space to identify the global maximum. The fitness of each particle in the PSO is evaluated using a fitness function and a velocity is assigned to it. Based on this velocity, the particle searches the optimal solution following the current optimal particle. The PSO is initialized with the generation of random particles and each particle is assigned two "best" values. The first value is the most optimal solution (fittest) the particle has achieved known as the "pbest." The second value is the fittest solution obtained by any particle in the population termed as "gbest," this is what the particle tracks in the search space. The particle velocity and position are updated based on (15.24), also a $v_{\max}$ parameter is defined by the user to prevent the particle to achieve a velocity beyond this point.

$$v\,[] = v\,[] + (c1 \times rand() \times (pbest\,[] - present\,[])) + (c2 \times rand() \times (gbest\,[] - present\,[])) \tag{15.24}$$

$$present\,[] = present\,[] + v\,[] \tag{15.25}$$

where present [] and the particle velocity are the current particle solutions, rand () is a random number between 0 and 1, and c1 and c2 are learning factors usually considered as 2. The pseudo code of the algorithm is as follows:

```
Generate random population
For each particle
    Assign particle velocity and best values
End
While iteration is not equal to maximum Do
    For each particle
        Calculate fitness values
        If fitness value is better than pbest, update pbest
    End
    Choose the particle with the best fitness and assign it to gbest
    For each particle
        Update velocity using (24)
        Update position using (25)
    End
End While
```

15.4.1.2.2 Artificial Bee Colony

The ABC optimization algorithm was developed by Karaboga and Basturk in 2007 based on the honey bee hive [20]. In ABC, the population generated is divided into three groups: employed, onlookers, and scout bees. The entire search space is divided into two halves employed bees and onlooker bees. Each food source (possible solution) is assigned an employed bee and an onlooker bee. The amount of nectar in each food source (fitness) is evaluated. Initially, the food source of employed bees is evaluated. If the fitness of the food source is poor it is abandoned and renamed as a scout bee. The algorithm learns from the fitness test of employed bees and based on that the fitness of onlooker bees is evaluated after all employed bee's fitness have been evaluated. The total number of possible solutions is equal to the number of employed bees and onlooker bee, whereas the total number of rejected solutions is equal to the number of scout bees. The pseudo code of the mathematical adaption of the algorithm is presented as follows:

Initialize the population
Divide the population into groups
While** maximum iteration not reached **Do
Produce a solution for employed bees
    Evaluate the employed bees
Perform the selection of fit bees
Assign non-fit bees as scout bees
    Produce new solution for onlooker bees based on the learning process of employed bees
    Perform the selection of fit bees
Assign non-fit bees as scout bees
Assign a new solution for all scout bees based on the current learning
    Memorize the best solution achieved so far
End While

15.4.1.2.3 Fish Swarm Algorithm

The FSA was proposed by Li and is based on the fish swarm in search of food [21]. For the optimization process, each fish represents a candidate in the school (population). Position $X_{i,}$ represents the fish and food satisfaction (FS) factor represents the fitness of each fish. The prime aim of the algorithm is to minimize FS. Fishes usually swim close to each other in a swamp to prevent predator, therefore, the distance between each fish is represented by the Euclidean distance $d_{ji} = X_i – X_i$. The position of the fish located in the center of the swarm is represented by X_c, which is used to identify the attributes of the entire group. The prime aim of each fish is to attain a position, where FS is maximum (i.e., food source is sufficient) without overcrowding the swarm. If the fish finds a greater concentration of food ($FC_j < FC_i$), it changes its position i to j in steps of X_i to X_{i+1}. This change of position takes place only if the fish within visual range of the current fish ($X_{\min}$) finds a greater food source. The three major factors involved in FSA are the visual distance ($X_{\min}$), the number of fish in the swarm (n_f), and crowd factor. The pseudo code of the FSA is as follows:

Initialize fish swarm
    While** optimal condition not achieved **Do
        ***For** all fish*
            Find min FC in X_{min} range
            ***If** found update position*
            Else
            Find new swarm
            ***If** found update position*
            Reduce X_{min} to prevent rejection
            End If
        End For

Update fish position
End while
Get optimal solution

15.4.1.2.4 Firefly Algorithm

FA was proposed by Yang in 2009 and it is based on the flashing behavior of the fireflies [22]. Each firefly represents the agent in the search space. These agents communicate among them self by flashing lights. A firefly glowing brighter than the other fireflies attracts them toward itself. The brightness of the firefly is modelled to diminish as the distance increases. The attenuation in the brightness of the light is governed by absorption coefficient γ, defined by the user. The idea behind this optimization is that each firefly follows the firefly brighter than itself with consideration of attenuation of brightness with distance. The degree of attractiveness is inversely proportional to the distance between the fireflies. The brightness of the firefly represents the fitness, which is evaluated by the fitness function. The adaptive pseudo code of the algorithm is as follows:
Formulate the objective function
Generate the initial population
Calculate the light intensity (I) of each fly (determined by fitness function)
While *optimal constraint not met* **Do For** *all fireflies*
If $I_j > I_i$
Move firefly I in the direction of j
End If
Attraction attenuates with distance (r) based on exp(-rγ)
Update intensity of all fireflies
End For
Rank fireflies based on the intensity of brightness
Check for most optimal solution
End while

15.4.1.2.5 Bacterial Foraging Algorithm

Bacterial foraging algorithm (BFA) was proposed by Passino in 2002 based on the foraging behavior of bacterium *Escherichia coli* [23]. In this technique, the bacteria (all possible solution) find food source, which gives the maximum nutrients (optimal solution). Communication in this technique between bacteria is through signals. The foraging process between bacteria is governed by two processes: communication signal and energy consumed from nutrients. The foraging process takes place in two steps: tumbling and swimming (chemotaxis). In the tumbling process, the bacteria change its orientation, whereas in the swimming process, the bacteria try to move toward a place having more nutrients per unit of energy. After foraging of all bacteria, the best half of the population is kept for reproduction, whereas the rest are rejected. To prevent the algorithm from being stuck at local minima, few bacteria are displayed at random location where they reproduce and the search for the optimal solution is continued. The adaptive pseudo code of the algorithm is as follows:
Generate population
Define objective function
Calculate the fitness of bacteria
While *global optimal solution not found* **Do**

```
    For each chemotaxis of bacteria
        Calculate the fitness
        Perform tumbling
        Move in the direction of tumbling
        Calculate step size
        Swim
    End For
    While chemotaxis is performed by bacteria Do
        Reproduction
    End While
    Perform dispersal
    Perform elimination
    End For
End While
```

15.4.1.2.6 Ant Colony Optimization

ACO was introduced by Dorgio and Gambardella in 1197. It is modelled on the action of ant seeking a path between the ant colony and food source [24]. In ACO, the most optimal path is found between the ant colony and the various food sources (possible solution). The ants lay down pheromones on its path while exploring the search space. Depending on the quality of pheromones and ant's current position, the other ants follow the path laid in the next iteration. Following parameters need to be defined by the user for the execution of the ACO:

- Type of the optimal solution required, based on which the ants will make a probabilistic incremental transition to the path.
- A function η to judge the quality of the components to be added to the current partial solution.
- Rules to modify the pheromones and the extent to which each pheromone can be modified in a single step (τ).
- A transition rule based on η and τ used to construct the solution in the proceeding iteration.

The pseudo code of ACO is as follows:

```
Generate the construction graphs
Initialize the pheromones and ant's values
While optimization criteria not satisfied Do
    For all ants
        k ants are placed at the starting node
        For problem size
            Choose the next stage based on probabilistic transition
        End For
        Update the intensity of pheromones
        Compare the obtained solution with the best solution achieved
    so far
    End For
End While
```

15.4.1.2.7 Cuckoo Search Optimization (CSO)

Cuckoo search optimization (CSO) algorithm was proposed by Yang and Deb in 2009. It is based on the process of cuckoos, which lay eggs in the nest of host birds [25]. If the host birds identify the alien egg, it is thrown out and the nest is destroyed (i.e., poor solution). The basic rules of cuckoo search algorithm are as follows:

- Only one egg is laid by each cuckoo, which is randomly dumbed in a nest.
- The nests having high-quality eggs (solution) are retained, whereas the nests with poor quality of the nests are dismissed.
- The identification of alien eggs by the host is based on probability, whereas the number of host nest is fixed (number of the possible solution).
- The main objective of the CSO is to attain the high-quality eggs in the nest by replacing the poor-quality eggs with better-quality eggs.

The pseudo code of CSO is as follows:

Define objective function
Generate all the host nests (the population)
While** optimization criteria not satisfied **Do
Randomly select a cuckoo to lay an egg (generate solution)
  Evaluate the fitness of the egg (solution)
  Randomly choose a host nest
  ***If** the fitness of selected host egg is high*
    Reject the cuckoo's eggs and build a new nest
  Else
    Update the eggs in the host nest with the cuckoo's eggs
  End If
Rank the best solutions and identify the current best solution
End While
Obtain the global optimal solution

15.4.1.2.8 Bat Algorithm

It is a new novel evolutionary optimization technique introduced by Yang in 2010. It is based on echolocation behavior of bats [26]. Bats use echolocation capacity to sense distance and distinguish between food and prey. There are three characteristics associated with each bat (solution): velocity, frequency, and loudness of echo. The distinguishing between food and prey is done based on the frequency and loudness of the echo. These characteristics are employed to obtain the global optimal solution in bat algorithm (BA). The algorithm is initialized by the generation of bats (all possible solutions) and assigning starting position to them. The loudness and frequency of each bat are assigned randomly. After each iteration, the frequency and loudness of each bat are updated to find a better solution (identification of prey is the solution). The prime aim is to move all the bats from its current location to the global optimal position after each iteration. The algorithm ends when all the bats have achieved the global optimal solution. BA has the following pseudo code:

Define objective function
Generate the population of bats
Define the frequency and loudness parameter form each bat
While** all bats at optimal solution **Do

Generate new solution randomly
If *the frequency of random solution > than the current solution*
Select the new frequency parameter and generate a new solution based on it
End If
If *loudness at the current frequency < current solution*
Accept solution
Increase frequency and reduce loudness
End If
Find the best solution by ranking the bats
End while

Merits and demerits of GSOA techniques are summarized in Table 15.2 [27].

Table 15.2. Summary of various GSOA techniques.

GSOA Technique	Operators	Merits	Demerits
GA	Crossover, mutation, inversion, and selection	• Works well for complex and large search space. • No mathematical analysis required	• The tendency to converge towards a local optimal solution • Not suitable for optimization problems which are constraint specific
ES	Mutation, selection, and recombination	• Strategy parameters are self-adapted	• Requires high computational time and power
EP	Mutation and selection	• Suited for complex search algorithms. • Robust in nature	• Due to lack of cross over the rate of convergence is low. (Mutation is a slow process to locate the optimal solution)
PSO	Initializer and evaluator	• Less dependent on the nature of the objective function. • Less dependent on the initial points set. • High-efficiency rate	• Lacks mathematical foundation, based on a stochastic search algorithm

(*Continued*)

Table 15.2. Summary of various GSOA techniques. (*Continued*)

GSOA Technique	Operators	Merits	Demerits
ABC	Reproduction, replacement, and selection	• The inherent decision mechanism helps locate the areas in search space where the possible solution may lie • Has fewer control parameters • Easy to implement, requires basic logical and mathematical operations	• Takes time to escape if stuck at local minima
FSA	Swarm, follow, and search	• Can be applied for optimization of complex nonlinear problems. • Has a high convergence rate due to fewer parametrical adjustments	• Lacks the ability to learn from the updating of other group members (intergroup communication is absent)
FA	Flashing and behavior	• Based on brightness automatic subdivision takes place • Can easily handle multimodal problems	• Tendency to stuck at local minima. • Lacks recording of the solution obtained, which may lead to loss of optimal solution
BFA	Reproduction, chemotaxis, dispersion, and elimination	• Always converges at global minima	• A high number of parameters must be defined. • Highly affected by the step size of the tumbling operator
ACO	Pheromones, update, and measure	• Premature convergence is avoided • The process is fast as multiple parallel paths are surveyed simultaneously	• Adaptability for various optimization process is a little difficult
CSO	Search and random selection	• Fast convergence time.	• The algorithm may get stuck if a prudent probability is not used to locate host nests.
BA	Echolocation and random selection	• Low computational time	• Convergence strongly affected with switching time between exploitation and exploration stages

15.4.2 Artificial Neural Network (ANN)

It is a machine learning approach based on the working of a human brain. This technique, when applied for optimization of HES, has the capability of adaptive power management. This is especially important when renewable sources such as PV and wind are used, which are very dynamic in nature. The basic working of ANN is analogous to a biological neuron, whose function is to take input from the external world, process it, and provide the desired output. A simple neural network is depicted in Fig. 15.5. There are primarily three layers to a neural network: the input layer, the output layer, and the hidden layer. The neuron receives the signal by interfacing its input layer to the real world. After processing the data based on adaptive weights, the output is provided to the output layer. Hidden layer is formed by the neurons, which are hidden between the input layer and the output layer. A signal is also provided from the output layer to the input layer called as feedback. The interaction between the output layer neurons also takes place, which is known as competition. Each neuron is linked to other neurons through some weights. Weights are initially initialized randomly by the user but through the learning process, the weights are updated to achieve native adaptability to handle any input. There are two types of learning: supervised learning and unsupervised learning. In supervised learning, the output generated for a given input is already known, whereas in unsupervised learning the network must make sense of the input on its own. The weights are updated through the training process. The rate at which the ANN learns is the learning rate. A learning rate must be prudently chosen by the user, as slow learning rate takes a longer time and high learning rate may not be able to differentiate between fine differences in the input. The merits of this technique are that it can adapt to implement complex nonlinear optimization problems and requires less formal numerical training. The demerits of this technique are that it requires very large computational power.

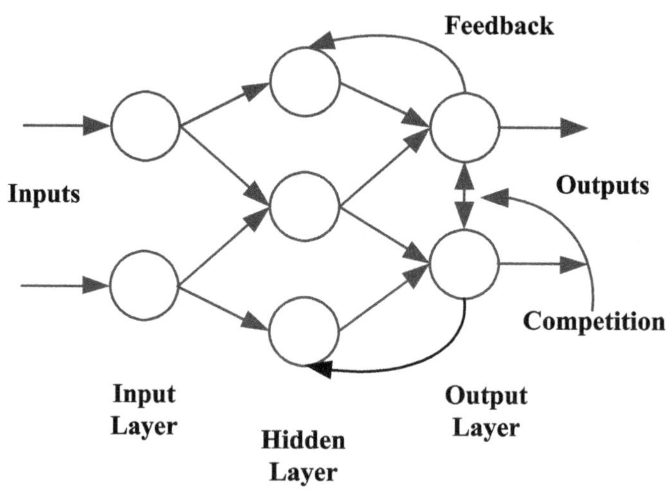

Figure 15.5. Simple architecture of a neural network.

15.4.3 Fuzzy Logic (FL)

Fuzzy logic (FL) was introduced by Lotfi A. Zadeh in 1965. It is based on the concept of multivalued logic, where intermittent values can be defined between yes/no, true/false, high/low, hot/cold, and so on. [28]. Universe of discourse provides all the elements of the fuzzy set. The membership function expresses the membership of each element with respect to a set. The membership function $\mu_A(x)$ expresses the membership value of x in set A. A fuzzy system basically consists of four components: fuzzifier, fuzzy interface, defuzzyfier, and fuzzy rule base. The basic steps involved for implementation of FL along with the fuzzy system model is shown in Fig. 15.6. The merits of FL algorithm are: easy learning process, ability to solve nonlinearities in the problem, applications for problems where the process model is not well defined. The demerits of the FL algorithm are the accuracy of the optimal solution is not high and depending on the rule base the computational time varies significantly.

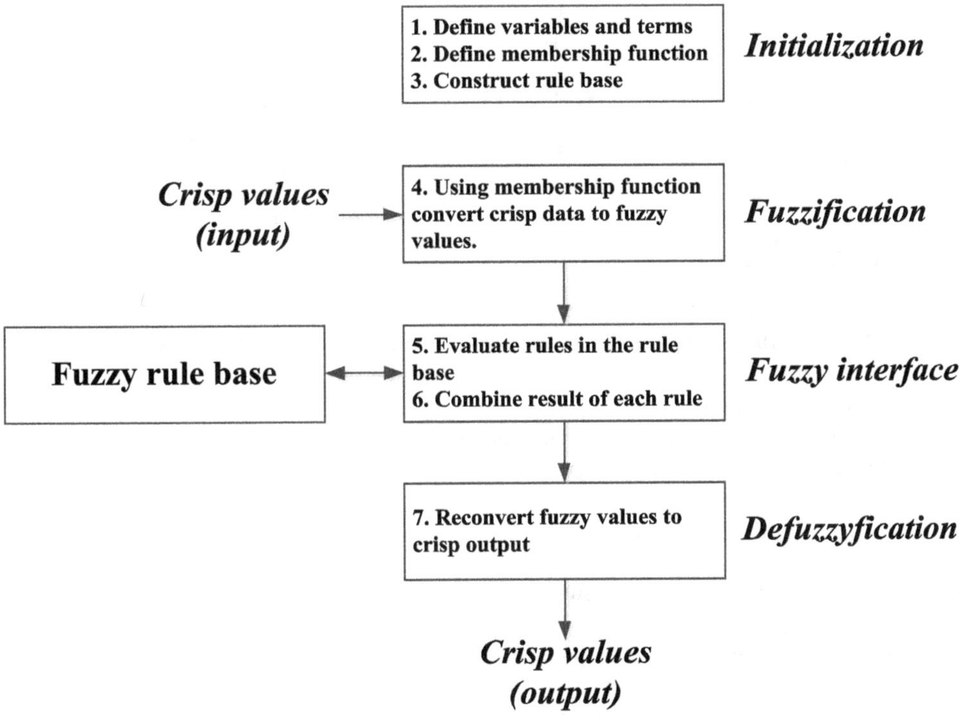

Figure 15.6. Fuzzy system model with FL algorithm steps.

15.4.4 Conditioning Approximate Reasoning (CAR)

It is a very popular tool for modelling uncertainty in system modelling, which is a feature of renewable energy systems. This method depends on the interdependencies in between states. The two most conditioning approximate reasoning (CAR) types are Markov models and Bayesian networks.

15.4.4.1 Hidden Markov Models (HMM)

Hidden Markov model (HMM) is based on the formulation of probabilistic models of problem space. Using conceptual toolkit complex models are presented in pictorial form. HMM is restricted to discrete time process ideally suited for optimization of HES. The discrete time process in HMM is defined by $\{X_k, Y_k\}_k \geq 0$, where $\{X_k\}$ is a node of the Markov chain dependent on the node number k and $\{Y_k\}$ is a conditional node dependent on $\{X_k\}$. The structure of HMM is shown in Fig. 15.7 [29].

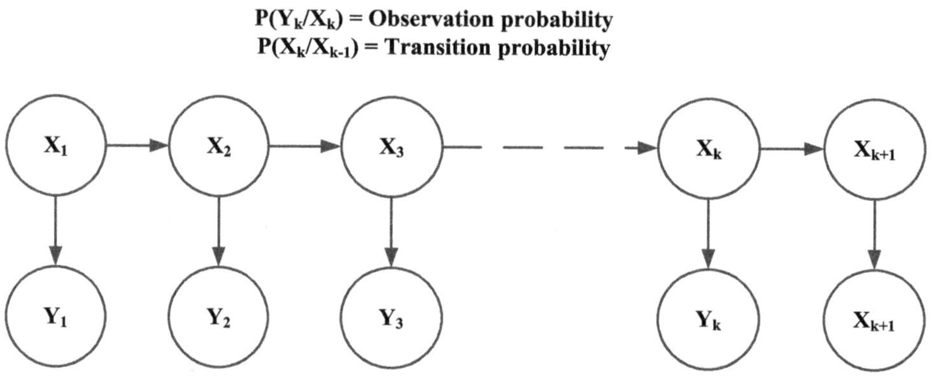

Figure 15.7. Structure of HMM.

The HMM models consist of five parameters that define the model as follows:

- Number of states in the model, N
- Number of observational symbols per state, M
- The state transitional probability, A
- The observational probability, B
- The initial state distribution π

The HMM algorithm works as follows:

1. Choose an initial state from the distribution π, say q_t
2. Initialize time of analysis, t = 1
3. Choose one of the observational symbols of the current state (physical output of the system)
4. Transit to new state q_{t+1} using the above chosen path
5. Increment t = t + 1
6. If t < T (time period of analysis), go to step 3
7. Terminate

The prime feature of HMM is that it can represent the interdependency between various components with good accuracy. The drawback of HMM is that time spent at a state is not counted, which can lead to nonrealization of the solution or high time consumption for complex systems.

15.4.4.2 Bayesian Belief Networks (BBN)

Probabilistic dependency of various variables among themselves is graphically represented in Bayesian belief network (BNN) [30]. The BNN also consists of nodes and arrows, where a node represents the variables in the dependency model and arrow represents the relationship between the connected nodes. BBN consists of structure S encoding the interdependency of variable X and a set P, which consists the probability distribution of dependency of each variable. An arrow from one node to another represents the direct dependency with a child–parent relationship. To get the relationship between random variables, conditional probability is defined with each node: $P = \left\{\underline{p}\left(X_1/Y_1\right), \ldots, p\left(X_1/Y_1\right)\right\}$, where Y_i is the parent set of X_i. For example, if there is a direct link between X_i and X_j, then X_i is the parent of X_j, hence X_j also belongs to Y_i. For a node A, the parents $B_{1 \text{ to } n}$, a conditional probability table is constructed given by $P(A/B_1, B_2, \ldots, B_n)$. The belief linked to each node is only updated if the belief of any of the states is directly connected to that node changes. The merits of the BNN technique is very efficient and has a simple representation of interdependency of uncertainties of the real world and gives a better understanding of the system dynamics due to the change in one or several dependent parameters. Considerable knowledge of the system is required for accurate modelling of system in BNN, as insufficient knowledge may lead to false results.

15.4.5 Software-Based Techniques

There are several software tools available for optimization of HES [31]. Some of the most common and frequently used software tools are HOMER, HYBRID 2, RETScreen, and iHOGA. The schematics of these softwares are shown in Fig. 15.8, where the prerequisites and the final output of each software are mentioned. These software techniques are not flawless, each having their own merits and demerits as summarized in Table 15.3.

15.4.6 Mathematical Iterative Approach

15.4.6.1 Power Pinch Point Analysis (PoPA)

PoPA is inspired from the pinch analysis used in the optimization of material production and thermal energy analysis [32,33]. This technique is applicable for optimization of the energy storage unit of the HES. The analysis is based on the concept of obtaining a grand composite curve (GCC) of the charge storage in the BESS and modifying it to obtain the size of BESS and the minimum initial charge requirement. A typical GCC of a BESS in HES is shown in Fig. 15.9. The steps involved in performing successful PoPA are as follows:

1. Input generation and load data
2. Plot storage curve of the BESS based on the generation and load requirement so as to completely meet the load requirement
3. Find the minimum value of the charge storage from the GCC. For Fig. 15.9, this point is "C," where minimum energy is Q_{min} and t_{ref} is the pinch point
4. Creating a new GCC by adding the Q_{min} at $t = 0$, mentioned as shifted curve in Fig. 15.9
5. Obtain the maximum value and initial value of the shifted charge curve. These values are Q_{min} and Q_{max2} for Fig. 15.9

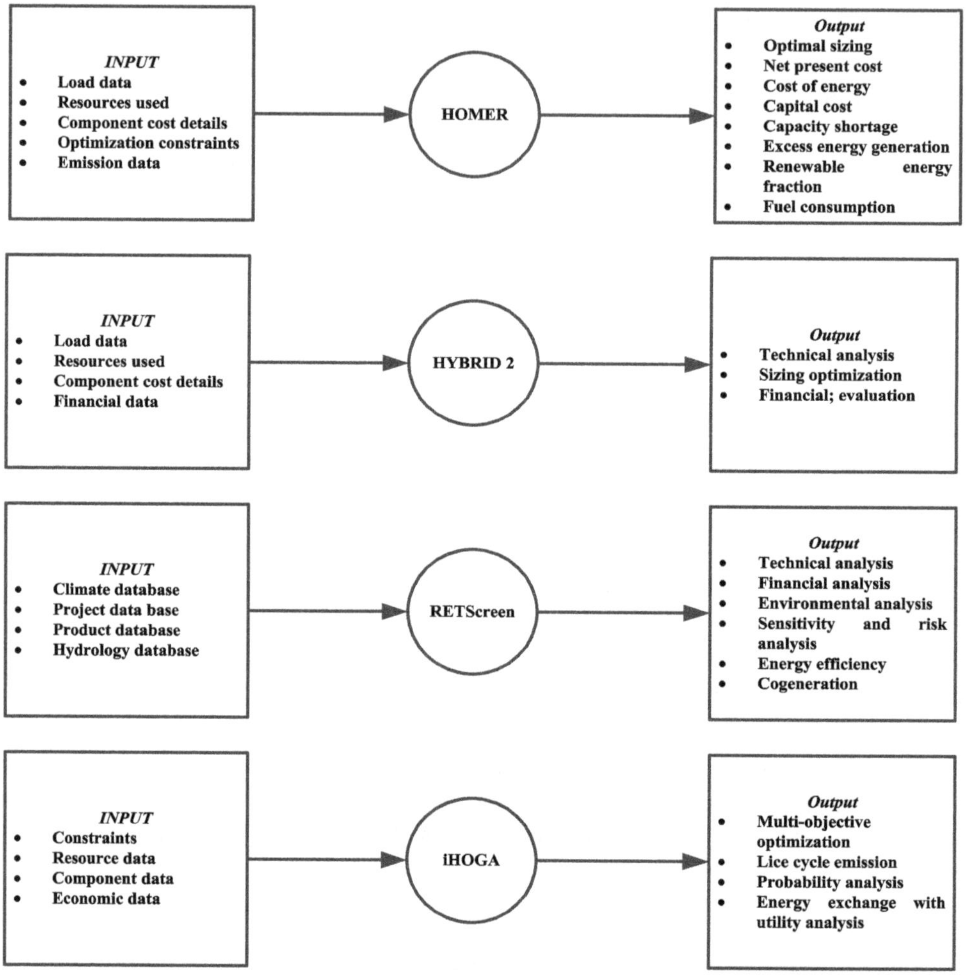

Figure 15.8. Schematic representation of popular software.

Table 15.3. Comparison of optimization software tools.

Tools	Developer	Analysis Performed	Systems Modelled	Merits	Demerits
HOMER	National Renewable Energy Laboratory, United States	Economic, Technical and Environmental	PVS, WECS, DG, BESS, Hydro, BG	• Graphical representation of results is well presented. • Can handle data with hourly step size.	• The software optimization code is hidden. • Cannot handle nonlinear problems. • Daily average data cannot be imported.
HYBRID 2	University of Massachusetts, United States	Technical and Environmental	PVS, WECS, DG, BESS, TS	• Multiple options of the type of loads. • Dispatching options available in detail.	• Not compatible with windows. • It is a little patchy.
RETScreen	Ministry of Natural Resources, Canada	Economic, Technical and Environmental	PVS, WECS, BESS	• Reliable metrological database. • Financial analysis forms the main strength	• Time series data cannot be imported. • Limited option to represent processed data in graphs.
iHOGA	University of Zaragoza, Spain	Economic, Technical and Environmental	PVS, WECS, DG, BESS, Hydro	• Low computational time required.	• Limited analysis options available

PVS-PV System, WECS—Wind Energy Conversion System, DG—Diesel Generator, BESS—Battery Energy Conversion System, TS—Thermal System

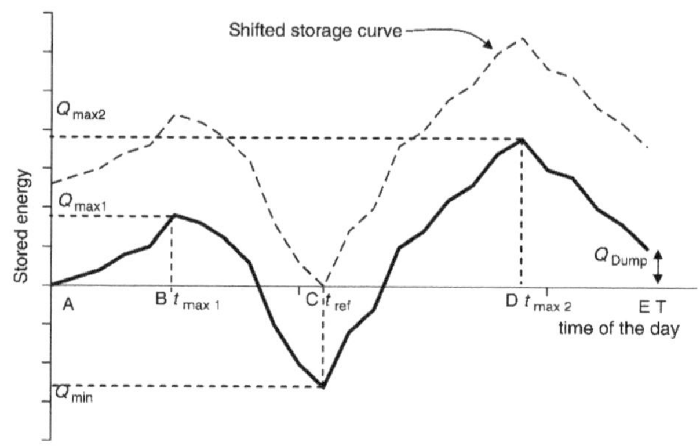

Figure 15.9. Typical GCC of BESS in a HES [32].

6. Make sure the GCC lies in the positive region (for no loss of power analysis)
7. The BESS capacity can be evaluated using Q_{max2} and minimum initial charge of the BESS is Q_{min}

It is a very easy method for obtaining the BESS size and understand the working of the generation shortage and excess depending on the load demand. However, the disadvantage of this method is that it can only optimize the energy storage unit of the HES.

15.4.6.2 Reformed Electric System Cascade Analysis (RESCA)

This method was proposed by Ho et al. [34] as ESCA, which was later improved upon to form reformed electric system cascade analysis (RESCA) [35]. The demerits of PoPA are addressed in RESCA, which also incorporates the optimization of energy-generating units taking into the consideration the mode of operation of the HES. There are various factors on which the optimization of the energy-generating units can take place, for example, final excess energy (FEE), renewable energy fraction (REF), energy generation ratio (EGR), system cost, and LPSP. The steps involved in the implementation of RESCA are as follows [36]:

1. Input generation, load, resources, and equipment characteristics data
2. Set optimization constraints
3. Make an initial assumption on the rating of energy-generating units
4. Compute the energy generation in each time step for the time period of analysis
5. Compute net energy excess of deficient based on load and energy generation
6. Check for optimization constraint satisfaction, if not satisfied update the system size in priority of the energy catering capacity
7. Repeat steps 4 to 6
8. Once constraint is satisfied the size of generating units are optimized. Now create the GCC of the BESS
9. Perform PoPA to get the BESS size
10. Optimization process completed

The credibility of the RESCA method is its ease of application of the process and also through the graphical representation of the process data, the system designer knows about the system performance and gets the idea as to why a particular solution is obtained. The merits of RECA are yet to be tested for environmental criteria.

15.5 Conclusion

This chapter has presented the various criteria, constraints, and techniques required for optimization of HES. The need of multi-criteria optimization of HES is highlighted in the chapter based on which formulation of optimization constraints are presented along with their merits and demerits pertaining to each constraint. The chapter has presented an exhaustive study of the various optimization techniques currently in use and provides a brief on their implementation. The merits and demerits of these techniques have been articulately presented to enhance the reader comprehensibility.

15.6 References

1. The International Renewable Energy Agency (IRENA) renewable power generation capacity statistics: https://www.irena.org/publications/2021/March/Renewable-Capacity-Statistics-2021
2. Energy Information Administration (US), *International Energy Outlook 2013 with Projections to 2040.*, US Energy Information Administration, 2013.
3. F. Díaz-González, A. Sumper, O. Gomis-Bellmunt, and R. Villafáfila-Robles, "A review of energy storage technologies for wind power applications," *Renewable and Sustainable Energy Reviews*, vol. 16, no. 4, pp. 2154–2171, May 2012.
4. A. Kumar, A. R. Singh, Y. Deng, X. He, P. Kumar, and R. C. Bansal, "Multiyear load growth based techno-financial evaluation of a microgrid for an academic institution," *IEEE Access*, vol. 6, pp. 37533–37555, 2018.
5. R. Singh, and R. C. Bansal, "Optimization of an autonomous hybrid renewable energy system using reformed electric system cascade analysis," *IEEE Transactions on Industrial Informatics*, vol. 15, pp. 399–409, 2018.
6. R. Hosseinalizadeh, H. Shakouri G, M. S. Amalnick, and P. Taghipour, "Economic sizing of a hybrid (PV–WT–FC) renewable energy system (HRES) for stand-alone usages by an optimization-simulation model: Case study of Iran," *Renewable and Sustainable Energy Reviews*, vol. 54, pp. 139–150, Feb. 2016.
7. G. M. Masters, *Renewable and Efficient Electric Power Systems*, John Wiley & Sons, New Jersey, 2013.
8. M. K. Deshmukh, and S. S. Deshmukh, "Modeling of hybrid renewable energy systems," *Renewable and Sustainable Energy Reviews*, vol. 12, no. 1, pp. 235–249, Jan. 2008.
9. M. Abdelaziz, Mohamed, and A. M. Eltamaly, "Modeling of hybrid renewable energy system," in *Modeling and Simulation of Smart Grid Integrated with Hybrid Renewable Energy Systems*, vol. 121, M. Abdelaziz Mohamed, and A. M. Eltamaly (eds.), pp. 11–21, Springer International Publishing, Cham, 2018.
10. A. Kumar, A. R. Singh, Y. Deng, X. He, P. Kumar, and R. C. Bansal, "A novel methodological framework for the design of sustainable rural microgrid for developing nations," *IEEE Access*, vol. 6, pp. 24925–24951, 2018.
11. R. Singh, and R. C. Bansal, "Review of HRESs based on storage options, system architecture and optimisation criteria and methodologies," *IET Renewable Power Generation*, vol. 12, no. 7, pp. 747–760, May 2018.

12. M. D. A. Al-falahi, S. D. G. Jayasinghe, and H. Enshaei, "A review on recent size optimization methodologies for standalone solar and wind hybrid renewable energy system," *Energy Conversion and Management*, vol. 143, pp. 252–274, Jul. 2017.
13. E. L. V. Eriksson, and E. M. Gray, "Optimization and integration of hybrid renewable energy hydrogen fuel cell energy systems—A critical review," *Applied Energy*, vol. 202, pp. 348–364, Sep. 2017.
14. R. Singh, R. C. Bansal, and A. R. Singh, "Optimization of an isolated photo-voltaic generating unit with battery energy storage system using electric system cascade analysis," *Electric Power Systems Research*, vol. 164, pp. 188–200, Nov. 2018.
15. F. J. Varela, and P. Bourgine, *Toward a Practice of Autonomous Systems: Proceedings of the First European Conference on Artificial Life*, MIT Press, 1992.
16. J. K. Kaldellis, D. Zafirakis, and E. Kondili, "Optimum autonomous stand-alone photovoltaic system design on the basis of energy pay-back analysis," *Energy*, vol. 34, no. 9, pp. 1187–1198, Sep. 2009.
17. T. Baeck, D. B. Fogel, and Z. Michalewicz, *Evolution strategies*, vol. 1, Taylor Francis, New York, 2000.
18. D. B. Fogei, L. J. Fogel, and V. W. Porto, "Evolving neural networks," *Biological Cybemetics 63*, vol. 6, pp. 487–493, 1990.
19. R. Eberhart, and J. Kennedy, "A new optimizer using particle swarm theory," Proceedings of the Sixth International Symposium on Micro Machine and Human Science, 1995. MHS '95, pp. 39–43, Nagoya, 1995.
20. D. Karaboga, and B. Basturk, "A powerful and efficient algorithm for numerical function optimization: Artificial bee colony (ABC) algorithm," *Journal of Global Optimization*, vol. 39, no. 3, pp. 459–471, Nov. 2007.
21. Li, Xiao lei, and Ji-xin Qian, "Studies on artificial fish swarm optimization algorithm based on decomposition and coordination techniques," *Journal of Circuits System*, vol. 1, no. 1, pp. 1–6, 2003.
22. X.-S. Yang, "Firefly Algorithms for Multimodal Optimization," *Journal of Circuits System*, vol. 1, no. 6, pp. 169–178, Oct. 2009.
23. N. Rajasekar, N. Krishna Kumar, and R. Venugopalan, "Bacterial foraging algorithm based solar PV parameter estimation," *Solar Energy*, vol. 97, pp. 255–265, Nov. 2013.
24. A. Abraham, S. Das, and S. Roy, "Swarm intelligence algorithms for data clustering," in *Soft Computing for Knowledge Discovery and Data Mining*, pp. 279–313, Springer, Boston, 2008.
25. G.-G. Wang, A. H. Gandomi, X.-S. Yand, and A. H. Alavi, "A new hybrid method based on krill herd and cuckoo search for global optimisation tasks," *International Journal of Bio-Inspired Computation*, vol. 5, no. 1, pp. 286–299, 2016.
26. X.-S. Yang, "A new metaheuristic bat-inspired algorithm," in *Nature Inspired Cooperative Strategies for Optimization (NICSO 2010)*, J. Gonzlez, D. Pelta, C. Cruz, G. Terrazas, and N. Krasnogor (eds.), pp. 65–74, Springer, Berlin, Heidelberg, 2010.
27. S. Sumathi, L. Kumar, and P. Surekha, *Computational Intelligence Paradigms for Optimization Problems Using MATLAB/SIMULINK*, vol. 1, CRC Press, Taylor Francis Group, Boca Raton, 2015.
28. R. E. Bellman, and L. A. Zadeh, "Decision-making in a fuzzy environment," *Management Science*, vol. 17, no. 4, p. B-141, 1970.
29. S. R. Eddy, "Hidden Markov models," *Current Opinion in Structural Biology*, vol. 6, no. 3, pp. 361–365, Jun. 1996.
30. S. Deb, D. Ghosh, and D. K. Mohanta, "Reliability analysis of PV cell, wind turbine and diesel generator by using Bayesian network," 2016 International Conference on Electrical, Electronics, and Optimization Techniques (ICEEOT), 2016, pp. 2714–2719.
31. S. Sinha, and S. S. Chandel, "Review of software tools for hybrid renewable energy systems," *Renewable and Sustainable Energy Reviews*, vol. 32, pp. 192–205, Apr. 2014.

32. S. Bandyopadhyay, "Design and optimization of isolated energy systems through pinch analysis," *Asia-Pacific Journal of Chemical Engineering*, vol. 6, no. 3, pp. 518–526, May 2011.
33. N. E. Mohammad Rozali, S. R. Wan Alwi, Z. A. Manan, and J. J. Klemeš, "Peak-off-peak load shifting for hybrid power systems based on Power Pinch Analysis," *Energy*, vol. 90, Part 1, pp. 128–136, Oct. 2015.
34. W. S. Ho, M. Z. W. M. Tohid, H. Hashim, and Z. A. Muis, "Electric System Cascade Analysis (ESCA): Solar PV system," *International Journal of Electrical Power & Energy Systems*, vol. 54, pp. 481–486, Jan. 2014.
35. R. Singh, R. C. Bansal, and N. Tiwari, "Optimization and comparison of autonomous renewable energy system based on ESCA technique," Presented at the IEEE International WIE Conference on Electrical and Computer Engineering, pp. 1–5, Dehradun, India, 2017.
36. R. Singh, R. C. Bansal, and A. R. Singh, "Multi-objective optimization of hybrid renewable energy system using reformed electric system cascade analysis for islanding and grid connected modes of operation," *IEEE Access*, vol. 6, pp. 47332–47354, 2018.

CHAPTER 16

Techno-Economic Evaluation of Hybrid Renewable Energy Sources for a University Campus

Saheed Lekan Gbadamosi[1], Aminat Oyeleke[2], Nnamdi I. Nwulu[1], and O. M. Babatunde[3]

[1]*Department of Electrical and Electronic Engineering Science, University of Johannesburg, Auckland Park, South Africa. gbadamosiadeolu@gmail.com; nnwulu@uj.ac.za*
[2]*Department of Electrical, Electronic and Computer Engineering, Afe Babalola University Ado-Ekiti, Ado-Ekiti, Nigeria. oyelekeaminat@rocketmail.com*
[3]*Department of Electrical and Electronic Engineering, University of Lagos, Lagos, Nigeria. olubayobabatunde@gmail.com*

Abstract The fluctuation in fossil fuel prices and environmental pollution associated with conventional power generation has led to an increased adoption of renewable energy sources among university campuses. The purpose of this study is to optimally size an off-grid hybrid renewable energy system for a university campus in Nigeria. Furthermore, this study seeks to conduct a techno-economic evaluation of the proposed system using the HOMER software. Sensitivity analysis was carried out based on power generated, economic feasibility, and environmental considerations of the hybrid renewable energy system. Obtained results reveal that the proposed system leads to significant reduction of pollutants and an investment return during the fourth year of a 25-year investment lifetime. There is also cost savings of approximately $14,279,601 with a 31% return on investment and a positive net profit value of $17.4 million.

Keywords: Hybrid renewable energy, Technical and economic analysis, University, Optimal sizing, Electricity, Emission, Cost.

16.1 Introduction

Globally, university campuses are becoming leading consumers of energy due to their large building sizes and the continuous need to supply the electricity, heating, and cooling loads of the laboratories, hostels, and libraries for research [1]. Climate change threats and energy security concerns are pushing universities toward clean and efficient energy generation systems. Renewable energy sources (RESs) have been considered as alternative options for sustainable and reliable energy owing to their resilience and emissions reduction ability [2,3]. More importantly, RES especially hybrid energy systems can be considered as the best options for developing countries such as Nigeria, where citizens face lack of adequate power supply with demand superseding the electricity generation thereby crippling the economy of the country [4]. Among the various RESs, wind and solar are getting attention in the energy sector owing to their viability, maturity, environmental friendly, and clean nature [5,6].

As the development of renewable energy technologies rises, there is the concern about the intermittent and unpredictable nature of wind and solar energy [7]. Therefore, in order to ensure continuous electricity supply and avoid significant demand variations, utilization of energy storage is highly essential. The deployment of more than one RES together with storage system increases the power system reliability and efficiency [8]. Energy storage have been widely deployed in renewable energy systems due to their fast response capability, adequate power delivery, and ability to meet energy demand in off-grid hybrid renewable energy systems [9–11]. Therefore, storage systems have become an inseparable part of hybrid RE systems in ensuring energy security and improving power quality. In [12], a combination of battery and supercapacitor was employed as hybrid energy storage for transmitting solar power in a grid-connected photovoltaic (PV) system. Also, in [13], a supercapacitor and battery storage system was employed for microgrid system. In [14], a battery and hydrogen storage system for a residential electrical load through an independent hybrid energy system was considered.

The literatures present some works that have been done on hybrid renewable energy systems. In [15], techno-economic evaluation for an off-grid hybrid energy system for rural healthcare systems in Nigeria regions was investigated. HOMER software was employed for the optimization of different selected regions and the simulated result shows that the hybrid system is encouraging due to the quality of renewable energy potentials in Nigeria. In [16], a thermo-economic grid-connected hybrid energy system coupled with storage is presented. The results show a remarkable mitigation of operating costs and improved self-consumed energy. In [17], a control strategy for grid-connected hybrid energy systems to reduce CO_2 emissions and increase RES penetration is developed. The results show that hybrid system support grid balancing and provider financial gains. In [18], technical and economic analysis of a hybrid energy system using metaheuristic optimization approach for a radio station was performed. Obtained results show the financial viability of the proposed system. In [19], a comprehensive assessment of grid-connected PV power system for small residential areas in China is presented. The techno-economic performance was analysed using HOMER software, and the results obtained reveal the feasibility of PV in the selected locations with decrease in CO_2 emissions. In [20], the potential of grid-connected hybrid microgrid for local inhabitants in Pakistan is investigated. The techno-economic assessment was optimized using HOMER software. In [21], the authors proposed an optimal configuration of a grid-connected RE system in Oman. The technical and economic simulation was done on MATLAB, and the result reveal that the investment returns in PV system is very promising in Oman.

So far, only a few works have considered employing hybrid RE systems for powering universities in Nigeria. This research direction is pertinent due to the frequent power outages in the country. In [22], a comparative cost assessment of hybrid RE with diesel and battery system as an energy source for an administrative university building in Nigeria was evaluated. The results indicate that solar power in addition to battery and diesel is more economical for optimal system design to meet the site demand compared to electricity supplied from diesel generators. However, this work considered a small portion of the university's load profile, and emissions such as sulfur dioxide, nitrogen oxide, and unburned carbon were not considered. In [23], the reduction of CO_2 emissions discharged from fossil fuel via recycling of bulbs in different hostels within a university premises in Nigeria was considered. However, hybrid renewable energy system is not considered in this study.

From the gap identified in the literature, this study focuses on designing an effective hybrid renewable energy system for Afe Babalola University in Ado-Ekiti, Nigeria. This study is motivated by the desire to assess the technical, economic, and emissions reduction, as well as feasibility of an off-grid hybrid RE system using HOMER software. The key contributions of the research are

- Optimal system design and analysis of the proposed off-grid hybrid RE system at the university campus.
- Determination of economic and technical performance of the proposed off-grid hybrid RE system.
- Comparative evaluation of emissions (carbon dioxide, sulfur dioxide, unburned hydrocarbon, particulate matter, carbon monoxide, and nitrogen oxide) from the grid-connected network and the proposed hybrid energy systems.

The organization of this chapter is as follows: Section 16.2 presents the background of the study; Section 16.3 describes the system components use for modelling the hybrid renewable energy system; Section 16.4 presents the simulation results and the discussion is based on system performance indicator for the technical, economic, and environment; and Section 16.5 presents the conclusion.

16.2 Background of the Study

16.2.1 Description

Afe Babalola University Ado-Ekiti (ABUAD) is a private university situated in Ado-Ekiti, Ekiti State, Nigeria, established in 2009. It is situated on 130 hectares of land at an altitude of 70°36′ N and longitude of 50°13′ E. The university has five colleges: a postgraduate school, conference halls, teaching hospital, student hostels, staff quarters, university farm, and other auxiliary services such as laundry, cafeteria, a water processing plant, and a bakery.

16.2.2 Load Profile

In this chapter, we investigate the daily load profile of the university campus over a period of 5 years as obtained from the works department. Figures 16.1 and 16.2 present the annual load profile and load curve for the university campus. The average hourly load consumption for a year is computed and presented in Fig. 16.3. The energy flow of the existing university campus power system network is presented in Fig. 16.4, showing the grid and the diesel generators supplying the university and farm load (Table 16.1).

16.2.3 The Proposed Hybrid System

In this study, the proposed hybrid RE system is structured to consist of two RESs: wind and solar in addition to battery and generator for system back-up in order to monitor adequacy of the supply. The schematic diagram of the energy flow of the proposed system is presented in Fig. 16.5. The proposed system is independent of the grid supply, having its supply from RESs and generator. The hybrid RE system and battery bank are essential to ensure good reliability and higher economic feasibility for the off-grid system.

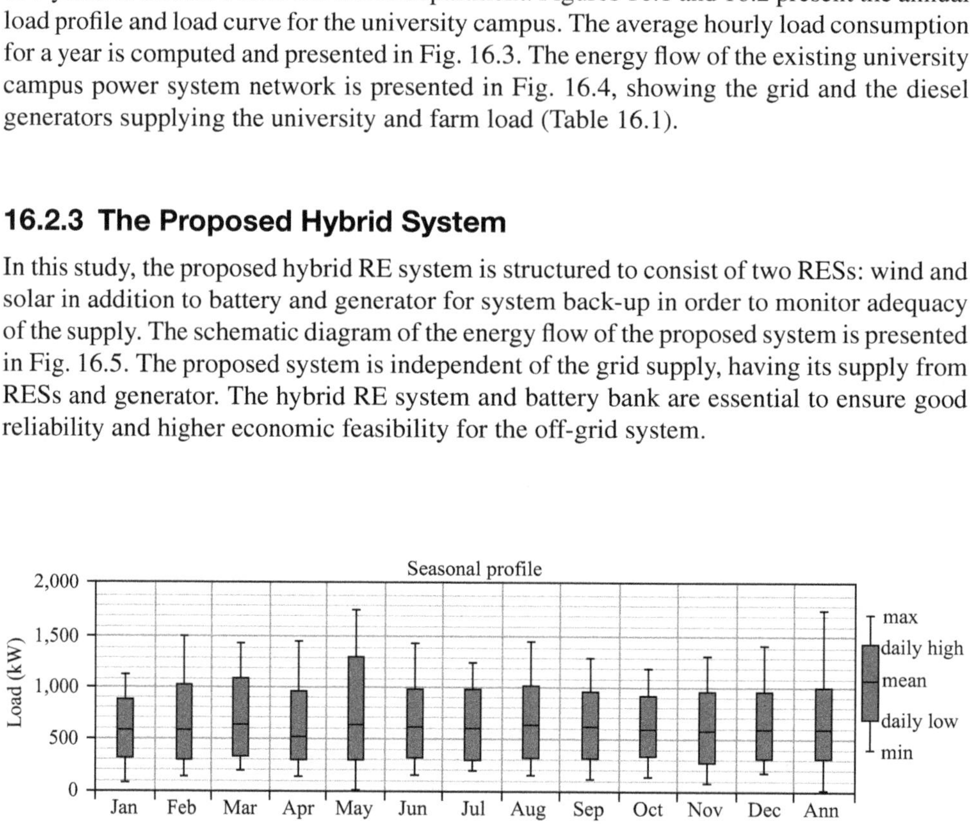

Figure 16.1. Yearly load profile at the university campus.

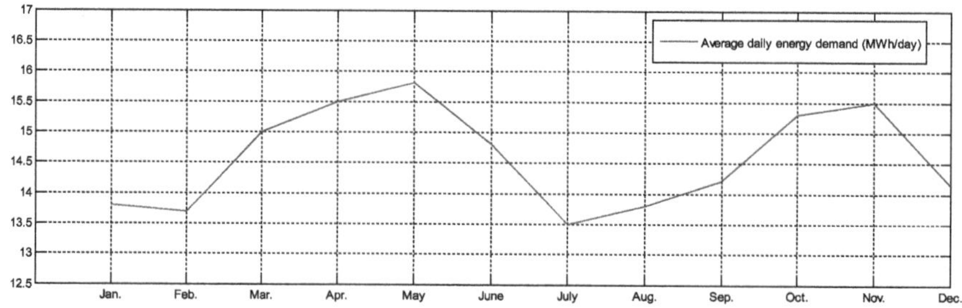

Figure 16.2. Average annual load curve for the university campus.

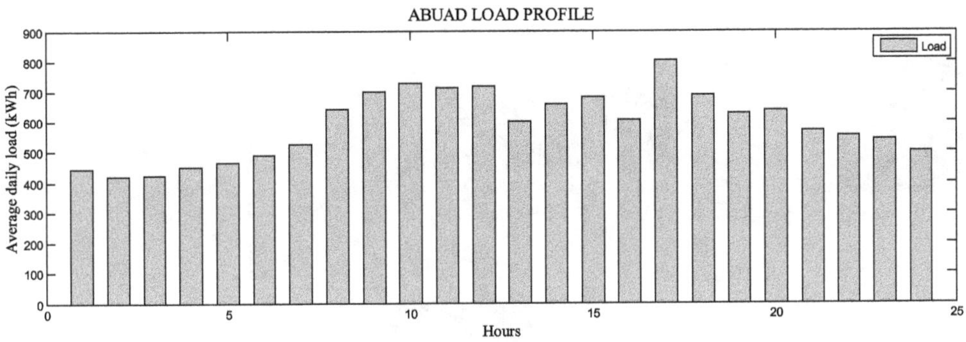

Figure 16.3. The average daily load profile for university campus.

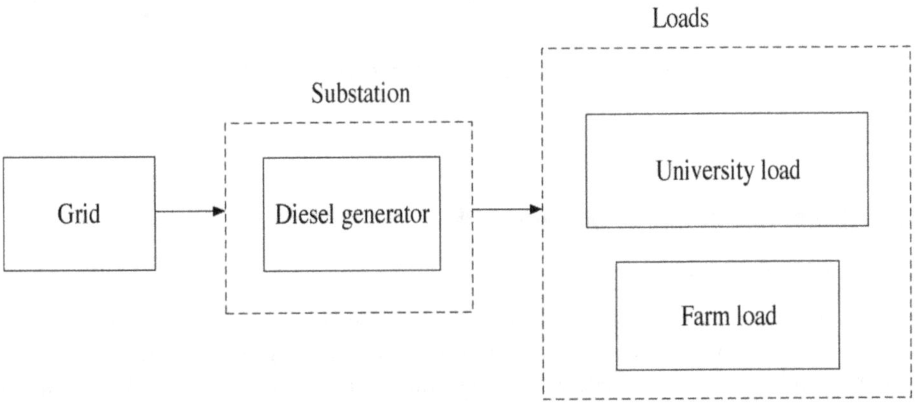

Figure 16.4. Energy flow of the grid-connected system of the existing network of the university.

Table 16.1. Estimated average energy demand [24].

Months	Average Daily Demand (MWh/day)	Months	Average Daily Demand (MWh/day)
Jan.	13.8	July	13.4
Feb.	13.7	Aug.	14.2
Mar.	15	Sept.	14.5
Apr.	15.5	Oct.	15.3
May	15.8	Nov.	15.7
June	14.8	Dec.	14.15

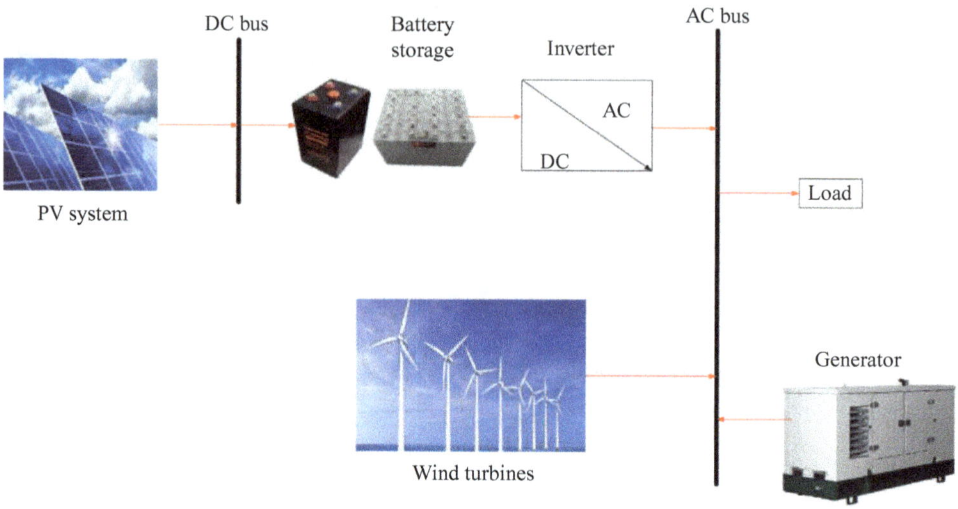

Figure 16.5. Energy flow of the proposed hybrid system for the university.

16.3 System Components of Hybrid RES

16.3.1 Wind Energy

A wind turbine extracts available kinetic energy from the wind and converts it into electrical energy by rotating the blades of its turbine [20]. The location of the wind farm depends on the availability of wind speed, which determines the feasibility of the design project. This is essential for the proper estimation of annual energy production from the proposed wind turbine installation. Figure 16.6 presents the wind speed for the university campus. The output power from the wind turbine is nonlinear owing to the uncertain nature of wind speed. The wind turbine hourly power output can be depicted as [25]

$$P = \begin{cases} 0 & 0 = v = v_{ci} \\ \dfrac{(v - v_{ci})}{(v_r - v_{ci})} P_r & v_{ci} = v = v_r \\ P_r & v_r = v = v_{co} \\ 0 & v_{co} = v \end{cases} \tag{16.1}$$

where P_r represents the rated power in (kW), v_{ci} and v_{co} are the cut-in and cut-out speed in (m/s), respectively, and v is the rated speed in (m/s).

The necessary part of wind energy designing is the appropriate selection of wind turbine. In this study, Fuhrlander 100 (FL100) is used and the technical specification of the wind turbine is summarized in Table 16.2.

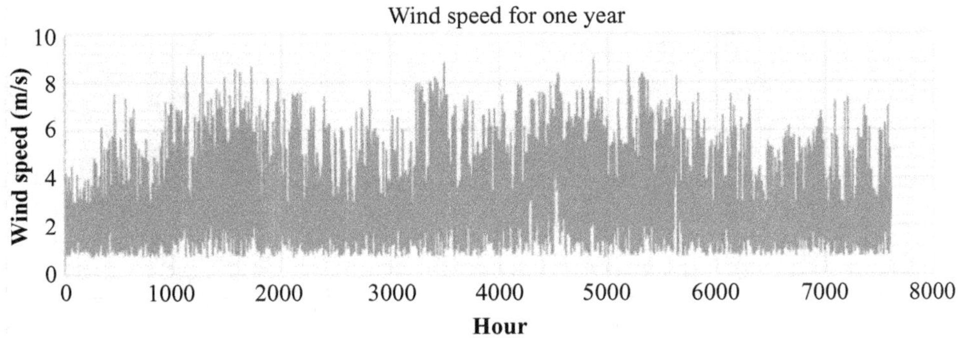

Figure 16.6. Wind speed for the university campus.

Table 16.2. Technical parameter of wind turbine.

Parameters	Data
Rated power	100 kW
Max power	65 kW
Start wind speed	2.5 m/s
Rated wind speed	9 m/s
Survival wind speed	59.5 m/s
Rotor diameter	22.3 m
Swept area	390.4 m^2
Efficiency	90%

16.3.2 Solar Energy

Solar energy describes the electricity produced from solar radiation through conversion of sunlight into electrical energy by tapping light photons to obtain electrons. This process generates DC current, which can be amplified with DC–DC converters and then inverted to deliver the required AC power. The solar PV consists of several series and parallel connections of PV modules to produce the desired power output. The ambient temperature of solar irradiance and system configuration determines the output power of solar energy, which is a function of PV module locations. Figure 16.7 reveals the average solar radiation available for the site location in the university campus. The power output of solar energy is depicted as [26]

$$P_{pv} = V * I * n_m * FF \tag{16.2}$$

$$FF = \frac{V^{max} * I^{max}}{V^{oc} * I^{sc}} \tag{16.3}$$

$$I = s(t) * (I^{sc} * K * (T_{ct} - 25)) \tag{16.4}$$

$$V = V^{oc} + K_v * T_{ct} \tag{16.5}$$

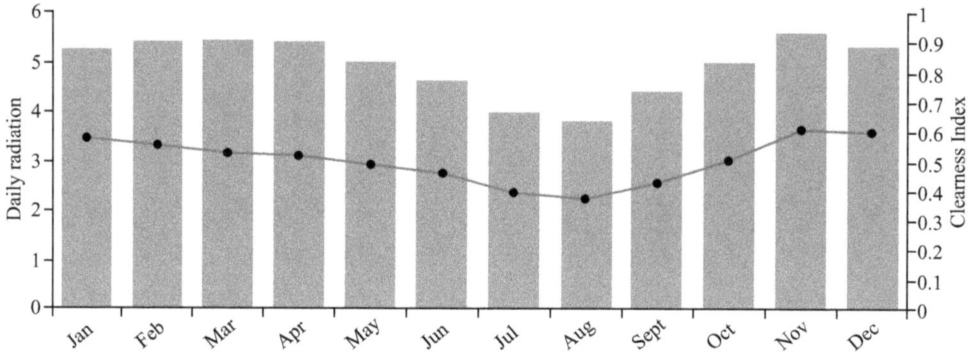

Figure 16.7. Monthly average solar radiation at university campus.

Table 16.3. Technical parameter of PV panel.

Electrical Data	STC	NOTC
Peak power watts—P_{max} (Wp)	300	218
Power output tolerance—P_{max} (%)	0/+3	0/+3
Maximal power voltage—V_{mp} (V)	36.9	33.3
Maximal power current—*Impp* (A)	8.13	6.55
Open-circuit voltage—V_{oc} (V)	45.3	41.3
Short-circuit current—I_{sc} (A)	8.6	7.04
Module efficiency—$\acute{\eta}$ (%)	15.5	15.5

$$T_{ct} = T_{at} + s(t) * \left(\frac{NOCT - 25}{0.8}\right), \tag{16.6}$$

where P_{pv} represents power output of the PV in (W), V is output voltage (V), I is terminal current (A), FF is fill factor, V^{max} and I^{max} are the voltage (V) and current (A) at peak power, I^{sc} is the short-circuit current (A), V^{oc} is open circuit voltage (V), K is current temperature coefficient (mA/°C), T_{ct} is cell temperature (°C), K_v is coefficient of voltage temperature (mV/°C), T_{at} is ambient temperature (°C), NOCT is nominal cell temperature (°C), and $s(t)$ is the random irradiance.

The solar radiation is available abundantly in the university campus (ABUAD) with solar per day insolation of approximately 3500 Wh/m^2/day. The specification of the selected PV module is summarized in Table 16.3.

16.3.3 Battery System

In this chapter, a battery bank is employed for system backup known as energy storage system owing to the intermittent characteristics of RES so as to maintain a steady supply of electricity. A 4500-Ah, 48-V lithium-ion battery (Li-Ion) is chosen and the technical specification of this battery is presented in Table 16.4. The battery bank assists the hybrid energy system by storing electricity, sustaining a steady voltage profile and dissipating the stored

Table 16.4. Technical parameter of battery bank.

Parameters	Data
Battery	4500 Ah, 48 V Li-ion
Battery loss	15%
Depth of discharge	40%
Batteries in series connection	13
Batteries in parallel connection	11
Total batteries required	143
Efficiency	90%
Lifetime	8 years
Costs	$300/kWh

energy with high efficiency. The battery banks are placed in series and parallel connections to give the desirable nominal DC operating voltage and ensure optimal capacity for energy storage. The battery banks in series and parallel connections are given as follows:

$$N_B^s = \frac{V_{bus}^{dc}}{V_B^{nom}} \tag{16.7}$$

$$N_B^p = \frac{L}{V_{bus}^{dc}} \tag{16.8}$$

$$N_B = N_B^s * N_B^p, \tag{16.9}$$

where N_B represents the required number of battery, N_B^s is the battery in series connection, N_B^p is the battery in parallel connection, L is the load, V_B^{nom} is the battery nominal voltage, and V_{bus}^{dc} is the DC bus voltage.

16.3.4 Inverter

Inverter is an important device for hybrid energy systems especially when considering an off-grid system. They are required for balancing energy flow between AC and DC devices. Inverters have the feature of ensuring safety and stability of the system. ABB PVS800 1 MW inverter is adopted and the technical specification is summarized in Table 16.5.

16.4 Simulation Results and Discussions

The simulation and optimization of the proposed off-grid hybrid RE systems are modelled with Hybrid Optimization of Multiple Energy Resources (HOMER) software [16]. Many system configurations are available in HOMER software, grid-connected diesel option as in the existing power system network, and proposed Wind-PV-Diesel option for the hybrid energy system was considered in this study. This software is employed to optimize and analyse the hybrid systems to achieve optimal results of decision variables for the system. This software makes the required weather data available by stating the longitude and

Table 16.5. Technical parameter of inverter.

Parameters	Data
Model	ABB PVS800
Rating	1MW
Efficiency	95%
Input (DC)	
Maximum input power	2 × 1200 kW
DC voltage range, MPP	600V–850 V
Maximum DC current	2 × 1710 A
Required MPP trackers	2
Output (AC)	
Nominal power output	2 × 1000 kW
Power at cos $\phi = 0.95$	2 × 950 kW
Maximum AC output power	2 × 1200
Nominal current	2 × 1445 A
Nominal voltage output	400 V
Frequency	50/60 Hz
Power factor compensation	Yes

latitude of the desired locations via connecting through the NASA atmospheric Science Data Centre. Figure 16.8 depicts the flowchart for the operational and economic assessment of the proposed hybrid energy system.

16.4.1 System Optimization

The simulation and optimization of the proposed off-grid hybrid renewable energy systems are modelled using HOMER software [16]. Many system configurations are available in HOMER software, grid-connected diesel option as in the existing power system network, and proposed wind-PV-diesel option for the hybrid energy system was considered in this study. This software is employed to optimize and analyze the hybrid system to obtain an optimal value of system decision variables. HOMER software examines the sources combination to meet the required load and locates the hybrid energy system that technically achieving the desired targets with less economic values using the cost parameters. This software makes the required weather data available by stating the longitude and latitude of the desired locations via connecting through the NASA atmospheric Science Data Centre. Figure 16.8 depicts the flowchart for the operational and economic analysis of the proposed hybrid energy system.

16.4.2 Performance Indicators

In this section, the simulated results based on the grid-connected system and proposed hybrid energy system are compared and assessed based on the following performance indicator: technical analysis, economic analysis, and environmental analysis.

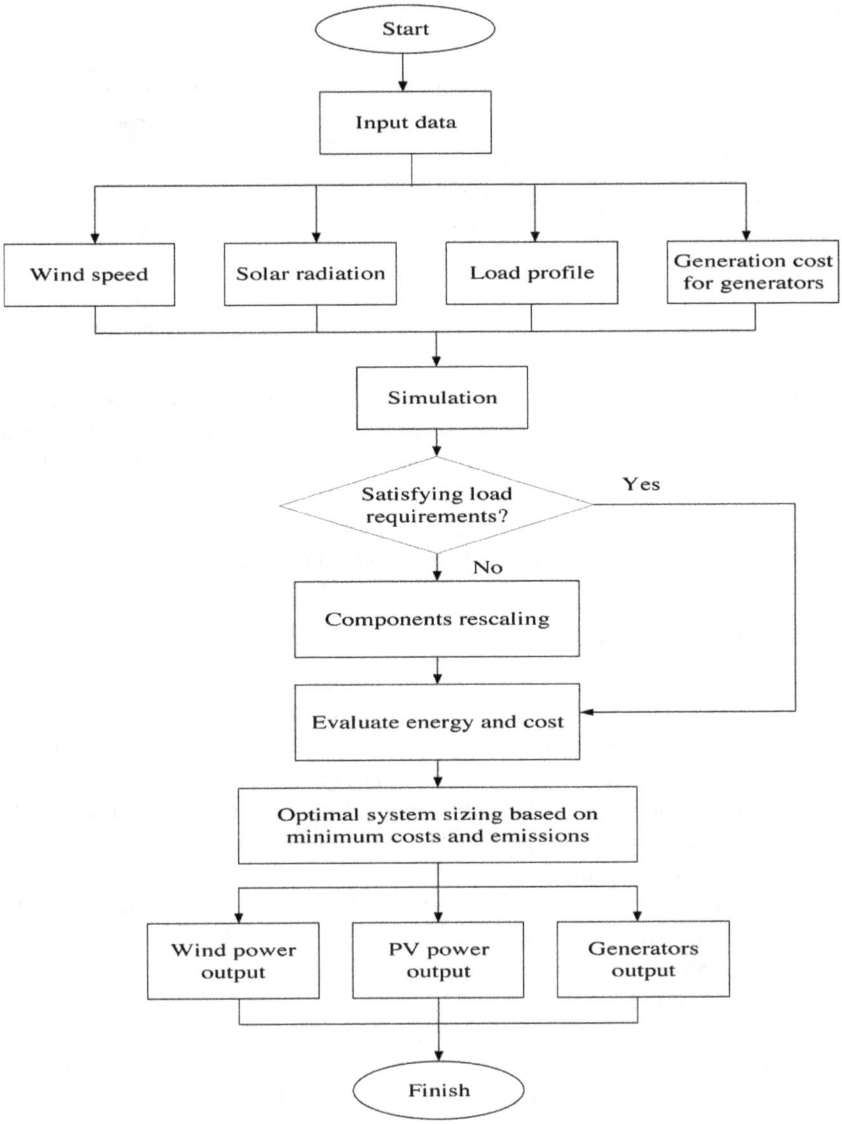

Figure 16.8. Flowchart for operational and life-cycle cost analysis of hybrid renewable energy system.

16.4.2.1 Technical Analysis

Operational performance is an essential criterion for determining the effectiveness of energy systems for energy users. Table 16.6 summarizes power generated by each component in the hybrid RE system. The wind–PV–diesel hybrid system has an annual energy production of 5,615,373 kWh/year. Figure 16.9 shows the percentage contributions from each of the sources.

Table 16.6. Electricity output by different components in the hybrid energy system.

Components	Energy Produced (kWh/yr)	Percentage of Energy Produced
Wind turbines	476724	8
PV system	1514697	27
Generator 1	1847403	33
Generator 2	1776549	32
Total	5615373	100

The wind represents 8% of the total energy produced, whereas PV accounts for 27% of the energy produced. The insufficient energy produced by wind and PV is complemented by the diesel generators. It is observed that wind energy is fairly low due to low wind potential experienced in the university location, whereas the PV experienced a great proportion of supply owing to favorable solar radiation at the region. The average power produced by the components of the system on monthly basis is depicted in Fig. 16.10.

16.4.2.2 Economic Analysis

To accurately analyze the hybrid energy system output, it is necessary to pursue an economic performance of the system. Figure 16.11 presents the annual cash flow for some components of the hybrid RE system. The negative sign is an indication that the system experienced profits from wind and PV generation despite the high initial investment cost, replacement cost, and salvage values. Table 16.7 summarizes the cash summary of available generators in the current system network of the university, and Fig. 16.12 depicts an account of the nominal cash flow for 25 years. The cost of running the diesel generators for the planning period of 25 years is approximately $19,288,744 and the cost of supply from the grid for the duration of 10 h daily from utility is approximately $12,786,875. Therefore, the total cost of electricity generation for the university campus considering the existing power system network is $32,075,619 (Table 16.8).

The existing grid-connected diesel electricity generation depicted in Fig. 16.13 shows that the cost of fuelling the generators took the major share of 98% of the total cost of electricity generation in the university campus owing to higher price of diesel. Figure 16.14 presents the cost summary of proposed hybrid energy system in the university campus with initial investment cost accounting for 91% and 94% share of the total cost of wind and PV system, whereas the operation and maintenance cost are relatively smaller than the generators in the hybrid energy system. The investment cost is larger for wind and PV in the first year because of the installation of all components required for project. The replacement cost constituted the larger share of the battery cost owing to its short lifetime and also the investment cost constituted the larger share of converter cost.

The energy production cost for the proposed hybrid RE system for the planning period is $14,279,601, which is lower than the estimated cost of electricity generation through the existing grid-connected diesel energy system in the university campus, which is

Table 16.7. The cash summary of the generators in the university.

Components	Capital ($)	Replacement ($)	O&M ($)	Fuel ($)	Salvage ($)	Total Costs ($)
Generator 1	7079	21,688	30,201	701,623	−115	760,475
Generator 2	53,115	160,232	171,061	5,372,730	−2991	5,754,146
Generator 3	39,836	101,471	113,109	3,552,701	−387	3,806,730
Generator 4	28,351	95,088	91,274	2,704, 316	−5791	2,913,240
Generator 5	30,164	90,249	114, 596	3,160,851	−2319	2,292,542
Generator 6	20,984	70,021	105,067	2,469,091	−4555	2,660,608
System	179,529	538,750	625,308	17,961,310	−16, 159	19, 288,744

Table 16.8. Cost analysis of the proposed hybrid system.

Components	Capital ($)	Replacement ($)	O&M ($)	Fuel ($)	Salvage ($)	Total ($)
PV system	2,352,000	0	153,400	0	0	2,505,401
Wind turbine	1,000,000	0	102,267	0	0	1,102,267
Battery	56,632	218,015	279,956	6,299,918	−12,799	6,841,721
Converter	106,230	159,102	199,367	6,088,659	−1526	6,551,832
Generator 1	42,900	396,741	146,242	0	−4998	580,885
Generator 2	150,000	0	63,917	0	0	213,917
System	3,707,762	773,859	945,148	12,388,575	−19,324	17,796,018

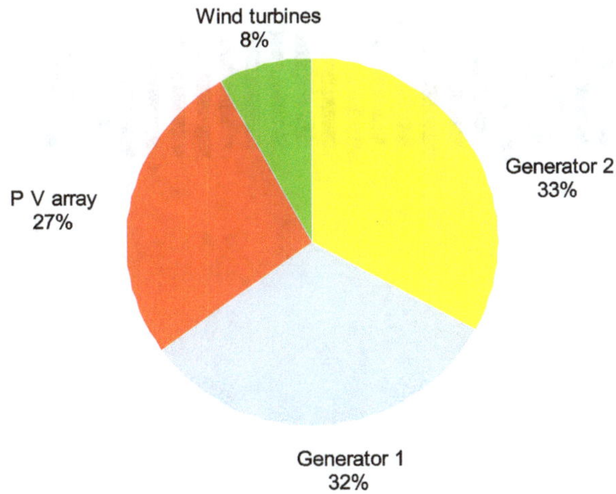

Figure 16.9. Electricity output by different components of the hybrid energy system.

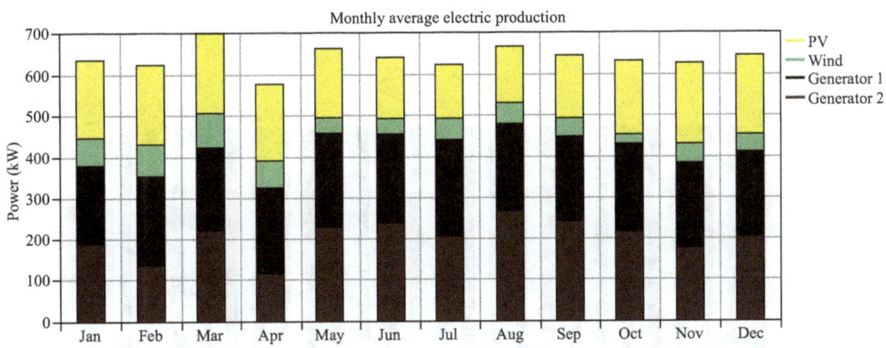

Figure 16.10. Monthly electricity output by different components.

\$32,075,619 as presented in Fig. 16.15. The net profit value is \$17.4 million, which shows that the project investment returns occur in the fourth year of the 25-year planning period rendering the hybrid energy system more economical than the existing network in the university. The payback period is a key parameter that reveals the time period the investors recover their initial investment.

16.4.2.3 Environmental Analysis

Here, the environmental assessment of electricity generation from the existing network and proposed hybrid energy system are evaluated. Table 16.9 summarizes the air pollutant emissions, which includes carbon dioxide, nitrogen oxide, sulfur dioxide, unburned hydrocarbon, carbon monoxide, and particulate matter from the existing grid-connected diesel supply and the proposed hybrid RE system. It is observed that these air pollutant emissions are lower for the proposed hybrid RE system compared with the existing grid-connected

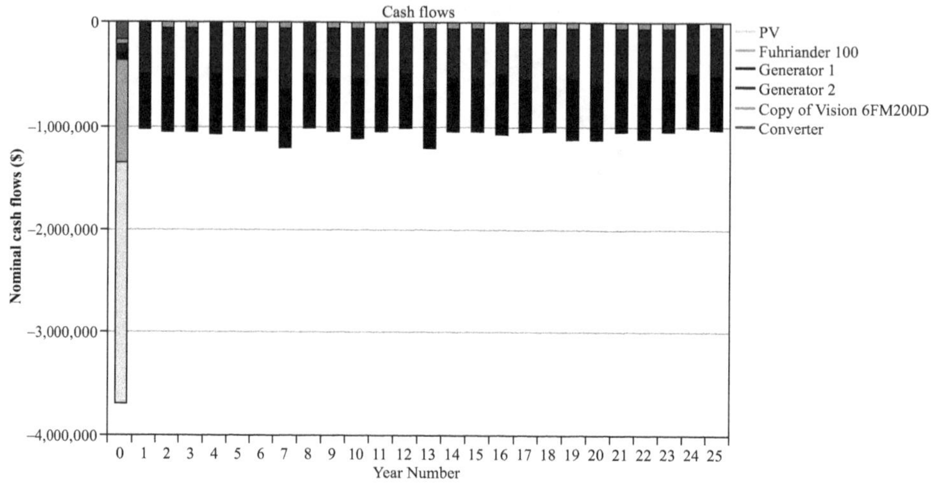

Figure 16.11. Annual cash flow of each components in the hybrid system for 25 years.

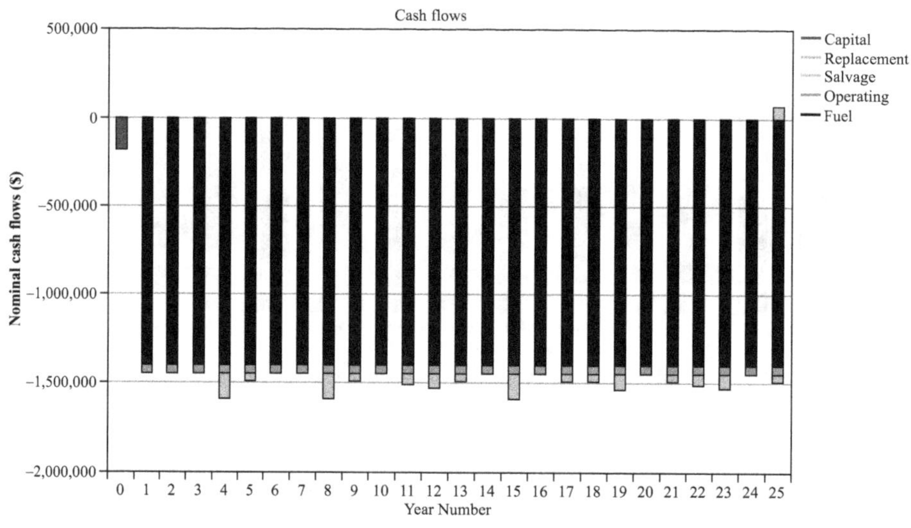

Figure 16.12. Cash flow of the generators for 25 years.

diesel system as depicted in Figs. 16.16 and 16.17. This is because the RESs' penetration significantly reduces the air pollutant emission to some extent.

16.5 Conclusion

In this chapter, the operational and economic feasibility of hybrid RE system in a university campus in Nigeria was investigated. The availability of solar radiation and wind speed of the selected location was obtained from NASA meteorology database. Through simulation,

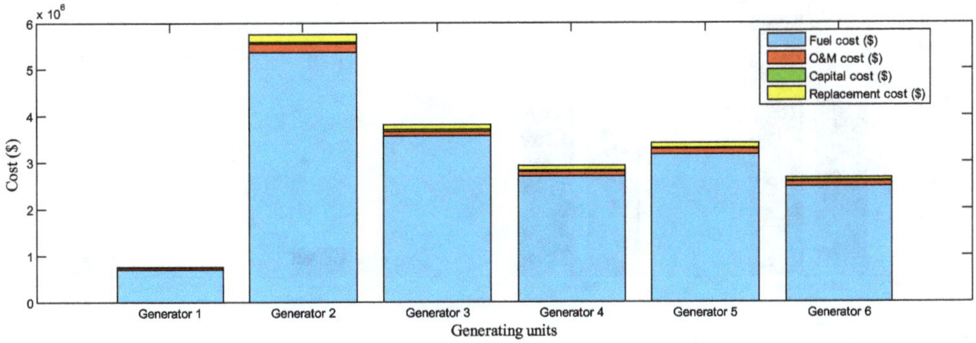

Figure 16.13. The cost analysis of the existing power system network in the university.

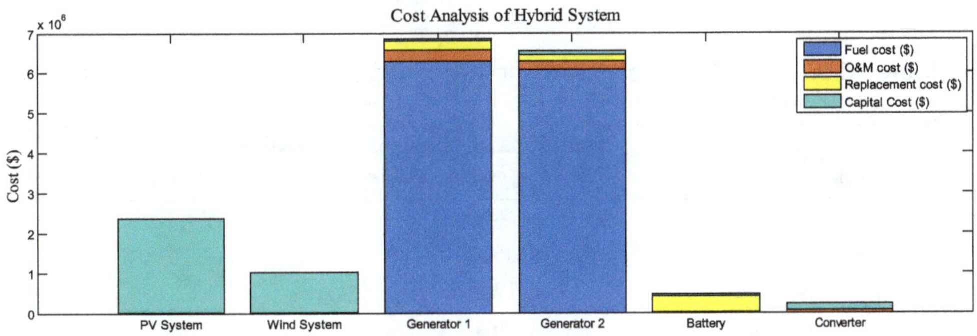

Figure 16.14. The cost analysis of the proposed hybrid system.

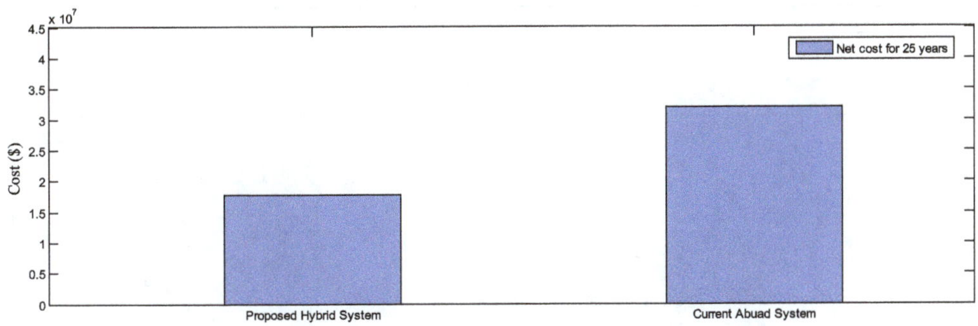

Figure 16.15. Systems net investment cost.

the optimal system configuration, costs, and environmental effects were effectively analyzed. The existing grid-connected diesel system and the proposed hybrid RE system were modelled and optimized using HOMER software to obtain an optimal energy sizing with efficient economic solution. The obtained results show that the hybrid renewable energy system is more economical in the university campus layout owing to the favorable wind and solar potential in the location. The hybrid energy system offers an equity payback during the fourth year of its 25-year investment lifetime with a 31% return on investment.The air

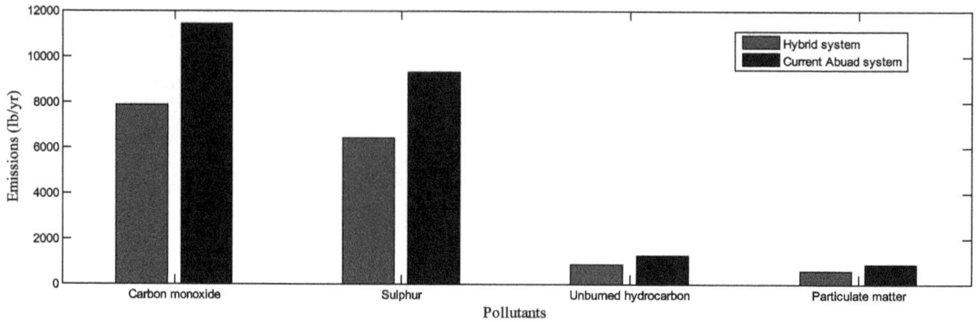

Figure 16.16. Air pollutant emissions in the existing and proposed hybrid energy system.

Table 16.9. Air pollutant emissions in proposed hybrid and current system of the university.

Pollutants	Emissions of Current ABUAD System (Ib/yr)	Emissions of Hybrid System (Ib/yr)
Carbon dioxide	4,624,961	3,190,006
Carbon monoxide	11,416	7874
Unburned hydrocarbon	1265	872
Particulate matter	861	594
Sulfur dioxide	9288	6406
Nitrogen oxide	101,866	70,261

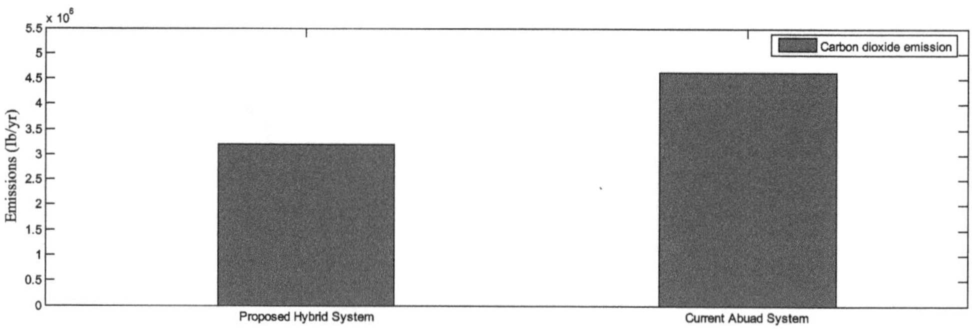

Figure 16.17. Carbon dioxide in the existing and proposed hybrid energy system.

pollutant emissions were effectively reduced with the proposed hybrid RE system owing to the presence of RESs. This is an indication that the proposed hybrid RE system is an economically feasible solution to meet the pressing demand of low-emission energy system with optimum costs of the university.

16.6 References

1. M. S. Gul, and S. Patidar, "Understanding the energy consumption and occupancy of a multi-purpose academic building," *Energy Buildings*, vol. 87, pp. 155–165, 2015.
2. M. Husein, and I. Chung, "Optimal design and financial feasibility of a university campus microgrid considering renewable energy incentives," *Applied Energy*, vol. 225, pp. 273–289, 2018.
3. S. L. Gbadamosi, N. I. Nwulu, and Y. Sun, "Harmonic and power loss minimization in power systems incorporating renewable energy sources and locational marginal pricing," *Journal of Renewable and Sustainable Energy*, vol. 10, pp. 1–24, 2018.
4. A. T. Brimmo, A. Sodiq, S. Sofela, and I. Kolo, "Sustainable energy development in Nigeria: Wind, hydropower, geothermal," *Renewable and Sustainable Energy Reviews*, vol. 74, pp. 474–490, 2016.
5. S. L. Gbadamosi, N. I. Nwulu, and Y. Sun, "Multi-objective optimisation for composite generation and transmission expansion planning considering offshore wind power and feed-in tariffs," *IET Renewable Power Generation*, vol. 12, no. 14, pp. 1687–1697, 2018.
6. Saheed Lekan Gbadamosi and Nnamdi I. Nwulu, "Optimal planning of renewable energy systems for power loss reduction in transmission expansion planning", *Journal of Engineering, Design and Technology*, 18 (2), pp 61–80, 10.1108/JEDT-11-2019-0291.
7. S. L. Gbadamosi, N. I. Nwulu, and Y. Sun, "Optimal Scheduling in Offshore Wind Farms Considering Harmonic Losses and Locational Marginal Pricing," IEEE PES-IAS 5th Power Africa Conf., pp. 658–663, 2018.
8. A. Al-shara, A. Z. Sahin, T. Ayar, and B. S. Yilbas, "Techno-economic analysis and optimization of solar and wind energy systems for power generation and hydrogen production in Saudi Arabia," *Renewable and Sustainable Energy Reviews*, vol. 69, pp. 33–49, 2016.
9. Y. Yang, S. Bremner, C. Menictas, and M. Kay, "Battery energy storage system size determination in renewable energy systems: A review," *Renewable and Sustainable Energy Reviews*, vol. 91, pp. 109–125, 2018.
10. Azizat O. Gbadegesin, Yanxia Sun, Nnamdi I. Nwulu, Saheed Lekan Gbadamosi, "Optimising fuel efficiency for renewable energy-based systems incorporating hybrid energy storage", 2019 11th International Conference on Applied Energy, 12-15 August 2019.
11. Saheed Lekan Gbadamosi and Nnamdi I. Nwulu, "A dynamic optimal expansion planning integrating renewable energy sources", 11th International Conference on Applied Energy, 12-15 August 2019.
12. P. Kumar, S. Roy, H. B. Karayaka, Y. Yan, and Y. Alqudah, "Investigations into best cost battery-supercapacitor hybrid energy storage system for a utility scale PV array," *Journal of Energy Storage*, vol. 22, pp. 50–59, 2019.
13. J. Li, B. Cornelusse, P. Vanderbemden, and D. Ernst, "A SC / battery Hybrid Energy Storage System in the Microgrid Assessing the feasibility of using the heat demand-outdoor temperature function for a long-term district heat demand forecast," *Energy Procedia*, vol. 142, pp. 3697–3702, 2016.
14. W. Zhang, A. Maleki, M. A. Rosen, and J. Liu, "Optimization with a simulated annealing algorithm of a hybrid system for renewable energy including battery and hydrogen storage," *Energy*, vol. 163, pp. 191–207, 2018.
15. L. Olatomiwa, S. Mekhilef, and O. S. Ohunakin, "Hybrid renewable power supply for rural health clinics (RHC) in six geo-political zones of Nigeria," *Sustainable Energy Technologies and Assessments*, vol. 13, pp. 1–12, 2016.
16. A. Buonomano, F. Calise, M. Dentice, and M. Vicidomini, "A hybrid renewable system based on wind and solar energy coupled with an electrical storage: Dynamic simulation and economic assessment," *Energy*, vol. 155, pp. 174–189, 2018.

17. L. Bartolucci, S. Cordiner, V. Mulone, and J. L. Rossi, "Electrical power and energy systems hybrid renewable energy systems for household ancillary services," *International Journal of Electrical Power & Energy Systems*, vol. 107, pp. 282–297, 2019.
18. M. Das, M. Anil, K. Singh, and A. Biswas, "Techno-economic optimization of an off-grid hybrid renewable energy system using metaheuristic optimization approaches – Case of a radio transmitter station in India," *Energy Conversion and Management*, vol. 185, pp. 339–352, 2019.
19. C. Li, D. Zhou, and Y. Zheng, "Techno-economic comparative study of grid-connected PV power systems in five climate zones, China," *Energy*, vol. 165, pp. 1352–1369, 2018.
20. J. Ahmad, M. Imran, A. Khalid, et al., "Techno economic analysis of a wind-photovoltaic-biomass hybrid renewable energy system for rural electrification: A case study of Kallar Kahar," *Energy*, vol. 148, pp. 208–234, 2018.
21. H. A. Kazem, M. H. Albadi, A. H. A. Al-waeli, A. H. Al-busaidi, and M. T. Chaichan, "Case studies in thermal engineering techno-economic feasibility analysis of 1 MW photovoltaic grid connected system in Oman," *Case Studies in Thermal Engineering*, vol. 10, pp. 131–141, 2016.
22. B. O. Ariyo, M. F. Akorede, I. O. A. Omeiza, S. A. Y. Amuda, and S. A. Oladeji, "Optimisation analysis of a stand-alone hybrid energy system for the senate building, university of Ilorin, Nigeria," *Journal of Building Engineering*, vol. 19, pp. 285–294, 2018.
23. S. M. Abolarin et al., "A collective approach to reducing carbon dioxide emission: A case study of four University of Lagos Halls of residence," *Energy Buildings*, vol. 61, pp. 318–322, 2013.
24. Annual Technical Report on System Operation of Power Generation Unit, Issued by Afe Babalola University 2017.
25. M. Y. Ramandi, K. Afshar, A. S. Gazafroudi, and N. Bigdeli, "Electrical power and energy systems reliability and economic evaluation of demand side management programming in wind integrated power systems," *International Journal of Electrical Power & Energy Systems*, vol. 78, pp. 258–268, 2016.
26. B. M. Radhakrishnan, and D. Srinivasan, "A multi-agent based distributed energy management scheme for smart grid applications," *Energy*, vol. 103, pp. 192–204, 2016.

CHAPTER 17

Optimal Distributed Resources Allocation in Contemporary Distribution Systems

Neeraj Kanwar[1], Nikhil Gupta[2], and K. R. Niazi[2]

[1]*Manipal University Jaipur, Jaipur, India. nk12.mnit@gmail.com*
[2]*Malaviya National Institute of Technology, Jaipur, India. nikhil2007_mnit@yahoo.com; krn152001@yahoo.co.in*

Abstract The origin of many power system issues are typically based on the electrical distribution systems as they are the tail ends of electric power systems. Whatever is embedded in distribution systems, its impact percolates in the whole power system. Different key technologies are available for distribution system performance improvement such as optimal distributed resources (DRs) allocation and network reconfiguration. The amount of actual benefit of these technologies depends on the modelling of load profile of the distribution system. The system load profile should be modelled by considering ground realities of distribution system such as allocation of dedicated feeders to different types of customers and load profile pattern of different category of customers. In this chapter, the DR allocation problem in radial distribution systems is modelled by considering a more realistic load profile of the system. Also, the consequences of ignoring load diversity among distribution buses on performance of distribution system are presented.

Keywords: Commercial customers, Distributed resources, Distribution systems, Energy loss, Industrial customers, Load modeling, Meta-heuristic technique, Network reconfiguration, Residential customers, Shunt capacitors, Smart grid, Teaching–learning-based optimization.

17.1 Introduction

The present distribution systems now become smart to achieve larger socio-economic and other non-tangible benefits. Smart grids are designed to achieve different goals such as lesser carbon footprints, better asset utilization, improved energy efficiency, reliability, and security and power quality. The passive distribution systems will be gradually transformed into active distribution systems having widespread deployment of distributed resources (DRs). Though this transition requires a paradigm shift in both planning and operations of distribution systems. The construction of next-generation active distribution networks requires the exploitation of existing infrastructure, use of new technologies of generation, and changes in operational practices [1]. The integration of distributed or dispersed generations (DGs), shunt capacitors (SCs), and network reconfiguration (NR) are three important technologies for realizing smart distribution systems. These key technologies may be synchronized together to get better solutions so that contemporary distribution systems can achieve optimum performance. Feeders in a distribution system normally have a mixture of different types of customers such as industrial, commercial, and residential [2]. Also each type of customer has its distinct load profile. Therefore, these ground realities of distribution systems should be considered with a good degree of accuracy, otherwise counterproductive results so obtained may jeopardize the planning and operation of distribution systems. This chapter addresses the simultaneous optimal allocation of DRs such as DGs and SCs in the view of NR to reduce annual energy losses and to enhance node voltage profiles of distribution systems. The main emphasis of this chapter is to show the importance of more practical formulations for these optimization problems considering realistic operational issues and realities of modern distribution systems.

The chapter is organized as follows: The introduction about theme of the chapter is explained in Section 17.1. The literature about optimal allocation of DRs and their simultaneous allocation are discussed in Section 17.2. Coordinated approach of DR allocation and NR is presented in Section 17.3. Annual load profile considering load diversity among customers is presented in Section 17.4. Problem formulation for DR allocation and NR is presented in Section 17.5. Simulation results are dealt with in Section 17.6, and the analysis of results is discussed in Section 17.7. Finally, the summary of the chapter is presented in Section 17.8.

17.2 Optimal Allocation of DRs

The passive distribution networks are currently being transformed into active distribution networks by integrating more and more DGs to achieve the objectives of smart distribution systems. DRs such as DGs and SCs can independently regulate active and reactive power flow among distribution feeders and reduce their flow from source to the location of these resources [2]. Therefore, performance of distribution systems can be effectively regulated when they are placed at strategic locations in distribution networks. From long back, researchers were attracted toward the optimal allocation of SCs for power factor correction; however, in the present scenarios, they are still important as the performance of modern distribution systems can be enhanced by adopting suitable strategy of placing several types of DRs. The following sections present a brief literature review about the optimal DRs allocation, that is, SCs and DGs, their simultaneous allocation, and a combined approach for DR allocation and NR in contemporary distribution system.

17.2.1 Optimal Placement of SCs

Reactive power compensation in distribution systems using SCs is typically an old classical problem of power systems. The capacitors must be allocating optimally, otherwise line losses may increase and develop over voltages during light load hours. The optimal capacitor placement problem involves the determination of their optimal number, sizing, and siting. Some pioneering works reported may include classical techniques [3,4], mathematical programming techniques [5,6], analytical methods [7–9], and numerical programming methods [6,10]. Many researchers are attracted toward heuristic methods [11,12], which are simple, fast, and easy to understand but usually converge to suboptimal solution. The evolutionary and swarm-based intelligence techniques have become the center of attraction for researchers from the beginning of this century as their performance does not depend on the type and shape of objective function, or the number of objectives or problem constraints employed, and are potentially very strong to get global or near global solution. Different objectives and constraints, which are generally considered in literature to solve optimal capacitor placement problem, are listed in Table 17.1.

DGs are becoming very common in distribution systems to achieve several techno-economic as well as social objectives. A wealth of literature is available to discuss optimal DG placement in distribution systems, which is briefly described in the Section 17.2.2.

17.2.2 Optimal Placement of DGs

The potential promises of the smart grid include reliability and power quality enhancement, peak system demand reduction, transmission congestion cost reduction, energy efficiency enhancement, better asset utilization, higher penetration of renewables, and so on [13]. Further, the active losses cannot be reduced in distribution systems using SCs alone as active power is provided by the swing bus [14]. Therefore, to supply active loads of the distribution system, local generation should be privileged by DGs. DGs refer to small generating units typically connected to the utility grid in parallel near load centers and can satisfy these objectives. DG penetration has been increased significantly during the last decade owing to advancement in several generation technologies, deregulation of power systems, environmental and construction issues related to new transmission lines, and so on [15].

The problem of optimal DG allocation involves exploring the optimal number and siting and sizing of DG units to achieve certain objectives while satisfying several network operation constraints. It has been found that strategically allocated DG units can yield several benefits, whereas improper allocation and/or operation, in general, may reduce benefits

Table 17.1. Objectives and constraints for optimal capacitor placement

Objectives	Constraints
1. Minimizing peak active power loss/line losses/cost of power losses **2.** Minimizing capacitor cost **3.** Minimizing energy loss cost **4.** Capacity release/maximizing marginal loading of feeders **5.** Minimizing node voltage deviation/enhance voltage stability **6.** Minimizing cost of reliability **7.** Minimizing switching operations **8.** Minimizing total harmonic distortion	**1.** Active and reactive power flow balance equations/power balance constraints **2.** Voltage limit constraints **3.** Reactive compensation limit **4.** Maximum total compensation **5.** Overall system power factor **6.** Number and size of SCs **7.** Total harmonic distortion of voltage **8.** Total active power loss on all branches **9.** Branch active power limit/power source limit/apparent power flow limit **10.** Determination of type of capacitors **11.** Selection of a unique bank per node/impossibility to locate capacitors at certain nodes **12.** Maximum limit of branch current/line current/line capacity limit **13.** Maximum unbalance factor/distortion indices **14.** Capacitor switching transients

and even jeopardize system operation [16]. A lot of research work is available to solve optimal DG allocation problem by considering various technical, economical, or techno-economic objectives while considering different constraints as summarized in Table 17.2. The problem is solved using a variety of solution techniques that can be briefly mentioned here. Among classical methods, the most efficient are the nonlinear programming [17,18], sequential quadratic programming [19], ordinal optimization methods [20], and so on. On the other hand, exhaustive search methods such as brute force technique and dynamic programming [21] guarantee to find the global optimum, but they have the curse of dimensionality so are unfit for modern distribution systems. Several mathematical optimization techniques, such as gradient search, Hereford ranch algorithm, decision theory approach, sequential quadratic programming, linear programming, primal-dual interior-point method, mixed-integer nonlinear programming, and so on, have been addressed to solve this problem [22]. However, the major difficulty with these methods is that they provide only one solution, which may not be feasible for DG placement owing to strategic location, political reasons, and so on. Analytical methods [23,24] are simple to understand and implement, and quickly executed, but may provide impractical solution as they are based on certain unrealistic assumptions. Though population-based metaheuristics suffer from high computational burden, this is not a serious limitation for DG placement applications [25]. Therefore, the swarm- and evolutionary-based metaheuristic techniques have gained much attention of the researchers in solving DG allocation problems efficiently.

Table 17.2. Objectives and constraints for DG penetration

Objectives	Constraints
1. Minimizing total/active power loss	**1.** Power flow equality constraints/P and Q mismatch equations
2. Minimizing energy loss/cost of energy loss/maximizing apparent power export	**2.** Node voltage limits
3. Minimizing system average interruption duration index (SAIDI) and frequency index (SAIFI)/interruption/reliability cost	**3.** Capacity limit of lines/transformers
4. Minimizing investment, O&M cost of DG/reinforcement cost of the distribution system	**4.** Total harmonic distortion limit
5. Improving voltage stability margin	**5.** Reliability index SAIDI
6. Maximizing system investment deferral incentives	**6.** Distribution substation capacity limit
7. Maximizing profit/benefit to cost ratio	**7.** Power generation limits of DG
8. Maximizing capacity adequacy cost/identifying spare capacity to accommodate DGs	**8.** Protection coordination limits
9. Maximizing voltage limit loadability/enhance line loadability	**9.** DG power factor limits
10. Minimizing system upgrade cost/line capacity release	**10.** DG annual operation time limit
11. Maximizing the DG penetration level/rating of DGs	**11.** Only one DG can be installed in one installation position
12. Improving node voltage profile	**12.** Discrete size of DG units
13. Minimizing total harmonic distortion	**13.** DG loss constraints/active power losses of branches
14. Minimizing purchase cost of energy	**14.** Number and size of DG units/total installed DG capacity at each node/reactive power flow flowing back to source
15. Maximizing security margin/minimizing short-circuit current	**15.** Line thermal limits
16. Minimizing number of DG units	**16.** Short-circuit limits
17. Minimizing distributed loads/minimizing load curtailment	**17.** Exclusion of right of the way buses
18. Minimizing environmental cost	**18.** Generator bus constraints
	19. Power flow limits/power source limit
	20. Penetration level of DGs

17.2.3 Simultaneous Allocation of SCs and DGs

In existing distribution systems, there is a certain limit for DG penetration as most of the existing DGs operate at unity power factor control mode. This limit could be increased if sufficient and coordinated reactive support is available by deploying SCs. The optimal generation of active and reactive power from these DRs fruitfully affects power transactions with the grid, thus regulating power flows within the system. Optimal capacitor placement regulates reactive power flow, whereas the active power flow is regulated by optimal DG placement. Therefore, SCs and DGs can independently act and regulate the power flow in the system, however, their optimal allocation is not independent; the presence of one component may affect the optimal allocation of the other and vice versa. Moreover, they facilitate effective utilization and life extension of existing infrastructure [13] and also contribute toward self-adequacy of the system. In this view, the simultaneous allocation of these DRs needs an investigation to extract optimum benefits and also to avoid counter-productive consequences. Recently, some researchers [14,26–30] addressed the problem of simultaneous optimal allocation SCs and DGs using different approaches and have shown promising common impact of these components on distribution system performance. The problem was solved to optimize one or more objectives related to loss minimization, node voltage enhancement, reliability improvement, and so on, and is solved using analytical, mathematical, or heuristic optimization techniques.

17.3 Optimal DR Allocation and NR

DR is one of the new system planning options, which should be investigated in combination with the other traditional distribution system planning options such as NR [31]. NR is a well-known operational strategy that is used to achieve high performance of distribution systems, so it may also be employed in conjunction with optimal allocation of DRs. Smart grid facilitates integrated solutions for optimal allocation of DRs and NR for coexistence of these strategies of different horizons to achieve better efficiency and power supply of good quality [13]. Such coordinated strategies can benefit network owner as well as its users and other stake holders. However, the feasibility of DG investment versus other traditional planning options should be evaluated [25].

NR is a process that was initially invented to restore supply in the distribution system by isolating the faulty section. This is achieved by altering the status of the sectionalizing switches (normally close switches) and tie switches (normally open switches). Later on, NR becomes potential strategy of radial distribution systems to optimize vital objectives such as feeder load balancing, feeder power loss minimization, node voltage enhancement, reliability improvement, and so on [32]. Merlin and Back were the first to propose the idea of NR in 1975. Since then, extensive research work has been carried out to address reconfiguration problem using several objectives such as loss minimization, voltage profile enhancement, and reliability enhancement using diverse optimization techniques, namely exhaustive algorithms, heuristic algorithms, mathematical programming, analytical methods, AI techniques, and hybrid approaches.

In [16,33–36], NR in conjunction with the optimal allocation of DRs is employed and is found to be a very effective strategy to enhance system performance, whereas some researchers [35] suggested joint optimization by considering DR allocation along with NR.

However, the methodology is not realistic because optimal DG allocation may not remain optimal with the variation in network topology. Hung et al. [16] compared several strategies for DR allocation with NR where they concluded that DR allocation problem should be solved after NR. However, this is unrealistic as the planning problem should be followed by any operational strategy.

The modelling of load profile is vital whenever dealing with optimization problem of distribution system of planning or operation horizon. An incorrect modelling may produce counterproductive results. Typical distribution systems have diverse customer base, which are being divided into several groups on account of their specific load pattern, tariff structures, or social obligations. These groups may acquire strategic location in a distribution network, therefore, usually being fed through dedicated distribution buses and feeders. The specific loading among distribution buses becomes time variant, but never varies in unison owing to diversity in usage pattern of different group of customers. These realistic concerns should be addressed while formulating load profile of distribution systems.

17.4 Annual Load Profile Modelling Considering Load Diversity among Customers

The individual load profile modelling of distribution system plays an important role in deciding annual load profile of the distribution system. Therefore, it also affects DR allocation problem. In general, DR allocation problem is solved by considering piecewise multiple-level load profile [37,38]. This type of load modelling is simplest but not realistic. It ignores actual load diversity, which occurs due to load class mix of different types of customers that are connected to different buses. This load diversity among customers should be considered while modelling annual load profile of the system. It will generate more practical solution for any planning and operation problem. The consideration of load diversity decides actual shape of the annual load profile including peak demand, its duration, and also node voltage profiles of the system. The customer load demand varies with ambient conditions such as temperature and air pressure. It also faces seasonal variations. The daily load profile of distribution feeders can be aggregated for a given season by considering hourly variation in load demand to reduce the simulation time. Individual load profile of three different types of customers as residential, commercial, and industrial are shown in Fig. 17.1 [34]. The figure also shows aggregated daily load profile of the system having various load levels due to diversities attributed to different categories of customers. In a similar manner, the combined load profile for other types of seasons may be generated with an appropriate multiplying factor [33]. Finally, annual load profile can be developed by integrating load profile obtained for different seasons of the year.

The DR allocation and NR problems can be formulated as given in Section 17.5.

17.5 Problem Formulation

The DR allocation problem is formulated to maximize reduction in annual energy loss and to improve voltage profiles of system nodes. These objectives are further optimized using optimal NR. The solution of DR allocation problem determines the optimal number, sites, and sizing of DGs and SCs. These DRs are assumed to be controllable (tuneable), so that their optimal dispatches are determined for each operating state of the system to achieve optimum objectives.

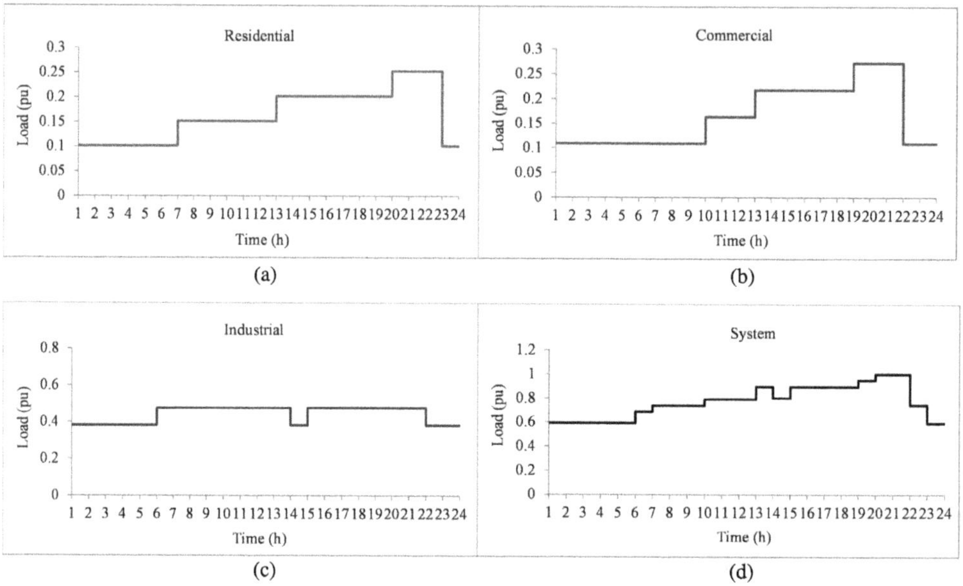

Figure 17.1. Daily load profiles of (a) residential customers, (b) commercial customers, (c) industrial customers, and (d) distribution system for spring/fall season.

17.5.1 Optimal DR Allocation

The optimal DR allocation problem is expressed as maximization problem:

$$\text{Obj. Function } 1 = E_{Loss}^{base} - E_{Loss}^{DR} \tag{17.1}$$

where E_{Loss}^{base} and E_{Loss}^{DR} represent annual energy losses for base case and after DR placement. Energy losses can be calculated using Eq. (17.2):

$$E_{Loss} = \sum_{s=1}^{N_s} D_s \left(\sum_{j=1}^{L} H_{j,s} \left(\sum_{res=1}^{N_{res}} R_{res} \frac{P_{res,j,s}^2 + Q_{res,j,s}^2}{|V_{res,j,s}|^2} + \sum_{ind=1}^{N_{ind}} R_{ind} \frac{P_{ind,j,s}^2 + Q_{ind,j,s}^2}{|V_{ind,j,s}|^2} + \sum_{com=1}^{N_{com}} R_{com} \frac{P_{com,j,s}^2 + Q_{com,j,s}^2}{|V_{com,j,s}|^2} \right) \right) \tag{17.2}$$

Objective function 1 is solved using following constraints [33]:

17.5.1.1 Power Flow Equations

The power import from utility grid should be balanced with the system load demand and losses. Different power flows are given in (17.3)–(17.7):

$$P_{n+1} = P_n - R_n \frac{P_n^2 + Q_n^2}{V_n^2} - p_{n+1}; \ \forall n \in \{N_{res}, N_{ind}, N_{com}\} \tag{17.3}$$

$$Q_{n+1} = Q_n - X_n \frac{P_n^2 + Q_n^2}{V_n^2} - q_{n+1};\ \forall n \in \{N_{res}, N_{ind}, N_{com}\} \tag{17.4}$$

$$V_{n+1}^2 = V_n^2 - 2\left(R_n P_n + X_n Q_n\right) + \left(R_n^2 + X_n^2\right) \frac{P_n^2 + Q_n^2}{V_n^2};\ \forall n \in \{N_{res}, N_{ind}, N_{com}\} \tag{17.5}$$

$$p_{n+1} = p_{n+1}^L - p_{n+1}^{\mathrm{DG}};\ \forall n \in \{N_{res}, N_{ind}, N_{com}\} \tag{17.6}$$

$$q_{n+1} = q_{n+1}^L - q_{n+1}^{\mathrm{SC}};\ \forall n \in \{N_{res}, N_{ind}, N_{com}\} \tag{17.7}$$

17.5.1.2 Node Voltage Constraint

All node voltages $V_{res,j,s}$ of the residential nodes should remain in their minimum and maximum bounds using conventional hard node voltage constraint as shown in (17.8). Similarly, node voltage constraint for industrial and commercial nodes can also be defined.

$$V_{\min} \leq V_{res,j,s} \leq V_{\max};\ \forall r \in N_{res}, \forall j \in L \tag{17.8}$$

17.5.1.3 Feeder Current Constraint

The flow of current in each distribution line should remain below or equal to the rated ampacity. This imposes feeder current constraints as given in (17.9).

$$I_n \leq I_n^{\max};\ \forall n \in \{N_{res}, N_{ind}, N_{com}\} \tag{17.9}$$

17.5.1.4 Bus Compensation Limit

Each bus must have total active and reactive power injection within their permissible limits as shown in (17.10) and (17.11).

$$p_{n,\min}^{DG} \leq p_n^{DG} \leq p_{n,\max}^{DG};\ \forall n \in \{N_{res}, N_{ind}, N_{com}\} \tag{17.10}$$

$$q_{n,\min}^{\mathrm{SC}} \leq q_n^{\mathrm{SC}} \leq q_{n,\max}^{\mathrm{SC}};\ \forall n \in \{N_{res}, N_{ind}, N_{com}\} \tag{17.11}$$

17.5.1.5 System Compensation Limit

The total injection of active and reactive power at optimal nodes must be less than nominal system active and reactive power demand, respectively. For *loc* candidate locations, the active and reactive system compensation limit can be calculated using (17.12) and (17.13).

$$\sum_{n=1}^{loc} p_n^{DG} \leq p_D;\ \forall n \in \{N_{res}, N_{ind}, N_{com}\} \tag{17.12}$$

$$\sum_{n=1}^{loc} q_n^{\mathrm{SC}} \leq q_D;\ \forall n \in \{N_{res}, N_{ind}, N_{com}\} \tag{17.13}$$

The repetition of candidate sites for DGs and SCs can be prohibited using Eqs. (17.14) and (17.15), respectively. DGs and SCs are modelled using (17.16) and (17.17) due to their commercial availability in discrete sizes.

$$loc_a^{DG} \neq loc_b^{DG};\ a,b \in \{N_{res}, N_{ind}, N_{com}\} \tag{17.14}$$

$$loc_a^{SC} \neq loc_b^{SC};\ a,b \in \{N_{res}, N_{ind}, N_{com}\} \tag{17.15}$$

$$p_n^{DG} \leq K_d P_d;\ K_d = 0, 1, 2, 3, ..., N_{DG} \text{ and } \forall n \in \{N_{res}, N_{ind}, N_{com}\} \tag{17.16}$$

$$q_n^{SC} \leq K_b Q_b;\ K_b = 0, 1, 2, 3, ..., N_{SC} \text{ and } \forall n \in \{N_{res}, N_{ind}, N_{com}\} \tag{17.17}$$

The solution of Eq. (17.1) provides optimal allocation of DRs in terms of their optimal number, sizing, and siting. Next, this equation is optimized to define the optimal dispatches of these installed DRs at different load levels. Therefore, Eq. (17.1) is solved at each load level separately. The optimal siting obtained in optimal solution is kept same, but sizing limit is restricted by solution provided by the optimal solution.

17.5.2 Network Reconfiguration

The NR problem of distribution systems is solved to minimize feeder power loss and is formulated in (17.18).

Minimize,

$$\begin{aligned} P_{Loss} = \sum_{s=1}^{N_s} \sum_{j=1}^{L} \Bigg(\sum_{res=1}^{N_{res}} R_r \frac{P_{res,j,s}^2 + Q_{res,j,s}^2}{|V_{res,j,s}|^2} + \sum_{ind=1}^{N_{ind}} R_i \frac{P_{ind,j,s}^2 + Q_{ind,j,s}^2}{|V_{ind,j,s}|^2} \\ + \sum_{com=1}^{N_{com}} R_c \frac{P_{com,j,s}^2 + Q_{com,j,s}^2}{|V_{com,j,s}|^2} \Bigg) \end{aligned} \tag{17.18}$$

To solve reconfiguration problem, network topology should be radial without any closed path. Therefore, the radiality constraint for the zth radial topology is defined in [34] as

$$\Phi_j(z) = 0 \tag{17.19}$$

The other problem constraints used are same as defined for DR allocation. While solving the NR problem, the radial topology constraint can be handled using any method available in literature such as in [39].

17.6 Application Results

Simulation results are obtained on a well-known standard IEEE 33-bus distribution system [40]. The brief about system data is summarized in Table 17.3. Detailed system data can be referred from [40]. A total of 27 load levels are obtained in aggregated load profile of system for 1 year considering load diversity and dedicated feeders. The load factors and corresponding load duration can be referred from [34]. The load levels are taken as 1–9 for spring/fall, 10–18 for winter, and 19–27 for summer seasons, respectively [33].

Optimization is done using a modified version of metaheuristic technique, that is, teaching–learning-based optimization (TLBO) [26] for the following four different cases [33]:

Case 1: Base system after optimal DR allocation
Case 2: Base system after optimal DR dispatches
Case 3: Case-1 after NR
Case 4: Case-2 after NR

Case 1 represents the base case after optimal DR allocation. The results are summarized in Table 17.4. Optimal dispatches of DRs at each load level are obtained using this solution known as Case 2. Cases 3 and 4 represent the NR after optimal DR allocation and NR after optimal dispatches allocation, respectively. Optimal dispatches obtained in Case 2 and optimal configuration obtained in Cases 3 and 4 are summarized in Table 17.5.

Network performance is obtained using optimal dispatches at each load level. The power loss obtained at each load level is shown in Fig. 17.2. It can be seen that power losses are considerably reduced among all cases with respect to Base Case. Moreover, power losses are further reduced in Cases 3 and 4 with respect to Cases 1 and 2. This shows the utility of NR after DR allocation. This fact can be observed at each load level. However, power losses are not reduced so much during off-peak hours of the winter season (load levels 10–13) in Cases 1 and 3 as compared to Cases 2 and 4. This shows the importance of DG output power control during off-peak hours. Therefore, Cases 2 and 4 are generating promising results in this regard. It shows that system performance is improved with optimal control of DR sizing with each load level.

The impact of DR power control and optimal dispatches can also be seen on minimum node voltage obtained at each load level as shown in Fig. 17.3. It can be observed that voltage profile is improved at each load level. This is true for all seasons. However, the

Table 17.3. Brief system data of 33-bus test distribution system [34]

Particular	Value
Line voltage (kV)	12.66
Nominal active system demand (kW)	3715
Nominal reactive system demand (kVA)	2300
Base system configuration	33–37
Residential feeders	1–15
Industrial feeders	22–29
Commercial feeders	16–21, 30–33

Table 17.4. Optimal solutions of DRs for Case 1

Location (DG Capacity in kW)	Location (SC Capacity in kVAr)
14(494), 24(954), 30(722)	12(300), 25(300), 30(600)
TC: 2170	TC:1200

TC: Total capacity

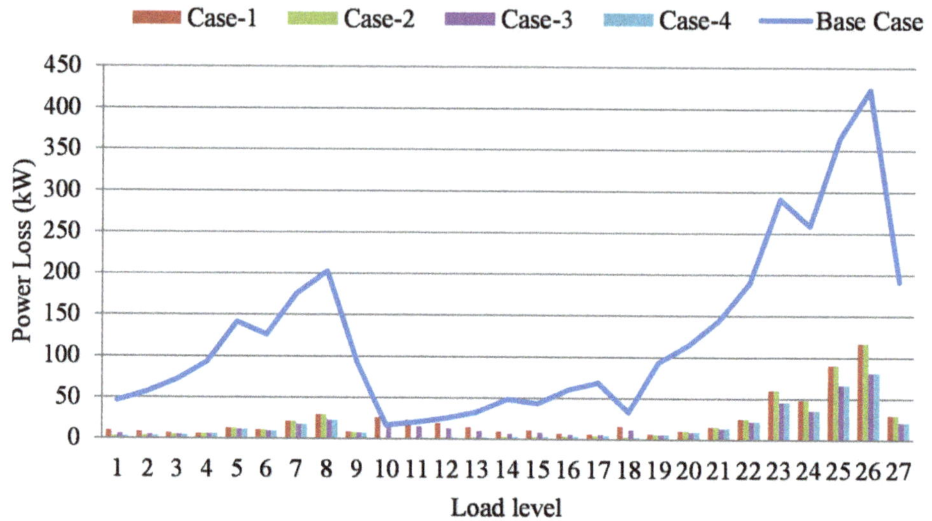

Figure 17.2. Power losses at different load levels.

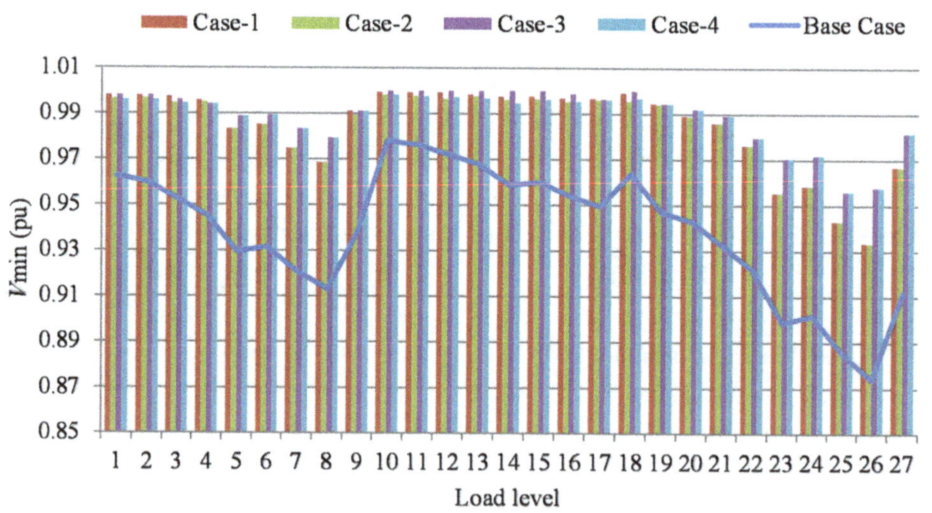

Figure 17.3. Minimum node voltage obtained at different load levels.

minimum voltage among all nodes remains below 0.94 pu. Precisely, it can be notified at load level 26 in Cases 1 and 2. This load level shows the peak load condition of the system. It is noteworthy that voltage profiles are further improved using NR in Cases 3 and 4 to such an extent that all node voltages are maintained above 0.94 pu even during peak load hours. Therefore, integrated approach of DR allocation and NR can provide promising results.

Table 17.5. Solution obtained for Cases 2, 3, and 4

Load Level	Optimal Dispatches of SCs (in kVAr)/DGs (in kW)	Optimal Configuration Case 3/Case 4
1	200, 300, 400/308, 800, 502	3, 6, 9, 17, 33/7, 8, 9, 25, 36
2	200, 300, 500/313, 949, 539	6, 9, 17, 25, 33/7, 8, 9, 26, 36
3	200, 300, 500/426, 954, 584	7, 9, 17, 25, 33/7, 9, 21, 25, 36
4	300, 300, 600/458, 954, 712	7, 21, 25, 34, 36/7, 8, 9, 25, 36
5	300, 300, 600/494, 954, 722	7, 8, 9, 26, 35/7, 8, 9, 26, 35
6	300, 300, 600/494, 921, 722	7, 8, 9, 25, 35/7, 8, 9, 17, 26
7	300, 300, 600/494, 946, 722	7, 8, 9, 28, 35/7, 9, 27, 34, 35
8	300, 300, 600/494, 954, 722	7, 8, 9, 17, 28/7, 8, 9, 17, 28
9	300, 300, 500/494, 910, 675	7, 8, 9, 15, 37/7, 8, 9, 15, 37
10	100, 200, 300/186, 476, 300	6, 11, 15, 33, 37/7, 9, 17, 21, 37
11	100, 200, 300/188, 593, 324	3, 6, 11, 15, 33/7, 9, 17, 21, 37
12	100, 200, 300/244, 602, 353	3, 7, 10, 16, 33/7, 9, 21, 25, 37
13	200, 200, 400/276, 617, 420	3, 7, 10, 17, 33/7, 8, 9, 25, 36
14	200, 200, 500/360, 615, 526	4, 7, 9, 17, 33/7, 8, 9, 25, 36
15	200, 200, 500/356, 525, 503	3, 9, 16, 33, 37/33, 34, 35, 36, 37
16	200, 200, 600/386, 618, 596	5, 7, 9, 17, 33/33, 34, 35, 36, 37
17	300, 200, 600/455, 633, 629	7, 9, 16, 33, 37/33, 34, 35, 36, 37
18	200, 200, 300/369, 522, 387	4, 8, 10,16, 33/7, 9, 16, 21, 37
19	300, 300, 600/437, 943, 719	7, 21, 26, 32, 34/7, 8, 9, 26, 36
20	300, 300, 600/485, 946, 722	6, 27, 32, 34, 35/6, 27, 32, 34, 35
21	300, 300, 600/494, 954, 722	7, 8, 9, 27, 36/7, 8, 9, 27, 36
22	300, 300, 600/494, 953, 722	7, 8, 28, 32, 34/7, 8, 28, 32, 34
23	300, 300, 600/494, 954, 722	7, 9, 28, 34, 36/7, 9, 28, 34, 36
24	300, 300, 600/494, 954, 722	7, 9, 28, 34, 36/7, 9, 28, 34, 36
25	300, 300, 600/494, 954, 722	7, 10, 28, 32, 34/7, 10, 28, 32, 34
26	300, 300, 600/494, 954, 722	7, 10, 28, 34, 36/7, 10, 28, 34, 36
27	300, 300, 600/494, 954, 722	7, 8, 9, 15, 26/7, 8, 9, 15, 26

Figure 17.4 compares annual energy losses, percentage annual energy losses, and minimum node voltage obtained for various cases. It can be seen that it is Case 4 that is giving best results among all other cases. It provides a significant reduction in annual energy loss of about 89%, although maintaining node voltage of minimum value of 0.9561 pu during the year. The contribution of NR for energy loss reduction and voltage profile enhancement seems to be marginal. But, these enhancements are significant while considering the original state of the distribution system being well compensated.

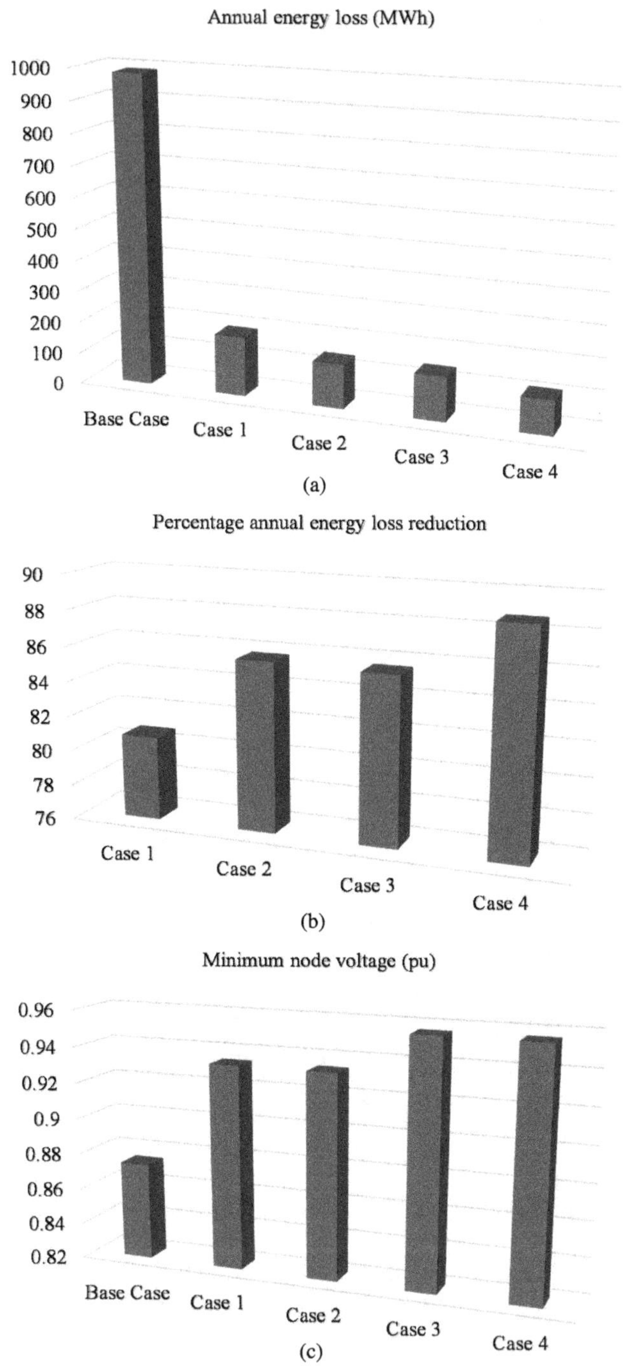

Figure 17.4. System performance for different cases.

17.7 Result Analysis

17.7.1 Integrated Approach for DR Allocation and NR

The proposed integrated approach of DR allocation and NR has shown promising results. The desired objectives have been continuously improved by installing DRs in the distribution network, by regulating the generation of power with changing load conditions and then optimally reconfiguring the distribution network. Thus, allocation of DRs at few optimal locations can control energy losses during peak and off-peak hours. Therefore, this approach can be beneficial for electric utilities to improve their profits and further system expansion to cope up increase in load demand.

17.7.2 Importance of Considering Load Diversity

Several works have been reported where the three-level piece wise linearization method is used to model annual load profile. Dedicated feeders assigned to various customers as well as the load diversity exist due to different types of customers is ignored in general. The impact of dedicated feeders and load diversity can be seen while comparing the network performance with and without considering load diversity. Network performance is first calculated considering three load levels that represent light, nominal, and peak loading condition. Load variation during light and peak load condition is taken as ±40% than nominal loading, respectively. The load durations for light, nominal, and peak loading is considered as 2000, 5260, and 1500 h, respectively. The best result is obtained after 100 trial runs. Table 17.6 presents optimal SC and DG dispatches and also optimal network configuration obtained at each load level [33]. Network performance is calculated after allocation

Table 17.6. Optimal solution for DR allocation and optimal NR without load diversity

	Optimal Location (Optimal Sizing)		
Load Level	**Siting (Sizing in kVAr) for SC**	**Siting (Sizing in kW) for DG**	**Optimal Network Configuration**
Light	14(200), 24(300), 30(600) TC: 1100	14(449), 24(638), 30(624) TC: 1711	7, 9, 17, 35, 37
Nominal	14(300), 24(600), 30(900) TC: 1800	14(740), 24(1059), 30(1038) TC: 2837	7, 9, 17, 25, 35
Peak	14(300), 24(600), 30(900) TC: 1800	14(740), 24(1059), 30(1038) TC: 2837	7, 9, 28, 34, 36

TC: Total capacity

of obtained optimal dispatches at each load level. The power losses of about 3.94, 11.39, and 44.52 MW can be obtained at light, nominal, and peak load conditions, respectively. Similarly, minimum node voltage of about 0.9950, 0.9908, and 0.9712 pu can be achieved at light, nominal, and peak load conditions, respectively.

A comparison analysis of considering load diversity is summarized in Table 17.7. The table shows that the results obtained without considering load diversity are optimistic. It generates a better system performance in terms of annual energy loss reduction and minimum node voltage. However, the system performance will be affected if these optimistic results are implemented on a network having actual loading profile. It may hamper the planning and operation of the distribution network by generating serious errors.

The table also shows that annual energy losses of base case condition differ while load diversity is considered. It compares energy loss and node voltage in both conditions. The losses are quite less on account of load diversity existing among customers. Figure 17.5 presents the network performance when the results obtained in Case 4 are implemented in a system having realistic load profile.

Table 17.7. Comparison analysis of considering load diversity

Particulars	Load Profile	Base Case	Proposed Approach
Annual energy losses (MWh)	With load diversity	984.63	111.12
	Without load diversity	1837.77	134.58
Annual energy loss reduction (%)	With load diversity	–	88.71
	Without load diversity	–	92.67
Minimum node voltage (pu)	With load diversity	0.8740	0.9561
	Without load diversity	0.8740	0.9712

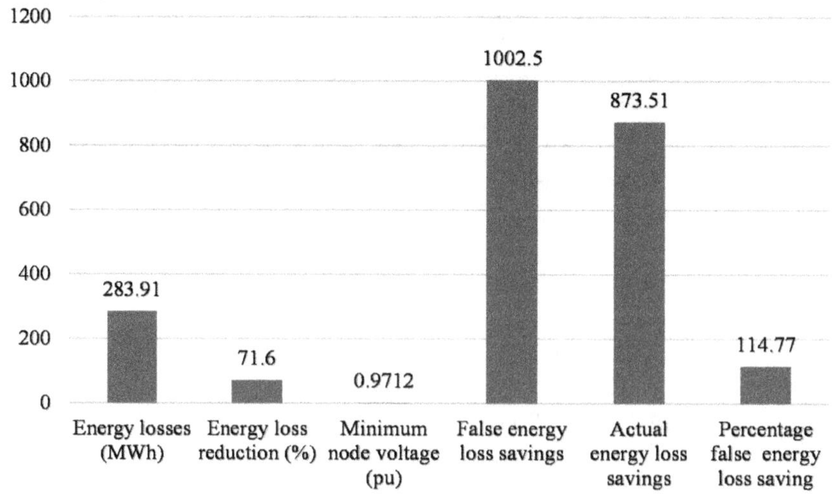

Figure 17.5. Network performance while implementing erroneous solution for DR allocation.

The comparison of Fig. 17.5 with Table 17.7 reveals that the reduction in annual energy losses is decreased from 92.67% to 71.60%. In addition, the table presents an energy loss savings of 1002.5 MWh, which is a false indication. However, actual saving in the network is 873.51 MWh. Therefore, the false indication of saving for annual energy loss is as high as 115% of the actual savings as shown in Fig. 17.6. These calculations are shown in Fig. 17.6. It happens because the diversity existing among customers' load demand affects annual profile of system loading, peak load demand, and also its total duration. Further, it is interesting that this substantial amount of false annual energy loss savings occurred while installing DRs having about 150% overcapacity. Although these overcapacities of DRs contribute to further enhance the node voltage profiles, as shown in Fig. 17.5. However, this volume of overrated optimal capacities of DRs along with the substantial false savings in annual energy losses may hamper planning and operation of distribution systems.

17.7.3 Importance of Considering Dedicated Feeders on Node Voltage Profile

The impact of considering dedicated feeders on system node voltage profile can be investigated by implementing the optimal solution obtained without considering load diversity on a system that actually have load diversity and dedicated feeders. Three major load levels are considered for comparison as shown in Fig. 17.7. The figure shows that node voltage profiles are either comparable or better when considering three load levels than actual load profile. However, these results are achieved at the cost of substantial higher optimal sizing of DRs. It is noteworthy that the proposed integrated approach of DR allocation and NR not only enhances feeder node voltage profiles, but also reduces their wide variations with varying load conditions as NR introduces some load diversity along distribution feeders.

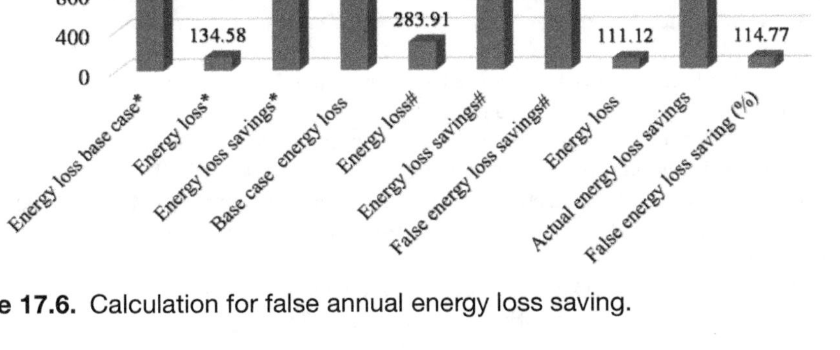

Figure 17.6. Calculation for false annual energy loss saving.

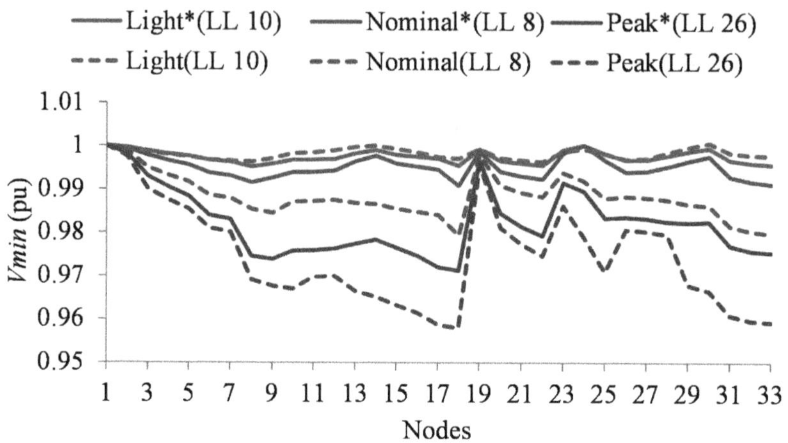

Figure 17.7. Impact of load diversity on node voltage profiles.

17.8 Summary

The chapter presents impacts of load diversity and dedicated feeders on distribution system performance. Optimal DR allocation problem is considered to show the diversity impact on system performance. The system performance is evaluated in terms of improvement in loss reduction and node voltage profile. It can be seen that when load diversity and dedicated feeders are ignored, a better system performance is obtained. Load diversity consideration provides a more practical solution; however, system performance degrades by considering these reality issues. The ignorance of these reality issues also gives a false indication of saving for annual energy loss and it happens at 150% of overcapacity of DRs. Therefore, the load diversity plays a crucial role and its ignorance can give an optimistic performance of distribution system and overcapacities of optimal DRs. The consideration of dedicated feeders to various types of customers spoils load diversity along the distribution feeders. However, NR enhances it to some extent thus preventing wide variations in node voltage profiles during the year. However, a proper coordination among distribution company (DISCO) and DG operators (DGOs) is desired to furnish very promising results while implementing the proposed integrated approach.

17.9 References

1. N. Kanwar, N. Gupta, K. R. Niazi, and A. Swarnkar, "Optimal distributed resource planning for microgrids under uncertain environment," *IET Renewable Power Generation*, vol. 12, no. 2, pp. 244–251, 2018.
2. A. F. Zobaa, and R. C. Bansal, *Handbook of Renewable Energy Technology*, 1st ed., World Scientific Publishing Co Pte Ltd, Singapore, 2011.
3. N. E. Chang, "Generalized equations on loss reduction with shunt capacitors," *IEEE Transactions on Power Apparatus and Systems*, vol. 91, no. 5, pp. 2189–2195, 1972.
4. J. J. Grainger, and S. H. Lee, "Optimum size and location of shunt capacitors for reduction of losses on distribution feeders," *IEEE Transactions on Power Apparatus and Systems*, vol. 100, no. 3, pp. 1105–1118, 1981).
5. M. E. Baran, and F. F. Wu, "Optimal capacitor placement on radial distribution systems," *IEEE Transactions on Power Delivery*, vol. 4, no. 1, pp. 725–734, 1989.

6. R. Baldick, F. F. Wu, "Efficient integer optimization algorithms for optimal coordination of capacitors and regulators," *IEEE Transactions on Power Systems*, vol. 5, no. 3, pp. 805–812, 1990.
7. S. H. Lee, and J. J. Grainger, "Optimum placement of fixed and switched capacitors on primary distribution feeders," *IEEE Transactions on Power Apparatus and Systems*, vol. 100, no. 1, pp. 345–352, 1981.
8. J. J. Grainger, and S. Civanlar, "Volt/Var control on distribution systems with lateral branches using shunt capacitors and voltage regulators, Part I: The overall problem," *IEEE Transactions on Power Apparatus and Systems*, vol. 104, no. 11, pp. 3278–3283, 1985.
9. S. Civanlar, and J. J. Grainger, "Volt/Var control on distribution systems with lateral branches using shunt capacitors and voltage regulators, Part III: The numerical results," *IEEE Transactions on Power Apparatus and Systems*, vol. 104, no. 11, pp. 3291–3297, 1985.
10. M. Ponnavaikko, and K. S. P. Rao, "Optimal choice of fixed and switched shunt capacitors on radial distributors by the method of local variations," *IEEE Transactions on Power Apparatus and Systems*, vol. 102, no. 6, pp. 1607–1615, 1983.
11. M. Chis, M. M. A. Salama, and S. Jayaram, "Capacitor placement in distribution systems using heuristic search strategies," *IEE Proceedings—Generation, Transmission and Distribution*, vol. 144, no. 2, pp. 225–230, 1997.
12. M. R. Raju, K. V. S. R. Murthy, and K. Ravindra, "Direct search algorithm for capacitive compensation in radial distribution systems," *International Journal of Electrical Power & Energy Systems*, vol. 42, pp. 24–30, 2012.
13. M. E. El-Hawary, "The smart grid—state-of-the-art and future trends," *Electric Power Components and Systems*, vol. 42, no. 3–4, pp. 239–250, 2014.
14. R. A. Hooshmand, and H. Mohkami, "New optimal placement of capacitors and dispersed generators using bacterial foraging oriented by particle swarm optimization algorithm in distribution systems," *Electrical Engineering*, vol. 93, pp. 43–53, 2011.
15. Z. Moravej, and A. Akhlaghi, "A novel approach based on cuckoo search for DG allocation in distribution network," *International Journal of Electrical Power & Energy Systems*, vol. 44, pp. 672–679, 2013.
16. D. Q. Hung, N. Mithulananthan, and R. C. Bansal, A combined practical approach for distribution system loss reduction, *International Journal of Ambient Energy*, vol. 36, no. 3, pp. 123–131, 2015.
17. L. F. Ochoa, and G. P. Harrison, "Minimizing energy losses: optimal accommodation and smart operation of renewable distributed generation," *IEEE Transaction on Power System*, vol. 26, no. 1, pp. 198–205, 2011.
18. Y. M. Atwa, and E. F. El-Saadany, "Probabilistic approach for optimal allocation of wind-based distributed generation in distribution systems," *IET Renewable Power Generation*, vol. 5, no. 1, pp. 79–88, 2011.
19. M. F. AlHajri, M. R. AlRashidi, and M. E. El-Hawary, "Improved sequential quadratic programming approach for optimal distribution generation deployments via stability and sensitivity analyses," *Electric Power Components and Systems*, vol. 38, pp. 1595–1614, 2010.
20. R. A. Jabr, and B. C. Pal, "Ordinal optimisation approach for locating and sizing of distributed generation," *IET Generation, Transmission & Distribution*, vol. 3, no. 8, pp. 713–723, 2009.
21. D. Singh, R. K. Mirsa, and D. Singh, "Effect of load models in distributed generation planning," *IEEE Transaction on Power System*, vol. 22, no. 4, pp. 2204–2212, 2007.
22. D. K. Khatod, V. Pant, and J. Sharma, "Evolutionary programming based optimal placement of renewable distributed generators," *IEEE Transaction on Power System*, vol. 28, no. 2, pp. 683–695, 2013.
23. D. Q. Hung, N. Mithulananthan, and R. C. Bansal, "Analytical expressions for DG allocation in primary distribution networks," *IEEE Transactions on Energy Conversion*, vol. 25, no. 3, pp. 814–820, 2010.

24. D. Q. Hung, and N. Mithulananthan, "Multiple distributed generators placement in primary distribution networks for loss reduction," *IEEE Transactions on Industrial Electronics*, vol. 60, no. 4, pp. 1700–1708, 2013.
25. P. S. Georgilakis, and Hatziargyriou, "Optimal distributed generation placement in power distribution networks: Models, methods, and future research," *IEEE Transactions on Power Systems*, vol. 28, no. 3, pp. 3420–3428, 2013.
26. N. Kanwar, N. Gupta, K. R. Niazi, and A. Swarnkar, "Simultaneous allocation of distributed resources using improved teaching learning based optimization," *Energy Conversion and Management*, vol. 103, pp. 387–400, 2015.
27. N. Kanwar, N. Gupta, K. R. Niazi, A. Swarnkar, and R. C. Bansal, "Simultaneous allocation of distributed energy resource using improved particle swarm optimization," *Applied Energy*, vol. 185, pp. 1684–1693, 2017.
28. N. Kanwar, N. Gupta, K. R. Niazi, and A. Swarnkar, "Improved meta-heuristic techniques for simultaneous capacitor and DG allocation in radial distribution networks," *International Journal of Electrical Power & Energy Systems*, vol. 73, pp. 653–664, 2015.
29. H. R. Esmaeilian, O. Darijany, and M. Mohammadian, "Optimal placement and sizing of DG units and capacitors simultaneously in radial distribution networks based on the voltage stability security margin," *Turkısh Journal of Electrıcal Engıneerıng Computer Scıences*, pp. 1–14, 2014.
30. A. Zeinalzadeh, Y. Mohammadi, and M. H. Moradi, "Optimal multi objective placement and sizing of multiple DGs and shunt capacitor banks simultaneously considering load uncertainty via MOPSO approach," *International Journal of Electrical Power & Energy Systems*, vol. 67, pp. 336–349, 2015.
31. R. C. Bansal, *Handbook of Distributed Generation: Electric Power Technologies, Economics and Environmental Impacts*, 1st ed., Springer, Switzerland, 2017.
32. S. Tan, J. X. Xu, and S. K. Panda, "Optimization of distribution network incorporating distributed generators: An integrated approach," *IEEE Transactions on Power Systems*, vol. 28, no. 3, pp. 2421–2432, 2013.
33. N. Kanwar, N. Gupta, K. R. Niazi, and A. Swarnkar, "An integrated approach for distributed resource allocation and network reconfiguration considering load diversity among customers," *Sustainable Energy, Grids and Networks*, vol. 7, pp. 37–46, 2016.
34. N. Kanwar, N. Gupta, K. R. Niazi, and A. Swarnkar, "Optimal distributed generation allocation in radial distribution systems considering customer-wise dedicated feeders and load patterns," *Journal of Modern Power and Clean Energy,* vol. 3, no. 4, pp. 475–484, 2015.
35. R. S. Rao, K. Ravindra, K. Satish, and S. V. L. Narasimham, "Power loss minimization in distribution System using network reconfiguration in the presence of distributed generation," *IEEE Transactions on Power Systems*, vol. 28, no. 1, pp. 317–325, 2013.
36. T. T. Nguyen, A. V. Truong, and T. A. Phung, "A novel method based on adaptive cuckoo search for optimal network reconfiguration and distributed generation allocation in distribution network," *International Journal of Electrical Power & Energy Systems*, vol. 78, pp. 801–815, 2016.
37. D. Das, "Optimal placement of capacitors in radial distribution system using a Fuzzy-GA method," *International Journal of Electrical Power & Energy Systems*, vol. 30, no. 6–7, pp. 361–367, 2008.
38. D. Kaur, and J. Sharma, "Multiperiod shunt capacitor allocation in radial distribution systems," *International Journal of Electrical Power & Energy Systems*, vol. 52, pp. 247–253, 2013.
39. N. Gupta, A. Swarnkar, and K. R. Niazi, "Reconfiguration of distribution systems for real power loss minimization using adaptive particle swarm optimization," *Electric Power Components and Systems*, vol. 39, pp. 317–330, 2011.
40. M. E. Baran, and F. Wu, "Network reconfiguration in distribution system for loss reduction and load balancing," *IEEE Transactions on Power Delivery*, vol. 4, no. 2, pp. 1401–1407, 1989.

CHAPTER 18

Multiobjective Approach for Optimal Placement of DG Units at Optimum Power Factor Along with Shunt Capacitors for Distribution Systems

Srinivasa Rao Gampa[1] and D. Das[2]

[1]*Department of Electrical and Electronics Engineering, Gudlavalleru Engineering College, Andhra Pradesh, India. gsr_gsrinu@yahoo.co.in*
[2]*Department of Electrical Engineering, Indian Institute of Technology, Kharagpur, India. ddas@ee.iitkgp.ernet.in*

Abstract Distributed generations (DGs) are defined as electric power generation units integrated with the distribution networks. DGs became inherent integral components of modern power distribution systems and played a key role in supporting to meet increasing load demand. The DGs that are mainly contributing to electric grid are nondisplaceable renewable energy sources such as solar power and wind power. However, dispatchable energy sources such as biomass are also helping to meet the customer-side electricity demand. The reactive power demands in distribution system can be compensated by shunt capacitors. The distribution system performance can be improved in terms of reducing real power losses and improving voltage profile by appropriately placing DGs and shunt capacitors. Hence, it is important to study and develop techniques for optimum sizing and placement of DGs and shunt capacitors in distribution systems. Modern DGs based on doubly fed induction generator (DFIG)-based wind power systems and solar photovoltaic systems are capable of operating at lagging power factor. Therefore, it is also important to identify the optimum power factor at which DGs can supply reactive power.

In this chapter, the sensitivity analysis-based approach and simultaneous optimum allocation-based methods for optimum sizing and placement of DGs and shunt capacitors are discussed. The optimum sizing of DGs at optimum power factor and shunt capacitor sizing at optimum locations are obtained using genetic algorithm considering the objectives of minimizing burden on substation, reducing real and reactive power losses, improving voltage profile while maintaining branch current constraints. Simulation results are demonstrated considering distribution networks from the literature to show the effect of DGs and shunt capacitors incorporation on the performance improvement of the distribution system.

Keywords: Distributed generation, Shunt capacitors, Sensitivity analysis, Multiobjective approach, Simultaneous optimization, Performance indices.

18.1 Introduction

The distributed generation (DG) mainly depends on the renewable energy sources such as wind power, photovoltaic, solar thermal systems, and biomass, and in small scale also considers energy sources such as combined heat and power (CHP), gas turbine, and diesel generators. Placement of DGs at inappropriate locations results in greater power losses and causes magnitude reduction in voltage profile. Hence, the selection of optimum location and size of DG is essential for minimizing power losses and improving voltage profile before it is integrated to distribution system [1–3]. The traditional practice of power engineers is to incorporate shunt capacitors (SCs) for reactive power compensation in distribution system for better voltage profile of the distribution system. In the present days, doubly fed induction generator (DFIG)-based wind power generators and voltage source inverter (VSI)-based solar photovoltaic systems can supply both real and reactive power and can operate at variable power factors. Hence, it is important to identify the optimum power factors at which DGs will operate when they are located at optimum locations identified [4]. In this chapter, a genetic algorithm (GA)-based multiobjective-based methodology is proposed for simultaneously identifying optimum locations and sizing for DGs operating at optimum power factor and SCs.

18.2 Performance Indices of Distribution System

The DGs and SCs are incorporated into the distribution system mainly to reduce the power supply burden on the substation and to improve the performance of the distribution system. The performance of the distribution system can be improved by considering the following performance indices:

(i) ***Substation apparent power index (SSAPI)*:** The active and reactive power delivered by the substation can be optimized by simultaneously considering the reduction apparent power delivered by the substation.

$$SSAPI = \frac{SSAP^{DGSC}}{SSAP^{Base}} \tag{18.1}$$

where $SSAP^{DGSC}$ is the total substation apparent power considering DGs and SCs, and $SSAP^{Base}$ is the total substation apparent power for base case without considering DGs and SCs in the distribution network.

(ii) ***Real power loss index (RPLI):*** The RPLI is the ratio of real power loss of distribution system with DGs and SCs to the real power loss at base case [5].

The RPLI is defined as

$$RPLI = \frac{RPL^{DGSC}}{RPL^{Base}} \tag{18.2}$$

where RPL^{DGSC} and RPL^{Base} are the total real power loss of distribution system considering DGs and SCs and for base case, respectively.

(iii) ***Reactive power loss index (XPLI):*** The *XPLI* is the ratio of reactive power loss of distribution system with DGs and SCs to the reactive power loss at base case. The *RPLI* is defined as

$$XPLI = \frac{XPL^{DGSC}}{XPL^{Base}}, \tag{18.3}$$

where XPL^{DGSC} is the total reactive power loss with DGs and SCs installation, and is XPL^{Base} the total reactive power loss for base case without DGs and SCs in the distribution network.

(iv) ***Branch current-carrying capacity limit (BCI_i):*** BCI_i is defined as the ratio of branch current (I_i) to its current-carrying capacity (IC_i).

Now we define branch current capacity ratio as

$$BCI_i = \frac{I_i^{DGSC}}{IC_i}, \text{ for } i = 1, 2, 3, ..., NB - 1. \tag{18.4}$$

Now we define branch current capacity index considered for the total distribution system as

$$BCIT = \frac{1}{NB-1} \sum_{i=1}^{NB-1} BCI_i, \tag{18.5}$$

where NB is the total number of nodes of the distribution system; I_i^{DGSC} is the magnitude of current in "*i*th" branch due to the placement of DGs and SCs in the distribution network; and IC_i is the maximum current-carrying capacity of branch "*i*."

(v) ***Voltage deviation index (VDIT):*** The sum of the squares of the distribution system voltage magnitude deviations from the minimum voltage limit ($V_{\min}$) and the maximum voltage limit ($V_{\max}$) are calculated at all the nodes and *VDIT* is taken as the overall summation. The voltage at substation is not considered because it is a constant value and specified as unit per unit value. The *VDIT* can be described by the following equation [5]:

$$VDIT = \frac{\sum_{i=2}^{NB} \left\{ \left(V_i^{DGSC} - V_{\min}\right)^2 + \left(V_i^{DGSC} - V_{\max}\right)^2 \right\}}{\sum_{i=2}^{NB} \left\{ \left(V_i^{Base} - V_{\min}\right)^2 + \left(V_i^{Base} - V_{\max}\right)^2 \right\}} \tag{18.6}$$

In Eq. (18.6), V_i^{Base} and V_i^{DGSC} are the ith node voltage magnitude with base case and with DGs and SCs incorporation, respectively. In this work $V_{\min} = 0.95$ pu and $V_{\max} = 1.05$ pu are minimum and maximum limits considered.

18.3 Multiobjective Function Formation

A multiobjective function can be formed combining all the performance indices as described by the following equation.

$$J = W_1\, SSAPI + W_2\, RPLI + W_3\, XPLI + W_4\, BCIT + W_5\, VDIT \tag{18.7}$$

$$\sum_{i=1}^{5} W_i = 1 \tag{18.8}$$

The multiobjective function can be minimized subject to the operational constraints for obtaining the optimum sizing of DGs with optimum power factor along with optimum SCs. The weighting factors W_1–W_5 can be selected depending on the importance given to particular index. In this work, all the performance indices functions are given equal importance.

18.4 Operational Constraints on the Distribution System

The multiobjective function (Eq. (18.7)) is minimized subject to various operational constraints to satisfy the technical requirements of the distribution system for obtaining optimal sizing of DGs and SC units.

The constraints can be discussed as follows:

(i) DG penetration limit: In the current work, the DG penetration at the optimum nodes selected is limited to predefined fractional part of the total real power load of the distribution system.

$$0 \le P_{DG_i} \le P_{DG}^{\max} \tag{18.9}$$

where P_{DG_i} and $P_{DG}^{\max}$ are DG sizing at ith optimum node and maximum limit on DG sizing.

(ii) Capacitor reactive power injection limit: In the current work, the reactive power injection limits at the optimum nodes by the capacitor placement is limited to a predefined fractional part of the total reactive power load of the distribution system.

$$0 \le QC_i \le QC^{\max} \tag{18.10}$$

where QC_i and $QC^{\max}$ are SC sizing at ith optimum node and maximum limit on SC sizing.

(iii) BCI_i: The branch currents should be limited to maximum current-carrying capacity of the branches due to incorporation of DG and SC units.

$$I_i \le IC_i \quad \text{for } i = 1, 2, 3 \ldots NB - 1. \tag{18.11}$$

(iv) Bus voltage limits: The voltage at each bus must lie within the specified limits

$$V_{\min} \le V_i \le V_{\max} \quad \text{for } i = 1, 2, 3 \ldots NB. \tag{18.12}$$

18.5 Modeling of DG Units and SCs

In this work for modeling of DG units and SCs, PQ load modeling generally used for load flow analysis is considered. The active power supplied by the lagging power factor is modeled as negative active power load and the reactive power supplied by the lagging power factor DG units is modeled as negative reactive power load. The SC sizing is modeled as negative reactive power load for load flow analysis. The mathematical models for lagging power factor DG units and SCs can be represented by the following equations [5–10]:

$$P_{L_i}^{DG} = P_{L_i} - P_{DG_i} \tag{18.13}$$

$$Q_{L_i}^{DG} = Q_{L_i} - (\tan(\cos^{-1}(PF_{DG_i})P_{DG_i})) \tag{18.14}$$

$$Q_{L_j}^{SC} = Q_{L_j} - QC_j, \tag{18.15}$$

where

P_{L_i} = Active power load at ith node
Q_{L_i} = Reactive power load at ith node
P_{DG_i} = DG sizing at ith node
$P_{L_i}^{DG}$ = Active power load modification with DG consideration
$Q_{L_i}^{DG}$ = Reactive power load modification with DG consideration
PF_{DG_i} = Power factor of DG units at ith node
$Q_{L_i}^{SC} = Q_{L_i} - QC_j$
Q_{L_j} = Reactive power load at jth node
QC_j = SC sizing at jth node
$Q_{L_j}^{SC}$ = Reactive power load modification with QC consideration

18.6 Selection of Optimum Nodes for the Placement of DG Units

Solar photovoltaic and wind generator renewable energy sources are nondispatchable and intermittent in nature [11,12]. Hence along with the solar and wind energy DG units, the integration of dispatchable renewable energy resources such as biomass generators will be used to supply continuous and reliable power to the customer [13,14]. In the practical scenario, the selection of locations for the installation of solar photovoltaic, biomass, and wind DG systems in the distribution system depends on urbanity, that is, the availability of adequate space and the abundance of the energy available from the nature [3]. The forecasted load demand and the technical and economical benefits will influence the utilities to choose the size and locations of the DG units. The sensitivity analysis [15,16] and evolutionary computing techniques [5] can be used for identifying the optimum nodes among the available convenient locations for DG units placement.

18.7 Selection of Optimum Nodes for the Placement of SCs

SCs usage is a traditional choice of the utilities to inject reactive power into the radial distribution systems. SCs placement reduces real power losses, improves the substation power factor and voltage profile. The loss sensitivity analysis is used to identify the best nodes with highest real power loss reduction based on ranking system obtained by compensating the total reactive load when a capacitor is installed at every node. The ranking of the nodes is given based on descending order of real power loss reduction and the top-ranked nodes are selected for SC installation. The voltage instability in power distribution systems is mainly due to insufficient reactive power supply under disturbance or critical loading conditions and sometimes it may spread into transmission system and can cause severe blackouts. Therefore, the voltage stability should be given considerable attention for improving the stability of the operating conditions of the distribution system. Hence in the current work, the sensitivity analysis combining the loss sensitivity and voltage sensitivity is proposed for optimal allocation of SCs in a distribution network [17,18].

18.8 Methods of Optimal Selection of Nodes for DGs and SCs

The selection of optimum nodes for the placement of DGs and SCs can be done using mainly sensitivity analysis approach and simultaneous selection of optimum nodes along with optimum sizing of DGs and SCs [15,16,19]. In both the methodologies, the nodes that are not practically favorable for the placement of DGs and SCs should be excluded from the list. In the present case, it is assumed that all the nodes are available for the placement of DGs and SCs.

18.8.1 Sensitivity Analysis Method

The performance of the distribution system can be improved in terms of loss reduction and voltage profile improvement by suitably selecting optimum nodes for the placement of DGs and SCs considering both real power loss sensitivity and voltage sensitivity. Hence, a simple sensitivity index [16] combining loss and voltage sensitivities is described by the Eq. (18.16) is used for identifying the best nodes for incorporating DG units and SCs in the distribution network.

$$S_k = \frac{\Delta V_k^{\max}}{Ploss_k} \tag{18.16}$$

where

$$\Delta V_k^{max} = \max\left(|\overset{\sim DG/SC}{V_{m,k}}| - |\overset{\sim base}{V_{m,k}}| \right), \text{ for } m = 2, 3, \ldots, NB \tag{18.17}$$

$\Delta V_k^{\max}$ = Maximum change in voltage magnitude due to active power injection by unity power factor DG or reactive power injection by SC at node k.

$Ploss_k$ = Active power loss of the network due to active power injection by unity power factor DG or reactive power injection by SC at node k.

$\overset{\sim DG/SC}{V_{m,k}}$ = Voltage of node "m" before the placement of DG or SC at node k.

$\overset{\sim base}{V_{m,k}}$ = Voltage of node "m" after the placement of DG or shunt capacitor at node k.

For DG placement, considering from node 2 onward at every node active power and for capacitor placement, reactive power is injected and the sensitivity indices are calculated from Eq. (18.16). The nodes will have maximum sensitivity indices with minimum power loss and maximum improvement in voltage magnitude.

For the present analysis, 11-kV, 51-node radial distribution system [5] and a 12.66-kV, 69-node [20] radial distribution network are considered.

The order of ranking of the distribution system nodes can be obtained by the sensitivity index S_k described by Eq. (18.16) for the placement of DGs and SCs. At each node, 20% of total real power load and 20% of total reactive power load is injected independently and from Eq. (18.16) value of S_k is computed. The bar graphs of the Sensitivity Index (S_k) at each node for the placement of DGs for 51-node and 69-node distribution are shown in Figs. 18.1 and 18.2, respectively.

The order of ranking of nodes is given according to the descending order of S_k and the ranking of nodes for both 51- and 69-node distribution systems for the placement of DGs and SCs are summarized in Table 18.1.

In general, solar and wind are widely used renewable energy resources as DG resources in the distribution system. Because the solar and wind are intermittent in nature and solar power is not available at night to compensate the deficiency in power supply, the biomass DG is used as a third resource for supplying reliable power to the consumer [5]. Hence, the authors selected nodes with top three ranks for DG units placement and an equal number of SC units are considered to supply the reactive power. From Table 18.1, it is observed that in the case of DGs nodes 16, 15, and 45 are the optimal nodes and 16, 15, and 14 are the optimal nodes in the case of SCs nodes. In the case of 69-node distribution network, 65, 64, and 63 are the optimal nodes for the placement of both SCs and DGs units.

18.8.2 Simultaneous Optimization Method

In simultaneous optimization method, the optimum nodes for DG placement and optimum nodes for SC placement are simultaneously obtained along with optimum sizing of DG units with optimum power factor and SCs. The solution of simultaneous optimum placement of DG units and SCs can be obtained using evolutionary computing techniques such as GAs,

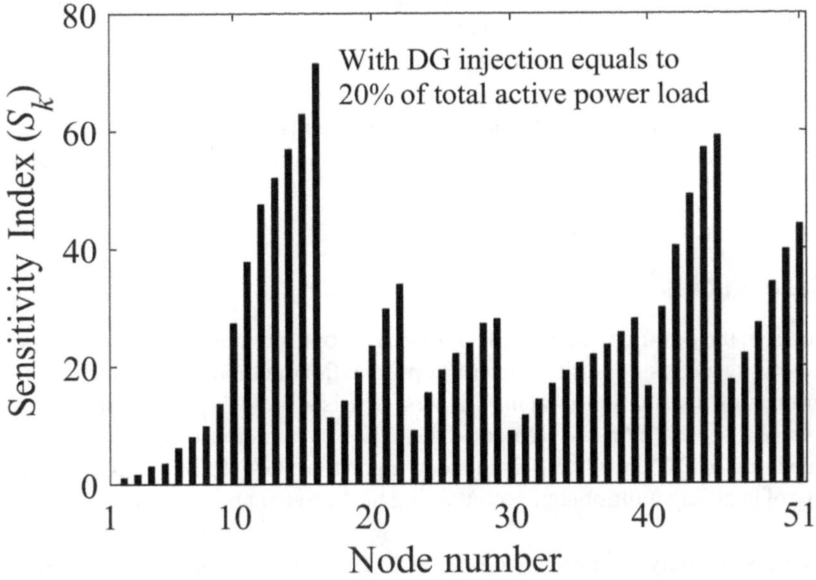

Figure 18.1. Sensitivity index S_k with active DG power injection of 51-node network.

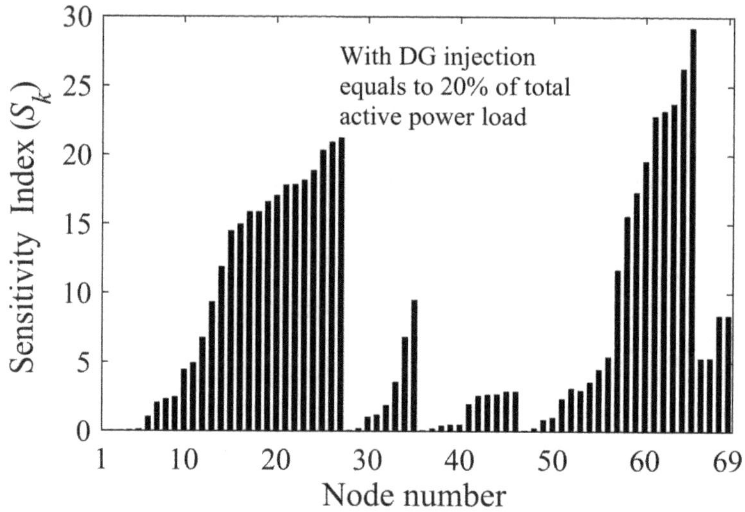

Figure 18.2. Sensitivity index S_k with active DG power injection of 69-node network.

Table 18.1. Ranking of nodes using sensitivity analysis

	51-Node Distribution System		69-Node Distribution System	
Ranking of Nodes	**Nodes for DG Injection**	**Nodes for QC Injection**	**Nodes for DG Injection**	**Nodes for QC Injection**
1	16	16	65	65
2	15	15	64	64
3	45	14	63	63

particle swarm optimization, and ant colony optimization by minimizing the multiobjective function described by the performance indices.

In this work, GA is used because of its ease to handle both integer and noninteger variables simultaneously for obtaining optimum nodes for DG units and SC allocation. The optimum nodes identified for DG units and SC placements by optimizing the multiobjective function described by Eq. (18.7) along with their optimum sizes using GA are summarized in Table 18.2.

18.9 Case Studies

In this section, the optimum sizing of DG units with optimum power factor along with SCs using sensitivity analysis and simultaneous optimization methods are analyzed considering two examples of distribution system. The first one is 11-kV, 51-node distribution system and the second one is 12.66-kV, 69-node distribution system. In the present case studies for optimum sizing of DG units and SCs, genetic algorithm (GA) [5,21] is used because of its capability of handling multiobjective containing both integer and real values simultaneously [15,16].

In the present analysis, three locations for DGs placement and three locations for SCs placement are considered. In the case of sensitivity analysis method, the optimum nodes

Table 18.2. Optimum nodes using simultaneous optimization method.

	51-Node Distribution System		69-Node Distribution System	
Optimum Nodes	Nodes for DG Injection	Nodes for QC Injection	Nodes for DG Injection	Nodes for QC Injection
1	42	20	62	50
2	35	25	16	68
3	12	47	64	60

Table 18.3. Optimum sizing of DGs and SCs for 51 bus.

Sensitivity Analysis Method	Simultaneous Optimization Method
$P_{DG_{16}} = 173.86$ kW	$P_{DG_{42}} = 486.81$ kW
$PF_{DG_{16}} = 0.9969$ lag pf	$PF_{DG_{42}} = 0.9177$ lag pf
$P_{DG_{15}} = 484.87$ kW	$P_{DG_{35}} = 469.42$ kW
$PF_{DG_{15}} = 0.9995$ lag pf	$PF_{DG_{35}} = 0.8685$ lag pf
$P_{DG_{45}} = 492.60$ kW	$P_{DG_{12}} = 490.67$ kW
$PF_{DG_{45}} = 0.8499$ lag pf	$PF_{DG_{12}} = 0.8794$ lag pf
$QC_{16} = 75.0$ kVAr	$QC_{20} = 150$ kVAr
$QC_{15} = 100.00$ kVAr	$QC_{25} = 300$ kVAr
$QC_{14} = 275.0$ kVAr	$QC_{47} = 300$ kVAr

for placement of DGs and SCs are selected using sensitivity analysis method. Hence, three parameters for DGs sizing, three parameters for optimum power factor, and three parameters for SCs sizing are to be found using GA. In the case of simultaneous optimization technique in addition to the above nine parameters, six more parameters, which are three optimum nodes for DGs placement and three optimum nodes for SCs placement, are to be identified. In the present GA parameters, crossover probability and mutation probability are taken as 0.95 and 0.001 and the population size and number of generations is taken as 200. The optimum sizing of DG units along with their power factor and SC sizing is obtained by optimizing the multiobjective function described by using Eq. (18.7) using GA. In the multiobjective function, all the weighting factors are considered of equal value.

The optimum sizing of DGs and SCs for the two approaches considered is summarized in Table 18.3 for 51 bus system and in Table 18.4 for 69 bus system for both sensitivity analysis and simultaneous optimization method.

From Tables 18.3 and 18.4, it can be observed that more DG and SC can be accommodated at the optimum nodes identified by the simultaneous analysis approach compared to the optimum nodes identified by the sensitivity analysis approach.

The performance of the distribution system with both sensitivity analysis approach and simultaneous optimization approach for 51-distribution system is summarized in Table 18.5 and for 69-distribution system is summarized in Table 18.6.

From Tables 18.5 and 18.6, it can be observed that the apparent power delivered by the substation can be reduced significantly with the incorporation of DG units operating at optimum power factor along with SCs. Hence, the active power load and reactive power

Table 18.4. Optimum sizing of DGs and SCs for 69 bus.

Sensitivity Analysis Method	Simultaneous Optimization Approach
$P_{DG_{65}} = 393.62$ kW	$P_{DG_{62}} = 748.47$ kW
$PF_{DG_{65}} = 0.9829$ lag pf	$PF_{DG_{62}} = 0.8611$ lag pf
$P_{DG_{64}} = 760.4$ kW	$P_{DG_{16}} = 736.54$ kW
$PF_{DG_{64}} = 0.9812$ lag pf	$PF_{DG_{16}} = 0.9591$ lag pf
$P_{DG_{63}} = 760.4$ kW	$P_{DG_{64}} = 754.44$ kW
$PF_{DG_{63}} = 0.8499$ lag pf	$PF_{DG_{64}} = 0.8794$ lag pf
$QC_{65} = 75$ kVAr	$QC_{50} = 525$ kVAr
$QC_{64} = 125$ kVAr	$QC_{68} = 325$ kVAr
$QC_{63} = 525$ kVAr	$QC_{60} = 425$ kVAr

Table 18.5. Performance of 51-node distribution system

51 Bus	Base Case	Sensitivity Analysis	Simultaneous Optimization-Based Method
S/S apparent power (kVA)	3089.67	1596.37	1045.06
S/S active power (kW)	2592.56	1371.23	1041.37
S/S reactive power (kVAR)	1680.68	817.39	87.67
Real power loss (kW)	129.56	59.56	25.27
Reactive power loss (kVAR)	111.68	31.79	13.01
Voltage (V (pu)) (minimum)	0. 9081(V_{min}^{16})	0. 9677(V_{min}^{39})	0. 9789(V_{min}^{22})

Table 18.6. Performance of 69-node distribution system

69 Bus	Base Case	Sensitivity Analysis	Simultaneous Optimization-Based Method
S/S apparent power (kVA)	4903.08	2316.31	1614.17
S/S active power (kW)	4027.19	1921.07	1573.58
S/S reactive power (kVAR)	2796.77	1294.14	359.72
Real power loss (kW)	225.00	33.29	10.84
Reactive power loss (kVAR)	102.17	19.25	7.89
Voltage (V (pu)) (minimum)	0. 9092(V_{min}^{65})	0. 9733(V_{min}^{27})	0. 9935(V_{min}^{61})

load requirements can be met with the help of DG units and SCs and accordingly the burden on substation can be reduced. The performance of the distribution system can be improved in terms of real power loss reduction, reactive power loss reduction, and voltage profile improvement. It can be observed that the performance of the distribution system is much better with the allocation of DG units at optimum power factor along with SCs when placed at the optimum nodes identified using simultaneous optimization technique compared to sensitivity analysis approach.

The performance improvement of distribution system in terms of percentage reduction values with respect to base case are summarized in Tables 18.7 and 18.8 for 51-node and 69-node distribution systems. From Tables 18.7 and 18.8, it can be observed that the apparent power reduction at substation, the real power, the reactive power loss reductions, the real power, and the reactive power loss reductions in terms of percentage values are quite good in the case of simultaneous optimization technique compared to sensitivity analysis approach.

Table 18.7. Performance improvement of 51-node distribution system

51-Node Distribution System	Sensitivity Analysis	Simultaneous Optimization-Based Method
S/S apparent power supply reduction (%)	48.33	66.18
S/S active power supply reduction (%)	47.11	59.83
S/S reactive power supply reduction (%)	51.37	94.78
Real power loss reduction (%)	54.03	80.50
Reactive power loss reduction (%)	71.53	88.35

Table 18.8. Performance improvement of 69-node distribution system

69-Node Distribution System	Sensitivity Analysis	Simultaneous Optimization-Based Method
S/S apparent power supply reduction (%)	52.76	67.08
S/S active power supply reduction (%)	52.30	60.93
S/S reactive power supply reduction (%)	53.72	87.14
Real power loss reduction (%)	85.20	95.18
Reactive power loss reduction (%)	81.16	92.27

From Figs. 18.3 and 18.4, it can be observed that the branch current-carrying capacity and voltage profile are improved significantly with the incorporation of DG units at optimum power factor and SCs and the improvement is much better in the case of simultaneous optimization technique.

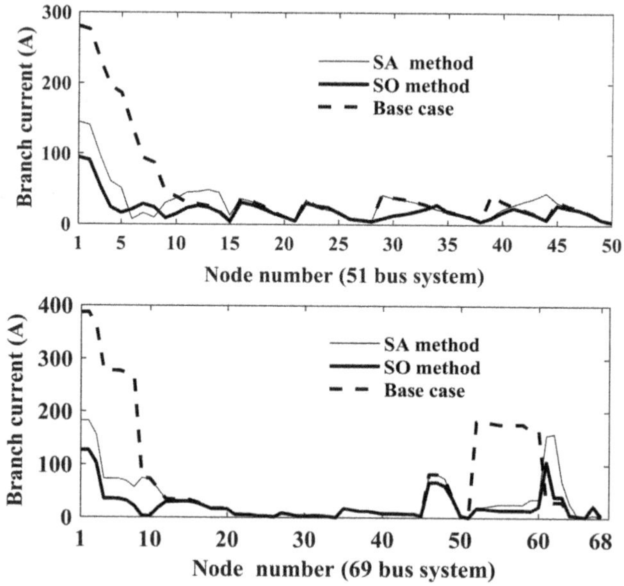

Figure 18.3. Branch current-carrying capacity improvement.

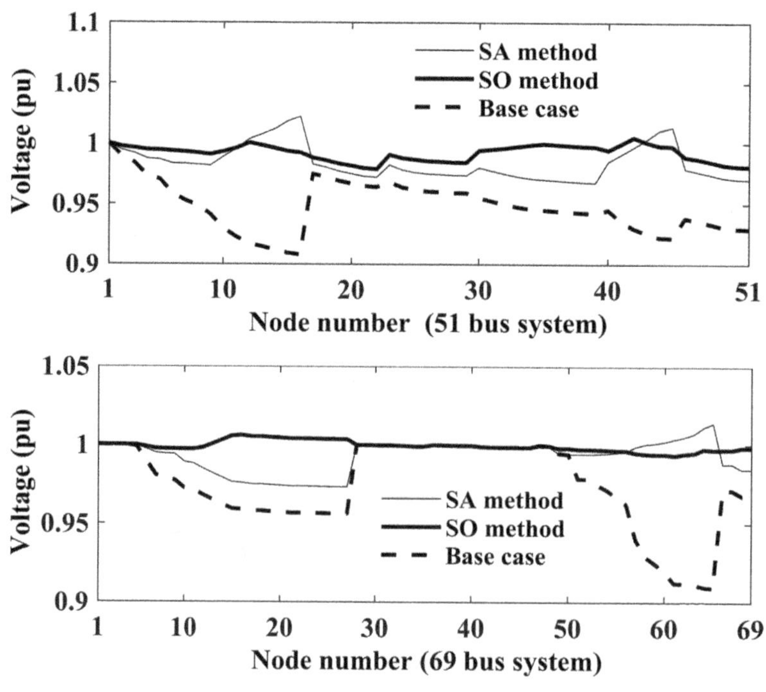

Figure 18.4. Voltage profile improvement.

18.10 Summary

In this work, GA-based multiobjective methodology is proposed for simultaneous placement of DGs at optimum power factor along with SCs in distribution systems for performance improvement of the distribution system. The objectives considered in the present work aimed at reduction of apparent power at substation, that is, real and reactive power supplies at substation, reduction of the real and reactive power loss of the distribution system, improving branch current capacity, and voltage profile of the distribution system. The distribution system performance is analyzed considering the combination of DG units with optimum power factor along with SCs. From the result analysis, it can be said that the distribution system performance is much better in the case of simultaneous optimization technique compared to sensitivity analysis-based optimal placement of DGs and SCs. The active and reactive power supply needs of the distributions system can be effectively met with combination of SCs with DG units operating at optimum power factor. The real power loss reduction, reactive power loss reduction, voltage profile improvement, branch current capacity improvement, and voltage stability improvement can be achieved much better with the combination of lagging pf DGs and SCs using GA-based simultaneous optimization technique.

18.11 References

1. D. Q. Hung, N. Mithulananthan, and R. C. Bansal, "Analytical expressions for DG allocation in primary distribution networks," *IEEE Transactions on Energy Conversion*, vol. 25, no. 3, pp. 814–820, 2010.
2. H. D. Mathur, N. C. Hien, N. Mithulananthan, D. Joshi, R. C. Bansal, *Distributed Generation: A Power System Perspective, Handbook of Renewable Energy Technology,* pp. 563–585, World Scientific Publishing Co. Pvt Ltd, Singapore, 2011.
3. S. A. Arefifar, Y. A. I. Mohamed, T. H. M. El-Fouly, "Supply-adequacy-based optimal construction of micro grids in smart distribution systems," *IEEE Transactions on Smart Grids*, vol. 3, no. 3, pp. 1491–1502, 2012.
4. A. Ellis, R. Nelson, E. Von Engeln, et al., "Reactive power performance requirements for wind and solar plants," Proceedings of IEEE PES General Meeting, pp. 1–8, San Diego, 2012.
5. S. R. Gampa, and D. Das, "Optimum placement and sizing of DGs considering average hourly variations of load," *International Journal of Electrical Power & Energy Systems*, vol. 66, pp. 25–40, 2015.
6. D. Q. Hung, N. Mithulananthan, and R. C. Bansal, "Analytical strategies for renewable distributed generation integration considering energy loss minimization," *Applied Energy*, vol. 105, pp. 75–85, 2013.
7. R. S. Al Abri, E. F. El-Saadany, and Y. M. Atwa, "Optimal placement and sizing method to improve the voltage stability margin in a distribution system using distributed generation," *IEEE Transactions on Power Systems*, vol. 28, no. 1, pp. 326–334, 2013.
8. D. Q. Hung, N. Mithulananthan, and K. Y. Lee, "Optimal placement of dispatchable and nondispatchable renewable DG units in distribution networks for minimizing energy loss," *International Journal of Electrical Power and Energy Systems*, vol. 55, pp. 179–186, 2014.
9. M. M. Aman, G. B. Jasmon, A. H. A. Bakar, and H. Mokhlis, "A new approach for optimum simultaneous multi-DG distributed generation units placement and sizing based on maximization of system loadability using HPSO (hybrid particle swarm optimization) algorithm," *Energy*, vol. 66, no. 1, pp. 202–215, 2014.

10. N. Kanwar, N. Gupta, K. R. Niazi, A. Swarnkar, and R. C. Bansal, "Multi-objective optimal DG allocation in distribution networks using bat algorithm," 3rd Southern African Solar Energy Conference, pp. 11–13, South Africa, 2015.
11. M. N. Kabir, Y. Mishra, and R. C. Bansal, "Probabilistic load flow for distribution systems with uncertain PV generation," *Applied Energy*, vol. 163, pp. 343–351, 2016.
12. A. F. Zobaa, and R. C. Bansal, (Eds.) *Handbook of Renewable Energy Technology*, World Scientific Publishers, Singapore, 2011.
13. T. Adefarati, and R. C. Bansal, "Integration of renewable distributed generators into the distribution system: A review," *IET—Renewable Power Generation*, vol. 10, no. 7, pp. 873–884, 2016.
14. R. C. Bansal (Ed.), *Handbook of Distributed Generation: Electric Power Technologies, Economics and Environmental Impacts,* Springer, New York, 2017.
15. S. R. Gampa, and D. Das, "Multiobjective approach for reconfiguration of distribution systems with distributed generations," *Electric Power Components and Systems*, vol. 45, no. 15, pp. 1678–1690, 2017.
16. S. R. Gampa, and D. Das, "Fuzzy GA based multiobjective methodology for DGs and shunt capacitors placement in distribution networks," Proceedings of the National Power Systems Conference, NIT, Trichy, Dec. 14–16, 2018.
17. S. R. Gampa, and D. Das, "Optimum placement of shunt capacitors in a radial distribution system," *International Journal of Electrical Power & Energy Systems*, vol. 77, pp. 314–326, 2016.
18. D. Das, "Optimal placement of capacitors in radial distribution system using a Fuzzy-GA method," *International Journal of Electrical Power and Energy Systems*, vol. 30, no. 6–7, pp. 361–367, 2008.
19. M. Kumawat, N. Gupta, N. Jain, and R. C. Bansal, "Swarm-intelligence-based optimal planning of distributed generators in distribution network for minimizing energy loss," *Electric Power Components and Systems*, vol. 45, no. 6, pp. 589–600, 2017.
20. M. Chakraborty, and D. Das, "Voltage stability analysis of radial distribution networks," *International Journal of Electrical Power Energy Systems*, vol. 23, no. 2, pp. 129–135, 2001.
21. D. E. Goldberg, *Genetic Algorithms in Search, Optimization and Learning*, 1st edition, Addison-Wesley Longman Publishing Co., Boston, 1989.

CHAPTER 19

Control and Stability of Microgrids

Farhad Shahnia

Discipline of Engineering and Energy, Murdoch University, Perth, Australia.
F.Shahnia@murdoch.edu.au

Abstract This chapter briefly discusses the various control principles and techniques for the primary and secondary control of microgrids. More specifically, proper active and reactive power-sharing among the distributed energy resources is discussed under the primary control aspect, whereas dynamic power-sharing ratio modification and regulating the voltage and frequency are discussed under the secondary control. A modified control technique is also discussed for the BESs in comparison to DERs. The advantages and limitations of each suggested technique are discussed and demonstrated by some simulation-based examples. The chapter has also discussed the small signal stability of microgrids, focusing on the most important states of the system, which have a dominating factor in making the system unstable.

Keywords: Microgrid, Primary controller, Secondary controller, Power sharing, Droop curve, Small signal stability, Distributed energy resource, Battery energy storage system.

19.1 Introduction

The rising electricity demand and the inevitability of reliability improvement and cost reduction are motivating the application of distributed energy resources (DERs) within power systems in place of expanding distribution systems [1, 2].

Microgrids are referred to as a set of DERs, battery energy storage units (BESs), and electrical loads (see Fig.19.1). To deliver power with high quality and reliability, the microgrid has to perform as a controlled system that reacts punctually to demand variations [3]. Microgrids have been studied and suggested in many technical reports for various applications such as supplying electricity to remote towns without utility feeder [4–7], supplying the demand of an island [8], reducing the levelized cost of energy production [9, 10], and off-grid remote area greenhouses [11].

Microgrids can function in a grid-connected and islanded (autonomous) modes throughout planned or unplanned outages of the utility grid because of network maintenance or faults [12]. Microgrid's frequency and voltage are enforced by the grid when operating under the grid-connected regime, thus, the DERs function at their rated capacities or according to maximum power point tracking scheme, under a constant power strategy [13]. However, in the microgrid's islanded mode of operation, the load demand is shared by the DERs if their overall capacity is more than the loads' demand. As such, the DERs are controlled under a frequency and voltage droop mechanism [14]. Furthermore, the droop mechanism can lead to power-sharing among multiple DERs similar to that among conventional generators of a power system. The frequency and voltage droop can also be replaced by a voltage and angle droop in microgrids [15]. To improve the system response, some modified droop controls are also proposed [16] or an arctan power frequency droop has replaced conventional droop technique in [17] to improve the microgrid's small signal stability. Lately, adaptive and intelligent droop controls [18, 19] are also used to eliminate droop control's dependency on network line characteristics. In [20], the droop control has been replaced in the microgrid with a potential function to fine-tune the common set-points for power-sharing among DERs [20]. After assigning the references for the DERs, their converters may operate under voltage or current-controlled strategies [21–23] to track the desired references. In microgrids, DERs are coupled to the microgrid's lines via coupling impedances and are regulated to inject the preferred power into the system.

Islanding detection is another important issue in microgrids, which refers to microgrid's isolation from the grid. Various islanding detection techniques, introduced in [24], can be

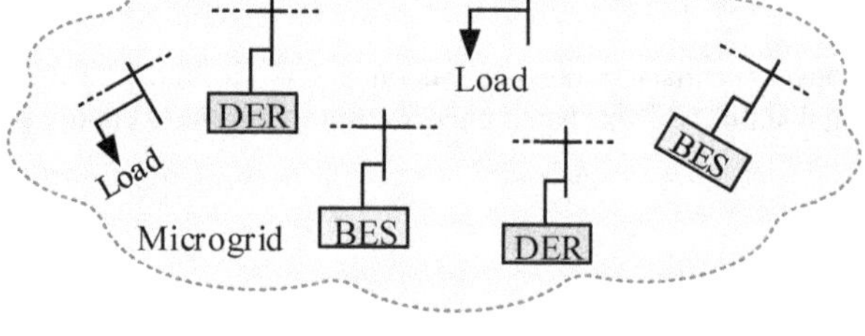

Figure 19.1. Sample representation of a microgrid, composed of DERs, BESs, and loads.

used. On the other hand, a communication link can transfer the status of the circuit breaker to the converters of DERs (an ideal method because it does not have any none-detection zones that exist in other islanding detection methods). In [25–28], the possibility of employing a ZigBee-based communication technology in microgrids has been discussed and its impact on the system performance has been evaluated.

Resynchronization is the other key concern in microgrids, which refers to the microgrid's reconnection to the grid or a DER's reconnection to the microgrid [29]. Resynchronization is not required when reconnecting a current-controlled DER to the microgrid; however, proper resynchronization is essential when the DER runs in voltage-controlled strategy. Reconnection has to occur only when the voltage magnitude and phase difference on either side of the circuit breaker are almost zero or less than a minuscule predetermined level. This is because inappropriate reconnection can lead to large current oscillations that have the capability to damage the system's assets or even cause instability. In [30, 31], various resynchronization approaches has been reviewed.

The following sections present an overview of the control perspectives and stability issues of three-phase AC microgrids. The interested readers are referred to [32–34] for DC microgrids and [35–37] for hybrid AC–DC microgrids. Also, in [38, 39] three-phase microgrids with unequal distribution of single-phase DERs among the phases have been focused, whereas in [40, 41], the techniques to integrate wind energy to microgrids have been discussed in detail.

19.2 Microgrid Control

A three-level hierarchical controller is vital for a microgrid's appropriate operation within a distribution network, as discussed below [42, 43].

19.2.1 The Primary Controller

Primary controller, which is also the lowest level of control, composes of outer- and inner-loop controls. The inner-loop control is in charge of appropriately turning on/off the switches within the converter to facilitate an appropriate tracking of the preferred voltage or current reference at the output of the DER's converter. The local voltage and current measurements at the DER's converter output, as well as the references defined by the outer-loop control, are used in this controller. The references defined by the outer-loop control facilitate the suitable level of power that the DERs inject to the microgrid. This controller produces suitable references for the inner-loop control, which changes depending on the microgrid's grid-connected and autonomous modes of operation.

19.2.2 The Secondary Controller

The secondary controller is the middle level of control and acts as microgrid's central controller in autonomous mode and determines the reference of the voltage magnitude and angle or frequency by monitoring the frequency and voltage in the network when necessary. It operates in a slower frame versus the primary controller.

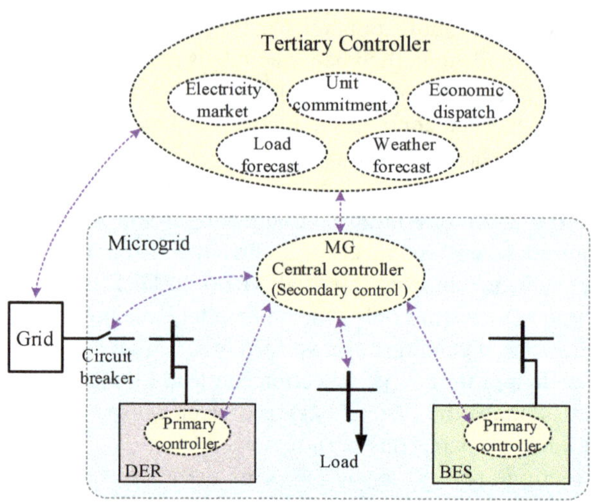

Figure 19.2 Representation of the hierarchical controller for microgrids.

19.2.3 The Tertiary Controller

The tertiary controller is the network's highest level of control and interconnects with the microgrids' central controllers as well as the circuit breakers and the protective relays within the system. Overall, this controller can use demand response, load-forecasting, and electricity market data to realize an optimal power flow within the microgrids and the system [44]. Moreover, this controller can employ the data transmitted from the circuit breakers to determine the system's overall status and configuration when composed of several neighboring microgrids [45, 46]. Under such a condition, an agent coordinating the microgrids has to determine the necessity of the microgrids' coupling and those circuit breakers that should open/close for transmitting suitable activation command signals. Moreover, the isolation of the coupled microgrids from each other will be determined by this controller. Figure 19.2 illustrates schematically this control system for the microgrids.

The concept of tertiary control is beyond the scope of this chapter; however, as discussed above, it can be used for the formation of interconnected microgrids to each other via back-to-back converters [47] or conventional circuit-breakers [48–51], which need proper synchronization [52, 53]. In [54–56], the mechanism to detect the necessity to form a system of coupled microgrids have been discussed, which has then be extended to be accompanied by decision-making tools [57–59] or optimization-based techniques [60–70], under two general approaches of corrective or preventive.

19.3 Control of DERs

Suppose DERs have a voltage source converter (VSC) and an output passive filter as shown in Fig. 19.3(a). The equivalent, single-phase representation of this is shown in Fig. 19.3(b) and has a Thevenin equivalent circuit (i.e., V_{Th} and Z_{Th}) as shown in Fig. 19.3(c). From

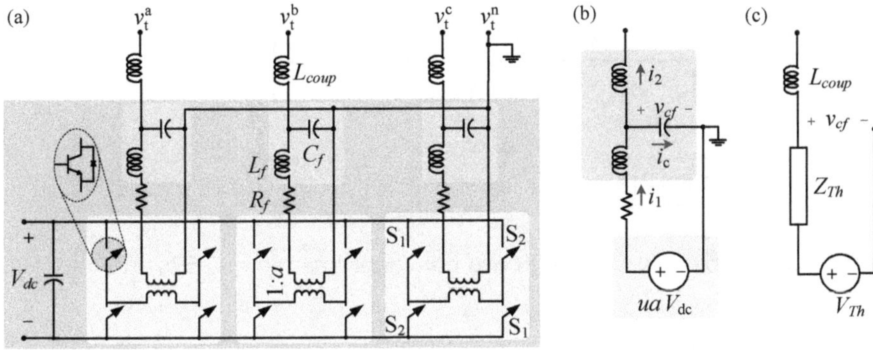

Figure 19.3 (a) Assumed three-phase VSC and passive output filter system for the DERs and BESs, (b) Its single-phase equivalent circuit, (c) Its simplified Thevenin equivalent circuit.

this figure, the active and reactive powers (p, q) supplied by DER-i to its point of common coupling (PCC) can be expressed as in [71].

$$p_i = \frac{\left(|V_{T,i}|\,|V_{Th,i}|\cos\varphi_i - |V_{Th,i}|^2\right)\cos\theta_i + |V_{T,i}|\,|V_{Th,i}|\sin\varphi_i\sin\theta_i}{|Z_{Th,i} + Z_{coup,i}|}$$

$$q_i = \frac{\left(|V_{T,i}|\,|V_{Th,i}|\cos\varphi_i - |V_{Th,i}|^2\right)\sin\theta_i - |V_{T,i}|\,|V_{Th,i}|\sin\varphi_i\cos\theta_i}{|Z_{Th,i} + Z_{coup,i}|} \tag{19.1}$$

where

$$\varphi_i = \delta_{Th,i} - \delta_{T,i}$$
$$\theta_i = \angle\left(Z_{Th,i} + Z_{coup,i}\right)$$

in which V_T is the voltage at the PCC, Z_{coup} is the coupling impedance at the DER's output, and $V= |V| < \delta$ denotes $v(t)$'s phasor notation. The coupling impedance is more inductive (i.e., $Z_{coup} \approx j\omega_{coup}$), whereas Z_{Th} has a minuscule magnitude and is also inductive at 50 Hz (i.e., $Z_{Th} \approx j\omega_{conv}$) [43]. Thus, in Fig. 19.3(c), $V_{Th} \approx V_{cf}$ in which V_{cf} denotes the voltage across capacitor C_f. Therefore, (19.1) can be represented by

$$p_i = \frac{|V_{T,i}|\,|V_{cf,i}|\,\sin\left(\delta_{cf,i} - \delta_{T,i}\right)}{\omega L_{conv,i} + \omega L_{coup,i}}$$

$$q_i = \frac{|V_{T,i}|\,|V_{cf,i}|\cos\left(\delta_{cf,i} - \delta_{T,i}\right) - |V_{T,i}|^2}{\omega L_{conv,i} + \omega L_{coup,i}} \tag{19.2}$$

The average output active and reactive powers (P, Q) injected by a DER can be determined from p and q and with the help of a low-pass filter [72].

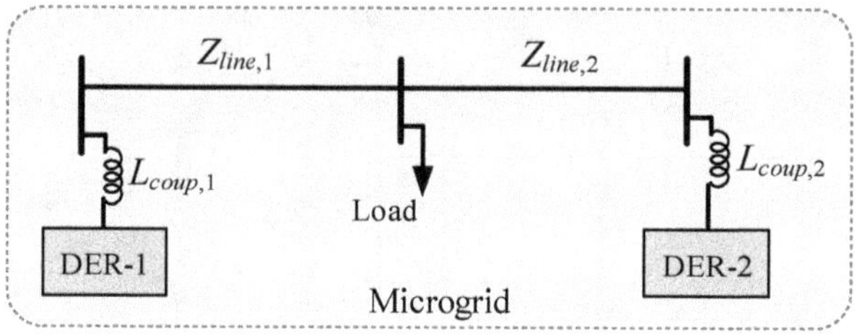

Figure 19.4 A sample microgrid with two converter-interfaced DERs and one load.

A voltage-controlled strategy is used in this chapter, which monitors the voltage at the DER's PCC to control the converter's output voltage of V_{cf} such that the preferred active and reactive powers get supplied. Hence, the voltage magnitude and angles at the PCC are instantaneously measured and employed in (19.2) to determine the reference of the converter's output voltage of $V_{cf,ref}$ for the preferred powers. Later, the calculated reference voltage is applied to the inner-loop control to produce the on/off signals for the switches within the VSC.

Now, consider the microgrid of Fig.19.4 with two converter-coupled DERs that are supplying a load. Assume that DER-1 is coupled to the load via a feeder with an impedance of $Z_{line,1}$, whereas DER-2 is coupled via another feeder with an impedance of $Z_{line,2}$. Assume a highly inductive feeder in which $Z_{line} \approx j\omega\ _{line}$. Denoting the voltage at the load's PCC by V_{load}, the active and reactive powers that the load receives from each DER can be given by

$$p_i = \frac{|V_{load,i}|\,|V_{cf,i}|\sin\left(\delta_{cf,i} - \delta_{load,i}\right)}{\omega L_{conv,i} + \omega L_{coup,i} + \omega L_{line,i}}$$

$$q_i = \frac{|V_{load,i}|\,|V_{cf,i}|\cos\left(\delta_{cf,i} - \delta_{load,i}\right) - |V_{load,i}|^2}{\omega L_{conv,i} + \omega L_{coup,i} + \omega L_{line,i}} \quad (19.3)$$

The active power and reactive power are decoupled (because the feeder is highly inductive). Furthermore, the dissimilarity of the voltage angle between V_{cf} and V_{load} is minuscule. Therefore, a DC load flow analysis technique can be employed from which the average output active power that the load receives from each DER becomes

$$P_i = \frac{|V_{load,i}|\,|V_{cf,i}|\left(\delta_{cf,i} - \delta_{load}\right)}{\omega L_{conv,i} + \omega L_{coup,i} + \omega L_{line,i}} \quad (19.4)$$

Thus, the angle of the voltage across DER-i's output capacitor becomes

$$\delta_{cf,i} = P_i\left(\Gamma_{conv,i} + \Gamma_{coup,i} + \Gamma_{line,i}\right) + \delta_{load} \quad (19.5)$$

where

$$\Gamma_i = \frac{\omega L_i}{|V_{cf,i}|\,|V_{load}|}$$

For achieving a decentralized power-sharing among multiple DERs within a microgrid, the DERs output's voltage magnitude and angle should be modified under the droop control as [19]

$$\delta_{cf,i} = \delta_{rated,i} - m_i \left[\frac{X_{line}}{Z_{line}}\left(P_i - P_{rated,i}\right) - \frac{R_{line}}{Z_{line}}\left(Q_i - Q_{rated,i}\right)\right]$$
$$|V_{cf,i}| = V_{rated,i} - n_i \left[\frac{R_{line}}{Z_{line}}\left(P_i - P_{rated,i}\right(+ \frac{X_{line}}{Z_{line}}\left(Q_i - Q_{rated,i}\right(\right] \tag{19.6}$$

where $_{rated}$ and V_{rated} are the DER's rated voltage angle and magnitude when it injects rated active and reactive powers of P_{rated} and Q_{rated}, whereas m and n are, respectively, the active power-angle and reactive power-voltage droop coefficients in rad/W and V/VAr. Equation (19.6) can be updated as

$$\delta_{cf,i} = \delta_{rated,i} - m_i\left(P_i - P_{rated,i}\right)$$
$$|V_{cf,i}| = V_{rated,i} - n_i\left(Q_i - Q_{rated,i}\right) \tag{19.7}$$

Assuming that the microgrid's lines are highly inductive, which will be adjusted via the outer-loop control. Hence, the outer-loop controller defines the preferred V_{cf} for a DER as

$$v_{\text{ref,a}}(t) = \sqrt{2}|V_{cf,i}|\sin(2\pi f_{rated}t + \delta_{cf,i})$$
$$v_{\text{ref,b}}(t) = \sqrt{2}|V_{cf,i}|\sin(2\pi f_{rated}t + \delta_{cf,i} - 120°) \tag{19.8}$$
$$v_{\text{ref,c}}(t) = \sqrt{2}|V_{cf,i}|\sin(2\pi f_{rated}t + \delta_{cf,i} + 120°)$$

Fig. 19.5 shows the VSC's control block diagram. Note that the reference generation for i_c and i_2, the linear quadratic regulator gain calculation, and hysteresis function are described in detail in [16] and are not repeated here. Also, more detailed info on the VSC structure and control are provided in [73, 74].

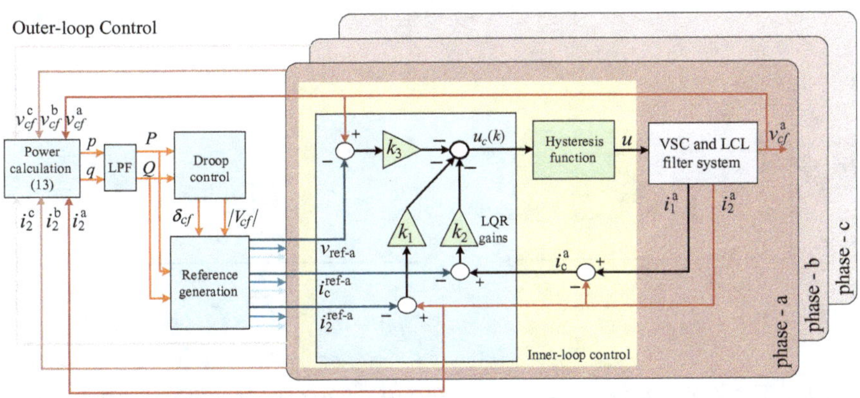

Figure 19.5 The closed-loop controller of a DER.

When using a frequency droop instead of the angle droop in the microgrid, the outer-loop control can be expressed by [75] as

$$\frac{f = f_{rated,i} - m_i' \left[\frac{X_{line}}{Z_{line}}\left(p_i - p_{rated,i}\right) - \frac{R_{line}}{Z_{line}}\left(Q_i - Q_{rated,i}\right)\right]}{\left|V_{cf,i}\right| = V_{rated,i-n_i}\left[\frac{R_{line}}{Z_{line}}\left(p_i - p_{rated,i}\right) - \frac{X_{line}}{Z_{line}}\left(Q_i - Q_{rated,i}\right)\right]} \tag{19.9}$$

$$\begin{aligned} f &= f_{rated,i} - m_i'\left(P_i - P_{rated,i}\right) \\ |V_{cf,i}| &= V_{rated,i} - n_i\left(Q_i - Q_{rated,i}\right) \end{aligned} \tag{19.10}$$

$$\begin{aligned} v_{\mathrm{ref,a}}(t) &= \sqrt{2}|V_{cf,i}|\sin(2\pi f t) \\ v_{\mathrm{ref,b}}(t) &= \sqrt{2}|V_{cf,i}|\sin(2\pi f t - 120°) \\ v_{\mathrm{ref,c}}(t) &= \sqrt{2}|V_{cf,i}|\sin(2\pi f t + 120°) \end{aligned} \tag{19.11}$$

instead of (19.6)–(19.8), where f_{rated} is the microgrid's rated frequency and m' (in Hz/W) is the droop coefficient. Figure 19.6 shows this schematically.

Assume that the microgrid's frequency decreases by $\Delta\omega$ when a DER's output active power rises from zero to the rated (nominal) level. Thus, the DER's active power-angle droop coefficient can be given by [76].

$$m_i = \frac{\Delta\omega}{P_{rated,i}} \tag{19.12}$$

Assuming $\Delta\omega$ the same and fixed for DERs i and j with dissimilar rating yields

$$\frac{m_i}{m_j} = \frac{P_{rated,\,j}}{P_{rated,\,i}} \tag{19.13}$$

Likewise, presume that the microgrid's voltage decreases by ΔV when a DER's output reactive power rises from zero to the rated level. As such, the DER's reactive power-voltage droop coefficient can be given in [76] as

$$n_i = \frac{\Delta V}{Q_{rated,i}} \tag{19.14}$$

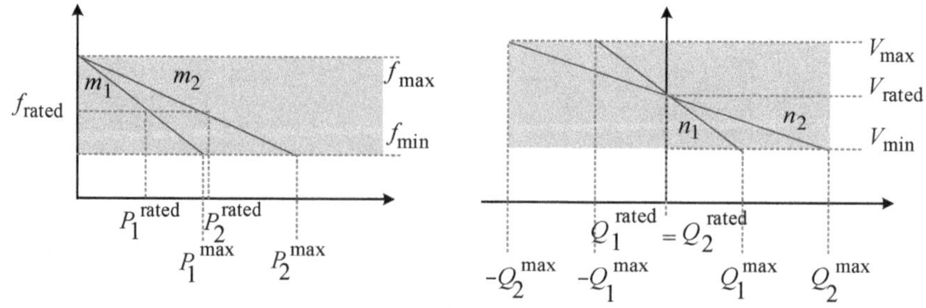

Figure 19.6 The active power-frequency and reactive power-voltage droop curves for two DERs.

Assuming ΔV the same and fixed for DERs i and j with dissimilar rating yields.

$$\frac{n_i}{n_j} = \frac{Q_{rated,j}}{Q_{rated,i}} \tag{19.15}$$

Suppose DERs i and j within a steady-state microgrid have an equal δ_{rated}. Therefore, from (19.7) and (19.13),

$$\begin{gathered}\delta_{cf,i} - \delta_{cf,j} = \\ (\delta_{rated,i} - \delta_{rated,j}) - m_i(P_i - P_{rated,i}) + m_j(P_j - P_{rated,j}) = m_jP_j - m_iP_i\end{gathered} \tag{19.16}$$

Replacing$_{cf,i}$ and $_{cf,j}$ from (19.5) in (19.16) yields

$$P_i(\Gamma_{conv,i} + \Gamma_{coup,i} + \Gamma_{line,i}) - P_j(\Gamma_{conv,j} + \Gamma_{coup,j} + \Gamma_{line,j}) = m_jP_j - m_iP_i \tag{19.17}$$

Therefore, the relationship of the active power injected by two DERs becomes

$$\frac{P_j}{P_i} = \frac{\Gamma_{conv,i} + \Gamma_{coup,i} + \Gamma_{line,i} + m_i}{\Gamma_{conv,j} + \Gamma_{coup,j} + \Gamma_{line,j} + m_j} \tag{19.18}$$

Equation (19.18) confirms that a DER's output active power is reciprocal to total Γ at observed at its output. $\Gamma_{\text{line,}}\Gamma_{\text{coup}}$ and Γ_{conv} depend on the inductances between the load and the DER. Because includes the square of the voltage magnitude in the denominator, we have

$$\Gamma_{conv,i} << \Gamma_{line,i} << \Gamma_{coup,i} << m_i \tag{19.19}$$

Thus, (19.18) can be updated as

$$\frac{P_j}{P_i} \approx \frac{m_i}{m_j} \tag{19.20}$$

Likewise, the relationship of the reactive powers injected by two DERs becomes

$$\frac{Q_j}{Q_i} \approx \frac{n_i}{n_j} \tag{19.21}$$

According to the above notion, from (19.13), (19.20), (19.15), and (19.21), the relationship of the output active and reactive power of two DERs within the microgrid is equal to the proportion of their power ratings.

It is preferred to have a small difference in the voltage angles at either side of the coupling inductances (i.e., $\delta_{cf} - \delta_{PCC}$) [76] to lie at the linear part P-δ characteristic of (19.2). Likewise, it is preferred to observe a small voltage magnitude difference at either side of the coupling inductances (i.e., $|V_{cf}| - |V_{PCC}|$) [77]. To realize this, the coupling inductances can be selected as

$$\frac{L_{coup,i}}{L_{coup,j}} = \frac{P_{rated,j}}{P_{rated,i}} = \frac{Q_{rated,j}}{Q_{rated,i}} \tag{19.22}$$

Hence, a suitable power-sharing between two DERs within a microgrid requires [43]

$$\begin{gathered}\frac{P_j}{P_i} \approx \frac{m_i}{m_j} = \frac{L_{coup,i}}{L_{coup,j}} = \frac{P_{rated,j}}{P_{rated,i}} \\ \frac{Q_j}{Q_i} \approx \frac{n_i}{n_j} = \frac{L_{coup,i}}{L_{coup,j}} = \frac{Q_{rated,j}}{Q_{rated,i}}\end{gathered} \tag{19.23}$$

Consider the microgrid of Fig.19.4 with DER-1 and DER-2 and the technical data of Table 19.1. It is preferred to realize the power ratios of DER-1:DER-2 as 2:1. The microgrid is presumed at the steady-state condition with a total demand of 0.41 pu. This load rises to 1 pu at t = 0.5 s, then reduces to 0.53 pu at t = 1 s. The microgrid's load further decreases to 0.17 pu at t = 1.5 s. The ratio of the output active powers between the DERs is upheld as desired throughout all load changes, as shown in Fig.19.7(a). The total electrical demand of the microgrid is shown in Fig.19.7(b). As seen from Fig.19.7(c–d), the microgrid's frequency and voltage stay within the permissible ranges throughout the study. For the assumed microgrid, V_{rated} = 1 pu and f_{rated} = 50 Hz, as shown in Fig.19.7(e–f).

Table 19.1. Technical data of the network parameter

Network	400 Vrms line-line, 50 Hz, 1 pu = 6 kVA
Line	$R = 0.2\ \Omega$, $L = 10$ mH
Load	Constant-impedance type
VSCs and filters structure and control parameters	$R_f = 0.1\ \Omega$, $L_f = 0.36$ mH, $C_f = 50\ \mu$F, $V_{dc} = 350$ V, $a = 2k_1 = 6.1802$, $k_2 = 2.2968$, $k_3 = 22.968$
DER-1	$L_{coup} = 1.36$ mH, $m = 1.5708$ rad/kW, $n = 9$ V/kVAr
DER-2	$L_{coup} = 2.72$ mH, $m = 3.1416$ rad/kW, $n = 18$ V/kVAr

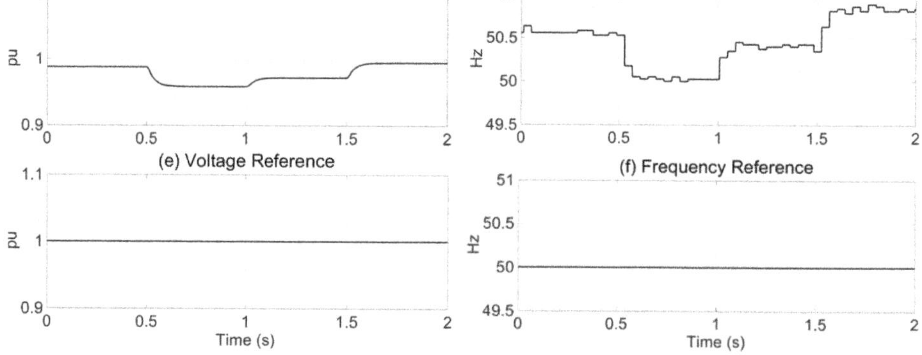

Figure 19.7 Results for the assumed microgrid.

19.4 Power-Sharing among DERs

Because usually $L_{line} \ll L_{coup}$ and $L_{conv} \ll L_{coup}$, the active and reactive power that a DER injects depends on its coupling inductance. Therefore, the output active power of the DERs in Figure 19.4 can be expressed from (19.3) as

$$p_1 \approx \frac{|V_{load}|\,|V_{cf,1}|\sin\delta_{cf,1}}{\omega L_{coup,1}}$$
$$p_2 \approx \frac{|V_{load}|\,|V_{cf,2}|\sin\delta_{cf,2}}{\omega L_{coup,2}} \tag{19.24}$$

Three techniques are deliberated below that can control the ratio of power-sharing among DERs [78].

19.4.1 Technique-1

The relationship of the average active and reactive power between two DERs can be expressed from (19.24) in the form of

$$\frac{P_1}{P_2} = \frac{L_{coup,2}}{L_{coup,1}}\frac{|V_{cf,1}|\sin\delta_{cf,1}}{|V_{cf,2}|\sin\delta_{cf,2}}$$
$$\frac{Q_1}{Q_2} = \frac{L_{coup,2}}{L_{coup,1}}\frac{|V_{cf,1}|\cos\delta_{cf,1} - |V_{load}|}{|V_{cf,2}|\cos\delta_{cf,2} - |V_{load}|} \tag{19.25}$$

Presume that the preferred ratio of active power between the DERs is $P_1/P_2 = k_p > 0$; that is, [3, 79]

$$\begin{cases} \dfrac{P_1}{P_2} = k_p \\ \dfrac{L_{coup,2}}{L_{coup,1}} = k_p \end{cases} \Rightarrow |V_{cf,1}|\sin\delta_{cf,1} = |V_{cf,2}|\sin\delta_{cf,2} \tag{19.26}$$

Also, suppose the ratio of their reactive power is $Q_1/Q_2 = k_q = k_p$. For simplicity, it is thought in [3, 79] that

$$\begin{cases} \dfrac{Q_1}{Q_2} = k_q = k_p \\ \dfrac{L_{coup,2}}{L_{coup,1}} = k_p \end{cases} \Rightarrow |V_{cf,1}|\cos\delta_{cf,1} = |V_{cf,2}|\cos\delta_{cf,2} \tag{19.27}$$

19.4.2 Technique-2

In general, the ratio of the coupling inductances of two DERs can be $k' > 0$, that is,

$$\begin{cases} \dfrac{P_1}{P_2} = k_p \\ \dfrac{L_{coup,2}}{L_{coup,1}} = k' \end{cases} \Rightarrow |V_{cf,1}|\sin\delta_{cf,1} = \frac{k_p}{k'}|V_{cf,2}|\sin\delta_{cf,2} = k''|V_{cf,2}|\sin\delta_{cf,2} \tag{19.28}$$

where $k'' = k_p/k'$. In such a condition, the DERs' output voltages correlate by a factor of k'', whereas the ratio of their reactive powers is

$$\begin{cases} \frac{Q_1}{Q_2} = k_q \\ \frac{L_{coup,2}}{L_{coup,1}} = k' \end{cases} \Rightarrow \quad |V_{cf,1}|\cos\delta_{cf,1} - |V_{load}|^2 = \frac{k_q}{k'}\left(|V_{cf,2}|\cos\delta_{cf,2} - |V_{load}|^2\right) \quad (19.29)$$

In general, $k_p \neq k_q$; hence, from (19.28) and (19.29), it can be seen that although the output active power of two DERs are equal to k_p, the ratio of their output reactive powers will not be k_p. Thus, the coupling inductances can have any ratio if only maintaining the ratio of the active powers between the DERs is intended.

19.4.3 Technique-3

Technique-2's key disadvantage is that the proportion of the DERs' output reactive power is different from that of their active powers because of the mismatch of the proportion of their coupling inductances. To improve this, two internal inductances can be applied to the DERs by altering the controller of their VSCs to balance the equivalent inductance with the proportion of the preferred output active and reactive powers. This modification will cause the voltage appearing at each VSC's output to be slightly less than the original level. The primary objective is determining a suitable voltage drop across the internal balancing inductance (L_{conv}) in such a way that the preferred active and reactive power ratio is realized at the DERs' outputs regardless of the proportion of the installed coupling inductances. Therefore, it is prefered that

$$\begin{aligned} \frac{P_1}{P_2} &= \frac{L_{coup,2} + L_{conv,2}}{L_{coup,1} + L_{conv,1}} = \frac{m_2}{m_{1\}}} \\ \frac{Q_1}{Q_2} &= \frac{L_{coup,2} + L_{conv,2}}{L_{coup,1} + L_{conv,1}} = \frac{n_2}{n_{1\}}} \end{aligned} \quad (19.30)$$

Assuming $P_1 > P_2$ and hence,$L_{coup,2} + L_{conv,2} > L_{coup,1} + L_{conv,1}$, one can allocate a very small $L_{conv,1}$ and then determine $L_{conv,2}$ to satisfy (19.30) without causing instability for the system.

After imposing this balancing inductance, the phase voltage at each DER's output (v_{ref}) is defined by [80, 81]

$$v_{ref} = v_{droop} - v_{conv} \quad (19.31)$$

where $v_{droop} = |V_{cf}| < \delta_{cf}$ is the initial voltage determined by the droop equation (19.7), whereas v_{conv} denotes the voltage drop across the internal balancing inductance. Expressing (19.31) in $\alpha\beta$ coordinates in the form of

$$\begin{aligned} |V_{ref-\alpha}| &= |V_{cf}|\cos\delta_{cf} + \omega L_{conv}|I|\sin\theta \\ |V_{ref-\beta}| &= |V_{cf}|\sin\delta_{cf} - \omega L_{conv}|I|\cos\theta \end{aligned} \quad (19.32)$$

where $\omega = 2\pi f_{rated}$, whereas $|I| < \theta$ denotes the current flowing through the coupling inductance in phase-a. After transforming (19.32) into the polar coordination, the voltage reference at the VSC's output can be expressed as

$$\begin{aligned} v_{ref,a}(t) &= \sqrt{2}|V_{ref}| \sin(\omega t + \theta_{ref}) \\ v_{ref,b}(t) &= \sqrt{2}|V_{ref}| \sin(\omega t + \theta_{ref} - 120°) \\ v_{ref,c}(t) &= \sqrt{2}|V_{ref}| \sin(\omega t + \theta_{ref} + 120°) \end{aligned} \tag{19.33}$$

where

$$|V_{ref}| = \sqrt{|V_{ref-\alpha}|^2 + |V_{ref-\beta}|^2}$$
$$\theta_{ref} = \tan^{-1} \frac{|V_{ref-\beta}|}{|V_{ref-\alpha}|}$$

This voltage will then be employed for the VSC's other two phases with a ± 120 phase shift.

Consider the microgrid of Fig.19.4 composed of DER-1 and DER-2 and the technical parameters of Table 19.2. The microgrid is autonomous and at the steady-state condition with a total demand of 0.88 pu. The microgrid's load decreases to 0.77 pu at t = 1.5 s, whereas it increases to 0.91 pu at t = 3 s. Fig.19.8(a) illustrates the microgrid's active and reactive demand.

It is prefered to maintain the proportion of the output active and reactive power between DER-1 and 2 as $P_1/P_2 = Q_1/Q_2 = 2$. Thus, the proportion of their coupling inductances are $L_{coup,1}/L_{coup,2} = 0.5$, whereas that of their droop coefficients are $m_1/m_2 = n_1/n_2 = 0.5$. These assumptions satisfy (19.26). Fig.19.8(b) and (c) shows that the output active and reactive power ratio between the DERs is the same as the preferred level by the help of the designed droop coefficients and coupling inductances.

Now, assume that both DERs have equal coupling inductances. Fig.19.9(a) shows that the DERs keep the proportion of their output active powers nearly the same as the expected level during the load variations by properly adjusting their droop coefficients. Nevertheless, Fig.19.9(b) shows that the proportion of their reactive powers is 0.58, which is different from that of their active powers.

Now, let us assume that the internal balancing inductance is imposed to both DERs. Assuming the coupling inductance of both DERs as 30 mH, the internal balancing inductance of DER-1 (i.e., the DER with a larger output power) is thought to be small (e.g., 0.5 mH) based on which the corresponding internal balancing inductance of DER-2 is defined to be 31.4 mH from (19.30). The impact of imposing the internal balancing inductances in realizing the preferred proportion of both active and reactive powers among the DERs is shown in Figs.19.10(a) and (b).

19.5 Control of BESs

BESs can be employed for various reasons in microgrids; however, one of the key applications is contributing to the frequency regulation [82–85]. Employing the abovementioned droop control technique for BESs has some drawbacks. When a conventional droop control is employed, the output power of BESs becomes directly proportional to their power rating,

Table 19.2. Technical data of the network parameter

Network	400 Vrms line-line, 50 Hz, 1 pu = 5.8 kVA
Line	$R = 0.2\Omega$, L =6.4 mH
Load	Constant-impedance type
VSCs and filters structure and control parameters	$R_f = 0.1\ \Omega$, $L_f = 0.36$ mH, $C_f = 50\ \mu$F, $V_{dc} = 350$ V, $a = 2k_1 = 6.1802$, $k_2 = 2.2968$, $k_3 = 22.968$
DER-1	$L_{coup} = 30$ m H$m = 1.5708$ rad/kW, $n = 9$ V/kVAr
DER-2	$L_{coup} = 60$ mH (in Technique-1), $L_{coup} = 30$ mH (in Technique-2 and 3) $m = 3.1416$ rad/kW, $n = 18$ V/kVAr

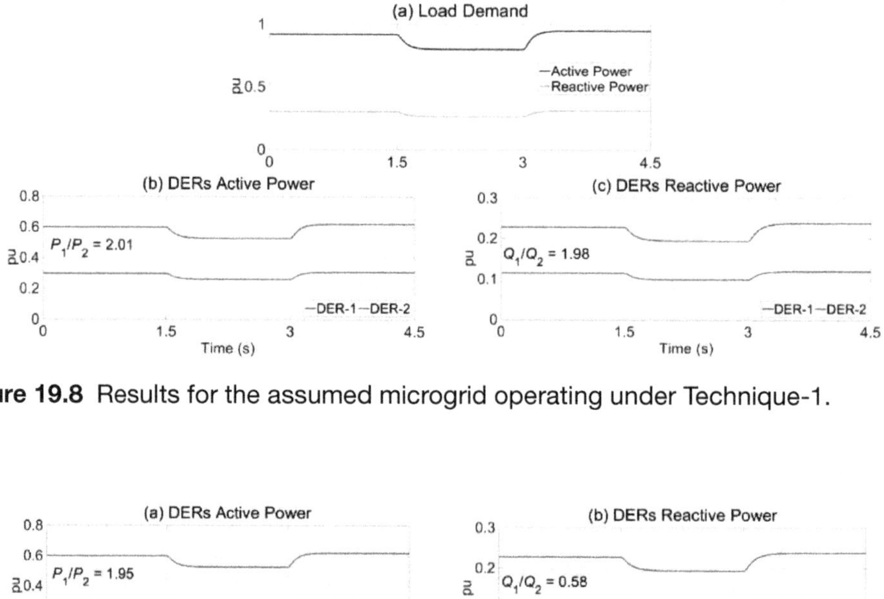

Figure 19.8 Results for the assumed microgrid operating under Technique-1.

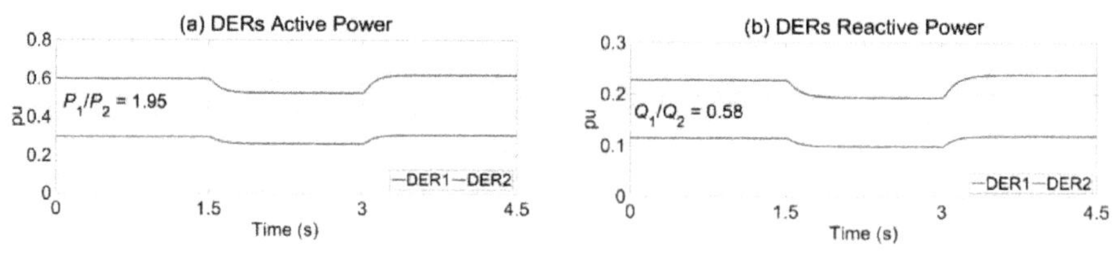

Figure 19.9 Results for the assumed microgrid operating under Technique-2.

whereas the available energy stored in them is ignored. Thus, the duration that a BES discharges and supplies the microgrid demand is not managed. Hence, a BES that has a lower state of charge (SoC) will be discharged faster than the other BESs [86, 87].

When the microgrid consists of BES in addition to DERs, a different power-sharing technique can be utilized. Considering that the energy stored in a BES reduces as it continues to discharge, the SoC of BESs can be employed in place of the BES power capacity in

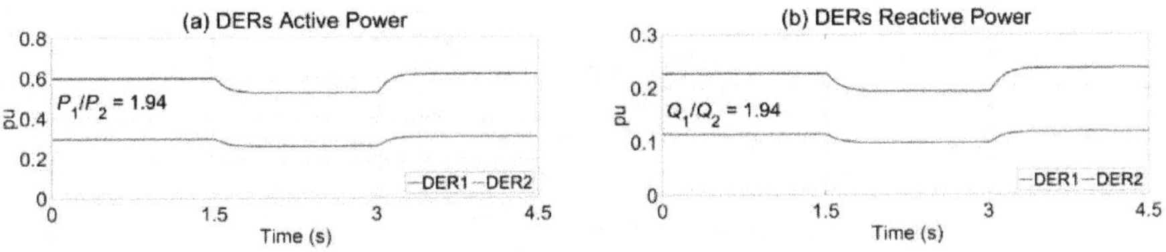

Figure 19.10 Results for the assumed microgrid operating under Technique-3.

(19.23). Therefore, the energy stored in the BES is estimated at any given time by

$$E_{\mathrm{BES}}(t) = E_{\mathrm{BES}}^{\mathrm{initial}} - \int P_{\mathrm{BES}}(t).dt \tag{19.34}$$

in which $E_{\mathrm{BES}}^{\mathrm{initial}}$ represents the BES's initial stored energy, whereas P_{BES} represents its average output active power, which is measured at the output of its VSC. Hence, the BES's SoC is expressed as

$$SoC_{\mathrm{BES}}(t) = \frac{E_{\mathrm{BES}}(t)}{E_{\mathrm{BES}}^{\max}} \times 100 \tag{19.35}$$

where $SoC_{\mathrm{BES}}(t)$ and $E_{\mathrm{BES}}^{\max}$ denote the BES's instantaneous SoC and energy capacity. To control the BES' output power based on the SoC, its P-δ droop coefficient in (19.7) can be dynamically varied based on the SoC. Hence, the BES with lower SoC supplies a lower amount of active power. This prevents a quick discharge of a BES that has a lower SoC. Fig.19.11 shows this approach schematically from which it can be seen that with the reduction of the BES's SoC, the BES's operating point relocates from point A to B; therefore, the output active power of the BES reduces.

As the SoC varies instantaneously, it cannot be employed directly in the droop control. Hence, a discretized SoC ($SoC_{\mathrm{BES}}^{\mathrm{d}}$) in 10% steps can be employed as [86]

$$SoC_{\mathrm{BES}}^{\mathrm{d}}(t) = \begin{cases} 0.9 & 90\% \le SoC \le 100\% \\ 0.8 & 80\% \le SoC < 90\% \\ \vdots & \\ 0.2 & 20\% \le SoC < 30\% \end{cases} \tag{19.36}$$

The BES will continue discharging until SoC_{BES} reduces to its lowest allowed value. Assuming a BES's largest depth of discharge as 70%, the BES stops discharging when SoC_{BES} reduces to 30% of $E_{\mathrm{BES}}^{\max}$.

From (19.36), the SoC of each BES in the microgrid will be updated continuously. Each BES's P-δ droop coefficient is then determined by

$$\frac{P_{\mathrm{BES}}}{P_0} = \frac{SoC_{\mathrm{BES}}^{\mathrm{d}}}{SoC_0} = \frac{m_0}{m_{\mathrm{BES}}} \tag{19.37}$$

in which m_0 denotes a global droop coefficient designed for P_0 and $SoC_0 = 1$. From (19.37), every BES's droop coefficient will be updated from

$$m_{\mathrm{BES}}(t) = \frac{m_0}{SoC_{\mathrm{BES}}^{\mathrm{d}}(t)} \tag{19.38}$$

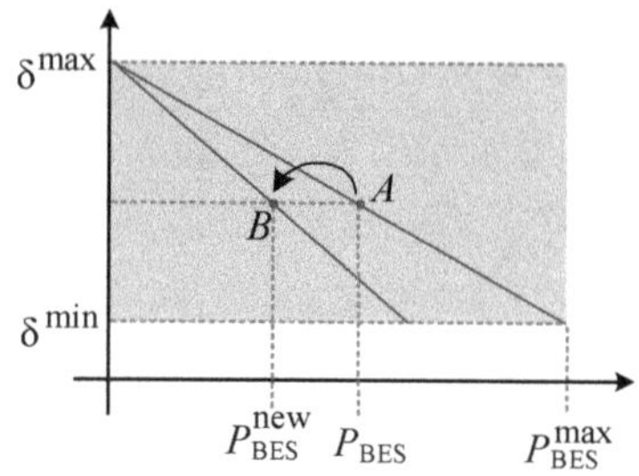

Figure 19.11 BES's P-δ droop coefficient modification.

Hence, for a microgrid with two BESs, the power ratio equation (19.23) will be described by

$$\frac{P_{\text{BES1}}}{P_{\text{BES2}}} = \frac{SoC_{\text{BES1}}}{SoC_{\text{BES2}}} = \frac{m_{\text{BES2}}}{m_{\text{BES1}}} \tag{19.39}$$

The weakness of this approach is that a BES's output power will change with respect to the reduction of the output power of another BES. To prevent this problem, the droop coefficients of the DERs can be updated to regulate their output powers as soon as the output power of a BES reduces. Fig.19.12 illustrates this approach schematically. In this figure, when a BES's SoC reduces and its operating point relocates from point A to B, the DER's operating point also relocates from point C to D. Thus, the DER starts to deliver the BES's reduced power. Hence, a BES's output power only relates to its own SoC and does not get affected by the SoC of other BESs. For this, each DER's droop coefficients is defined by

$$m_{\text{DER}}(t) = m_0 \prod_{j=1}^{N} SoC^{\text{d}}_{\text{BES}-j}(t) \tag{19.40}$$

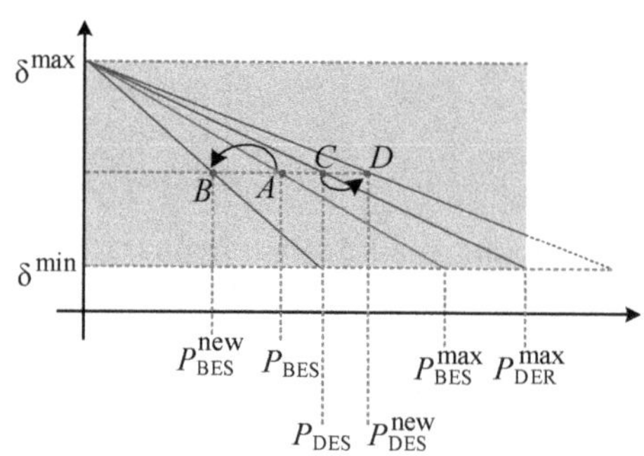

Figure 19.12 P-δ droop coefficient change for a DER and BES together.

in which N is the BES number in the microgrid and $SoC^{\text{d}}_{\text{BES}-j}$ accounts for the SoC of the jth BES. The control block diagram for each converter-interfaced DER and BES is shown in Fig.19.13.

Consider the microgrid of Fig.19.14 with two DERs and two BESs and with the technical data of Table 19.3. The maximum power capacity of the DERs (P_1 and P_4) are assumed

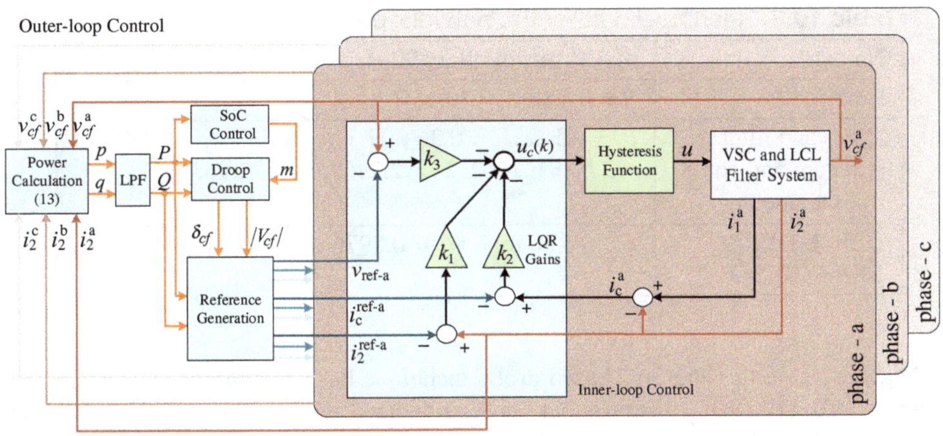

Figure 19.13 The control system of the VSC and output passive filter for a converter-interfaced DER.

to be $P_1/P_4 = 2$, whereas the BESs' overall energy capacity (P_2 and P_3) are assumed to be $P_2/P_3 = 1$. Initially, P_1/P_2 is expected to be unity. First, suppose the SoC-based droop technique is employed for the BESs using (19.39), whereas the control of the DERs relies on conventional droop technique. Fig.19.15(a) shows that the load demand is kept constant in this case. Fig.19.15(b) shows that with the reduction of the BESs' SoC, they reduce their output active power; for example, at $t = 4.8$ s, BES-2's SoC reduces below 90%; thus, its P-δ droop coefficient increases. As such, its output active power (i.e., P_3) decreases. This reduction is picked up by the DERs as well as BES-1. The SoC of the BESs is shown in Fig.19.15(c) and illustrates that for the considered period, the SoC of BES-1 and -2 reduce to 43% and 61%.

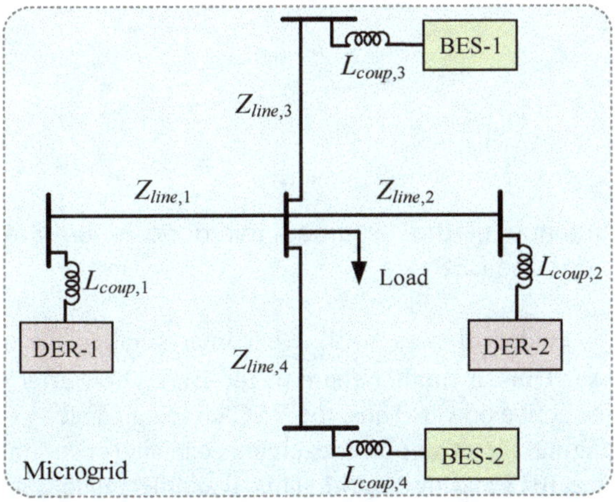

Figure 19.14 Assumed microgrid with two DER and two BESs.

Table 19.3. Technical data of the network parameter

Network	400 Vrms line-line, 50 Hz
Line	$R = 0.5\ \Omega$, $L = 3.2$ mH
VSCs and filters structure and control parameters	$R_f = 0.1\ \Omega$, $L_f = 0.36$ mH, $C_f = 50\ \mu$F, $V_{dc} = 350$ V, $a = 2$ $k_1 = 6.1802$, $k_2 = 2.2968$, $k_3 = 22.968$
DER-1, DER-2, BES-1, BES-2	$L_{coup} = 30$ mH, $m_o = 0.8976$ rad/kW, $P_o = 3$ kW

Now, presume that the SoC-based droop technique has been employed for the DERs using (19.40). Fig.19.16(a) shows that the load demand remains unchanged. In addition, the BES produces less active power when its SoC decreases. However, the BES's reduced active power is addressed solely by the DERs, as shown in Fig.19.16(b). The BESs' SoCs are shown in Fig.19.16(c) and demonstrate that in the considered period, the SoC of BES-1 and BES-2 reduces to 52% and 75%, which shows an improvement versus the above case.

The previous study is repeated supposing a 16% load decrease and a 25% load increase, respectively at t = 6 and 15 s, as shown in Fig.19.17(a). During this period, the BESs' droop coefficients are not changed. Fig.19.17(b) shows that the output power of all DERs and BESs vary based on the load changes and according to their desired power ratios. Fig.19.17(c) demonstrates the BESs' SoC in this period.

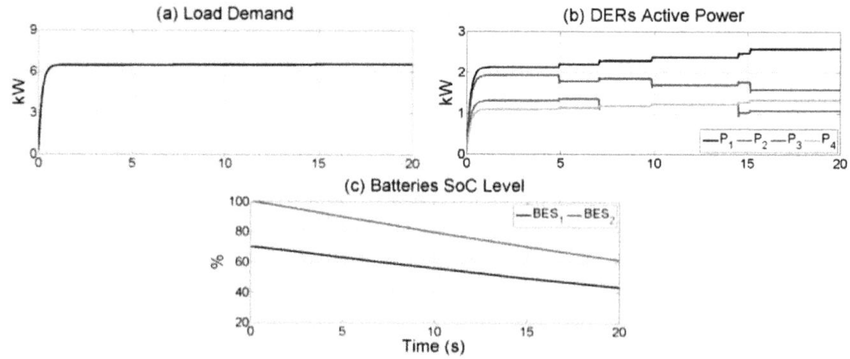

Figure 19.15 Performance of the assumed microgrid when SoC-based droop is applied to the BESs.

From (19.39), it is evident that as SoC^{d}_{BES} becomes smaller, the output active power of BES-i will decrease. Thus, a smaller share of the BES's converter's capacity (S^{cap}_{BES}) is used for the flow of the active power. Thus, the VSC's vacant capacity can be employed for reactive power exchanging. Reactive power exchange can also occur on top of active power exchanged between the BES and microgrid. This available capacity in BES-i's converter (Q^{av}_{BES}) is updated over time from

$$Q^{av}_{BES} = \sqrt{\left(S^{cap}_{BES}\right)^2 - \left(P_{BES}\right)^2} \tag{19.41}$$

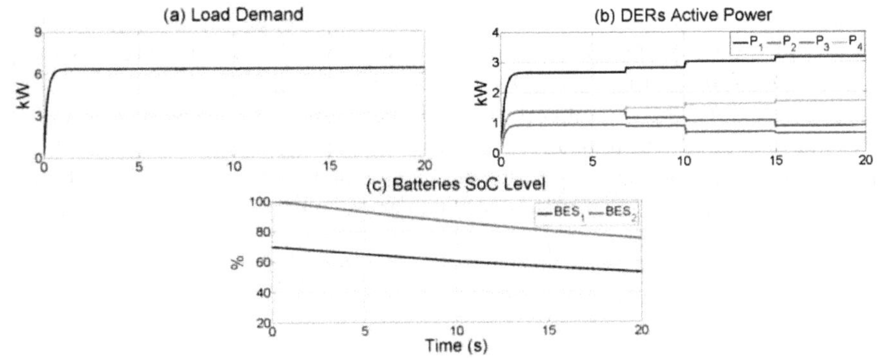

Figure 19.16 Performance of the assumed microgrid when SoC-based droop is applied to the BES and DERs.

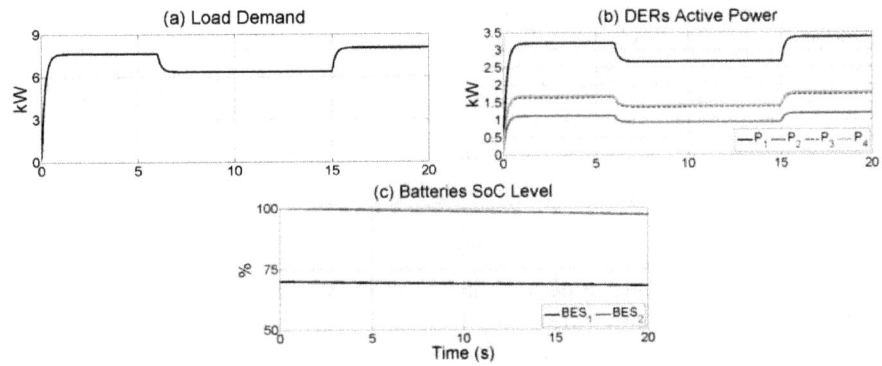

Figure 19.17 Performance of the assumed microgrid with load changes.

It is later employed to determine the BES's V-Q droop coefficient (n_{BES}) in the form of

$$n_{\mathrm{BES}} = \frac{\Delta V}{Q_{\mathrm{BES}}^{\mathrm{av}}} \tag{19.42}$$

Note that the stored energy within the BES and its SoC do not get affected by exchanging reactive power between the converter and the microgrid.

19.6 Microgrid Central Controller

The central controller of the microgrid interconnects with the distribution network's tertiary controller and the DERs' primary controllers. Its key obligation is defining the set-points and references of the DERs' primary controller, which are used later to determine each DER's output active and reactive power.

As discussed before, each DER not only has to supply a percentage of the microgrid's demand, but also should regulate the frequency and voltage in the microgrid when operating in autonomous mode. In addition, the central controller is in charge of checking these parameters. Therefore, the central controller will determine new set-points for the DERs' primary controllers and will transmit them this information if the microgrid's voltage and frequency drops below or increases above the acceptable thresholds. This procedure has to occur quicker than the predefined operation time of the microgrid's under- or overvoltage and frequency protective systems.

The microgrid's central controller is composed of multiple modules, which are discussed below [75].

19.6.1 Dynamic Power Ratio Adjustment

The distribution network's tertiary controller determines the preferred ratio of the output power of the DERs within the microgrid when operating under both grid-connected and autonomous modes. Therefore, it can change this ratio dynamically and pass this data to the DERs' primary controllers via the microgrid's central controller. As an example, suppose the relationship of the output active powers of DER-1 and DER-2 of Fig. 19.4 is k_{p1} at $t = t_1$. Therefore, $P_{\text{rated},2}/P_{\text{rated},1} = m_1/m_2 = k_{p1}$ according to (19.23). Thus, from (19.10),

$$f_1^{t_1} = f_{\text{rated}} - m_1\left(P_1^{t_1} - P_{\text{rated},1}\right) \tag{19.43}$$

If the tertiary controller modifies this ratio to k_{p2} at $t = t_2$, from (19.10),

$$f_1^{t_2} = f_{\text{rate}} - m_1\left(P_1^{t_2} - P_{\text{rated},1}\right) \tag{19.44}$$

Assuming the load demand is same in t_1 and t_2, the DERs' total active power generation will remain unchanged at these two instants. Thus,

$$P_1^{t_1} + P_2^{t_1} = P_1^{t_2} + P_2^{t_2} \tag{19.45}$$

Equation (19.45) can be simplified by replacing the power ratios as

$$\begin{aligned} & P_1^{t_1} + k_{p1}P_1^{t_1} = P_1^{t_2} + k_{p2}P_1^{t_2} \\ \Rightarrow \quad & P_1^{t_1}(1 + k_{p1}) = P_1^{t_2}(1 + k_{p2}) \\ \Rightarrow \quad & P_1^{t_2} = P_1^{t_1}\frac{1 + k_{p1}}{1 + k_{p2}} \end{aligned} \tag{19.46}$$

From (19.43), (19.44), and (19.46), the variation of the microgrid's frequency because of this ratio change is

$$\Delta f^t = f_1^{t_2} - f_1^{t_1} = m_1\left(P_1^{t_1} - P_1^{t_2}\right) = P_1^{t_1}m_1\left(1 - \frac{1 + k_{p1}}{1 + k_{p2}}\right) = \frac{k_{p2} - k_{p1}}{1 + k_{p2}}m_1P_1^{t_1} \tag{19.47}$$

This deviation is acceptable if the microgrid's new frequency stays between $f_{\max}$ and $f_{\min}$. Thus,

$$\begin{aligned} & f_{\min} \le f_1^{t_1} + \Delta f^t \le f_{\max} \\ & \Rightarrow f_{\min} - f_1^{t_1} \le \Delta f^t \le f_{\max} - f_1^{t_1} \\ & \Rightarrow f_{\min} - f_1^{t_1} \le \frac{k_{p2} - k_{p1}}{1 + k_{p2}}m_1P_1^{t_1} \le f_{\max} - f_1^{t_1} \end{aligned} \tag{19.48}$$

Assuming

$$\begin{cases} H_1 = \dfrac{f_{\min} - f_1^{t_1}}{m_1 P_1^{t_1}} \\ H_2 = \dfrac{f_{\max} - f_1^{t_1}}{m_1 P_1^{t_1}} \end{cases}$$

to avoid the observance of an unacceptable frequency within the microgrid, k_{p2} must satisfy

$$\begin{cases} k_{p2} > 0 \\ k_{p2} \le \dfrac{k_{p1} + H_1}{1 - H_1} \\ k_{p2} \le \dfrac{k_{p1} + H_2}{1 - H_2} \end{cases} \tag{19.49}$$

Another analysis can be performed for the microgrid's voltage magnitude. Suppose the relationship of the output reactive powers of DER-1 and DER-2 of Figure 19.4 is k_{q1} that at $t = t_1$. Hence, $Q_{\text{rated},2}/Q_{\text{rated},1} = n_1/n_2 = k_{q1}$ according to (19.23). In such a condition, from (19.10),

$$|V_1|^{t_1} = V_{\text{rated}} - n_1 \left(Q_1^{t_1} - Q_{\text{rated},1}\right) \tag{19.50}$$

If at $t = t_2$, the tertiary controller changes this ratio to k_{q2}, from (19.10),

$$|V_1|^{t_2} = V_{\text{rated}} - n_1 \left(Q_1^{t_2} - Q_{\text{rated},1}\right) \tag{19.51}$$

Assuming the load demand is equal at t_1 and t_2, the DERs' total reactive power generation remains unchanged at these instants. Thus,

$$Q_1^{t_1} + Q_2^{t_1} = Q_1^{t_2} + Q_2^{t_2} \tag{19.52}$$

Equation (19.52) can be simplified by replacing the power ratios as

$$\begin{aligned} & Q_1^{t_1} + k_{q1} Q_1^{t_1} = Q_1^{t_2} + k_{q2} Q_1^{t_2} \\ \Rightarrow\ & Q_1^{t_1}(1 + k_{q1}) = Q_1^{t_2}(1 + k_{q2}) \\ \Rightarrow\ & Q_1^{t_2} = Q_1^{t_1} \frac{1 + k_{q1}}{1 + k_{q2}} \end{aligned} \tag{19.53}$$

From (19.50)–(19.52), the variation of the voltage because of this ratio change is

$$\Delta |V_1|^t = |V_1|^{t_2} - |V_1|^{t_1} = n_1 (Q_1^{t_1} - Q_1^{t_2}) = Q_1^{t_1} n_1 (1 - \frac{1 + k_{q1}}{1 + k_{q2}}) = \frac{k_{q2} - k_{q1}}{1 + k_{q2}} n_1 Q_1^{t_1} \tag{19.54}$$

This deviation is acceptable if the microgrid's new voltage magnitude stays within $|V|_{\min}$ and $|V|_{\max}$. Hence,

$$\begin{aligned} & |V|_{\min} \le |V_1|^{t_1} + \Delta |V|^t \le |V|_{\max} \\ \Rightarrow\ & |V|_{\min} - |V_1|^{t_1} \le \Delta |V|^t \le |V|_{\max} - |V_1|^{t_1} \\ \Rightarrow\ & |V|_{\min} - |V_1|^{t_1} \le \frac{k_{q2} - k_{q1}}{1 + k_{q2}} n_1 Q_1^{t_1} \le |V|_{\max} - |V_1|^{t_1} \end{aligned} \tag{19.55}$$

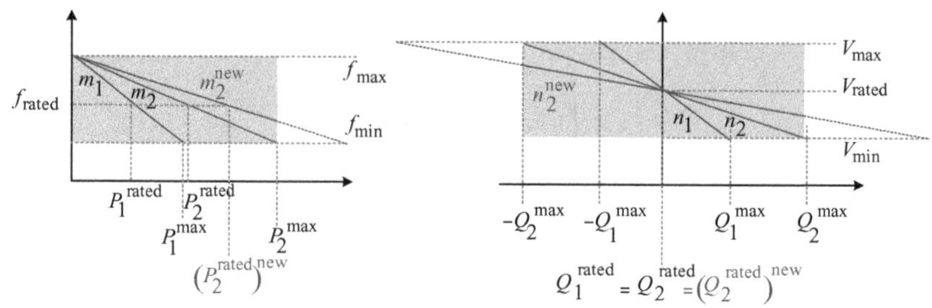

Figure 19.18 Schematic illustration of the adjustment of the droop curve to facilitate power ratio change.

Assuming

$$\begin{cases} G_1 = \dfrac{|V|_{\min} - |V_1|^{t_1}}{n_1 Q_1^{t_1}} \\ G_2 = \dfrac{|V|_{\max} - |V_1|^{t_1}}{n_1 Q_1^{t_1}} \end{cases}$$

to avoid an unacceptable voltage magnitude within the microgrid, k_{q2} must satisfy

$$\begin{cases} k_{q2} > 0 \\ k_{q2} \leq \dfrac{k_{q1} + G_1}{1 - G_1} \\ k_{q2} \leq \dfrac{k_{q1} + G_2}{1 - G_2} \end{cases} \tag{19.56}$$

DERs' f_{rated} and V_{rated} have to be updated accordingly to uphold the microgrid's frequency and voltage magnitude within the acceptable ranges if k_{p2} leads to a violation in them. The principle of the dynamic power ratio update is illustrated schematically in Fig.19.18.

Consider the microgrid of Fig. 19.4 with two DERs at the steady-state condition. Both DERs are connected via a coupling inductance of 6.8 mH. As shown in Figcomp 19.19(a), the ratio of their output active power is preferred to be 2:1 initially. Thus, as shown in Figcomp19.19(b), the ratio of their droop coefficients are determined as $m_{\text{DER-1}}/m_{\text{DER-2}}$ = 1:2. Moreover, $P_{rated,1}/P_{rated,2} = 2$. The tertiary controller alters the ratio of their output active powers to 1:1 and transmits this data to the microgrid's central controller, which changes the ratio of their droop coefficients and rated active powers as $m_{\text{DER-1}}/m_{\text{DER-2}}$ = 1 and $P_{rated,1}/P_{rated,2} = 1$ and then transmits this information to their primary controllers. Likewise, the tertiary controller determines this ratio to be as 1:3 at $t = 1$ s. Therefore, the microgrid's central controller modifies the ratio of their droop coefficients and rated active powers as $m_{\text{DER-1}}/m_{\text{DER-2}} = 3$ and $P_{rated,1}/P_{rated,2} = 1/3$ and transmits this information to their primary controllers. As a result, the preferred ratio of the active powers is observed for the DERs as shown in Figure 19.19(a).

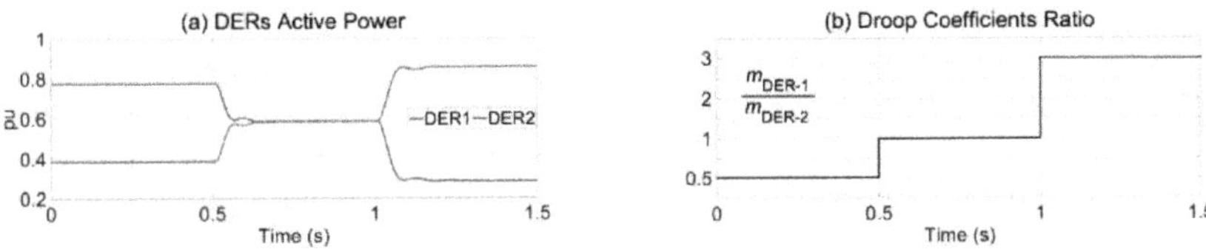

Figure 19.19 Results for the assumed microgrid with variable ratios of power between two DERs: (a) DERS active power and (b) droop coefficients ratio.

19.6.2 Droop Curve Adjustment

For a microgrid operating under the autonomous mode, its central controller has to act accordingly to increase or reduce the rated levels of the droop curve if and when the microgrid's frequency or the voltage magnitude drops below or rises above the permissible limits. It is expected that the microgrid's voltage magnitude and frequency are adjusted to the preferred levels. As such, it is necessary to determine the variation of the microgrid's frequency from its preferred level ($\Delta\omega$) as well as a new maximum frequency as

$$f_{\max}^{new} = f_{\max}^{old} + \Delta f = f_{\max}^{old} + \left(f_{desired} - f\right) \tag{19.57}$$

Likewise, it is necessary to determine the variation of the microgrid's voltage magnitude from its preferred level (ΔV) as well as a new rated voltage as

$$V_{\text{rated}}^{new} = V_{\text{rated}}^{old} + \Delta V = V_{\text{rated}}^{old} + \left(V_{desired} - |V|\right) \tag{19.58}$$

as schematically illustrated in Figure 19.20.

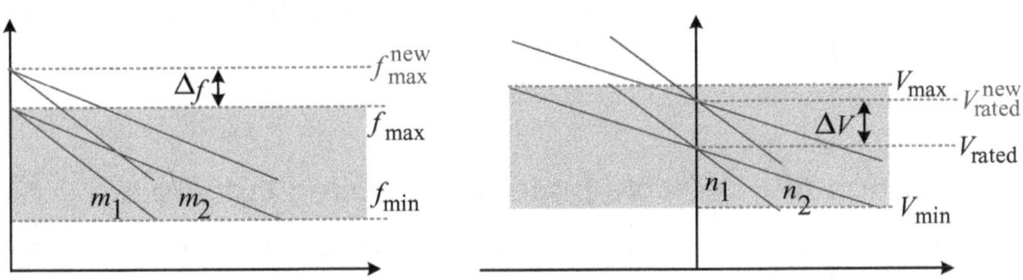

Figure 19.20 Adjustment of the droop curve settings to avoid unacceptable violation in microgrid's frequency and voltage magnitude.

Consider the microgrid of Figure 19.4 at the steady-state condition with a total demand of 0.15 pu. Figure 19.21(a) shows that the ratio of the DERs' output active powers is kept as 2:1, as preferred by the microgrid's central controller. As seen in Figure 19.21(b), the load increases to 0.4 pu at $t = 0.5$ s and consequently, the microgrid's voltage drops to 0.89 pu (below the permissible level of 0.9 pu), as shown in Fig. 19.21(c), whereas as shown in Fig. 19.21(d), the microgrid's frequency remains within the allowed limits of 49 to 51 Hz. As was discussed earlier, the microgrid's central controller, which operates slower than the

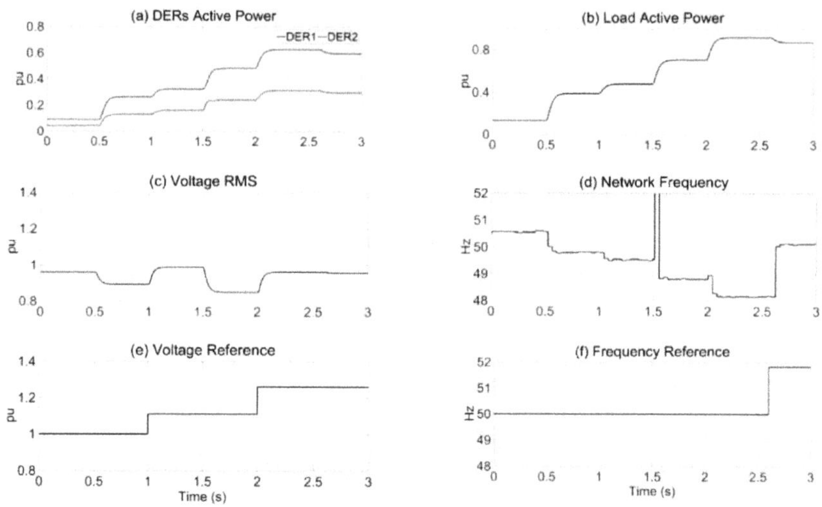

Figure 19.21 Results for the assumed microgrid after droop curve adjustments: (a) DERs active power; (b) load active power; (c) voltage RMS; (d) network frequency; (e) voltage reference; and (f) frequency reference.

DERs' primary controllers, updates V_{rated} to 1.1 pu for both DERs at t = 1 s, as shown in Figure 19.21(e). As a result, the microgrid's voltage increases to 1 pu. The load increases again at t = 2 s, thereby the microgrid's frequency reduces below the permissible level. Therefore, as shown in Figure 19.21(f), the central controller adjusts f_{rated} to 51.9 Hz at t = 2.6 s, which leads to the frequency to restore to 50 Hz t = 2.6 s. As the loads were assumed as constant-impedance type, their active power consumption slightly increases after each increase of V_{rated} or f_{rated}.

19.6.3 Selection of a Suitable Internal Balancing Inductance

Assume a microgrid in which two DERs are sharing the load based on their capacities. Hence, they function with droop coefficients of m_1, m_2, n_1, and n_2. Once the microgrid's central controller receives a new preferred relationship for the DERs, it proceeds to determine a new set of ratio for their droop coefficients while satisfying (19.23). The variation of the droop coefficient for a DER is same as the change of its virtual (assumed) maximum active and reactive power capacities; thereby, a DER only injects a power lower than the minimum of its virtual and actual capacities. In [78], it has been shown that this change will result in changing the ratio of the active power between the DERs, whereas it does not satisfy the change in the ratio of their reactive powers because the latter depends on the DERs' coupling inductances. As discussed before, an internal balancing inductance can be employed by the controller of the DERs' converters to address this problem. For this, the microgrid's central controller first defines an optimal internal balancing inductance for the DER, which has to generate the highest power [80], based on which the suitable internal balancing inductances for the other DERs will then be calculated from (19.30). Figure 19.22(a) shows the flowchart of selecting the internal balancing inductance, whereas the flowchart of selecting the optimal internal balancing inductance for the DER with the largest power generation is shown in Figure 19.22(b). To select the optimum L_{conv}, an objective function can be formulated as

$$OF = \alpha\,\varepsilon_{\mathrm{P}} + \beta\,\varepsilon_{\mathrm{Q}} \tag{19.59}$$

to determine the difference of the power ratios for various values of L_{conv} in which

$$\begin{aligned} \varepsilon_{\mathrm{P}} &= 1 - \frac{P_{\text{ratio}}^{\text{actual}}}{P_{\text{ratio}}^{\text{desired}}} \\ \varepsilon_{\mathrm{Q}} &= 1 - \frac{Q_{\text{ratio}}^{\text{actual}}}{Q_{\text{ratio}}^{\text{desired}}} \end{aligned} \tag{19.60}$$

and α, β (e.g., $\alpha = \beta = 0.5$) are weighting coefficients. It is vital that the selected L_{conv}s do not cause instability for the microgrid. Thus, a small signal stability analysis has to be conducted as well, as denoted within Fig. 19.22(b). Figure 19.23 shows the closed-loop control for the DERs' converters operating under this mechanism.

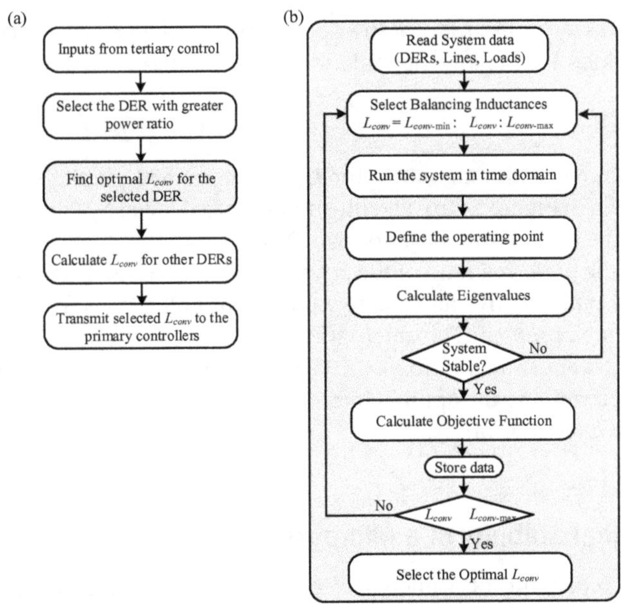

Figure 19.22 Flowchart to select (a) internal balancing inductances for all DERs and (b) Optimal internal balancing inductances.

Consider the microgrid of Figure 19.4 in which the power ratio of DER-1:DER-2 varies from 2 to 10. Figure 19.24(a) illustrates ε_{P} and ε_{Q} for various L_{conv}s from which it can be seen that ε_{Q} decreases significantly as L_{conv} increases, whereas ε_{P} observes a negligible

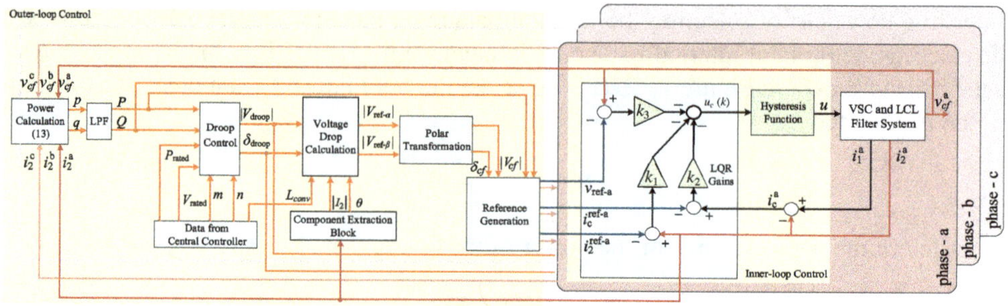

Figure 19.23 The closed-loop control block diagram of the VSC and output passive filter of the DERs operating under dynamic power-sharing ratio.

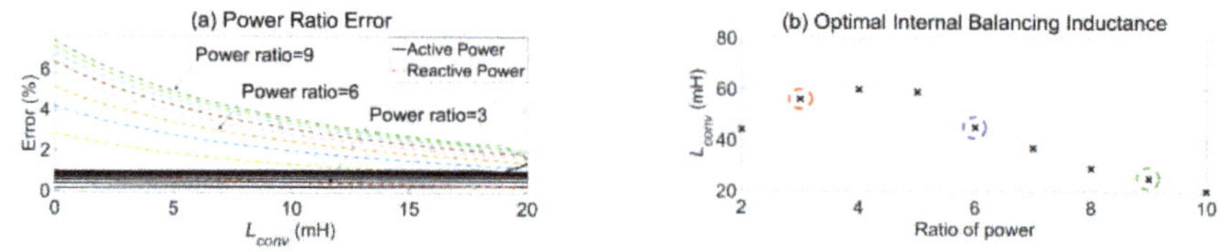

Figure 19.24 (a) Error of the ratio of the active and reactive power and (b) the optimal values for the internal balancing inductance.

increase. The different optimum values for L_{conv}, determined via the flowchart of Figure 19.22(b), are shown in Figure 24(b), and illustrates that a decrease of the optimal L_{conv} leads to an increase in the ratio of the power of two DERs.

Now, assume the preferred ratio of the powers of DER-1:DER-2 as 3:1. Figure 19.25(a) and (b) shows the output active and reactive powers of both DERs from which $\varepsilon_P = 0.3\%$ and $\varepsilon_Q = 0.002\%$ are observed. This microgrid has also been examined for ratios of 6:1 and 9:1 and their results are shown in Figure 19.25(c–f). The studies show that ε_P and ε_Q are below 1% after determining the optimal internal balancing inductances and imposing them to the DERs' converter controllers.

19.7 Small Signal Stability of a Microgrid

The small signal modelling of a microgrid composed of converter-interfaced DERs and constant-impedance type loads is presented in this section. The state-space presentation of DER-i, with the VSC and the output passive filter structure of Fig. 19.3(c), can be expressed in the direct-quadrant domain as [88, 89]

$$\Delta\dot{\mathbf{x}}^i_{\mathrm{DER}} = \mathbf{A}^i\Delta\mathbf{x}^i_{\mathrm{DER}} + \mathbf{B}^i\Delta\mathbf{u}^i_{\mathrm{DER}} + \mathbf{C}^i\Delta\mathbf{v}^i_{\mathrm{DER}} \tag{19.61}$$

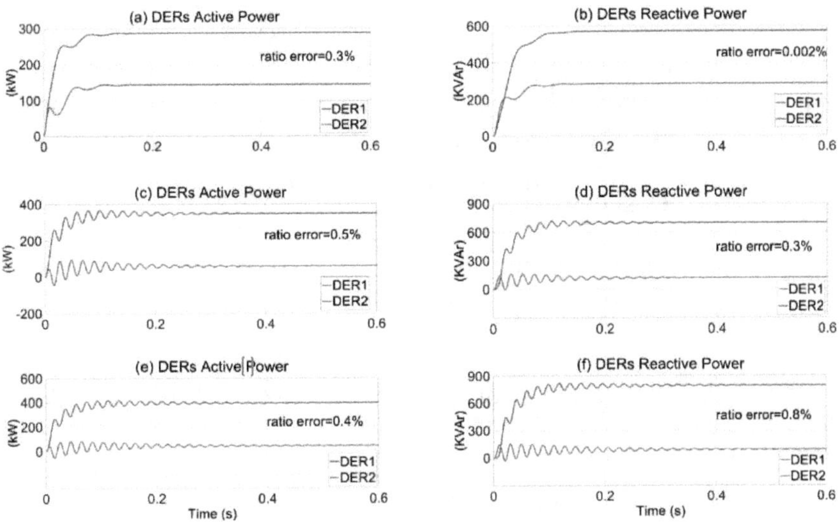

Figure 19.25 The output active (a, c, e) and reactive (b, d, f) powers of the DERs in the assumed microgrid for various ratios of DER1:DER-2.

as expanded in

$$\Delta\begin{bmatrix} i_1^{\mathrm{d}} \\ i_1^{\mathrm{d}} \\ i_2^{\mathrm{d}} \\ i_2^{\mathrm{q}} \\ v_{\mathrm{c}}^{\mathrm{d}} \\ v_{\mathrm{c}}^{\mathrm{q}} \end{bmatrix}^{\bullet} = \begin{bmatrix} -R_{\mathrm{f}}/L_{\mathrm{f}} & \omega & 0 & 0 & -1/L_{\mathrm{f}} & 0 \\ -\omega & -R_{\mathrm{f}}/L_{\mathrm{f}} & 0 & 0 & 0 & -1/L_{\mathrm{f}} \\ 0 & 0 & 0 & \omega & 1/L_{\mathrm{coup}} & 0 \\ 0 & 0 & -\omega & 0 & 0 & 1/L_{\mathrm{coup}} \\ 1/C_{\mathrm{f}} & 0 & -1/C_{\mathrm{f}} & 0 & 0 & \omega \\ 0 & 1/C_{\mathrm{f}} & 0 & -1/C_{\mathrm{f}} & -\omega & 0 \end{bmatrix} \Delta\begin{bmatrix} i_1^{\mathrm{d}} \\ i_1^{\mathrm{d}} \\ i_2^{\mathrm{d}} \\ i_2^{\mathrm{q}} \\ v_{\mathrm{c}}^{\mathrm{d}} \\ v_{\mathrm{c}}^{\mathrm{q}} \end{bmatrix} + \frac{aV_{\mathrm{dc}}}{L_{\mathrm{f}}}\begin{bmatrix} 1 & 0 \\ 0 & 1 \\ 0 & 0 \\ 0 & 0 \\ 0 & 0 \\ 0 & 0 \end{bmatrix} .\Delta\begin{bmatrix} u^{\mathrm{d}} \\ u^{\mathrm{q}} \end{bmatrix} - \frac{1}{L_{\mathrm{T}}}\begin{bmatrix} 0 & 0 \\ 0 & 0 \\ 1 & 0 \\ 0 & 1 \\ 0 & 0 \\ 0 & 0 \end{bmatrix} .\Delta\begin{bmatrix} v_{\mathrm{t}}^{\mathrm{d}} \\ v_{\mathrm{t}}^{\mathrm{q}} \end{bmatrix} \tag{19.62}$$

Suppose a flawless tracking of the references. Then, u_{c} can be characterized by u. Hence, the state-feedback control can be expressed in the direct-quadratic frame as

$$\Delta\mathbf{u}_{DER}^{i} = -\mathbf{K}^{i}[\Delta\mathbf{Z}_{DER} - (\Delta\mathbf{Z}_{DER}^{i})^{ref}] = \mathbf{G}^{i}\Delta\mathbf{x}_{DER}^{i} + \mathbf{H}^{i}(\Delta\mathbf{Z}_{DER}^{i})^{ref} \tag{19.63}$$

where

$$\mathbf{Z}_{\mathrm{DER}} = \left[i_2^{\mathrm{d}} i_2^{\mathrm{q}} i_{\mathrm{c}}^{\mathrm{d}} i_{\mathrm{c}}^{\mathrm{q}} v_{\mathrm{c}}^{\mathrm{d}} v_{\mathrm{c}}^{\mathrm{q}}\right]^T$$
$$\mathbf{K} = \begin{bmatrix} k_1 & k_1 & k_2 & k_2 & k_3 & k_3 \end{bmatrix}$$
$$\mathbf{G} = \begin{bmatrix} -k_2 & 0 & k_2 - k_1 & 0 & -k_3 & 0 \\ 0 & -k_2 & 0 & k_2 - k_1 & 0 & -k_3 \end{bmatrix}$$
$$\mathbf{H} = \begin{bmatrix} k_1 & 0 & k_2 & 0 & k_3 & 0 \\ 0 & k_1 & 0 & k_2 & 0 & k_3 \end{bmatrix}$$

From (19.61) and (19.63), the state-space equation of DER-i's VSC and its output passive filter system along with its inner-loop control becomes

$$\Delta\dot{\mathbf{x}}^i_{\mathrm{DER}} = \mathbf{A}^i_{\mathrm{DER}}\,\Delta\mathbf{x}^i_{DER} + \mathbf{B}^i_{\mathrm{DER}}(\Delta\mathbf{Z}^i_{\mathrm{DER}})^{ref} + \mathbf{C}^i_{\mathrm{DER}}\Delta\mathbf{v}^i_{\mathrm{DER}} \tag{19.64}$$

where

$$\mathbf{A}^i_{\mathrm{DER}} = \mathbf{A}^i + \mathbf{B}^i\mathbf{G}^i$$
$$\mathbf{B}^i_{\mathrm{DER}} = \mathbf{B}^i\mathbf{H}^i$$

As the BESs are assumed to have the same VSC and the output passive filter and closed-loop control systems, BES-i's state-space equation becomes

$$\Delta\dot{\mathbf{x}}^i_{\mathrm{BES}} = \mathbf{A}^i_{\mathrm{BES}}\,\Delta\mathbf{x}^i_{\mathrm{BES}} + \mathbf{B}^i_{\mathrm{BES}}(\Delta\mathbf{Z}^i_{\mathrm{BES}})^{ref} + \mathbf{C}^i_{\mathrm{BES}}\Delta\mathbf{v}^i_{\mathrm{BES}} \tag{19.65}$$

Note that if another structure or controller is used for the DERs' or BESs' VSCs, the relevant state-space presentation should be used.

Although a droop-based load or a load managed under demand side management can be assumed in microgrids [90, 91], the above studies considered the microgrid's loads as constant-impedance RL type. Thus, load-j's state-space equation when connected to microgrid's bus-jj is given as [88, 89]

$$\Delta\dot{\mathbf{x}}^j_{load} = \mathbf{A}^j_{load}\,\Delta\mathbf{x}^j_{load} + \mathbf{C}^j_{load}\Delta\mathbf{v}^{jj}_{load} \tag{19.66}$$

as expanded in

$$\Delta\begin{bmatrix} i^{\mathrm{d}} \\ i^{\mathrm{q}} \end{bmatrix}^{\bullet} = \begin{bmatrix} -\dfrac{R_{load}}{L_{load}} & \omega \\ -\omega & -\dfrac{R_{load}}{L_{load}} \end{bmatrix} .\Delta\begin{bmatrix} i^{\mathrm{d}} \\ i^{\mathrm{q}} \end{bmatrix} + \frac{1}{L_{load}}\mathbf{I}_{2\times 2}.\Delta\begin{bmatrix} v_{\mathrm{t}}^{\mathrm{d}} \\ v_{\mathrm{t}}^{\mathrm{q}} \end{bmatrix} \tag{19.67}$$

where $I_{2\times2}$ denotes order-2 identity matrix.

Likewise, the lines are assumed as constant-impedance RL type branches. Hence, line-k's state-space equation, when connected between bus kk and ll, is represented by [88, 89]

$$\Delta\dot{\mathbf{x}}_{line}^{k} = \mathbf{A}_{line}^{k}\,\Delta\mathbf{x}_{line}^{k} + \mathbf{C}_{line}^{k}\,\Delta\mathbf{v}_{MG} \tag{19.68}$$

as detailed in

$$\Delta\begin{bmatrix} i^{d} \\ i^{q} \end{bmatrix}^{\bullet} = \begin{bmatrix} -\dfrac{R_{line}}{L_{line}} & \omega \\ -\omega & -\dfrac{R_{line}}{L_{line}} \end{bmatrix} .\Delta\begin{bmatrix} i^{d} \\ i^{q} \end{bmatrix} + \frac{1}{L_{line}}\begin{bmatrix} \overbrace{\mathbf{O}_{2\times2}\ldots\mathbf{O}_{2\times2}}^{\text{bus}(1)\text{ to bus}(kk-1)}\ \overbrace{\mathbf{I}_{2\times2}}^{\text{bus}(kk)}\ \overbrace{\mathbf{O}_{2\times2}\ldots\mathbf{O}_{2\times2}}^{\text{bus}(kk+1)\text{ to bus}(ll-1)} \\ \overbrace{-\mathbf{I}_{2\times2}}^{\text{bus}(ll)}\ \overbrace{\mathbf{O}_{2\times2}\ldots\mathbf{O}_{2\times2}}^{\text{bus}(ll+1)\text{ to bus}(N_{\text{bus}})} \end{bmatrix} \times \Delta\begin{bmatrix} \overbrace{(v_{t}^{d})^{1}\,(v_{t}^{q})^{1}}^{\text{bus}(1)}\ldots\overbrace{(v_{t}^{d})^{kk-1}\,(v_{t}^{q})^{kk-1}}^{\text{bus}(kk-1)}\overbrace{(v_{t}^{d})^{kk}\,(v_{t}^{q})^{kk}}^{\text{bus}(kk)}\overbrace{(v_{t}^{d})^{kk+1}\,(v_{t}^{q})^{kk+1}}^{\text{bus}(kk+1)} \\ \ldots\overbrace{(v_{t}^{d})^{ll-1}\,(v_{t}^{q})^{ll-1}}^{\text{bus}(ll-1)}\overbrace{(v_{t}^{d})^{ll}\,(v_{t}^{q})^{ll}}^{\text{bus}(ll)}\overbrace{(v_{t}^{d})^{ll+1}\,(v_{t}^{q})^{ll+1}}^{\text{bus}(ll+1)} \\ \ldots\overbrace{(v_{t}^{d})^{N_{bus}}\,(v_{t}^{q})^{N_{bus}}}^{\text{bus}(N_{\text{bus}})} \end{bmatrix}^{T} \tag{19.69}$$

where $\boldsymbol{O}_{2\times2}$ represents a matrix of zeros.

From (19.64) to (19.66) and (19.68), the microgrid's overall state-space presentation becomes

$$\Delta\dot{\mathbf{x}}_{\text{MG}}^{1} = \mathbf{A}_{\text{MG}}^{1}\Delta\mathbf{x}_{\text{MG}}^{1} + \mathbf{B}_{\text{MG}}^{1}(\Delta\mathbf{Z}_{\text{MG}}^{1})^{\text{ref}} + \mathbf{C}_{\text{MG}}^{1}\,\Delta\mathbf{v}_{MG} \tag{19.70}$$

where

$$\mathbf{x}_{\mathrm{MG}}^{1}=\left[\overbrace{\mathbf{x}_{\mathrm{DER}}^{1}\cdots\mathbf{x}_{\mathrm{DER}}^{N_{\mathrm{DER}}}}^{\text{DERs}}\ \overbrace{\mathbf{x}_{\mathrm{BES}}^{1}\cdots\mathbf{x}_{\mathrm{BES}}^{N_{\mathrm{BES}}}}^{\text{BES Systems}}\ \overbrace{\mathbf{x}_{\mathrm{load}}^{1}\cdots\mathbf{x}_{\mathrm{load}}^{N_{\mathrm{load}}}}^{\text{Loads}}\ \overbrace{\mathbf{x}_{\mathrm{line}}^{1}\cdots\mathbf{x}_{\mathrm{line}}^{N_{\mathrm{line}}}}^{\text{Lines}}\right]^{T}$$

$$\mathbf{Z}_{\mathrm{MG}}^{1}=\left[\overbrace{\mathbf{Z}_{\mathrm{DER}}^{1}\cdots\mathbf{Z}_{\mathrm{DER}}^{N_{\mathrm{DER}}}}^{\text{DERs}}\ \overbrace{\mathbf{Z}_{\mathrm{BES}}^{1}\cdots\mathbf{Z}_{\mathrm{BES}}^{N_{\mathrm{BES}}}}^{\text{BES Systems}}\right]^{T}$$

$$\mathbf{A}_{\mathrm{MG}}^{1}=\mathrm{diag}\left(\overbrace{\mathbf{A}_{\mathrm{DER}}^{1},\ldots,\mathbf{A}_{\mathrm{DER}}^{N_{\mathrm{DER}}}}^{\text{DERs}},\overbrace{\mathbf{A}_{\mathrm{BES}}^{1},\ldots,\mathbf{A}_{\mathrm{BES}}^{N_{\mathrm{BES}}}}^{\text{BES Systems}},\overbrace{\mathbf{A}_{\mathrm{load}}^{1},\ldots,\mathbf{A}_{\mathrm{load}}^{N_{\mathrm{load}}}}^{\text{Loads}},\overbrace{\mathbf{A}_{\mathrm{line}}^{1},\ldots,\mathbf{A}_{\mathrm{line}}^{N_{\mathrm{line}}}}^{\text{Lines}}\right)$$

$$\mathbf{B}_{\mathrm{MG}}^{1}=\left[\begin{array}{c}\mathrm{diag}\left(\overbrace{\mathbf{B}_{\mathrm{DER}}^{1},\ldots,\mathbf{B}_{\mathrm{DER}}^{N_{\mathrm{DER}}}}^{\text{DERs}},\overbrace{\mathbf{B}_{\mathrm{BES}}^{1},\ldots,\mathbf{B}_{\mathrm{BES}}^{N_{\mathrm{BES}}}}^{\text{BES Systems}}\right)\\\hline\mathbf{O}_{2(N_{\mathrm{load}}+N_{\mathrm{line}})\times 6(N_{\mathrm{DER}}+N_{\mathrm{BES}})}\end{array}\right]$$

$$\mathbf{C}_{\mathrm{MG}}^{1}=\left[\begin{array}{c}\mathrm{diag}\left(\overbrace{\mathbf{C}_{\mathrm{DER}}^{1},\ldots,\mathbf{C}_{\mathrm{DER}}^{N_{\mathrm{DER}}}}^{\text{DERs}},\overbrace{\mathbf{C}_{\mathrm{BES}}^{1},\ldots,\mathbf{C}_{\mathrm{BES}}^{N_{\mathrm{BES}}}}^{\text{BES Systems}},\overbrace{\mathbf{C}_{\mathrm{load}}^{1},\ldots,\mathbf{C}_{\mathrm{load}}^{N_{\mathrm{load}}}}^{\text{Loads}}\right)\\\hline\mathbf{C}_{\mathrm{line}}^{1}\\\vdots\\\mathbf{C}_{\mathrm{line}}^{N_{\mathrm{line}}}\end{array}\right]$$

in which diag(x,y,…) denotes a diagonal matrix with x,y,… at its diagonal elements.

A DER's instantaneous output active and reactive powers can be given in direct-quadratic domain by [88, 89]

$$\begin{bmatrix}p\\q\end{bmatrix}=\frac{3}{2}\begin{bmatrix}i_{2}^{\mathrm{d}} & i_{2}^{\mathrm{q}}\\-i_{2}^{\mathrm{q}} & i_{2}^{\mathrm{d}}\end{bmatrix}\begin{bmatrix}v_{\mathrm{c}}^{\mathrm{d}}\\v_{\mathrm{c}}^{\mathrm{q}}\end{bmatrix}\tag{19.71}$$

Suppose a low-pass filter filters frequencies above$_{\mathrm{c}}$. Then, the linearized average active and reactive powers around DER-i's equilibrium (i.e., operating) point (denoted by subscript $_{\mathrm{o}}$) become

$$\Delta\dot{\mathbf{p}}_{\mathrm{DER}}^{i}=\mathbf{A}_{\mathrm{P-DER}}^{i}\,\Delta\mathbf{p}_{\mathrm{DER}}^{i}+\mathbf{B}_{\mathrm{P-DER}}^{i}\,\Delta\mathbf{x}_{\mathrm{DER}}^{i}\tag{19.72}$$

$$\Delta\dot{\mathbf{q}}_{\mathrm{DER}}^{i}=\mathbf{A}_{\mathrm{Q-DER}}^{i}\,\Delta\mathbf{q}_{\mathrm{DER}}^{i}+\mathbf{B}_{\mathrm{Q-DER}}^{i}\,\Delta\mathbf{x}_{\mathrm{DER}}^{i}\tag{19.73}$$

as expanded in

$$\dot{P} = -\omega_c P + \frac{3}{2}\omega_c \left[0 \; 0 \; (v_c^d)_o \; (v_c^q)_o \; (i_2^d)_o \; (i_2^q)_o \right] \times \left[i_1^d \; i_1^q \; i_2^d \; i_2^q \; v_c^d \; v_c^q \right]^T$$
$$\dot{Q} = -\omega_c Q + \frac{3}{2}\omega_c \left[0 \; 0 \; (v_c^q)_o \; -(v_c^d)_o \; -(i_2^q)_o \; (i_2^d)_o \right] \times \left[i_1^d \; i_1^q \; i_2^d \; i_2^q \; v_c^d \; v_c^q \right]^T \tag{19.74}$$

It is to be noted that (19.72) and (19.73) are applicable for BES-i. Thus,

$$\Delta \dot{\mathbf{p}}_{BES}^i = \mathbf{A}_{\text{P-BES}}^i \, \Delta \mathbf{p}_{BES}^i + \mathbf{B}_{\text{P-BES}}^i \, \Delta \mathbf{x}_{BES}^i \tag{19.75}$$

$$\Delta \dot{\mathbf{q}}_{BES}^i = \mathbf{A}_{\text{Q-BES}}^i \, \Delta \mathbf{q}_{BES}^i + \mathbf{B}_{\text{Q-BES}}^i \, \Delta \mathbf{x}_{BES}^i \tag{19.76}$$

In addition, the voltage angle at the output of a DER's VSC can be computed from the converter's angular frequency of $\omega = 2f$ versus a common angular frequency of $\omega_{\text{com}} = m_{\text{com}} P_{\text{com}}$ as [88, 89]

$$\delta = \int (\omega - \omega_{\text{com}})\, dt \tag{19.77}$$

in which $_{\text{com}}$ is the angular frequency of one of the converters (e.g., $_1$). Differentiating (19.77) and linearizing it,

$$\Delta\dot{\delta} = \Delta\omega - \Delta\omega_{\text{com}} \tag{19.78}$$

On the other hand, from (19.10), one can get

$$\omega = \omega^{\max} - m'' P \tag{19.79}$$

in which $m'' = 2\pi\, m'$ (in rad/W). Now, differentiating (19.79) and linearizing it,

$$\Delta\dot{\omega} = -m''\Delta P \tag{19.80}$$

Replacing (19.80) in (19.78) yields

$$\Delta\dot{\delta}_{\text{DER}}^i = \mathbf{A}_{\delta\text{-DER}}^i \, \Delta \mathbf{p}_{DER}^i + \mathbf{D}_{\delta\text{-DER}}^i P_{\text{com}} \tag{19.81}$$

as expanded in

$$\Delta\dot{\delta} = -m''\Delta P + m''_{\text{com}} \Delta P_{\text{com}} \tag{19.82}$$

Again, it is to be noted that (19.81) is valid for BES-i. Thus,

$$\Delta\dot{\delta}_{BES}^i = \mathbf{A}_{\delta\text{-BES}}^i \, \Delta \mathbf{p}_{BES}^i + \mathbf{D}_{\delta\text{-BES}}^i P_{\text{com}} \tag{19.83}$$

Moreover, because a BES's SoC is calculated as [92]

$$SoC = SoC^{\text{int}} - \frac{\int p dt}{E^{\text{cap}}} \tag{19.84}$$

where p, SoC^{int}, and E^{cap} correspondingly denote the BES's instantaneous output active power, initial SoC, and storage capacity. Linearizing (19.84), one can get

$$\Delta \mathbf{S\dot{o}C}_{BES}^i = \mathbf{B}_{\text{SoC-BES}}^i \, \Delta \mathbf{x}_{BES}^i \tag{19.85}$$

as expanded in

$$S\dot{o}C = \frac{-3}{2E^{\mathrm{cap}}}\left[0\ 0\ (v_c^d)_o\ (v_c^q)_o\ (i_2^d)_o\ (i_2^q)_o\right] \times \left[i_1^d\ i_1^q\ i_2^d\ i_2^q\ v_c^d\ v_c^q\right]^T \tag{19.86}$$

Combining (19.72), (19.73), (19.75), (19.76), (19.78), (19.82), and (19.84) yields

$$\Delta\dot{\mathbf{x}}_{MG}^2 = \mathbf{A}'^{\,2}_{MG}\Delta\mathbf{x}_{MG}^2 + \mathbf{B}_{MG}^2\,\Delta\mathbf{x}_{MG}^1 + \mathbf{D}_{MG}^2\,P_{\mathrm{com}} \tag{19.87}$$

in which

$$\mathbf{x}_{MG}^2 = \left[\overbrace{\mathbf{p}_{\mathrm{DER}}^1 \cdots \mathbf{p}_{\mathrm{DER}}^{N_{\mathrm{DER}}}}^{\text{DERs}}\ \overbrace{\mathbf{p}_{\mathrm{BES}}^1 \cdots \mathbf{p}_{\mathrm{BES}}^{N_{\mathrm{BES}}}}^{\text{BES Systems}}\ \overbrace{\mathbf{q}_{\mathrm{DER}}^1 \cdots \mathbf{q}_{\mathrm{DER}}^{N_{\mathrm{DER}}}}^{\text{DERs}}\ \overbrace{\mathbf{q}_{\mathrm{BES}}^1 \cdots \mathbf{q}_{\mathrm{BES}}^{N_{\mathrm{BES}}}}^{\text{BES Systems}}\ \overbrace{\delta_{\mathrm{DER}}^1 \cdots \delta_{\mathrm{DER}}^{N_{\mathrm{DER}}}}^{\text{DERs}}\ \overbrace{\delta_{\mathrm{BES}}^1 \cdots \delta_{\mathrm{BES}}^{N_{\mathrm{BES}}}}^{\text{BES Systems}}\ \overbrace{\mathbf{SoC}_{\mathrm{BES}}^1 \cdots \mathbf{SoC}_{\mathrm{BES}}^{N_{\mathrm{BES}}}}^{\text{BES Systems}}\right]^T$$

$$\mathbf{A}'^{\,2}_{MG} = \mathrm{diag}\left(\overbrace{\mathbf{A}_{\mathrm{P-DER}}^1, \ldots, \mathbf{A}_{\mathrm{P-DER}}^{N_{\mathrm{DER}}}}^{\text{DERs}}, \overbrace{\mathbf{A}_{\mathrm{P-BES}}^1, \ldots, \mathbf{A}_{\mathrm{P-BES}}^{N_{\mathrm{BES}}}}^{\text{BES Systems}}, \overbrace{\mathbf{A}_{\mathrm{Q-DER}}^1, \ldots, \mathbf{A}_{\mathrm{Q-DER}}^{N_{\mathrm{DER}}}}^{\text{DERs}}, \overbrace{\mathbf{A}_{\mathrm{Q-BES}}^1, \ldots, \mathbf{A}_{\mathrm{Q-BES}}^{N_{\mathrm{BES}}}}^{\text{BES Systems}}, \overbrace{\mathbf{A}_{\delta\mathrm{-DER}}^1, \ldots, \mathbf{A}_{\delta\mathrm{-DER}}^{N_{\mathrm{DER}}}}^{\text{DERs}}, \overbrace{\mathbf{A}_{\delta\mathrm{-BES}}^1, \ldots, \mathbf{A}_{\delta\mathrm{-BES}}^{N_{\mathrm{BES}}}}^{\text{BES Systems}}, \overbrace{\mathbf{O}_{N_{\mathrm{BES}}\times N_{\mathrm{BES}}}}^{\text{BES Systems}}\right)$$

$$\mathbf{B}_{MG}^2 = \mathrm{diag}\left(\overbrace{\mathbf{B}_{\mathrm{P-DER}}^1, \ldots, \mathbf{B}_{\mathrm{P-DER}}^{N_{\mathrm{DER}}}}^{\text{DERs}}, \overbrace{\mathbf{B}_{\mathrm{P-BES}}^1, \ldots, \mathbf{B}_{\mathrm{P-BES}}^{N_{\mathrm{BES}}}}^{\text{BES Systems}}, \overbrace{\mathbf{B}_{\mathrm{Q-DER}}^1, \ldots, \mathbf{B}_{\mathrm{Q-DER}}^{N_{\mathrm{DER}}}}^{\text{DERs}}, \overbrace{\mathbf{B}_{\mathrm{Q-BES}}^1, \ldots, \mathbf{B}_{\mathrm{Q-BES}}^{N_{\mathrm{BES}}}}^{\text{BES Systems}}, \overbrace{\mathbf{O}_{(N_{\mathrm{DER}}+N_{\mathrm{BES}})\times 6(N_{\mathrm{DER}}+N_{\mathrm{BES}})}}^{\text{DERs \& BES Systems}}, \overbrace{\mathbf{B}_{\mathrm{SoC-BES}}^1, \ldots, \mathbf{B}_{\mathrm{SoC-BES}}^{N_{\mathrm{BES}}}}^{\text{BES Systems}}\right)$$

$$\mathbf{D}_{MG}^2 = \left[\mathbf{O}_{2(N_{\mathrm{DER}}+N_{\mathrm{BES}})\times 1}\ \overbrace{\mathbf{D}_{\delta\mathrm{-DER}}^1 \cdots \mathbf{D}_{\delta\mathrm{-DER}}^{N_{\mathrm{DER}}}}^{\text{DERs}}\ \overbrace{\mathbf{D}_{\delta\mathrm{-BES}}^1 \cdots \mathbf{D}_{\delta\mathrm{-BES}}^{N_{\mathrm{BES}}}}^{\text{BES Systems}}\ \mathbf{O}_{N_{\mathrm{BES}}\times 1}\right]^T$$

Because P_1 is assumed to be P_{com} in the above discussions and is the first element in $\mathbf{x}_{MG}^2$, thereby, $\mathbf{D}_{MG}^2$ has to be added to the first column of $\mathbf{A}'^{\,2}_{MG}$ to form $\mathbf{A}_{MG}^2$. Thus, (19.85) changes to

$$\Delta\dot{\mathbf{x}}_{MG}^2 = \mathbf{A}_{MG}^2\Delta\mathbf{x}_{MG}^2 + \mathbf{B}_{MG}^2\,\Delta\mathbf{x}_{MG}^1 \tag{19.88}$$

Transforming the time-domain references to the common direct-quadratic domain and linearizing them, $(\Delta\mathbf{Z}^i_{\mathrm{DER}})^{\mathrm{ref}}$ becomes [89]

$$(\Delta\mathbf{Z}^i_{\mathrm{DER}})^{\mathrm{ref}} = \mathbf{M}^i_{\mathrm{DER}}\Delta\begin{bmatrix}\mathbf{p}^i_{DER}\\ \mathbf{q}^i_{DER}\\ \delta^i_{DER}\end{bmatrix} \tag{19.89}$$

where

$$\mathbf{M}^i_{\mathrm{DER}} = \begin{bmatrix}\gamma_{11} & \gamma_{12} & \gamma_{13}\\ \gamma_{21} & \gamma_{22} & \gamma_{23}\\ 0 & -C_f n\omega\cos\delta_{\mathrm{o}} & -C_f V_{\mathrm{cmo}}\omega\sin\delta_{\mathrm{o}}\\ 0 & -C_f n\omega\sin\delta_{\mathrm{o}} & C_f V_{\mathrm{cmo}}\omega\cos\delta_{\mathrm{o}}\\ 0 & -n\sin\delta_{\mathrm{o}} & V_{\mathrm{cmo}}\cos\delta_{\mathrm{o}}\\ 0 & n\cos\delta_{\mathrm{o}} & V_{\mathrm{cmo}}\sin\delta_{\mathrm{o}}\end{bmatrix}$$

in which

$$\begin{cases}\gamma_{11} = 2\sin\delta_{\mathrm{o}}/3V_{\mathrm{cmo}}\\ \gamma_{12} = -2\left[V_{\mathrm{cmo}}\cos\delta_{\mathrm{o}} + n\left(Q_{\mathrm{o}}\cos\delta_{\mathrm{o}} - P_{\mathrm{o}}\sin\delta_{\mathrm{o}}\right)\right]/3V^2_{\mathrm{cmo}}\\ \gamma_{13} = 2\left(Q_{\mathrm{o}}\sin\delta_{\mathrm{o}} + P_{\mathrm{o}}\cos\delta_{\mathrm{o}}\right)/3V_{\mathrm{cmo}}\\ \gamma_{21} = -2\cos\delta_{\mathrm{o}}/3V_{\mathrm{cmo}}\\ \gamma_{22} = -2\left[V_{\mathrm{cmo}}\sin\delta_{\mathrm{o}} + n\left(Q_{\mathrm{o}}\sin\delta_{\mathrm{o}} + P_{\mathrm{o}}\cos\delta_{\mathrm{o}}\right)\right]/3V^2_{\mathrm{cmo}}\\ \gamma_{23} = -2\left(Q_{\mathrm{o}}\cos\delta_{\mathrm{o}} - P_{\mathrm{o}}\sin\delta_{\mathrm{o}}\right)/3V_{\mathrm{cmo}}\end{cases}$$

In a similar manner, $(\Delta\mathbf{Z}^i_{\mathrm{BES}})^{\mathrm{ref}}$ for the VSC and output passive filter of BES-i becomes

$$(\Delta\mathbf{z}^i_{\mathrm{BES}})^{\mathrm{ref}} = \mathbf{M}^i_{\mathrm{BES}}\,\Delta\begin{bmatrix}\mathbf{p}^i_{\mathrm{BES}}\\ \mathbf{q}^i_{\mathrm{BES}}\\ \delta^i_{\mathrm{BES}}\end{bmatrix} \tag{19.90}$$

Let us reshape $\mathbf{M}^i_{\mathrm{DER}}$ and $\mathbf{M}^i_{\mathrm{BES}}$ of (19.89) and (19.90) as

$$\begin{aligned}\mathbf{M}^i_{\mathrm{DER}} &= \left[\mathbf{M}^i_{1-\mathrm{DER}}\;\; \mathbf{M}^i_{2-\mathrm{DER}}\;\; \mathbf{M}^i_{3-\mathrm{DER}}\right]\\ \mathbf{M}^i_{\mathrm{BES}} &= \left[\mathbf{M}^i_{1-\mathrm{BES}}\;\; \mathbf{M}^i_{2-\mathrm{BES}}\;\; \mathbf{M}^i_{3-\mathrm{BES}}\right]\end{aligned} \tag{19.91}$$

Merging (19.89) and (19.90), $(\Delta\mathbf{Z}_{\mathrm{MG}})^{\mathrm{ref}}$ can be given as a function of $\mathbf{x}^2_{\mathrm{MG}}$ as

$$(\Delta\mathbf{Z}_{\mathrm{MG}})^{\mathrm{ref}} = \mathbf{M}\,\Delta\mathbf{x}^2_{\mathrm{MG}} \tag{19.92}$$

where

$$\mathbf{M} = \begin{bmatrix} \mathrm{diag}\left(\overbrace{\mathbf{M}^{1}_{1-\mathrm{DER}}, \ldots, \mathbf{M}^{N_{\mathrm{DER}}}_{1-\mathrm{DER}}}^{\mathrm{DERs}}\right) & \mathrm{diag}\left(\overbrace{\mathbf{M}^{1}_{1-\mathrm{BES}}, \ldots, \mathbf{M}^{N_{\mathrm{BES}}}_{1-\mathrm{BES}}}^{\mathrm{BES\ Systems}}\right) \\ \mathrm{diag}\left(\overbrace{\mathbf{M}^{1}_{2-\mathrm{DER}}, \ldots, \mathbf{M}^{N_{\mathrm{DER}}}_{2-\mathrm{DER}}}^{\mathrm{DERs}}\right) & \mathrm{diag}\left(\overbrace{\mathbf{M}^{1}_{2-\mathrm{BES}}, \ldots, \mathbf{M}^{N_{\mathrm{BES}}}_{2-\mathrm{BES}}}^{\mathrm{BES\ Systems}}\right) \\ \mathrm{diag}\left(\overbrace{\mathbf{M}^{1}_{3-\mathrm{DER}}, \ldots, \mathbf{M}^{N_{\mathrm{DER}}}_{3-\mathrm{DER}}}^{\mathrm{DERs}}\right) & \mathrm{diag}\left(\overbrace{\mathbf{M}^{1}_{3-\mathrm{BES}}, \ldots, \mathbf{M}^{N_{\mathrm{BES}}}_{3-\mathrm{BES}}}^{\mathrm{BES\ Systems}}\right) \end{bmatrix}$$

Replacing (19.92) in (19.70) yields

$$\Delta\dot{\mathbf{x}}^{1}_{\mathrm{MG}} = \mathbf{A}^{1}_{\mathrm{MG}}\Delta\mathbf{x}^{1}_{\mathrm{MG}} + \mathbf{B}^{1}_{\mathrm{MG}}\mathbf{M}\,\Delta\mathbf{x}^{2}_{\mathrm{MG}} + \mathbf{C}^{1}_{\mathrm{MG}}\,\Delta\mathbf{v}_{MG} \tag{19.93}$$

Combining (19.88) and (19.93), the microgrid's full linearized state-space presentation becomes

$$\dot{\mathbf{x}}_{\mathrm{MG}} = \mathbf{A}_{\mathrm{MG}}\mathbf{x}_{\mathrm{MG}} + \mathbf{C}_{\mathrm{MG}}\mathbf{v}_{\mathrm{MG}} \tag{19.94}$$

in which

$$\mathbf{x}_{\mathrm{MG}} = \left[\mathbf{x}^{1}_{\mathrm{MG}}\ \mathbf{x}^{2}_{\mathrm{MG}}\right]^{T}$$

$$\mathbf{A}_{\mathrm{MG}} = \begin{bmatrix} \mathbf{A}^{1}_{\mathrm{MG}} & \mathbf{B}^{1}_{\mathrm{MG}}\mathbf{M} \\ \mathbf{B}^{2}_{\mathrm{MG}} & \mathbf{A}^{2}_{\mathrm{MG}} \end{bmatrix}$$

$$\mathbf{C}_{\mathrm{MG}} = \begin{bmatrix} \mathbf{C}^{1}_{\mathrm{MG}} \\ \mathbf{O}_{(3N_{\mathrm{DER}}+4N_{\mathrm{BES}})\times 2N_{\mathrm{bus}}} \end{bmatrix}$$

Presuming a virtual resistor of $R_{\mathrm{v}} = 10^4$ between microgrid's every bus and the ground, bus-*jj*'s voltage can be calculated from [88, 89].

$$\Delta\mathbf{v}^{jj}_{\mathrm{bus}} = \Delta\begin{bmatrix} v^{\mathrm{d}} \\ v^{\mathrm{q}} \end{bmatrix}^{jj}_{\mathrm{bus}} = R_{\mathrm{v}} \times \left\{ \begin{matrix} \sum_{j=1}^{N_{\mathrm{DER}}} \mu^{\mathrm{DER}-j}\Delta\begin{bmatrix} i^{\mathrm{d}}_{2} \\ i^{\mathrm{q}}_{2} \end{bmatrix}^{j}_{\mathrm{DER}} + \sum_{j=1}^{N_{\mathrm{BES}}} \mu^{\mathrm{BES}-j}\Delta\begin{bmatrix} i^{\mathrm{d}}_{2} \\ i^{\mathrm{q}}_{2} \end{bmatrix}^{j}_{\mathrm{BES}} \\ + \sum_{j=1}^{N_{\mathrm{load}}} \mu^{\mathrm{load}-j}\Delta\begin{bmatrix} i^{\mathrm{d}} \\ i^{\mathrm{q}} \end{bmatrix}^{j}_{\mathrm{load}} + \sum_{k=1}^{N_{\mathrm{line}}} \mu^{\mathrm{line}-k}\Delta\begin{bmatrix} i^{\mathrm{d}} \\ i^{\mathrm{q}} \end{bmatrix}^{k}_{\mathrm{line}} \end{matrix} \right\} \tag{19.95}$$

where

$$\mu^{\text{DER}-j} = \begin{cases} \mathbf{I}_{2\times2} & \text{for DER} - j \text{ connected to bus} - jj \\ \mathbf{O}_{2\times2} & \text{otherwise} \end{cases}$$

$$\mu^{\text{BES}-j} = \begin{cases} \mathbf{I}_{2\times2} & \text{for BES} - j \text{ connected to bus} - jj \text{ and discharging} \\ -\mathbf{I}_{2\times2} & \text{for BES} - j \text{ connected to bus} - jj \text{ and charging} \\ \mathbf{O}_{2\times2} & \text{otherwise} \end{cases}$$

$$\mu^{\text{load}-j} = \begin{cases} -\mathbf{I}_{2\times2} & \text{for load} - j \text{ connected to bus} - jj \\ \mathbf{O}_{2\times2} & \text{otherwise} \end{cases}$$

$$\mu^{\text{line}-k} = \begin{cases} \mathbf{I}_{2\times2} & \text{for line} - k \text{ entering bus} - jj \\ -\mathbf{I}_{2\times2} & \text{for BES} - k \text{ exiting bus} - jj \\ \mathbf{O}_{2\times2} & \text{otherwise} \end{cases}$$

By calculating (19.94) for all buses, v_{MG} can be given as a function of x_{MG} as

$$\Delta\mathbf{v}_{\text{MG}} = \mathbf{N}_{\text{MG}}\,\mathbf{x}_{\text{MG}} \tag{19.96}$$

in which

$$\mathbf{N}_{\text{MG}} = \left[\mathbf{N}_{\text{bus}}^{1} \ldots \mathbf{N}_{\text{bus}}^{N_{\text{bus}}}\right]^{T}$$

whereas

$$\mathbf{N}_{\text{bus}}^{jj} = \left[\begin{array}{c} \overbrace{\overbrace{\mathbf{O}_{2\times2}\ \mu^{\text{DER}-1}\ \mathbf{O}_{2\times2}}^{\text{DER}-1^{\text{MG}-i}} \ldots \overbrace{\mathbf{O}_{2\times2}\ \mu^{\text{DER}-N_{BES}}\ \mathbf{O}_{2\times2}}^{\text{DER}-N_{BES}}}^{\text{DERs}} \\ \overbrace{\overbrace{\mathbf{O}_{2\times2}\ \mu^{\text{BES}-1}\ \mathbf{O}_{2\times2}}^{\text{BES}-1} \ldots \overbrace{\mathbf{O}_{2\times2}\ \mu^{\text{BES}-N_{BES}}\ \mathbf{O}_{2\times2}}^{\text{BES}-N_{BES}}}^{\text{BES Systems}} \\ \overbrace{\mu^{\text{load}-1} \ldots \mu^{\text{load}-N_{load}}}^{\text{loads}}\ \overbrace{\mu^{\text{line}-1} \ldots \mu^{\text{line}-N_{line}}}^{\text{lines}}\ \mathbf{O}_{2\times(3\,N_{DER}+4\,N_{BES}(} \end{array}\right]$$

Replacing (19.96) in (19.94),

$$\dot{\mathbf{x}}_{\text{MG}} = \mathbf{A}_{\text{MG}}^{\text{H}}\,\mathbf{x}_{\text{MG}} \tag{19.97}$$

where

$$\mathbf{A}_{\text{MG}}^{\text{H}} = \mathbf{A}_{\text{MG}} + \mathbf{C}_{\text{MG}}\mathbf{N}_{\text{MG}}$$

from which the microgrid's eigenvalues ($_{\text{MG}}$) are the roots of [93]

$$\det(\lambda_{\text{MG}}\mathbf{I} - \mathbf{A}_{\text{MG}}^{\text{H}}) = 0 \tag{19.98}$$

in which det(.) is the determinant function and **I** has the same order of $\mathbf{A}_{\text{MG}}^{\text{H}}$. The contribution of each state of $\mathbf{x}_{\text{MG}}$ on each eigenvalue can be defined from the participation factor matrix (**PF**) formulated by [94]

$$\mathbf{PF} = \left(\Gamma^{-1}\right)^{T} \circ \Gamma \tag{19.99}$$

in which denotes the right eigenvectors matrix of $\mathbf{A}_{\mathrm{MG}}^{\mathrm{H}}$ and $\circ$ denotes the Hadamard production, whereas $\mathbf{PF}(x,y)$ shows how much eigenvalue y is impacted by state x. The states with the biggest contributors are the effective states of that eigenvalue and their normalized weightings can be computed from

$$||\mathbf{PF}(x, y)|| = \frac{|\mathbf{PF}(x, y)|}{\sum_{\mu} |\mathbf{PF}(\mu, y)|} \tag{19.100}$$

in which μ denotes all states of eigenvalue y.

Each mode's damping ratio, undamped natural frequency, period and number of cycles during which its amplitude is damped by 50%, respectively denoted by , $_n$, T, and $NoC_{50\%}$ are determined as [95]

$$\begin{aligned} \zeta &= 1/\sqrt{1 + (\upsilon/\sigma)^2} \\ \omega_{\mathrm{n}} &= -\sigma/\zeta \\ T &= 2\pi/\omega_n\sqrt{1 - \zeta^2} \\ NoC_{50\%} &= ln(2)\sqrt{1 - \zeta^2}/2\pi\zeta \end{aligned} \tag{19.101}$$

in which ln(.) denotes the natural logarithm function.

A sensitivity analysis can then be used to determine the boundary of the droop coefficients for which the microgrid is stable by varying droop coefficients m and n, respectively, from 0.1 m_{o} and 0.1 n_{o} to 10 m_{o} and 10 n_{o} in which m_{o} and n_{o} denote the operating point's droop coefficients. The first m and n for which the microgrid becomes unstable are denoted by m_{critical} and n_{critical}. When all DERs have equal nominal powers, m_{critical} and n_{critical} are the same for all DERs. Nevertheless, they will be different when DERs have dissimilar rated powers. Hence, stability margin indices of f_{critical} and V_{critical} of

$$\begin{aligned} \Delta f_{\mathrm{critical}} &= \frac{m_{\mathrm{critical}-i} \times P_{\mathrm{DER}-i}^{\mathrm{cap}}}{f_{\mathrm{nominal}}} \\ \Delta V_{\mathrm{critical}} &= \frac{n_{\mathrm{critical}-i} \times Q_{\mathrm{DER}-i}^{\mathrm{cap}}}{V_{\mathrm{nominal}}} \end{aligned} \tag{19.102}$$

can be used, which are constant for a microgrid.

The eigenanalysis-based small signal stability is used in [96–98] to assess the stability margins of some sample microgrids. Consider MG-1, MG-2, and MG-3 of Fig. 19.26 with the technical parameters of Table 19.4 and operating point droop coefficients of m_{o} = 0.0333 Hz/kW and n_{o} = 2.3 V/kVAr. Figure 19.27(a)and (b) illustrate MG-1's dominant eigenvalues' trajectory for m and n variations, whereas and , respectively, denote those at the operating and marginal stability points. This trajectory is composed of one complex-conjugate pair of critical eigenvalues ($_{1,1'}$), with characteristics of Table 19.5. For MG-1, f_{critical} = 19.4468% and V_{critical} = 7.7102%. Figure 19.27(c) depicts the participation factor of various states on the critical eigenvalue and shows that $_1$ is affected mostly by the output current states of the DERs (i.e., $(i_2)_{\text{DER-1}}$ and $(i_2)_{\text{DER-2}}$), whereas the rest of the microgrid's states have less than 4% impact altogether.

A similar study is performed for MG-2 and MG-3 of Figure 19.26 and the trajectory of their dominant eigenvalues' and participation factors are shown in Figure 19.28 and 19.29, whereas their characteristics are summarized in Tables 19.6–19.8.

The interested readers are referred to [99–104], which have discussed in detail the various aspects of stability analysis of microgrids. In the above study, it is assumed that the uncertainty of the ambient conditions does not affect the output DC voltage of the DERs [105]. The above study can be repeated to consider how the eigenvalues and stability of the microgrids will change with respect to ambient uncertainties.

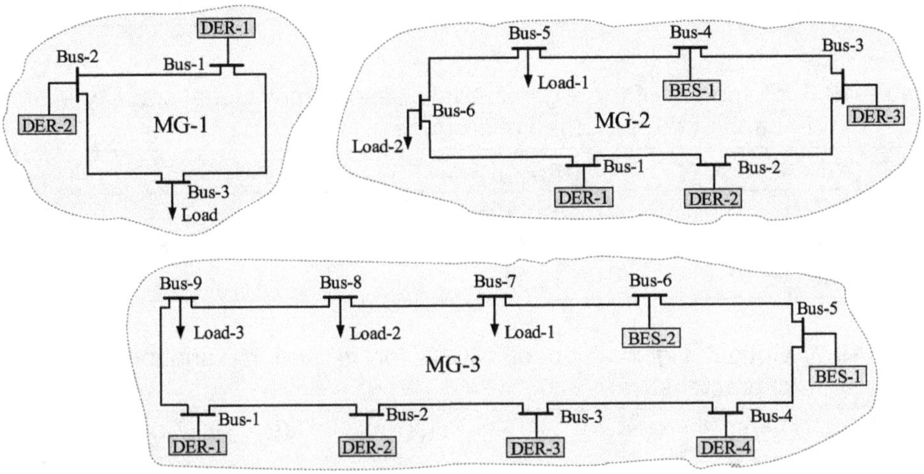

Figure 19.26 Assumed microgrids.

Table 19.4. Technical data of the network parameter

Microgrid parameters	V_{base} = 400 Vrms line-line, V^{max} = 241.5 V, V^{min} = 218.50 V $f_{nominal}$ = 50 Hz, f^{max} = 50.5 Hz, f^{min} = 49.5 Hz Z_{load} = 50.57 + j 16.6504 /phase, Z_{line} = 0.1 + j 1 /phase
VSCs and filters structure and control parameters	R_f = 0.1 Ω, L_f = 0.36 mH, C_f = 50 μF, V_{dc} = 350 V, a = 2,k_1 = 2.2968, k_2 = 6.1802, k_3 = 22.968, $_c$ = 31.4159 rad
DER parameters	P_{rated} = 3 kW, L_{coup} = 13.6 mH

Table 19.5. Critical eigenvalues of MG-1 for *m* and *n* variation and their characteristics

Critical eigen values	σ (1/s)	υ (rad/s)	ζ (%)	ω_n (rad)	T (s)	$NoC_{50\%}$	Critical for
$\lambda_{1,1'}$	−8.408	309.839	2.71	309.953	0.0203	4.0649	*m, n*

Table 19.6. Summary of the eigenanalysis-based small signal stability evaluation of the assumed microgrids

System	$\Delta f_{\text{critical}}(\%)$	$\Delta V_{\text{critical}}(\%)$
MG-1	19.4468	7.7102
MG-2	17.3083	6.5085
MG-3	12.3984	6.2142

Table 19.7. Critical eigenvalues of MG-2 for *m* and *n* variation and their characteristics

Critical eigen values	σ (1/s)	υ (rad/s)	ζ (%)	ω_n (rad)	T (s)	$NoC_{50\%}$	Critical for
$\lambda_{1,1'}$	−14.699	309.887	4.74	310.236	0.0203	2.325	*m, n*
$\lambda_{2,2'}$	−8.624	309.895	2.78	310.015	0.0203	3.963	*m, n*
$\lambda_{3,3'}$	−5.966	309.967	1.92	310.024	0.0203	5.731	*m, n*
$\lambda_{4,4'}$	−314.159	314.159	70.71	444.288	0.0200	0.110	*m*

Table 19.8. Critical eigenvalues of MG-3 for *m* and *n* variation and their characteristics

Critical eigen values	σ (1/s)	υ (rad/s)	ζ (%)	ω_n (rad)	T (s)	$NoC_{50\%}$	Critical for
$\lambda_{1,1'}$	−22.772	309.933	7.33	310.769	0.0203	1.501	*m, n*
$\lambda_{2,2'}$	−12.617	309.948	4.07	310.205	0.0203	2.709	*m, n*
$\lambda_{3,3'}$	−7.746	309.913	2.50	310.009	0.0203	4.413	*m, n*
$\lambda_{4,4'}$	−3.059	310.093	0.99	310.108	0.0203	11.180	*m, n*
$\lambda_{5,5'}$	−6.941	309.883	2.24	309.961	0.0203	4.925	*m, n*
$\lambda_{6,6'}$	−314.159	314.159	70.71	444.288	0.0200	0.110	*m, n*
$\lambda_{7,7'}$	−15.551	44.959	32.69	47.572	0.1398	0.318	*m*

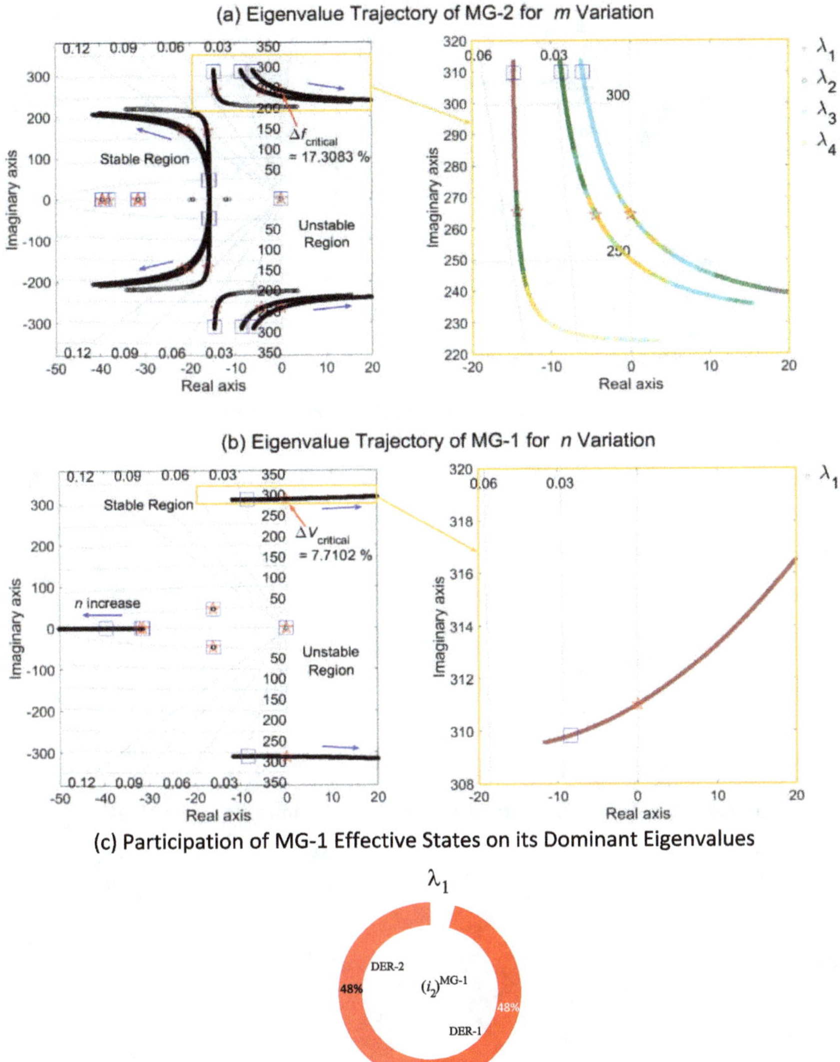

Figure 19.27 (a) Eigenvalue trajectory of MG-1 for m variation, (b) eigenvalue trajectory of MG-1 for n variation, and (c) participation factors of MG-1 effective states on its dominant eigenvalues.

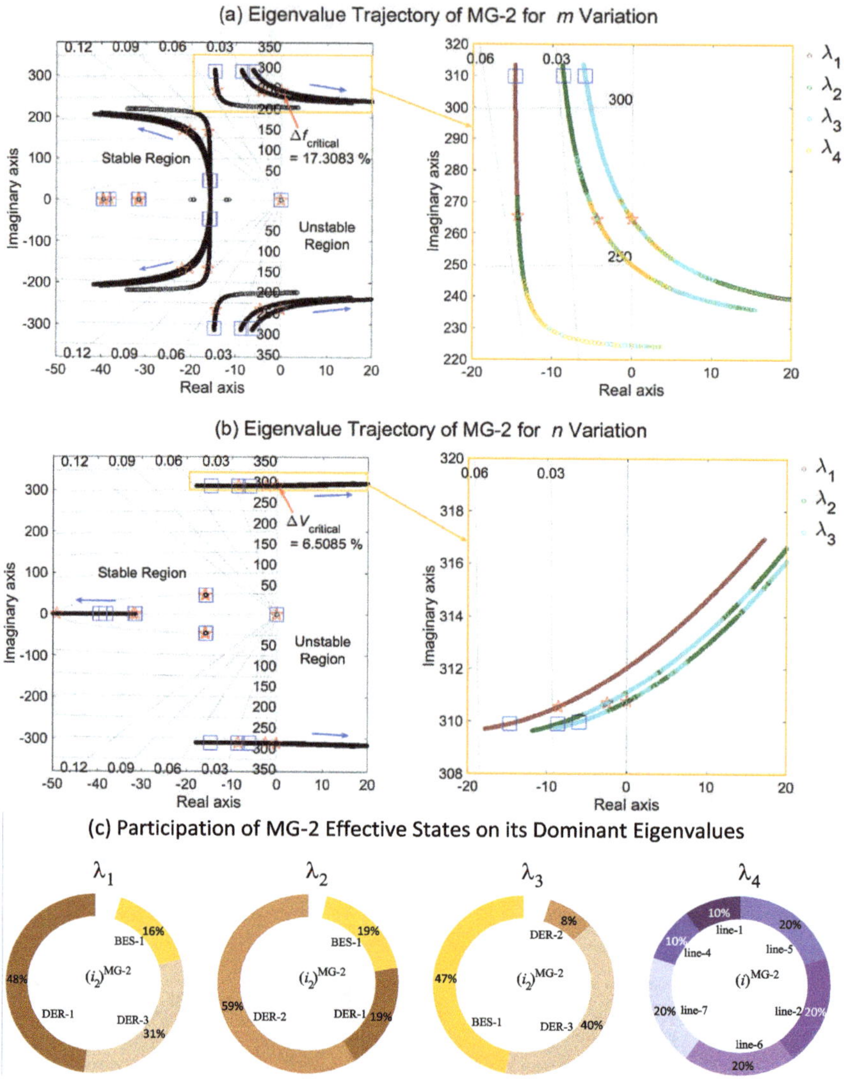

Figure 19.28 (a) Eigenvalue trajectory of MG-2 for m variation, (b) eigenvalue trajectory of MG-2 for n variation, and (c) participation factors of MG-2 effective states on its dominant eigenvalues.

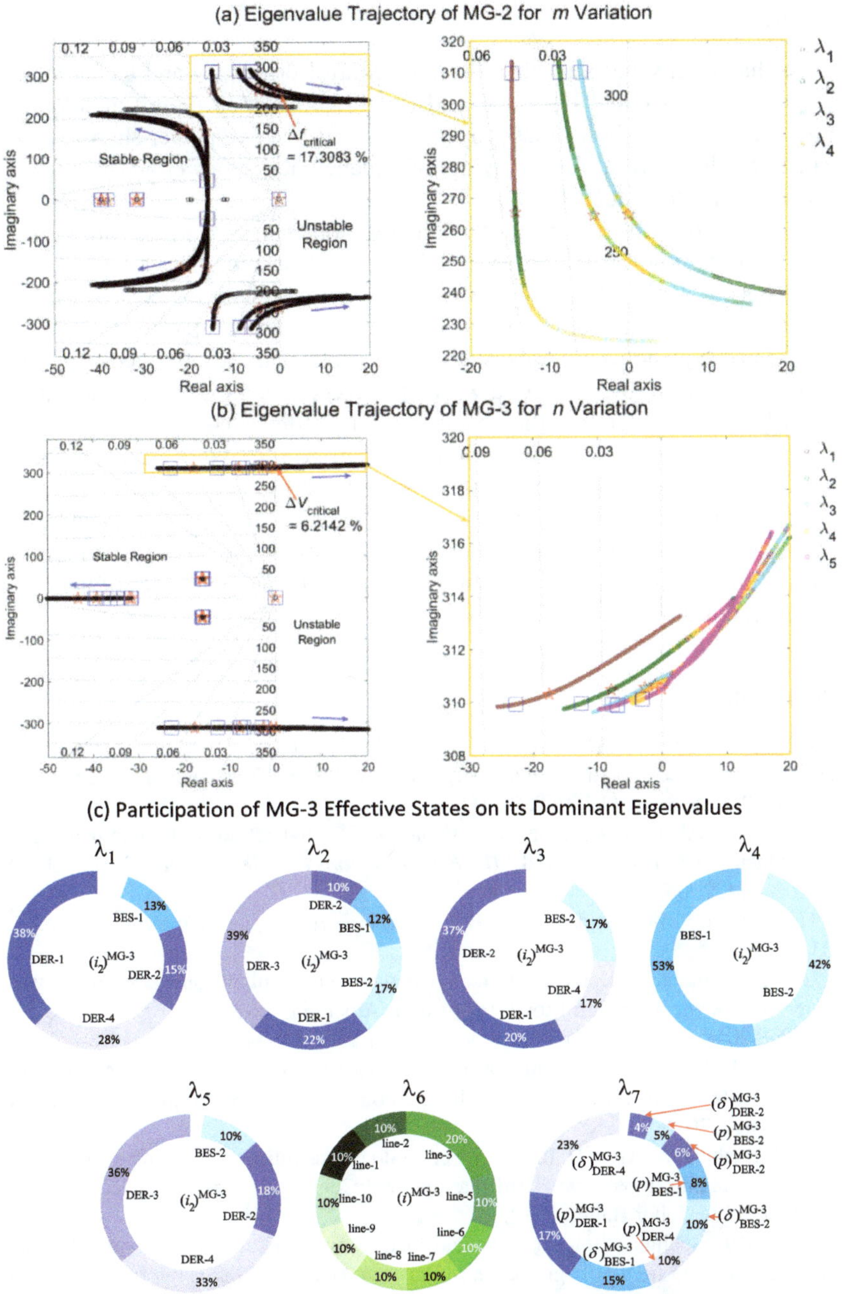

Figure 19.29 (a) Eigenvalue trajectory of MG-3 for m variation, (b) eigenvalue trajectory of MG-3 for n variation, and (c) participation factors of MG-3 effective states on its dominant eigenvalues.

19.8 Summary

This chapter has briefly introduced the various control principles and techniques for the primary and secondary control of microgrids, highlighting their main functionalities and key features. It has been shown that proper active and reactive power-sharing among the DERs can be easily realized by employing a frequency-active power and voltage-reactive power droop control technique.

The chapter also showed that by dynamically modifying the slop of the droop lines, the ratio of the output power among two DERs can be adjusted under a secondary control technique. This can be helpful in minimizing the energy generation cost by allowing the cheaper source to generate more power, when available. The chapter further showed that the voltage and frequency set-points of the droop lines can be modified to regulate the voltage and frequency in the microgrid within the acceptable range if they deviate beyond the allowed limits. Furthermore, the chapter has suggested a coordinated way to charge or discharge battery energy storages within the microgrid to avoid a quicker charging or discharging of any battery regardless of their capacity and state of charge.

The chapter has also discussed the small signal stability of microgrids, and highlighted that the output current states of the converters of the sources as well as the current states of the lines are usually the most dominating states in critical eigenvalues of the microgrid, which can affect its stability.

19.9 References

1. B. Kroposki, C. Pink, R. DeBlasio, H. Thomas, M. Simões, and P. K. Sen, "Benefits of power electronic interfaces for distributed energy systems," *IEEE Transactions on Energy Conversion*, vol. 25, no. 3, pp. 901–908, 2010.
2. T. Senjyu, T. Nakaji, K. Uezato, and T. Funabashi, "A hybrid power system using alternative energy facilities in isolated island," *IEEE Transactions on Energy Conversion.*, vol. 20, no. 2, pp. 406–414, 2005.
3. R. H. Lasseter, "MicroGrids," *IEEE Power Engineering Society Winter Meeting*, vol. 1, pp. 305–308, 2002.
4. L. Ali, and F. Shahnia, "Determination of an economically-suitable and sustainable standalone power system for an off-grid town in Western Australia," *Renewable Energy*, vol. 106, pp. 243–254, 2017.
5. L. Ali, and F. Shahnia, "Selection of a suitable standalone power supply system for an off-grid town in Western Australia," IEEE 16th International Conference on Environment and Electrical, Florence, 2016.
6. L. Ali, F. Shahnia, A. Arefi, H. Iu, and T. Fernando, "Feasibility analysis of a sustainable system for an Australian remote town," 3rd International Conference on Power Generation Systems and Renewable, Johor Bahru, 2017.
7. L. Ali, and F. Shahnia, "Impact of annual load growth on selecting the suitable sustainable standalone system for an off-grid town in Western Australia," IEEE International Conference on Power System Technology (Powercon), pp. 1–5, Wollongong, 2016
8. F. Shahnia, M. Moghbel, A. Arefi, G. Shafiullah, M. Anda, and A. Vahidnia, "Levelized cost of energy and cash flow for a hybrid solar-wind-diesel microgrid on Rottnest island," 27th Australasian Universities Power Engineering Conference (AUPEC), pp. 1–6, Melbourne, 2017.
9. R. Fornarelli, F. Shahnia, M. Anda, P.A. Bahri, and G. Ho, "Selecting an economically suitable and sustainable solution for a renewable energy-powered water desalination system: A rural Australian case study," *Desalination*, vol. 435, pp. 128–139, 2018.
10. F. Shahnia, R. Fornarelli, M. Anda, and P. A. Bahri, "A renewable energy-driven water treatment system in regional Western Australia," 27th Australasian Universities Power Engineering Conference (AUPEC), pp. 1–6, Melbourne, 2017.

11. N. Shekarchi, and F. Shahnia, “A comprehensive review of solar-driven desalination technologies for off-grid greenhouses,” *International Journal of Energy Research*, vol. 43, pp. 1357–1386, 2019.
12. N. Hatziargyriou, H. Asano, R. Iravani, and C. Marnay, “Microgrids,” *IEEE Power and Energy Magazine*, vol. 5, no. 4, pp. 78–94, 2007.
13. J. A. P. Lopes, C. L. Moreira, and A. G. Madureira, “Defining control strategies for Microgrids islanded operation,” *IEEE Transactions on Power Systems,* vol. 21, no. 2, pp. 916–924, 2006.
14. M. C. Chandorkar, D. M. Divan, and R. Adapa, “Control of parallel connected inverters in standalone AC supply systems,” *IEEE Transactions on Industrial Application*, vol. 29, no. 1, pp. 136–143, 1993.
15. R. Majumder, A. Ghosh, G. Ledwich, and F. Zare, “Angle droop versus frequency droop in a voltage source converter based autonomous microgrid,” IEEE Power Engineering Society General Meeting, pp. 1–8, Calgary, 2009.
16. R. Majumder, F. Shahnia, A. Ghosh, G. Ledwich, M. Wishart, and F. Zare, “Operation and control of a microgrid containing inertial and non–inertial micro sources,” IEEE Region 10 Conference (TENCON), pp. 1–6, Singapore, 2009.
17. C. N. Rowe, T. J. Summers, R. E. Betz, D. J. Cornforth, and T. G. Moore, “Arctan power–frequency droop for improved Microgrid stability,” *IEEE Transactions on Power Electronics*, vol. 28, no. 8, pp. 3747–3759, 2013.
18. E. Rokrok, and M. E. H. Golshan, “Adaptive voltage droop scheme for voltage source converters in an islanded multibus microgrid,” *IET Generation Transmission & Distribution*, vol. 4, no. 5, pp. 562–578, 2010.
19. H. Bevrani, and S. Shokoohi, “An intelligent droop control for simultaneous voltage and frequency regulation in islanded microgrids,” *IEEE Transactions on Smart Grid*, vol. 4, no. 3, pp. 1505–1513, 2013.
20. M. J. Sanjari, and G. B. Gharehpetian, “Small signal stability based fuzzy potential function proposal for secondary frequency and voltage control of islanded microgrid,” *Electric Power Components and Systems*, vol. 41, no. 5, pp. 485–499, 2013.
21. R. Majumder, B. Chaudhuri, A. Ghosh, R. Majumder, G. Ledwich, and F. Zare, “Improvement of stability and load sharing in an autonomous microgrid using supplementary droop control loop,” *IEEE Transactions on Power Systems*, vol. 25, no. 2, pp. 796–808, 2010.
22. R. P. S. Chandrasena, F. Shahnia, S. Rajakaruna, and A. Ghosh, “Control, operation and power sharing among parallel converter-interfaced DERs in a microgrid in the presence of unbalanced and harmonic loads,” 23rd Australasian Universities Power Engineering Conference (AUPEC), pp. 1–6, Hobart, 2013.
23. F. Katiraei, and M. R. Iravani, “Power management strategies for a microgrid with multiple distributed generation units,” *IEEE Transactions on Power Systems*, vol. 21, no. 4, pp. 1821–1831, 2006.
24. R. Teodorescu, M. Liserre, and P. Rodriguez, *Grid Converters for Photovoltaic and Wind Power Systems*, Wiley, 2011.
25. M. A. Setiawan, F. Shahnia, A. Ghosh, and S. Rajakaruna, “Proposing a new algorithm for defining the shortest distance among ZigBee-based communication devices in microgrids,” 24th Australasian Universities Power Engineering Conference (AUPEC), pp. 1–6, Perth, 2014.
26. M. A. Setiawan, F. Shahnia, S. Rajakaruna, and A. Ghosh, “ZigBee-based communication system for data transfer within future microgrids,” *IEEE Transactions on Smart Grid*, vol. 6, no. 5, pp. 2343–2355, 2015.
27. M. A. Setiawan, F. Shahnia, A. Ghosh, and S. Rajakaruna, “Developing the ZigBee based data payload coding for data communication in microgrids,” 24th Australasian Universities Power Engineering Conference (AUPEC), pp. 1–6, Perth, 2014.

28. M. A. Setiawan, F. Shahnia, R. P. S. Chandrasena, and A. Ghosh, "Data communication network and its delay effect on the dynamic operation of distributed generation units in a microgrid," 6th IEEE PES Asia-Pacific Power and Energy Engineering Conference (APPEEC), pp. 1–6, Hong Kong, 2014.
29. F. Blaabjerg, R. Teodorescu, M. Liserre, and A. V. Timbus, "Overview of control and grid synchronization for distributed power generation systems," *IEEE Transactions Industrial Electronics*, vol. 53, no. 5, pp. 1398–1409, 2006.
30. J. Rocabert, G. Azevedo, I. Candela, R. Teoderescu, P. Rodriguez, and I. E. Otadui, "Microgrid connection management based on an intelligent connection agent," IEEE 36th Annual Conf. on Industrial Electronics (IECON), pp. 3028–3033, Glendale, 2010.
31. T. L. Vandoorn, B. Meersman, J. D. M. De Kooning, and L. Vandevelde, "Transition from islanded to grid-connected mode of microgrids with voltage-based droop control," *IEEE Transactions on Power Systems*, vol. 28, no. 3, pp. 2545–2553, 2013.
32. M. A. Setiawan, A. Abu-Siada, and F. Shahnia, "A new technique for simultaneous load current sharing and voltage regulation in dc Microgrids," *IEEE Transactions on Industrial Informatics*, vol. 14, no. 4, pp. 1403–1414, 2017.
33. M. A. Setiawan, A. Abu-Siada, and F. Shahnia, "Voltage regulation in dc microgrids with various circuit configurations," 2nd IEEE Annual Southern Power Electronics Conference (SPEC), pp. 1–6, Auckland, 2016.
34. M. A. Setiawan, A. Abu-Siada, and F. Shahnia, "New technique for power sharing in dc microgrids," 2nd IEEE Annual Southern Power Electronics Conference (SPEC), pp. 1–6, Auckland, 2016.
35. R. P. S. Chandrasena, F. Shahnia, A. Ghosh, and S. Rajakaruna, "Operation and control of a hybrid ac-dc nanogrid for future community houses," 24th Australasian Universities Power Engineering Conference (AUPEC), pp. 1–6, Perth, 2014.
36. F. Shahnia, "Hybrid nanogrid systems for future small communities," in *Sustainable Development for Energy Systems*, B. Azzopardi (ed.), pp. 19–38, Springer, 2017.
37. R. P. S. Chandrasena, F. Shahnia, S. Rajakaruna, and A. Ghosh, "Dynamic operation and control of a hybrid nanogrid system for future community houses," *IET Generation, Transmission & Distribution*, vol. 9, no. 11, pp. 1168–1178, 2015.
38. R. P. S. Chandrasena, F. Shahnia, S. Rajakaruna, and A. Ghosh, "Operation and control of three phase microgrids consisting of single-phase DERs," IEEE 8th International Conference on Industrial and Information Systems (ICIIS), Peradeniya, 2013.
39. F. Shahnia, and R. P. S Chandrasena, "A three-phase community microgrid comprised of single-phase energy resources with an uneven scattering amongst phases," *International Journal of Electrical Power & Energy Systems*, vol. 84, pp. 267–283, 2017.
40. M. Goyal, Y. Fan, A. Ghosh, and F. Shahnia, "Techniques for a wind energy system integration with an islanded microgrid," *International Journal of Emerging Electric Power Systems*, vol. 17, no. 2, pp. 191–203, 2016.
41. Y. Fan, M. Goyal, A. Ghosh, and F. Shahnia, "Integration of wind energy conversion system with microgrid and utility," 24th Australasian Universities Power Engineering Conference (AUPEC), pp. 1–6, Perth, 2014.
42. J. M. Guerrero, J. C. Vasquez, J. Matas, L. G. de Vicuna, and M. Castilla, "Hierarchical control of droop-controlled ac and dc microgrids- A general approach toward standardization," *IEEE Transactions Industrial Electronics*, vol. 58, no. 1, pp. 158–172, 2011.
43. F. Shahnia, R. P. S. Chandrasena, S. Rajakaruna, and A. Ghosh, "Primary control level of parallel distributed energy resources converters in system of multiple interconnected autonomous microgrids within self-healing networks," *IET Generation, Transmission & Distribution*, vol. 8, no. 2, pp. 203–222, 2014.
44. F. Katiraei, R. Iravani, N. Hatziargyriou and A. Dimeas, "Microgrids management," *IEEE Power Energy Magazine*, vol. 6, no. 3, pp. 54–65, 2008.

45. E. Pashajavid, F. Shahnia, and A. Ghosh, "Development of a self-healing strategy for resilience enhancement of islanded microgrids," *IEEE Transactions on Smart Grid*, vol. 8, pp. 868–880, 2017. doi:10.1109/TSG.2015.2477601
46. E. Pashajavid, F. Shahnia, and A. Ghosh, "Overload management of autonomous microgrids," 11th IEEE International Conference on Power Electronics and Drive Systems (PEDS), pp. 1–6, Sydney, 2015.
47. J. Susanto, F. Shahnia, A. Ghosh, and S. Rajakaruna, "Interconnected microgrids via back-to-back converters for dynamic frequency support," 24th Australasian Universities Power Engineering Conference (AUPEC), pp. 1–6, Perth, 2014.
48. F. Shahnia, R. P. S. Chandrasena, S. Rajakaruna, and A. Ghosh, "Interconnected autonomous microgrids in smart grids with self-healing capability," in *Renewable Energy Integration, Challenges* and Solutions, J. Hossain and M. A. Mahmud (eds.), pp. 347–381, Springer, 2014.
49. F. Shahnia, "Interconnected microgrid systems for remote areas," in *Sustainable Development for Energy Systems*, B. Azzopardi (eds.), pp. 39–63, Springer, 2017.
50. M. Batool, S. M. Islam, and F. Shahnia, "Master control unit based power exchange strategy for interconnected microgrids," 27th Australasian Universities Power Engineering Conference (AUPEC), pp. 1–6, Melbourne, 2017.
51. F. Shahnia, R. P. S. Chandrasena, S. Rajakaruna, and A. Ghosh, "Autonomous operation of multiple interconnected microgrids with self-healing capability," IEEE Power and Energy Society General Meeting, pp. 1–5, Vancouver, 2013.
52. F. Shahnia, and S. Bourbour, "A practical and intelligent technique for coupling multiple neighboring microgrids at the synchronization stage," *Sustainable Energy, Grids and Networks*, vol. 11, pp. 13–25, 2017.
53. S. Bourbour, and F. Shahnia, "A suitable mechanism for the interconnection phase of temporary coupling of adjacent microgrids," IEEE PES Innovative Smart Grid Technologies Asian Conference (ISGT-Asia), pp. 1–6, Melbourne, 2016.
54. E. Pashajavid, F. Shahnia, and A. Ghosh, "Provisional internal and external power exchange to support remote sustainable microgrids in the course of power deficiency," *IET Generation, Transmission & Distribution*, vol. 11, no. 1, pp. 246–260, 2017.
55. E. Pashajavid, F. Shahnia, and A. Ghosh, "Overloading conditions management in remote networks by coupling neighboring microgrids," 50th International Universities Power Engineering Conference (UPEC), pp. 1–6, Stoke-on-Trent, 2015.
56. E. Pashajavid, F. Shahnia, and A. Ghosh, "A decentralized strategy to remedy the power deficiency in remote area microgrids," 50th International Universities Power Engineering Conference (UPEC), pp. 1–6, Stoke-on-Trent, 2015.
57. F. Shahnia, S. Bourbour, and A. Ghosh, "Coupling neighboring microgrids for overload management based on dynamic multi-criteria decision-making," *IEEE Transactions on Smart Grid*, vol. 8, no. 2, pp. 969–983, 2017.
58. S. Bourbour, F. Shahnia, and A. Ghosh, "Selection of a suitable microgrid to couple with an overloaded neighboring microgrid based on decision making," 47th IEEE North American Power Symposium (NAPS), pp. 1–6, Charlotte, 2015.
59. S. Bourbour, and F. Shahnia, "Impact of the weightings of different criteria in selecting the suitable microgrids when forming a system of coupled microgrids," IEEE PES Innovative Smart Grid Technologies Asian Conference (ISGT-Asia), pp. 1–6, Melbourne, 2016.
60. F. Shahnia, and A. Arefi, "Defining the suitable adjacent microgrids to form a temporary system of coupled microgrids," IEEE Region 10 Conference (Tencon), pp. 1–4, Singapore, 2016.
61. A. Arefi, and F. Shahnia, "Tertiary controller-based optimal voltage and frequency management technique for multi-microgrid systems of large remote towns, *IEEE Transactions on Smart Grid*, vol. 9, no. 6, pp. 5962–5974, 2018.

62. M. A. Shoeb, F. Shahnia, G. Shafiullah, and X. Su, "Sensitivity of prediction error on the performance of a preventive controller for microgrids," 20th IEEE International Conference on Industrial Technology (ICIT), pp. 1–6, Melbourne, 2019.
63. M. A. Shoeb, F. Shahnia, and G. Shafiullah, "Optimal coupling of multiple microgrid clusters," 9th IEEE PES Innovative Smart Grid Technologies Asian Conference (ISGT-Asia), pp. 1–6, Chengdu, 2019.
64. M. A. Shoeb, F. Shahnia, and G. Shafiullah, "A multilayer and event-triggered voltage and frequency management technique for microgrid's central controller considering operational and sustainability aspects," *IEEE Transactions on Smart Grid*, vol. 10, pp. 5136–5151, 2018. doi:10.1109/TSG.2018.2877999
65. M. A. Shoeb, F. Shahnia, and G. Shafiullah, "A multilayer optimization scheme to retain the voltage and frequency in standalone microgrids," 7th IEEE PES Innovative Smart Grid Technologies Asian Conference (ISGT-Asia), pp. 1–6, Auckland, 2017.
66. M. A. Shoeb, F. Shahnia, and G. Shafiullah, "A multilayer preventive control to regulate voltage and frequency in autonomous microgrids," 28th Australasian Universities Power Engineering Conference (AUPEC), pp. 1–6, Auckland, 2018.
67. M. Batool, S. M. Islam, and F. Shahnia, "Power transaction management amongst coupled microgrids in remote areas," 7th IEEE PES Innovative Smart Grid Technologies Asian Conference (ISGT-Asia), pp. 1–6, Auckland, 2017.
68. M. Batool, F. Shahnia, and S. M. Islam, "Multi-level supervisory emergency control for the operation of remote area microgrid clusters," *Journal of Modern Power Systems and Clean Energy*, vol. 7, pp. 1210–1228, 2019. doi:10.1007/s40565-018-0481-6
69. T. Jamal, T. Urmee, G. Shafiullah, and F. Shahnia, "Using experts' opinions and multi-criteria decision analysis to determine the weighing of criteria employed in planning remote area microgrids," IEEE International Conference on Green Energy for Sustainable Development (ICUE), Phuket, 2018.
70. M. Batool, F. Shahnia, and S. M. Islam, "The impact of scaled fitness functions on a floating-point genetic algorithm to optimize the operation of standalone microgrids," *IET Renewable Power Generation*, vol. 13, pp. 1280–1290, 2019. doi: 10.1049/iet-rpg.2018.5519
71. J. C. Vasquez, R. A. Mastromauro, J. M. Guerrero, and M. Liserre, "Voltage support provided by a droop-controlled multifunctional inverter," *IEEE Transactions on Industrial Electronics,* vol. 56, no. 11, pp. 4510–4519, 2009.
72. F. Shahnia, "Impact of low pass filters of the droop control on converter-interfaced DERs of an islanded microgrid," IEEE International Conference on Industrial Technology (ICIT), pp. 1–6, Taipei, 2016.
73. R. P. S. Chandrasena, F. Barbieri, F. Shahnia, A. Ghosh, and S. Rajakaruna, "Developing the guidelines for fabrication of laboratory prototype voltage source converters," 24th Australasian Universities Power Engineering Conference (AUPEC), pp. 1–8, Perth, 2014.
74. F. Barbieri, R. P. S. Chandrasena, F. Shahnia, S. Rajakaruna, and A. Ghosh, "Application notes and recommendations on using TMS320F28335 digital signal processor to control voltage source converters," 24th Australasian Universities Power Engineering Conference (AUPEC), pp. 1–7, Perth, 2014.
75. R. P. S. Chandrasena, F. Shahnia, A. Ghosh, and S. Rajakaruna, "Secondary control in microgrids for dynamic power sharing and voltage/frequency adjustment," 24th Australasian Universities Power Engineering Conference (AUPEC), pp. 1–6, Perth, 2014.
76. R. H. Lasseter, and P. Piagi, *Control and design of microgrid components*, Final project report, Power Systems Engineering Research Center, University of Wisconsin-Madison, 2006.
77. A. M. Salamah, S. J. Finney, and B. W. Williams, "Autonomous controller for improved dynamic performance of AC grid, parallel–connected, single–phase inverters," *IET Generation Transmission & Distribution*, vol. 2, no. 2, pp. 209–218, Mar. 2008.

78. T. H. Mehr, F. Shahnia, R. P. S. Chandrasena, and A. Ghosh, "Different power sharing techniques for converter-interfaced DERs in an autonomous microgrid," 6th IEEE PES Asia-Pacific Power and Energy Engineering Conference (APPEEC), pp. 1–6, Hong Kong, 2014.
79. F. Shahnia, R. Majumder, A. Ghosh, G. Ledwich, and F. Zare, "Operation and control of a hybrid microgrid containing unbalanced and nonlinear loads," *Electric Power System Research*, vol. 80, no. 8, pp. 954–965, 2010.
80. T. H. Mehr, F. Shahnia, R. P. S. Chandrasena, and A. Ghosh, "Dynamic power sharing control among converter-interfaced DERs in an autonomous microgrid," IEEE PowerTech Conference, pp. 1–6, Eindhoven, 2015.
81. J. M. Guerrero, N. Berbel, J. Matas, J. L. Sosa, and L. G. de Vicuna, "Droop control method with virtual output impedance for parallel operation of uninterruptible power supply systems in a microgrid," IEEE 22nd Applied Power Electronics Conference (APEC), pp. 1126–1132, Anaheim, 2007.
82. F. Shahnia, "Semi-decentralized charging and discharging control of floating batteries in microgrids," 2nd IEEE Annual Southern Power Electronics Conference (SPEC), pp. 1–6, Auckland, 2016.
83. T. H. Mehr, A. Ghosh, and F. Shahnia, "Cooperative control of battery energy storage systems in microgrids," *International Journal of Electrical Power & Energy Systems*, vol. 87, pp. 109–120, 2017.
84. M. Goyal, A. Ghosh, and F. Shahnia, "Overload prevention in an autonomous microgrid using battery storage units," IEEE Power and Energy Society General Meeting, pp. 1–5, Washington DC, 2014.
85. M. Goyal, A. Ghosh, and F. Shahnia, "Distributed battery storage units for overload prevention in an islanded microgrid," 24th Australasian Universities Power Engineering Conference (AUPEC), pp. 1–6, Perth, 2014.
86. T. H. Mehr, F. Shahnia, and A. Ghosh, "Power sharing control of batteries within autonomous microgrids based on their state of charge," 25th Australasian Universities Power Engineering Conference (AUPEC), pp. 1–6, Wollongong, 2015.
87. T. Morstyn, B. Hredzak, and V. G. Agelidis, "Distributed cooperative control of microgrid storage," *IEEE Transactions on Power Systems*, vol. 30, no. 5, pp. 2780–2789, 2015.
88. N. Pogaku, M. Prodanovic, and T. C. Green, "Modeling, analysis and testing of autonomous operation of an inverter-based microgrid," *IEEE Transactions Power Electronics*, vol. 22, no. 2, pp. 613–625, 2007.
89. E. Pashajavid, F. Shahnia, and A. Ghosh, "Interconnection of two neighboring autonomous microgrids based on small signal analysis," 9th IEEE International Conference on Power Electronics (ICPE-ECCA Asia), pp. 213–220, Seoul, 2015.
90. A. Ashourpouri, A. Ghosh, and F. Shahnia, "Droop-based demand dispatch schedule in smart grid," 25th Australasian Universities Power Engineering Conference (AUPEC), pp. 1–6, Wollongong, 2015.
91. Y. Astriani, G. Shafiullah, and F. Shahnia, "Additional controls to enhance the active power management within islanded microgrids," *Energy Procedia*, vol. 158, pp. 2780–2786, 2019.
92. Y. S. Kim, E. S. Kim, and S. I. Moon, "Frequency and voltage control strategy of standalone microgrids with high penetration of intermittent renewable systems," *IEEE Transactions on Power Systems*, vol. 31, no. 1, pp. 718–728, 2016.
93. A. Tewari, *Modern Control Design*, Wiley, 2002.
94. P. Kundur, *Power System Stability and Control*, McGraw-Hill, 1994.
95. S. G. Kelly, *Mechanical Vibrations*, Cengage Learning, 2013.
96. F. Shahnia, "Stability and eigenanalysis of a sustainable remote area microgrid with a transforming structure," *Sustainable Energy, Grids and Networks*, vol. 8, pp. 37–50, 2016.

97. F. Shahnia, "Eigenvalues and their participation factors for the system of coupled adjacent microgrids," 2nd IEEE Annual Southern Power Electronics Conf. (SPEC), pp. 1–6, Auckland, 2016.
98. F. Shahnia, "Stability of a sustainable remote area microgrid," IEEE Region 10 Conference (Tencon), pp. 1–4, Singapore, 2016.
99. J. Susanto, F. Shahnia, and A. Arefi, "Effects of network characteristics and topology on the stability of converter-dominated microgrids," 27th Australasian Universities Power Engineering Conference (AUPEC), pp. 1–6, Melbourne, Australia, Nov 2017.
100. A. Kunwar, F. Shahnia, and R. C. Bansal, "Impact of the capacity and number of inertial and non-inertial distributed energy resources within a microgrid on its stability margins," 27th IEEE International Symposium on Industrial Electronics (ISIE), pp. 1–6, Cairns, 2018.
101. F. Shahnia, and A. Arefi, "Eigenanalysis-based small signal stability of the system of coupled sustainable microgrids," *International Journal of Electrical Power & Energy Systems*, vol. 91, pp. 42–60, 2017.
102. F. Shahnia, J. Susanto, and M. A. Mahmud, "Evaluating the stability impact of the interconnecting tie-line between coupled neighboring microgrids," IEEE Power Electronics, Drives and Energy Systems Conference (PEDES), pp. 1–6, Chennai, 2018.
103. J. Susanto, F. Shahnia, and D. Ludwig, "A framework to technically evaluate integration of utility-scale photovoltaic plants to weak power distribution systems," *Applied Energy*, vol. 231, pp. 207–221, 2018.
104. J. Susanto, F. Shahnia, and A. Arefi, "Small signal model load flow initialisation for converter-dominated microgrids," IEEE International Conference on Power Systems Technology (Powercon), pp. 1–6, Wollongong, 2016.
105. M. Batool, S. M. Islam, and F. Shahnia, "Stochastic modeling of the output power of photovoltaic generators in various weather conditions," 26th Australasian Universities Power Engineering Conference (AUPEC), pp. 1–5, Brisbane, 2016.

Section IV: Miscellaneous: Power Quality, Smart Grid

CHAPTER 20

Power Quality Problems and Solutions in Renewable Energy

S. P. Dubey[1] and S. P. Makhija[2]

[1]*Electrical Engineering Department, ICFAI University, Raipur, Chhattisgarh, India.*
spdubeyfee@gmail.com
[2]*Electrical Engineering Department, ITM University, Raipur, India.*
satyaprakashm@itmuniversity.org

Abstract Renewable energy sources connected in distribution systems utilizing power electronics devices to interface lead to various power quality problems. This chapter presents a review on power quality issues associated with the grid-connected renewable energy systems and mitigation techniques. To mitigate the power quality issues, an effective role is played by power electronic devices and custom power devices such as active power filters (APFs) and flexible AC transmission systems (FACTs). This chapter also discusses IEC and IEEE standards for grid-connected renewable energy system.

Keywords: Power Quality, Power Electronics, Renewable energy, Grid, Problems, Solutions.

ANSI	:	American National Standards Institute
APF	:	Active Power Filter
ASTM	:	American Society for Testing and Materials
AWEA	:	American Wind Energy Association
CENELEC	:	European Committee for Electro technical Standardization
DFIG	:	doubly fed induction generator
DVR	:	Dynamic Voltage Restorer
FACTS	:	Flexible AC Transmission Systems
HAPF	:	Hybrid Active Power Filter
HPF	:	Hybrid Power Filter
IEC	:	International Electro-technical Commission
IGBT	:	Insulated Gate Bipolar Transistor
NIST	:	National Institute of Standards and Technology
PCC	:	Power Control Centre
PF	:	Passive Filters
PWM	:	Pulse Width Modulation
SAC	:	Services Accounting Code
SEMI	:	Semiconductor Equipment and Materials International
STATCOM	:	Static Compensators
SVC	:	Static VAR Compensator
THD	:	Total Harmonic Distortion
VAR	:	Volt Ampere Reactive
VSC	:	Voltage Source Controlled

20.1 Introduction

The world has been continuously moving toward higher per capita power consumption for last two decades especially. This is due to increasing living standards of people and increasing population in majorly developing countries. As per the World Energy Outlook report 2018 [1], every year the electrical energy demand is increasing continuously by 2.5%. To match this increase in demand, worldwide electrical energy generation, which was 20 trillion kWh in 2010, will rise to almost 42 trillion kWh in 2040 [2]. Looking at world's current focus on reduction of global warming, renewable energy sources, especially solar photovoltaic (PV) systems and wind energy systems, are increasingly being installed with a clear vision by various countries. In an analysis for feasibility of renewable energy for five major countries, that is, China, India, Russia, United Kingdom, and United States, it has been observed that the area required for the installation of large-scale solar and wind energy plants in order to meet half of their electricity requirements will spread in only a fractional portion of the present land and close-by off-shore area in these countries [3]. Apart from all favorable conditions, there are some issues related to quality of power output from these renewable energy sources especially when connected to grid. These issues are due to power electronic circuitry, sudden load variations, and inconsistent renewable energy input, that is, solar radiations and wind speed, throughout a year. The power quality issues include voltage dips and drops, voltage swells and rises, and harmonics [4, 5]. Power quality issues closely affect the economics of power-generating plants, the consuming units, and thus the whole power system. The details of power quality issues, its economics, and ways to mitigate these issues are discussed in the following sections.

20.2 Power Electronic Structures of Grid-Connected PV and Wind Systems

Renewable energy available in the form of solar radiations and wind speed is required to be converted into electrical energy for usefulness. PV cells and wind generators along with various power electronics equipment are employed for this purpose. Power electronics equipment are required for conditioning of power output from PV cells and wind generators.

20.2.1 Grid-Connected PV System

The DC output voltage of PV modules is brought at required level using boost converters followed by conversion to AC through inverter. This AC voltage is matched with grid voltage level by the use of transformer. Figure 20.1 shows the schematic diagram of grid-connected PV system [6]. The DC power generated by PV arrays is delivered by boost converter to inverter at a voltage higher than the input voltage. The variation and voltage control can be obtained by varying the duty cycle of a switch present in the circuitry of boost converter. A boost converter operates in two modes. In first mode when switch is closed, it is in charging mode. When the switch is open, it is in second mode, that is, discharging mode of operation. The inverters are used for solar PV integration in AC grid. The selection of inverter is an important factor, which may have an impact on cost, reliability, and overall system efficiency. The basic categories of the inverter used in grid-connected PV system are central, micro, and string inverters. From these, microinverters have lowest power ratings with the most compact size.

The advantages of these inverters are higher reliability, possible to monitor individual panels, and easy to increase the system size as the inverters are connected to individual PV module to convert DC power into AC power. The overall performance of system is not affected if some or any panel is affected by any obstruction or natural shading. A number of modules are connected to string inverters; these modules are joined in cascade to construct a string of module. To achieve optimal efficiency of string inverters, all the panels must have same characteristics and operating conditions. Central inverters are suitable for higher power ratings (up to megawatts) and used for commercial purposes. To generate high voltage without voltage amplification, the inverters are cascaded. To be capable of withstanding a large amount of power, the strings of inverters are connected in parallel. In central inverter, single inverters are used to interface with grid.

20.2.2 Grid-Connected Wind Turbine System

Normally, wind turbine system comprises a doubly fed induction generator (DFIG) used for conversion of wind energy to electrical output. In the model shown in Fig. 20.2, two converters have been used together one after the other.

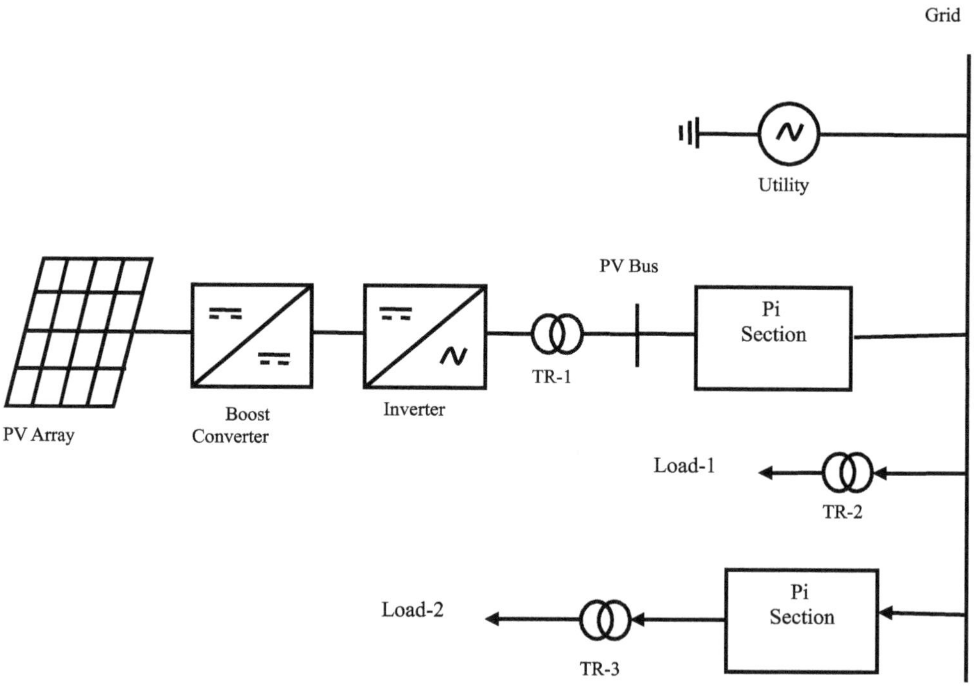

Figure 20.1. Schematic diagram of grid-connected PV system.

Grid parameters are fed to the grid side converter for controlling and regulating the DC-link voltage at a specified reference level. The objective of other converter on the rotor side is to track maximum available power and ensure its availability at generator output. Thus, energy flow is bidirectional between the grid and the rotor of the induction generator [7].

The controls used in wind energy systems are more sophisticated compared to the PV system. To control the operation of DFIG, The following four control blocks are used [8]:

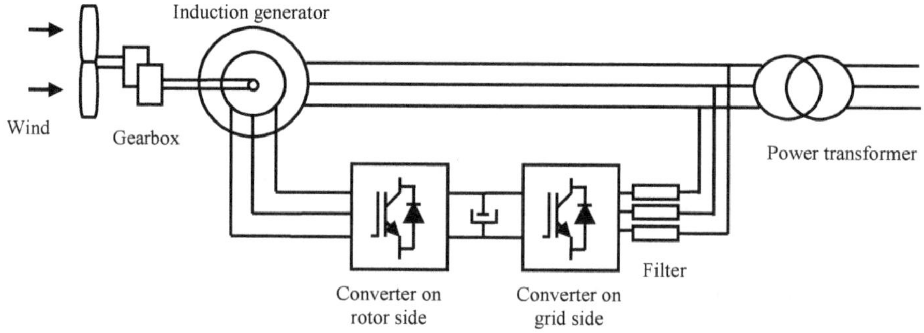

Figure 20.2 Schematic diagram of grid-connected wind turbine system.

- Voltage of capacitor is kept fixed by grid voltage source controller (VSC), wherein capacitor voltage is controlled with real component of current and the terminal voltage is regulated using quadrature component.
- The grid VSC controls and rotor VSC controls are similar. The d and q rotor currents are independently decoupled for ensuring control of the rotor excitation and electrical torque.
- A constant capacitor voltage is maintained by grid VSC. It is regulated by current, wherein capacitor voltage is controlled by real component and terminal voltage is adjusted by quadrature component.
- A protective shunt resistor is used as crow bar control between the two converters along with a switch in series.

Power converter interfaces wind generator and grid. The power converter must fulfill the requirements of both the sides. Power converter must have the capability to control wind generator output voltage and frequency and comply with grid codes. Power converter must be able to control the reactive power and quick response for active power. Various power converter topologies are

(i) Soft starter
(ii) Diode rectifier
(iii) Back-to-back converters
(iv) Multilevel converters
(v) Matrix converters
(vi) Resonant converters
(vii) B-4 converter

20.3 Need of Power Quality Studies

It is expected that the supply voltages at each point in any power system (distribution as well as transmission) should be sinusoidal and should have constant and equal magnitudes, fixed frequency, equal-phase displacement of 120° among the three phases, and above all there should be no zero-sequence voltage. However, because of nonlinear loads connected to the distribution system, the supply voltages no longer are found to exhibit these characteristics. It may be pointed out that the good power supply voltage quality implies both reliability and quality [9, 10]. The most critically affected areas are the process industries and IT sectors. Any power disturbance may cause huge financial loss with loss of productivity and competitiveness. A continuous process industry and the information technology services are the mostly affected areas. A sudden power quality issue may result in lower production, loss of competitiveness, and associated financial problems. It is evident that necessary steps should be taken for acquiring appropriate power quality level.

20.3.1 Power Quality Issues in Grid Integration of Renewable Energy System

(A) Solar PV system

The solar radiation energy conversion into electricity output of PV systems is a function of radiation intensity and the shadow effect due to clouds. The efficiency of components of PV systems, that is, modules, inverters, controllers, and filters, affects the overall performance of PV system. As per study by Golovanov et al. [11], low-voltage distribution grids are affected by even small duration fluctuations in radiations and cloud shadow effect in the case of high PV penetration. Thus, it is very important that proper monitoring of voltage and power flow be ensured. The quality issues in voltage and power flow can be solved by the use of super capacitors. Although this may scale up the cost of PV system by around 20%.

(B) Wind energy system

The integration of wind power plants to the utility system after adopting grid code were led by American Wind Energy Association (AWEA). According to IEC-61400-21, the rules have been defined at the distribution network for grid operation of wind power systems. As per the rules, for maintaining electric system stability, the control of voltage and frequency along with fault ride through capability must be provided by new wind farms.

20.3.2 International Standards

To maintain power quality at an acceptable level, various agencies have specified the limits on harmonics contents in the power supply. These limits have been proposed for harmonic distortion level in voltages and currents at different frequencies. Commonly known guidelines are the IEEE-519 [12] in North America and the IEC-1000 Standard (formerly IEC 555) [13] prepared by the International Electrical Commission (in effect since 1996). Various organizations, such as ANSI, NIST (National Institute of Standards and Technology), ASTM, IEEE, SEMI, SAC, CENELEC, and IEC, are working in renewable energy generation systems, but the IEC is the most important international body with regard to renewable energy system-related standardization. The activities in standardization of quality parameters in renewable systems have been increased considerably through Technical Committee 82 (TC82). For PV systems, wind energy systems, and batteries, TC82, TC88, and TC21 are responsible, respectively.

The working team of Technical Committee 82 (Photovoltaic) of IEC (International Electro-technical Commission) develops the international standards:

- **(i)** IEC TS 63157 ED1: These standards are for quality assurance of power conversion equipments used in photovoltaic systems
- **(ii)** IEC TS 62910 ED2: It defines test procedure for low-voltage ride-through measurements for utility-interconnected photovoltaic inverters

The working team of Technical Committee-88 (Wind) of IEC (International Electro-technical Commission) develops the international standards:

1. IEC 61400-12: Standards defining wind turbine performance
2. IEC 61400-13: Defines measuring procedure in determining the power behavior in wind turbine
3. IEC 61400-21: Guidelines for generating system of wind turbine
4. IEC 61400-3-7: Guidelines for fluctuating load's emissions limit assessment.

20.3.3 Parameters of Power Quality

Different types of disturbances affect voltage and frequency of the electrical power supply. In such situations, continuous mentoring of these parameters is required. Various specific parameters are associated with voltage, frequency, and noise level leading to poor power quality. Such specific parameters are detailed in the following paragraphs:

(i) Voltage sags and swells: For any load, voltage magnitude in terms of its root mean square (RMS) value should be of specified and required level. Voltage sag is a phenomenon in which magnitude of RMS voltage gets reduced, for durations of 0.5 cycles to 1 minute, between 10% and 90% of the nominal RMS voltage at the power frequency [14], as can be seen in Fig. 20.3(a). The voltage sag occurs due to sudden increase in current flowing in transmission or distribution lines. This sudden increase in current is the result of connection of heavy loads or faults in the system. Sudden increase in current causes higher voltage drop in lines resulting in reduced voltage at the load ends. This happens mostly in parallel feeders.

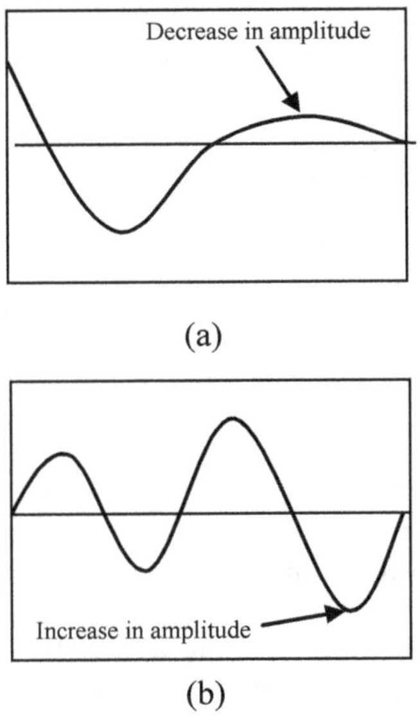

Figure 20.3 (a) Voltage sag and (b) voltage swell [15].

The voltage swell is the result of sudden decrease of heavy load such as stoppage of large motors. The current gets decreased suddenly causing reduced voltage drop and increased voltage at the load side as can be seen in Fig. 20.3(b).

(ii) Harmonic distortion: In a power system, voltage and current waveforms are desired to be pure sinusoidal with a constant frequency. But, due to several reasons, the waveforms of voltage and current get distorted after superimposition of additional higher frequency voltage and current waveforms on original waveforms. The frequencies of these additional waveforms are twice, thrice, or even more times than that of fundamental frequency of original waveform. This is called harmonic distortion of voltage or current waveforms and is shown in Fig. 20.4(a). Harmonic distortion results due to presence of different types of nonlinear loads and components such as arc furnaces, rectifiers/inverters, welding machines, switched mode power supplies, and data-processing equipment [16]. Total harmonic distortion (THD) is the term that describes the content of harmonic distortion in a waveform. The value of THD is calculated using Eq. (20.1),

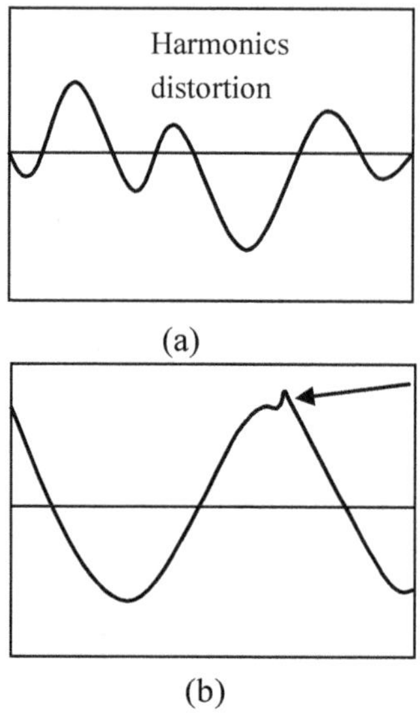

Figure 20.4 (a) Harmonic distortion and (b) voltage spike [15].

$$\%\text{THD} = \sqrt{I(D_1)^2 + I(D_2)^2 + \ldots} \tag{20.1}$$

where $I(D_n)$ = nth harmonic magnitude as a percentage of the fundamental (individual distortion)

A perfectly sinusoidal wave contains zero THD. As per IEEE standards, 5.0% THD is the commonly maximum allowable limit for distribution circuits.

(iii) Voltage spike: Voltage spike is accounted if voltage increases for a very small time in the range of few microseconds to few milliseconds as shown in Figure 20.4(b). Lightening, switching of capacitors, and disconnection of very heavy loads are the causes of voltage spikes. Sometimes it results in electromagnetic interference along with damages in electronic equipment and insulating materials.

(iv) Voltage interruptions: As per IEEE standards, a voltage interruption is considered when the generator's terminal voltage becomes lower than 10% of its nominal supply voltage for a short or long duration. Interruptions occur normally due to malfunctioning of protection devices [17]. A short-duration interruption occurs for few milliseconds to 1 or 2 s as shown in Fig. 20.5(a), whereas a long-duration voltage interruption persists for more than 1 or 2 s as shown in Fig. 20.5(b).

(v) Voltage unbalance: Three-phase voltages are called balanced if their RMS magnitudes are same and they have 120° phase difference between any two phases. If these conditions do not hold, the three voltages become unbalanced as shown in Fig.

20.6(a). Unbalancing occurs due to different load in three phases. The three-phase equipment is most affected by unbalancing situation.

(vi) Noise: When high-frequency signals get superimposed on power system voltage or current waveforms, the original waveforms get distorted as shown in Fig. 20.6(b). This distortion is called noise in waveforms. Electromagnetic interference, radiations due to furnaces and welding machines, and so on are the sources of noise [18]. Noise may cause malfunctioning of sensitive equipment and data-processing errors.

(vii) Flickers and fluctuations: As the wind velocity is never constant in magnitude and direction, the output power of wind turbines continuously fluctuates. The terminal voltage of grid-connected wind farm fluctuates more due to load variations. When voltage fluctuation propagates into the power system at load end, the illumination level of lighting loads also varies very fast. This phenomenon is called flickers.

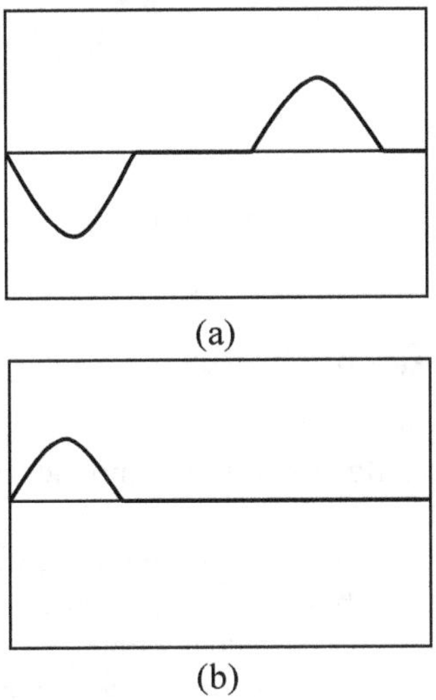

Figure 20.5 (a) Short-duration voltage interruption and (b) long-duration voltage interruption [15].

(viii) Frequency deviations: Frequency of power supply is a very critical parameter in power system. Even a deviation of more than 2% may cause damage to sensitive equipment. It is quite challenging to keep frequency at nominal value under variable load conditions.

(ix) Power factor: Power factor is a parameter that quantifies the percentage of efficient use of input power by a load. It varies in the range of 0–1. The higher the power factor, the lower is the input power for given output power. Mathematically, the power factor at the load is the result of dividing average power by apparent power. Power factor gets affected due to harmonic distortion. Filter circuits are used to rectify the effect of higher harmonics content on power factor. In PV and wind energy systems, the inconsistent solar radiations and wind speed are responsible for poor power factor at various power and voltage unbalance conditions.

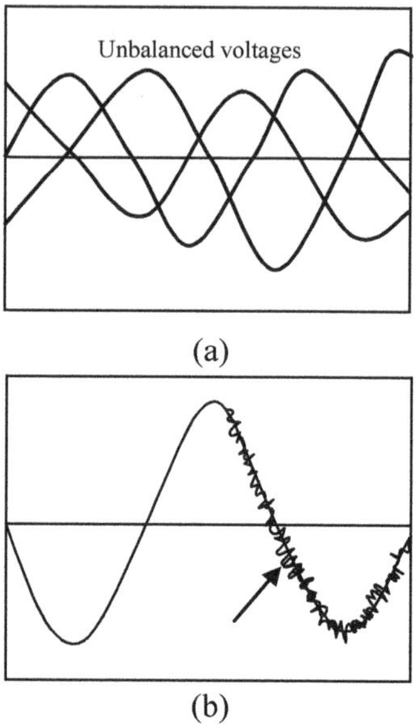

Figure 20.6 (a) Voltage unbalance and (b) noise [15].

20.4 Economics of Power Quality Issues

As discussed earlier, poor power quality is measured by accounting various parameters. Each parameter directly or indirectly affects the economics of power generation, transmission, and distribution.

The extent of effect of power quality on various costs depends upon type of industry, type of equipment, and market conditions. Poor power quality may result in the peculiar operation of equipment and thus may cause heavy financial loss to industries and individuals. The percentage financial losses due to different power quality issues are shown in Fig. 20.7 [20]. The evaluated costs due to poor power quality may be categorized in two ways, that is, direct and indirect costs.

20.4.1 Direct Costs

The costs that are directly affected by power quality are termed direct costs. Poor power quality may result in burning of motor windings, fusing of lighting appliances, bursting of air compressors, and pumping loads. Apart from this, personal computing devices' components get damaged due to power quality issues. Poor power quality makes many equipment work with lower efficiency. This results in higher input energy consumption and thereby higher utility bills. Lower efficiency causes overheating of equipment and thus reduced life. The effects of power quality vary from equipment to equipment. Few critical factors affect the economics associated with different equipment.

The type, duration, and magnitude of power quality disturbance directly affect the extent of damage to equipment. The efficiency of electrical motors gets reduced due to core and copper losses. The core losses that are of two types, that is, hysteresis losses and eddy current losses, depend on frequency of power supply. Harmonic distortion, which is due to higher multiples of fundamental frequencies, gives rise to higher core losses and hence reduces efficiency. Higher the duration and magnitude will further increase the adverse effect on efficiency and hence cost of operation of industrial unit.

Sensitivity of equipment to a disturbance is another parameter affecting the cost parameters. Higher the sensitivity higher will be the loss to equipment due to disturbance. Higher frequency of occurrence of a disturbance drastically reduces the life of equipment.

Equipment's age decides its efficiency and tolerance to a power quality issue. The dielectric property of insulating materials used in electrical machines goes on reducing with age. Such equipment gets damaged and sometimes undesired breakdowns occur in a continuous process industry resulting in huge loss to that unit in terms of cost.

The impact of power quality on equipment is affected by their location also. Because control circuits have low current levels and effect of harmonics is higher on such circuits. If power cable runs in parallel to control cables, then the level of distortion depends upon frequency of harmonics and the distance between the cables [16].

20.4.2 Indirect Costs

The costs that are difficult to evaluate are called indirect costs. Taking example of a continuous process industry, there are some costs that affect the process time, material flow, and product quality and thus not easy to evaluate [21]. Sometimes even 1 h process stoppage due to power quality issues may result in loss of thousands/lacks of dollars. The costs incurred in prevention of power quality issues also come under indirect costs.

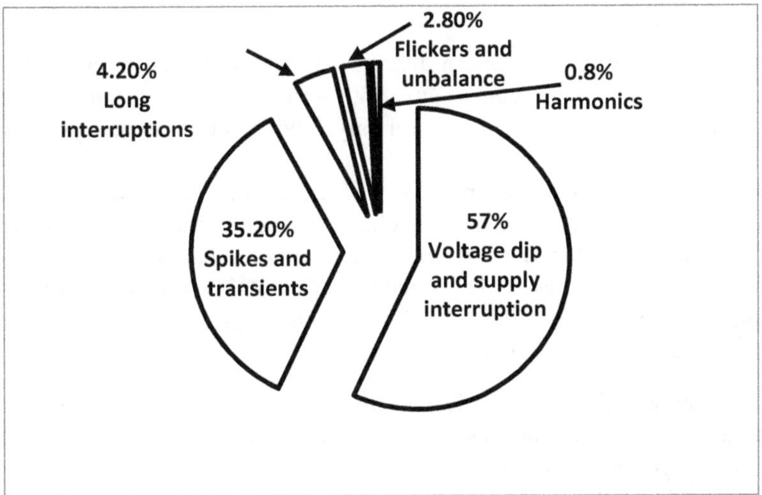

Figure 20.7. Percentage cost of power quality issues [19].

20.5 Mitigation of Power Quality Problems

The scientists, researchers, and power companies are continuously working toward finding solution of power quality problems. Various power quality issues have been discussed in Section 20.4. The world population is increasingly using nonlinear loads, which create further higher power quality issues such as harmonics. Flexible AC transmission system (FACTS) devices, such as active power filters (APFs) and static compensators (STATCOMs), are the technologies that are preferred for solving power quality issues in industrial and commercial fields [22].

In the initial stage, passive filters (PF) were used to solve harmonic grid problems. The limitations of PFs have been overcome by APFs, the second-stage filters providing an effective solution to harmonic problems. Further, the issues of shunt APFs are solved using hybrid power filter (HPF) devices. HPFs have both PFs and shunt APFs in combination. In another step, unified power quality conditioners (UPQCs) have been developed to compensate the issues of HPFs [23]. FACTS devices such as distribution STATCOM (DSTATCOM), dynamic voltage restorer (DVR), static volt–ampere reactive (VAR), and compensator (SVC) are used to solve the power quality problems such as poor power factor, load imbalanced conditions, and load harmonics.

Current and voltage are the two types of harmonic sources based on impedance. A smoothing inductor and a smoothing capacitor in a diode rectifier characterize a current harmonic source and a voltage harmonic source, respectively. Some IEEE international standards are referred by various equipment manufacturers, industrial users, and utility companies. The limitations and guidelines for equipment manufacturers and power supply companies are provided under IEC 61000-3-2. The harmonic control practices and requirements are recommended under IEEE-510 [24].

Harmonic contents in grid-connected systems can be mitigated using different methods. These methods include PFs, APFs, and STATCOMs for solving harmonic power quality issues. The following sections discuss these methods in detail.

20.5.1 Shunt PFs

The solution to power quality issues in distribution networks was initially given by PFs, which was very economical. PFs consist of various series–parallel combinations of damping resistors, inductors, and capacitors. A commonly used PF configuration is shown in Fig. 20.8 [25].

The fundamental source impedance influences each operation. The values of inductances and capacitances decide the filtering characteristic. To track the desired harmonics, the working of the filter is tuned as per the order of required harmonic.

To increase the effectiveness of harmonic reduction and compensation of reactive current component, normally the coupling of PFs is with a thyristor-controlled reactor. Some limitations associated with PFs are as follows [26]:

- Only one component (from harmonic and fundamental) may pass through PFs
- For each harmonic component, PFs require a separate filter
- The filtering range of PFs is limited
- PFs get overloaded and saturated when harmonic currents are large
- PFs have limited compensation, large size, and tuning issues as they eliminate only those frequencies to which they are tuned.

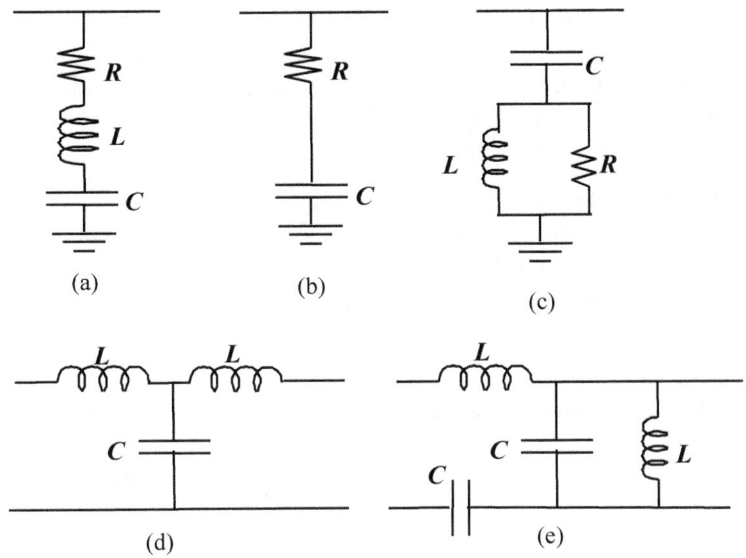

Figure 20.8. PF structures: (a) single-tuned; (b) first-order high-pass; (c) second-order high-pass; (d) LCL; and (e) LLCC filters.

20.5.2 Active Power Filters

(A) Shunt and series APFs

The limitations of PFs are solved by using shunt APFs. The APFs have better compensation characteristics, which are achieved by passive-energy storage devices, such as inductors and capacitors and an active switching device. It includes compensation for current and voltage harmonics, compensation for consumers due to current imbalance, and compensation for utilities due to voltage imbalance.

APFs work in order to mitigate sudden voltage distortion, frequency variation, neutral current, reactive power, voltage notch, changing line impedance, and transient disturbance. Power factor of current and voltage also improves fusing APFs in medium-power systems [27]. Figure 20.9(a) shows the configuration of shunt APF. The shunt APFs have fundamental compensation principle of removing current harmonics produced by critical loads. Normally, the installations of APFs are done near a nonlinear load and in shunt position. This is for the purpose of compensating the consequences of harmonic nonlinearity. The series active power filter is connected in series to the distribution system as shown in Fig. 20.9(b). The series active power filter acts as controlled voltage source and injects voltage component in series with the supply voltage to compensate voltage disturbances such as sags and swells.

(B) Hybrid active power filters

To reduce the unwanted issues of passive and stand-alone APFs, the combination of these, that is, hybrid filter, was developed [28]. The topologies of HAPF comprise both active and passive-tuned filters. The rating of hybrid active filter can be reduced by improvements in compensation characteristics of the PFs. Figure 20.10(a) and (b) shows the configurations of HAPFs. The HAPFs have the merits of both PF and APF.

The objective behind the development of HAPF is to reduce APF duty. With reduced APF duty, the HAPF can be employed in high voltage and capacity grids. HAPFs have proved their better performance in decreasing harmonics along with reactive power compensation with a low capacity of APF.

The APFs have few practical limitations as follows [29]:

- Initially the cost of installation is high.
- The controlling of filter is very complicated because of higher harmonics owing to complex design and control structure.
- The APF gives a design trade-off with high-power rating system demand and rapid dynamic current response.

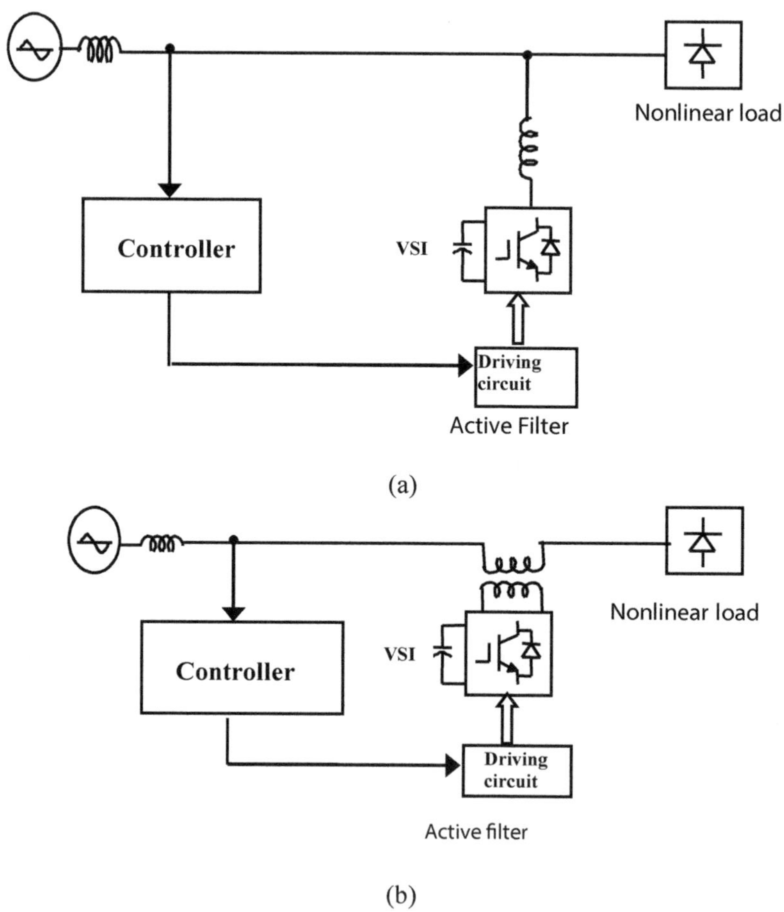

Figure 20.9 APF: (a) shunt APF and (b) series APF.

The APFs have few practical limitations as follows [29]: (1) Initially the cost of installation is high. (2) The controlling of filter is very complicated because of higher harmonics owing to complex design and control structure. (3) The APF gives a design trade-off with high-power rating system demand and rapid dynamic current response.

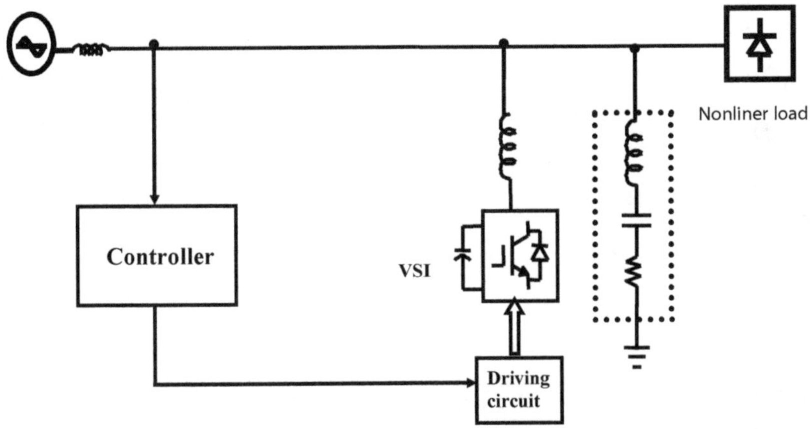

(a)

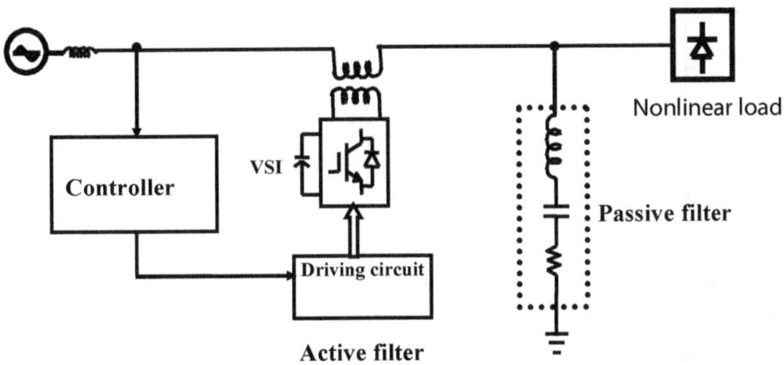

(b)

Figure 20.10 HAPF: (a) PF with shunt APF and (b) PF with series APF.

The selection of shunt, series, and configuration of HPAF depend on requirements. Table 20.1 summarizes the selection of APFs for different applications.

20.5.3 STATCOM

In modern transmission system, the requirements of FACTS devices are increasing rapidly. The FACTS devices are of two types: SVCs and static synchronous compensators (STATCOMs). The primary objective of connecting the FACTS devices in transmission line is to provide a precise, fast responding, and variable quantity of reactive power to the system. This can be obtained by adjustment of polarity along with magnitude of reactive component of current. STATCOMs connected in shunt with power control center (PCC) are used to provide voltage support and active and reactive power as presented in Fig. 20.11. Additionally, STATCOMs are utilized for dynamic reactive power compensation (i.e., dynamic

Table 20.1. Selection of shunt, series, and hybrid APFs for different applications.

Sl. No.	Application	Type of Filter			
		Series Active Filter	Shunt Active Filter	Hybrid Filter (Active series and Passive Shunt)	Hybrid Filter (Active Series and Active Shunt)
1	Voltage flicker reduction	√	√		√
2	Voltage harmonic compensation	√		√	√
3	Improving voltage regulation	√	√	√	√
4	Removing voltage sags	√	√	√	√
5	Current harmonic compensation		√	√	√
6	Reactive power compensation		√	√	√
7	Improving load balancing		√		
8	Neutral current compensation		√	√	
9	2 + 3	√			√
10	1 + 4 + 5 + 6			√	√
11	1 + 2 + 3 + 4	√			√
12	5 + 6		√	√	√
13	2 + 6			√	√
14	5 + 6 + 7 + 8		√		
15	5 + 7		√		
16	5 + 6 + 7		√		√
17	5 + 7 + 8		√		

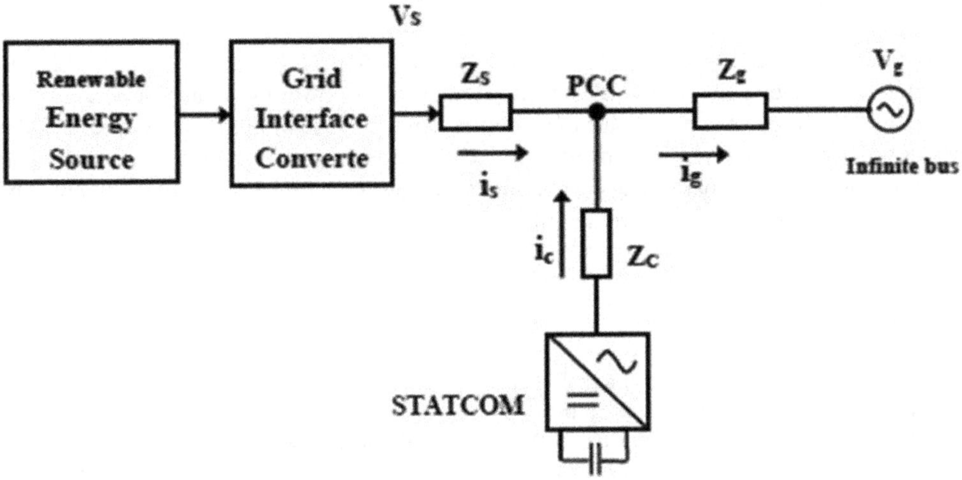

Figure 20.11. STATCOMs connected in shunt with PCC.

power factor correction), reduction of voltage fluctuations in the power supply, which will ensure prevention of equipment's damage and minimize the plant's running costs [30].

(A) Multilevel PV-STATCOM applications in grid-connected systems

The function of PV-STATCOM systems is to control active and reactive power flows through the converter and reduce the constraints associated with conventional PV inverters. In multilevel PV-STATCOM, the cascaded H-bridge inverter is an advanced configuration for high solar power applications. In this system, no additional DC source is required for STATCOM devices. When solar energy is available during daytime, PV inverter functions as STATCOM providing active power optimally, and when radiations are low it provides reactive power.

A PV-STATCOM along with active power filtering is capable of handling power quality issues such as in the PV system [27]. A PV-active filter-STATCOM functions in order to mitigate harmonic distortion and continual control of reactive power for grid applications.

(B) Wind turbine STATCOM (WT-STATCOM) applications in grid-connected systems

The STATCOM used in wind turbine applications consists of IGBT-equipped and PWM-controlled major component of modular VSC. The battery energy storage system (BESS) can be employed as an energy storage element for voltage regulation. The BESS keeps the DC condenser voltage constant, which helps in fast injection or consumption of reactive power to stabilize. When there is a variation in the power system, the power fluctuations with the load and discharge process can be controlled using BESS. The battery is connected in shunt with the DC capacitor of the STATCOM [27].

The advance classifications of APFs and STATCOMs are based on the test results of research based on power conversion techniques, semiconverter devices, control techniques, and passive power components.

20.6 Summary

Today, the renewable energy penetration has been continuously increasing in the world. The integration of renewable system power into grid requires power electronic devices. These power electronic devices inject harmonic components into the system that deteriorates the power quality. The requirements of power electronic converters to integrate the renewable system, power quality parameters, economics of power quality issue, and mitigation of power quality problems are addressed in this chapter.

20.7 References

1. World Energy Outlook 2018-International Energy Agency, International Energy Agency (IEA), France, 2018. Available Online: https://www.iea.org/weo2018/
2. S. P. Makhija, and S. P. Dubey, "Feasibility analysis of distributed solar electricity penetration in cement manufacturing plants," *International Journal of Ambient Energy*, vol. 17, pp. 218–227, Feb. 2019.
3. M. Asif, and T. Muneer, "Energy supply, its demand and security issues for developed and emerging economies," *Renewable and Sustainable Energy Reviews*, vol. 11, pp. 1388–1413, Sep. 2007.
4. L. Gidwani, H. Tiwari, and R. C. Bansal, "Improving power quality of wind energy conversion system with unconventional power electronic interface," *International Journal of Electrical Power & Energy Systems*, vol. 44, pp. 445–453, Jan. 2013.
5. P. Thakur, A. K. Singh, and R. C. Bansal, "Novel way for classification and type detection of voltage sag," *IET Generation, Transmission & Distribution*, vol. 7, pp. 398–404, Jun. 2013.
6. J. Sreedevi, N. Ashwin, and M. N. Raju, "A study on grid connected PV system," National Power Systems Conference (NPSC), Indian Institute of Technology, Bhubaneshwar, Odisha, India, pp. 1–6, Dec. 2016.
7. F. E. Taveiros, L. S. Barros, and F. B. Costa, "Back-to-back converter state-feedback control of DFIG (doubly-fed induction generator)-based wind turbines," *Energy*, vol. 89, pp. 896–906, Sep. 2015.
8. A. Hariri, and M. O. Faruque, "Impacts of distributed generation on power quality," *IEEE North American Power Symposium (NAPS)*, vol. 7, pp. 1–6, Nov. 2014.
9. E. Hossain, S. P. Selim Ay, and I. Khan, "Analysis and mitigation of power quality issuesin distributed generation systems using custom power devices," *IEEE Access*, vol. 6, pp. 16816–16833, Mar. 2018.
10. K. Thangaraj, and S. Gopalasamy, "Power quality analysis and enhancement of grid connected solar energy system," *Scientific Research Publishing, Circuits and Systems*, vol. 7, pp. 1954–1961, Jun. 2016.
11. N. Golovanov, G. C. Lazaroiu, M. Roscia, and D. Zaninelli, "Power quality assessment in small scale renewable energy sources supplying distribution systems," *Energies,* vol. 6, pp. 634–645, Jan. 2013.
12. A. Dejanović, M. Babić, and M. Tomašević, "Measuring electric power quality of renewable energy sources," *International Journal of Electrical and Computer Engineering Systems*, vol. 21, no.2, pp. 81–88, 2011.
13. S. W. Mohod, and M. V. Aware, "Power quality and grid code issues in wind energy conversion system," INTECH Creative Commons Attribution License, 2013. Available online: http://creativecommons.org/licenses/by/3.0.
14. R. Bansal, *Handbook of Distributed Generation: Electric Power Technologies, Economics and Environmental Impacts*, Springer, 2017, [Online ISBN- 978-3-319-51343-0]
15. A. De Almeida, L. Moreira, and J. Delgado, "Power quality problems and new solutions," *International Conference on Renewable Energies and Power Quality*, 3: pp. 1–45, Apr. 2003.

16. J. C. Cebrian, N. Kagan, and J. V. Milanovic, "Probabilistic estimation of distribution network performance with respect to voltage sags and interruptions considering network protection setting—Part I: The methodology," *IEEE Transactions on Power Delivery*, vol. 33, pp. 42–51, Dec. 2016.
17. A. Kalair, N. Abas, A. R. Kalair, Z. Saleem, and N. Khan, "Review of harmonic analysis, modeling and mitigation techniques," *Renewable and Sustainable Energy Reviews*, vol. 78, pp. 1152–1187, Oct. 2017.
18. F. A. Borges, R. A. Fernandes, I. N. Silva, and C. B. Silva, "Feature extraction and power quality disturbances classification using smart meters signals," *IEEE Transactions on Industrial Informatics*, vol. 12, pp. 824–833, Apr. 2016.
19. "Economic Aspects of Power Quality." Central Electricity Regulatory Commission (CERC), New Delhi, India, 2018. Available Online: http://shodhganga.inflibnet.ac.in/bitstream/10603/16056/10 /10_chapter%203.pdf
20. F. H. Gandoman, A. Ahmadi, A. M. Sharaf, P. Siano, J. Pou, B. Hredzak, and V. G. Agelidis, "Review of FACTS technologies and applications for power quality in smart grids with renewable energy systems," *Renewable and Sustainable Energy Reviews*, vol. 82, pp. 502–514, Feb. 2018.
21. E. J. Coster, J. M. Myrzik, B. Kruimer, and W. L. Kling, "Integration issues of distributed generation in distribution grids," *Proceedings of the IEEE*, vol. 99, pp. 28–39, Jan. 2011.
22. R. Targosz, and J. Manson, *Cost of Poor Power Quality. de Handbook of Power Quality*, West Sussex, John Wiley & Sons Ltd., vol. 18, pp. 545–594, May 2008.
23. R. Sirjani, J. Rezaee, and A. Jordehi, "Optimal placement and sizing of distribution static compensator (D-STATCOM) in electric distribution networks: A review," *Renewable and Sustainable Energy Reviews*, vol. 77, pp. 688–694, Sep. 2017.
24. P. Mahela, O. Gafoor, and A. Shaik, "Topological aspects of power quality improvement techniques: A comprehensive overview," *Renewable and Sustainable Energy Reviews*, vol. 58, pp. 1129–1142, May 2016.
25. P. Kanjiya, V. Khadkikar, and H. H. Zeineldin, "Optimal control of shunt active power filter to meet IEEE Std. 519 current harmonic constraints under nonideal supply condition," *IEEE Transactions of Industrial Electronics*, vol. 62, pp. 724–734, Jul. 2014.
26. M. Buyuk, A. Tan, M. Tumay, and K. C. Bayındır, "Topologies, generalized designs, passive and active damping methods of switching ripple filters for voltage source inverter: A comprehensive review," *Renewable and Sustainable Energy Reviews*, vol. 62, pp. 46–69, Sep. 2016.
27. W. Tareen, M. Aamir, S. Mekhilef, M. Nakaoka, M. Seyedmahmoudian, B. Horan, M. Memon, and N. Baig, "Mitigation of power quality issues due to high penetration of renewable energy sources in electric grid systems using three-phase APF/STATCOM technologies: A review," *Energies*, vol. 11, pp. 1–41, Jun. 2018.
28. Z. Salam, T. P. Cheng, A. Jusoh, "A hybrid active power filter connected to a photovoltaic array," *Journal - The Institution of Engineers, Malaysia*, vol. 69, pp. 45–52, Mar. 2008.
29. R. Mahendran, M. Rajasekar, P. Swadeeswaran, and M. Vignesh, "Renewable energy source compensator for power quality issues using active power filter in grid," *IJAREEIE*, vol. 6, pp. 1986–1993, Mar. 2017.
30. B. Singh, R. Saha, A. Chandra, and K. Al-Haddad, "Static synchronous compensators (STATCOM): A review," *IET Power Electron*, vol. 2, pp. 297–324, Jul. 2009.

CHAPTER 21

Optimal Economic Incentive for Renewable Energy in Composite Expansion Planning

Saheed Lekan Gbadamosi and Nnamdi I. Nwulu

Department of Electrical and Electronic Engineering Science, University of Johannesburg, Auckland Park, South Africa. gbadamosiadeolu@gmail.com; nnwulu@uj.ac.za

Abstract In this study, we evaluate economic incentive schemes for improving the effective utilization of renewable energy sources in composite expansion planning. The economic incentive schemes investigated are investment tax credits and feed in tariffs. The optimal expansion planning strategy is achieved by minimizing the investment cost for the renewable energy sources, minimizing the operational cost and emissions of the fossil fuel generators, as well as maximizing the incentive schemes while satisfying all the necessary constraints. The developed mathematical model is solved using CPLEX 12.6.3 solver implemented in Advanced Interactive Multidimensional Modelling System (AIMMS) and is tested on two practical case studies: IEEE 24-bus test system and real-world data of the Nigerian power system. Sensitivity analyses were investigated on the system performance of the proposed model and the results give an insight on the benefit of incentive schemes on effective utilization of renewable energy sources. Investment tax credit was discovered to be more effective when compared to feed-in tariff in composite expansion planning strategy by enhancing more utilization of renewable energy sources, improving voltage profile, as well as reducing fossil fuel emissions.

Keywords: feed-in-tariff, composite expansion planning, renewable energy sources, economic incentives, investment tax credit

Nomenclature

Set

B	:	number of buses
I	:	number of fossil fuel generators
$\mathbb{R}$	:	number of renewable energy generators
J	:	number of loads
L	:	number of transmission lines

Indices

b	:	buses
i	:	fossil fuel generators
$\mathfrak{R}$	:	renewable energy generators
l	:	transmission lines
rl	:	receiving end of transmission lines
sl	:	sending end of transmission lines
j	:	demand
k	:	generation option of renewable energy generators

Parameters

Ω^l	:	line susceptance L
y^i	:	operational cost of fossil fuel generators i
α^i	:	emissions of the fossil fuel generators i
$P^l_{\max}$	:	maximum transmittable power capacity of line L
$x^{\mathfrak{R}}$	:	annualized investment cost of RE generating unit $\mathfrak{R}$

(Continued)

$\overline{P^{\Re}}$	:	upper bound of RE generating unit $\Re$ capacity
$P^{\Re k}$	:	capacity of generation for investment choice k of RE generating unit $\Re$
P^j	:	power demand of load j
P^i_{max}	:	peak generating capacity of fossil fuel generators i
P^i_{min}	:	least generating capacity of fossil fuel generators i
$m * [^i], n^i$and o^i	:	generation cost coefficients of fossil fuel generators i
$a * [^i], b^i$and c^i	:	emission coefficients of fossil fuel generators i
e_a	:	weighting function
∂	:	duration factor of operation

Variables

$z^{\Re k}$	:	decision variable for RE generating unit $\Re$ of option k.
P^i	:	power generated by fossil fuel generators i
$P^{\Re}$	:	power generated by renewable energy generators $\Re$.
$P^{\Re}_{max}$	:	capacity of renewable energy generators $\Re$.
P^l	:	power flow through transmission line L
E	:	incentives applied (feed-in tariff or ITC)
ϕ_b	:	voltage angle at bus b

21.1 Introduction

There is great concern about climate change across the globe, which serves as a threat to the world's economy and ecosystem [1]. The role of energy is vital owing to its contributions to greenhouse gas emissions, because fossil fuel remains the major source of electricity generation in the world. Electricity generation and its related activities are responsible for 36% of global warming across the world [2]. To ensure proper decarbonization of the electricity grid and meeting the growing energy demand due to an increase in the world energy consumption, renewable energy sources (RESs) serve as an alternative source of energy. RES plays a crucial role in the development of sustainable electrification in the world because of their potential in reducing the impact of fossil fuels while ensuring energy security [3]. Examples of RESs include wind, solar, geothermal, tidal wave, biomass, and hydro. Among these sources, wind and solar are gradually finding place in the electricity market owing to their clean nature, maturity, and ability to generate in large quantities [4].

Several governmental policies have assisted in fostering the utilization of RES in the energy sector via various economic incentive schemes. These incentive schemes can be either production- or investment-based incentives. Examples of production-based incentives include feed-in tariffs (FITs), obligatory quotas, and production tax credit, whereas investment-based incentives are investment tax credits, interest-free loans, and VAT exemption. These incentive schemes are essential for building RES investor trust, reducing management costs, guaranteeing low production costs, boosting RES investment, and making RES competitive in the electricity market [5]. In 2009, the Canadian government adopted the FIT program as a way of boosting RES investment and also encouraging green economy [6]. Similarly, Germany's FIT program established by the renewable energy act in 2000 recorded 43,657 MW of RE added to German Grid within 10 years of the program [7]. Some other countries that adopted the FIT plan include Austria, Cyprus, Czech Republic, Denmark, Greece, Hungary, Ireland, Kenya, Uganda, and Croatia [8]. Investment tax credits are practiced by Belgium, Canada, France, Austria, Italy, Mexico, Portugal, Sweden, United States, Germany, Greece, the Netherlands, and India. VAT exemption are done in Estonia, Finland, France, Germany, Hungary, Korea, Malta, Poland, and the United Kingdom [8].

Generation and transmission expansion planning is highly essential to determine how to effectively expand the existing network with incorporation of RESs to optimally serve the system load while decarbonizing the global economy and improving energy security [4,9,10]. Several optimization methods have been developed to achieve an optimal expansion planning in the literature. These methods are broadly categorized as classical, heuristic, and metaheuristic optimization methods. The classical optimization technique includes Benders decomposition (BD), branch and bound, linear programming (LP), and mixed integer linear programming (MILP). Heuristic technique includes genetic algorithm (GA), differential evolution (DE), tabu search (TS), and simulated annealing (SA). Examples of metaheuristic optimization techniques are ant colony optimization (ACO), particle swarm optimization (PSO), gravitational search algorithm (GSA), and exchange market algorithm (EMA) [9].

21.2 Renewable Energy Sources

RES is experiencing tremendous growth in the world owing to its ability to compete with conventional generating systems based on rated capacity, electricity generation cost, and ability to meet the growing demand [11]. In this chapter, wind and solar energies are considered because of their unique benefits such as ready availability and relatively mature technologies.

21.2.1 Wind Energy

Wind is a natural resource that is readily available and free in nature. It is the most promising source of alternative energy. These potentials have fostered the development of wind turbines and other auxiliary equipment. Wind turbine extracts available kinetic energy from wind and converts it to electrical energy by rotating the turbine blades [12]. Wind energy potential is highly influenced by wind speed estimation, wind farm site selection, and operation scheme. Hence, site of the wind farm depends on the wind speed availability. The power output from wind turbine is nonlinear owing to the uncertain nature of wind speed. The relationship between output power of wind turbine and wind speed is expressed in [13] as

$$P = \begin{cases} 0 & 0 \le v \le v_{ci} \\ \dfrac{(v - v_{ci})}{(v_r - v_{ci})} P_r & v_{ci} \le v \le v_r \\ P_r & v_r \le v \le v_{co} \\ 0 & v_{co} \le v \end{cases} \tag{21.1}$$

where P_r is the rated power in (kW), v_{ci} is the cut in speed in (m/s), v_{co} is the cut out speed in (m/s), and v is the rated speed in (m/s).

21.2.2 Solar Energy

Solar energy describes the electricity generated through conversion of sunlight into electrical energy by tapping light photons to obtain electrons. This process generates DC current, which can be amplified with DC–DC converters and then inverted to deliver required AC power. The solar PV comprises several PV modules in series and parallel connection to produce the desired power output. The ambient temperature of solar irradiance and system configuration determines the power output of solar PV, which is a function of solar PV locations. The output power of solar PV is depicted in [14].

$$P_{pv} = V * I * n_m * FF \tag{21.2}$$

$$FF = \frac{V^{\max} * I^{\max}}{V^{oc} * I^{sc}} \tag{21.3}$$

$$I = s(t) * (I^{sc} * K * (T_{ct} - 25)) \tag{21.4}$$

$$V = V^{oc} + K_v * T_{ct} \tag{21.5}$$

$$T_{ct} = T_{at} + s(t) * \left(\frac{NOCT - 25}{0.8}\right) \tag{21.6}$$

where P_{pv} is output power of the PV in (W), V is output voltage (V), I is terminal current (A), FF is fill factor, V^{max} and I^{max} are the voltage (V) and current (A) at peak power, respectively, V^{oc} is open circuit voltage (V), I^{sc} is the short-circuit current (A), K is coefficient of current temperature (mA/°C), T_{ct} is cell temperature (°C), K_v is coefficient of voltage temperature (mV/°C), T_{at} is ambient temperature (°C), $NOCT$ is nominal operating cell temperature (°C), and $s(t)$ is the random irradiance.

21.3 Incentives Schemes for RES

Incentive system is a policy mechanism in the form of financial support designed to achieve RES targets as well as mitigate environmental impacts and heavy dependence on fossil fuel [15]. The aim is to support the technological development of RES for the purpose of making RES competitive with fossil fuel. The incentive schemes can be categorized as production- and investment-based incentives, which are shown in Fig. 21.1.

21.3.1 Production-Based Incentives

Production-based incentives are also known as performance-based incentives. They provide support based on the energy generated by renewable energy system. This system ensures payment is based on the actual system performance to ensure adequate project quality. It can be further classified into FIT, production tax credit, and quota system, which are discussed in the following sections.

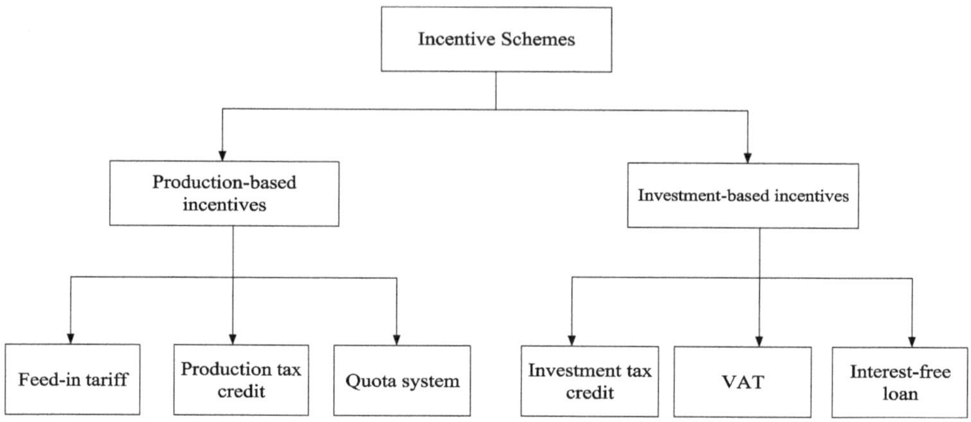

Figure 21.1. Classification of economic incentive schemes.

21.3.1.1 Feed-in Tariff

FIT is a standard support policy that provides RES provider with guaranteed payment for electricity generated from RES that is injected into the power grid [16]. This incentive plan rewards all energy fed into the grid regardless of RES provider (prosumer or RES farm). Policy maker decides the tariff plan to attract RES investment and also to control the rate of RES integration into the grid. Mathematically, it can be expressed as

$$E = \varepsilon \left(\sum_{\mathfrak{R}=1}^{\mathbb{R}} P^{\mathfrak{R}} \right) \tag{21.7}$$

where E is the incentive scheme, ε is the FIT, and $P^{\mathfrak{R}}$ is the power generated from RES.

21.3.1.2 Production Tax Credit

This type of incentive plan provides the RES investors with yearly tax credit based on the total amount of energy injected into the network and it is mostly used by investors with adequate tax capacity. This scheme does not have long-term assurance; therefore, investors do not see it as being reliable [1].

21.3.1.3 Quota System

Quota system is also known as renewable portfolio standards (RPSs) and this requires that a certain amount of energy should be provided by RES. The RES provider receives a green certificate for any unit of generated clean energy. This system relies on the private market to consume the generated energy. Here, government does not provide any price guarantee, but rather support to establish quotas for promoting RES growth [17].

21.3.2 Investment-Based Incentives

Investment-based incentive provides RES investors and financial supporters with the opportunity to offset the investment against their own personal income tax. The purpose is to grant financial relief so as to attract RES investors.

21.3.2.1 Investment Tax Credit

This incentive plan allows RES investors or project owners to be eligible to 30% tax credit claim for RES investment against liability of residential, commercial, and utility. This is one of the most important incentive policies that support financial relief to investors and also quick recovery of investment by the investors. It continues to drive growth in the industry and job creation across the country.

$$E = \beta\left(\sum_{\mathfrak{R}}^{\mathbb{R}} x^{\mathfrak{R}}\right) \tag{21.8}$$

21.3.2.2 VAT Exemption

This incentive plan helps to increase the attractiveness for investors in energy sector. It exempts the purchase, import, or production of equipment used in the construction of renewable energy projects.

21.3.2.3 Interest-Free Loan

This incentive scheme involves some governmental agencies providing interest-free loans for investors to purchase renewable energy equipment serving as a financial reliable.

21.4 Mathematical Modelling

The mathematical formulation is introduced in this section for the optimal sizing of grid-connected RESs. The objective function in Eq. (21.9) minimizes the investment cost of RES, objective function in Eq. (21.10) minimizes operational costs of fossil fuel generators, and objective function in Eq. (21.11) minimizes the emissions from fossil fuel generators, whereas objective function in Eq. (21.12) maximizes economic incentives. The incentive system is applied for two different schemes and evaluated based on their effectiveness: (1) FIT and (2) investment tax credit.

21.4.1 Objective Function

- Investment cost for RESs

$$Min\left[e_1 \sum_{\mathfrak{R}=1}^{\mathbb{R}} x^{\mathfrak{R}} P_{\max}^{\mathfrak{R}}\right] \tag{21.9}$$

- Operational cost for fossil fuel generators

$$Min\left[e_2\left(\partial \sum_{i=1}^{I} y^i P^i\right)\right] \tag{21.10}$$

- Emissions from fossil fuel generators

$$Min\left[e_3\left(\partial\sum_{i=1}^{I}\alpha^i P^i\right)\right] \tag{21.11}$$

- Economic incentives for RESs

$$Max\left[e_4\sum_{\Re}^{\mathbb{R}}E\right] \tag{21.12}$$

Quadratic function for fossil fuel costs and emissions are established, the operational fuel cost function in Eq. (21.13) represents the costs in relation to its output power ($/h) and also the total emissions in relation to its power output of the fossil fuel generators (Ib/h).

$$y^i P^i = m^i + n^i P^i + o^i (P^i)^2 \tag{21.13}$$

$$\alpha^i P^i = a^i + b^i P^i + c^i (P^i)^2 \tag{21.14}$$

21.4.2 Model Constraints

- Weighting factor constraints
  To obtain a single solution for the four objective functions, a weighting factor as in Eq. (21.15) is necessary to make the objectives comparable, where e_a is a nonnegative weighting factor as stated in Eq. (21.16) [18,19].

$$\sum_{a=1}^{A} e_a = 1 \tag{21.15}$$

$$e_a \geq 0 \forall a \tag{21.16}$$

- Constraints for RESs capacity
  Renewable energy-generating capacity constraints (21.17) and (21.19) enforce limits on the capacity of RES-generating unit to be built over the planning period, whereas constraints (21.20) and (21.21) are binary decision variable ensuring that a new RES-generating unit can be constructed only once.

$$\sum_{\Re=1}^{\mathbb{R}} x^{\Re} P^{\Re}_{\max} \leq B^{\Re} \forall \Re \tag{21.17}$$

$$0 \leq P^{\Re}_{\max} \leq \overline{P^{\Re}} \forall \Re \tag{21.18}$$

$$P^{\Re}_{\max} = \sum_{\Re=1}^{\mathbb{R}} z^{\Re k} p^{\Re k} \forall \Re \tag{21.19}$$

$$\sum_{k=1}^{K} z^{\Re k} = 1 \forall \Re \tag{21.20}$$

$$z^{\Re k} \in (0, 1) \forall \Re, k \tag{21.21}$$

- Power balance constraints

  Equation (21.22) is a power balance constraint that ensures that at any available bus, the summation of power from fossil fuel generators i, renewable energy generators $\mathfrak{R}$, and the power flowing into and out of the available lines equal total load.

$$\sum_{i=1}^{I} P^i + \sum_{\mathfrak{R}=1}^{\mathbb{R}} P^{\mathfrak{R}} - \left(\sum_{l|sl}^{L} P^l + \sum_{l|rl}^{L} P^l \right) = \sum_{j=1}^{J} P^j \forall B \tag{21.22}$$

- Transmission lines limit

  The transmission line limit ensures that power flowing on the lines must not exceed its recommended limits.

$$P^l = \Omega^l \left(\phi_{sl} - \phi_{rl} \right) \forall L \tag{21.23}$$

$$-P^l_{\max} \leq P^l \leq P^l_{\max} \forall l \tag{21.24}$$

- Generation limits for fossil fuel and RESs generators

  Constraints (21.25) and (21.26) impose that the fossil fuel and renewable energy generators, respectively, does not exceed their limits.

$$P^i_{\min} \leq P^i \leq P^i_{\max} \forall i \tag{21.25}$$

$$0 \leq P^{\mathfrak{R}} \leq P^{\mathfrak{R}}_{\max} \forall \mathfrak{R} \tag{21.26}$$

- Angle limit

  Constraint (21.27) ensures that at any bus, the voltage angle does not exceed its limit.

$$-\Pi \leq \phi_b \leq \Pi \forall B \tag{21.27}$$

21.5 Numerical Test

To confirm the efficiency of the developed model, the MILP problem is implemented on two test systems: IEEE 24-bus test system and the real-world data of Nigeria power network. The model analysis is carried out for three different cases with some level of RES integration. Case 1 is the system network without economic incentive, Case 2 is the system network with inclusion of FIT, and Case 3 is the system network with inclusion of investment tax credit. Investment cost for the new RES-generating unit is considered to be \$1.2M/MW [20] and the FIT plan considered for the power generated from RES is estimated to be \$23/MWh [21]. The MILP model is solved through the CPLEX 12.6.3 solver embedded in AIMMS platform. AIMMS is a numerical solvers and modelling language designed for large-scale optimization problems and advanced planning systems [22]. The illustration for the flowchart of the study model is depicted in Fig. 21.2.

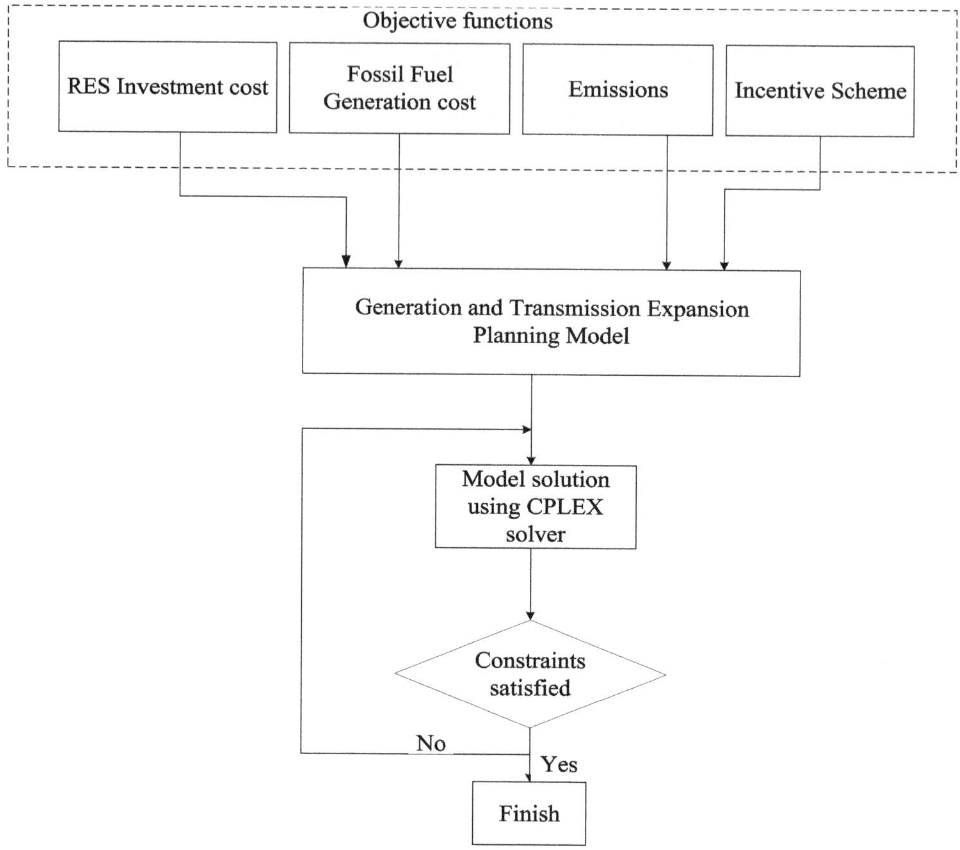

Figure 21.2. Flowchart for generation and transmission expansion planning model.

21.5.1 IEEE 24-Bus Test System

The proposed model is validated on IEEE 24-bus test system is as shown in Fig. 21.3. It consists of 24 buses with 3405 MW of fossil fuel-installed capacity and a load demand of 3200 MW. The proposed RES-generating units are installed on seven different buses with an installed capacity of 2700 MW.

21.5.2 48-Bus System of Nigeria Power Network

The applicability of the proposed model in a larger and reliable system is verified using real-world data of generation and demand of 48-bus system of Nigeria power network as shown in Fig. 21.4. Nigeria is endowed with abundant natural resources such as crude oil, biomass, coal hydro, natural gas, wave and tidal, solar radiation, and wind [23]. Presently, Nigeria generates electricity at commercial scale from only four energy sources, which are hydro, coal, natural gas, and oil. To harness the potential of these resources to generate adequate power, several studies have been done by researchers especially on the characterization of wind and solar speed and pattern to recognize the reasonable locations for wind and solar power generation (Table 21.1).

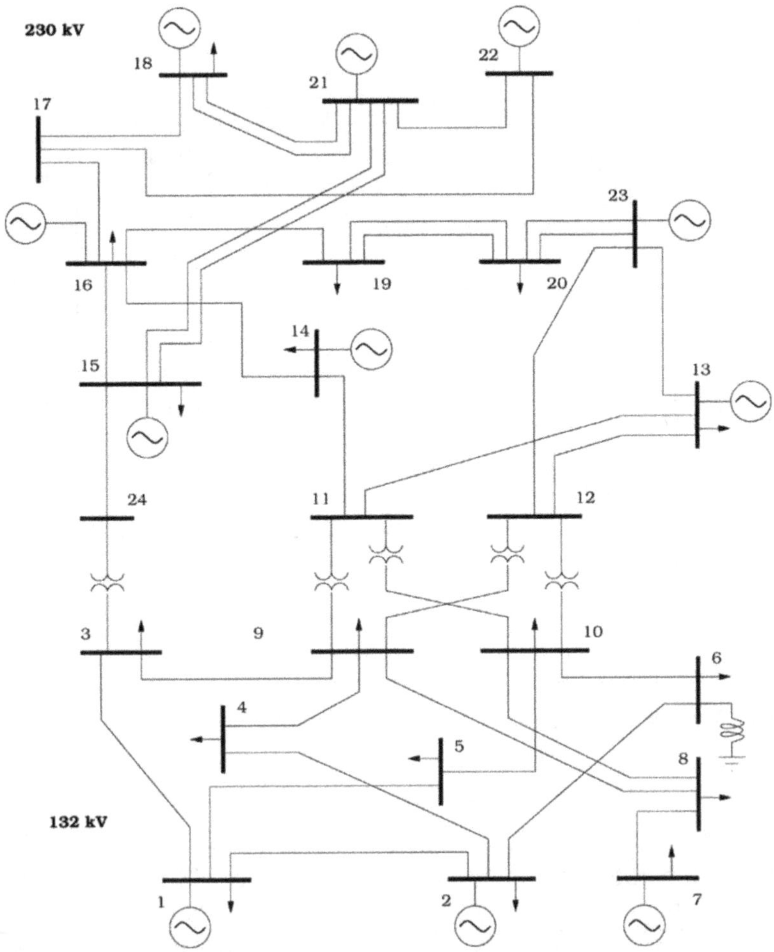

Figure 21.3. IEEE 24-bus system network [4].

In this study, some locations have been selected for installation of RESs based on the recommendation from some scholars who have carried out rigorous research [24–27]. The northern part of the country has been considered for installation of four solar farms because of the intensity of solar radiation in the region, whereas the southern part (coastal area) are map for installation of 12 wind farms. The network comprises 3 hydro and 18 fossil fuel generators, 35 loads, and 81 transmission lines as shown in Table 21.1. The load and transmission parameters are provided in Appendices (Table 21.10 and 21.11).

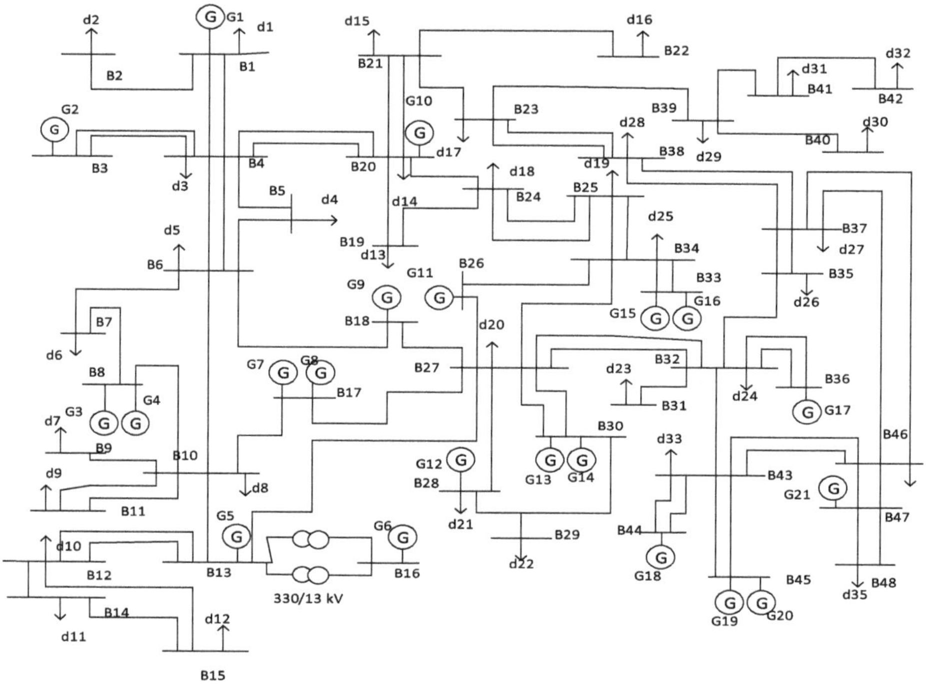

Figure 21.4. The 48-bus system of Nigeria power network.

Table 21.1. Generator parameters and location of 48-bus Nigeria power network [28]

Generators	Bus	Stations	Available Capacity (MW)	Transient X (pu)	Synchronous X (pu)
G1	B1	Kainji G.S	368	1.612	1.612
G2	B3	Jebba G.S	374	1.56	1.56
G3	B8	Olorunsogo NIPP	160	0.74	0.74
G4	B8	Olorunsogo G.S	200.7	1.184	1.184
G5	B13	Egbin G.S	820	1.522	1.522
G6	B16	AES	85	1.197	1.197
G7	B17	Omotosho G.S	150	1.184	1.184

(Continued)

Table 21.1. Generator parameters and location of 48-bus Nigeria power network [28] (*Continued*)

G8	B17	Omotosho NIPP	105.7	0.592	0.592
G9	B18	Ihovbor NIPP	129.4	0.2	0.2
G10	B20	Shiroro G.S	300	0.8	0.8
G11	B26	Azura G.S	256	0.8	0.8
G12	B28	Delta G.S	526	5.52	5.52
G13	B30	Sapele G.S	84	0.532	0.532
G14	B30	Sapele NIPP	143.9	0.8	0.8
G15	B33	Geregu G.S	152	0.444	0.444
G16	B33	Geregu NIPP	200	0.444	0.444
G17	B36	Okpai G.S	363	0.399	0.399
G18	B44	Alaoji G.S	45	0.587	0.587
G19	B45	Afam IV-V	125	2.527	2.527
G20	B45	Afam VI	426	0.532	0.532
G21	B47	Odukpani G.S	273.9	0.352	0.352

21.6 Simulation Results

Here, sensitivity analysis is carried out to determine the relevance and compatibility of different incentive schemes for RES on generation and transmission expansion planning. This will aid a central planner in making investment decisions for the system to meet the required energy demand. The simulated results presented from the expansion planning are analyzed based on three cases of the incentive schemes.

Case 1: Composite expansion planning with no economic incentive. In this case, only the three objectives in Eqs. (21.9)–(21.11) are minimized.

Case 2: Composite expansion planning with FIT. In this case, an FIT of \$23/MWh is payable to RES investors for the renewable energy injected to power grid.

Case 3: Composite expansion planning with investment tax credit. In this case, 30% tax credit claim on RES investment for initiating RES growth as well as stimulating renewable energy investment.

21.6.1 IEEE 24-Bus System

The IEEE 24-bus test system [4] has a generating capacity of 3405 MW from 33 fossil fuel-generating units. The system peak load is 2850 MW from 17 different loads as depicted in Fig. 21.5.

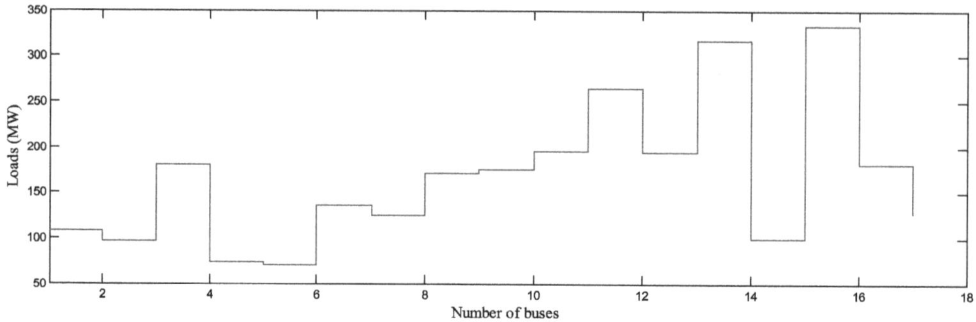

Figure 21.5. Load profile of IEEE 24-bus system.

The simulated results obtained from power generated from fossil fuel units by all the three cases are furnished in Table 21.2. The results show operating capacity of all the fossil fuel generators after the optimization process. The locations and size of the proposed RES-generating units connected to the IEEE 24-bus test system to operate effectively are summarized in Table 21.3. The simulated results for the RES-generating units after optimization for all three cases can also be obtained in Table 21.3.

The results of the simulation of the power flowing on transmission lines for all the three cases in IEEE 24-bus test system are summarized in Table 21.4. The obtained results show a total of 37 transmission lines are optimized and the power transmittable on these lines are 2561.96, 2554.96, and 2251.30 MW for Cases 1, 2, and 3, respectively. This is an indication that incentive schemes available in Cases 2 and 3 ensure installation of RES-generating units at the proposed locations thereby reducing the amount of transmittable power on the lines. Table 21.5 gives the characteristic performance of IEEE 24-bus test system for

Table 21.2. Optimal power generated from fossil fuel based on three case studies for IEEE 24-bus system.

Buses	Installed Capacity (MW)	Operating Capacity (MW)		
		Case 1	Case 2	Case 3
B1	192	152	132	112
B2	192	52	152	152
B7	300	256.25	127.77	125
B13	591	170.03	4.56	3.81
B15	215	155	155	105
B16	155	155	95	55
B18	400	333.06	318.4	291.44
B21	400	321.66	300.3	182.24
B22	300	300	300	300
B23	660	660	640	628.51

Table 21.3. Optimal power generated from RES based on three case studies for IEEE 24-bus system

Buses	Proposed Capacity (MW)	Operating Capacity (MW)		
		Case 1	Case 2	Case 3
B1	600	45	35	85
B6	250	90	115	130
B10	300	100	165	210
B13	450	–	50	70
B15	300	–	80	110
B19	550	60	80	160
B24	250	–	60	130

all the three cases. The cases with incentives show improvement in their system performance, this indicate that the optimal values of characteristic parameter improve the system performance.

Table 21.4. Optimal power flowing on lines based on three case studies on IEEE 24-bus system

S/N	Between Buses	Case 1	Case 2	Case 3
1	1–2	19.11	18.24	5.87
2	1–5	69.89	70.76	53.13
3	2–4	53.70	37.31	55.84
4	2–6	20.41	35.93	5.04
5	3–9	30.25	8.50	19.74
6	3–24	210.25	188.50	190.74
7	4–9	20.30	36.69	18.16
8	5–10	1.11	0.24	17.87
9	6–10	0.59	10.07	0.96
10	7–8	2.77	131.25	9
11	8–9	70.12	37.19	59.53
12	8–10	98.11	2.56	111.47
13	9–11	91.48	86.39	84.76
14	9–12	143.69	153.99	148.19
15	10–11	49.33	69.84	31.33
16	10–12	100.48	138.04	83.98
17	11–13	33.73	75.16	40.98

(Continued)

Table 21.4. Optimal power flowing on lines based on three case studies on IEEE 24-bus system (*Continued*)

S/N	Between Buses	Case 1	Case 2	Case 3
18	11–14	107.08	81.06	75.11
19	12–13		48.76	32
20	12–23	244.17	243.27	200.17
21	13–23	244.17	218.89	232.17
22	14–16	301.08	275.06	269.11
23	15–16	132.93	95.56	108.54
24	15–21	222.59	223.03	190.14
25	15–21	222.59	223.03	190.14
26	15–24	210.25	188.50	169.74
27	16–17	190.49	175.66	160.39
28	16–19	77.35	51.16	54.82
29	17–18	42.00	31.35	15.22
30	17–22	148.49	144.31	145.17
31	18–21	18.32	15.65	28.39
32	18–21	18.32	15.65	28.39
33	19–20	21.83	34.92	33.09
34	19–20	21.83	34.92	33.09
35	20–23	85.83	98.92	97.09
36	20–23	85.83	98.92	97.09
37	21–22	151.51	155.69	154.83

Table 21.5. Performance of the three case studies on IEEE 24 bus test system

Characteristics	Case 1	Case 2	Case 3
Objective function ($\$10^6$)	169	125	120
Total generation cost ($\$10^6$)	457	330	324
Total power generated from fossil fuel (MW)	2555	2225.03	1955
Total power generated from RES (MW)	295	585	895
Total power transmitted on the lines (MW)	2561.96	2554.96	2251.30
Total emission (10^6 Ib)	130	123	105

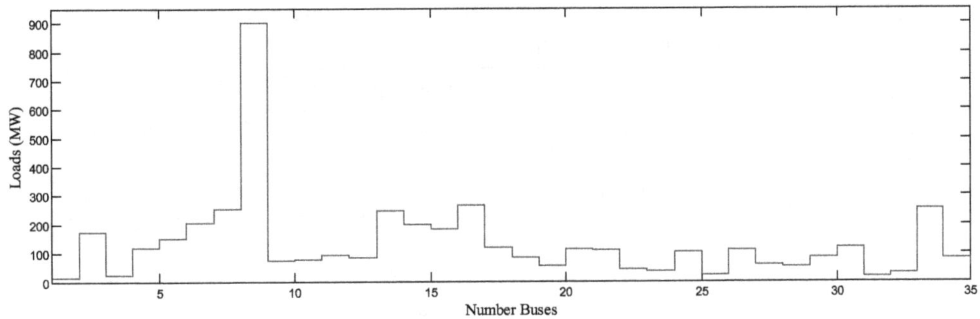

Figure 21.6. Load profile of 48-bus Nigeria power system.

21.6.2 48-Bus System of Nigeria Power Network

A 330-kV practical transmission network of Nigeria power system is employed as a second case study. The system consists of 5327 MW capacity from 21 generators and a total peak load of 4619.37 MW from 35 locations as shown in Fig. 21.6. From Fig. 21.6, it is seen that Ikeja west load in Lagos state accounted for approximately 20% of the total load on the grid. Lagos state is the business hub of the nation, resulting in demand for higher load to meet residential, commercial, and industrial requirements.

Table 21.6 summarizes generator locations and their respective capacity. The generator capacity used in this study is obtained from the average capacities obtained over the period of 5 years, because most of the generators are not operating at the installed capacity owing to old age or unavailability of gas [21].

The results obtained from the simulation are summarized in Table 21.6 showing the operating capacity of the generators for the three cases. G15 and G16 were not in use due to the presence of RES-generating units in the system. Table 21.7 summarizes the RES-generating unit locations and results obtained for operating power for 4 solar and 12 wind farms for the three cases. The power transmittable on the transmission lines for the three cases is summarized in Table 21.8, and they are 4353.57, 4225.59, and 4158.22 MW for Cases 1, 2, and 3, respectively. Table 21.9 summarizes the system performance for Nigeria power systems without incentive, with FIT, and with investment tax credit.

Table 21.6. Optimal power generated from fossil fuel based on three case studies for 48-bus Nigeria power system

Generators	Buses	Installed Capacity (MW)	Operating Capacity (MW)		
			Case 1	Case 2	Case 3
G1	B1	368	368	368	368
G2	B3	374	374	374	374
G3	B8	160	98.4	98.4	65.75
G4	B8	200.7	98.4	98.4	65.75

(Continued)

Table 21.6. Optimal power generated from fossil fuel based on three case studies for 48-bus Nigeria power system (*Continued*)

Generators	Buses	Installed Capacity (MW)	Operating Capacity (MW)		
			Case 1	Case 2	Case 3
G5	B13	820	27.65	27.51	22.3
G6	B16	85	73.18	72.8	59.04
G7	B17	150	52.41	102.14	50.27
G8	B17	105.7	51.85	101.58	49.73
G9	B18	130	104.83	104.83	70.05
G10	B20	300	300	300	300
G11	B26	256	34.64	34.46	27.93
G12	B28	526	53.73	53.45	43.32
G13	B30	84	61.91	61.59	49.91
G14	B30	144	64.12	63.78	51.71
G15	B33	152	50	–	–
G16	B33	200	50	–	–
G17	B36	363	62.15	91.82	37.4
G18	B44	45	45	45	45
G19	B45	125	55.48	55.48	37.9
G20	B45	426	102.22	102.22	69.8
G21	B47	274	31.5	31.5	19

Table 21.7. Optimal power generated from RES based on three case studies for 48-bus Nigeria power network

Types of RES	Buses	Proposed Capacity (MW)	Operating Capacity (MW)		
			Case 1	Case 2	Case 3
Solar 1	B2	200	175	175	175
OFW 1	B9	450	350	350	450
OFW 2	B12	250	137.5	75	250
OFW 4	B15	250	–	175	–
OFW 5	B17	250	112.5	150	250
Solar 2	B21	300	50	300	300
OFW 6	B23	250	185	250	250
OFW 7	B28	300	250	300	300

(*Continued*)

Table 21.7. Optimal power generated from RES based on three case studies for 48-bus Nigeria power network (*Continued*)

Types of RES	Buses	Proposed Capacity (MW)	Operating Capacity (MW)		
			Case 1	Case 2	Case 3
OFW 8	B30	350	100	–	50
OFW 9	B36	300	–	157.5	212.5
OFW 10	B38	250	187.5	175	112.5
Solar 3	B39	300	112.5	200	212.5
Solar 4	B42	300	112.5	175	37.5
OFW 11	B45	450	37.5	150	100
OFW 12	B47	250	50	300	112.5

Table 21.8. Optimal power flowing on the lines based on three case studies for 48-bus Nigeria power network

S/N	Between Buses	Case 1	Case 2	Case 3
1	1–4	151.37	151.37	151.37
2	1–4	201.63	201.63	201.63
3	3–4	149.51	149.51	149.51
4	3–4	224.49	224.49	224.49
5	4–5	203.26	203.26	203.26
6	4–6	332.16	332.16	332.16
7	4–6	166.08	166.08	166.08
8	5–6	83.26	83.26	83.26
9	6–7	205.00	205.00	205.00
10	6–10	330.93	330.93	296.15
11	6–18	104.83	104.83	70.05
12	8–10	196.72	196.72	131.50
13	9–10	97.00	97.00	197.00
14	10–11	74.65	74.65	74.65
15	10–17	350.00	350.00	350.00
16	12–13	–	–	5.00
17	12–13	7.50	7.50	2.50
18	12–14	57.54	4.00	91.30
19	12–15	8.96	–	87.70
20	13–16	73.18	72.80	59.04

(*Continued*)

Table 21.8. Optimal power flowing on the lines based on three case studies for 48-bus Nigeria power network (*Continued*)

S/N	Between Buses	Case 1	Case 2	Case 3
21	13–27	93.33	92.80	73.84
22	14–15	36.46	90.00	2.70
23	17–27	4.26	3.72	–
24	19–20	63.00	–	–
25	19–24	184.00	247.00	247.00
26	20–21	37.00	–	–
27	20–21	–	34.50	22.00
28	20–24	–	65.50	78.00
29	21–22	265.00	265.00	265.00
30	21–23	127.00	114.50	127.00
31	23–39	3.00	15.50	3.00
32	24–25	269.40	266.90	254.40
33	25–34	324.40	321.90	309.40
34	26–27	34.64	–	–
35	26–34	–	34.46	27.93
36	27–28	243.73	243.45	233.32
37	27–30	–	83.48	109.76
38	27–30	84.14	–	–
39	27–32	0.28	0.60	–
40	27–34	347.31	310.35	304.43
41	29–30	41.75	41.75	41.75
42	31–32	35.82	35.82	35.82
43	32–35	110.00	110.00	110.00
44	32–36	249.54	249.22	249.82
45	33–34	–	0.12	0.19
46	33–34	0.24	0.12	–
47	37–38	58.50	121.00	58.50
48	37–46	0.50	63.00	–
49	39–40	120.00	120.00	120.00
50	39–41	10.50	10.50	10.50
51	41–42	7.50	7.50	7.50
52	43–44	22.50	45.00	22.50
53	43–44	22.50	–	22.50
54	43–45	207.70	207.70	207.70

(*Continued*)

Table 21.8. Optimal power flowing on the lines based on three case studies for 48-bus Nigeria power network (*Continued*)

S/N	Between Buses	Case 1	Case 2	Case 3
55	46–47	79.50	–	79.50
56	46–47	–	17.00	–
57	47–48	52.00	–	–
58	47–48	–	52.00	52.00

Table 21.9. Performance of 48-bus Nigeria power network

Characteristics	Case 1	Case 2	Case 3
Objective function ($\$10^6$)	38,097.76	28,472.93	21,282.40
Total emission (10^6 Ib)	244.42	224.15	203.08
Total generation cost ($\$10^6$)	22,778.25	22,688.69	12,573.19
Total power transmitted on lines (MW)	4353.57	4225.59	4158.22
Total power generated from fossil fuel (MW)	2609.47	2186.96	1806.86
Total power generated from RES (MW)	1910	2432.5	2812.5

21.7 Discussion of Results

The obtained results will be examined and discussed based on the two case studies in this section. The optimization results are analyzed along three directions, which state how sensitive the outcomes with and without incentive schemes of renewable energy is to some system parameters and performance.

Figs. 21.7 and 21.8 present the optimal power generated from fossil fuel-generating units of IEEE 24-bus system and 48-bus Nigeria power system. It is clearly demonstrated from the figures that the optimal power obtained decreases with integration of incentives to improve the usage of RES-generating units. More importantly, the RES-generating units with investment tax credit have less active fossil fuel-generating units as compared to FIT incentive plan.

Observing Fig. 21.9 and Table 21.7, by comparing these three figures, we can see how the power systems perform without incentive and how differently each incentive scheme performs according to the effective utilization of RES. From the simulated results, it is obvious that Case 3 outperforms all other two cases and Case 2 also outperforms Case 1. The worst of the three is Case 1 as regard to RES utilization, because there is no return for RES investment.

Figures 21.10 and 21.11 clearly show comparison of voltage profile for each bus of all three cases of the two case studies. There is a significant improvement in bus voltage profile in Case 3 as compared to other two cases, whereas Case 2 is fairly better as compared to Case 1. This shows that the increase in utilization of RES owing to incentive schemes

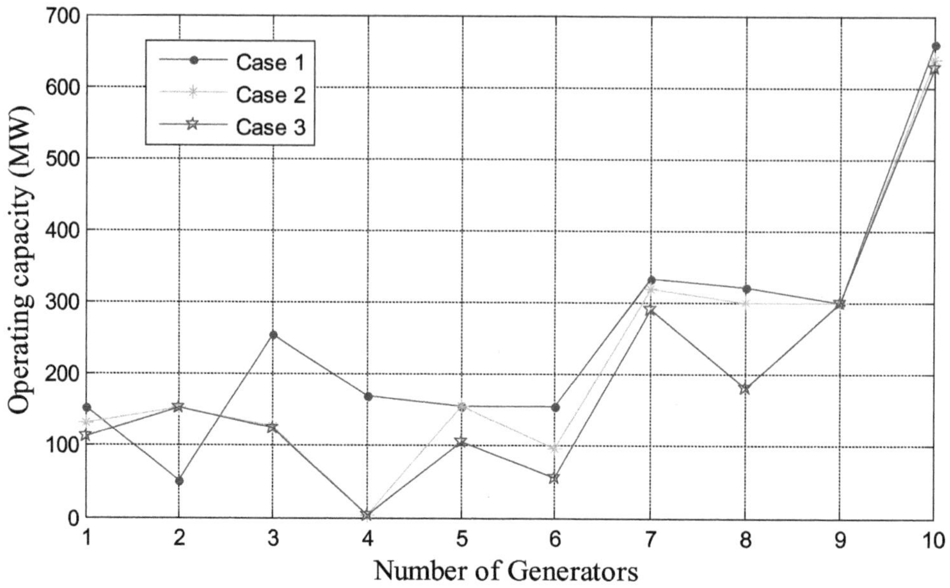

Figure 21.7. Optimal power generated from fossil fuel generators of IEEE24-bus system.

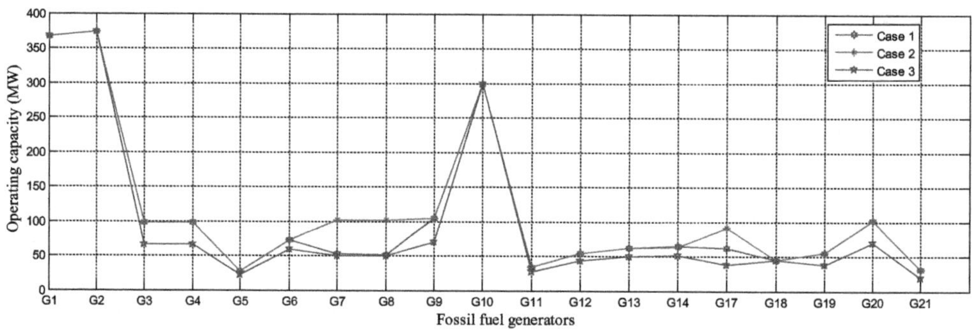

Figure 21.8. Optimal power generated from generators for Nigeria power system.

enhanced the system performance of voltage profile. The performance comparison of the system on three cases is depicted in Figs. 21.12 and 21.13. It is clearly seen that the cost incurred with integration of RES in Case 1 is higher than other two cases, because there is no return on investing in RES. This high cost is also followed with higher emissions, higher generation costs, and higher transmittable power on transmission lines as compared to other two cases with incentive schemes. Case 3 has lower emissions as compared with others because of more utilization of RES and lesser fossil fuel power in the system. This happens because the success of emissions in the power systems is only possible with increase in utilization of RES and reducing energy injected from fossil fuel-generating units. The

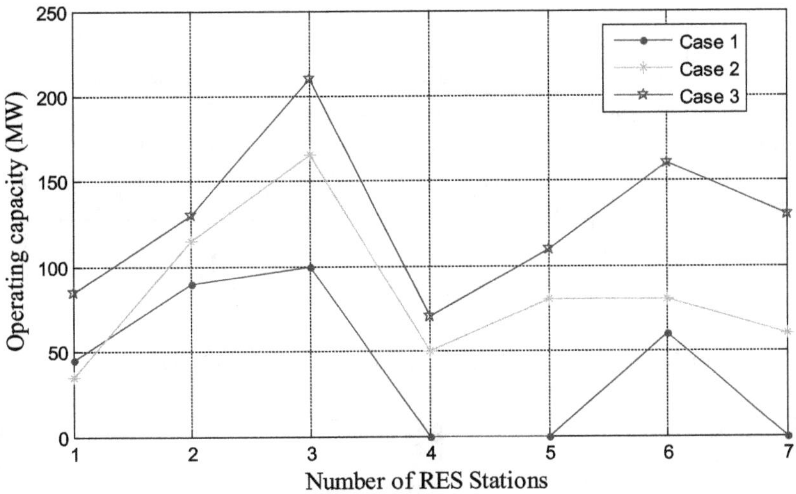

Figure 21.9. Optimal power generated from proposed RES for IEEE 24-bus test system.

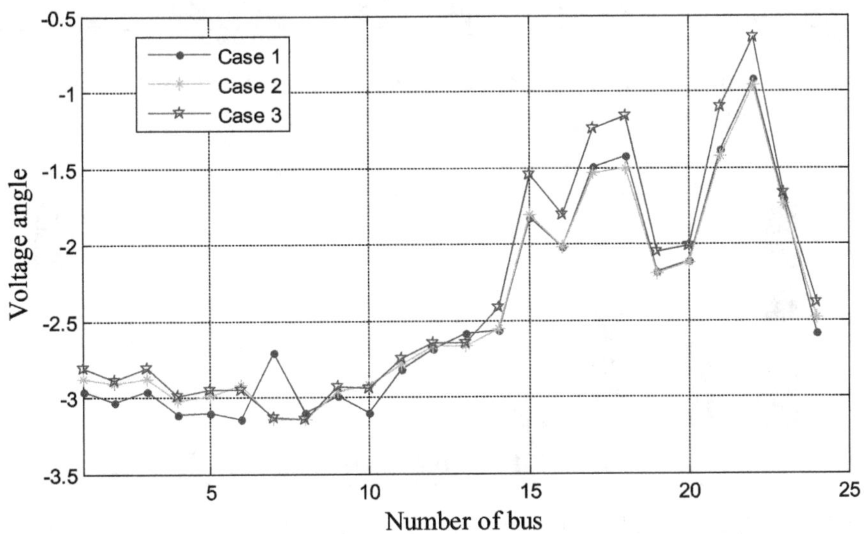

Figure 21.10. Variation of bus voltage angle of IEEE 24-bus system.

obtained results in IEEE 24-bus system shows 10%, 21%, and 31% of RES utilization for Cases 1, 2, and 3, respectively, whereas that of Nigeria power system is given as 40%, 51%, and 60% of RES utilization for all the three cases, respectively. This indicates that the more the improvement in investment return, the more the growth of RES in power systems.

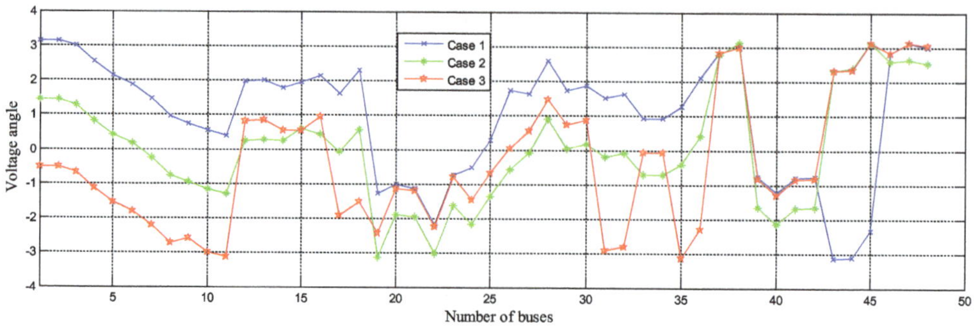

Figure 21.11. Variation of bus voltage angle of Nigeria power system.

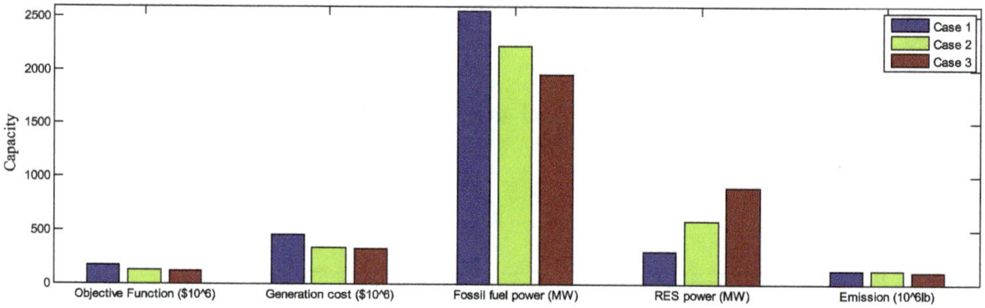

Figure 21.12. Optimal system performance with integration of RES of IEEE 24-bus test system.

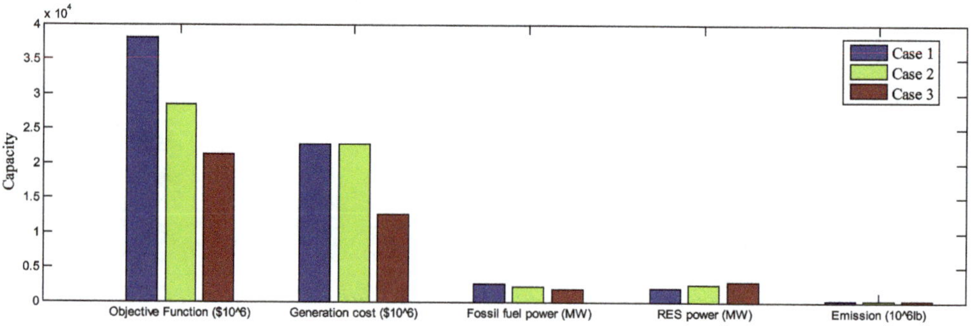

Figure 21.13 Optimal system performance with integration of RES of 48-bus Nigeria power system.

21.8 Conclusion

This chapter has developed a multi-objective optimization model to evaluate and optimize various incentive schemes for renewable energy in composite expansion planning with the view of improving effective utilization of RESs. The effectiveness of the proposed model was implemented on IEEE 24-bus system and real-world data of Nigeria power system. It can be deduced from the simulated results that the system performance is based on the effectiveness of incentive schemes. Case 3 was found to be more effective in terms of RES utilization, emission reductions, generation cost minimization, and voltage profile improvement. The worst performer is Case 1, because there is no return on RES investment in the system. This will assist power system central planner in making investment decisions for the system to meet the required energy demand.

21.9 References

1. H. Ayca, and N. Simsek, "Recent incentives for renewable energy in Turkey," *Energy Policy*, vol. 63, pp. 521–530, 2013.
2. A. Bhuvanesh, S. T. Jaya Christa, S. Kannan, and M. Karuppasamy Pandiyan, "Aiming towards pollution free future by high penetration of renewable energy sources in electricity generation expansion planning," *Futures*, July, 2018.
3. S. L. Gbadamosi, N. I. Nwulu, and Y. Sun, "Harmonic and power loss minimization in power systems incorporating renewable energy sources and locational marginal pricing," *Journal of Renewable and Sustainable Energy*, vol. 10, pp. 1–24, 2018.
4. S. L. Gbadamosi, N. I. Nwulu, and Y. Sun, "Multi-objective optimisation for composite generation and transmission expansion planning considering offshore wind power and feed-in tariffs," *IET Renewable Power Generation*, vol. 12, no. 14, pp. 1687–1697, 2018.
5. C. Roldán, A. A. Sánchez de la Nieta, R. García-Bertrand, and R. Mínguez, "Robust dynamic transmission and renewable generation expansion planning: Walking towards sustainable systems," *International Journal of Electrical Power & Energy Systems*, vol. 96, pp. 52–63, 2018.
6. A. Yatchew, and A. Baziliauskas, "Ontario feed-in-tariff programs," *Energy Policy*, vol. 39, no. 7, pp. 3885–3893, 2011.
7. L. Alagappan, R. Orans, and C. K. Woo, "What drives renewable energy development?" *Energy Policy*, vol. 39, no. 9, pp. 5099–5104, 2011.
8. A. Pyrgou, A. Kylili, and P. A. Fokaides, "The future of the Feed-in Tariff (FiT) scheme in Europe: The case of photovoltaics," *Energy Policy*, vol. 95, pp. 94–102, 2016.
9. T. S. Kishore, and S. K. Singal, "Optimal economic planning of power transmission lines: A review," *Renewable and Sustainable Energy Reviews*, vol. 39, pp. 949–974, 2014.
10. S. L. Gbadamosi and N. I. Nwulu, "A dynamic optimal expansion planning integrating renewable energy sources," 2019 11th International Conference on Applied Energy, ICAE2019, 12-15 August 2019.
11. S. L. Gbadamosi and N. I. Nwulu, "Optimal planning of renewable energy systems for power loss reduction in transmission expansion planning," *Journal of Engineering, Design and Technology*, 18 (2), pp 61–80, 10.1108/JEDT-11-2019-0291.
12. A. S. Aliyu, J. O. Dada, and I. K. Adam, "Current status and future prospects of renewable energy in Nigeria," *Renewable and Sustainable Energy Reviews*, vol. 48, pp. 336–346, 2015.
13. M. Y. Ramandi, K. Afshar, A. S. Gazafroudi, and N. Bigdeli, "Electrical power and energy systems reliability and economic evaluation of demand side management programming in wind integrated power systems," *International Journal of Electrical Power & Energy Systems*, vol. 78, pp. 258–268, 2016.
14. B. M. Radhakrishnan, and D. Srinivasan, "A multi-agent based distributed energy management scheme for smart grid applications," *Energy*, vol. 103, pp. 192–204, 2016.
15. M. Ozcan, "Assessment of renewable energy incentive system from investors' perspective," *Renewable Energy*, vol. 71, pp. 425–432, 2014.
16. M. Mihaylov, R. Roxana, L. Arco, S. Jurado, N. Avellana, and A. Now, "Comparing stakeholder incentives across state-of-the-art renewable support mechanisms," *Renewable Energy*, vol. 131, pp. 689–699, 2019.
17. G. Aquila, E. De Oliveira, A. Rodrigo, D. Queiroz, P. Rotela, and M. Nunes, "An overview of incentive policies for the expansion of renewable energy generation in electricity power systems

and the Brazilian experience," *Renewable and Sustainable Energy Reviews*, vol. 70, pp. 1090–1098, 2017.

18. N. I. Nwulu, and X. Xia, "Optimal dispatch for a microgrid incorporating renewables and demand response," *Renewable Energy*, vol. 101, pp. 16–28, 2017.
19. S. L. Gbadamosi, N. I. Nwulu, and Y. Sun, "Optimal Scheduling in Offshore Wind Farms Considering Harmonic Losses and Locational Marginal Pricing," 2018 IEEE PES-IAS 5th Power Africa Conf., pp. 658–663, 2018.
20. M. Bilgili, A. Yasar, and E. Simsek, "Offshore wind power development in Europe and its comparison with onshore counterpart," *Renewable and Sustainable Energy Reviews*, vol. 15, no. 2, pp. 905–915, 2011.
21. S. Kota, S. B. Bayne, and S. Nimmagadda, "Offshore wind energy: A comparative analysis of UK, USA and India," *Renewable and Sustainable Energy Reviews*, vol. 41, no. 5, pp. 685–694, 2015.
22. J. Bisschop, *AIMMS Optimization Modeling*, 2017 published by Lulu.com.
23. S. L. Gbadamosi, and A. Melodi, "Effects of steel plants with three-phase induction furnaces on power distribution quality of the existing 33 kV network in Nigeria," *Advances in Science and Technology Research Journal*, vol. 9, no. 27, pp. 1–10, 2015.
24. D. O. Kolawole, O. S. Fayomi, U. K. Efemwenkieki, et al., "Viability of hydroelectricity in Nigeria and the future prospect," *Energy Procedia*, vol. 157, pp. 871–878, 2019.
25. A. T. Brimmo, A. Sodiq, S. Sofela, and I. Kolo, "Sustainable energy development in Nigeria: Wind, hydropower, geothermal," *Renewable and Sustainable Energy Reviews*, vol. 74, pp. 474–490, 2017.
26. M. Shaaban, and J. O. Petinrin, "Renewable energy potentials in Nigeria: Meeting rural energy needs," *Renewable and Sustainable Energy Reviews*, vol. 29, pp. 72–84, 2014.
27. U. B. Akuru, I. E. Onukwube, O. I. Okoro, and E. S. Obe, "Towards 100 % renewable energy in Nigeria," *Renewable and Sustainable Energy Reviews*, vol. 71, pp. 943–953, 2017.
28. Annual Technical Report on Grid System Operation, Issued by National Control Centre 2017.

Appendices

Table 21.10. Load parameters and locations of 48-bus Nigeria power system [28]

Load	Bus	Stations	Demand
d1	B1	Kainji	15
d2	B2	B'Kebbi	175
d3	B4	Jebba	25.5
d4	B5	Ganmo	120
d5	B6	Osogbo	150.4
d6	B7	Ayede	205
d7	B9	Sakete	253
d8	B10	Ikeja West	900
d9	B11	Akangba	74.65
d10	B12	Aja	78.5
d11	B14	Alagbon	94
d12	B15	Lekki	85
d13	B19	Katampe	247
d14	B20	Shiroro	200
d15	B21	Kaduna	184
d16	B22	Kano	265
d17	B23	Jos	120
d18	B24	Gwagwalada	85.4
d19	B25	Lokoja	55
d20	B27	Benin	112.5
d21	B28	Effurun	110
d22	B29	Aladja	41.75
d23	B31	Asaba	35.82
d24	B32	Onisha	104
d25	B34	Ajaokuta	23.15
d26	B35	New Haven	110
d27	B37	Ugwuaji	58
d28	B38	Makurdi	54
d29	B39	Gombe	85
d30	B40	Yola	120
d31	B41	Damaturu	18
d32	B42	Maiduguri	30
d33	B43	Alaoji T.S	252.7
d34	B46	Ikot_Ekpene	80
d35	B48	Adiabor	52

Table 21.11. Transmission line parameters and their respective distance [28]

S/N	Between Buses	Distances (km)	R (?)	X (?)
1	1–2	310	12.09	102.61
2	1–4	81	3.16	26.81
3	1–4	81	3.16	26.81
4	3–4	8	0.31	2.65
5	3–4	8	0.31	2.65
6	4–5	78	2.51	26.72
7	4–6	157	6.12	51.97
8	4–6	157	6.12	51.97
9	4–20	244	9.52	80.76
10	4–20	244	9.52	80.76
11	5–6	104	3.86	31.02
12	6–7	119	4.09	34.7
13	6–10	235	11.54	97.98
14	6–18	241	9.79	83.08
15	7–8	73	2.46	25.64
16	8–10	105	3.91	33.12
17	9–10	25	2.03	19.8
18	10–11	18	0.7	5.96
19	10–11	18	0.7	5.96
20	10–13	62	2.42	20.52
21	10–17	160	6.02	54.72
22	12–13	14	0.57	4.68
23	12–13	14	0.57	4.68
24	12–14	6	0.21	2.87
25	12–15	10	0.4	3.2
26	13–16	8	0.31	2.65
27	13–16	8	0.31	2.65
28	13–27	210	10.92	92.68
29	14–15	8	0.31	2.65
30	17–27	120	4.24	35.2
31	18–27	10	0.4	3.2
32	19–20	144	5.62	47.66
33	19–24	60	2.42	20.52
34	20–21	95	3.74	31.78
35	20–21	95	3.74	31.78
36	20–24	144	5.62	47.06

(*Continued*)

Table 21.11. Transmission line parameters and their respective distance [28] (*Continued*)

S/N	Between Buses	Distances (km)	R (?)	X (?)
37	21–22	230	8.97	76.13
38	21–23	197	7.68	65.21
39	23–38	285	10.57	93.59
40	23–38	285	10.57	93.59
41	23–39	265	10.34	87.72
42	24–25	107	4.01	28.15
43	24–25	107	4.01	28.15
44	25–34	35	2.08	20.6
45	25–34	35	2.08	20.6
46	26–27	23	2.03	19.8
47	26–34	145	5.63	76.8
48	27–28	107	4.01	28.15
49	27–30	50	1.95	16.55
50	27–30	50	1.95	16.55
51	27–32	137	5.34	45.35
52	27–32	137	5.34	45.35
53	27–34	195	7.61	64.55
54	28–29	30	1.17	9.93
55	29–30	33	2.46	20.85
56	31–32	27	2.05	20.15
57	32–35	96	3.74	31.78
58	32–36	56	2.18	18.54
59	32–36	56	2.18	18.54
60	32–43	138	6.01	50.97
61	33–34	7	0.3	2.6
62	33–34	7	0.3	2.6
63	35–37	8	0.31	2.65
64	35–37	8	0.31	2.65
65	37–38	136	5.93	48.45
66	37–38	136	5.93	48.45
67	37–46	143	5.6	76.1
68	37–46	143	5.6	76.1
69	39–40	240	9.79	83.08
70	39–41	160	6.02	54.72
71	41–42	260	10.26	85.32
72	43–44	7.5	0.3	2.65
73	43–44	7.5	0.3	2.65

(*Continued*)

Table 21.11. Transmission line parameters and their respective distance [28] (*Continued*)

S/N	Between Buses	Distances (km)	R (?)	X (?)
74	43–45	25	0.98	8.28
75	43–45	25	0.98	8.28
76	43–46	38	1.05	10.26
77	43–46	38	1.05	10.26
78	46–47	8	0.31	2.65
79	46–47	8	0.31	2.65
80	47–48	10	0.4	3.2
81	47–48	10	0.4	3.2

CHAPTER 22

Demand Dispatch for Renewable Energy Integration in Prosumer Smart Grids

Uyikumhe Damisa and Nnamdi I. Nwulu

Department of Electrical and Electronic Engineering Science, University of Johannesburg, Johannesburg, South Africa. uyikumhe@yahoo.com; nnwulu@uj.ac.za

Abstract With high penetration of intermittent renewable energy sources, the modern power grid is becoming increasingly difficult to manage due to the erratic nature of these sources. Supply and demand balance, which is crucial to power system stability, is thus more challenging to sustain. The situation is exacerbated by an increase in prosumers, who consume electricity and sometimes send energy back to the grid. With regard to maintaining this balance, direct control of loads, known as demand dispatch, and conventional dispatch of generation are complementary. Hence, a coordinated dispatch of both supply and demand resources is presented in this chapter. We develop a mathematical model of the operational dynamics of a prosumer smart grid and formulate the problem of power balance as a constrained optimization problem whose solution minimizes the operating cost of the grid and achieves power balance by implementing demand dispatch. Our formulation is solved with Advanced Interactive Multidimensional Modeling System (AIMMS) outer approximation module using data for Ontario's grid network. The performance of our formulation is tested and results obtained authenticate the effectiveness of demand dispatch strategies in achieving supply and demand dispatch in prosumer smart grids. Furthermore, we show that demand dispatch curtails the adverse effects of the randomness of renewable energy generation.

Keywords: Demand dispatch, Economic dispatch, AIMMS, Renewable energy-assisted.

22.1 Introduction

The modern power system is relatively more difficult to manage due to high penetration of intermittent renewable generation. The power from a solar system varies with time of day and may rise/drop abruptly as a result of passing clouds. Wind power, on the other hand, varies with wind speed and may also rise/drop sharply when the speed rises above cut-out speed. Power sent intermittently to the grid, as a result of excess power from renewable generation in prosumer's premises, further increases the randomness of grid power. This high variability of supply, coupled with the unpredictable nature of demand, makes it challenging to balance power supply and demand. The importance of this balance in a power system cannot be overemphasized; it is crucial for system stability. For many decades, this problem of balancing power supply and demand has been captured under the well-known "economic dispatch" problem. Economic dispatch is usually modelled as an optimization problem whose solution gives optimal generation schedule for generating plants; it deals with dispatch of only supply resources—generators. In modern power systems, more sophisticated methods are required. A relatively new approach, termed "demand dispatch," was proposed in [1]. It can be defined as *"An operating model used by grid operators to dispatch "behind-the-meter" resources in both directions—increasing and decreasing load as viewed at the system level—as a complement to supply (generation) dispatch to more effectively optimize grid operations"* [2]. Demand response and demand dispatch may seem similar if viewed from the perspective of the devices involved; however, these are not exactly the same. Although the former entails curbing peak demand, demand dispatch involves constantly adjusting loads to achieve power balance and economic operation of the system. Furthermore, unlike demand response, the system operator has direct control over flexible loads in demand dispatch. Demand dispatch (together with supply dispatch) has a greater capacity than economic dispatch (supply dispatch alone) to manage the stochastic behavior of renewable energy sources because demand can be adjusted in response to fluctuating supply. Also, the capacities of backup power supply and energy storage required to ensure system reliability decrease significantly with demand dispatch. Demand dispatch has become viable due to advances in grid technology and control and communication capabilities of the smart grid.

Economic dispatch has been investigated for many years and successfully implemented in traditional power systems. Over the years, several variants of the economic dispatch model for thermal generators have been studied. The authors of [3] presented a simple model, which assumed a smooth fuel cost curve and typical constraints, but the authors of [4] included an effect caused by the steam admission valve of a turbine, known as the valve point effect. This made the resulting fuel cost curve nonsmooth. Other effects that can be represented in the economic dispatch model are generator ramp rate limits and prohibited zones of operation. These were included in the model developed in [5]. In addition to the economics of generation, utilities are faced with the challenge of curbing emission produced by thermal units during power generation. Various methods were presented to tackle this problem in [6–8]. A more practical form of economic dispatch is the dynamic economic dispatch formulation, which is presented in [9].

Deregulation of the power sector and implementation of favorable renewable energy policies have led to unprecedented growth in distributed generation. This brings with it various technical challenges with regard to appropriate technologies to integrate these systems into the grid, and strategies to optimally manage them. Power systems optimization and planning remain essential to the operation of renewable energy-assisted hybrid systems; hence, new optimization models have been developed by researchers over the years. The authors of [10] proposed a decentralized power dispatch algorithm that minimizes a user-defined cost that can include weighted power generation cost and distribution loss. The emphasis was on dispatching from multiple power sources for improved grid efficiency. The authors of [11] proposed a direct search method for solution to the economic dispatch problem of a microgrid. In [12], the existing microgrid energy management system architectures were reviewed, and the advantage of each stated. Also, a centralized energy management system architecture for a microgrid, made up of a multistage economic load dispatch block and a unit commitment block, was proposed. The authors of [13] carried out a survey on stochastic modelling and optimization tools for microgrid planning and operation. From the survey, it was observed that most of the works were based on Monte Carlo simulation, and potential stochastic modelling and optimization tools such as stochastic game and stochastic inventory theory were mentioned. Several other dispatch approaches have been proposed by researchers to make the best use of renewable sources and other conventional microsources in microgrids, but the focus has been largely on the generation side.

The authors of [14] incorporated demand response in the optimal dispatch of a microgrid consisting of wind, solar, and diesel power generators. Simulations show that optimality is introduced to both the demand and supply sides of the grid when energy management incorporates demand response. With the aim of minimizing the overall cost of a facility's electricity usage, the authors of [15] presented a method for determining the optimal schedule of controllable loads. The test system studied was made up of a utility, uncontrollable loads, and buildings with heating, ventilation and air conditioning (HVAC) chillers as controllable loads. The work in [15] was extended in [16] by incorporating the delayed response times of controllable loads and using a nonlinear model to represent HVAC loads. In [17], based on a statistical model of plug-in electric vehicles' charging demand, smart charging patterns were developed and used to carry out demand dispatch. In [18], deferrable appliances were optimally scheduled in response to grid electricity price, thereby minimizing total cost of energy consumption. The authors of [19] achieved power supply and demand balance primarily by scheduling elastic loads. The authors of [20] presented a mathematical formulation for the demand dispatch of a microgrid connected to a main grid via a distribution transformer. Dispatchable loads including electric vehicles, washing machines, and dish washers were considered in the setup. Comparison of results obtained for scenarios with and without demand dispatch showed that demand dispatch reduces microgrid operating cost and curtails distribution transformer overloading. In [21], a two-stage operational planning scheme for a microgrid was proposed. In the scheme, via demand response programs, demand resources are involved in energy and reserve scheduling. The microgrid studied comprises wind and solar energy resources, fuel cell, microturbine, battery, and industrial and residential loads. Its operational cost is seen to reduce with the involvement of demand-side resources. The use of demand-side resources (such as flexible loads and batteries) for the provision of grid ancillary services has also been explored in the literature, thereby corroborating the employability of these resources in demand dispatch. The authors of [22] presented a report on the provision of contingency reserve using demand response. The system reserve capacity of a stand-alone power system in Australia was augmented using load control, in the work presented in [23]. The combined economic and emissions dispatch problem of a microgrid was investigated in [24], and battery storage and deferrable loads were used to augment the reserve capacity provided by conventional generators. After studies carried out on the use of solar batteries to deliver primary control reserve, the authors of [25] stated that solar batteries are capable of performing their primary obligation while supplying primary control reserve. The planning of batteries for participation in the combined day-ahead, reserve, and regulation markets was investigated in [26]. The authors of [27] presented a strategy for using batteries to provide frequency containment reserve. Batteries have a quick response time [28], hence they are suitable for providing ancillary services [29]. From the study reported in [30], electric vehicles are also capable of providing grid ancillary services. In this chapter, the mathematical model of a smart grid is presented and the problem of power balancing is formulated as a constrained optimization problem whose solution maintains power balance in the system at all times. Our formulation makes use of demand dispatch to minimize operating cost and maintain power balance in the grid. The rest of this chapter is organized as follows: the smart grid component models are presented in Section 22.2. Our problem formulation is detailed in Section 22.3 and a brief methodology is given in Section 22.4. Results obtained are shown in Section 22.5 and discussed in Section 22.6. The chapter concludes with Section 22.7.

22.2 Smart Grid

22.2.1 Architecture

Figure 22.1 is a diagrammatic representation of the smart grid model adopted in this chapter. It is made up of a conventional generator, a wind farm, a solar farm, a utility storage unit, and buildings (prosumers) with onsite solar generators. Figure 22.2 shows possible direction of power flow between elements of the grid. Power flow between prosumers and the grid is possible in both directions, but not concurrently. Each prosumer has a portion of its load under the direct control of the system operator who determines when the loads would be served. The system operator is also committed to meeting the energy requirements of these loads within the window specified by prosumers. In this chapter, these loads are referred to as flexible loads. Energy management is the sole responsibility of the system operator and this is carried out centrally in time slots $t \epsilon [1, 2, \ldots, T]$.

22.2.2 Nomenclature

t, T Index and number of time slots

g, G Index and number of conventional generators

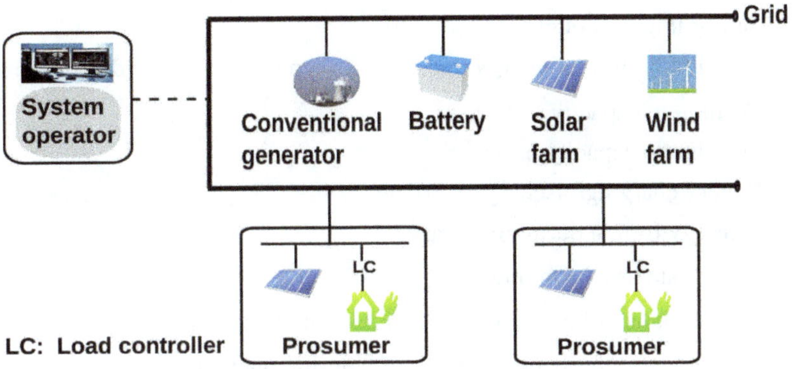

Figure 22.1. Smart grid architecture.

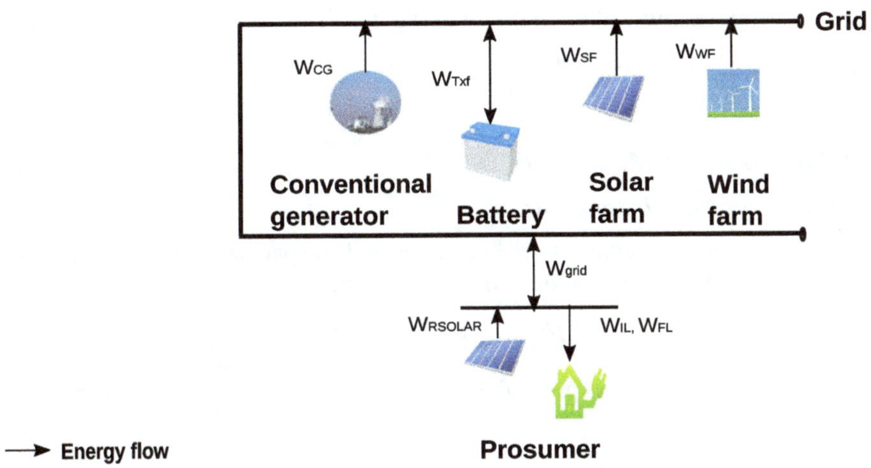

Figure 22.2. Energy flows.

b, B	Index and number of buildings/prosumers
Δt	Period of a time slot
W^t_{SF}	Solar farm output during interval t
W^t_{WF}	Wind farm output during interval t
$W^t_{Rsolar,b}$	Output of rooftop solar in b during interval t
$W^t_{Il,b}$	Required inflexible load energy in b during interval t
$W_{Fl,b}$	Energy supply to flexible load in b per time slot
$W^{total}_{Fl,b}$	Total energy required by flexible load in b
$P^{min}_{CG,g}$	Minimum power output of g
$P^{max}_{CG,g}$	Maximum power output of g
r	Ramping coefficient of conventional generators
a_g, b_g	Fuel cost parameters of g
$C^t_{CG,g}$	Fuel cost of g

C_s^t Unit selling cost of energy (\$/kWh)
W_{Txf}^{max} Maximum limit of energy transfer between battery and grid per time slot
W_{Batt}^{min} Minimum capacity of battery
W_{Batt}^{max} Maximum capacity of battery
η Battery charging/discharging efficiency
$W_{CG,g}^t$ Energy from g during interval t
$S_{CG,g}^t$ On/off status of g in interval t
$S_{Fl,b}^t$ On/off status of flexible load in b in interval t
$W_{grid,b}^t$ Energy transfer between grid and b during interval t
$P_{CG,g}^t$ Power output of g in interval t
W_{Txf}^t Energy to/from battery during interval t
W_{Batt}^t Energy content of battery in interval t

22.2.3 Component Modelling

22.2.3.1 Conventional Generator

$P_{CG,g}^t$, the power output of conventional generator g, is dispatchable, that is, it can be controlled. Due to its physical limitations, the power output from a generator is constrained as

$$P_{CG,g}^{min} \leq P_{CG,g}^t \leq P_{CG,g}^{max} \tag{22.1}$$

Its cost function, $C_{CG,g}^t$, is expressed as

$$C_{CG,g}^t = a_g\left(P_{CG,g}^t\right)^2 + b_g P_{CG,g}^t \tag{22.2}$$

Due to ramping limitations of conventional generators, (22.3) and (22.4) are included [31]:

$$P_{CG,g}^t - P_{CG,g}^{t-1} \leq r * P_{CG,g}^{max} \tag{22.3}$$

$$P_{CG,g}^{t-1} - P_{CG,g}^t \leq r * P_{CG,g}^{max} \tag{22.4}$$

22.2.3.2 Renewable Sources

Given that a photovoltaic (PV) generator has efficiency of η_{pv}, array area of A_{pv} and hourly incident solar irradiance of I_{pv}, its output energy per hour, E_{pv}, is given as [32]

$$\eta_{pv} A_{pv} I_{pv} \tag{22.5}$$

Given that a wind turbine has power coefficient of $C_p C_p$, rotor swept area of A_w, efficiency of η_w, and generator efficiency of η_g, its power output, P_w, when wind velocity is V_w, and air density factor is ρ_a, is given in (22.6) [33].

$$P_w = \eta_w * \eta_g * 0.5 * \rho_a * C_p * A_w * V_w^3 \tag{22.6}$$

22.2.3.3 Battery

W^t_{Batt} is constrained due to the physical limitations of the battery by

$$W^{min}_{Batt} \leq W^t_{Batt} \leq W^{max}_{Batt} \tag{22.7}$$

The relationship between the energy content of a battery in adjacent time slots is shown as

$$W^t_{Batt} = W^{t-1}_{Batt} + \eta \left(W^t_{Txf} \right) \tag{22.8}$$

W^t_{Txf} can take both positive and negative values. A positive value would imply that the battery is being charged and a negative value means it is being discharged. It is const rained as

$$-W^{max}_{Txf} \leq W^t_{Txf} \leq W^{max}_{Txf} \tag{22.9}$$

To account for battery inefficiency, (22.10) is introduced [34].

$$-W^{t-1}_{Batt} \geq -W^t_{Txf} \tag{22.10}$$

22.2.3.4 Building/Prosumer Load

Each building has two load categories: flexible and inflexible. Although the energy requirement of inflexible loads must be supplied on request, that of flexible loads can be delayed or even interrupted as situation demands.

Constraint (22.11) ensures that flexible loads are sufficiently served within the window specified.

$$\sum\nolimits_{t=start}^{end} \left(S^t_{Fl,b} * W_{Fl,b} \right) = W^{total}_{Fl,b} \tag{22.11}$$

Outside this window, flexible loads must not be served, hence the need for (22.12) [34].

$$S^t_{Fl,b} = 0 \quad \text{for all end} < t < \text{start} \tag{22.12}$$

For control of the frequency of interruption of service to flexible loads, (22.13) is enforced.

$$\sum\nolimits_{t=start}^{end-1} \left(S^t_{Fl,b} * S^{t+1}_{Fl,b} \right) \geq \frac{W^{total}_{Fl,b}}{W_{Fl,b}} - 2 \tag{22.13}$$

The foregoing constraint permits only one interruption per serving window; however, it can be modified to allow for more interruptions.

From Fig. 22.2, the energy balance equation at each prosumers's node is deduced, and shown in (22.14).

$$W^t_{Rsolar,b} - W^t_{grid,b} - W^t_{Il,b} - W_{Fl,b} * S^t_{Dl,b} = 0 \tag{22.14}$$

$W^t_{grid,b}$ is a free variable, that is, it can take positive or negative values. A positive value would indicate that energy is flowing to the grid and a negative value would mean that energy is flowing out of the grid (to prosumer).

22.2.4 Energy Balance

For system stability, there must be balance between supply and demand, hence the need for the energy balance equation

$$W^t_{SF} + W^t_{WF} + \sum_{g=1}^{G} \left(S^t_{CG,g} * W^t_{CG,g}\right) - W^t_{Txf} + \sum_{g=1}^{S} W^t_{grid,b} = 0 \tag{22.15}$$

The aggregate energy being generated by conventional generators per time slot is represented by the third term of the energy balance equation above and the last term represents the net transfer of energy between the grid and prosumers per time slot.

22.3 Problem Formulation

With the smart grid component models developed above, a constrained optimization problem whose solution balances power supply and demand, and minimizes operating cost, is formulated. Our objective function represents the cost metric of the system operator, and is shown as

$$Min \sum_{t=1}^{T} \sum_{g=1}^{G} \left(S^t_{CG,g} * C^t_{CG,g}\right) + \sum_{t=1}^{T} \sum_{b=1}^{B} \left(C^t_s * W^t_{grid,b}\right) \tag{22.16}$$

The terms in the expression above represent total cost of conventional generation and net cost of energy transfer between the grid and buildings. Constraints (22.17) through (22.30) are enforced.

$$S^t_{CG,g} \epsilon \; [0, 1] \quad \forall \, t \, \epsilon \, [1, \mathrm{T}] \tag{22.17}$$

$$S^t_{Dl,s} \epsilon \; [0, 1] \quad \forall \, t \, \epsilon \;\; 1, \mathrm{T}] \tag{22.18}$$

$$W^t_{SF} + W^t_{WF} + \sum_{g=1}^{G} \left(S^t_{CG,g} * W^t_{CG,g}\right) + \sum_{b=1}^{B} W^t_{grid,b} = 0 \, \forall \, t \, \epsilon \; [1, T] \tag{22.19}$$

$$W^t_{Rsolar,b} - W^t_{grid,b} - W^t_{Il,b} - W_{Fl,b} * S^t_{Fl,b} = 0 \quad \forall \, \mathrm{b} \, \epsilon \; [1, B] \, \forall \, t \, \epsilon \; [1, T] \tag{22.20}$$

$$\sum_{t=start}^{end} \left(S^t_{Fl,b} * W_{Fl,b}\right) = W^{total}_{Fl,b} \tag{22.21}$$

$$\sum_{t=start}^{end-1} \left(S^t_{Fl,b} * S^{t+1}_{Fl,b}\right) \geq \frac{W^{total}_{Fl,b}}{W_{Fl,b}} - 2 \, \forall \, \mathrm{b} \, \epsilon \; [1, B] \tag{22.22}$$

$$S^t_{Fl,b} = 0 \, \forall \, \mathrm{end} \; < t < \mathrm{start} \tag{22.23}$$

$$P^{min}_{CG,g} \leq P^t_{CG,g} \leq P^{max}_{CG,g} \forall \, g \, \epsilon \; [1, \mathrm{G}] \; \forall \, t \, \epsilon \; [1, T] \tag{22.24}$$

$$P^t_{CG,g} - P^{t-1}_{CG,g} \leq r * P^{max}_{CG,g} \forall \, \mathrm{g} \, \epsilon \; [1, \mathrm{G}] \; \forall \, \mathrm{t} \, \epsilon \; (1, \mathrm{T}) \tag{22.25}$$

$$P^{t-1}_{CG,g} - P^t_{CG,g} \leq r * P^{max}_{CG,g} \forall \, \mathrm{g} \, \epsilon \; [1, \mathrm{G}] \; \forall \, \mathrm{t} \, \epsilon \; (1, \mathrm{T}) \tag{22.26}$$

$$W^t_{Batt} = W^{t-1}_{Batt} + \left(W^t_{Txf}\right) \forall \text{b} \epsilon \ [1, B] \ \forall t \epsilon \ [1, T] \tag{22.27}$$

$$W^{min}_{Batt} \leq W^t_{Batt} \leq W^{max}_{Batt} \forall b \varepsilon \ [1, B] \ \forall t \varepsilon \ [1, T] \tag{22.28}$$

$$-W^{max}_{Txf} \leq W^t_{Txf} \leq W^{max}_{Txf} \forall \text{b} \varepsilon \ [1, B] \ \forall t \varepsilon \ [1, T] \tag{22.29}$$

$$\text{-}W^{t-1}_{Batt} \geq -W^t_{Txf} \forall \text{ b} \epsilon \ [1, B] \ \forall t \varepsilon \ [1, T] \tag{22.30}$$

The products of decision variables in the formulation above are replaced with new variables, which are constrained appropriately [35]. $S^t_{CG,g} * C^t_{CG,g}$, $S^t_{CG,g} * W^t_{CG,g}$, and $S^t_{Fl,b} * S^{t+1}_{Fl,b}$ are replaced with SC^t_g, SW^t_g and SS^t_s, respectively, and the resulting formulation is (22.31) through (22.57).

$$Min \sum_{t=1}^{T} \sum_{g=1}^{G} \left(S^t_{CG,g} * C^t_{CG,g}\right) + \sum_{t=1}^{T} \sum_{b=1}^{B} \left(C^t_s * W^t_{grid,b}\right) \tag{22.31}$$

Subject to :

$$SC^t_g \leq UC^t_g * S^t_{CG,g} \forall \text{ g } \epsilon \ [1, \text{G}] \ \forall \text{t} \epsilon \ [1, \text{T}] \tag{22.32}$$

$$SC^t_g \geq LC^t_g * S^t_{CG,g} \forall \text{ g } \epsilon \ [1, \text{G}] \ \forall \text{t} \epsilon \ [1, \text{T}] \tag{22.33}$$

$$SC^t_g \leq C^t_{CG,g} - LC^t_g \left(1 - S^t_{CG,g}\right) \forall \text{ g } \epsilon \ [1, \text{G}] \ \forall \text{t} \epsilon \ [1, \text{T}] \tag{22.34}$$

$$SC^t_g \geq C^t_{CG,g} - UC^t_g \left(1 - S^t_{CG,g}\right) \forall \text{ g } \epsilon \ [1, \text{G}] \ \forall \text{t} \epsilon \ [1, \text{T}] \tag{22.35}$$

$UC^t_g LC^t_g$ are the maximum and minimum limits, respectively, of conventional generation cost.

$$S^t_{CG,g} \epsilon \ [0, 1] \ \forall t \epsilon \ [1, T] \tag{22.36}$$

$$S^t_{Dl,s} \epsilon \ [0, 1] \ \forall t \epsilon \ [1, T] (20) \tag{22.37}$$

$$W^t_{SF} + W^t_{WF} + \sum_{g=1}^{G} \left(SW^t_g\right) + \sum_{b=1}^{B} W^t_{grid,b} = 0 \forall \text{ t } \epsilon \ [1, \text{T}] \tag{22.38}$$

$$SW^t_g \leq UE^t_g * S^t_{CG,g} \forall \text{ g } \epsilon \ [1, \text{G}] \ \forall \text{ t } \epsilon [1, \text{T}] \tag{22.39}$$

$$SW^t_g \geq LE^t_g * S^t_{CG,g} \forall \text{ g } \epsilon \ [1, \text{G}] \ \forall \text{ t } \epsilon [1, \text{T}] \tag{22.40}$$

$$SW^t_g \leq W^t_{CG,g} - LE^t_g \left(1 - S^t_{CG,g}\right) \forall \text{ g } \epsilon \ [1, \text{G}] \ \forall \text{ t } \epsilon [1, \text{T}] \tag{22.41}$$

$$SW^t_g \geq W^t_{CG,g} - UE^t_g \left(1 - S^t_{CG,g}\right) \forall \text{ g } \epsilon \ [1, \text{G}] \ \forall \text{ t } \epsilon [1, \text{T}] \tag{22.42}$$

$UE^{t\,t}_{gg}$ are the maximum and minimum limits, respectively, of conventional generator output.

$$W^t_{Rsolar,b} - W^t_{grid,b} - W^t_{Il,b} - W_{Fl,b} * S^t_{Fl,b} = 0\,\forall\, \mathrm{b} \,\epsilon\, [1, \mathrm{B}]\, \forall\, \mathrm{t} \,\epsilon\, [1, \mathrm{T}] \tag{22.43}$$

$$\sum\nolimits_{t=start}^{end} \left(S^t_{Fl,b} * W_{Fl,b} \right) = W^{total}_{Fl,b} \tag{22.44}$$

$$\sum\nolimits_{t=start}^{end-1} \left(SS^t_s \right) \geq \frac{W^{total}_{Fl,b}}{W_{Fl,b}} - 2\,\forall\, \mathrm{b} \,\epsilon\, [1, \mathrm{B}] \tag{22.45}$$

$$SS^t_s \leq S^t_{Fl,b} \forall\, \mathrm{b} \,\epsilon\, [1, \mathrm{B}]\, \forall\, \mathrm{t} \,\epsilon\, [\mathrm{start}, \mathrm{end} - 1] \tag{22.46}$$

$$SS^t_s \leq S^{t+1}_{Fl,b}\, \forall\, \mathrm{b} \,\epsilon\, [1, \mathrm{B}]\, \forall\, \mathrm{t} \,\epsilon\, [\mathrm{start}, \mathrm{end} - 1] \tag{22.47}$$

$$SS^t_s \geq S^t_{Fl,b} + S^{t+1}_{Fl,b} - 1\,\forall\, \mathrm{b} \,\epsilon\, [1, \mathrm{B}]\, \forall\, \mathrm{t} \,\epsilon\, [\mathrm{start}, \mathrm{end} - 1] \tag{22.48}$$

$$SS^t_s \,\epsilon\, [0, 1]\ \ \forall\, \mathrm{b} \,\epsilon\, [1, \mathrm{B}]\, \forall\, \mathrm{t} \,\epsilon\, [\mathrm{start}, \mathrm{end} - 1] \tag{22.49}$$

$$S^t_{Fl,b} = 0\,\forall\, \mathrm{end} < \mathrm{t} < \mathrm{start} \tag{22.50}$$

$$P^{min}_{CG,g} \leq P^t_{CG,g} \leq P^{max}_{CG,g} \forall\, \mathrm{g} \,\epsilon\, [1, \mathrm{G}]\, \forall\, \mathrm{t} \,\epsilon\, [1, \mathrm{T}] \tag{22.51}$$

$$P^t_{CG,g} - P^{t-1}_{CG,g} \leq r * P^{max}_{CG,g}\, \forall\, \mathrm{g} \,\epsilon\, [1, \mathrm{G}]\, \forall\, \mathrm{t} \,\epsilon\, (1, \mathrm{T}) \tag{22.52}$$

$$P^{t-1}_{CG,g} - P^t_{CG,g} \leq r * P^{max}_{CG,g}\, \forall\, \mathrm{g} \,\epsilon\, [1, \mathrm{G}]\, \forall\, \mathrm{t} \,\epsilon\, (1, \mathrm{T}) \tag{22.53}$$

$$W^t_{Batt} = W^{t-1}_{Batt} + \eta \left(W^t_{Txf} \right) \forall\, \mathrm{b} \,\epsilon\, [1, \mathrm{B}]\, \forall\, \mathrm{t} \,\epsilon\, [1, \mathrm{T}] \tag{22.54}$$

$$W^{min}_{Batt} \leq W^t_{Batt} \leq W^{max}_{Batt} \forall\, \mathrm{b} \,\epsilon\, [1, \mathrm{B}]\, \forall\, \mathrm{t} \,\epsilon\, [1, \mathrm{T}] \tag{22.55}$$

$$-W^{max}_{Txf} \leq W^t_{Txf} \leq W^{max}_{Txf} \forall\, \mathrm{b} \,\epsilon\, [1, \mathrm{B}]\, \forall\, \mathrm{t} \,\epsilon\, [1, \mathrm{T}] \tag{22.56}$$

$$\eta W^{t-1}_{Batt} \geq -W^t_{Txf} \forall\, \mathrm{b} \,\epsilon\, [1, \mathrm{B}]\, \forall\, \mathrm{t} \,\epsilon\, [1, \mathrm{T}] \tag{22.57}$$

22.4 Methodology

For this study, a module in AIMMS (a mathematical modelling language integrated with solvers, which has been used successfully in [36] and [37]), which was developed based on the outer approximation method presented in [38], is employed to solve our formulation. The module, known as the AIMMS outer approximation module (AOA), makes use of CONOPT and CPLEX solvers to solve a sequence of nonlinear programming (NLP) and mixed integer programming (MIP) models, respectively. The basic AOA algorithm is as follows [39]:

- With integer variables converted to continuous variables within bounds, the original problem is solved as a NLP problem.
- New linear constraints obtained from linearizing around the optimal solution are combined with the original linear constraints of the problem to form a master mixed integer linear programming problem.
- This master problem is solved, and the integer part of the solution obtained is temporarily fixed, whereas the original mixed integer nonlinear problem with fixed integer variables is solved as a nonlinear problem.
- Again, the new linear constraints obtained from linearizing about the optimal solution are added to the master mixed integer linear programming problem. To avoid cycling, the previously found integer solution is cut off by adding one or more constraints.
- The last two steps above are repeated until the master mixed integer linear programming problem becomes infeasible or a termination criterion is satisfied.

22.5 Numerical Tests

22.5.1 Experimental Setup

Table 22.1 summarizes the ratings of the renewable energy generators in the smart grid and Table 22.2 summarizes the utility battery parameters. The smart grid setup consists of a conventional generator and 30 buildings (each having a solar system installed, flexible, and inflexible loads). For each building, $W_{Dl,s}$ is taken to be 1 kWh and $W_{Dl,s}^{total}$ is given in Table 22.3. A total of 24 time slots are considered, each spanning a 1-h period, summing up to a time horizon of 1 day. $C_{CG,g}^{t} = a_g\left(P_{CG,g}^{t}\right)^2 + b_g P_{CG,g}^{t}$, where a_g and b_g are summarized in Table 22.4 [24]. Wind, solar, electricity cost, and demand data are summarized in Table 22.5 [29]. Table 22.5 gives the rating of flexible load in each of the 30 buildings used in this study.

Table 22.1. Renewable generator ratings.

	Maximum Power Output (kW)
Solar farm	27
Wind farm	50
Rooftop solar	1.5

Table 22.2. Battery parameters.

	W_{Batt}^{min}(kWh)	W_{Batt}^{max}(kWh)	W_{Txf}^{max}(kWh)	
Battery	5	150	50	0.95

22.5.2 Results

22.5.2.1 Base Case

Shown in this section are results obtained from simulation using the base case data presented in Section 22.5.1. Figure 22.3 shows the optimal power schedule obtained. The net undispatchable supply shown in the key of Fig. 22.3 is the sum of solar farm output, wind farm output and total rooftop solar output, minus total inflexible load demand, per time slot. For the purpose of clarity and easy comprehension of the figure, the following sign conventions for energy flow to/from battery and energy supply to flexible loads are adopted: Positive energy to/from battery indicates that battery supplies energy in that time slot and vice versa; negative energy supply to flexible load means that flexible load is being served in that time slot. Figure 22.4 displays the conventional generator output per time slot with and without demand dispatch. A summary of the results for the base case scenario is summarized in Table 22.6.

22.5.2.2 Effect of Battery Capacity on Optimal Power Schedule

The effect of battery capacity on optimal power schedule is investigated by varying the battery capacity from 50 to 250 kWh in steps of 50 kWh. Optimal schedules obtained are presented in Figs. 22.5–22.8.

22.5.2.3 Effect of Battery Capacity on System Operator's Cost

Battery capacity is varied from 50 to 250 kWh, and its effect on the system operator's cost is displayed in Fig. 22.9. Two scenarios are shown: with and without demand dispatch.

22.5.2.4 Effect of Demand Dispatch on Conventional Generator Output

The conventional generator output per time slot, with and without demand dispatch, for battery capacities 50–250 kWh are displayed in Figs. 22.10–22.13.

22.5.2.5 Effect of Size of Flexible Load Serving Window of System Operator's Cost Metric Value

The number of time slots within which flexible load energy requirements must be met (flexible load serving window) is varied from 18 to 24 time slots and its effect on system operator's cost is shown in Fig. 22.14.

22.5.2.6 Effect of Battery Capacity on Energy Flow to and fro Battery

Figure 22.15 shows the energy flow to and from the battery for battery capacities from 50 to 250 kWh.

Table 22.3. Flexible load energy requirement per building.

Building	**1**	**2**	**3**	**4**	**5**	**6**	**7**	**8**	**9**	**10**	**11**	**12**	**13**	**14**	**15**
(*kWh*	6	8	16	11	8	14	5	11	6	10	9	6	8	10	9
Building	16	17	18	19	20	21	22	23	24	25	26	27	28	29	30
(*kWh*	7	14	5	16	6	10	13	17	10	5	12	18	15	15	17

Table 22.4. Conventional generator parameters.

	$P_{CG,g}^{min}$(kW)	$P_{CG,g}^{max}$(kW)	a_g(\$/kW^2h	b_g(\$/kWh)	r
Generator	8	45	0.003	0.25	0.2

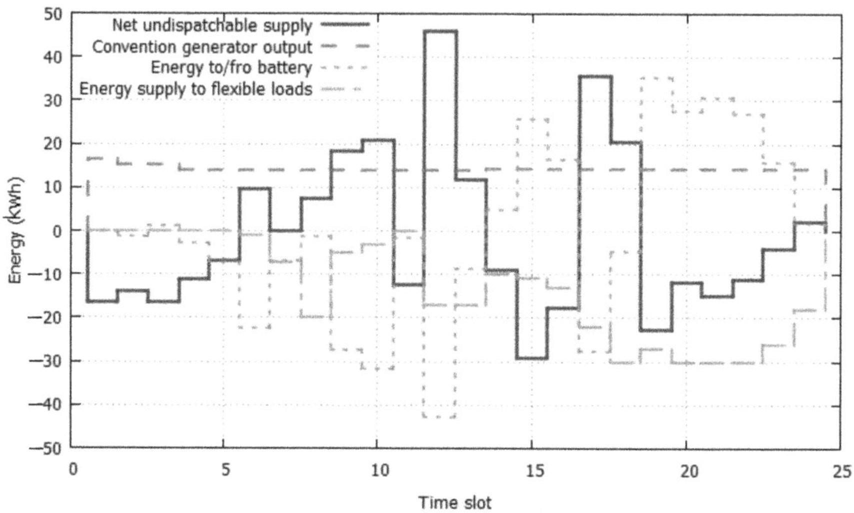

Figure 22.3. Optimal power schedule for base case (battery capacity of 150 kWh).

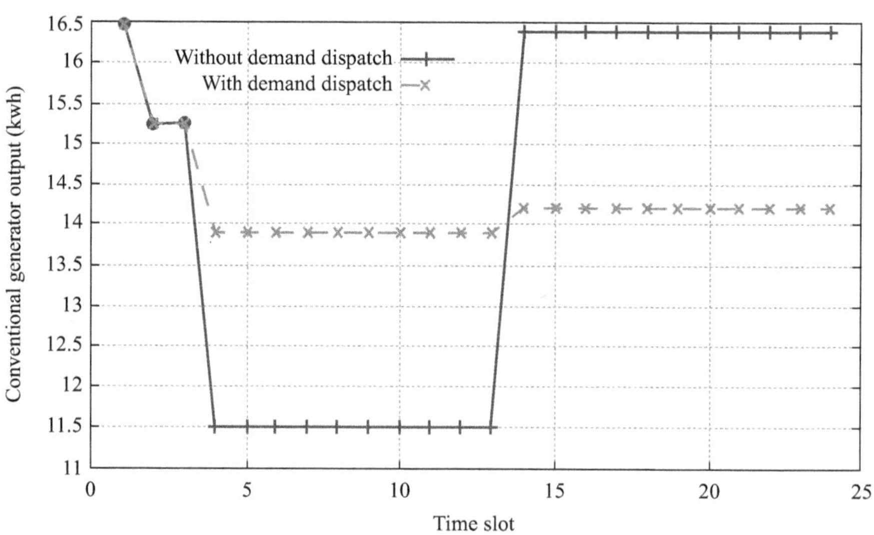

Figure 22.4. Conventional generator output per time slot for base case (battery capacity of 150 kWh).

Table 22.5. Ontario data (scaled down).

Time slot	W^t_{SF}(kWh)	W^t_{WF}(kWh)	$W^t_{Rsolar,s}$(kWh)	$W^t_{IL,s}$ (kWh)	C^t_s (\$/kWh)
1	0.0	22.15	0.00000	1.2874	0.00273
2	0.0	23.40	0.00000	1.2445	-0.0005
3	0.0	19.95	0.00000	1.2170	-0.00208
4	0.0	25.05	0.00000	1.2079	-0.00229
5	0.0	29.60	0.00000	1.2137	-0.00229
6	0.1	47.40	0.00566	1.2695	0.00092
7	0.9	39.00	0.05094	1.3803	0.00317
8	5.3	38.10	0.30000	1.5005	0.01097
9	12.3	29.45	0.69623	1.4781	0.00271
10	17.1	18.85	0.96792	1.4751	0.00250
11	7.0	12.15	0.39623	1.4443	-0.00002
12	26.5	18.20	1.50000	1.4562	0.00952
13	16.0	11.50	0.90566	1.4254	0.00614
14	8.2	11.00	0.46415	1.4070	0.00286
15	1.9	8.05	0.10750	1.4091	0.00594
16	5.9	9.10	0.33396	1.4204	0.00602
17	22.8	17.10	1.29057	1.4345	0.00763
18	16.6	19.55	0.93962	1.4622	0.01455
19	3.5	12.35	0.19811	1.4778	0.00634
20	2.0	28.50	0.11321	1.5226	0.02346
21	0.1	32.60	0.00566	1.5941	0.01888
22	0.0	36.15	0.00000	1.5757	0.01927
23	0.0	39.65	0.00000	1.4530	0.00936
24	0.0	42.10	0.00000	1.3354	0.00699

Table 22.6. Summary of results for base case.

Quantity	Value
Total Solar farm output (kWh)	146.2
Total wind farm output (kWh)	590.95
Total rooftop solar output (kWh)	248.26
Total energy supplied to inflexible loads (kWh)	1010.75
Total conventional generation (kWh)	342.34
Total energy supplied to flexible loads (kWh)	317
System operator cost without demand dispatch ($)	92.05
System operator cost with demand dispatch ($)	91.47

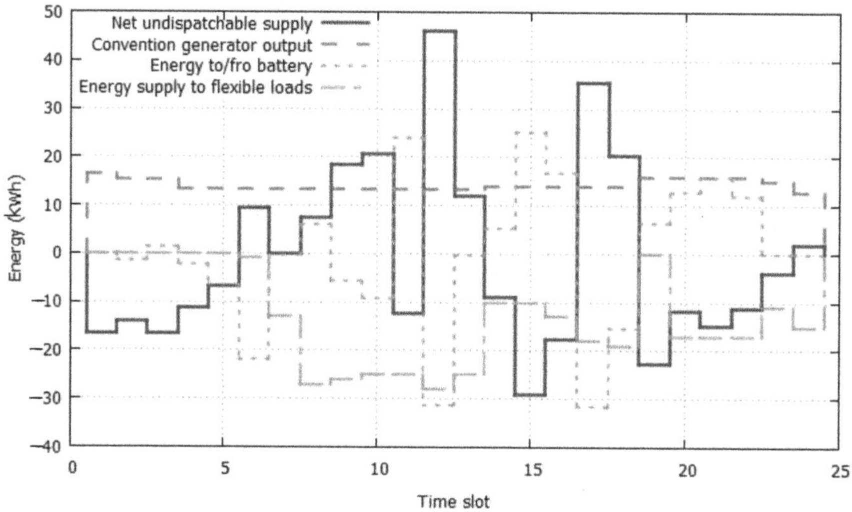

Figure 22.5. Optimal power schedule (battery capacity of 50 kWh).

22.6 Discussion

The proposed smart grid operational model is such that all renewable generation is used up, to either serve loads or charge the utility battery. This can be observed in the optimal power schedule shown in Fig. 22.3. The net undispatchable supply is positive in time slots 6, 8–10, 11, 12, 17, 18, and 24, that is, in these time slots, the total output from the solar and wind farms and rooftop solar systems exceeds the sum of inflexible loads. This excess renewable energy is used to serve flexible loads and/or to charge the utility battery. By doing this, no renewable energy is lost, and the fluctuating nature of solar and wind power is better managed. When renewable generation is in short supply, for example, in time slots 1–5 and 19–23, the conventional generator (sometimes together with the utility battery) meets the excess demand requirements, thereby ensuring system reliability. Although slight variations in conventional generator output are noticed within the first four time slots in

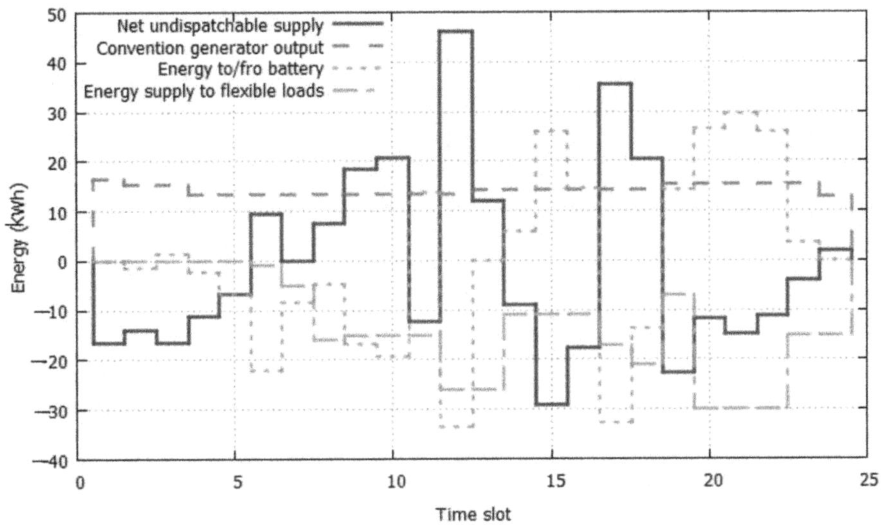

Figure 22.6. Optimal power schedule (battery capacity of 100 kWh).

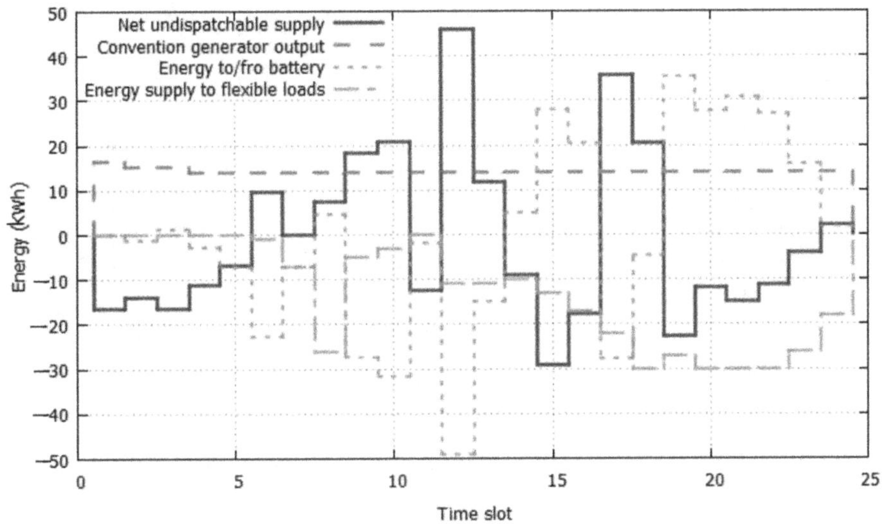

Figure 22.7. Optimal power schedule (battery capacity of 200 kWh).

Fig. 22.3, relative stability is observed in subsequent time slots. This stability results due to the utility battery's ability to absorb surplus energy and supply energy deficits. Therefore, the use of utility-size battery storage can curtail the adverse effects of renewable energy variations. In Fig. 22.4, the role of demand dispatch in ensuring relative stability of conventional generator output can be observed. Demand dispatch is also seen to have a positive impact on the system operator's cost as displayed in Table 22.6. A close look at Figs. 22.5–22.8 gives an interesting insight with regard to the effect of utility battery size on the output of the conventional generator. The stability of the output is seen to improve as the battery

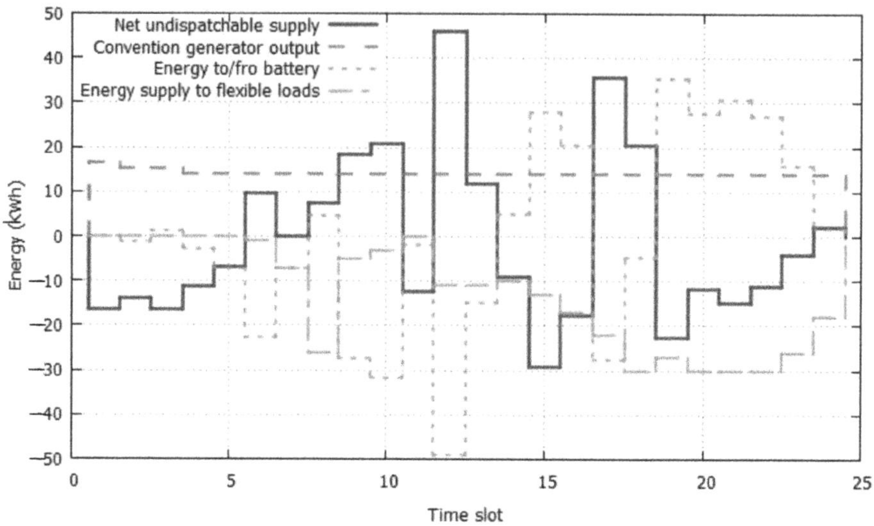

Figure 22.8 Optimal power schedule (battery capacity of 250 kWh).

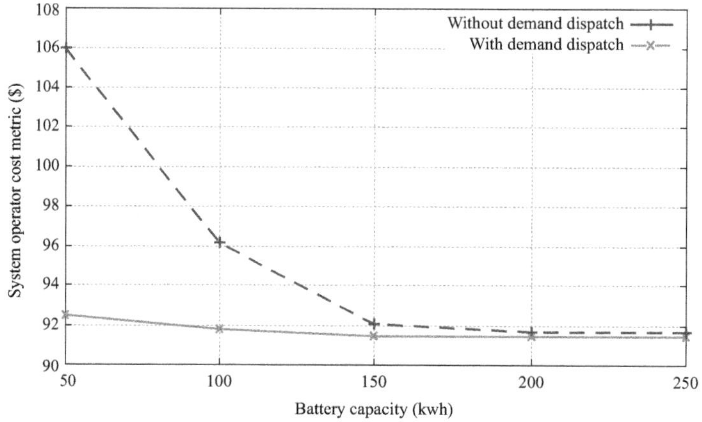

Figure 22.9. System operator cost with respect to battery capacity.

size increases, albeit not indefinitely; this improvement stops at a battery size of 200 kWh. No improvement is seen when battery size is increased from 200 kWh (Fig. 22.7) to 250 kWh (Fig. 22.8). Besides the economic benefit of demand dispatch to the system operator, depicted in Fig. 22.9, it can also be observed that the system operator's costs, with and without demand dispatch, approach equality as the battery size increases. This implies that sufficient battery capacity can play the role of demand dispatch; however, the economic implication of procuring large storage facilities is a concern. In addition to the stabilizing effect of increasing battery size on generator output, Figs. 22.10–22.13 show that until a certain battery capacity is reached, demand dispatch complements the battery to stabilize conventional generator output. In our model, flexible loads must be served within specified time slots (serving windows). Figure 22.14 shows the effect of the size of this window on the

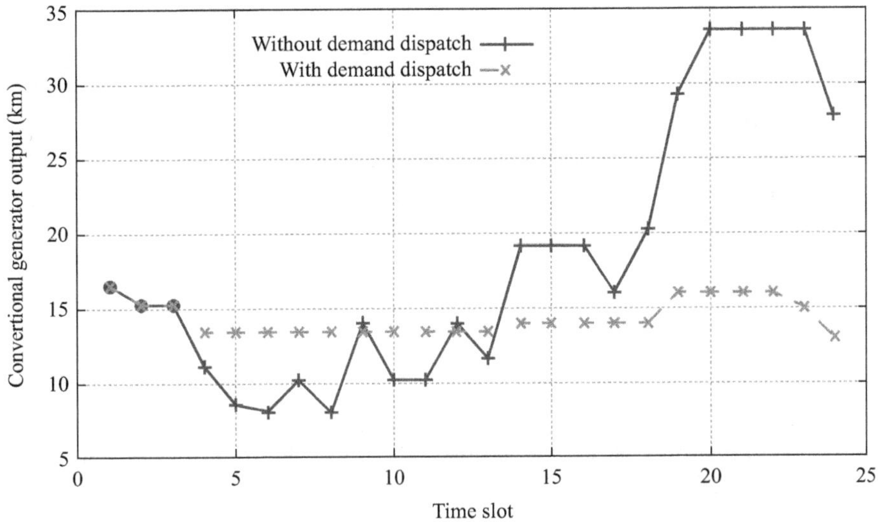

Figure 22.10. Conventional generator output per time slot (battery capacity of 50 kWh).

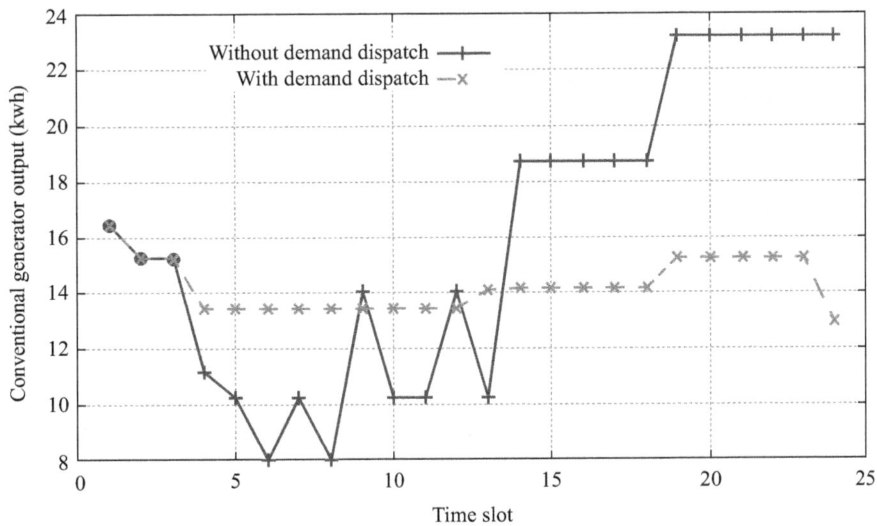

Figure 22.11. Conventional generator output per time slot (battery capacity of 100 kWh).

system operator's cost. An increase from 18 to 19 time slots caused a sharp drop in operating cost, however, no cost reductions were recorded despite further increase in flexible load serving window. This means that, although the serving windows became wider, more optimal flexible load schedules were not discovered. Figure 22.15 shows that the energy to and from the battery increases with battery capacity until a certain capacity is reached. Hence,

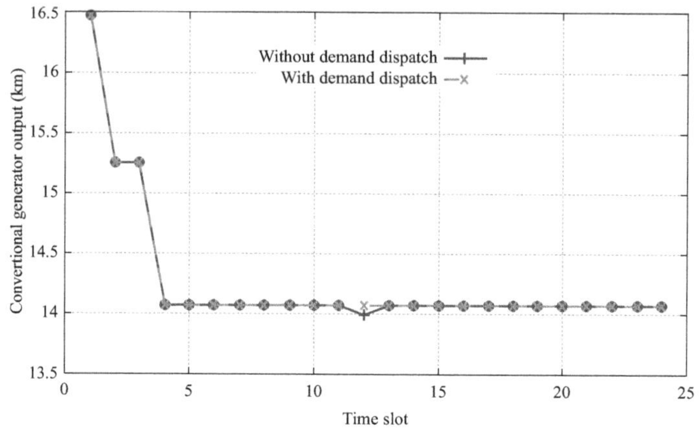

Figure 22.12. Conventional generator output per time slot (battery capacity of 200 kWh).

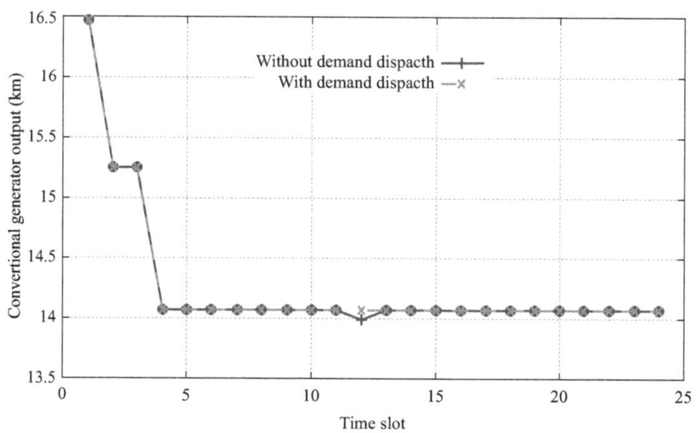

Figure 22.13. Conventional generator output per time slot (battery capacity of 250 kWh).

more energy is lost with increase in battery capacity due to transmission losses and battery charging/discharging inefficiency. To curtail this, the cost of these losses should feature in the objective function (system operator's cost) of the formulation. It is also shown in Fig. 22.15 that the total energy to the battery is equal to the amount out of the battery over the time horizon considered. That is, the solution algorithm ensures that virtually all the energy absorbed by the battery within the time horizon is used up by the end of the time horizon. This is because it is cheaper to use up all the energy transferred to the battery than generate more.

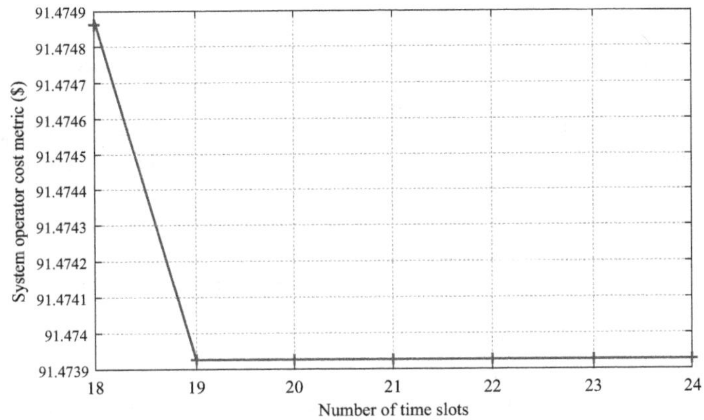

Figure 22.14. System operator cost with respect to size of flexible load serving window.

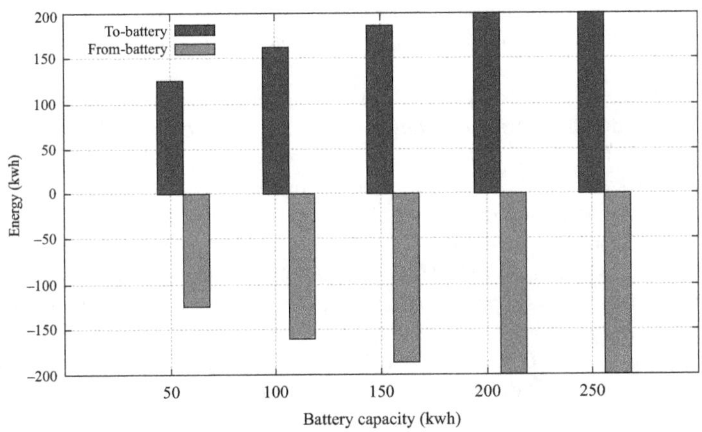

Figure 22.15. Total energy to/from battery.

22.7 Conclusion

In this chapter, the problem of supply and demand balance in a prosumer smart grid was investigated. The mathematical models for the components of the smart grid architecture studied were developed, and the problem of power balance was formulated as a constrained optimization problem whose solution ensures power balance in the grid at all times. Results revealed that (1) demand dispatch has economic benefits for the system operator (2); the energy to and from the battery increases with battery capacity until a certain capacity is reached (3); the system operator's cost drops with increase in flexible load serving window, only up to a certain extent; and (4) a utility-size battery complements demand dispatch to reduce fluctuations in conventional generator output thereby reducing wear and tear of generator parts.

To further the work presented in this chapter, a stochastic version of our formulation, which accounts for the unpredictable nature of wind power, solar power, and demand may be investigated.

22.8 References

1. A. Brooks, E. Lu, D. Reicher, C. Spirakis, and B. Weihl, "Demand Dispatch," IEEE Power Energy Magazine, vol. 8, no. 3, pp. 20–29, May 2010.
2. J. Goellner, M. Prica, J. Miller, S. Pullins, J. Westerman, J. Harmon, and D. Coen, *Demand Dispatch—Intelligent Demand for a More Efficient Grid*, US National Energy Technology Lab, 2011.
3. L. Chopra, and R. Kaur, "Economic load dispatch using simple and refined genetic algorithm," *International Journal Advanced Engineering Technology*, vol. 5, no. 1, pp. 584–590, 2012
4. I. Musirin, N. H. F. Ismail, M. R. Kalil, M. K. Idris, T. K. A. Rahman, and M. R. Adzman, *Ant Colony Optimization (ACO) Technique in Economic Power Dispatch Problems*, Springer, Dordrecht, pp. 191–203, 2009.
5. H. Nezamabadi-pour, S. Yazdani, M. M. Farsangi, and M. Neyestani, "A solution to an economic dispatch problem By A Fuzzy adaptive genetic algorithm," University Sistan Baluchestan, vol. 8, no. 3, pp. 1–21, Oct. 2011.
6. R. M. Kumar, T. Bavisetti, and K. Kiran, "Optimization of combined economic and emission dispatch problem: A comparative study for 30 bus systems," *Journal Electrical Electronic Engineering (IOSR-JEEE)*, vol. 2, pp. 37–3, 2012
7. M. Garg, "A survey on environmental economic load dispatch using Lagrange multiplier method", *International Journal of Electronics & Communication Technology*, vol. 3, pp. 43–46, 2012.
8. S. Dinu, I. Odagescu, and M. Moise, "Environmental economic dispatch optimization using a modified genetic algorithm," *International Journal Computer Applications*, vol. 20, no. 2, pp. 7–14, Apr. 2011.
9. S. Rajasomashekar, and P. Aravindhababu, "Biogeography based optimization strategy for dynamic economic emission dispatch", *International Journal Engineering Science and Technology*, vol. 5(2), pp. 419–429, 2013.
10. C. Ahn, H. Peng, C. Ahn, and H. Peng, "Decentralized and real-time power dispatch control for an islanded microgrid supported by distributed power sources," *Energies*, vol. 6, no. 12, pp. 6439–6454, Dec. 2013.
11. W. T. Huang, K. C. Yao, C. C. Wu, W. T. Huang, K. C. Yao, and C. C. Wu, "Using the direct search method for optimal dispatch of distributed generation in a medium-voltage microgrid," *Energies*, vol. 7, no. 12, pp. 8355–8373, Dec. 2014.
12. D. E. Olivares, C. A. Canizares, and M. Kazerani, "A centralized optimal energy management system for microgrids," IEEE Power and Energy Society General Meeting, Detroit, Michigan, USA, pp. 1–6, 2011.
13. H. Liang, W. Zhuang, H. Liang, and W. Zhuang, "Stochastic modeling and optimization in a microgrid: A survey," *Energies*, vol. 7, no. 4, pp. 2027–2050, Mar. 2014.
14. N. I. Nwulu, and X. Xia, "Optimal dispatch for a microgrid incorporating renewables and demand response," *Renew Energy*, vol. 101, pp. 16–28, Feb. 2017.
15. J. Berardino, C. O. Nwankpa, and K. Miu, "Economic demand dispatch of controllable building electric loads for demand response," IEEE Trondheim PowerTech, Trondheim, Norway, pp. 1–6, 2011.
16. J. Berardino, and C. O. Nwankpa, "Economic demand dispatch of controllable building electrical loads incorporating delayed response times," IEEE PES Innovative Smart Grid Technologies Conference (ISGT), Washington, DC, USA, pp. 1–6, 2013.
17. T. Wu, G. Wu, Z. Bao, Q. Yang, and W. Yan, "Demand dispatch of smart charging for plug-in electric vehicles," International Conference on Control Engineering and Communication Technology, Shenyang, China, pp. 803–806, 2012.

18. S. Chen, N. B. Shroff, and P. Sinha, "Heterogeneous delay tolerant task scheduling and energy management in the smart grid with renewable energy," *IEEE Journal Selected Areas Communications*, vol. 31, no. 7, pp. 1258–1267, Jul. 2013.
19. P. He, M. Li, L. Zhao, B. Venkatesh, and H. Li, "Water-filling exact solutions for load balancing of smart power grid systems," *IEEE Transactions on Smart Grid*, vol. 9, no. 2, pp. 1397–1407, Mar. 2018.
20. F. D. Farimani, and H. R. Mashhadi, "Modeling and implementation of demand dispatch approach in a smart micro-grid," Integral Methods in Science and Engineering, Cham: Springer International Publishing, pp. 129–141, 2015.
21. A. Zakariazadeh, and S. Jadid, "Smart microgrid operational planning considering multiple demand response programs," *Journal of Renewable and Sustainable Energy*, vol. 6, no. 1, p. 013134, Jan. 2014.
22. M. Motalleb, M. Thornton, E. Reihani, and R. Ghorbani, "A nascent market for contingency reserve services using demand response," *Applied Energy*, vol. 179, pp. 985–995, Oct. 2016.
23. D. Nikolic, M. Negnevitsky, and M. De Groot, "Fast demand response as spinning reserve in microgrids," Mediterranean Conference on Power Generation, Transmission, Distribution and Energy Conversion (MedPower 2016), pp. 75–75, 2016.
24. U. Damisa, N. I. Nwulu, and Y. Sun, "A mathematical formulation of the joint economic and emission dispatch problem of a renewable energy-assisted prosumer microgrid," *African Journal Science Technology Innovation Development*, vol. 11(3), pp. 359–367, 2019.
25. F. Braam, L. M. Diazgranados, R. Hollinger, B. Engel, G. Bopp, and T. Erge, "Distributed solar battery systems providing primary control reserve," *IET Renewable Power Generation*, vol. 10, no. 1, pp. 63–70, Jan. 2016.
26. M. Kazemi, H. Zareipour, N. Amjady, W. D. Rosehart, and M. Ehsan, "Operation scheduling of battery storage systems in joint energy and ancillary services markets," *IEEE Transactions on Sustainable Energy*, vol. 8, no. 4, pp. 1726–1735, Oct. 2017.
27. T. Thien, D. Schweer, D. vom Stein, A. Moser, and D. U. Sauer, "Real-world operating strategy and sensitivity analysis of frequency containment reserve provision with battery energy storage systems in the German market," *Journal of Energy Storage*, vol. 13, pp. 143–163, Oct. 2017.
28. J. Fleer, and P. Stenzel, "Impact analysis of different operation strategies for battery energy storage systems providing primary control reserve," *Journal of Energy Storage*, vol. 8, pp. 320–338, Nov. 2016.
29. D. Kottick, M. Blau, and D. Edelstein, "Battery energy storage for frequency regulation in an island power system," *IEEE Transactions on Energy Conversion*, vol. 8, no. 3, pp. 455–459, 1993.
30. M. Alipour, B. Mohammadi-Ivatloo, M. Moradi-Dalvand, and K. Zare, "Stochastic scheduling of aggregators of plug-in electric vehicles for participation in energy and ancillary service markets," *Energy*, vol. 118, pp. 1168–1179, Jan. 2017.
31. R. Pal, C. Chelmis, M. Frincu, and V. Prasanna, "MATCH for the prosumer smart grid the algorithmics of real-time power balance," *IEEE Transactions on Parallel Distributed Systems*, vol. 27, no. 12, pp. 3532–3546, Dec. 2016.
32. H. Tazvinga, X. Xia, and J. Zhang, "Minimum cost solution of photovoltaic–diesel–battery hybrid power systems for remote consumers," *Solar Energy*, vol. 96, pp. 292–299, Oct. 2013.
33. S. Ashok, "Optimised model for community-based hybrid energy system," *Renew Energy*, vol. 32, no. 7, pp. 1155–1164, Jun. 2007.
34. S. Salinas, M. Li, P. Li, and Y. Fu, "Dynamic energy management for the smart grid with distributed energy resources," *IEEE Transactions on Smart Grid*, vol. 4, no. 4, pp. 2139–2151, Dec. 2013.
35. Johannes Bisschop, "AIMMS Optimization Modeling", Google Books, 2006." Available Online: https://books.google.co.za/books?hl=en&lr=&id=oHR8vSZFMvgC&oi=fnd&pg=PR

5&dq=AIMMS+optimization+modeling&ots=Cj1mRMIpHn&sig=vVYtKwxU-j1wDgL9P6pV0ro4xuw#v=onepage&q=AIMMS optimization modeling&f=false

36. U. Damisa, N. I. Nwulu, and Y. Sun, "Microgrid energy and reserve management incorporating prosumer behind-the-meter resources," *IET Renewable Power Generation*, vol. 12, no. 8, pp. 910–919, Jun. 2018.
37. U. Damisa, N. I. Nwulu, and Y. Sun, "A robust energy and reserve dispatch model for prosumer microgrids incorporating demand response aggregators," *Journal of Renewable and Sustainable Energy*, vol. 10, no. 5, p. 055301, Sep. 2018.
38. M. A. Duran, and I. E. Grossmann, "An outer-approximation algorithm for a class of mixed-integer nonlinear programs," *Math Program*, vol. 36, no. 3, pp. 307–339, Oct. 1986.
39. M. Hunting, "The AIMMS Outer Approximation Algorithm for MINLP," Paragon Decision Technology, Haarlem, 2011.

Index

A

B

C

D

E

F

G

H

I

J

K

L

M

N

O

P

Q

R

S

T

U

V

W

Z

www.ingramcontent.com/pod-product-compliance
Lightning Source LLC
LaVergne TN
LVHW081351011026
849161LV00013BA/2788